S0-DUU-377

OCEAN YEARBOOK 2

OCEAN YEARBOOK 2
BOARD OF EDITORS

Francis Auburn
University of Western Australia
John Bardach
The East-West Center
Frank Barnaby
Stockholm International Peace Research Institute
Thomas S. Busha
Inter-Governmental Maritime Consultative Organization
S. J. Holt
(UK)
Yves LaPrairie
Groupement Interprofessionnel pour L'Exploitation des Océans
Arvid Pardo
University of Southern California
Lord Ritchie-Calder
The House of Lords
Mario Ruivo
Intergovernmental Oceanographic Commission
Warren S. Wooster
University of Washington

INTERNATIONAL OCEAN INSTITUTE
BOARD OF TRUSTEES

H. S. Amerasinghe *(Chairman)*
President, United Nations Conference on the Law of the Sea
Zakaria Ben Mustapha
Ministère de l'Agriculture, Tunisia
Elisabeth Mann Borgese
Dalhousie University
Jorge Castañeda
Secretary of Foreign Affairs, Mexico
Edwin J. Borg Costanzi
University of Southhampton
Gunnar Myrdal
University of Stockholm
Aurelio Peccei
President, Club of Rome
Jan Pronk
M.P., The Netherlands
Roger Revelle
University of California, San Diego
Mario Ruivo
Secretary, Intergovernmental Oceanographic Commission
Hernan Santa Cruz
United Nations Development Programme
Anton Vratuŝa
President, Socialist Republic of Slovenia, Yugoslavia
Layachi Yaker
Ambassador of Algeria to the USSR

INTERNATIONAL OCEAN INSTITUTE
PLANNING COUNCIL

Elisabeth Mann Borgese *(Chairman)*
Dalhousie University
Silviu Brucan
University of Bucharest
Maxwell Bruce, Q.C.
(Canada)
Thomas S. Busha
Inter-Governmental Maritime Consultative Organization
Edwin J. Borg Costanzi
University of Southhampton
Peter Dohrn
Mediterranean Association for Marine Biology and Oceanology, Italy
R. J. Dupuy
College de France
Reynaldo Galindo Pohl
(El Salvador)
J. King Gordon
International Development Research Centre, Canada
G. L. Kesteven
(Australia)
Anatoly L. Kolodkin
UNCTAD, Geneva
Frank LaQue
(U.S.A.)
Arvid Pardo
University of Southern California
Jacques Piccard
Fondation pour l'Étude et la Protection de la Mer et des Lacs, Switzerland
Christopher Pinto
(Sri Lanka)
Sir Egerton Richardson
(Jamaica)
Lord Ritchie-Calder
The House of Lords
Father Peter Serracino-Inglott
Scuola Beate Angelico, Milan
Bhagwat Singh
United Nations Secretariat
Jun Ui
Tokyo University
V. K. S. Varadan
Director-General, Geological Survey of India
Jan van Ettinger
Director, RIO Foundation, The Netherlands
Joseph Warioba
Attorney-General, Tanzania
Alexander Yankov
Ambassador of Bulgaria to the United Nations

OCEAN YEARBOOK 2

Sponsored by the
International Ocean Institute

Edited by
Elisabeth Mann Borgese and
Norton Ginsburg

Assistant Editor, Daniel Dzurek

The University of Chicago Press

Chicago and London

The University of Chicago Press, Chicago 60637
The University of Chicago Press, Ltd., London

© 1980 by the University of Chicago
All rights reserved. Published 1980
Printed in the United States of America

International Standard Book Number: 0-226-06603-7
Library of Congress Catalog Card Number: 79-642855

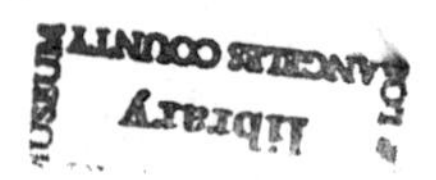

COPYING BEYOND FAIR USE. The code on the first page of an article in this volume indicates the copyright owner's consent that copies of the article may be made beyond those permitted by Sections 107 or 108 of the U.S. Copyright Law provided that copies are made only for personal or internal use, or for the personal or internal use of specific clients and provided that the copier pay the stated per-copy fee through the Copyright Clearance Center Inc. To request permission for other kinds of copying, such as copying for general distribution, for advertising or promotional purposes, for creating new collective works, or for resale, kindly write to the publisher. If no code appears on the first page of an article, permission to reprint may be obtained only from the author.

Contents

Regional Developments

Appendices

Acknowledgments

We wish to thank FAO, ECOR, ILO, IMCO, IOC, SCOR, SIPRI, UNEP, WHO, and WMO for their cooperation; and the General Service Foundation, the government of the Netherlands (Minister for Development and Cooperation), the Gulbenkian Foundation, and the RIO Foundation for financial and other support which has made the *Ocean Yearbooks* possible. Last, but by no means least, we wish to thank the Department of Geography at the University of Chicago for providing office facilities for the editors.

THE EDITORS

The Year in Perspective

The Editors

The evolving law of the sea symbolizes a paradigmatic shift in the nature of international jurisprudence. For the first time, the entire international community will, in the near future, be bound by a set of rules concerning the conservancy of most of the surface of its planet. Not merely rights, not merely obligations, but also management responsibilities are being incorporated in the new jurisprudential system. To be sure, earlier models involving such responsibilities have appeared, concerned with outer space, airwaves, and Antarctica, among others, but none has dealt with so complex an array of issues with which all nations are deeply and immediately concerned. The new law of the sea itself is unprecedentedly complex, straddling as it does matters not only of conventional jurisprudence but also of economics, geography, oceanography, marine biology, industrial management, and, of course, international politics. Few individuals could command knowledge about all these matters. Certainly, very few laymen are likely to; but even those charged with the task of creating this new instrument of hemiglobal management have difficulty finding their way through the conceptual and factual forest that fuels the vast lawmaking effort.

The objective of the *Ocean Yearbook*s is to present an integrated view of man's activities in and relating to the oceans, to analyze trends, and to present them in their interaction. In fulfilling this function, it will provide a guide, a source of reliable generalizations, a reservoir of facts for those, representing, we hope, all the countries of the world, who will have to ratify the forthcoming Convention, whenever it appears, and for those responsible world citizens—in government, business, and academia—who share an abiding interest in the future of humankind.

As the themes of volume 1 are taken up and further developed in this volume, a profile of work in progress appears to take shape and a clearer structure is emerging. It is hoped that volume 2 moves somewhat closer to the achievement of an integrated view of the emerging maritime global order, although such a view remains an illusive goal, for it keeps changing as the components change and as perceptions of events, trends, and priorities keep changing.

In the introductory pages, volume 1 projected a number of major trends. It may be of interest to look at these again and to reexamine them in the light of the events reported in this volume 2.

© 1980 by The University of Chicago. 0-226-06603-7/80/0017$01.00

1. The overall slow increase in the production of living resources continues, albeit with a pronounced shift from overexploited or "collapsed" fisheries to "unconventional" species and exotic regions. Whereas volume 1 predicted a drop in production due, in part, to the jurisdictional changes in the oceans, it is remarkable how slight, thus far, has been the international impact of the new political boundaries on the production and distribution of living resources. Distant-water fishing nations have adjusted surprisingly quickly to the widespread establishment of extended fisheries or exclusive economic zones, of which there are already over 60. Distant-water fishing nations have made their agreements, largely in the form of joint ventures, with coastal states, and their production, on the whole, has remained constant or risen. Rich coastal nations, with effective marine-management capabilities, have profited most from the newly acquired fishing zones. Canada, for example, has seen a dramatic increase of over 11 percent in its Atlantic and 29 percent in its Pacific fisheries between 1976 and 1978, largely as a consequence of the establishment of its efficiently managed 200-mile fisheries zone. The greatest increase in the Atlantic is due to the introduction of a new fishery, squid. In the poorer countries, on the other hand, local consumption of fish shows no sign of increase, and the beneficial impact of extended Exclusive Economic Zones on the nutrition of local population, on employment, and on coastal development is yet to come (see the article by S. J. Holt and C. Vanderbilt).

2. Following last year's analysis of oil production and consumption, the present volume presents a survey of the geology of oil (see the article by Gaskell and Simpson). Further confounding the pseudoscientific assumptions underlying present negotiations on the limits of the outer continental margin, the authors point out that in many cases "the lesson to be remembered, when applying North Sea experience to other parts of the world, is that coastlines can also be geological boundaries and that land geology will not necessarily continue out to shallow water." The geological evidence they cited confirms that oil is a finite resource with a remaining useful life of some 30 years.[1] Unless alternative energy sources are exploited and new technologies developed, there may be a shortfall in world energy supplies by the year 2000 of as much as 40 percent. In the development of these alternative sources and new technologies the oceans will play an important role.

3. The world economic recession of the late seventies has affected ocean development in various ways. There can be no doubt, for instance, that the

1. The Tenth World Petroleum Congress, meeting in Bucharest, Romania, in 1979, came to considerably more optimistic conclusions. Oil and gas reserves north of the Arctic Circle, according to these forecasts, might match those of the Middle East. These include huge gas and oil fields in Soviet Siberia and lesser reservoirs in the Beaufort Sea off Canada and Alaska. Vast untapped resources lie in deep waters, including the international area. Experience in drilling off Thailand, Surinam, and elsewhere has proved that "deep water is not so hostile if you disregard high cost." Even pessimists agreed that probably not even half of the known reservoirs have been tapped.

depression of the world nickel market, more than the intricacy of legal and institutional problems, has slowed down negotiations concerning the International Seabed Authority, while at the same time slackening the pressure of unilateral legislation leading to seabed mineral exploitation. In the somewhat longer term—given the long lead times necessary—the effects on the global shipping industry are striking. For example, U.S. shipping companies were reported to be receiving $290 million annually in federal operating subsidies. By March 1978, one London ship-broking firm estimated that over one-third of world tanker tonnage was laid up (see the article by Susan Strange and Christopher Cragg).[2] An aspect of the industry generally neglected in the discussions on shipping and its role in the relations between the developed and the developing countries is the insurance business in this sector. The first Third-World Insurance Congress was held in the Philippines in October 1977 and produced some startling figures. It was estimated that in 1975 only 6.5 percent of the total volume of world premium income originated from all the developing countries put together. Only 16 percent of insurance companies operated in developing areas at all, and 61 percent of those operating in Asia were foreign owned. Regional cooperation among developing countries could, in this sector as in others, alter these ratios; but the task of adapting the multibillion dollar shipping insurance business to the challenges arising from new technologies and the magnitude of ecological risks and pollution liabilities is an arduous one and will take time and require structural changes in the industry itself.

4. Two important UN conferences, where marine sciences and technologies were to play, or should have played, an important role, were held in 1979. One was the World Climate Conference (WCC), held in Geneva, February 12–13, 1979, and organized by the World Meteorological Organization (WMO) in collaboration with FAO, Unesco, UNEP, WHO, ICSU, and IIASA. The crucial importance of the oceans in meteorology is stressed in the report in Appendix A in this volume, "Meteorology and Ocean Affairs." The conference dealt in particular with the interactions between climate variability and the development and management of freshwater resources and marine resources, including fisheries. It is possible that in a shorter term than hitherto anticipated climatic changes will affect agricultural systems in such a way as to create greater pressures on aquatic sources of food production than has been true in the past. Forthcoming climatic changes, however slight, also may contribute to

2. According to 1978 OECD statistics, the developing countries have fared better than the developed countries, and better also than flag-of-convenience countries. The goal for the Second Development Decade (1970–80) of 10 percent of world shipping for developing countries was practically reached by 1978. The OECD countries, which in 1964 accounted for 73 percent of world shipping, decreased to just above 50 percent. The Socialist countries of Eastern Europe continued their slow growth (4.5 percent). Their percentage in world tonnage remained constant. Greece, with a growth rate of 15 percent, showed the greatest increase and now possesses the third largest shipping fleet in the world, trailing Japan by 5 million tons.

the transformation of "hunting" fisheries into systems of "husbanding" and of the marginal gathering of seaweeds into the large-scale scientific cultivation of seaweeds and algae for food, industry, and pharmaceutical purposes (see the article by Akio Miura).

The second conference was the UN Conference on Science and Technology for Development (UNCSTD) held in Vienna, August 1979. Here concern with the oceans was conspicuously absent, in spite of the fact that the marine sciences are essential not only to understanding the evolution of our planet but also to the development of marine resources (for international cooperation in marine sciences and technology, see the article by Warren Wooster), and that the development of marine resources is becoming an integral and, in many cases, essential part of development in general. The conference merely addressed a series of topics drawn from the seemingly obsolescent development literature of the 1960s and early 1970s. Due partly to the sectoral structure of the UN system, partly to a variety of political reasons, the oceans play no readily discernible role in the evolution and promulgation of these strategies. Even now, the proportion of development funds spent on the transfer of marine scientific knowledge and technologies has been negligible. The recent establishment of a $7 million fund and $100 million program, under FAO auspices, to facilitate the transfer of fisheries technologies, although pointing in the right direction, can scarcely correct the imbalance. It is unfortunate that the occasion UNCSTD could have provided to further this objective has been wasted.

5. The expansion of national jurisdictions in ocean space has had important effects, readily apparent to many but not intended or foreseen by those idealists who, during the early phase of the Law of the Sea Conference in Caracas (1974), hailed the concept of the Exclusive Economic Zone as an instrument of distributive justice and a harbinger of a New International Economic Order. It is in the Pacific in particular and its marginal seas that the establishment of economic zones and archipelagic waters around tiny and in many cases dependent islands, as well as around archipelagic states like Indonesia, the Philippines, and Fiji, is bringing 40 percent of this largest of all oceans under national jurisdiction—to the dismay of those, also idealists, who have cherished and nurtured the concept of ocean space and resources as a Common Heritage of Mankind. Here one observes the seemingly inexorable advance of marine territorialism as a *primus modus operandi* in international affairs. A basic motive force is the belief that the seas constitute an unlimited source of potential wealth; for the island countries newly come into being in the southwestern Pacific, and even those with longer histories in Southeast Asia, this golden objective outweighs all other considerations. In the South China Seas, the territory of which is already virtually exhausted by the overlapping and conflicting jursdictional claims of its riparian bounding states, the smell of oil permeates the council chambers of governments; but the national prides of all riparians, old like China or new like the Philippines, also seem at stake in the heat of international disputes over the islets and waters of the sea (see the paper

by Choon-ho Park). In the eastern China seas similar issues prevail, muted somewhat by the greater sophistication in international affairs of the antagonists, though exacerbated by the importance of the military dimension (see Bruce Larkin's essay). In the western Pacific too great blocks of oceanic territory are coming under claim, largely under the initiative of the newly independent insular and archipelagic states, but, ironically, the chief benefactor of this process appears rather to be the United States which, though long an opponent of the EEZ, will, through its Micronesian possessions, obtain control over up to 4 million square miles of ocean space. The implications of these developments are by no means clear, but they certainly will have major economic (as, e.g., regarding seabed mining) and strategic consequences, not least the intensification of territorial conflicts wherever they might occur and the need to resolve them (see J. R. V. Prescott's discussion in this volume).

6. The expansion of maritime jurisdictional claims coincides with the militarization of the oceans, which in turn is associated with the need for surveillance of the vast expanses of ocean that are now coming under national jurisdiction (see the articles contributed by staff members of SIPRI in this issue). The most obvious development in recent years has been the enormous expansion of the nuclear-powered submarine fleets of the Great Powers and their immediate allies. This proliferation has cast the confrontation of the United States and the Soviet Union in global terms markedly different from those of, say, 20 years ago. One obvious consequence of their increasing reliance on nuclear-powered submarines has been the elevation of the international straits issue to a new level of priority in international deliberations. At the same time, the promulgation of EEZs and fisheries zones of similar size has been associated with the extremely rapid development of so-called light naval forces, composed chiefly of highly sophisticated Fast Patrol Boats and others at the frigate level and smaller. Parenthetically, these units are highly efficient deterrents of the larger naval vessels of the major powers' navies at the local scale and help even out, at least for purposes of defense, the inequalities between the great navies and the smaller ones, such as those of China and Indonesia. Ironically, they also materially contribute to the export trades of the wealthier countries vis-à-vis the poorer ones, for most such vessels come from the naval shipyards of the developed states.

7. As was noted in volume 1, the extension of national jurisdictions, by a curious dialectical process, increases the need for international cooperation; witness the South China Sea once more. What had been left unmanaged in the past must now be managed, and purely national management is in many cases frustrated by the overlapping of political and ecological boundaries and by the high costs of exploitative and managerial technologies and infrastructures, which are beyond the means of most small, poor, newly independent countries. These factors should encourage regional cooperation and the development of a new international economic order in the oceans based on that cooperation, but this healthy development is slowed down by the widespread divergence of

interests between countries within given regions and, strikingly in some cases, by the differently perceived views and interests of the very poor countries and the less poor ones in those regions. As the "poorest of the poor" fall behind their neighbors, a "development gap" thus appears to be widening among the developing countries themselves (see the article authored by George Kent).

In spite of these trends, the regionalization of marine affairs and the resolution of regional differences in national ocean policies under the evolving Law of the Sea continue to occupy an unfortunately low priority in the established scheme of things, even though the Informal Composite Negotiating Text frequently refers to "appropriate regional bodies" as the means of dealing with many difficult issues that cannot be resolved on a global scale. Already existing international associations of countries (e.g., the Association of Southeast Asian Nations and the Organization of American States) have shown little inclination, let alone capability, to assume responsibility for management decisions and implementation relating to the oceans at the regional level.

Nevertheless, a marine-centered, ecologically oriented regionalism, complementing and overlapping with the more traditional, land-based political and economic regionalism, is in the making. African scientists, legal experts, and political leaders, gathered in December 1978 in Yaoundé, Cameroon, for Pacem in Maribus IX, gave expression to this trend, stressing the need to articulate the emerging law of the sea in regional terms, to adapt it to regional situations, and to develop it for the benefit of the people in specified localized circumstances (see the report of Pacem in Maribus IX).

Remarkable, in this connection, is the sea-centered regional network emerging from the developments fostered by UNEP in cooperation with a multitude of national governments and intergovernmental and nongovernmental organizations (see the report by Peter Thacher and Nikki Meith). The Regional Seas Program of UNEP is exemplary in that it begins to build an institutional infrastructure for the implementation of Parts XII, XIII, and XIV of the proposed Law of the Sea Convention (see pp. 646–60 in *Ocean Yearbook 1*) long before it comes into force, or even is agreed upon. With all the difficulties it is necessarily encountering, UNEP's Regional Strategy is a successful example for the interweaving of international and national activities, sea based, land based, and outer space based. It demonstrates the need for integrating the management of traditional uses of the sea (fishing, navigation, mining, and tourism) with new and "unconventional" ones. Raising its sights far above the horizon of the Third UN Conference on the Law of the Sea, UNEP's Blue Plan for the Mediterranean constitutes an important beginning in investigating the international regional requirements of mariculture and in advancing its development at the national and international level; and mariculture, it can safely be predicted, is going to play a rapidly increasing role in the world fisheries strategy. UNEP's Regional Program—far ahead of UNCLOS—is turning its attention to the new "alternative" technologies to produce renewable energy from the seas and oceans, such as Ocean Thermal Energy Conversion

(OTEC). The program is remarkable also in that it stresses the importance of water management in a comprehensive sense, including both seawater and freshwater in their multiple uses. In this emphasis, too, UNEP's program extends beyond both the UN Conference on Water (Rio de la Plata, 1977), which ignores seawater, and UNCLOS, the scope of which does not include freshwater. Water management as an integrated concept, IOI research has stressed on other occasions,[3] could instead be a basic instrument for the integration of land-based and sea-based regional economic planning—in other words, for the integration of marine resources and ocean management into national and regional development strategies. Here, then, is a challenge for the remaining years of this century: integrated water management, a tool essential for development, beneficial for the further evolution of the law of the sea, and conducive to the building of a New International Economic Order.

8. Although national claims, at the Law of the Sea Conference and unilaterally, continue to erode the concept of the Common Heritage of Mankind in a territorial sense, the concept shows resilience in other respects. In this volume a case is made for including archeological treasures on the continental shelves and in international waters in the category of common-heritage resources (see the article by George Bass). A wreck, it is argued, does not necessarily belong to the country in whose waters it lies. There are at present no international agreements to cover ancient sites discovered on the outer continental shelf or in international waters. At present, finder is keeper. The ICNT contains an article (Art. 149) providing that "all objects of an archeological and historical nature found in the Area shall be preserved or disposed of for the benefit of the international community as a whole, particular regard being paid to the preferential rights of the State or country of origin, or the State of cultural origin, or the State of historical and archeological origin." Thus, archeological and historical objects, the "common cultural heritage" of many nations, appear to be included in the category of nonliving resources of the international area which are the Common Heritage of Mankind. Considering, however, the uncertain status of the boundaries of the international area,[4] many questions remain unsolved.

9. Groping steps toward expanding the idea of the Common Heritage of Mankind have also been taken in two other directions. Considering the inextricable links between resource management and technology—resources become identifiable, meaningful, and exploitable only as technologies become available—it is understandable that developing countries are beginning to press for the inclusion of technological knowledge in the category of common

3. Arvid Pardo and Elisabeth Mann Borgese, "Marine Resources, Ocean Management, and International Development Strategy for the Eighties and Beyond," IOI occasional paper no. 7, Malta, 1980 (in abbreviated form, in *IFDA Dossier,* November 1979).

4. Ibid.

heritage. The Arusha Symposium of African States (January 30–February 4, 1978), in preparation for UNCSTD, proposed that "it must be accepted universally, for a start, that technological knowledge is the Common Heritage of Mankind." No action was taken on this proposal at UNCSTD, but patent laws and the ownership of intellectual property are in general disarray, and a new order for science and technology for development, including marine science and technology, may have to be based in some ways on the common-heritage principle. At the same time, developing countries in the Group of 77 are beginning to direct their attention to Antarctica and its vast living and nonliving resources and to call for the application of a common-heritage regime to that area. Action in the United Nations General Assembly may be forthcoming.

In one respect, however, the concept has been consummated and enshrined in a consensus document of the United Nations. The 1979 conference of the UN Committee on the Peaceful Uses of Outer Space agreed that the resources of the moon and other celestial bodies are a Common Heritage of Mankind, subject to international management and benefit sharing, just as the resources of the deep seabed are professed to be. The economic exploitation of the resources of the moon and other celestial bodies may not take place for decades; but in the meantime, the principle of common heritage could be usefully applied to resources produced or processed in outer space. Satellite-based factories, taking advantage of the lack of gravity (weightlessness) for the processing of certain materials (e.g., silicon, of growing importance in the expanding microelectronics industry), are already on the drawing boards. They could be placed under a common-heritage regime involving an international system of management. Idealism lives on!

10. A detailed analysis of the Informal Composite Negotiating Text (ICNT) and the ongoing work of UNCLOS was presented in volume 1.[5] In the meantime, a revised text (ICNT/Rev. 1) has been issued (for the most important changes, see Appendix B, pp. 458–515), and discussions on it have begun. The new text constitutes a significant technical improvement on the previous one, but it remains to be seen whether or not it can provide a basis for consensus. Since work is still in progress, and likely to be concluded next year, a detailed analysis of the revised text will be deferred to volume 3.

Still, viewing matters with detachment from the frustrations and intricacies of the continuing UNCLOS negotiations, we conclude on an optimistic note. The seed, sown by Malta in 1967, is about to bear fruit. The great congress of nations, called in the wake of Malta's initiative to determine the future of two-thirds of the earth, has indeed begun to transform the structure of the world order and the international relations that inform it. Traditional concepts of sovereignty and property will, it is certain, never be the same.

5. Arvid Pardo, "The Evolving Law of the Sea: A Critique of the Informal Composite Negotiating Text (1977)," *Ocean Yearbook 1,* ed. Elisabeth Mann Borgese and Norton Ginsburg (Chicago: University of Chicago Press, 1978), pp. 9–34.

Living Resources

Marine Fisheries[1]

S. J. Holt
Food and Agriculture Organization, Rome

C. Vanderbilt
International Ocean Institute, Malta

THE DATA BASE

The chapter on marine fisheries in *Ocean Yearbook 1* surveyed the results of 25 years of marine fisheries development,[2] mainly on the basis of data compiled and published by the Food and Agriculture Organization of the United Nations (FAO) up to and including the year 1975. Some provisional fisheries statistics for 1976 also were taken into account. In that chapter, on page 39, 1976 was viewed as "the last year in which the greater part of fisheries throughout the world were conducted under a legal regime characterized by the principle of freedom of fishing on the high seas," and "a baseline against which to compare future developments under another regime, involving new international arrangements and perhaps new technologies."

At the time of writing, data for 1976 are the most recent published by FAO, in its *Yearbook of Fishery Statistics,* volumes 42 and 43 (1976). Incomplete data are available for 1977, so some appraisal of recent events can be attempted, but missing data include those for Japan and the USSR, the two countries with the biggest marine catches; global conclusions will therefore be tentative. The recent data submissions by countries also include some corrections for 1976 and earlier years, and these have here been taken into account.

In 1977 Holt also prepared a review which included a detailed analysis of

1. Opinions expressed in this review are those of the authors and do not necessarily represent policies of the organizations by which they are employed. We wish to acknowledge particular debts for helpful suggestions and constructive criticism, particularly with respect to the section on nutrition, to Mr. H. Watzinger, Dr. G. L. Kesteven, Mr. G. Saetersdal, Prof. Claudia Carr, Mr. M. Robinson, and Ms. A. Crispoldi. Colleagues in the FAO Fisheries Department, particularly Mr. E. F. Akyüz and Mr. D. Gertenbach and their staffs, have, as always, gone to much trouble to make statistical and other data available to us promptly and to explain some of the pitfalls in interpreting them.

2. Sidney Holt, "Marine Fisheries," in *Ocean Yearbook 1,* ed. Elisabeth Mann Borgese and Norton Ginsburg (Chicago: University of Chicago Press, 1978), pp. 38–83.

© 1980 by The University of Chicago. 0-226-06603-7/80/0002$01.00

the marine contribution to world food supplies.[3] The nutritional and dietary basis for that review derived in large part from the fourth edition of the *Food Balance Sheets,* published by FAO in 1971 and containing information from 132 countries for the period 1964–66. In 1977 FAO published *Provisional Food Balance Sheets* for the years 1972–74 based on information from 162 countries; the more recent data are used in the present review.

The most recent data published on the monetary values of marine catches are for 1973, after which collection ceased. In *Ocean Yearbook 1* the base year 1971 was adopted for discussion of values of catches, as at that time the 1973 data had not been fully analyzed. The FAO will resume as of 1978 the collection of data on landed values, and we have accordingly decided to postpone further detailed consideration of the subject until *Ocean Yearbook 3* is prepared. However, a general appraisal of the 1973 data is given below.

In *Ocean Yearbook 1* catches of whales were included in world totals, since these were a substantial part of the world catch by weight until the early 1960s. They now contribute a tiny and still declining fraction of the catch, and the greater part of their contribution is to nonedible products since the sperm whale became the species most sought after. Whales have therefore been excluded from the tables in the present review.

PRODUCTION

By Weight

The total harvest of marine fishes, shellfish, and seaweeds fell from 1974 to 1975, as it had once previously done from 1971 to 1972. Whereas this latter fall was due to a collapse of the stocks of anchoveta (*Engraulis ringens*) fished by Peru and Chile, the more recent decline could not be attributed only to a repetition of that cause. Revised data show the fall in production from 1974 to 1975 (excluding anchoveta) was less than originally thought. Nevertheless, the previous continuous growth had faltered. In 1976 the general growth rate of recent years was resumed, and both total catches and catches excluding anchoveta reached their highest levels ever (table 1). Present indications are that catches excluding anchoveta continued to increase in 1977.[4] This seems to have happened notwithstanding some expressions of fears that disruptions to global fishing patterns resulting from widespread declarations of national jurisdiction

3. S. J. Holt, "A Contribution to Discussion of a New Economic Order, with Reference to the Living Resources of the Ocean," offset (Rome: FAO, 1977).

4. Provisional 1977 total (including inland) as of October 10, 1978, is 72.88 million metric tons (m.t.), compared with 73.96 for 1976 (revised total); but excluding anchoveta is 72.07 compared with 69.67 (3.4 percent increase). These figures include FAO "estimates"; if these are excluded, the increase rate becomes 4 percent.

TABLE 1.—WORLD MARINE CATCHES

	Fish, Shellfish, and Seaweeds				
		Excluding Anchoveta		Excluding Anchoveta and Capelin	
Year	Total m.t.	m.t.	Annual Increase (%)	m.t.	Annual Increase (%)
1970	61.4	48.3	7.4	46.8	6.2
1971	61.4	50.2	3.9	48.6	3.8
1972	56.8	52.0	3.6	50.2	3.3
1973	57.4	55.7	7.1	53.6	6.8
1974	61.0	57.1	2.5	55.2	3.0
1975	60.3	57.0	−.2	54.8	−.7
1976	64.1	59.8	4.9	56.4	2.9
1977	62.7	61.9	3.5	57.9	2.7

NOTE.—m.t. = millions of metric tons in all tables.

over 200–mile zones would lead to a significant, even possibly a substantial, fall in the world catch. At least one of the major fishing countries most likely to be affected by extensions of jurisdiction by others—Japan—maintained its marine catch; a provisional figure (available in June 1978) is 10.49 million metric tons (m.t.), compared with 10.42 m.t. in 1976. The USSR catch is, however, believed to have fallen.

In 1977 the second collapse within a decade of the anchoveta fishery was confirmed, with a regulated catch of only 0.81 m.t. compared with 4.30 m.t. in 1976.

Of the increase of 2.8 m.t. from 1975 to 1976, 1.2 million can be attributed to a continuing increase in catches of capelin (*Mallotus villosus*) in the North Atlantic (see table 2). This species is, like the anchoveta, taken almost exclusively for reduction to fish meal, and catches of it have been rising very rapidly over the past decade. This growth tends to obscure other events in the pattern of world sea fisheries, and it is useful also to exclude it from certain summaries, as in the last column of table 1. Similarly, of the increase of 2.1 m.t. from 1976 to 1977, 0.8 m.t. is accounted for by the increase in the capelin catch, which in the latter year exceeded 4 m.t.

A very few fish species form about one-quarter of the world catch, excluding anchoveta, and between one-quarter and one-third of the total marine catch, as shown in table 3. These are species catches each of which have at some time reached 1 m.t. With the sole exception of the anchoveta they are taken overwhelmingly by developed countries, and high proportions of most of them are reduced to meal and oil.

Table 4 shows the catches by economic groups of countries (cf. table 17 in *Ocean Yearbook 1,* p. 70). Short-term trends are distorted by the absence of recent data from China and the other "centrally planned developing" countries and by

TABLE 2.—FISHERIES FOR SPECIES OF WHICH ANNUAL CATCHES HAVE REACHED OR EXCEEDED 1 m.t.*

Species†	Area	Date Reached .5 m.t.	Peak m.t.‡	1976 Catch	Date Fell to ½ Peak	Date Fell to .5 m.t.
Anchoveta *(Engraulis ringens)*	S. E. Pacific	1957	13.1 (1970)	4.3	1972	1977
Alaska pollack *(Theragra chalcogramma)*	N. Pacific	1960	5.0 (1975)	5.1	...	...
Herring *(Clupea harengus)*	N. Atlantic	Before 1910	4.1 (1965)	1.2	1972	1977?
Cod *(Gadus morhua)*	N. Atlantic	1910 or earlier	3.7 (1969)	2.4	1973	...
Capelin *(Mallotus villosus)*	N. Atlantic	1966	2.2 (1975)	3.4	...	...
Blue whale *(Balaenoptera musculus)*	World§	1927	1.8 (1933)	.0	1938	1952
Fin whale *(B. physalus)*	World§	1926	1.5 (1938)	<.1	1965	1966
Mackerel *(Scomber scombrus)*	N. Atlantic	1966	1.1 (1975)	1.1	...	...
Japanese pilchard *(Sardinops melanosticta)*	N. Pacific	1975	1.1 (1976)	1.1	...	...
Sprat *(Sprattus sprattus)*	N. Atlantic	1973	1.0 (1975)	.9	...	...

*An update of table 12 in S. J. Holt, "A Contribution to Discussion of a New Economic Order, with Reference to the Living Resources of the Ocean," offset (Rome: FAO, 1977).

†Species are arranged in descending order of peak catches, by weight. The order by an index of total value would be quite different, with the list headed by cod and the two baleen whales, followed by herring and Alaska pollack, then mackerel, capelin, and the other clupeoids at the bottom.

‡Date in parentheses.

§Mainly Southern Ocean.

TABLE 3.—CONTRIBUTION OF THE FISH SPECIES LISTED IN TABLE 2 TO WORLD CATCHES, BY WEIGHT (m.t.)

Year	Ancho-veta	Alaska Pollack	Herring	Cod	Capelin	Mack-erel	Japanese Pilchard	Sprat	Total of 8 Species % of World Catch
1970	13.1	3.1	2.3	3.1	1.5	.7	...	.2	39
1971	11.2	3.6	2.1	2.9	1.6	.8	.1	.3	37
1972	4.8	4.2	1.9	2.7	2.0	.8	.1	.3	30
1973	1.7	4.6	2.0	2.5	2.1	1.0	.3	.5	26
1974	4.0	4.9	1.6	2.8	1.9	1.0	.4	.7	29
1975	3.3	5.0	1.5	2.4	2.2	1.1	.5	1.0	29
1076	4.3	5.1	1.2	2.4	3.4	1.1	1.1	.9	31
1977	.8	...	...	...	...	...	...	...	...

the adoption by FAO of a constant total for this group of 3.96 m.t. for each of the past 5 yr (see table 4,†). In table 5, therefore, this economic group has been omitted from totals and percentages and in calculations of rates of change. Table 5 illustrates the fact that apart from the faltering in 1975 the fisheries of both market economy groups increased over the period. In the developed centrally planned group the growth was continuous. In both developed groups

TABLE 4.—1976 MARINE CATCHES, BY ECONOMIC GROUPINGS OF COUNTRIES

	m.t.	% of Total Catch	% 1974 Human Population*
Developed m.e.	27.1	43.5	19.3
Developing m.e.	20.6	33.1	48.8
Total m.e.	47.7	76.6	68.1
Developed c.p.e.	10.6	17.0	9.3
Developing c.p.e.	4.0†	6.4	22.6
Total c.p.e.	14.6	23.4	31.9
Developed countries	37.7	60.5	28.6
Developing countries	24.6	39.5	71.4
Total world	62.2‡	100.0	100.0

NOTE.—m.e. = market economies; c.p.e. = centrally planned economies on all tables.

*3,875 million in 1974, about 4,000 million in 1976.

†This value is an FAO estimate unchanged since 1973. It is now presumably an underestimate. It affects all sums and calculated growth rates in which it occurs.

‡The discrepancy between this total and the value 64.1 in table 1 is due to a small percentage of catches not being allocated to economic class in FAO basic data.

TABLE 5.—WORLD MARINE CATCHES, BY ECONOMIC GROUPINGS OF COUNTRIES

	Catches (m.t.)				Rate of Annual Increase (%)		
	1973	1974	1975	1976	1973–74	1974–75	1975–76
A. All Marine Species							
Market economies	43.0 (83.0)	45.8 (82.6)	44.2 (81.0)	47.7 (81.8)	6.59	−3.53	7.86
Developed	26.4 (50.9)	26.3 (47.3)	25.6 (46.9)	27.1 (46.4)	−.54	−2.59	5.79
Developing	16.6 (32.0)	19.6 (35.2)	18.6 (34.1)	20.6 (35.4)	17.85	−4.79	11.29
Centrally planned economies, developed	8.8 (17.0)	9.7 (17.4)	10.4 (19.1)	10.6 (18.2)	9.40	7.30	2.13
Developed countries	35.2 (68.0)	35.9 (64.8)	36.0 (65.9)	37.7 (64.6)	1.95	.07	4.73
All countries	51.8 (100)	55.5 (100)	54.6 (100)	58.3 (100)	7.04	−1.64	6.77
B. Excluding Anchoveta and Capelin							
Market economies	39.5 (82.2)	40.3 (81.3)	39.4 (80.3)	40.9 (80.9)	2.08	−2.39	4.00
Developed	24.6 (51.2)	24.7 (49.9)	24.1 (49.1)	24.6 (46.4)	.50	−2.72	2.25
Developing	14.9 (31.0)	15.6 (31.4)	15.3 (31.2)	16.3 (32.3)	4.70	−1.85	6.75
Centrally planned economies, developed	8.6 (17.8)	9.3 (18.7)	9.6 (19.7)	9.7 (19.1)	8.32	3.83	.46
Developed countries	33.2 (69.0)	34.0 (68.6)	33.7 (68.8)	34.3 (67.7)	2.52	−.93	1.74
All countries	48.1 (100)	49.6 (100)	49.0 (100)	50.6 (100)	3.20	−1.22	3.31

NOTE.—Percent of total shown in parentheses.

much of the growth was due to increases in capelin catches, and in the developing market economy group to the beginning of recovery of the anchoveta fishery. The small relative changes in the relative contributions by these country groups to the total are illustrated, which brings out the slight but rather steady improvement in the performance by the developing countries. The last three columns in this series show the variability of the annual rate of growth but indicate also that the average over the 4-yr period, excluding anchoveta and capelin, is about 1.7–1.8 percent per annum, which is below the rate of human population increase, then 1.9 percent annually but now possibly somewhat slower (about 1.7 percent).

In considering trends in catches by the developed countries, account must also be taken of the special importance of two gadoid species—Alaska pollack and cod. Catches of these species are very large but have not been changing very much over recent years, although the distribution of the catch of pollack between groups of countries has changed substantially (table 6). Such changes obscure other features of the global fisheries pattern. Thus the average growth rate over the 4-yr period achieved by the group of developed centrally planned countries of 4.2 percent annually (excluding capelin) is reduced to 1.7 percent if pollack is excluded. Likewise, the growth rates by the developed market economies are particularly affected by the decline of the Japanese catch of Alaska pollack and of European catches of Atlantic herring.

Total changes reported in FAO *Yearbooks of Fishery Statistics* include "estimates" of the catches by countries that do not report in time. Each such estimate is, in fact, nearly always a repetition of the previous year's catch, and that may in turn be a figure repeated from an earlier year. Thus any rate of change calculated from successive annual totals contains the assumption that countries which did not report did not change. The effect of this is to dampen the estimated degree of change, whether an increase or a reduction was observed. In the 1976 FAO *Yearbook,* 115 out of 227 "country" entries are such estimates of

TABLE 6.—RECENT CATCHES OF ALASKA POLLACK, ATLANTIC COD, AND ATLANTIC HERRING (m.t.)

	1973	1974	1975	1976
Alaska pollack:				
Developed m.e. (Japan)	3.02	2.86	2.68	2.45
Developing m.e. (Korea Rep.)	.26	.30	.39	.45
Developed c.p.e. (USSR)	1.34	1.76	1.96	2.09
Atlantic cod:				
Developed m.e.	1.86	1.94	1.75	1.78
Developed c.p.e.	.68	.87	.68	.60
Atlantic herring:				
Developed m.e.	1.48	1.13	1.09	.87
Developed c.p.e.	.50	.44	.43	.31

nonreported figures which are repeated from 1975 to 1976. These accounted for 14.6 m.t. of the world total (marine plus inland) of 73.5 m.t. given for 1976 and 15.4 m.t. of 69.9 m.t. reported for 1975 (i.e., 21 percent of the total). Similarly, 67 countries did not report in both 1974 and 1975, and they accounted for 20 percent of the catches in those years. These are not negligible proportions, and we have therefore examined the consequences with respect to the estimates of the rates of change in marine catches. The revised estimates, based only on the catches of reporting countries, are a 1.5 percent decrease (instead of 1.3 percent) from 1974 to 1975 and 7.7 percent increase (instead of 6.1 percent) from 1975 to 1976. These revisions, if applied to the total catches, contain, of course, the assumption that the catches by nonreporting countries changed from one year to the next by the same percentage as those by the reporting countries; such an assumption is perhaps slightly more realistic than an assumption of no change. This is borne out by the fact that when, in subsequent FAO *Yearbooks*, estimates are replaced by late arrived figures for some countries these data, grouped, tend to follow the world trend. Applying this assumption to the data since 1970 and assuming the 1970 global figure of 48.0 m.t. (excluding anchoveta) is correct, we arrive at an estimate of the 1976 marine catch of about 62 m.t. instead of 58.8. This represents an average annual growth rate of 4.3 percent over the 6-yr period.

By Value

As mentioned earlier, the most recent data available refer to 1973 catches. Table 7D of the Appendices to this volume lists the 51 countries whose marine catches reached 100,000 metric tons in 1973 or soon thereafter; they are ranked in the order of the 1973 catches. The data on landed value (first sale) per metric ton caught are obtained by dividing the total tonnage by the total reported value where both are given by FAO (20 countries). These data refer to marine *plus inland* catches. Where data were incomplete in 1973, values per ton have been extrapolated from previous years or estimated by comparing countries with similar economic structure, with similar fishing industries, and located in the same region, following a study by Alan Marriot. From this series have been calculated the values of marine plus inland catches, totaling US$19,300 million. The marine catches by these countries totalled 85.8 percent of their marine plus inland catches, so a rough estimate of the value of their marine catches is US$16,600 million. Their marine catches were 94.2 percent of the weight of the world marine catch in 1973, so a rough estimate of the value of the total marine catch is US$17,600 million. This estimate is somewhat higher than figures given in *Ocean Yearbook 1* (US$11,000 million in 1971 and US$15,000 million in 1974); it may, however, be exaggerated by not having taken account of the fact that, more often than not, the value per ton of inland fish caught in a country is higher than that of the marine fish.

This table also shows the value of aquatic catches per capita population and as a percentage of the Gross National Product (GNP) of each country. It is well known that this latter figure is not very reliable, and the percentages should therefore be regarded only as giving a broad indication of the relative contributions of fishing to the national economies.

The 1973 data for the 51 countries whose aquatic catches exceeded 100,000 metric tons permit an appraisal of relative values of fisheries by economic groups. Table 7 shows the breakdown of catch weights and values and comparable human population and GNP data.

By Species Groups

In table 5 of *Ocean Yearbook 1* an analysis was given of changes in catch composition over a 25-yr period. Obviously, major changes cannot take place in the short space of 1 or 2 yr except those brought about by the growth or decline of certain major fisheries, such as for anchoveta or capelin. This latter is in fact the cause of the recent increase, in an absolute and relative sense, in catches of "jacks, mullets, sauries, etc." (Appendix table 8D), into which category FAO has

TABLE 7.—CATCH VALUES BY ECONOMIC GROUPINGS OF COUNTRIES
(1973, Percentages of Totals)

	Catch by			
	Weight	Value	Population	GNP
Developed m.e.	42.5	40.8	20.8	69.4
North America	6.0	6.2	7.1	39.2
Western Europe	17.3	16.2	9.2	21.1
Oceania and other	19.2	18.4	4.5	9.1
Developing m.e.	28.5	22.8	43.5	10.0
Africa	4.1	2.2	4.4	.7
Latin America	7.5	3.3	7.1	4.5
Asia and Far East	16.0	16.4	32.6	4.4
Near East and other	.9	.7	.4	.5
Total m.e.	71.0	63.6	64.3	79.4
Developed c.p.e.	15.3	17.4	9.5	15.7
Developing c.p.e.	13.7	19.0	26.2	4.9
Total c.p.e.	28.9	36.4	35.7	20.6
Developed countries	57.8	58.2	30.3	85.1
Developing countries	42.2	41.8	69.7	14.9
Total world	100.0	100.0	100.0	100.0

now decided to include capelin instead of with the mainly diadromous "salmons, trouts, and smelts, etc."

By Major Fishing Areas

Appendix table 2D shows the recent trends in distribution of catches by ocean regions. All ocean totals show declines from 1974 to 1975 and increases from 1975 to 1976, but some areas *within* oceans have divergent trends—for example while the catch from the Northeast Atlantic has been increasing continuously, that from the Northwest Atlantic has been declining.

DISPOSITION OF CATCHES

In appraising the contribution of marine fish and shellfish catches to human nutrition, a primary consideration is the extent to which catches are eaten fresh, processed for direct human consumption, and reduced for the production of meal to be used in livestock feeds and for oil, most of which enters human diets as a variety of fatty and oily products. The proportion of the world catch which is reduced to meal and oil increased continuously from 14 percent in 1948 to 42.4 percent in 1970. This latter was the peak year of the anchoveta catch, practically all of which was, and still is, reduced. Thereafter the proportion declined somewhat, though irregularly (table 8, col. 1). This decline has been claimed as a result of efforts made throughout the world to better utilize catches

TABLE 8.—MARINE CATCH OF FISHES, SHELLFISH, AND SEAWEEDS REDUCED TO MEAL AND OIL (%)*

		All Species Except		
Year	All Species (1)	Anchoveta (2)	Anchoveta and Capelin (3)	All Fishes Except Anchoveta and Capelin (4)
1965	36.0	23.1	23.0	26.3
1970	42.4	28.0	25.6	29.3
1971	41.2	28.7	26.1	30.2
1972	35.0	30.4	26.7	30.5
1973	31.7	30.3	27.2	31.0
1974	34.6	30.1	28.0	31.9
1975	34.6	31.3	28.4	33.1
1976	35.5	31.1	26.8	30.7
1977	32.8	31.9	27.2	31.3

*Includes small quantities used for miscellaneous purposes other than human consumption.

for direct human consumption and to give priority to stimulation of fishing specifically for human consumption, and especially by developing countries. Unfortunately, the statistics do not support this claim; the overall declining tendency in recent years is eliminated if the anchoveta is excluded from the data, and in fact a contrary trend becomes evident (col. 2). Only in 1976 is the trend reversed, and it continues again in 1977. The proportions reduced in later years are substantially affected also by the capelin catch. If this is also excluded (col. 3) we see that the trend reversal in 1976 is more marked, but in 1977 the proportion reduced is back again to the 1973 level.

It is not easy to identify the reasons for the trend reversals. It could be that certain industries based on major species which are used partly for consumption and partly for reduction, in changing proportions, have altered.

The trend of increasing utilization of catches for meal and oil production has occurred in three of the four economic categories of countries for which some data are available (table 9).

In the mode of processing the catches which are utilized for direct human consumption the previous trends or increase in the percentage of the catch which is frozen and a compensatory decline in the percentage marketed fresh have been reversed in 1977. The percentage cured continues gradually to decline. The figures given in Appendix table 3D are, however, for both marine and inland catches; the data base does not permit separation of the two categories.

In *Ocean Yearbook 1* attention was given mainly to sea fish and marine products as contributions to protein supplies, directly or indirectly. However, the sea as a significant source of edible oil, and hence as assimilable and storable calories in human diets, should not be ignored. In 1950 the catches of baleen whales, from which 0.4 m.t. of oil were produced, accounted for 60 percent of all edible oils of marine origin. By 1975 baleen-oil production was down to less than one-tenth of this figure and accounted for only 3 percent of edible marine oils. Apart from the changing contribution of whale oil, most edible marine oil now comes from the bodies of oily fish species which are taken primarily for reduction to meal (table 10). These fishes, excluding anchoveta, yield from 0.23 to 0.30 metric tons of oil for every ton of meal. Anchoveta oil yield is variable

TABLE 9.—MARINE CATCH (excluding Anchoveta) REDUCED TO MEAL AND OIL (%)

	Market Economies		Centrally Planned Economies
Year	Developed	Developing	Developed
1961–63	27	10	6
1975	40	22	14

TABLE 10.—FISH OIL AND MEAL PRODUCTION (Thousand Metric Tons)

	Body oils		Meals		
Year	Anchoveta	Other Species	White Fish	Anchoveta	Other Oily Fishes
1960	51	340	222	581	1,178
1965	135	593	305	1,355	1,952
1970	323	647	461	2,397	2,606
1971	446	649	476	2,124	2,732
1972	227	640	482	964	2,809
1973	41	725	527	401	3,016
1974	212	748	496	914	3,020
1975	213	757	478	732	3,241
1976	109	833	493	921	3,432
1977	106	879	495	182	3,733

but generally lower, in the range of 0.10–0.24 tons oil per ton of meal. Excluding anchoveta meal, from 85 to 90 percent of total fish meal is made from oily fishes and the rest from "white" fishes and offal, with a slight trend of increase in the proportion from oily fish over the last 2 decades.

The total export value of fish body oils relative to the value of fish meal is generally about the same as the yield ratio because, although there are considerable and apparently independent year-to-year fluctuations in prices, the average price received for oil and for meal from oily fishes is about the same (e.g., about US$350 per metric ton in 1973–75).

Production of and trade in fish oil are dominated by a very few countries in each economic group and in each region. In the period 1972–74 the percentages of oil production were as follows: African countries, 8.7; North America, 11.9; South America, 18.2; Asia, 16.1; Europe, 34.0; Oceania, 0.6; and USSR, 15.1. However, of the African production 72 percent was by South Africa and Namibia; of North American, 82 percent by the United States; of South American, 89 percent by Peru (which accounted for 86 percent of *all* developing country production); of Asian, 99 percent by Japan; of European, 55 percent by Norway and a further 30 percent by three other Nordic countries. Likewise, with respect to fish-oil trade, imports to European countries accounted for 92 percent of all recorded imports, and of this 21 percent was into the United Kingdom, 25 percent into the Federal Republic of Germany, and 19 percent into the Netherlands. In all, between 55 and 60 percent of marine-oil production enters international trade. Much of this trade is among developed countries with market economies, but a substantial amount is between that group and the countries with developing market economies, with a *net* flow from the latter to the former group.

MARINE PRODUCTS IN HUMAN NUTRITION

Data are not yet available to update the information on 1970 marine fish supplies and consumption given in *Ocean Yearbook 1* (summarized in its table 21, p. 75). Here we limit ourselves to a different approach, through the new *Food Balance Sheets.* Data from these for 1970 permit comparison with the above summary, and those for 1974 give the most up-to-date information and a basis for examination of trends. First, however, it may be useful briefly to review the general nutritional background to such studies.

The World Bank has recently presented evidence that 75 percent of the people in developing countries with market economies—that is, 1,100 million people—had, in 1965, calorie-deficient diets; 840 million had deficits exceeding 10 percent of the threshold level.[5] This situation existed despite the fact that there was no caloric deficit in the world as a whole and that in these developing countries the average deficit was not more than 7 percent, equivalent to only 4 percent of world production of cereals at that time. The calorie deficits which existed then, and which continued despite efforts to increase food production, occurred because of inequities in distribution between developed and developing countries; between developing countries themselves, even within each major region; and especially within countries, between the relatively rich and relatively poor people. By 1970 the overall "calorie gap" in the developing countries had been reduced to 4 percent, but as a result of a sequence of climatically caused setbacks the average situation had deteriorated again by 1975.

As far as we know, more recent data have not yet been analyzed, but the World Bank study gave projections to 1975 (and also to 1990). Calorie deficits per capita are highly correlated with income; how much people in poor countries eat is determined by what they can afford. For these projections alternative growth rates of per capita income were assumed, but no change in the pattern of income distribution. A further assumption, of doubtful validity, was that population growth rates of income groups in each country would be the same as in the country as a whole. The projections were not, however, very sensitive to this assumption over the period of extrapolation. The observed increase in per capita calorie consumption in the period 1965–70 roughly corresponded with the highest of the three income growth rates assumed in the World Bank study. This gave the most optimistic predictions of the present and future extent of malnutrition, and we here look at the consequences of these. They indicate that the *average* calorie deficit in the developing countries with market economies would have been 2.3 percent in 1975 and would have been eliminated by 1990 (in fact a 5 percent *surplus* was predicted by that date). When the slowest of the

5. S. Reutlinger and M. Selowsky, "Malnutrition and Poverty: Magnitude and Policy Options," Occasional Paper, World Bank staff, 1976.

three average income growth rates is assumed, the deficits predicted were 5.1 and 2.3 percent, respectively. The conclusions from these projections were that "the absolute number of people suffering from undernutrition would be higher, but their proportion of the total world population would decline." However, the percentage per capita deficit *among the undernourished* (i.e., among those income groups with a deficit) would decline only from 18.4 percent in 1965 to between 18.1 and 13.5 percent by 1975 and to between 18.4 and 5.1 percent by 1990, according to the slow and fast income-increase assumptions. Were income distribution to remain unchanged, therefore, "extremely high rates of growth in the demand for, and supply of, food would be required to achieve *per capita* growth rates of food consumption that would eliminate calorie deficits in the lowest income classes."[6] It should be said that all the conclusions cited so far are rather insensitive to the precise levels taken as average calorie requirements per capita in various regions.

Protein

Until about 1970 the theory providing the focus for international action with respect to food and agriculture was that there was a serious shortfall of protein consumption, especially in the developing countries. It has since become clear, however, that protein deficiency is for the most part the indirect result of inadequate *energy* intake, which results in the body catabolizing some of that protein which is eaten. It has been affirmed that a diet in which 5 percent of the calories come from good-quality protein would practically always satisfy the individual's protein needs, whether he be a young child or an adult, provided that his total energy intake meets requirements.[7]

Supplies of total protein in the developing countries have exceeded requirements in every developing region (table 11); for the developing market economies as a whole the percentage of excess was 88 in 1965, 90 in 1970, and 87 in 1974.

There are no analyses for protein deficiency comparable with the World Bank study of calorie deficiency, but Sukhatme, following a similar approach, concluded that of the population of developing countries a slightly higher proportion failed in 1970 to satisfy their protein needs for health than failed to satisfy their calorie needs. However, very few of the people who did not satisfy their protein needs satisfied their calorie needs, and the proportion of people consuming inadequate protein in absolute sense (i.e., whose diet would still be

6. Ibid.

7. P. V. Sukhatme, "Human Calorie and Protein Needs and How Far They Are Satisfied Today," in *Resources and Population: Proceedings,* Eugenics Society, 9th Symposium, ed. B. Benjamin et al. (London: Academic Press, 1973), pp. 25–43.

TABLE 11.—DAILY PER CAPITA FOOD REQUIREMENTS AND SUPPLIES

	% of 1974 World Population	Per Capita Requirements*		Supplies as % of Requirements							
				Calories				Protein			
		Calories	Protein (g)	1962	1965	1970	1974	1962	1965	1970	1974
All developing market economies	48.7	2,287	29.5	93	96	97	97	184	188	190	187
Asia	27.6	2,210	28.0	91	94	94	93	175	179	179	175
Middle East	4.7	2,450	35.0	93	96	98	100	188	190	192	196
Africa	8.3	2,350	32.0	88	90	91	90	162	165	168	164
Latin America	8.1	2,390	29.0	101	104	105	107	221	225	227	224
Developed regions	28.7	2,560	30.0	118	121	123	132	288	302	307	327
All regions	77.4	2,390	29.7	102	105	107	111	227	232	237	237

* The differences in *per capita* requirements between regions derive from consideration of human biomass per capita (varying with age structure of the population) and climatic differences. These *average* "requirements" therefore take into account, for example, the differing real requirements of children and adults.

protein deficient even if they consumed more nonprotein calories) was very small.

Sukhatme concluded with the observation that "it appears that people with high and rising incomes find it difficult not to eat more, especially tasty animal foods, when they can afford to do so."[8] The adverse nutritional and health consequences of this behavior in affluent societies have recently been brought vigorously to our attention by the World-Watch Institute pointing out that "undernutrition and overnutrition have similar consequences for the individual: reduced life expectancy, increased susceptibility to disease, and reduced productivity; and the number of people afflicted by the modern plague of overnutrition is approaching that suffering from undernutrition."[9] The authors of this study, and most reviewers of it, have tended to focus on the incidence of overnutrition in the richer countries, but the conclusions apply also to the smaller numbers of relatively affluent people in some of the poorer countries. It is ironic that the skewed distributions of income in most of the developing countries are leading, as average income rises, to increasing overnutrition toward the upper end of the range, while malnutrition at the lower end is only very slowly being reduced!

In 1974, only 21 percent of the protein consumed in the developing regions was of animal origin, as against 56 percent in the developed regions. Between 1962 and 1975 this proportion had increased slightly (from 20 and 49 percent, respectively). The total protein intake per capita, as well as that of the animal protein, had increased, from 53.5 to 54.2 g per day in the developing regions and from 90.8 to 98.1 g per day in the developed regions. Table 12 shows that *(a)*

TABLE 12.—DAILY PROTEIN CONSUMPTION PER CAPITA (% Annual Increase)

	1962–70		1970–74	
	Total	Animal	Total	Animal
Developed m.e.	.6	1.6	.3	.7
Developing m.e.	.3	.9	–.4	.0
All m.e.	.4	.9	–.2	.2
Developed c.p.e.	1.0	2.9	.7	2.6
Developing c.p.e.	1.4	1.9	.9	.4
All c.p.e.	1.3	2.2	.8	1.7
Developed countries	.7	2.1	.4	1.3
Developing countries	.5	1.2	–.1	.1
World	.5	1.2	.0	.4

8. Ibid.

9. E. Eckholm and F. Record, *The Two Faces of Malnutrition,* Worldwatch Papers (Washington, D.C.: Worldwatch Institute, 1976).

animal protein consumption per capita is continually increasing in all economic groups of countries except, recently, in the developing countries with market economies, in which the total protein consumption also has slightly declined; *(b)* for all groups and both periods, except for the developing centrally planned economies, the rate of increase in animal protein consumption exceeded that of total protein; *(c)* for all groups the rate of increase in both total and animal protein was less in the later period than the earlier period; *(d)* in all cases the rates of increase in developed countries were higher than in developing countries, except for the increase in total protein consumption in developing countries with centrally planned economies as compared with developed countries with centrally planned economies; *(e)* in all cases the rates of increase in countries with centrally planned economies were higher than in market economies, developed or developing, over the same period. Now, although the rate of increase everywhere has slowed, there is clearly no cause yet for immediate concern that total protein consumption, which in all areas exceeds physiological need on the average, will not keep pace with population growth, particularly with respect to animal protein.

At this point, it is worthwhile to look again at the differences between various groups of developing countries (table 11). Evidently average protein consumption relative to requirements, while in excess in all areas, is closely correlated with calorie consumption relative to requirements. The relative numbers of undernourished people in each area cannot, however, be deduced directly from the average relative calorie intakes because the skewness of income, and calorie consumption as a function of income differs among the different developing regions (table 13). Unfortunately, available data do not permit calculations of overall protein undernutrition by regions, but the incidence of the deficiency diseases, marasmus and kwashiorkor, is greatest in the region where a high proportion of the population has a calorie deficit and where the protein surplus is smallest—that is, in Africa. Even this interpretation has, however, become complicated by the fact that whereas these have been regarded as clinical manifestations of protein-calorie and protein deficiencies, respectively, it has recently been suggested that other facts in addition to diet,

TABLE 13.—STATISTICS OF UNDERNOURISHMENT (1965)

Region	Estimated % of Population with Calorie Consumption Less than Mean Level of Region	% of Population Undernourished (Calorie Deficit)
Asia and Far East	68	84–92
Middle East	56	66–71
Africa	71	75–84
Latin America	60	52–57

such as contamination of water supplies causing infections and parasitic infestations of the alimentary canal, also are important.[10]

Table 11, giving the standard daily per capita food requirements and estimated supplies, shows the considerable differences in the situation between regions and confirms that the food production on a global scale would meet the requirements for the present population if it were distributed equitably. However, the fact that everyone could be adequately fed if all food was distributed with the aim of meeting nutritional standards does not, of course, mean that increased production of food is not needed to solve the world food problems. There is a need to emphasize this point in particular with regard to fisheries precisely because protein supplies are *on the average* far in excess of the needs for this particular type of food. Food consumption, and particularly protein consumption, is not restricted to a nutrient motivation in man, but there are, as we all know, a number of underlying causes why people eat, what they eat, and how much they eat. It is unrealistic to imagine a world where people would restrict eating qualitatively and quantitatively to that which would meet the nutritional requirements. There is a strong correlation between income and diet, with greatly increasing consumption of animal protein with higher incomes. We conclude that, notwithstanding the apparent global adequacy of supplies, there is in fact a serious lack of food in the world today and that increased production both of calorie foods and protein is needed to alleviate the existing malnutrition. In addition comes the need to keep pace in food production with the inevitable population increase. When we consider the possibilities for increasing food production it is, however, important to relate these to their distributional effects and to take into account where in the world the increased amounts are primarily needed. The aim should be to increase self-sufficiency and self-reliance of countries and regions. Increased production which merely forms part of the industrialized countries' food industries and systems, and primarily only increases the availability of food to the already overfed part of the world or contributes further to the dependence of the less developed countries on the industrial world, should be given a very low priority in a global world food policy. We now look at some features of world fisheries against this background.

The nutritional value of fish is primarily in direct consumption as protein, whether fresh, frozen, or otherwise preserved, and in indirect protein consumption through conversion to fish meal which is then used as feed supplements for livestock, principally chickens, pigs and, recently, cultivated freshwater fishes such as trout. Fish, whether of freshwater or marine origin, has long been recognized as a superior source of animal protein. Taking the favorable amino-acid pattern into account, it has been estimated that 80–85

10. M. Muller, "New Evidence on Malnutrition," *New Scientist* (January 6, 1977), p. 4 (reporting work of the Dunn Nutritional Laboratory, Cambridge, discussed at a Symposium of the Royal Society of Tropical Medicine).

percent of the raw weight of fish can be utilized by human beings,[11] although protein-efficiency conversion expressed as a protein score (egg = 100) has been given as 75 percent for fish, compared with beef, 83 percent; pork, 36 percent; mutton, 94 percent; poultry, 95 percent.

Fish contributed directly 1 percent of the world calorie consumption but 4.4 percent of the protein consumption in 1974. There are, however, very considerable variations among categories of countries. In general the fish contribution to *animal* protein consumption is, in the developing countries, double that of the developed countries, but less as a fraction of the *total* protein.

The new *Food Balance Sheets* permit an evaluation of medium term trends in the direct contribution of fish and sea food to diet. Table 14 shows the general world change in the pattern of protein consumption. Thus fish are an increasing part of an increasing consumption. The increase in the percentage of total protein that is derived from animals, and of the total animal protein that is derived from fish, is found in practically all types of countries, but not to the same degree.

To appraise the role of *marine* products in nutrition it would be necessary to make two adjustments to table 14. First the "fish and sea food" category includes fish from inland waters, and the proportion of these in the totals of both production and consumption varies considerably among countries, although their proportions have not changed very much over time. Second, much of the fish production finds its way into consumption in the category of "animal protein other than fish"; the proportion varies between economic groupings

TABLE 14.—PROTEIN CONSUMPTION, BY ECONOMIC GROUPINGS OF COUNTRIES

	Animal as % of Total Protein			Fish and Sea Food as % of Animal Protein		
	1962	1970	1974	1962	1970	1974
Developed m.e.	53.5	57.9	58.8	12.3	12.0	12.3
Developing m.e.	19.8	20.7	21.0	15.1	18.4	19.3
All m.e.	30.7	31.8	31.8	14.2	16.5	17.3
Developed c.p.e.	39.8	46.2	49.8	12.8	13.2	14.2
Developing c.p.e.	19.2	19.4	19.0	25.7	33.3	33.6
All c.p.e.	28.4	30.4	31.6	17.2	20.8	21.2
Developed countries	49.0	54.1	55.9	12.5	12.4	13.0
Developing countries	19.6	20.3	20.4	18.7	23.3	23.8
World	32.6	34.2	34.8	14.5	16.2	16.7

NOTE.—World protein consumption in the respective years was 65.3, 68.6, and 68.7 g/day/capita. Fish and sea food contributed 4.7%, 5.5%, and 5.8% of the total protein consumed in these years.

11. F. W. Bell and E. R. Canterbury, *Aquaculture for the Developing Countries: A Feasibility Study* (Cambridge, Mass.: Ballinger, 1976).

TABLE 15.—CONTRIBUTION OF MARINE LIVING RESOURCES TO PROTEIN CONSUMPTION (1970)

Economic Category	Total Protein g/Capita/Day	% Marine Origin		% Protein of Marine Origin Consumed Directly
		of Total	of Animal	
Developed m.e.	94.0	9.0	15.8	75
Developing m.e.	55.0	3.4	16.6	85
All m.e.	66.6	5.7	15.9	79
Developed c.p.e.	101.5	5.8	12.6	90
Developing c.p.e.	60.4	2.7	13.9	98
All c.p.e.	72.6	4.1	13.3	92
Developed countries	96.4	8.0	15.0	79
Developing countries	56.8	3.2	16.0	87
World	68.6	5.2	15.1	80

both as to production and consumption and has increased over time. These adjustments can be made by updating and combining the information given in *Ocean Yearbook 1* (tables 21 and 22, pp. 75 and 77), taking 1970 as a common base date. In the earlier calculations it was supposed that 1 kg of fish reduced to meal contributed indirectly as much protein to human diets as 200 g of fish consumed directly. It now seems that the equivalent of 1 kg for reduction may be closer to 300–400 g of direct consumption; in the present calculations the equivalent 350 g has been assumed. The results are summarized in table 15. A remarkable feature of this table is the relative constancy, as between country types, of the figures for the percentage contribution of protein of marine origin to animal-protein consumption. It is necessary, however, to look beyond generalizations concerning groups of countries to the situations in particular ones, because there are large differences between countries *within* groups with respect to their degree of dependence on marine resources.

In table 16 are listed 88 countries and territories with populations over 100,000 selected from among all countries and territories listed in FAO statistical publications by the following criteria (applied to 1972–74 data):[12] (i) fish and sea foods contribute to diets 2 g or more protein per capita per day; and/or (ii) fish and sea foods contribute one-tenth or more of total protein and/or one-third or more of the animal protein in diets; and/or (iii) national marine fish production exceeds 200,000 metric tons annually; and/or (iv) the protein

12. An analysis of per capita consumption of sea food in each of 48 countries in which this contributed at least 1 g protein per day in 1970 was given in table 10 of Holt (n. 3 above). The new *Food Balance Sheets* show substantially different estimates of total, animal, and "fish and sea-food" (including fish caught in inland waters) consumption for 1970 as well as for recent years up to 1974. An updated analysis is now possible for 1972–74.

TABLE 16.—DIRECT CONTRIBUTION OF MARINE PRODUCTS TO PROTEIN CONSUMPTION COMPARED WITH PRODUCTION (1972–74 Average)

Country or Territory	Population (Millions)	Consumption: Total Protein (g/Capita/Day)	Consumption: Marine/Total Protein (%)	Consumption: Marine/Animal Protein (%)	Consumption: Marine Protein (g/Capita/Day)	Production: Marine Protein* (g/Capita/Day)	Production: Marine Catch (m.t.)
1. Maldives	.12	62.5 (52)	45 (1)	93 (1)	28.6 (1)	84 (4)	.034
2. Japan	108.3	85.5 (27)	25 (2)	52 (5)	21.0 (2)	26 (6)	10.43
3. Iceland	.2	113.7 (1)	12 (23)	15 (48)	14.1 (3)	1,111 (1)	.86
4. Portugal	8.7	93.6 (17)	15 (13)	37 (18)	14.1 (4)	14 (14)	.46
5. Singapore	2.2	74.8 (34)	18 (4)	36 (19)	13.6 (5)	2 (65)	.02
6. Hong Kong	4.1	78.6 (31)	15 (14)	27 (37)	12.1 (6)	9 (22)	.13
7. Granada	.1	56.6 (61)	20 (3)	42 (12)	11.5 (7)	5 (37)	.002
8. Fr. Polynesia	.12	70.5 (41)	16 (11)	34 (26)	11.4 (8)	6 (31)	.002
9. New Hebrides	.09	61.5 (54)	18 (5)	32 (27)	11.1 (9)	24 (8)	.008
10. Martinique	.35	72.3 (36)	15 (15)	29 (31)	11.1 (10)	3 (53)	.003
11. Brunei	.14	64.5 (48)	17 (6)	35 (22)	10.7 (11)	3 (54)	.002
12. Norway	4.0	91.9 (19)	12 (24)	18 (44)	10.6 (12)	203 (2)	2.94
13. Malaysia†	1.8	55.0 (65)	17 (7)	53 (6)	9.6 (13)	10 (19)	.06
14. Spain	34.6	91.1 (21)	10 (33)	21 (43)	9.5 (14)	12 (18)	1.53
15. Santa Lucia	.11	56.8 (59)	17 (8)	32 (28)	9.5 (15)	8 (23)	.002
16. Denmark	5.0	90.5 (22)	10 (34)	16 (47)	9.2 (16)	85 (3)	1.57
17. Guadaloupe	.34	71.6 (38)	13 (20)	29 (32)	9.2 (17)	4 (48)	.005
18. Korea, Rep.	32.6	72.0 (37)	13 (21)	69 (3)	9.0 (18)	14 (15)	1.68
19. Senegal	4.74	61.1 (55)	14 (16)	51 (7)	8.8 (19)	18 (11)	.31
20. Antigua	.07	53.5 (68)	16 (12)	27 (38)	8.6 (20)	3 (55)	.001
21. Ghana	9.3	51.7 (74)	2 (79)	6 (77)	8.6 (21)	6 (32)	.20
22. Jamaica	1.97	68.7 (44)	12 (25)	27 (39)	8.2 (22)	1 (78)	.01
23. Barbados	.24	79.7 (30)	10 (36)	18 (45)	8.2 (23)	5 (38)	.004
24. Macau	.28	56.8 (60)	14 (17)	26 (40)	8.1 (24)	10 (20)	.01

TABLE 16.—DIRECT CONTRIBUTION OF MARINE PRODUCTS TO PROTEIN CONSUMPTION COMPARED WITH PRODUCTION (1972–74 Average) (*Continued*)

Country or Territory	Population (Millions)	Consumption: Total Protein (g/Capita/Day)	Consumption: Marine/Total Protein (%)	Consumption: Marine/Animal Protein (%)	Consumption: Marine Protein (g/Capita/Day)	Production: Marine Protein* (g/Capita/Day)	Production: Marine Catch (m.t.)
25. Korea, Dem. Rep.	15.05	76.4 (33)	10 (35)	67 (4)	8.0 (25)	15 (12)	.80
26. Papua N.G.	2.59	47.7 (79)	17 (9)	43 (11)	7.9 (26)	4 (49)	.04
27. Philippines	41.6	45.5 (81)	17 (10)	46 (9)	7.7 (27)	7 (27)	1.13
28. Sweden	88.14	85.1 (28)	9 (38)	14 (50)	7.6 (28)	7 (28)	.21
29. Dominica	.07	56.2 (64)	13 (22)	28 (35)	7.5 (29)	2 (66)	.0005
30. Samoa	.15	52.1 (71)	14 (18)	35 (23)	7.5 (30)	2 (67)	.001
31. Ivory Coast	4.6	62.2 (53)	11 (29)	38 (17)	7.1 (31)	4 (50)	.07
32. Sierra Leone	2.8	49.7 (75)	14 (19)	71 (2)	7.1 (32)	6 (33)	.06
33. Fiji	.55	56.6 (62)	12 (26)	36 (20)	7.0 (33)	2 (68)	.005
34. Reunion	.48	67.5 (45)	10 (37)	22 (42)	6.5 (34)	1 (79)	.002
35. USSR	249.8	104.3 (5)	6 (46)	13 (53)	6.5 (35)	8 (24)	7.71
36. Vietnam	41.7	56.5 (63)	12 (27)	46 (10)	6.5 (36)	5 (39)	.83
37. Surinam	.40	53.4 (69)	11 (30)	28 (36)	5.8 (37)	3 (56)	.004
38. Yemen, Dem. Rep.	1.57	51.8 (73)	11 (31)	41 (14)	5.6 (38)	22 (9)	.13
39. German Dem. Rep.	17.1	94.7 (16)	6 (47)	10 (60)	5.5 (39)	5 (40)	.34
40. Poland	33.4	104.4 (4)	5 (55)	10 (61)	5.4 (40)	5 (41)	.58
41. Thailand	39.4	49.5 (76)	11 (32)	40 (15)	5.3 (41)	10 (21)	1.48
42. Neth. Antilles	.23	71.1 (39)	7 (44)	11 (59)	5.0 (42)	5 (57)	.001
43. Guyana	.76	54.4 (66)	9 (39)	23 (41)	5.0 (43)	7 (29)	.02
44. France	52.1	97.8 (15)	5 (56)	8 (67)	4.9 (44)	4 (51)	.81
45. Mauritius	.87	53.0 (70)	9 (40)	31 (29)	4.9 (45)	2 (69)	.007
46. Finland	4.66	93.5 (18)	5 (57)	8 (68)	4.7 (46)	5 (42)	.08
47. Solomon Is.	.18	39.8 (86)	12 (28)	40 (16)	4.6 (47)	14 (16)	.009
48. Cuba	9.1	70.1 (42)	7 (45)	15 (49)	4.6 (48)	5 (43)	.15
49. United Kingdom	56.2	91.9 (20)	5 (58)	8 (69)	4.4 (49)	5 (44)	1.12
50. Bahamas	.19	70.6 (40)	6 (48)	9 (62)	4.3 (50)	5 (45)	.003

51. Greece	8.9	101.2 (9)	4 (64)	9 (63)	4.2 (51)	3 (58)	.09
52. Belgium	10.1	98.9 (12)	4 (65)	7 (75)	3.9 (52)	1 (80)	.05
53. Angola	6.1	41.1 (83)	9 (41)	35 (24)	3.8 (53)	22 (10)	.49
54. Mauritania	1.23	62.8 (50)	6 (49)	12 (56)	3.8 (54)	6 (34)	.028
55. Gabon	.5	49.0 (77)	8 (43)	14 (51)	3.8 (55)	2 (70)	.1
56. Peru	14.5	60.6 (56)	6 (50)	17 (46)	3.7 (56)	71 (5)	3.73
57. Congo	1.3	40.4 (85)	9 (42)	31 (30)	3.6 (57)	3 (59)	.02
58. Italy	54.9	98.1 (14)	4 (66)	8 (70)	3.6 (58)	2 (71)	.40
59. S. Africa	23.4	77.8 (32)	5 (59)	12 (57)	3.6 (59)	8 (25)	.67
60. Netherlands	13.4	86.2 (26)	4 (67)	7 (76)	3.5 (60)	7 (30)	.034
61. Malta	.32	88.5 (24)	4 (68)	9 (64)	3.5 (61)	1 (81)	.001
62. Germany, Fed. Rep.	61.4	87.2 (25)	4 (69)	6 (78)	3.4 (62)	2 (72)	.46
63. Chile	9.9	73.6 (35)	4 (70)	13 (54)	1.1 (63)	25 (7)	.89
64. Gambia	.49	57.6 (57)	6 (51)	29 (33)	3.3 (64)	5 (46)	.008
65. Trinidad/Tob.	.99	64.8 (47)	5 (60)	13 (55)	3.3 (65)	1 (82)	.003
66. Panama	1.6	57.4 (58)	6 (52)	12 (58)	3.2 (66)	5 (13)	.09
67. Australia	13.1	99.3 (11)	3 (72)	5 (79)	3.2 (67)	3 (60)	.13
68. United States	210.4	104.6 (3)	3 (73)	4 (83)	3.1 (68)	3 (61)	2.64
69. New Zealand	3.0	106.8 (2)	3 (74)	4 (84)	2.9 (69)	6 (35)	.06
70. Ireland	3.05	104.3 (6)	3 (75)	5 (80)	2.9 (70)	8 (26)	.09
71. Cape Verde	.29	54.3 (67)	5 (61)	42 (13)	2.8 (71)	4 (52)	.005
72. Venezuela	11.5	62.6 (51)	4 (71)	9 (65)	2.8 (72)	3 (62)	.15
73. Canada	22.2	98.8 (13)	3 (76)	4 (85)	2.6 (73)	13 (17)	1.08
74. Indonesia	129.2	42.0 (82)	6 (53)	47 (8)	2.5 (74)	2 (73)	.89
75. Israel	3.2	101.9 (7)	2 (80)	5 (81)	2.5 (75)	1 (83)	.01
76. Sri Lanka	13.4	40.9 (84)	6 (54)	36 (21)	2.4 (76)	2 (74)	.10
77. Togo	2.13	52.1 (72)	5 (62)	35 (25)	2.4 (77)	1 (84)	.008
78. Switzerland	6.44	89.6 (23)	2 (81)	4 (86)	2.2 (78)	0 (86)	0
79. Bulgaria	8.6	101.1 (10)	2 (82)	5 (82)	1.9 (79)	3 (63)	.102
80. Comoros	.29	39.3 (87)	5 (63)	29 (34)	1.9 (80)	2 (75)	.002
81. Brazil	103.7	63.2 (49)	3 (77)	8 (71)	1.8 (81)	2 (76)	.60

TABLE 16.—DIRECT CONTRIBUTION OF MARINE PRODUCTS TO PROTEIN CONSUMPTION COMPARED WITH PRODUCTION (1972–74 Average) (*Continued*)

		Consumption				Production	
Country or Territory	Population (Millions)	Total Protein (g/Capita/Day)	Marine/Total Protein (%)	Marine/Animal Protein (%)	Marine Protein (g/Capita/Day)	Marine Protein* (g/Capita/Day)	Marine Catch (m.t.)
82. Ecuador	6.65	47.3 (80)	3 (78)	9 (66)	1.6 (82)	6 (36)	.15
83. Argentina	24.7	101.8 (8)	2 (86)	2 (87)	1.6 (83)	3 (64)	.27
84. Mexico	55.5	65.6 (46)	2 (83)	8 (72)	1.5 (84)	2 (77)	.45
85. Turkey	37.95	80.2 (29)	2 (84)	8 (73)	1.5 (85)	1 (85)	.19
86. Morocco	16.5	69.8 (43)	2 (85)	14 (52)	1.5 (86)	5 (47)	.32
87. India	584.0	48.5 (78)	1 (87)	8 (74)	.4 (87)	1 (87)	1.22
China‡	796.6	61.7	2	10	1.2	1	2.3

NOTE.—Numbers in parentheses are rankings within column.

*Calculated approximately as g protein/capita/day = annual catch in tons / 3.65 × population in thousands.

† Malaysia includes Sabah and Sarawak but not the peninsula itself, for which data are not available.

‡ Data for China are approximate estimates and so this country is not ranked with the others. In China (Island of Taiwan) fish and shellfish consumption in 1970 was 10 g/capita/day (total protein 68 g, 14% of total protein, 46% of animal protein).

content of marine catches equals or exceeds 2 g per capita per day. The countries and territories are listed in the order of the first criterion. Numbers in parentheses are the rankings of the figures in that column. The table therefore shows simultaneously several indicators of the importance of sea food and sea-fishing industries to countries. In interpreting the data on consumption, however, it must be remembered that they do not include the fish-meal contributions to supplies of nonmarine animal protein, which in some countries may approach the level of the direct contribution of fish and sea food. The countries and territories listed together contain 78 percent of the world population of 3,800 million (1972–74 average). The ranges of values are very great.

Table 17 shows the various indices of marine-product consumption and production summarized according to a range of values, giving the total population of countries and territories to which those ranges apply. These are derived from table 16. The countries and territories listed here account for 2,189 million out of a world population of 3,008 million (both excluding China) (i.e., 71 percent). In table 17 the grouped populations are given as percentages of 2,189 million. It is, of course, important to remember that these percentages do not reflect the actual numbers of people with the consumption levels indicated, but only the total population of countries in a particular range of average levels. Note that four of the five distributions are bimodal.

Table 18 gives the 1976 sea-fish and shellfish catches by country, the average annual percentage increase in catch over the period 1970–75, and the percentage change from 1975 to 1976. An indicator of national "performance" in recent years might be the difference between this increase rate and the popula-

TABLE 17.—FREQUENCY DISTRIBUTIONS OF INDICES OF MARINE CONSUMPTION AND PRODUCTION, BY POPULATION*

Consumption:						
Total protein (g/capita/day):						
Range	> 100	80–99.9	60–79.9	40–59.9	< 20	...
% population	25	19	14	42	...	...
Marine/total (%):						
Range	> 10	5–9.9	1–4.9	...	...	...
% population	17	30	53	...	...	...
Marine/animal (%):						
Range	> 50	40–49.9	30–39.9	20–29.9	10–19.9	0–9.9
% population	8	12	3	2	18	57
Marine protein (g/capita/day):						
Range	> 20	15–19.9	10–14.9	5–9.9	0–4.9	...
% population	5	0	17	23	71	...
Production:						
Marine protein (g/capita/day):						
% population	7	1	6	25	61	...

*2,189 million (excludes China).

TABLE 18.—CATCH LEVELS AND MARINE FISHERIES GROWTH PERFORMANCE RELATIVE TO POPULATION

Country	Catch Annual Change 1970–75 (%)	Marine Fisheries Performance 1970–75* (%)	Catch 1976 (m.t.)	Increase 1975–76 (%)
1. Japan	2.3	1.0	10.42	.9
2. USSR	6.9	5.9	9.36	4.1
3. Norway	−3.1	−3.9	3.44	34.7
4. USA	.2	−.8	2.93	9.6
5. Peru	−22.8	−25.7	4.34	26.1
6. China	1.9?	.2?	2.31?	?
7. Korea, Rep.	20.3	18.5	2.39	1.1
8. Spain	−.1	−1.2	1.47	−2.9
9. Thailand	−2.1	−5.1	1.46	5.2
10. Denmark	7.5	6.8	1.90	8.3
11. India	6.4	4.1	1.53	3.2
12. Philippines	6.2	3.2	1.32	4.8
13. United Kingdom	−2.1	−2.5	1.06	6.7
14. Canada	−6.2	−7.6	1.09	10.7
15. Iceland	6.3	5.1	.99	−.8
16. Indonesia	12.9	10.6	1.04	5.5
17. Vietnam	4.6	1.9	.84?	?
18. France	.6	−.2	.81	0
19. Korea, Dem. Rep.	?	?	.80?	?
20. S. Africa	−.9	−3.6	.64	.3?
21. Chile	−.9	−2.8	1.26	36.1
22. Brazil	6.1	3.2	.83?	13.6?
23. Poland	11.5	10.7	.73	−6.7
24. Angola	13.0	−14.4	.15?	?
25. Mexico	4.9	1.4	.55	15.2
26. Germany, Fed. Rep.	−6.5	−7.1	.44	2.8
27. Portugal	−5.9	−5.8	.34	−9.6
28. Malaysia	6.8	4.2	.51	8.9
29. Morocco	−3.3	−5.7	.28	31.9?
30. Italy	.2	−.5	.40	.3
31. German Dem. Rep.	3.2	3.3	.27	−26.5
32. Netherlands	3.0	1.9	.28	−19.0
33. Burma	2.7	.5	.37	3.7
34. Senegal	15.8	13.1	.35	−.6
35. Argentina	.5	−1.0	.27	27.1
36. Faroe Is.	6.6	5.7	.34	19.6
37. Sweden	−6.3	−6.9	.20	−2.9
38. Pakistan	2.2	−.7	.18	6.0
39. Ghana	8.5	5.9	.20	−8.0
40. Oman	2.0?	?	.20	−.5

TABLE 18.—CATCH LEVELS AND MARINE FISHERIES GROWTH PERFORMANCE RELATIVE TO POPULATION *(Continued)*

Country	Catch Annual Change 1970–75 (%)	Marine Fisheries Performance 1970–75* (%)	Catch 1976 (m.t.)	Increase 1975–76 (%)
41. Nigeria	−8.7	−11.2	.17	3.8
42. Turkey	7.5	4.9	.14?	−22.2?
43. Venezuela	3.9	.6	.14	−4.1
44. Cuba	9.1	7.3	.20?	17.4?
45. Ecuador	19.6	16.2	.22?	?
46. Yemen, Dem. Rep.	1.6	−1.3	.13?	?
47. Australia	−.1	−1.9	.11	5.6
48. Hong Kong	2.0	.1	.15	4.8
49. Bangladesh	?	?	.09?	?
50. Sri Lanka	5.2	3.2	.12	6.0
51. Bulgaria	12.2	11.6	.16	6.0
All	. . .	. . .	59.92	6.7
All, excl. "estimates"	. . .	. . .	53.29	6.9

*The difference between rate of change of catch and rate of population growth (average annual % rates 1970–75). Smaller fishing countries not included in the table that gained in production per capita, 1970–75: Papua New Guinea (15.6), Sierra Leone (15.3), Namibia (15.2), Uruguay (14.1), Puerto Rico (10.1), Tunisia (9.5), Tanzania (7.4), Ethiopia (7.1), Finland (6.0), Algeria (4.7), Panama (5.5), Yugoslavia (2.9), Greenland (1.5). Smaller fishing countries that neither lost nor gained more than 1% per capita per year: Ireland, Guyana, New Zealand, Liberia, Colombia, Malagasy Rep., Somalia. Smaller fishing countries which declined per capita: Cameroon (−1.2), Saudi Arabia (−1.2), Belgium (−2.0), Singapore (−2.2), Egypt (−3.4), Ivory Coast (−4.0), Iran (−4.5), Israel (−4.5), Zaire (−4.6), United Arab Emirates (−5.9), Maldive Is. (−6.1), Greece (−8.6), El Salvador (−8.7), Kuwait (−12.9), Kampuchea Dem. Rep. (−14.5), Mauritania (−18.4). Figures in parentheses are percentages. In some of these countries fish protein consumption was relatively important in 1970, but in many cases they relied on imports more oman on home production.

tion growth rate over the same period; this difference is the average rate of increase in production per capita. Small fishing countries which nevertheless performed well in increasing their catches are listed in the notes to table 18.

A large part of the world catch is taken by a very few countries. More than one-half of the 1976 catch was taken by six countries with, between them, 36 percent of the world population; three-quarters of the catch was taken by the leading 16 countries, having only 60 percent of that population.

It is important that the data in tables 17 and 18 are seen as representing a situation at a certain time which cannot necessarily be sustained. In several cases large catches, and large increase rates of catches, have been obtained at the expense of the stability of the resource, that is, their levels do not represent sustainable production.

During the decade 1965–74 the annual increase in world population was nearly 2 percent and in protein production of marine origin about 3 percent, so the per capita consumption of protein of marine origin has increased by nearly 10 percent over the period. However, neither the population growth nor the increase in fish *production* was the same in different country groups. In the last few years the developing countries' share of catch has increased slightly, but that of the developed centrally planned economies by much more, continuing a trend started in the 1960s (table 19).

So, whereas the per capita dietary protein production by developed countries with market economies had been falling, that by both developing countries and centrally planned countries had been rising rather fast. These increases had, however, very different consequences in the two groups of countries. That in the centrally planned countries corresponds with a large rise in *consumption* in those countries of protein of marine origin, whereas much of the increase by the developing countries has been exported. These exports, originally consisting largely of fish meal, now include increasing quantities of products for direct consumption, mainly in frozen form. The extent to which the foreign currency earnings from these exports are used to improve the nutrition of the people in those countries is an important question which we cannot go into here; it is sure, however, that this varies very much from one developing country to another.

For the world as a whole, not only the catch but also the catch per capita increased, but not as fast as did the real (i.e., corrected for U.S. dollar inflation) GNP per capita. The situation is very different in different economic groups. In developed market economies fisheries were declining relative to the total economic activity; they were increasing slightly in the developing countries and

TABLE 19.—GROWTH RATES OF MARINE PRODUCTION PER CAPITA AND GNP PER CAPITA BY ECONOMIC GROUPINGS OF COUNTRIES (%)

	Annual Increase 1965–74					
	Catch for Direct Consumption	Catch for Reduction	Total Equivalent Direct*	Population Growth	Catch per Capita†	GNP per Capita‡
Developed m.e.	.1	4.3	.3	.9	2.2	4.3
Developing m.e.	5.5	7.4	5.6	2.4	4.2	3.8
All m.e.	2.2	5.0	2.4	2.0	2.4	4.0
Developed c.p.e.	3.8	16.0	4.1	1.6	6.4	3.7
Developed countries . .	. . .	. . .	. . .	. . .	3.2	4.1
World	. . .	. . .	. . .	. . .	2.8	4.0

*Catch for direct consumption plus one-third of catch for reduction.
†Excludes anchoveta.
‡World Bank, 1976.

increasing rapidly in the developed countries with centrally planned economies.

Fats and Oils

Animals, including fish, contribute 18 percent of the world calorie supply, but there is a large difference between the relative importance of animal calories in developed countries, where they contribute over 30 percent of dietary calories, and in developing countries, where they contribute only 8 percent. Although attention has usually been focused on living marine resources as a source of protein, the fat in sea fishes and the oil extracted from them contribute significantly to fat supplies as a whole and hence to the calorific value of diets.

The world supply of fats, either as fatty substances intrinsic in whole foods or as fatty or oily products, is about 61 g per capita per day, of which rather more than half (56 percent) is of animal origin. However, the fat-calorie supply per capita in developed countries, from vegetable and animal sources, is about 3.5 times that in developing countries, and from animal sources alone about six times (table 20). Calories from fats constitute about 21 percent of the total

TABLE 20.—CONTRIBUTION OF FATS TO HUMAN NUTRITION (1974)

	Fat Supply (g/Capita/ Day)	Fat Cal.* / Total Cal. (%)	Fat Supply in or from Animals† (g/ Capita/Day)	Marine Oils as‡	
				% Total Fat	% Animal Fat
Developed m.e.	134	35	89 (66)	2.3	1.5
Developing m.e.	37	15	12 (34)	.3	.1
All m.e.	65	21	35 (53)	1.5	.8
Developed c.p.e.	109	27	78 (72)	1.2	.9
Developing c.p.e.	33	12	17 (52)	.01	.006
All c.p.e.	55	18	35 (63)	.7	.4
Developed countries	124	33	86 (68)	1.9	1.3
Developing countries	35	14	14 (39)	.2	.09
World	61	21	34 (56)	1.3	.7

*Calculated as 1 g fat yielding 8.8 cal.
†Percent of total shown in parentheses.
‡1972 data.

calorific content of the food reaching human consumers. However, they constitute a relatively much higher fraction (2.5 times) of the calorie supply to people in developed countries than to those in developing countries. Animal fats contribute 22 percent of the total calories in developed countries but only 5.5 percent in developing countries.

For the world as a whole 40 percent of the fat in the diet comes from fatty or oily extracted products of either vegetable or animal origin; the rest is contained within other foods including, of course, fresh and preserved fish. Of the total fat calories derived from marine fish or fish products just over half is consumed in the forms of fatty and oily products of marine origin. Oil products from fish contribute about 1.3 percent of the total fat calories in the diet and about 2.3 percent of the animal-fat calories.

Of the 46 m.t. of fat and oils produced annually in the period 1972–74, 40 million (87 percent) were edible types, though not all of this quantity actually entered the diets of humans or domesticated animals. Of the 40 m.t., 10 million (25 percent) was of animal origin, and of that amount 10 percent was of marine origin; 30 percent of the 40 m.t. figured in international trade statistics.

The distribution of total production of fats and oils is shown in table 21, together with data for daily production per capita (expressed in calorific terms) and fractions of vegetable and animal origin. There is a small net trade flow of these products toward the developed countries, which produce 60 percent but consume 63 percent of the total (table 22). The net flow in that direction is, however, largely due to a stronger net flow of vegetable-oil products toward the

TABLE 21.—PRODUCTION OF OIL AND FAT PRODUCTS, ALL SOURCES (Terrestrial and Marine) BY ECONOMIC GROUPINGS OF COUNTRIES (1972–74)

	Production* (m.t.)	Cal/Capita/Day	% of Animal Origin (Inc. Butter)
Developed m.e.	19.9 (43)	633	45
Developing m.e.	15.4 (33)	200	15
All m.e.	35.3 (76)	325	32
Developed c.p.e.	7.9 (17)	502	45
Developing c.p.e.	3.1 (7)	94	20
All c.p.e.	11.0 (23)	226	38
Developed countries	27.8 (60)	589	45
Developing countries	18.5 (40)	168	16
World	46.2 (100)	294	33

*Percent of total shown in parentheses.

TABLE 22.—SUPPLIES OF FAT AND OIL PRODUCTS FROM ALL SOURCES (Marine and Terrestrial) BY ECONOMIC GROUPINGS OF COUNTRIES*(1972–74)

				Marine Products as % of	
	All Sources (m.t.)	Animal (m.t.)	Animal All (%)	All Products	Animal Products (Incl. Butter)
Developed m.e.	22.0 (47)	8.3 (54)	38	3.9	10.2
Developing m.e.	13.8 (30)	3.1 (20)	23	.6	2.5
All m.e.	35.8 (77)	11.4 (74)	32	2.6	8.1
Developed c.p.e.	7.5 (16)	3.4 (22)	45	2.2	4.9
Developing c.p.e.	3.4 (7)	.7 (45)	21	.03	.1
All c.p.e.	10.9 (23)	4.1 (26)	37	1.5	4.1
Developed countries	29.5 (63)	11.7 (76)	39	3.4	8.7
Developing countries	17.2 (37)	3.8 (24)	22	.5	2.0
World†	46.7 (100)	15.5 (100)	33	2.3	7.1

* Percentages shown in parentheses.

† The figure for world supplies differs from that for world production, found in table 21 because of discrepancies in import/export figures.

developed countries (which produce 50 percent but consume 57 percent), which is only partially compensated by a net flow of animal fat and oil products *away* from the developed countries (which produce 82 percent of these and consume 75 percent). Even so, because of the differences in levels of production, this latter flow is not strong enough to prevent supplies of animal fat and oil products being three times higher in developed than in developing countries and seven times higher per capita.

Table 22 also shows the contributions of fatty and oily products of marine origin to these totals; they are, for the world as a whole, 2.3 percent of all oil/fat products, and 7.1 percent of all animal fat products. These fractions differ four- to six-fold as between developed and developing countries. Thus, the contribution of marine oils, 80 percent of which are produced by the developed countries, to the calorific content of the diet in developing countries, is minute. It will be seen from table 23 that, contrary to the overall trade flow in animal oil/fat products, the flow of marine oils is very strongly *toward* the developed countries, which produce nine times as much per capita as the developing countries but consume 22 times as much.

In summary, marine oils contribute overall about 0.3 of the calories in human diets; about 0.7 percent in developed countries but a negligible amount in developing countries. They contribute 1.3 percent of the *fat* calories in diets; 1.9 percent in that of developed countries and 0.3 percent in that of developing

TABLE 23.—MARINE OILS—ANNUAL PRODUCTION AND SUPPLY, 1972–74

	Production		Supply	
	Thousand Metric tons	Cal/ Capita/ Day*	Thousand Metric tons	Cal/ Capita/ Day
Developed m.e.	695 (65)	22	849 (78)	27
Developing m.e.	214 (20)	3	77 (7)	1
All m.e.	909 (85)	8	929 (85)	8
Developed c.p.e.	157 (15)	10	165 (15)	10
Developing c.p.e.	1 (.1)	.03	1 (.1)	.03
All c.p.e.	158 (15)	3	168 (15)	3
Developed countries	852 (80)	18	1,015 (93)	22
Developing countries	215 (20)	2	78 (7)	1
World	1,068 (100)	7	1,093 (100)	7

NOTE.—Percentages shown in parentheses.
* Calculated as 1g yielding 8.8 cal.

countries. As earnings, by developing countries, of the currencies of developed countries, they constitute about one-fifth of the value of exports of products from all fish taken for reduction purposes.

TRENDS IN INTERNATIONAL TRADE

Data provided to the Nineteenth Session of the FAO conference (November–December 1977) permit comparison of trade in fisheries products (overwhelmingly marine products) with trade in agricultural and forestry products. The total value of all such exports was, in 1976, US $108,000 million (an increase of 8 percent over 1975), of which 4.4 percent was contributed by fisheries products and 73.7 percent by agricultural products. The increase from 1975 to 1976 was very much greater in the case of fisheries than for agriculture (table 24). However, this relative increase in fisheries arises largely from increases in prices of fisheries products relative to agricultural products. Thus, on the one hand, the volume of trade in crop and livestock products increased by 8 percent while the unit value *declined* by 2 percent; on the other hand, the volume of trade in fisheries products increased only by 4 percent although the unit price *increased* by 9 percent.

Of the total value of exports of fisheries products in 1976, 63 percent was from developed countries with market economies, 32 percent from developing countries with market economies, and 5 percent from countries with centrally planned economies.

TABLE 24.—VALUE OF EXPORTS (% Change from 1975 to 1976)

	Fisheries	Agriculture	Forestry	All
Developed m.e.	25.7	4.2	18.4	7.9
Developing m.e.	27.3	11.8	43.1	14.8
All m.e.	26.2	6.6	21.5	9.9
All c.p.e.	−1.3	−6.0	7.4	−2.9
World	24.5	5.6	20.0	8.8

PROGNOSES

In 1974 and 1975 world aquatic catches of 70 m.t. were attained, of which 60 m.t. came from the sea. As noted in *Ocean Yearbook 1*, FAO reported in April 1977 that world (sea) catches were "probably capable of increase [only] by some 30/35 million tons under optimum conditions of exploitation," excluding possible development of large fisheries for unconventional resources—krill and the like.

The limit cited, of somewhat less than 100 m.t. at best, comes from analysis of the situation regarding the capacities of the natural resources to sustain larger catches. Other approaches have been taken to forecasting trends. One of these is to attempt to designate a "desirable target" figure for catches, to estimate the investment in production, processing, and distribution capacity, and in training manpower that might be required to reach such a target. One such published target is 130 m.t. by the year 2000,[13] and the calculated investment required was US $1,500 million annually from 1980 to 2000, totaling US $30,000 million. In this calculation account was not taken of the capacity of the resources to sustain such a catch level, or of the fact that as catches increase and the resources are depleted the cost of catching per ton increases at an accelerating rate. Nevertheless, these unrealistic figures have received wide publicity.

Another approach, followed by FAO with respect to agricultural commodities generally, has been to construct *demand projections.* The basic data for these are per capita consumption figures over a selected base period, derived from *Commodity Balance Sheets.* The demand for fish meal as such is not easily predicted; demand is for protein food, which is projected on estimated growth in livestock numbers and type which, in turn, is based on the demand for meat. The practice is then to assume that fish meal will continue to represent the same proportion of protein feed as it does now. Demands for *food-fish* are more directly calculated. Projections are based on *(a)* the projected increase in the human population (UN figures); *(b)* an assumption that *relative* prices of food

13. FAO, *Long-Term Targets for Fisheries Development* (Rome: FAO, 1977).

commodities are constant; and *(c)* projected changes in per capita income and a relationship between income and fish consumption derived from household budgetary surveys, assuming that this relationship will continue beyond the period covered by the survey. Clearly these assumptions may be invalidated by various changes, but analysis of their implications for demand growth, considered together with extrapolation of recent production trends, gives the best indications at present available of the likely strength in future of forces for fisheries development and hence possible trends.

The FAO has made such projects for fresh and frozen fish, processed fish, shellfish, and miscellaneous aquatic products. As with dietary data, trends in total aquatic and in marine production and demand are not distinguished. Table 25 summarizes such projections for the years 1980 and 2000. The world totals are to be compared with a total food-fish catch in 1976 of 51 m.t., of which 41 m.t. were of marine origin. The projection for 1985—the final target date set in 1970 under the Indicative World Plan—by this method would be 71 m.t.

The possible demand for 113 m.t. of food fish in the year 2000 goes beyond reasonable expectations of catches at that time, in the light of present knowledge about resource availability. The "gap" between calculated demand and possible catches would presumably be closed by a decrease in the demand resulting from relative increases in food-fish prices if the other assumptions held. Such increases might in turn make acceptance of restraining management measures, aimed at preventing overexploitation of the resources, more difficult, especially internationally. Such a feature has already been observed with respect to management of the capelin fishery. Further, given the degree to which food fish enters international trade, price increases are likely to reduce the consumption of fish in many developing countries and thus run counter to hopes for a "new international nutritional order" as well as for a new International Economic Order. Maintenance of healthy world fisheries and their contribution to nutrition and to dietary variety would seem to call for measures beyond the play of the economic factors, account of which is taken in making the demand projections.

Yet another approach to prognosis is projection forward of observed rates of increase in production. Simple extrapolations were made in 1968 by FAO of

TABLE 25.—FOOD-FISH DEMAND PROJECTIONS (m.t.)

	1980	2000
Developed m.e.	21 (35)	28 (25)
Developing m.e.	18 (30)	40 (35)
All m.e.	39 (65)	68 (60)
All c.p.e.	21 (35)	45 (40)
World	60 (100)	113 (100)

NOTE.—Percent of total shown in parentheses.

total catches based on continuation of an overall annual rate of increase of 7 percent noted to that date. Predictions, using data for the 20-yr period 1948–68, were for catches of "bony fish" of 80 m.t. by 1975 and up to 100 m.t. by 1980. Actual catches fell far short of such expectation, reaching only 49 m.t. by the mid-1970s. The major change between 1968 and 1975 was the unexpected collapse of the anchoveta fishery. In fact, the tremendous growth of the anchoveta fishery during the 1950–68 period (and continuing even to 1970) inflated the general increase rate which would otherwise have been 5 rather than nearly 7 percent. A 5 percent growth rate predicts a 1975 catch of 55 m.t., which is still somewhat too high. In fact, over the period 1950–68 the annual rate of increase in fish catch (excluding anchoveta) although variable, tended to decline, and a better extrapolation would attempt to take that factor into account. There are many ways of doing this.[14] A simple method assumes that the rate of annual increase in all marine catches (excluding anchoveta) will continue to fall as it has done for 2 decades. This leads to predictions of a maximum catch of about 110 m.t. at the turn of the century. The catch taken for reduction has been far more variable than that taken for consumption, and prediction of it is more hazardous. The pattern of change in consumption catches has been rather constant; extrapolation leads to a prediction that it will increase from the 1975 level of 38.3 m.t. to a maximum of between 43 and 55 m.t. by the turn of the century. This prediction could, of course, be upset by any technological and economic changes in the meantime which resulted in substantial quantities of the types of fish at present being caught primarily for reduction being consumed directly. Efforts to this end are being made. To the extent that such efforts may have been successful in recent years, they have in theory already been taken into account in the regressions on which the above predictions have been based. If, nevertheless, a major change occurs in future, the predicted percentage that reduced catch will be of consumed catch will be too high. However, the relations between the "reduced" and "consumed" catches are more complex than this. Some increase in reduction catch in recent years has come from exploitation of small species, such as capelin and sand eels, which were not previously caught at all, or at least not in large quantities. In other cases increases have been made *at the expense* of the consumption catches, either by taking the same species for both purposes or, worse, by taking large quantities of the young of the species which were originally only taken as adults for consumption. Future expansion of the fisheries for reduction purposes could follow either or both of these courses.

The move to several new species and groups might continue to take in lantern fishes, pelagic crabs, squid, and so on, but further expansion of direct competition with the consumption fisheries will further slow the growth of the latter, and even lead to an absolute decline in them. Vast expansion of the

14. See n. 3 above.

fishery for Antarctic krill would introduce a new factor, as great as or greater than the 1950s expansion of the anchoveta fishery.

In general, these predictions are not inconsistent with the calculations that have been made of the total potential of "conventional" resources and of expectation from some of the less conventional ones. They show the total catches, in the year 2000, still exceeding a projected human population growth rate of 1.6 percent annually. If, however, the trends of reduction and consumption catches continue as predicted, and assuming an overall 3:1 conversion for protein for human consumption obtained by feeding fish as meal through livestock, then the contribution of sea fish to protein supplies per capita would start falling in the period 1979–85. This suggests that energetic efforts to increase the efficiency with which "reduction" catches enter human diet, and to reduce wastage at all stages from catches to consumption, would be well justified. An attempt to counteract the declining increase rates by greatly increased investments in fishing power would be expected in the medium and long run merely to increase the rate of approach to the asymptotic level dictated by the availability of resources, and hence to bring nearer the time at which per capita production and consumption begin to fall as well as further increasing the cost per unit catch.

"WASTAGE" OF MARINE RESOURCES

If, as we have seen, it is likely that present catches of conventional types of sea food are approaching the natural limit of the ocean to sustain them, and if the real costs, however measured, of catching much more than at present must steadily rise, then a reappraisal is required of the orthodox route of fisheries development: ever more investment to intensify existing fisheries and to discover and exploit new fish stocks. This is basically the route followed in the past quarter-century. Such "developments" have included the consumption of a greater variety of species from traditional fishing grounds (such as the redfish, *Sebastes marinus*, from the traditional North Atlantic cod and haddock grounds and sei and minke whales from the blue and fin whale ground of the Southern Ocean); trawling in deeper waters, even beyond the continental shelf on to the continental slope; opening fisheries by vessels from Northern Hemisphere countries in the Southern Hemisphere, mainly off the coasts of developing countries; and the growth of the fish meal industry. Because these developments have necessitated technological changes and have usually involved increases in the size and cost of fishing units and operation at great distances from home bases and markets, the processing of catches has also undergone important changes. In industrialized fishing and whaling, the degree of utilization of the retained catch has steadily increased. Not only do filleting machines remove the edible part of the whole fish efficiently, but there is space and power available to utilize the inedible residues by reducing them to meal, and other forms. Some catches of species not familiar to, and therefore exciting resistance from, some "consumers" can be converted into a form more attractive than in their natural state, and marketed. At the same time, organization of some forms of industrial fishing has led to a high value being assigned to the hold space in fishing vessels and to the time required to sort catches of mixed species. When the mixture includes species such as shrimps with very high market values and "trash" with low values, then it seems inevitable under present economic conditions that large quantities of the latter would be wasted, and that is precisely what happens in shrimp fisheries in tropical and subtropical waters all over the world. For every ton of shrimp caught and frozen, several tons of trash fish (as well as vast quantities of sponges, rubble, and—in the eyes of the fishermen—organic "rubbish") are caught in trawls and discarded dead at sea. The same economic considerations prevent the utilization as meat of the tens of thousands of tons of dolphins still killed every year in purse-seine fisheries for tunas in the Eastern Tropical Pacific.

Other kinds of fisheries—and especially those employing purse seines—lead to the waste of considerable quantities of potential food for humans. If, as sometimes happens, a haul of a pelagic shoaling species such as mackerel is too large to be handled by the vessel or to be accommodated in the remaining hold space, the catch is "slipped," and none or only a small part of it taken aboard; the fish in the greater part which is slipped normally die from shock, suffocation, or mechanical damage while they were restrained in the net. Likewise, unwanted fish brought up in trawls from deep water die as a result of rapid pressure change, as well as from mechanical damage. Slipping is also practiced when the purse-seine haul is considered to contain too many of a species other than the main "target" species, or even of the wrong size-range of the target species.

It is difficult to put a figure on the total "wastage" of these kinds. It has been estimated that fish discards from shrimp fisheries alone now total about 4 m.t. annually, and that total deliberate discards of fish might reach 6 m.t., that is, more than 10 percent of the world retained catch. There are no estimates of the quantities slipped or otherwise killed by fishing gear and never brought on board vessels.

A bigger wastage in terms of human nutrition may be in the use of fish to feed livestock. If the conversion adopted here (3:1) is approximately correct, then the equivalent of 10 m.t. of sea-food catch is being lost annually, amounting to 17 percent of the total retained catch. Such calculations would justify very considerable efforts to use catches, which are at present reduced, for direct consumption, notwithstanding the technical and economic difficulties of doing this. It is possible that the conversion to protein from livestock is more efficient than assumed here; if that is so, present nutritional benefits from sea fisheries are greater than indicated. However, gross conversion efficiency is certainly not now higher than 50 percent at best, and less than this on the average, so that improvements would be nutritionally worthwhile even if the fish cannot conveniently be consumed directly. Some improvements can come from better choice of the livestock to be fed—in fact the highest efficiency is attained by using fish meal to raise aquatic animals such as rainbow trout.

Recognizing the special role of marine protein for many developing countries, and the immense potential of the "unconventional" resources such as krill, ocean squids, and lantern fishes, Saetersdal has recently proposed not only that much greater efforts should be made to incorporate fish protein concentrates (FPCs) into human diets directly, but that such products could be suitable technically and economically as high-grade food reserves for emergency use and in schemes for world food security.[15] Saetersdal also has argued that "a widespread production of fish powder would contribute to increased self-reliance of countries and regions in food production."

There are very considerable wastes at both ends of the production-consumption process. On the one hand the losses at capture and in primary processing, especially in the fish meal industry, are quite substantial—for example 10 percent or more of the anchoveta catch is lost in one way or another between setting of nets and feeding the bulk fish into the fish meal plants. This figure includes slipping of excessively large catches or parts of catches which would overload the vessel, losses through crushing of catches in holds (which are highest when holds are not completely filled) and at unloading. When anchoveta catches were very large indeed, as in the late 1960s and early 1970s, loss rates might have reached 25 percent, mainly through slipping some large catches but also because with so much raw material moving it was tempting to be careless with it. Since then the loss rates at most stages have declined, because large catches are now rare and new laws have ensured that meal plants now operate more efficiently and, in particular, utilize the "stickwater." Hold losses are however still substantial, and slipping to release catches which consist mainly of fish below legal minimum size of course continues, although it is said that good skippers can do this with little mortality to the released fish.

At the other end the wastage by the ultimate consumer appears to be increasing, particularly in the more affluent countries. Roy has reported, for example, that 7 percent of all home-produced and imported food in the United Kingdom finishes as kitchen and plate waste, at least partly as a result of the adoption of unnecessarily high aesthetic standards and the lowering of standards of domestic competence and practices.[16] It seems likely that the figure for foods of marine origin, which are relatively expensive as well as highly processed (frozen fillets and canned, mainly), is less than this; nevertheless the loss is probably substantial. In less affluent countries, and especially in the tropics, the important losses occur during storage and transport of products such as dried and salt- or smoke-cured fish, by bacterial action, and by depredation of

15. G. Saetersdal, "The Potential of Fish Protein Concentrates (FPC) in World Food Security," mimeographed (paper delivered to FAO/NORAD, Nordic Journalists Encounter, Oslo, December 1977).

16. R. Roy, *Wastage in the UK Food System* (London: Earth Resources Research, 1976).

insect and other pests. They are considerable, but we have not been able to locate data from which global losses could be deduced.

THE BOUNDS OF STOCKS AND THEIR EXPLOITATION

The biggest—and increasing—wastage as far as the production of sea foods is concerned is in the misuse of the resources themselves, and this waste could, in theory, be reduced with simultaneous *savings* in inputs.

In the 1930s it began to be realized that at least some stocks of sea fish were limited and that intensive fishing was affecting them. Salmon were a special case, depending rather obviously on the availability of spawning sites in rivers and on their freedom of access to those sites. By the late 1930s it was apparent that the Antarctic baleen whales were vulnerable to depletion, although the first international negotiations to regulate the catching of them appear to have been as much concerned with reducing the effects of economic competition between nations, and the impact of this on the market for whale oil, as with conserving the resources. Otherwise, the only stocks which it was feared were becoming "overfished" were bottom-living fishes, particularly flatfish (pleuronectids: halibuts, plaice, soles, etc.) in the North Pacific and the Northeast Atlantic. Such overfishing proved to be more or less reversible; if the intensity of fishing on these rather long-lived animals diminished for a few years—as it did during the two world wars—the stocks increased.

Some kinds of bottom- and near-bottom-living fishes such as the haddock, cod, and other gadoid species, tend to fluctuate in abundance much more than do the flatfish because their reproductive success varies considerably from year to year. Being more mobile and responding to shifts in environmental conditions, their accessibility is also variable. The combination of this with actual variation in abundance causes catches to fluctuate much more than is accounted for by changes in the amount of fishing effort, as determined by the number, power, and deployment of vessels. These fluctuations have made it rather more difficult to detect overfishing of these species, but few of them have escaped the consequences of the increase in fishing effort since 1945, and many of them, such as the haddock in the Northwest Atlantic, are now recognized as overfished.

The more or less obvious consequences of intensive fishing are a reduction in the catch per unit of fishing effort, accompanied by a decrease in the average size of individuals in catches because they have a lesser and lesser chance of growing to near full size before they are caught. Eventually the total catch fails to increase as effort increases, or it even begins to decline. In the typical case the effective biological reproduction of the stock in number is not noticeably affected, since most fishes, other than sharks and rays, are prolific producers of eggs and larvae. The "evolutionary strategy" that they have adopted—called technically an "*r*-strategy"—is one that is well suited to ensure survival under fluctuating environmental conditions. It involves very high levels or reproduction and, normally, very high levels of juvenile mortality from predation (including cannibalism by adults of the same species, in many cases) and limitations of food. The numbers surviving to be "recruited" into the exploited stock—that is, attaining the size at which they begin to be of interest for fisheries—are not closely correlated with the number of fertile eggs produced. Efforts to protect such species from overfishing therefore concentrate on ensuring better survival and growth of marketable fish. This means limiting the amount of fishing effort *and* also preventing the capture of small individuals, either by promulgating minimum legal sizes at which each species may be caught or retained or by technical means (such as adjusting the sizes of meshes of nets, entrances to traps, or hooks), so that smaller fish are either not taken by the fishing gear in the first place or can escape from it.

The big fisheries for pelagic (surface-living) species, such as the herring, have always been characterized by great year-to-year variation. This is, again, a result of both varying accessibility and varying abundance, but the latter variation in these cases may be especially high because these species tend to be smaller or to have shorter natural life spans than the demersal species. The large pelagic tunas are exceptions to this rule, and in fact several species of them obviously became affected by fishing rather early in the post–World War II fishery expansion.

For many years fishing did not seem to have a great effect on the stocks of small pelagic fishes, and what effects there might have been were obscured by their natural variability. The first collapse—of the Californian sardine fishery—was mysterious, and it was difficult to attribute the cause conclusively either to natural factors or to overfishing. The distinguishing feature of this collapse was that it was not reversible; when, for economic reasons, fishing stopped, the sardine did not recover. Since then other pelagic stocks have collapsed—the Northeast Atlantic herring and the anchoveta are prime examples—and it is unsure to what extent and how fast they will recover in the absence of intensive fishing. In the case of the herring a main contributor to the collapse undoubtedly has been the capture in recent years of large quantities of immature fish for fish meal, and it has meant the loss both of a fish-meal industry and of the original very large fishery for direct consumption. A feature of all such situations is that the eventual "ecocatastrophe" results from a combination of reduced spawning stock and unfavorable environmental conditions, these *r*-strategists having been reduced by man to levels at which their "strategy" is no longer effective. It has been suggested that the dynamics of these species, and of the ecosystems of which they are components, are such that if they are greatly reduced they may be replaced by other, perhaps similar, species which have not been under stress. Evidence for this happening in the sea is as yet inconclusive, but theory and observation converge in making it seem likely. In all the cases mentioned "overfishing" involves not only—or even mainly—excessively reducing the chances of survival of an individual to commercially valuable size, but also reducing the number of mature adults and the consequent production of eggs and larvae to such a degree that "recruitment" into the next generation is sharply reduced. This leads to a rapidly accelerated decline. Unless the animals concerned are extremely short lived—1 yr or 2 yr only—this decline may come as a surprise to the industry because of the natural delay between birth and recruitment and sexual maturity. The possibility of it happening can be inferred from observations of the abundance and age composition of the current stock; and these characteristics are now monitored in many fisheries. The likelihood of a decline can be judged from observations of the mortality of larval fish and juveniles; such monitoring is technically much more difficult, more costly, and rarely done. In any case, a good scientific basis for the prediction of survival from one generation to the next does not yet exist.

If the two kinds of regulatory measure—controlling size (and hence age) at first liability to capture and fishing effort—are not taken simultaneously, overfishing may not be prevented. This happened in the Northwest Atlantic, where the International Commission for the Northwest Atlantic Fisheries (ICNAF) sought for nearly 2 decades to regulate the meshes of trawl nets in very great detail, while any benefit thereby obtained was quickly removed by increases in the intensity of fishing. In other fisheries, the latter has been regulated, usually indirectly, by setting overall annual catch quotas. The most important result of this has been less to protect the resource than to intensify competition between national fleets and between fishing units within fleets, leading to extremely inefficient operations and to excessive investment in vessels and equipment in pursuit of competitive success. Another result was reduction in the lengths of fishing seasons.

In negotiating national shares of catches from a common resource it is apparent that even if the total permitted catch is agreed to be determined on "scientific" grounds—that is, in terms of what the limited resource can yield—nevertheless the highly political and competitive nature of the negotiation will ensure that attempts are made to apply strong political pressures to the process of determining the overall quotas and, *in extremis*, to distort the theoretically objective scientific assessments. Thus in this situation the decision as to the overall objective to be reached by conserving the resource is of an essentially political nature even if the arguments for a particular decision may wear a technical or scientific disguise.

In this connection it must be said that the concept of maximum sustainable yield (MSY) continues to maintain its place in current negotiation of a new law of the sea as it pertains to fishing. The present fashion is to speak of total allowable catch (TAC), a quantity which is vaguely related to a yield which is sustainable in the long run, and which, if it is not exactly the theoretical maximum, is nevertheless quite close to it. We suggest that the tenacity of the MSY concept, in one form or another, arises from its fundamentally ideological, and consequently its political, function. First, it is supposedly based on natural science, which makes it respectable, especially when economic science fails; it is also, ultimately, a tool of the scientifically and hence technologically stronger among competitors. Second, the ability to provide a maximum surplus is supposed to be a real and continuing property of nature, a reliable anchor for human enterprise in an unpredictable ocean. It might be more apt to say "willingness" rather than "ability" for nature to provide. It is then but a short step to the idea that not to accept this gift is a snub to nature, and a further short step to considering it is wrong to "waste" this gift of nature by leaving it in the sea. One can go still further down this slippery path. Nature itself would be better off if we were to accept the gift; a "thinned out" stock of fish might grow better, whales in a smaller-than-natural population would be fatter and bear healthier calves more frequently. The ocean will be improved if we exploit it "fully."

Thus the ideology is established. The political application of it, to resources which are accessible to many nations, is the corollary proposition that if fishermen from one nation do not take all the gift they should not hinder its acceptance by other nations, especially if those others can argue that they "need" fish, which now nearly always means, in practice, that some of their people can afford to "demand" more fish—in the economic sense.

The effect of applying this "principle" to high-seas fisheries has been to favor the technically advanced nations, and those who were already fishing, against newcomers, especially economically weak newcomers. Once a fishery has been established and grown to a fair size, the average abundance of the stock is reduced, and it becomes correspondingly more costly to take a unit catch from it. Any newcomer who must learn how to fish and where exactly the fish are, and establish markets and distribution arrangements, has the disadvantage of learning the business on an already depleted stock. The "newcomers," in the second half of the twentieth century, are the developing countries, and especially the recently independent ones. It has been argued that what they may lack in expertise and in capital for long-range or advanced fishing equipment is made up for by their proximity to some of the major ocean resources. This proximity has not yet been of much advantage to the West African countries in exploiting the rich central and southern Eastern Atlantic resources, nor to the Southern Hemisphere countries in their use of the Southern Ocean, although of course the costs of transit to and from operating bases and resource locations should eventually be more significant than hitherto in determining who benefits most.

The presence of large fishing fleets from industrial nations off the shores of developing countries has without doubt had a major influence on the decision of the latter group to opt for 200-mi exclusive economic zones (EEZ), with or without the approval of the UNCLOS, and to oppose the revolutionary idea that the living resources of the ocean, as well as the mineral resources of the deep seabed, are a "common heritage of mankind." The EEZ concept has in practice already been conceded, but it is still maintained that a TAC can in each case be identified and that whatever fraction of this the coastal state cannot or does not wish to take is in some mysterious way "surplus," and that therefore others who *do* wish to take the surplus should not be unreasonably impeded. We hear little of the fact that, if others *do* take the surplus, the costs per unit catch will be increased for all, coastal and noncoastal fishermen alike. Similarly, decisions concerning investment for economic "development" continue to be made in many instances on the basis of crude calculations which do not take into account the facts that when the number of vessels is increased the rate of catch per vessel will decline, and that the catch rate by boats in the existing fishery (often smaller boats operated by poorer fishermen) will also become depressed.[17] Any state which decided *not* to exploit a productive resource in its EEZ or, for example, to reserve it for light use in traditional ways by subsistence fishermen, would under the prevailing ideology stand accused of allowing to go to waste a precious resource, and moreover, a resource of food in what is supposedly a hungry world. It is ironic that such accusations would come especially from those who are *not* hungry!

If the fishery resources were, truly, self-contained within national EEZs, it could alternatively be argued that the coastal state has a full moral right to leave the resource unused and would be justified in deciding formally to do so. If moral justification were needed it could take the form of a wish to leave resources in a perfect state for future generations of mankind—especially the section of mankind that happens to live in the coastal state. In any case a moral argument can best be met with another moral argument: that a resource unused is said to be *wasted* is a distortion of the term, which means "to expend to no purpose or for inadequate result," and therefore in practice implies that something is actually moved or transformed, not merely that a natural process has been allowed to continue. Neither is an unfished resource wasted in the biological sense that biological production at that location is lost forever; at least that happens only on a limited scale and over a long time period as a result of nutrients sinking into the deep ocean sediments. In general, the nutrient materials are recycled in a practically closed system, and the absolute "loss" is small. The argument that the production is needed to feed a hungry world and that it is therefore a major sin to use legal or other means to withhold access to it has, we hope, been demolished earlier in this review. It can therefore equally be argued, counter to the "full utilization" philosophy, that it is unwise to interfere with a potential resource before it is really needed, and that not only is it wrong to reduce a resource to a level at which either its existence or its biological productivity are threatened, but it is also wrong to take the unmeasurable risk that such would happen as a result of seeking, while in a state of considerable ignorance, to maximize commercial production from it.

In fact very few fish stocks are self-contained within a single EEZ; on the contrary, most of them are shared between two or more zones, or between EEZs and the international areas. The existence of EEZs does now provide a legal and

17. J. Galtung, "Technology and Dependence, the Internal Logic of Excessive Modernization in a Fisheries Project in Kerala," *FAO Ceres* (September–October 1974), pp. 45–50.

political basis for negotiation where none existed before, but the continued lack of law and principle on the high seas is worrying. This means that in such cases where stocks in a coastal zone do in some way overlap the high seas, powerful fishing nations can, by intensive fishing outside the zone, reduce the availability of fish within the zone. It is equally true that a coastal state, by reducing habitat, or otherwise adversely affecting the inshore production of young fish, can cause depletion of stocks in adjacent areas, coastal and offshore.

In all cases there is now a need for agreed principles on which to found new international agreements. Since no universal rule or principle is yet formulated, it will be necessary to rely on continuing institutions rather than on the word of treaties, without implementing and monitoring mechanisms, to gradually establish wise practice. Until now, UNCLOS has given scant attention to the matter of fishery institutions or to the problems of fisheries in the international zone. While practical experience has been taking us away from the species-by-species approach to management, the UNCLOS text still classifies, supposedly pragmatically and realistically, species as "highly migratory," "sedentary," and so on, and prescribes different regimes for each. Furthermore, attention continues to be focused on the effects of fishing on the stocks, with a side glance at the little-known effects of pollution, while other human activities in the ocean are possibly having even greater effects than either.

We have become accustomed now to the idea that the continued existence of a *critical habitat* may determine the survival of a terrestrial species, of a marine mammal, or of certain specialized diadromous fish such as salmon. It is less well understood that many valuable coastal living fish and shellfish depend completely on the continued biological productivity of, and shelter provided by, mangroves, estuaries, sea-grass beds, and coastal wetlands generally. It is even less widely understood that the increasing removal of gravel from shallow seas, and of coral sand and rubble from tropical lagoons—mainly in order to cover hitherto productive land with concrete—is probably beginning to have serious effects on marine fish production.[18] Apart from the research needed to establish, and eventually to predict, such consequences, and the need to limit and otherwise regulate fishing through international machinery, there is clearly arising a need for a more integrated approach to the management of multiple uses of ocean space and the different kinds of resources contained therein. Thought is only just beginning to be given, at the political level, to the form such machinery might eventually take. Experience, from attempts at fishery management in isolation, suggests that success will depend on securing agreed procedures for equitable allocation of uses as well as defining overall limits of use, and on a critical scrutiny of proposed objectives for management to clarify in advance the political implications of the ideology propounded.

18. International Council for the Exploration of the Seas, "Report of the Working Group on Effects of Marine Sand and Gravel Extraction," in *Cooperative Research Report Number 46* (Copenhagen, 1975).

As regards fisheries, the present Conference on the Law of the Sea is much more concerned with appropriation of resources than with conserving or protecting them; more with the rights of states than with their responsibilities; more with the distribution of benefits among coastal states, developed and developing alike, than with global equity.[19] The degree to which the new order in ocean space will be a move in the direction of a New International Economic Order in the sense of one for the purpose of redistributing wealth from richer nations to poorer ones remains to be seen. Much depends on the skill with which the developing countries act in concert to secure benefits from others who wish to continue fishing near their coasts, and on the ways they use those short-term benefits to build skills, national institutions, and cooperative arrangements to permit wise long-term use of the resources taken into their custody. There need be no hurry for them "to utilize fully" those resources; earlier benefit from them will come from cautious use combined with a more equitable distribution of wealth, and hence of nonprotein staple foods, within as well as between countries—to ensure that the sea food which *is* consumed is physiologically well utilized and continues to contribute pleasure to eating.

DEVELOPMENT OF REGIONAL ARRANGEMENTS FOR FISHERIES

The biennium of 1977–78 has been a period of considerable activity with respect to the reshaping of arrangements for regional fisheries management. It is still too early to describe and evaluate the results of this process, but some general patterns can be discerned. For example, countries interested in fishing in the Northwest Atlantic, having abandoned the ICNAF, are negotiating a new body—the Northwest Atlantic Fishery Organization (NAFO). This body would have essentially a tripartite structure: one part through which coastal states of the region would cooperate, another involving also the noncoastal fishing countries and regulating high seas catches, and a third for the promotion and coordination of scientific research and the provision of collective advice to the governments concerned. The government of Canada submitted in June 1978 an outline of a similar approach to the revision of the International Convention on Whaling (1946), which members of the International Whaling Commission (IWC) and some other countries interested in whaling are attempting to renegotiate or to replace. In the Canadian approach, some species of cetaceans (whales and dolphins) would be dealt with under a general international agreement, irrespective of whether and when these migratory animals were, for some of their lives, within areas of national jurisdiction; other species would be dealt with by coastal states, singly or in cooperation, even if the animals spent

19. For a further discussion, see Arvid Pardo, "The Evolving Law of the Sea: A Critique of the Informal Composite Negotiating Text (1977)," in *Ocean Yearbook 1,* pp. 9–37.

part of their lives in waters *beyond* national jurisdiction. A common scientific body would provide advice for all purposes. This proposal is being studied by other governments.

In addition to changes in most regional and specialized bodies established under treaties, the many regional fisheries organizations established under the constitution of FAO are in a state of flux. Several are extending the scopes of their responsibilities, adjusting their boundaries of competence, and examining how they will respond to the new situation created by extensions of national jurisdiction. At the same time some of them have expressed interest in having a higher degree of autonomy while remaining within the overall framework provided by FAO.

The biennium has also seen a very rapid evolution in two areas of the ocean which had hitherto not been covered by fishery management bodies—the South Pacific and the Southern Ocean. Initiative with respect to the former area was taken in 1977 by the eighth South Pacific Forum, meeting at Port Moresby in October 1976, through a "Declaration on the Law of the Sea and a Regional Fisheries Agency." In pursuance of this initiative, meetings of representatives of states, with some participation by organizations and from dependent countries, have been held in Fiji, in November 1977 and June 1978, under the auspices of the South Pacific Bureau for Economic Cooperation (SPEC). A decision of principle had been taken to create a South Pacific Regional Fisheries Organization comprised of an agency and a conference. The South Pacific Forum adopted the convention at Honiara, Solomon Islands, in July 1979.[20] There are, however, several controversial issues, including the application of the convention to "highly migratory species," the membership of the organization, and its eventual competence. The original proposals were for a body in which virtually all authority rested with the independent coastal states; a body whose functions would include arranging for mutual assistance in such matters as surveillance and policing and for cooperation in harmonizing policies regarding licencing of fishing by noncoastal countries as well as agreements concerning fishing operations by any member in the area of jurisdiction of another. These aims to create a "strong" organization have been considerably modified over the past year. It seems likely that in this area also two or three linked agreements will be made to deal with cooperation among coastal states, cooperation between them and noncoastal fishing countries (principally Japan, Republic of Korea, and China [Taiwan]), and cooperation in scientific research and analysis. The eventual role of the South Pacific Commission (SPC), especially with respect to the latter, is unclear.[21]

An important feature of the discussion concerning the South Pacific fisheries is that there is very little "high-seas" area left between national claims

20. For the text, see pp. 575–78.

21. For an extended discussion of South Pacific fisheries, see the article by George Kent in this issue.

except for a region surrounded by Nauru, Kiribati, Tuvalu, Fiji, Solomon Islands, Papua New Guinea, and the southern tip of the Pacific Trust Territory. The claims are of enormous extent, both absolutely and relative to land areas and populations.[22] To this list must be added some part of the EEZ claims of Australia and New Zealand and a separated area around Easter Island of approximately 900,000 km^2.

By contrast, an even larger region of primary interest to a group of developed countries and to two developing countries of the Southern Hemisphere—Argentina and Chile—who are parties to the Antarctic Treaty, is the Southern Ocean.[23] Following an initiative at the ninth Consultative Meeting of Antarctic Treaty Powers (London, October–November 1977), two meetings of government representatives from parties to the treaty have been held—the first in Canberra in February, and the second in Buenos Aires in June 1978. These meetings have been for the purpose of drafting a convention, provisionally entitled a Convention for the Conservation of the Living Resources of the Southern Ocean, which would be open for adherence by treaty powers and by other countries engaged in exploitation of, or substantial research on, the living resources of the area. It was the declared intention of the powers to complete the drafting and open the convention for signature by the end of 1978. In this case the political issues are naturally somewhat different from those pertaining in the South Pacific. The northern limit of the area to be covered by the convention is a matter of debate, as also are the fundamental objectives of the convention, the scope of country participation to be permitted and the degree to which international organizations such as FAO may be involved in the discussions and the decisions to be taken. With respect to the objectives, some countries place great emphasis on the conservation and protection of the Antarctic marine ecosystem while others wish to exploit the "resources" there, especially the krill. The degree to which an expanded krill fishery will affect other elements of the system—especially the depleted whales—is a matter of much speculation and little scientific knowledge, as is the potential magnitude of the sustainable yield of krill. The FAO has now joined, through its Advisory Committee on Marine Resources Research, with SCAR and SCOR in the planning of the medium-term scientific research program on these resources, entitled BIOMASS.[24] The International Union for the Conservation of Nature and Natural Resources (IUCN), in an attempt to promote a rational ecosystem approach to management in an intensely politically charged sphere, has published proposals for the linking of a Southern Ocean convention with a revised international convention on whaling (or on the conservation of cetaceans) as well as with other relevant international arrangements.

22. Ibid., p. 349–51.

23. See G. L. Kesteven, "The Southern Ocean," *Ocean Yearbook 1,* pp. 467–99.

24. "SCAR/SCOR Group on the Living Resources of the Southern Ocean (SCOR Working Group 54)," *Ocean Yearbook 1,* p. 769–71.

One other initiative which should be reported is that by "Non-aligned and Other Developing Countries in the Sphere of Fishing." A first meeting, attended by representatives of 40 countries, in Havana, November 1977, established a Standing Commission[25] to identify areas of cooperation and to plan and coordinate activities concerning fisheries development, conservation, research, legislation, surveillance, aquaculture, and multinational enterprises.

25. Comprised of representatives of Angola, Cuba, Libya, Morocco, Somalia, Sri Lanka, and Vietnam.

Seaweed Cultivation: Present Practices and Potentials[1]

Akio Miura
Tokyo University of Fisheries

What is generally referred to as seaweed is, for the purposes of this paper, marine Rhodophyta, Phaeophyta, Chlorophyta, etc., that has multicellular systems. Seaweed is found in coastal ocean waters and plays a very important role in coastal ecology through the provision of food and adequate environmental conditions for spawning and the development of fish life.

From the ancient past through a long and varied history, the peoples of Asia have depended on and used a wide variety of seaweed for food by simply collecting it or cultivating it in great quantities. In the West, seaweed has been considered useless, and its utilization as a food is extremely rare. However, in Europe, from the end of the seventeenth century through to the eighteenth century, the industrial extraction of sodium carbonate from kelp for the production of glass was practiced by burning Phaeophyta under low temperatures. Also, potassium chloride and iodine were extracted as by-products, with such practices being continued until the beginning of the twentieth century, as the so-called kelp industry developed. Today, however, most of these products are obtained from salt, niter, and subterranean water containing natural gas, and the practice of producing soda and iodine from kelp has disappeared. After the decline of the kelp industry in Europe, the production of alginic acid from Phaeophyta, carageenan from Rhodophyta, and the production of agar were started and continue to this day. Beyond the use of seaweed as a raw material for industrial production, it has also been widely used for a long period of time in Europe as feed for domestic animals and as a fertilizer.

In looking at the future of seaweed cultivation, some interesting geographical patterns are suggested by the fact that seaweed cultivation for food developed in Japan, China, and Korea, whereas the use of seaweed as an industrial resource was developed in Europe and North America. It is also noteworthy for the future of seaweed production that Eucheuma, as the natural resource for the carageenan industry, has recently come into cultivation in the Philippines.

Let us now turn our attention to the future potential for seaweed cultivation in relation to the production and consumption of seaweed presently under cultivation.

1. Translated by Masaki Yokoyama and edited by Anthony Carter.

© 1980 by The University of Chicago. 0.226-06603-7/80/0002-0004$01.00

EDIBLE SEAWEED

Although seaweed foods generally contain little in the way of calories and the digestive ratio is rather low, they are rich in inorganic salts such as iodine and calcium. The average digestive ratio for the carbohydrates contained in seaweed is 67.7 percent—rather low when compared with other foods in general. However, seaweed is considered to have certain useful effects as a laxative. Among the various seaweed foods, the variety known as dried *nori* is considered to be unique in that it contains many important vitamins such as A, members of the B group, C, and D, plus substances effective in reducing blood cholesterol, and a protein content of 30 percent—about the percentage found in beans. Dried *nori* in sheet form contains from 36,000 to 50,000 international units of vitamin A per 100 g; thus two sheets of high-quality *nori* are equivalent to a half pound of butter or four chicken eggs. Vitamins B_2 and C also are contained in higher amounts than in ordinary foods.

The custom of eating seaweed is observed in various regions of the world, but in Europe the practice is limited only to a few coastal regions. In Japan, China, Korea, the Philippines, and Polynesia, people eat seaweed of many kinds in relatively large quantities. Especially in Japan, most seaweed production is intended for human consumption as a food, and as such seaweed is regarded as a very important marine product. Edible seaweed is produced by harvesting naturally growing varieties maintained under strict production environments or through intensive cultivation. These products are widely distributed as perishable processed foods. Almost all of the forms of seaweed that were consumed as food in ancient times by the Japanese are still utilized today as food. Above all, it is interesting to note that the consumption of *Porphyra, Undaria,* and *Laminaria* is still growing.

Let us look now at the actual circumstances relative to the production and consumption of *Porphyra, Undaria,* and *Laminaria* as important cultivated products of Japan's marine activity.

Porphyra (nori)

Porphyra belongs to the genus of Rhodophyta, which ranges widely over the earth, with many species found in cold-current areas. There are about 80 species known in the world today. Many of them grow in the tidal zone and appear in their luxuriant state in the summertime in cold-current areas and in the winter in warm-current areas in temperate and tropical climates. *Porphyra* plants are usually leafy, thin, and membranous and range from a few centimeters to about a meter in size (fig. 1). In general, the species found in cold-current areas are larger than those found in tropical areas.

When a *Porphyra* plant reaches maturity it gives off spores. These spores are sexually produced as a result of the fertilization between an egg and a

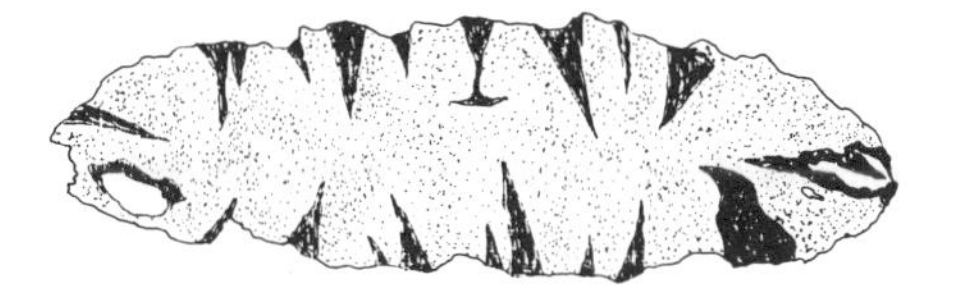

FIG. 1.—*Porphyra* plant

spermatium. The released spores germinate into microscopic moldlike filaments. These germlings usually bore into calcic materials such as clam and oyster shells, forming a filamentous network within the shell. When these filaments mature they give rise to asexually produced spores. These spores, when released, develop into the usual leafy *Porphyra* plants.

The cultivation of *Porphyra* is performed through the culture of the filament-inhabiting shells in such a way that the spores released from the filaments are collected on nursery nets, which are spread and placed horizontally in the sea to allow them to grow. A portion of the nursery nets is dried and stored in polyethylene bags under low temperatures. The nursery nets are spread and retained by props in shallow water or in floating rope frames maintained on floats in deeper water. These rope frames are fixed in place by anchor (fig. 2). Filament-inhabiting shells are cultured in February and March

FIG. 2.—*Porphyra* plant is grown in shallow water (at a depth of 4–5 m at flood tide) on horizontally deployed nets supported by poles. This is called floating cultivation.

and developed under control until September or October. The nursery nets are made between mid-September and early November, and their cultivation goes on from October to April with the *Porphyra* being harvested repeatedly from the same nets (fig. 3). When the harvest is degraded, the nets are replaced by those stored under low temperature. Good-harvesting varieties are isolated and widely cultivated in Japan. On the average, about 1,800 dried *nori* sheets (5.4 kg dried weight and 68 kg wet weight) are harvested per single unit of cultivation area (18 × 1.2 m) in one cultivation period from October to March. Under optimum conditions, harvests can reach five times the average amount. After the harvest has been completed, the *Porphyra* is cut or shredded into very small pieces and made into paper form and then dried into *nori* sheets. The size of a dried Japanese *nori* sheet is 19 × 20 cm and weighs about 3 g. The entire manufacturing process, from the harvest to the completed product, is mechanized. All *nori* sheets distributed on the open market are produced from cultivated *Porphyra*. Naturally occurring *Porphyra* is also collected, but it is not widely used as a commodity, for it is very difficult to make it into smooth, dried *nori* sheets because it is usually mixed with sand. The reason for turning the cultivated *Porphyra* into dried *nori* sheets is related to the fact that it is best preserved in that form. These sheets are used as food in various ways, the most

FIG. 3.—Harvesting *Porphyra* plants from the floating-cultivation system by means of a vacuum-cleaner-like harvester.

popular being to roast and dip it in soy sauce to be eaten with rice. The annual production of cultivated *Porphyra* in Japan was about 1 billion sheets (3,000 tons in dry weight) in 1945. Production then increased geometrically and reached a peak in 1973 of about 9.6 billion sheets (29,000 tons dry weight and 360,000 tons wet weight). After that the rate of return for *Porphyra*-cultivating families dropped due to imbalances in supply and demand which caused the market for dried *nori* sheets to stagnate in comparison with the increasing production costs. At the present time, there are about 30,000 families producing 7 billion sheets (21,000 tons in dry weight and 262,000 tons in wet weight) every year in 64,000 ha of sea cultivation area. As the annual production roughly corresponds with the annual consumption, the annual consumption is also estimated at 7 billion sheets, which means that a Japanese person consumes about 70 sheets per year. This figure should be considered as the maximum limit on consumption in Japan today. The monetary values from these production levels are approximately 140 billion yen, or 700 million U.S. dollars, and this level is about 7 percent of the total fisheries production in Japan.

The Japanese cultivation of *Porphyra* employs unique technologies which are not to be observed in the cultivation of other seaweeds, such as the artificial production of the nursery nets, the low-temperature storage of the nursery nets, the use of floating cultivation facilities, genetic improvement of cultivated *Porphyra*, and the mechanization of the harvesting and processing. These techniques were developed under the impetus of the exploding increase in demand for dried *nori* sheets after World War II. The diffusion of these techniques brought about the stability and increase in production that is to be found to the present day. Following are the basic reasons for the rapid expansion in the cultivation of *Porphyra*: (1) Consumption increased along with the general increase in the living standard, because Japanese, Koreans, and Chinese generally have a taste for *nori*. (2) *Nori* production became profitable with the rise in the market price caused by an insufficiency in supply relative to the increased demand for the product. (3) The expansion of the areas used for cultivation was caused by increased nutrient input into the coastal waters resulting from the rapid urbanization of coastal areas, and this fact worked together with technological advances to make the production of *nori* in deeper coastal waters possible.

In China and Korea, cultivated *nori* is produced in the same manner as in Japan, but the manner of cooking is a little different. The total production in China is not known, but in South Korea it is said to be 1 billion dried *nori* sheets (3,000 tons in dry weight). Until 1974 Japan was importing 460 million sheets annually from South Korea, but after 1975 such imports were stopped completely because of a loss in competitive price position. Certain quality standards were established for the importation of dried *nori* into Japan. In China and South Korea, as in Japan, cultivation which employs the use of artificial nursery nets is found, but in South Korea this practice is not very widely diffused. As a result, there are great fluctuations in production quality from year to year.

Undaria (wakame)

Undaria is a Phaeophyta closely related to *Laminaria (kombu)*. It is found only in the warm-current areas of Japan, the Korean peninsula, and China and nowhere else in the world. It is an annual plant which grows under the low-tide lines to a depth of about 15 m and luxuriates from spring to early summer, being washed away in the summertime. The plant is composed of a tree-branch-like holdfast, a compressed stipe, and a pinnatifid lamina. The length is 1–2 m (fig. 4). Except for the holdfast, the stipe and lamina are dried and salted to be used as food. It is cooked in various ways like vegetables and is commonly eaten in Japan.

In Japan, before 1960, the entire production of *Undaria* came from the collection of the naturally occurring plant. The production at that time was only 60,000 tons (wet weight). After 1965, large-scale production was started and expanded very rapidly. In 1976, the production of cultivated *Undaria* reached about 130,000 tons (wet weight). On the other hand, the collection of naturally occurring *Undaria* dropped to about 20,000 tons (wet weight). Naturally growing *Undaria* is collected in April and May, but the cultivated plant is harvested from January to April.

When it reaches maturity, *Undaria* gives off spores which produce microscopic filamentous plants. The filamentous plants produce *Undaria* through sexual reproduction. In order to cultivate *Undaria*, the seedlings are grown after planting the filamentous plants on twine about 3 mm in diameter. Then these twines with the seedlings are tied between ropes that are 1–3 cm in diameter. The usual *Undaria* plant must adhere to a solid substratum with the branch-like holdfast in order to grow, and if there is not enough adhesion it will not grow well. Thus thick ropes are used as a substratum for the cultivated *Undaria*. The ropes and the twine with the seedlings are placed horizontally and vertically in the water at 1 or 2 m in depth. In general, the spores are placed on twines in May and the filamentous plants are grown under controlled conditions until October. The cultivation in the sea starts in September or October

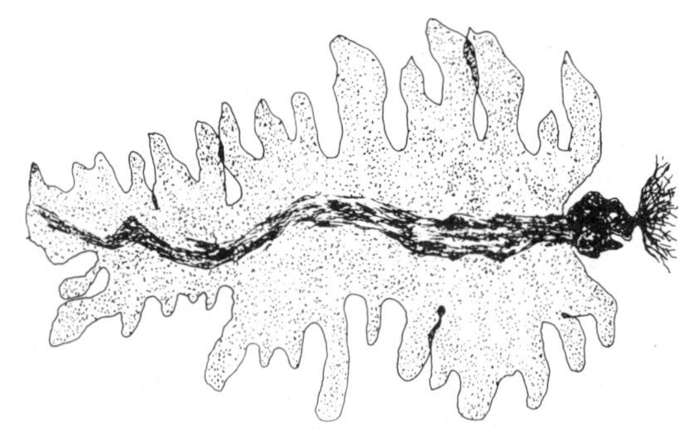

FIG. 4.—*Undaria* plant

when the water temperature drops below 20° C and lasts until the next April when the water temperature rises above 15° C. Harvest commences in January by individually cutting plants which have reached harvestable size. During 1 yr of cultivation (between October and April), an average of 5 kg (wet weight) or a maximum of 13 kg (wet weight) per meter of rope is harvested. *Undaria* is considered to be best if it is appropriately firm and shows a bright green color when boiled. In other words, it is important that the harvesting be completed before the plant gets too old. In Japan, 18,000 families produce 130,000 tons (wet weight) of *Undaria* on 130×10^6 m of a substratum rope retaining materials. This production value is 11.3 billion yen, or 57 million U.S. dollars.

While only the cultivated *Porphyra* plant is used for the production of dried *nori* sheets, in the case of *Undaria* this condition does not apply. In spite of this fact, however, the collection of naturally occurring *Undaria* has decreased, and at the present time the production of the cultivated *Undaria* is six times that of naturally occurring *Undaria*. The expansion in the demand for cultivated *Undaria* was led by a growth in consumption and the technical development of salt preservation. Also, planned production of a standard-quality product became possible with cultivation. Production through cultivation has the advantage of easier harvesting and a labor efficiency that is higher than that attainable with the collection of natural *Undaria*. In China and South Korea, *Undaria* is cultivated in the same manner as in Japan. The production amounts are not known, but a considerable amount of *Undaria* is imported from South Korea into Japan, creating an oversupply in Japan.

Laminaria (kombu)

Laminaria is a large Phaeophyta which ranges in cold-current areas both in the Northern and Southern Hemispheres. Three genera and 10 species of *Laminaria* are used for food in Japan, with the most typical species, *Laminaria japonica*, used for cultivation in Japan and China. It is a perennial marine plant in waters from 7 to 17 m in depth. It grows rapidly during the spring and summer months. After 2 or 3 yr it reaches maturity, gives off spores, and then is washed away. The body is composed of a tree-branch-like diverging holdfast, a cylindric stipe part, and a single strip of coriaceous lamina (fig. 5). The total length is from 2 to 6 m, and the width is 30 cm with the thickness of the lamina being about 3 mm. Only the lamina is used for food. After harvest it is dried, expanded to retain a good shape, sorted according to certain standards, and bundled as the finished product. The dried product is used as a raw material in various foods. It is finely cut or thinly shaved to produce various processed foodstuffs. *Laminaria* is harvested and used only when it reaches full growth and maturity. The naturally growing *Laminaria* is collected only after 2 yr during which the lamina attains the proper thickness. The *Laminaria japonica* has a wide and thick lamina of fine quality and is widely used for various kinds

FIG. 5.—*Laminaria* plant

of processed foodstuffs. Therefore, this species has the highest commercial value.

In Japan, almost all of the *Laminaria* is produced off the coast of Hokkaido and the northeastern Tohoku district of Honshū. The greater portion of the production comes from naturally occurring *Laminaria*. The annual collection of natural *Laminaria* in Japan amounts to about 160,000 tons (wet weight), with 6 percent of it being *Laminaria japonica* (about 10,000 tons, wet weight). Since 1970, following the example of China, the cultivation of *Laminaria* has also been undertaken in Japan, but not in Korea. The production of the cultivated variety is growing rapidly. *Laminaria japonica*, which is of the highest commercial value, is the main species under cultivation. Today, production through cultivation reaches about 20,000 tons (wet weight), which is twice the production of the naturally occurring plants. In China, the production of cultivated *Laminaria japonica* is estimated at 100,000 tons (wet weight). The total production is assumed to be used for food in China, but the manner in which it is processed often differs from that in Japan.

Laminaria has the same life cycle as *Undaria*. *Laminaria,* when it matures, gives off spores which produce microscopic filamentous plants. The filamentous plants, through sexual reproduction, produce *Laminaria* plants. Therefore, the most intensive part of cultivating *Laminaria* is the growing of the filamentous plants on a substratum of twines 3 mm in diameter. The twines with the seedlings are then tied on ropes of 1–3 cm in diameter. The ropes placed in water of from 2 to 7 m deep and allowed to grow. In Japan, a cultivation method which enables the *Laminaria* to grow in only 1 yr with the same or better quality than naturally grown plants has been developed and put into common use. This method is called forced cultivation.

With the forced-cultivation method the spores are collected to grow as seedlings in late August, and the seedlings grow up to 1 cm by the end of September or October, when the water temperature in the cultivation areas drops below 18° C. With wild plants, spores are given off from spring through fall, and the seedlings with a length of 1 cm do not appear until the next February.

In order to grow thick *Laminaria* of fine quality, it is necessary to maintain proper cultivation density. For this purpose, thinning to adjust for density is completed in February before the rapid-growth period. Afterward, as the plant grows, the ropes gain weight, necessitating the addition of more floats to sustain needed buoyancy. Then, after May when the lamina grows thicker, ropes are sustained at or near the surface level of the sea. The *Laminaria* is harvested from August to mid-September.

In 1976, 3,100 families produced 22,096 tons (wet weight) of cultivated *Laminaria* with 1,613 $\times$ 10^3 m of substratum ropes. The total value of that amount came to 4 billion yen, or 20 million U.S. dollars. The average annual production capacity per unit area is estimated to be about 14 kg (wet weight) per 10 m of substratum ropes. It is estimated that from 1933 to 1943 the annual consumption in Japan was around 100,000 tons. Today, 160,000 tons (wet weight) of natural *Laminaria* and 20,000 tons (wet weight) of cultivated *Laminaria*, totaling 180,000 tons, are consumed, showing a remarkable increase. The increase in consumption has resulted from improvements in living standards, as has been the case also for *Porphyra* and *Undaria*. In spite of the demand, the production of natural *Laminaria* of fine quality, such as the *Laminaria japonica,* has decreased. This provided the needed impetus for the cultivation of *Laminaria japonica*. It is not likely, however, that all production will be derived from cultivation. At the present time there is concern that an oversupply of cultivated *Laminaria* will cause the price to drop.

CULTIVATION OF SEAWEEDS FOR INDUSTRIAL MATERIALS

There are three major industrial materials which are derived from seaweeds. Carageenan, taken from some types of Rhodophyta such as *Chondrus, Eucheuma,* and *Gigartina,* solidifies casein and therefore has various uses in food processing—especially as a coagulator in dairy products. Agar-agar, or vegetable gelatin, extracted from *Gelidium, Gracilaria,* and the like, has several important properties. It solidifies at around 30° C, even in dilute solution, and liquifies at 80° C. Its high melting point, strong gelling power, high water-retaining capacity, and stability even in the presence of organic solvents and bacteria lead to many uses, such as in stabilizers for food processing, in antidessicants, etc. Alginic acid soda, from the Laminariales, Fucales, and Phaeophyta, slowly dissolves in water forming a sticky solution. It has many uses as a stabilizer for sticky foods and as a paste in dye prints.

All the seaweed sources of these materials, except the *Eucheuma* from which carageenan is extracted, are not generally cultivated but are collected from naturally growing plants under carefully controlled conditions.

Eucheuma

Eucheuma is a Rhodophyta that ranges from tropical to subtropical zones and grows on reefs in the lower tidal zones. It has a cartilaginous compressed or cylindric body branching irregularly (fig. 6). Branches are twined together, and the whole forms a large cluster at a height of from 10 to 25 cm. A fast-growing strain of *Eucheuma*, with extraordinary vegetative capacity, has been isolated and is cultivated. The regenerating piece of plant is fixed on a substratum of ropes or nets. These nets and ropes are placed horizontally just under the low-tide line and fixed to props. The plant is harvested four times each year. A certain amount of the regenerating plant piece must be left on the substratum. After the harvest is completed the plant is dried.

The standard-size operating unit is one family. Using 300 nylon ropes, each 10 m in length and 4 mm in diameter, in a 0.5-ha cultivation area with 25–30 regenerating plant pieces of 200 g weight attached to each rope, each piece grows to 3 kg in 90 days. Thus a family has a total production capacity of about 22,500 kg (wet weight). When dried the weight is reduced to about one-tenth; therefore, 2,250 kg of dried *Eucheuma* is produced. Deducting expenses for materials, the annual earnings for the four harvests is said to be 1,360 U.S. dollars (272,000 yen).

In the Philippines the production of cultivated *Eucheuma* was 0.6 tons (wet weight) in 1971 and then increased to 10,000 tons (wet weight) in 1974. On the other hand, the production of natural *Eucheuma* dropped from 534 tons (wet weight) to 75 tons (wet weight). In 1974, 2,249 families cultivated *Eucheuma*. The production of cultivated *Eucheuma* in the Philippines is estimated to be 3,500 tons (dry weight), but the total world demand is estimated at not over 4,000 tons a year. Most of the cultivated *Eucheuma* is exported to the United States as a raw material for carageenan. The total export earnings in 1970 were 170,000 U.S. dollars, and these increased to 4.5 million U.S. dollars in 1974.

FUTURE POTENTIALS IN SEAWEED PRODUCTION

In this survey I have dealt with the cultivation of *Porphyra, Undaria, Laminaria*, and *Eucheuma* because it is the basis of intense and well-organized industrial activities. *Porphyra, Undaria*, and *Laminaria* are used directly for food, whereas *Eucheuma* is used only as an industrial raw material. These various types of seaweed cultivation are supported by a high demand. While *Porphyra* has been cultivated for a long time, the farming of the other three goes back only to 1965. In Japan there still is an abundant natural production of *Undaria* and *Laminaria*, but cultivation is rapidly increasing because it enhances quality, facilitates standardization, and increases labor productivity.

In Japan, China, and South Korea, as well as in other parts of the world,

FIG. 6.—*Eucheuma* plant

there are many edible seaweeds which are not as yet cultivated, but as demand expands they will probably come under cultivation. When the great utility of seaweeds as sources of food is more readily understood and more information is available, the demand for seaweed in Western countries will increase even if the seaweed is not yet commonly eaten. The distinctive features of seaweed foods as "nonfattening" and "noncaloric" will lead to the study of these products in Western countries, where food is too rich in fat and protein. Already, in the United States, the use of seaweeds as condiments is gradually coming into practice. Also in Western Europe, many people have developed the custom of eating seaweed, and it is likely that production will increase in the near future. When the demand for carageenan, alginic acid, and agar-agar increases, the seaweeds which are the basis for these materials will also be increasingly cultivated. Among these, only carageenan is produced from a cultivated genus, *Eucheuma.* Based on present cultivation techniques, cultivation of seaweeds for alginic acid and agar-agar is not improbable. For such purposes plant breeding which will produce species with higher growth yields than the natural seaweeds will be an important subject for further study and research.

In the cases of *Porphyra, Laminaria,* and *Eucheuma,* the faster-growing strains are isolated and used. In the United States and Canada, mass cultivation of the fast-growing varieties of *Chondrus* and *Gigartina* in water tanks has been experimented with for the production of raw materials for carageenan. In Japan, construction of sea forests and seaweed farms has been the subject of experiments using large Phaeophyta in order to encourage the growth of coastal fish and shellfish resources. In such cases the experience and knowledge gained from seaweed cultivation should have wider applications.

Each one of the seaweeds described here—*Porphyra, Undaria, Laminaria,* and *Eucheuma*—has a different life cycle, and the cultivation techniques developed in relation to each of the life cycles suggest even greater potentials for seaweed cultivation in the future.

SELECTED BIBLIOGRAPHY

Deveau, L. M., and Castle, J. R. *The Industrial Development of Farmed Marine Algae: The Case History of Eucheuma in the Philippines and the U.S.A.* Report of the FAO Technical Conference on Aquaculture, Kyoto, Japan, FIR:AQ/Conf/E.56.

Doty, M. S. "Farming the Red Seaweed, *Eucheuma,* for Carageenans." *Micronesia* 9, no. 1 (1973): 59–73.

Druel, L. D. "Past, Present, and Future of the Seaweed Industry." *Underwater Journal and Information Bulletin* 4 (1972): 182–91.

Hasegawa, Y. "Progress of *Laminaria* Cultivation in Japan." *Journal of the Fisheries Research Board of Canada* 33, no. 4 (1976): 1002–6.

Miura, A. "*Porphyra* Cultivation in Japan." In *Advance of Phycology in Japan,* edited by J. Tokida and H. Hirose. Jena: Gustav Fischer, 1976.

Neish, I. C. *Culture of Algae and Seaweeds: Developments in the Culture of Algae and Seaweeds and the Future of the Industry.* Report of the FAO Technical Conference on Aquaculture, Kyoto, Japan, FIR:AQ/Conf/76/R.1.

Saito, Y. "*Undaria.*" In *Advance of Phycology in Japan,* edited by J. Tokida and H. Hirose. Jena: Gustav Fischer, 1976.

Oil: Two Billion B.C. – A.D. Two Thousand

T. F. Gaskell
Oil Industry International Exploration and Production Forum, London

S. J. R. Simpson
St. James's Research Group, London

FORMATION

The earth consists of an iron-nickel core and a surrounding mantle corresponding roughly in composition to the silicate mineral olivine ($FeMgSiO_4$). Natural carbon is only 0.034 percent of the earth,[1] while the carbon that is fundamental to our energy supplies (92 percent, including oil, natural gas, and solid fuels) has been formed as a consequence of events and processes subsequent to the establishment of the planet and derived from the recycling of once-living organisms.

The greatest concentration of carbon is in living cells (approximately one atom in every 10 compared to one in every 2,000 atoms in the crust of the earth and one in every 70,000 atoms of ocean water). Thus, it is in the accident of life itself and the earth's evolution that one might discover the origin of oil. Various radioactive-isotope methods of dating stony meteorites suggest that the earth is approximately 4.6 billion yr old:[2] the origin of life (anaerobic bacteria) has been put at 3.5 billion yr and the first presence of free oxygen at about 2 billion yr, from which period eukaryotic cells (algae) and the first multicellular plants and animals were formed.

Oil could be described as decayed marine organisms, the formation of which may have begun at this time in the biochemical evolution of life when the conditions for biodegradation were also established. It is possible, therefore, that oil's earliest formation would have been sometime in the late Precambrian epoch.

Oil is recoverable as a consequence of geological forces. Most notable of these are the processes of plate tectonics, in which the continental land masses, 20 mi or so thick, float at a rate of a few centimeters per year over the earth's

1. Richard E. Dickerson, "Chemical Evolution and the Origin of Life," *Scientific American* 239 (September 1978): 62–83.
2. Ibid.

© 1980 by The University of Chicago. 0-226-06603-7/80/0002-0014$01.00

mantle, giving rise to rock-layer faulting and fissures. "Continental drift" explains the distribution of living species, the consistency of continental topology, and the similarity between the rock formations of adjacent land masses (i.e., Africa and South America), and gives clues to the location of physiographic basins in which oil-bearing sedimentary rocks have formed. Continental drift has formed and reformed these physiographic basins in which rock sediments have accumulated, and created the geological conditions for the entrapment of petroleum reservoirs. It should be noted in this context that there is a regular pattern in the worldwide distribution of oilfields. Originally marine organisms were deposited in tropical zones which drifted to temperate and arctic positions (see fig. 1). This precondition for oil formation rules out many sedimentary areas whose original formation can be traced to nontropical zones.

The largest oil discoveries have been made in sediments of 65 million yr old or earlier (table 1).[3] The significance of these oil-producing horizons is not the relationship of depth to quantity but in the likelihood of conditions arising in which oil-bearing sand or limestone will be trapped beneath impermeable rock formations. Also, since the ocean sediments are not static and are subject to the continuous movements of the sea floor, older rock formations often overlie more recent sediments.

Two-thirds of the earth's surface is covered by oceans several miles deep. The topography of the ocean floor is quite uncharacteristic of the continents, and has geological structures equally different.[4] Drilling offshore, however, is carried out at depths of 200 m or so. Although this is often climatically hostile, 200 m is shallow compared to the 2-mi depths of most of the oceans. The deep oceans do not contain the accumulations of deposited material which make up the sedimentary rock sequences (up to 3,000 m) that have built up around continental shorelines, where offshore oil can be found (see fig. 2 and table 2).

Thus, oil and gas derive from organic plant and animal material accumulating in fine-grained rock sediments. The formation of petroleum has been described as part of the general process of sedimentation. Nearly all sedimentary rocks contain a small percentage of organic material from which petroleum could be formed. An isolated globule of oil in the pore spaces between the grains of a sandstone rock (fig. 3*A*) can only be affected by a few forces: (1) capillary attraction; (2) surface tension; (3) gravity—the effect of buoyancy, since oil is lighter than water; and (4) pressure differences due to the flow of water through the rock pores. Of these, the second acts so as to resist movement and the fourth will be of sufficient magnitude to overcome it if the length of the oil globule in the direction of movement is of the order of 1 m (fig. 3*B*). Most sediments ultimately reach the sea and are either dispersed or

3. British Petroleum Co., *Our Industry Petroleum, 1977* (London: British Petroleum Co., 1977), pp. 449–526, 527.

4. T. F. Gaskell, *Physics of the Earth* (London: Thames & Hudson, 1970), pp. 62, 59; and *Under the Deep Oceans* (London: Eyre & Spottiswoode, 1960), pp. 79–101.

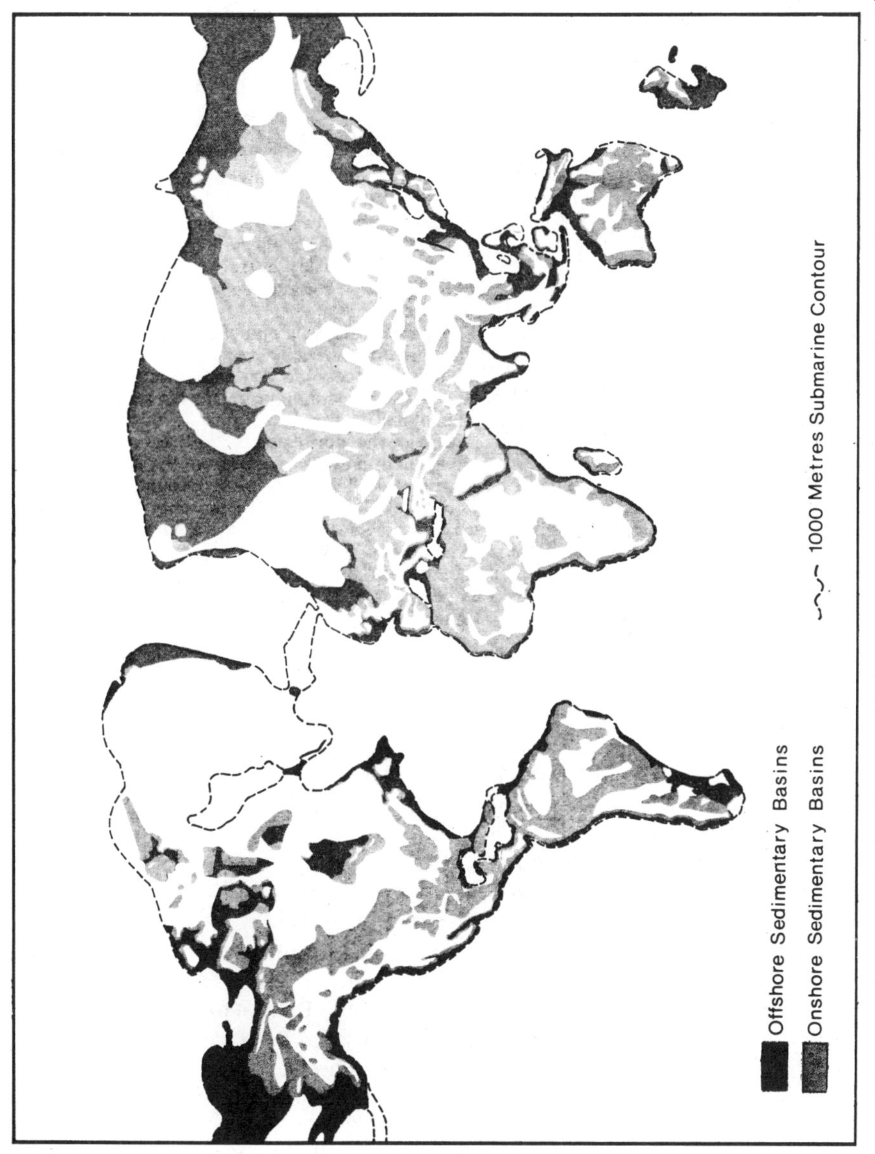

FIG. 1.—World sedimentary basins (taken with permission from *Our Industry Petroleum* [British Petroleum, 1977])

TABLE 1.—GEOLOGICAL/GEOGRAPHICAL DISTRIBUTION OF OIL RESERVES

Time Scale (Million Years) and Producing Horizon	Oil-producing Region
0–2:	
Pleistocene	Indonesia
Holocene	Japan
2–7 (Pliocene)	Bulgaria/Yugoslavia/Italy
	Brunei/Indonesia/Thailand
	U.S./Gulf Coast (offshore/coastal); California
7–26 (Miocene)	Austria/Czechoslovakia/Poland/Yugoslavia
	Colombia/Trinidad/Venezuela/Mexico
	Brunei/Burma/Indonesia/Taiwan/India
	Iran/Iraq/Turkey/Egypt
	Angola/Cabinda/Gabon/Nigeria
	Japan
26–38 (Oligocene)	Alsace/Bulgaria/Poland/Yugoslavia
	Colombia/Venezuela
	Iran/Iraq
	Libya
	India/Burma
	Nigeria
38–54 (Eocene)	Austria/Albania/Yugoslavia
	Colombia/Ecuador/Mexico/Peru/Venezuela
	Algeria/Egypt/Libya
	Iraq/Neutral Zone
	Angola/Cabinda/Gabon
	India/Pakistan
	Alaska
	Queensland (Australia)
54–65 (Paleocene)	Austria/U.K. (offshore)
	Colombia/Venezuela
	Libya
	Morocco
	Pakistan
65–136 (Cretaceous)	Austria/Aquitaine/Paris Basin/Netherlands/West Germany
	Argentina/Colombia/Bolivia/Brazil
	Chile/Ecuador/Mexico/Peru/Venezuela
	U.S. (Gulf Coast/East Texas/Rockies)
	Alberta/British Columbia/Alaska
	Cuba
	Abu Dhabi/Bahrain/Dubai/Kuwait/Neutral Zone
	Iraq/Oman/Saudi Arabia/Syria/Turkey
	Algeria/Egypt/Libya/Tunisia
	Angola/Cabinda/Gabon

TABLE 1. (*Continued*)

Time Scale (Million Years) and Producing Horizon	Oil-producing Region
	China
	Queensland (Australia)/New Zealand
136–95 (Jurassic)	U.K. (onshore)/Aquitaine/Paris Basin/U.K. (offshore)
	U.S. (Gulf Coast/Rockies)
	Alberta/Saskatchewan
	Mexico
	Argentina
	Abu Dhabi/Qatar/Saudi Arabia
	Morocco
	Pakistan
	Queensland (Australia)
	China
	USSR
195–225 (Triassic)	Italy/West Germany
	U.S. (Rockies)/Alberta/Alaska/British Columbia
	Argentina
	Algeria/Tunisia
	Iraq
	China
225–80 (Permian)	U.S. (West Texas/Mid-Continental/Rockies)
	East Germany
	Egypt
	USSR
	U.K. (North Sea)
280–345 (Carboniferous)	U.K. (onshore)/U.K. (North Sea)
	British Columbia/Alberta/Saskatchewan/Manitoba
	U.S. (Mid-Continental/Rockies)
	Bolivia/Argentina
	Mexico
	Alaska
	Algeria/Egypt
	USSR
345–95 (Devonian)	U.S. (Mid-Continental)
	U.K. (offshore)
	Alberta/Ontario/Northwest Territories
	Bolivia
	Algeria
	USSR
395–435 (Silurian)	U.S. (Mid-Continental)
435–500 (Ordovician)	U.S. (Mid-Continental)
	Algeria/Libya
	Queensland (Australia)

TABLE 1. (*Continued*)

Time Scale (Million Years) and Producing Horizon	Oil-producing Region
500–570 (Cambrian)	U.S. (Mid-Continental)
	Ontario
	Algeria/Libya
	Queensland (Australia)
570–4,500 (Pre-Cambrian)	Libya

SOURCE.—British Petroleum Co., *Our Industry Petroleum, 1977.*

accumulated. Most of the oil that is generated similarly disperses, so while oil itself is not an accidental phenomenon, its accumulation is subject to the geological variables described in this paper.

A sedimentary basin began with a subsidence of land and a subsequent invasion by the sea. Through a cycle of transgressive and regressive movements over millions of years and a consequent interchange of marine and terrestrial sediments and rock types, the geological conditions were established for the formation and accumulation of petroleum and other fossil fuels. Petroleum does not normally occur in significant quantities above 1,500 m below the surface. This depth puts an apparent upper limit on an oil-producing horizon to approximately the Lower Tertiary (Paleocene) to Upper Secondary era (Cretaceous) which in time represents some 50–100 million yr.

DISTRIBUTION

The earth does not offer uniform probabilities for the oil explorer. The areas where oil might be found, that is, in physiographic basins containing rock sediments, are clearly recognized (fig. 1). The shaded areas of the map thus far

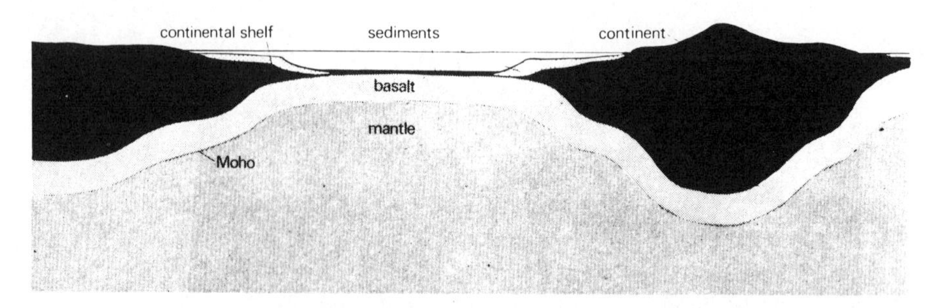

FIG. 2.—Relative thickness of the earth's crust under the continental land masses and under the ocean; also the relative thickness of sedimentary rocks between the continental shelf and deep ocean should be noted. Source.—T. F. Gaskell, *Physics of the Earth* (London: Thames & Hudson, 1970).

TABLE 2.—PRINCIPAL OFFSHORE SEDIMENTARY BASINS

Location	Producing Horizons
Bering Straits; Beaufort Sea (Alaska)	Cretaceous
Arctic Ocean (USSR, Canada)	Cretaceous
North Sea	Permian/Jurassic
Bass Strait (Australia)	Eocene
Northern Australia	Paleocene
Nigeria	Olig-Miocene
Malaysia/Indonesia/Australia/New Zealand	Miocene/Pliocene
Sea of Okhotsk (USSR)	Cretaceous
Angola/Cabinda/Congo/Gabon	Cretaceous
Baltic Sea	Cretaceous
Gulf of Mexico	Cretaceous
Mediterranean (Italy/Egypt/Tunisia)	Eocene/Miocene/U. Cretaceous/ Paleocene
Persian Gulf	Cretaceous/Jurassic
Argentina	U. Cretaceous/Jurassic
Caribbean	Miocene
Brazil	Cretaceous

NOTE.—In addition to delineated basins there are many other opportunities for offshore oil discovery along the continental shelves, where onshore sedimentary basins may extend oil-bearing rock formations.

have yielded about 30,000 oilfields, less than 8 percent of which (240) have reserves of more than 68×10^6 metric tons (MT) (500 million bbl). In hostile environments such as the North Sea, and further north into the Barents/Kara Seas, Beaufort Sea, or Gulf of Alaska, this level of reserve is below the minimum (800 million bbls) that can justify development costs at current oil prices, and while the purpose of this paper is not to discuss the economic aspects of oil or energy development, even with the growing value of fields which were hitherto considered uneconomic it is clear that oil, in relation to other energy sources, will not be a very attractive proposition.

Offshore Potentials

There is no reason to suppose that offshore oil would be more abundant than onshore reservoirs. However, since the aggregate of oil-bearing offshore areas is similar in size to oil-bearing land areas (50×10^6 km^2 and 66×10^6 km^2, respectively), it is not unreasonable to suppose that similar amounts of oil are present. The ocean represents 71 percent of the earth's surface—360×10^6 km^2; some 14 percent of this area is classified as containing potential oil-bearing sedimentary rock. The earth's land area, that is, the remaining 29 percent, represent some 201×10^9 km^2, of which approximately 33 percent contains potentially oil-bearing sedimentary rock formations.

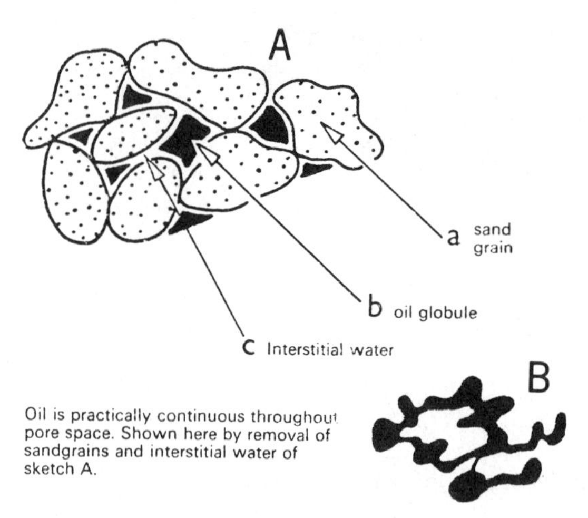

FIG. 3.—Diagram showing the position oil globules take up in the pores of a reservoir rock (taken with permission from *Our Industry Petroleum* [British Petroleum Co., 1970]).

Of the onshore fields found thus far, four contain a total of one-fifth (21 percent) of known reserves—an average of 4.5×10^9 MT each. One of these fields is in Venezuela and three are in the Arabian peninsula. The probability of finding similar fields offshore is considered to be remote in view of the prospecting already carried out. Nevertheless, improved detection techniques, applied in areas hitherto pronounced nonpetroliferous, may uncover economically feasible fields to sustain demand at the levels required (fig. 4) while alternative energy sources are being developed.

For example, recent developments in the Mexican Campeche Sound[5] (first discovered in 1972) have led some to suggest that 350×10^7 MT per annum will be produced by 1982. However, such reserves have not been proved, even though it is clear that Mexico's 1.6 million sq mi of sedimentary area gives her the potential of being a major oil producer for some time to come.[6]

The prolific Middle East oil-bearing region covers an area of approximately 400,000 sq mi and contributes some 36 percent of world annual production; it contains some 56 percent of world "published proved"[7] reserves. Another such find would not remove the urgency to develop alternative energy resources; even less so since any such find would be located in either a hostile environment or politically sensitive area, or both.

5. Frank E. Neiring, Jr., "A New Force in World Oil," *Petroleum Economist* (March 1979), pp. 105–13.

6. T. J. Stewart-Gordon, "Mexico's Oil—Myth, Fact and Future," *World Oil* (February 1, 1979).

7. American Petroleum Institute, *Oil and Gas Journal* (Worldwide Oil Issue) (December 26, 1977).

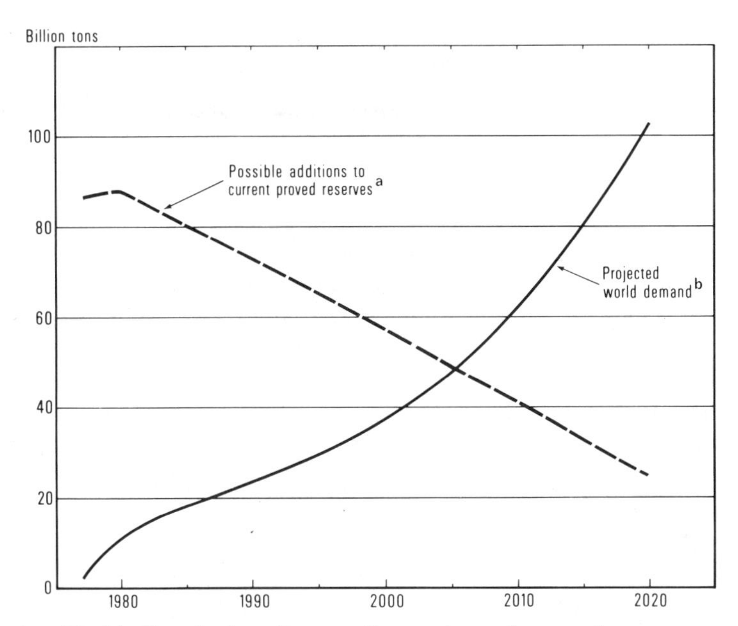

FIG. 4.—World oil outlook: *a,* By new discoveries and upward reassessment (recoverability) of existing fields, less consumption; 1977 reserves taken to be 88.6 billion tons, and a net depletion of 1.6 billion tons/yr from 1980 has been assumed. *b,* Assuming an increase in consumption of 5 percent annually from 1980 (1977 consumption taken to be 2.9 billion tons).

The North Sea is comparable in size to the Persian Gulf, and a comparable output has been implied.[8] The Persian Gulf offshore currently produces 1.6×10^8 MT per year from reservoirs of Olig-Mio-Cret-U. Jurassic horizons. Oil in the North Sea, on the other hand, is of older formation — Paleocene/Jurassic/Permian/Devonian. The most optimistic "estimates" for these fields put maximum annual production at 1×10^9 MT, which is some 60 percent more than the Persian Gulf. The feasibility of this proposition cannot be necessarily inferred on geological grounds, but if indeed the 27 offshore Persian Gulf fields whose average annual capacity is 6×10^7 MT were prorated to the more important North Sea fields, then an ultimate North Sea oil resource base, at least in mathemetical terms, could be in the region of 6×10^9 MT, which is about the equivalent of Kuwait (10 percent of world proved reserves) producing from Cretaceous horizons.

The world's largest fields are to be found in the rock layers associated with the Mesozoic period, most notably the vast Middle East fields (which produce from Cretaceous sands and limestones), and the major U.S., Mexican, North

8. Peter R. Odell, "Oil and Gas Exploration and Exploitation in the North Sea," in *Ocean Yearbook 1,* ed. Elisabeth Mann Borgese and Norton Ginsburg (Chicago: University of Chicago Press, 1978), p. 140.

African, and South American fields. It has been suggested that in the Cretaceous epoch there could have been an enormous expansion in the numbers of marine organisms, which could account in some part for this phenomenon.

Whereas some interesting oil discoveries have been made in formations as recent as 2–7 million yr (Pliocene), notably in California (which is regarded as being among the richest petroliferous regions known), it may be that these have resulted from a migration of oil formed in an earlier era and subsequently become entrapped in later faulted structures. As a corollary, the formation of oil may have occurred over one particular epoch (Cretaceous, for instance) and subsequently migrated "upward" and "downward," which would account for its distribution.

The recent Canadian discoveries in the Beaufort Sea, which are reported as comprising one of the world's biggest oilfields, are likely to be an extension of the Cretaceous reservoir in the Beaufort Basin. The time-lapse of the Precambrian era (constituting the first layer of sedimentary rock; some of the most ancient known are about 3.2 billion yr old) almost predates the earliest known oil formations of the Cambrian epoch (in the mid-continental United States, Ontario, North Africa, and Queensland, Australia). One field, however, the Nafoora-Augila in Libya, produces from a Precambrian horizon.

It is interesting to consider the geomorphology of Norway and Scotland, which is characterized by igneous rocks. These rocks, formed by cooling of molten material, do not contain oil. A simple extrapolation would suggest that the sea between Norway and Scotland also covers non-oil-bearing rocks. However, geological reconstruction indicates a breakthrough from the Atlantic more than 150 million yr ago, and a continuous deposition of sediment brought from the erosion of northwestern Europe. This has proved to be the case, and the oilfields that are being discovered in the North Sea are mostly situated near the middle line between Scotland and Norway where one would expect the sediments to be thickest. The lesson to be remembered, when applying North Sea experience to other parts of the world, is that geographical coastlines can also be geological boundaries and that land geology will not necessarily continue out to shallow water.[9]

The oilfield discoveries in the northern part of the North Sea have demonstrated that several different types of oil accumulation exist. The BP Forties field is in a sandstone which is about 70 million yr old. The Ekofisk group of fields, which lie mostly in Norwegian waters, are in older rocks of the Cretaceous (chalk) series; some of the larger fields, such as Brent and Ninian, east of the Shetlands, are in still older rocks of Jurassic age (180 million yr old).

The wide range of age of the reservoir rocks is encouraging for further discovery, because it may indicate a long period in which conditions were good for oil formation. On the other hand, oil is known to migrate from one rock

9. B. Cooper and T. F. Gaskell, *The Adventure of North Sea Oil* (London: Heinemann, 1976), pp. 45–61, Gaskell, *Under the Deep Oceans,* pp. 79–101.

layer to another, so that there may have been one particular period of abundant oil production in the past which has provided the source for all the reservoirs. The understanding of what has happened improves with each new discovery that is made, so that there is still reason to expect productive drilling of wildcat wells. Nevertheless, on the evidence available there is little basis for predicting any dramatic differences in the world's supply outlook.[10] It is more likely that some countries will become newly self-sufficient in oil provided that the resources are available to exploit their good fortune.

Seismic Surveys[11]

The seismic survey is used to provide detailed information of sediment structure against which a decision to drill exploratory wells is taken (see fig. 5). Seismic methodology cannot identify rock layers unambiguously because of the velocity range of individual rock types. Clays are usually within a 1.5–2.4 km/sec range, measured by the travel times of shock waves induced by explosive charges. Limestone has a similarly wide range, from 2.4 km/sec for chalk to 6.5 km/sec for dense massive limestone; granite rocks range from 5.0–6.0 km/sec, while basic volcanic rocks have been measured from 5.4–6.9 km/sec. Refracted waves can provide the general information about rock formations, thicknesses, etc. Reflection provides the detailed information of sediment structure.

Figure 6 shows several different possible paths for sound waves sent out by an explosion at *A*. The "direct" wave on the surface layer goes horizontally to the detector at *B* and will always arrive before the reflected wave *ABC*. The "refracted" wave traves along the path *ADEB*. Part of the path is down and up *AD* and *EB* in the top layer, but the main part is the horizontal section *DE* in the underneath layer. The wave traveling along *AD* is refracted at the boundary between the two layers. The sound wave is bent again at *E* to travel upward to the top layer and back to the detector at *B*, in a perfectly symmetrical manner. Although the path *ADEB* is longer than either that of the direct wave *AB* or that of the reflected wave-path *ACB*, it is not necessarily longer in travel time. If the horizontal distance *DE* is great enough, the time saved in the fast layer will more than make up for the down and up time in the slowest top layer.

10. Edward Symonds, "Offshore Oil and Gas," *Ocean Yearbook 1*, pp. 114–38. The point of our article is to illustrate that even highly significant new finds will not make any dramatic difference to the world's oil supply outlook, and we would refer the reader to standard predictions of energy utilization in the year 2000 and particularly to *Energy: Global Prospects, 1985–2000*, Workshop on Alternative Energy Strategies (New York: McGraw-Hill, 1977). See also the *Survey of Energy Resources* published by U.S. National Committee of the World Energy Conference, 1977.

11. British Petroleum Co., *Our Industry Petroleum, 1977*, p. 85; Gaskell, *Physics of the Earth*, p. 59.

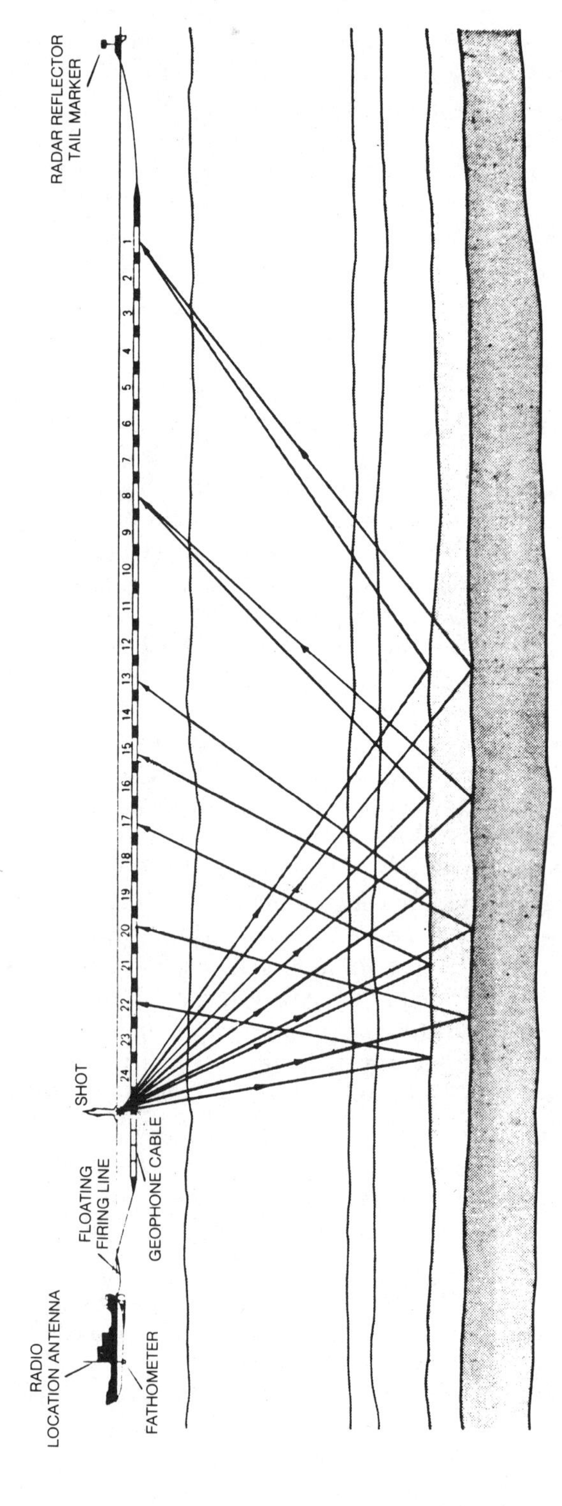

FIG. 5.—Marine reflection shooting (taken with permission from *Our Industry Petroleum* [British Petroleum Co., 1977])

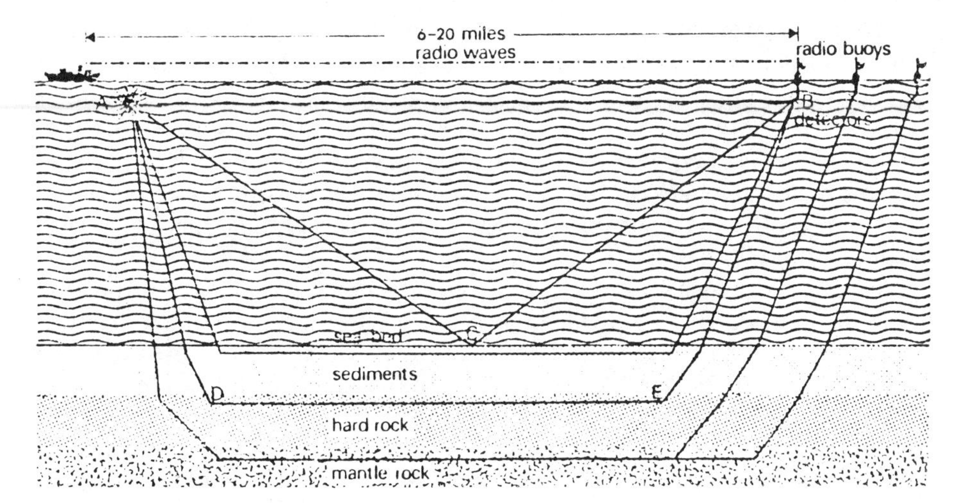

FIG. 6.—Seismic refraction. Source.—T. F. Gaskell, *Physics of the Earth* (London: Thames & Hudson, 1970).

The chances of finding oil in an area which could cover up to 200,000 sq mi of sedimentary rock formations by systematic drilling alone would be about 0.0005. Application of seismic techniques in association with drilling brings the probability of finding suitable oil-bearing structures to between 0.05 and 0.1.

DEPLETION

Recovery

The amount of oil that can be recovered from a well varies according to several highly significant characteristics of rock core: permeability, porosity, oil content, and connate (interstitial) water content. In addition, the properties of the oil itself and its pressure in the reservoir determine the feasibility of production.[12] One highly significant factor in the economic feasibility of a well is the specific gravity of the oil, expressed as °API (American Petroleum Institute)—that is, whether it is light (say, 24°–40° API) and under an effective water-gas drive mechanism, which might yield 80 percent of the total reservoir, or heavy, viscous oil (10°API) under a less effective drive mechanism, which may not yield as much as 10 percent of the oil in situ.

12. Obviously degrees of viscosity determine oil's utility and the economics of refining. However, we suggest that the reader refers to other sources for a determination of oil trade flows according to the different "cuts" which obtain from the enormous range of crudes produced. See, for example, James H. Gary and Glenn E. Handwerk, *Petroleum Refining, Technology and Economics* (New York: Dekker, 1975).

Nigerian oil is typical of the former, producing from Oligo-Miocene horizons. Venezuela, on the other hand, has an estimated 735×10^9 bbl of 8°–10° API in the Orinoco Basin, and the likelihood of production at this time is remote. But in Mexico, oil from Campeche already drilled is about 20 °API, similar to the Chicontepec basin. A recent Chevron find in the North Sea of similar-gravity oil makes production equally unlikely. In contrast, an attractive feature of most North Sea oil reserves is the generally high specific gravity, ranging from 36° to 44°API.

The rate at which oil can be produced depends upon the permeability of the rock—that is, the facility of the contained oil to migrate, which depends upon the size of the connecting channels between one rock pore and the next.

In addition to permeability of the rock, the other key factor in determining the development of an oil reservoir relates to the pressure of the oil and its association in this respect with gas and water, thus, its ability to flow. Gas and water provide the pressure for the oil to flow—as more oil is produced, natural pressure declines. At this stage it may be necessary to introduce secondary and tertiary recovery techniques that consist of pumping gas, water, or chemicals into the wellhead.

Clearly there is an economic advantage in allowing the oil to flow under natural pressures. However, frequently it is necessary to inhibit the natural flow in order to avoid a situation in which the gas or water drive mechanism overtakes the oil, thus necessitating earlier introduction of pressure maintenance. Middle East oil comes from highly permeable, cracked, and fissured rock; production thus far has relied upon the primary forces. In the North Sea the Forties oilfield has a low gas/oil ratio (expressed as cubic feet per barrel), and pressure maintenance has been anticipated for this and other North Sea reservoirs. By interesting comparison, Mexico's Reforma reservoirs are underpressured even though gas/oil ratios average about 1,500:1. There, waterdrive is ineffective at natural pressures, resulting in the necessity for water-injection programs.

About 10–25 percent of oil in a reservoir may be recovered unter natural pressure. Enhanced recovery techniques have increased production from some 25 percent of a reservoir in the 1940s (United States) to about 42 percent today. It would not be beyond technological ingenuity to improve this record, but the feasibility of such action could only be justified in the absence of alternative energy sources, where cost became immaterial. Statements of "ultimate" reserves must, therefore, be regarded with some caution.

Protecting the Ocean Environment

Between 2×10^6 and 5×10^6 MT of crude oil and its derivatives are released into the oceans each year, for one reason or another, from the sources shown in table 3. Thus about 0.14 percent of annual oil production ends up as detritus in

TABLE 3.—CRUDE OIL RELEASED INTO THE OCEANS

Source	Amount (Millions of Tons)	Percentage of Total
Tankers	1.25	38.0
Other shipping	.25	7.6
Industry effluent	1.71	52.0
Offshore exploration and production	.08	2.4
Total	3.3	100.0

SOURCE.—P. D. Wilmot and A. Slingerland, eds., "Technology Assessment and the Oceans," *Proceedings of the International Conference on Technology Assessment, Monaco, 26–30 October 1975* (Guildford: IPC Science & Technology Press, 1977).

the oceans. The worst effects of oil pollution are in estuaries and harbors and in coastal waters above the continental shelf. We can see that although the amount of oil "spilled" is not significant in global terms, its localized effects can be serious in the disruption of a local fishing industry or human amenity or upon local flora and fauna. Pollution control, therefore, is in practical terms largely a matter of regional responsibility to identify culprits and obtain compensation. By these means effective international agreements controlling maverick waste disposal might be established to protect the coastlines and territorial waters of those countries without the means of self-protection.

The probability of spills originating from blowouts, based upon present worldwide blowout statistics, is assumed to be in the range of 0.001 (one blowout in every 1,000 wells drilled) to 0.0015 (one blowout in every 667 wells drilled).[13] The most active area in the world has been the Gulf of Mexico, where offshore oil production started in earnest in the 1940s and where, for example in 1972, 17 percent of U. S. domestic crude oil originated. Although some mistakes have been made and accidental blowouts have occurred, there is no obvious ecological damage, for instance, along the Louisiana and Texas coasts. Offshore exploration and production are now taking place in many of the continental shelves of the world, and the lessons learned from the earlier activities are being applied, with suitable climatic and political modifications, to new areas.

We do not wish to understate the consequences of damage by oil pollution, particularly since it is difficult at present to predict the effects of changing levels on the growth and reproduction of marine life. On the other hand, it does appear that the priority is to avoid accidental (or deliberate) discharge of toxic materials and to control the discharge of materials which are known to cause ecological imbalance. In places such as Santa Barbara, California, and Cornwall, England (where the *Torrey Canyon* sank), spectacular oil spills have had no lasting effect. This also applies to the *Amoco Cadiz* spill (Brittany,

13. D. R. Blaikley et al., "'Sliktrak'—A Computer Simulation of Offshore Oil Spills, Clean-up, Effects and Associated Costs," *Proceedings of the 1977 Oil Spill Conference,* EPA/API/USCG (New Orleans, March 1977).

France), where it is reported that the provisional balance sheet indicates less damage to the marine environment than might have been expected.[14]

The volume of the oceans is 300 million cu mi compared with total world oil production of 1 cu mi/yr, so that, provided sufficient mixing takes place, the small percentage of oil that is spilled becomes sufficiently diluted for it to get lost by natural processes. Since oil is formed in a marine environment, there are always hydrocarbons of a petroleum-like nature associated with recent sediments on the seabed, and there do not appear to be any animal or plant chains of life that concentrate hydrocarbons, as occurs in the case of some metals. Research indicates, however, that shellfish contaminated by oil clean themselves in 2 weeks when the source of oil is removed.[15]

Evidence concerning the limited environmental effect of oil operations has been provided by a recent highly comprehensive study by the Gulf Universities Research Consortium. The objective of the study, known as the Offshore Ecology Investigation, or OEI, was to determine the effect of petroleum operations in the "Louisiana Oil Patch," the continental shelf area which has experienced about 25 yr of intensive petroleum exploration and production operations. If, indeed, exploration and production operations have an adverse effect upon the ecosystem, here is where one might expect the evidence of such effects. This is particularly true with regard to long-range effects which might be hidden by natural phenomena over the short term. That study indicated the following:

1. Natural phenomena completely dominate the characteristics, productivity, and general health of the ecosystem. These include seasonal changes in water quality and mass. The turbid layer arising from the Mississippi River, for instance, contributes far more to silting and sedimentation than does production and drilling activity.

2. The presence of offshore producing platforms and pipelines has an insignificant effect which, if anything, appears beneficial due to the reef effect of the structures increasing the productivity of basic nutrients in the vicinity.

3. Petroleum operations have not resulted in any significant accumulation of potentially toxic materials in either the sediment or water column in the vicinity of such operations.

4. No accumulation of hydrocarbons was found in the animal life in the area, and the accumulation of organic materials in sediments and beach sand was found to be of a low order and not ecologically significant.

14. C. Chassé, "The Ecological Impact on and near Shores of the *Amoco Cadiz* Oil Spill," *Marine Pollution Bulletin* 9 (November 1978): 298–301. Readers with a specific interest in oil and the environment should refer to *Proceedings of the 1979 Oil Spill Conference*, EPA/API/USCG (Los Angeles, March 19–22, 1979).

15. E. W. Mertens and R. C. Allred, "Impact of Oil Operation in the Aquatic Environment" (paper presented at the UNEP Industry Sector Seminar on Environmental Conservation in the Petroleum Industry, Paris, March 29–April 1, 1977).

Chronic pollution from land-based sources is a much more serious problem resulting in lasting damage to marine and saltmarsh habitants. Similarly, natural oil seeps of an estimated measure of 0.1×10^6 MT/yr contribute 1¼ times more than offshore operations such as Santa Barbara. Fortunately, gas chromatography and infrared spectroscopy can identify unambiguously most crude oils and seeps and remind us of our responsibility for good housekeeping.[16]

Demand and Consumption

Total world primary energy consumption in 1977 was equivalent to 6.7×10^9 MT of oil, of which oil itself accounted for some 44.4 percent.[17] In 1976 oil accounted for a similar proportion. Total world demand for primary energy increased by 3.5 percent in 1976–77, with the largest increases occurring in the USSR, China, Eastern Europe, the Middle East, and Africa, from where the largest future demand may be expected. However, in none of these regions is oil the dominant energy source (except the Middle East), and in China only some 16 percent of total energy demand is so far satisfied by oil. Since more than half the world's oil is contained within these areas, supplies to the rest of the world will be subject to price fluctuation, conservation policies, increased internal demand, disruption of supplies for political reasons, technological problems in the organization of large-scale production, and finance.

For oil to satisfy demand to the year 2000 an amount equivalent to current proved reserves (88.6 thousand million MT in 1977) would need to be discovered. In view of the annual 3 percent increase in world population and the similar expectation for growth in per capita income in the Third World in the coming decades, an annual 5 percent increase in demand for oil is a reasonable basis for forecasting. At this level of sustained consumption and because of the limits to the rate at which oil can be recovered, as well as the time lag between discovery and development, it can be predicted quite confidently that supplies will not only fail to meet demand but will run out. It is a canon of the oil industry that only 40 percent of an oil reservoir may be recovered. Even in the event that technology could be improved to squeeze a well dry, the additional 60 percent to current and anticipated reserves would not avoid a shortfall between supply and demand by the turn of the century.

Ultimate oil and gas resources under the oceans probably are in excess of

16. For a detailed account of the damages and threats from oil pollution, see Peter S. Thacher and Nikki Meith-Avcin, "The Oceans: Health and Prognosis," in *Ocean Yearbook 1*, pp. 294–301.

17. British Petroleum Co., *Statistical Review of the World Oil Industry, 1977* (London: British Petroleum Co., 1978).

estimates provided by national governments and the major oil companies. The implications that major oil producers are being irresponsible or self-seeking by understating the availability of recoverable resources of hydrocarbons[18] appear to us to be somewhat specious when even the most optimistic "estimates" are put into the context of present and growing oil consumption and the 4 percent per year growth in total energy demand (1970–77), all of which point to an absolute necessity for stimulated development of alternative, renewable, cheaper, cleaner, less politically sensitive, and domestically sited energy sources. Solid fuels, waterpower, and nuclear sources already provide some 40 percent of world energy. Obviously our total energy security relies upon the speed at which these resources can dominate oil.

CONCLUSION

This paper has presented some of the geological evidence that oil is a finite resource with a remaining useful life of some 30 yr, that is, from now until the year 2000. This fact should remind economists and politicians that even the most ingenious technological advances in oil development may not alter the picture very much, that unless alternative energy policies are vigorously pursued and implemented, there will be a shortfall in total energy by the year 2000 of as much as 40 percent (i.e., the amount presently contributed by oil to world energy consumption). This seems to be a somewhat pessimistic prediction in view of the already evident trend away from oil domination in Western Europe and Latin America; but the United States appears to be maintaining the proportion of oil consumed in line with natural increases in total energy consumption.

New discoveries of oil have been made at the rate of some 2×10^9 MT (2,000 million MT) per year over the last 5 yr. This represents some 66 percent of current world consumption (3×10^9 MT per year). It is not too likely that this rate of discovery will continue (although this is the basis for the predictions in fig. 4), even though offshore discoveries will make a significant contribution to reserves, underprospected Third World countries will be reexplored, and a large number of existing fields will be upwardly reappraised (in Mexico, for instance). However, there are no well-established geological estimates for the rate or volume of future additions to the present proved world reserves of about 90×10^9 MT. Similarly, however, there is no basis for disagreeing with speculative estimates of ultimate reserves which lie between 155×10^9 MT and 345×10^9 MT (within a range of probability declining from 90 percent to 10

18. Odell, pp. 147–57.

percent).[19] We can assume an understatement of official figures (which tend to be based upon what actually can be observed), but even if 300 × 10^9 MT were ultimately feasible (still producing a shortfall between supply and demand within the given period), political and economic constraints to development (i.e., conservation and cost) will not sustain consumption at current levels beyond the year 2000. For the purpose of introducing alternative energy, if for no other reason, it is prudent to regard published authoritative data as being closer to the top rather than the bottom line. Geological evidence strongly suggests that most, if not all, the prolific oil-bearing regions have been located, by either seismic testing or exploratory drilling.

Calculations of recoverable oil reserves (world "published proved")[20] are based upon 30 percent of the total oil in situ (based on technology currently available). There are constraints on the rate at which oil can be recovered. Even under optimum conditions of size, geology, and installed facilities of a field, the maximum that can be recovered in any one year, without reducing the amount that ultimately can be recovered, is about 10 percent. However, taking into account the different stages of development of the 30,000 known fields, a more realistic estimate of maximum annual production is nearer 6 percent of recoverable reserves.

In 1977 the remaining proved reserves were 88.6 × 10^9 MT. On the basis of the reserves-to-production (R/P) formula, maximum production capacity is currently 5.9 × 10^9 MT: actual production (1977) was 3.0 × 10^9; this is equivalent to an R/P ratio of 1:29, which is actually 3.4 percent of recoverable reserves, a proportion corresponding to world demand. Thus, current proved recoverable reserves are being depleted at a rate of some 3.4 percent. Allowing consumption to rise by 1–3 percent annually, these reserves will be exhausted before the year 2000. New finds of sufficient magnitude to sustain demand beyond the year 2000 would have to be made at twice the present discovery rate, but it can be seen that even this optimistic scenario could only maintain a situation in which oil would be consumed as fast as it could be produced.[21]

In terms of alternative natural resources, a vast mineral wealth held in solution in the earth's 300 million cu mi of seawater is awaiting recovery, but it is unlikely that any other minerals will be obtained from seawater until very cheap power becomes available. However, in perhaps 50 yr time, when nuclear power

19. B. A. Rahmer, "New Assessment of Resources," *Petroleum Economist* 46 (December 1979): 501. See also Michel T. Halbouty and John D. Moody, "World Ultimate Reserves of Crude Oil," *Proceedings of the World Petroleum Congress, 1979* (London and New York: Haydens, 1979), pp. 1–11; and *World Energy Resources, 1985–2020* (Guildford: IPC Science & Technology Press, 1978), pp. 1–48, for summaries of reports on resources, conservation, and demand to the Conservation Commission of the World Energy Conference.

20. *Oil and Gas Journal* (Worldwide Oil Issue) (December 26, 1977).

21. Authors' assumptions are based on currently available data as discussed herein.

should be taking over from fossil fuels, we shall probably be working down toward some of the elements that are present in lower concentrations. Titanium for example, is becoming popular for light-alloy work, and the sea's resources of 1,500 million tons may one day be tempting enough for an economic process to be developed. Uranium itself has been the subject of investigation by the British Atomic Energy Authority, which believes that some of the sea's 5,000 million tons could be harvested at only a few times the cost of mining and concentrating it on land. Supposing that disposal and safety problems, with the attendant political problems, could be solved, the entire world's present energy requirements—involving the consumption of 2,000 million tons of oil and 2,300 million tons of coal a year—could be met by modern nuclear plants fueled with about 3 million tons of uranium. So the quantity available in the sea provides a useful reserve for several centuries.

Transportation and Communication

The First Declaration of the Freedom of the Seas: The Rhodian Sea Law[1]

John Wilkinson

Although the maritime customs of the Mediterranean Sea appear to have a very long history,[2] the peculiar situation of the Aegean island of Rhodes during the Hellenistic Age gave rise to a certain number of maritime codes and practices which gradually assumed a binding character freely recognized by the seafaring states in the Mediterranean. Its principles were accepted by both Greeks and Romans, and meagre as our evidence is, it shows that the current maritime law of the Mediterranean—the rules which were known to every seaman and of which the Roman administration and the Roman jurists had to take account in building up their own maritime law—was commonly called the law of the Rhodians (*Lex Rhodia*).[3]

This implies that the Rhodians, in the period of their ascendancy in the Greek system, enforced on the seas a body of functional rules which attempted to sum up and perhaps explicitly codify all that Greek tradition had previously achieved in this field. The Rhodians showed their genius by accepting this framework. They collected, elaborated, and restated the maritime customs that they found in use, and they provided, for the first time in the history of international maritime relations, the conditions in which these customs could be enforced regularly and over a large area.[4]

Despite the lack of conclusive evidence as to the existence of a *written* body

1. Originally published as an appendix to the author's article, "A Tentative Program for the Simulation of the Historical 'Ecology' of the Mediterranean," in *Pacem in Maribus III: The Mediterranean Marine Environment and the Development of the Region,* ed. Norton Ginsburg, Sidney Holt, and William Murdoch (Malta: Royal University of Malta Press, for the International Ocean Institue, 1974), pp. 388–94. Reprinted here with permission of the International Ocean Institute.

2. See A. B. Bozeman, *Politics and Culture in International History* (Princeton, N.J.: Princeton University Press, 1960), pp. 110–11. The Code of Hammurabi contains probably the most ancient statement of shipping law in existence. We know little about the existence of such codes in Egypt, Minoan Crete, and Mesopotamia, and we can only make conjectures about the contents of the Phoenician records.

3. See M. Rostovtzeff, *The Social Economic History of the Hellenistic World* (Oxford: Clarendon Press, 1941), 2:688–89.

4. W. Ashburner, *The Rhodian Sea Law* (Oxford: Clarendon Press, 1909). This work provides an analysis of the manuscripts pertaining to the Rhodian Law.

of sea laws,[5] the Rhodian Code seems to have included regulations regarding copartnership, joint adventures, charter parties, and bills of lading. It established standards for the behavior of passengers while on shipboard, for the liability of commanders or seamen in cases of injuries to goods and carelessness or absence from duty; it also listed penalties for barratry (fraud or gross negligence), robbery of other ships, and careless collisions.[6] This maritime code dates from the third or second century B.C., and was apparently of great authority in the Mediterranean, its memory lasting down to the thirteenth century. But since the written form in which the sea law has reached posterity was supplied between the seventh and ninth centuries by Byzantine jurists, no one will ever know with certainty in what precise conformation these legal concepts were applied during the Hellenistic Age, under the Rhodian aegis. Of the many extant provisions, only the law of jettison (*lex de lactu*)[7] has been definitely identified as originating in Rhodes.

The significance of the Rhodian Sea Law, as well as the uses to which it has been applied in the contexts of different civilizations and particular legal systems, may perhaps be said to illustrate two general propositions: first, that the formal statutory expression of law does not always correspond with its existence in time; and, second, that the law itself tends to remain unaltered as long as the conditions that gave rise to it remain unaltered. A case in point is the Roman utilization of the Rhodian Sea Law: Cicero praised it for its humaneness; its adoption throughout the Roman Empire was ordered by Augustus; and its sovereignty was acknowledged by Rome's most outstanding Stoic emperor, Marcus Aurelius, when he responded to the appeal of a certain Eudaemon of Nicomedia, whose goods had been plundered when his ship had been wrecked: "I am the sovereign of the world, but the Rhodian law is sovereign wherever it does not run contrary to our statute law."[8] We next hear of the Rhodian Sea Law when the Emperor Justinian (483–565 A.D.) began restating the Roman civil code.[9] By this time, however, immemorial custom, Rhodian regulations, and Roman enactments had merged so completely that it

5. See in particular M. Rostovtzeff, "Rhodes, Delos and Hellenistic Commerce," in *The Cambridge Ancient History*, ed. S.A. Cook et al. (Cambridge: Cambridge University Press, 1930), 8, chap. 20: 690 ff.

6. See Ashburner, pp. cxii–cxv; and Bozeman, p. 111.

7. Rules regulating the circumstances in which goods can be thrown overboard to lighten a ship in distress.

8. W. S. Lindsay, *The History of Merchant Shipping and Ancient Commerce* (London: S. Low, Marston, Low & Searle, 1874–76), p. 183.

9. The most extensive compilation is attributed to Emperor Leo III *personally*. But this ascription, following the custom of the times, is certainly false. In addition, it contains much material only conventionally connected with the Rhodians themselves, a fact that demonstrates at least that the *cachet* of the name was invaluable, at least until the Mediterranean was turned into a Moslem sea. The Moslems understood sea-borne commerce almost exclusively and in every historical epoch in terms of pirates and corsairs. The Rhodian Sea Law is evidently useless with such a world view.

is difficult to distinguish the various components. The sea law was copied and handed down through the centuries in this form and apparently guided the nautical and commercial adventures of Europeans through the Middle Ages. Indeed, we find them embodied in the *Constitutum Usus* of the city of Pisa (published in 1160), as well as in the *"roles d'Oleron"* (Gascony), the Statues of Ragusa, and the practices of Venice, Genoa, and Amalfi. The authority of the Rhodian Law on jettison was even specifically edged by an English judge (in Burton v. English) in 1883.[10]

The Rhodian precedent must, however, be understood in historical, as well as political and economic, terms. The ascendancy of Rhodes, during the Hellenistic Age, underlines in particular a shift in the political and economic structure of the Hellenistic system. The Rhodian state, once a part of the Persian Empire, had been restored as a democracy by Alexander the Great. Once political union ("Synoecism") and constitutional reform had been effected in 407 B.C. by uniting three independent cities on the island, Rhodes became an independent and stable state, despite a short period of Persian dominance (397–388 B.C.).

From the beginnings of its history, Rhodes was first and foremost a commercial community. Above all else its strategic position between Egypt, Cyprus, the Syrian and Phoenician coast, and the world of the Greek cities made it an important intermediary between Greece and the Orient. By the third century B.C., Rhodes had replaced Athens as the commercial center of the Hellenistic world and was easily the richest Greek state. Thus, at the crossroads of commercial, strategic and political interests, equally accessible to the great and the secondary powers of three continents, the Rhodian state held a key position in the Hellenistic system of diplomacy and became one of the principal sea powers of the Mediterranean. All through the third century, and the first half of the second, Rhodes ruled the Mediterranean, not only by the power of the fleet, the extension of her trade as far west as the coasts of Gaul, Spain, and North Africa, and her public and private opulence,[11] but by her laws and regulations regarding marine matters. The extent of Rhodian prestige was attested to by the aid it received from the Mediterranean states during the catastrophic earthquakes of 225 and 222 B.C.[12]

The Rhodians realized early that their independence was inseparably connected with a policy of freedom of the seas. In this respect, Rhodes acted as "the protector of those who follow the sea."[13] She acted both as policeman of the

10. See Bozeman, pp. 112–13; and W. Cunningham, *An Essay on Western Civilization in Its Economic Aspects* (Cambridge: Cambridge University Press, 1913), 1:136.

11. In this connection, Seneca preceded Mr. Galbraith in commenting on Rome's "public squalor and private opulence." But even public squalor, after a certain level of "development," *always* becomes unbearably costly.

12. See Jules Toutain, *The Economic Life of the Ancient World* (New York: Barnes & Noble, 1951), p. 152.

13. Polybius *The Histories* 4.47.1.

Mediterranean and middleman of international trade, which implied a constant battle against piracy[14] and great power hegemony in the Mediterranean. Her foreign policy included attempts at collective security in the various leagues (i.e., the Delian League, the second Athenian confederacy), and after severing with Athens in 357 B.C., she became the nucleus for a new league of republics, as well as exercising a protectorate over the island of Delos.

Rhodian advice, mediation and arbitration were frequently sought after, such as in the feud between Byzantium and certain Celtic tribes, and in the dispute between Priene and Samos around 200 B.C.

The progressive rise of Rome in the Mediterranean put an end to the Rhodian ascendancy in the Mediterranean. After the Third Macedonian War, Rome set about to undermine Rhodian commerce by proclaiming Delos a free port. Rhodes became an ally of Rome on unfavorable terms and was annexed in 164 B.C. When the Roman Empire was divided in 395, Rhodes was assigned to the Eastern Roman (Byzantine) Empire.

The Rhodian Sea Law which resulted from this context appears then to be the result of several historical factors. It was not without historical precedents, but was uniquely combined with Greek traditions and maritime customs. It should not be isolated from the functional pursuits of the island of Rhodes during the Hellenistic era, nor should it be removed from the political and strategic conditions imposed upon the Rhodians through changes in the Hellenistic System.

Rhodes had, for a long time, "possessions" on the nearby mainland of the extreme Southwestern coast of Asia Minor. These were negligible in area. Nonetheless, a large quantum of Rhodian Sea Law appears to have arisen as a result of litigation between the Mother island and its mainland dependencies. The very existence of Rhodes itself as a political entity (the *Synoecism* of three independent cities referred to above) had similarly been the result of extensive litigation, that sometimes was exacerbated to armed skirmishes. The role of these litigations has yet to be examined; but the existence of *any* legal code known to history seems, without exception, to have involved more or less extensive courtroom processes.

One cannot emphasize too much the *exemplary* way in which Rhodes secured its status as the first league of nations. The Rhodian code was forced on no one with the exception of a few adjacent (and small) islands and mainland cities. The Mediterranean region recognized the code *voluntarily* because it represented that which could not be dispensed with—in any practical sense. Even when Rhodes itself had been leveled by severe earthquakes, the member "States" of the confederation built it up again by voluntary subscription. The last event, at least for the same reasons and to the same degree, is almost without

14. See H. M. Ormerod, *Piracy in the Ancient World* (Chicago: Argonaut, 1967), pp. 157 ff.

parallel in world history. The "idea" of Rhodes and its Sea Law was almost always in conflict with the "fact" of any actual Rhodian hegemony either with respect to naval dominion or trade. This sort of conflict between, say, the American need for a unified Europe and its actual existence is still an unsolved problem to which the Rhodian precedences, long vanished into history, are of extreme appositeness.

International Marine Insurance

Susan Strange and Christopher Cragg
London School of Economics and Political Science

On March 16, 1978, the *Amoco Cadiz* suffered a failure of steering gear and ran aground off Portsall on the rocky northern coast of Brittany. Flying the Liberian flag, owned by an American oil company, Amoco, and on charter, this supertanker or VLCC (very large crude carrier) released 220,000 tons of light crude oil onto French beaches, turning a popular tourist area into a slimy black wasteland. The ship's hull was insured for $12 million—about a third of the replacement cost.[1] Far greater were the third-party liabilities incurred by the accident. At the time, the cost of cleaning up the polluted areas and compensating fishermen, hotels, and others affected was variously estimated at between $60 and $100 million—far more than the $25.3 million of insurance funds available from international and industrial pollution compensation funds.[2] Later, suits brought in U.S. courts by the French government, local authorities, and tourist and fishing associations put the damage done at well over $500 million.

The case of the *Amoco Cadiz* serves to highlight some important facts about the current state of international marine insurance, the part it plays in the overall regime governing sea transport in the world economy, and about some fundamentally political issues which remain to be settled concerning the regulation of shipping, the legal assignment of liability for losses, and the distribution of profits and losses from insurance. How these issues are settled will be of great importance to all concerned—to shipowners and builders, to importers and exporters not only of oil but of many other cargoes, to banks and insurance enterprises as well as to governments and international organizations.

The most obvious fact to emerge from *Amoco Cadiz* was that a system of insurance originally designed to protect the owners of ships and their cargoes from bearing the whole burden of risks arising from the inevitable hazards of the sea was clearly quite inadequate to cope with the much greater risks of damage to third parties. The same point had been equally dramatically made

1. David Trash, "The Recession Chips away at the World Shipping Market," *Review* 180/4511 (May 19, 1978): 31–37. An earlier accident to another VLCC, the *Olympic Bravery,* on the same coast resulted in total loss. But the brand new ship was empty and the owners' insurance claims, at the bottom of the tanker market, were almost three times the price of an equivalent replacement.

2. Paul Taylor, "The Aftermath of Amoco Cadiz," *Financial Times* (July 6, 1978).

© 1980 by The University of Chicago. 0-226-06603-7/80/0002-0008$01.00

over a decade before when the *Torrey Canyon* slowly broke up in British waters in 1967, spilling less than half as much oil as the *Amoco Cadiz*. In the intervening decade, some attempts had been made to grapple with the problem. The oil companies had got together to organize funds—Tovalop and Cristal—for their mutual protection, governments had tried to tighten regulations, and international organizations had tried to draft new conventions. But the basic issues of responsibility for regulation and of who should share the costs of spreading it more widely were still open, the problem still unsolved.

Another fact, disputed by no one, was that although the source of pollution caused to third parties by such tanker accidents was usually only too obvious, those accidents were not the only or even the most important cause of environmental damage—just the culprits who got caught. Only a third of the oil in the sea comes from marine transport, only 5 percent from accidents to tankers.[3] The amount of damage done by any single accident, moreover, was apt to be determined more by luck than by human care or error. Changes of wind and tide and even the nature of the oil spilled on the water could multiply the harm done by quite a small tanker accident—or minimize that done by a much larger one. In October 1978, for instance, another tanker, the *Christos Bitas,* ran on the rocks in the Irish Sea; but good luck and unseasonably fair weather made it possible to pump most of the oil into other ships before it could spill and then to tow the ship out into the Atlantic Ocean and sink her. Far worse damage was done to the coasts of New England in December 1976 when a much smaller tanker, the *Argo Merchant,* broke up.

A third fact to emerge from all three of these tanker accidents was that there was a contributory element of negligence which might have been eliminated by more effective regulation.[4] The *Amoco Cadiz* spill would not have happened if the ship had not gone so imprudently close to a rocky coast—or if its dual steering systems had not both depended on the same power supply. The *Christos Bitas* had faulty radar but even so was wrecked on a calm, clear day in midafternoon when passing on the *wrong* side of a lighthouse. Not only was the *Argo Merchant* old and in bad condition, there was evidence that the radar and other navigational equipment it did have was not being properly used. And in yet another recent case, two sister ships, the *Venpet* and the *Venoil,* each of over 150,000 g.r.t. (gross registered tons) and both equipped with radar, managed to collide so badly off the coast of South Africa that $25 million–$30 million worth of damage was done to their hulls alone.

The political issue, though, is what better rules are needed, who should make them, and how are they to be enforced. Carrying out the rules and

3. Peter S. Thacher and Nikki Meith-Avcin, "The Oceans: Health and Prognosis," in *Ocean Yearbook 1,* ed. Elisabeth Mann Borgese and Norton Ginsburg (Chicago: University of Chicago Press, 1978), pp. 295–96.

4. Thomas Busha and James Dawson, "A Safe Voyage to a New World," ibid., pp. 217–29.

enforcing them entails costs to someone; so does bearing the expense of increased insurance. Already this can account for as much as 30 (or even in some cases 40) percent of the running costs of shipping. It seldom is less than 15 percent. If it is put much higher, should the burden be passed on in higher prices to all shippers, or only to some? What effect would this increased cost have on the future growth and nature of world trade? And, equally important, if the role of insurance is to grow in response to escalating risks, who is going to profit from the increased business? Where and how will the extra premiums be invested? Can the business and the capital be fairly shared between the great insurance enterprises of North America, Western Europe, and Japan and the newer, and usually smaller, insurance enterprises of the developing countries? Or will the increased use of insurance only deepen the perceived dependency of the latter on the great financial centers of the capitalist world? Whatever happens, one thing is certain. The outcome will be much affected by the current unsettled state of the insurance market and the precarious position of many operators in a business severely affected by inflation, by economic recession, and by the trend toward protectionism in economic policymaking—a situation in which there are already few profits to be shared among many fiercely competing interests.

These are difficult questions, but they can only be fully discussed against the broad background of the marine insurance business and the political as well as economic pressures upon it, for there are other respects in which the *Amoco Cadiz* and the other tankers that hit the world's headlines are not at all typical of the everyday business of marine insurance.

MARINE RISKS

Most of the claims on insurers, to begin with, are not for total losses nor even for what are technically called constructive total losses.[5] They are for repairs or for delays to ships and for losses of cargo by fire, theft, or all sorts of other causes. The threat to cargo from theft is well known, but its scale is not always appreciated. High-value commodities can simply disappear—like the 1,800-ton shipment of coffee valued at £5 million which was missing along with the 37 lorries carrying it *before* it reached Mombasa docks in 1977.[6] Fire can also cause damage to hulls and cargoes. Fire damage to the *Nedlloyd Hoorn* 2 mo before it was due to leave its original shipyard cost the insurers a cool $50 million. A fire in Antwerp docks early in 1977 produced claims amounting to $12 million to Lloyds cargo insurers alone, and the total cost was over $40 million.[7] Port delays

5. Constructive total loss occurs when a ship is so damaged by an insured peril that the cost of repairs would exceed her reported value.
6. ILU report in *Policy* 77 (February 1978): 127.
7. *Financial Times* (February 6, 1977).

not only add demurrage costs for shipowners but can cause huge losses of perishable goods—as in the famous pileup of cement cargoes outside Lagos in 1977. The Persian Gulf ports have been so congested that early the same year waiting time to unload at Shuwaikh, the main port of Kuwait, rose to over 3 mo.[8] Even containers, the ironclad solution to so many transport problems, are not immune to perils at sea or in dock. They can be broken up by bad handling, washed overboard, damaged by rain or seawater, contaminated, suffer refrigeration breakdown and/or internal sweating, or simply disappear in transit.[9] Usually the contents are well insured; a container-ship's cargo may be worth $50 million.

The possibility of some very large losses among a great many small ones makes the incidence of claims on insurers apt to vary considerably from year to year. For instance, 1976 and 1977 were both exceptionally bad years for insurers—in fact 1976 broke all postwar records (see fig. 1). In both years, there were losses of over 200 ships above 500 g.r.t. and aggregate tonnage losses over 1.2 million g.r.t. Yet in 1977 the largest single ship lost was an unremembered ore/bulk/oil (OBO) carrier, the Liberian registered *Exotic* of 70,337 tons. Nevertheless, over half the tonnage lost that year was accounted for by 26 tankers and 14 OBOs.[10]

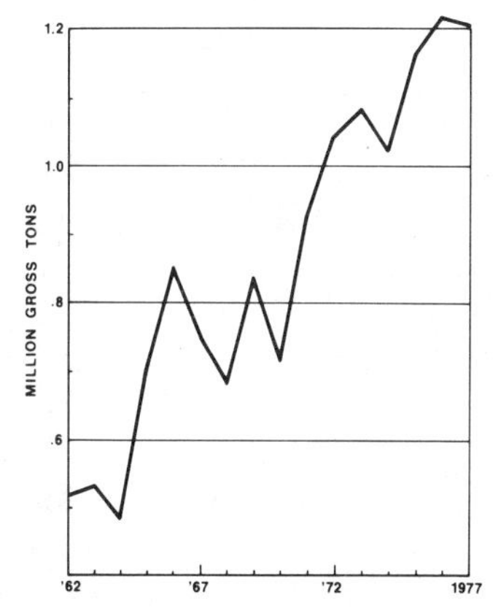

FIG. 1.—Annual losses by tonnage, 1962–77. Prepared by Cartographic Services, University of Chicago, based on *Annual Report*, Liverpool Underwriters' Association, March 1978.

8. "Port Congestion: A Solution in Sight?" *Kuwait Digest* 5 (October–December 1977): 815.

9. See, e.g., A. E. Mann, "Container Losses," Report to IUMI Council, reprinted in *Policy* 76 (suppl.; November 1977): xvi–xviii.

10. "Liverpool Underwriters' Association Annual Report 1978," *Policy* 77 (March 1978): 182–84.

It may be noted, however, that the proportion of ships lost to those afloat, calculated over the 5 yr 1972–77, averaged only 0.36 percent or 3½ tons lost for every 1,000 tons in use. Even in 1976 and 1977 the ships lost were only 4 percent in tonnage terms of the new tonnage launched in those years—an indication of the large and continuing increase in the world shipbuilding output at the very end of the boom years.

There are no comparable statistics relating to cargo losses. From 1960 to 1972, according to UNCTAD, international seaborne trade grew at an annual average rate of 6 percent, and total tonnage loaded more than doubled from 1,080 million tons to 2,861 million tons, while the measurement in ton miles more than quadrupled from 4,356 thousand million to 14,800 thousand million. More cargo was being taken further.[11] By 1975, the total tonnage loaded reached 3,175 million.[12] Thus, if the annual loss ratio of cargo resembled that on hulls, the total tonnage of cargo lost might be 11.5 million tons. (It must be emphasized, though, that this is only guesswork.)

In spite of the lack of figures, it is clear that the financial burden of cargo losses could not easily be borne by individual exporters or importers. The necessary cost of self-insurance would severely reduce the working capital of any manufacturer. Likewise, the cost of building and maintaining an oil rig in operation is sufficiently large that no single organization would conceivably do it and maintain the necessary investment to replace it should it burn up in a blowout. Individual ships also require large capital outlays for purchase, running, and repair. No shipping company has the reserves necessary to set aside the total value of its fleet in case of multiple losses. Most of them, in any case, are deeply in debt to the banking system; U.S. banks alone have outstanding loans to tanker owners of $3.5 billion.[13] Clearly no banker worth his assets could conceive of allowing one individual slice of collateral worth $50 million to float through the crowded sea lanes of the Persian Gulf or English Channel without some security against its loss. Bankers are in the business of giving loans, oil companies of finding oil, shipping companies of running ships, and exporters of exporting. None could double their investment base or diversify into nonrisk activities sufficiently to guard against fortuitous loss.

These are the burdens of the marine insurers, and the total level of world trade could not be anything like as high as it is without them. Given that there is a major accident involving a ship over 500 g.r.t. every 2 days, there simply could not be the present shipping tonnage if the complex, interconnected world of marine insurance did not exist.

11. UNCTAD, *Review of Maritime Transport, 1972–1973* (UNCTAD/B/C4/117), 1975, pp. 5–11.

12. United Nations, Department of Economics and Social Affairs, *World Statistics in Brief, 1977,* Statistical Papers, ser. 5, no. 2 (New York: U.N., 1977).

13. *Business Week* (May 1, 1978), p. 80.

BASIC PRINCIPLES AND DEVELOPMENT OF MARINE INSURANCE

The basic principles of marine insurance are simple and have a distinguished history. As the preamble to the Elizabethan Act of 1601 (written by Sir Francis Bacon) puts it, by the process of insurance "it commethe to passe that upon the losse or perishinge of any shippe there followethe not the undoing of any man, but the losse lighteth rather upon them that adventure not than those that doe adventure, whereby all merchantes speciallie the younger sorte are allured to adventure more willinglie and freely."[14]

This process of spreading the risks of seaborne trade is the essence of insurance and has such psychological and economic advantage that no Western maritime community since the Phoenicians has lacked a method to achieve it.

At its most basic, the spread of risk involved the payment of a premium by an individual shipowner to individual underwriters in return for an indemnity in the event of the loss of his ship and its cargo. The underwriter, by collecting a variety of premiums and investing them, obtained a profit depending upon the rate of premium charged, the number of claims, and the state of his investments. The rate of premium would be calculated according to the underwriter's knowledge of the ship, the route, the time of year of the voyage, and his past experience of the shipowner. In eighteenth-century London, an individual underwriter might take a proportion of the cost of indemnification, known as a "line" of perhaps four-fifths, or half, or a third, of the insured "risk." Effectively, the financial costs of any dangers of the voyage would be spread from shipowner to the underwriters.[15]

This involved considerable dangers for the latter, and a huge body of accepted practice and legal definitions of sea perils soon grew up surrounding marine insurance—conventions which are still reflected in the sometimes archaic language of insurance policies in use today. Although the simple transfer of risk from shipowner to underwriters was a great step forward, the spread of that risk was still inadequate. A rapid series of claims, despite the relative cheapness of seagoing vessels at the time, would force an insurer rapidly back on his financial reserves and eventually bankrupt him. Individual lines of 80 percent came to be thought of as very high.

While adequate expertise regarding types of ship, voyage, or cargo was some protection against this, the only real solution lay in a wider spread of risks within the insurance community itself. This wider spread was achieved in three ways. First, the underwriting groups which survived were those who spread risks evenly enough among themselves—notably those who started to meet

14. Quoted in V. Dover, *A Handbook of Marine Insurance*, 6th ed. (London: Witherby, 1962), p. 14.

15. H. A. L. Cockerell and E. Green, *The British Insurance Business 1547–1970* (London: Heinemann Educational Books, 1976), pp. 3–17.

regularly at Edward Lloyd's coffee house. Second, a large asset base could be constructed to withstand the larger claims if insurers operated as a joint-stock company with limited liability for shareholders. Third, shipowners themselves formed "mutual" insurance groups—the ancestors of the present P&I (protection and indemnity) clubs—which operated without profit purely to meet the needs of the shipping community. As a result a community of insurers emerged, dividing the risks into smaller lines and smoothing out the fluctuating accounts of each individual underwriter.

Other ways were devised to spread individual risks. It was recognized that insuring both hull and cargo together would involve high claims, so the accounts were separated. Since cargo and shipowners both shared the dangers to property implicit in the sea trade, a complex system of "general average" was created to establish that damage to either cargo or hull, but not both, was shared between insurers.[16]

Whereas for the purposes of gaining expertise and judging premiums it was useful to have underwriters who specialized in a particular area of claims experience (e.g., in the risks to vessels trading in the Baltic), bad weather or a decline of trade in that area could have severe repercussions if all the group's policies were written for such risks. In consequence, individual underwriting groups would try to maintain a balanced "portfolio" of lines of different types of vessel and voyage.

In addition, by the turn of the twentieth century insurers began to divide risks, not merely horizontally across the insurance community but vertically also, through reinsurance and even through the reinsurance of reinsurance, known as retrocession (see fig. 2). The result was a large interconnected community of insurers, each in competition for business yet also in cooperation, producing elaborate and frequently transnational chains of financial agreements.

There was a price to be paid for this diffusion of the financial risks of maritime transport. As the insurance business grew, no individual underwriting group or company could possibly find the manpower to check on the risks peculiar to each individual ship, to each voyage, and—sometimes more important—on the claims and safety record of the ship's owners. Nor could it draw out individual policy specifications, naming the precise legally defined peril, be it war, weather, piracy, or collision. In normal circumstances, the size of each individual line was by now less than 5 percent of the insured value.

Some underwriters, being more specialized in particular types of ship and voyage, became "lead" underwriters for those risks. They assessed and set the premium, and the rest of the insurance community accepted their judgment. The necessary expertise for individual risks was thus provided by one group

16. For an elementary guide to "general average," see H. A. Turner, *The Principles of Marine Insurance*, 6th ed. (London: Stone & Cox, 1971), pp. 69–80.

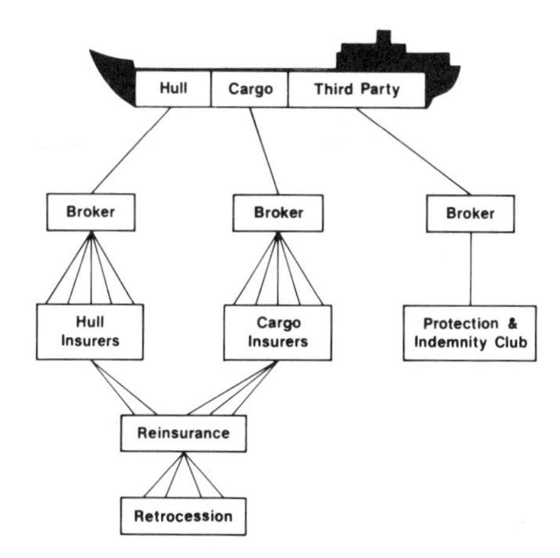

FIG. 2.—Spreading the risk. Prepared by Cartographic Services, University of Chicago

rather than the whole market, each lead nonetheless accepting reciprocal lines from other leads to maintain a balanced "portfolio."

Furthermore, it was also recognized that the simple measurement of premium on an ad hoc basis of one ship per voyage, with special legal consequences for deviation from an agreed route, was unnecessarily elaborate. Insurance could be "packaged," like sugar, and offered by the year; the area of voyage could be defined by merely setting limits to the latitudes sailed. Whole fleets could be insured together for convenience, while specific insurance policies for voyage and ship could still be made available to make the system flexible.

Since a complex series of legal definitions had evolved simultaneously with insurance, time and trouble could also be saved by codifying a common wording for insurance policies. In each insurance center a representative body of underwriters was organized to produce these agreed policy clauses (in Britain, the Institute of London Underwriters) and the resulting documents or "Institute clauses" defined the range of accepted perils. Thus a large risk, divided among many underwriters, could produce a single policy document.

Meanwhile, security for the policyholders and for the insurers was increased through the development of reinsurance. This was of three main kinds. Having followed the "lead," an individual underwriter might take up "facultative" reinsurance. By this means he would simply transfer part of his own line to another insurer, stating the nature of the specific risks. Alternatively, an "excess of loss" type of reinsurance would stipulate that if the claims on specific lines went above a certain figure the reinsurer would pay. A third system involved treaty agreements between underwriters so that the reinsurer took a proportion of the total lines of the direct insurer with the minimum enquiry as to the type of risks. Thus, while relying on the judgment of the direct insurer rather than on the risks themselves, the treaty reinsurer helped further reduce individual losses and gains.

Even without the necessary sophistication of long reinsurance chains, the division of risks between underwriters required the existence of an intermediary. While the insurer had every justification for scrutinizing the record of the shipowner, the scrutiny was also required in the other direction. Insurers could and did fail. These intermediaries, the brokers, had the task of collecting the original insurance business from the shipowner, finding the appropriate lead underwriter, and dividing up the risk among the rest of the community. Underwriters, being in competition for business with each other and unwilling to take a line on a whole ship, were not individually in a position to put together a total insurance package. Since underwriters and shipowners had opposed interests regarding the rate of premium, a market evolved in which brokers acted as regulators and middlemen. As market intermediaries, marine insurance brokers thus have a highly ambivalent relationship with underwriters—one of love and hate. A relationship furthermore that symbolizes the curious dichotomy, between competition and cooperation, which makes up the insurance market.

THE UNDERWRITING CYCLE

So far this explanatory account of the development of marine insurance has simplified the picture by carefully avoiding any mention of conditions in the market in order to emphasize the principle of the spread of risk and the technical expertise of underwriting. Yet in practice, as every underwriter or insurance broker is aware, the profits of marine insurance are not just the result of skill in the assessment of risk and the calculation of premium. Statistical experience is sometimes less important than guessing the state and direction of the market as a whole, subject as it is to the underwriting cycle (see fig. 3).

This cycle is quite simple in theory, but, due to the considerable range of influences upon it, it is extremely complex and difficult to predict in practice. Presuppose, for example, that the world shipping tonnage is expanding rapidly, as it did in the late 1960s and early 1970s. The pressure on existing insurers to underwrite larger lines on more ships will increase. Assuming that there is no expansion in existing insurance capacity, rates of premium will rise and so will profits. With enlarged profits, existing insurers will wish to expand their capacity and write more lines. Other general insurers will wish to enter the market. Institutions like Lloyds will expand their membership with more "names," and composite companies will expand their asset bases through reinvested profits and rights issues on the stock markets. As a result, "solvency ratios" or the ratio of premium income to capital assets will rise, allowing insurers to expand the amount they can underwrite.

At some point, however, the additional capacity of the market will reach equilibrium with the demand for insurance, and competition will begin to bring

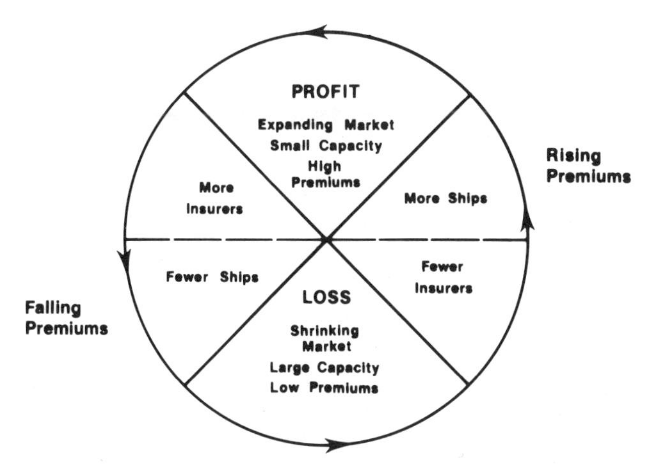

FIG. 3.—The underwriting cycle. Prepared by Cartographic Services, University of Chicago.

rates of premium down. This can happen as the result of a decline in tonnage to be insured through laying up, breakages, and a decline in shipbuilding as well as an increase in the supply side. On the cargo account, a general decline in world trade, such as followed the 1973 Arab oil embargo, can have the same effect. The result is falling underwriting profit. As competition grows the cooperative element in insurance decreases. Individual groups will start to cut rates rapidly and undermine the lead underwriter. After a certain point in rate cutting, it will simply become uneconomic to underwrite risks at all; underwriting profit will not be made no matter how well the risks are spread, and insurance groups will have to rely on investment income for overall profitability.

The consequence will be a sudden contraction of underwriting capacity as some insurers get out of the marine market, others limit their lines, and still others see a contraction of their solvency ratios due to a lack of profitability. Ship- and cargo owners will also realize that the cheapest rates do not always produce the best response to claims. As insurance capacity dwindles, the relationship of supply and demand will again change, and the cycle will come full circle.

Insurers will, of course, attempt to protect themselves by market agreements to fix rates. The Joint Hull Committee of the Institute of London attempts to do precisely this by imposing its collective will on the whole insurance community and forcing a common rejection of uneconomic rates.[17] In the long run, however, if the cycle is very deep, what ultimately counts is the total asset base and ability to invest even an inadequate premium. Insurance

17. George Sanders, "The Marine Problem: London Must Find the Answer," *Review* 180/4504 (February 10, 1978): 17–19.

companies collapse, not by making underwriting losses—that is a familiar part of the cycle—but by investment mistakes and disregard of the adequate solvency ratios defined by law.

Thus the general state of the business-and-investment cycle also affects the insurance industry, and the underwriting cycle tends to follow it.[18] Another factor may be the determination of one insurance center to gain business at the expense of another in an international rate-cutting war. The U.S. life insurance companies have in the past been accused of attempting to undermine the London marine market by setting attractive but "uneconomic" rates on hulls. With their huge investment income, they could afford to withstand a prolonged lack of underwriting profit in anticipation of damaging London's capacity.[19]

The position of the broker is also crucial. If there is a downward trend in the underwriting cycle and a limited number of brokers, the latter can exercise the privileges of semimonopoly, and detailed scrutiny of a ship and its owner's insurance record becomes a luxury. A recent insurance wrangle between Lloyds and the large brokers Willis, Fabers, and Dumas over a cargo of 301 Fiat cars "supposedly damaged by fire on board" illustrates this. Lloyds paid up on half the claim after it became clear that the cars were being sold at 80 percent of new value and were not therefore seriously damaged.[20] As both underwriters and brokers appreciate, in a falling market, if you want good risks, you have to take bad ones. The ratio between the two, however, depends on the size of the supplier, and, in consequence, most underwriters view the merger of broking houses with suspicion.[21]

Thus the insurance community provides a wide network of small-to-medium insurance companies and syndicates, which through cooperation provide a large underwriting capacity based upon the spread of risks through competition. This capacity ebbs and flows according to demand. The assessment of highly technical risks and the values at stake, the statistical measurement of claims experience, and a solid knowledge of the marine world set the margins for the rate. Market forces decide the rest.

PROBLEMS OF OVERCAPACITY

When the world's marine underwriters met at the Conference of the International Union of Marine Insurers (IUMI) in September 1978, most delegates were gloomy and talked of premiums "sliding into the abyss." The

18. Orio Giarini, "How Economic Cycles Affect the Insurance Industry," ibid., 180/4521 (October 6, 1978): 37–43.

19. John Gaselee, "Marine and Aviation Insurance Survey," *Financial Times* (June 19, 1974).

20. "MP Attacks Lloyds over Italian Claim Pressure," *Financial Times* (March 25, 1978).

21. "US Broker Chops Bid for Lloyd's Firm," ibid. (April 21, 1978).

reason for this general despondency—discounting the psychology of the individuals whose job it is to expect the worst—was the current relationship between the state of the world's shipping, sea trade, and underwriting overcapacity.[22] As the 1977 annual report of the Institute of London Underwriters (ILU) put it: "The over-capacity in shipping and shipbuilding finds a natural parallel in the international insurance market place, which is faced with a further decline in premium volume. Moreover, repair costs are still rising against lower ship values."[23]

This relationship, which produces the underwriting cycle, has been explained above. What have not been mentioned are the political aspects. For primarily nonmarket reasons, a decline in the need for insurance has recently coincided with an increase in capacity, thus deepening the negative part of the cycle. The fact is that governments have come to appreciate the importance of insurance in international trade and how it affects their place in the global economy. They have not been prepared to leave their country's marine insurance to the large transnational insurance enterprises of the Western industrialized countries, even if these are governed by purely market forces and not guilty (as it is sometimes suspected) of seeking oligopolistic profits for their services. National governments have preferred to develop insurance enterprises of their own and especially so in marine insurance, given the strategic and economic importance of maritime trade. Free trade being the philosophy of the economically powerful, protection of the native insurance market is widely considered vital nowadays in order to build up native underwriting capacity. It is this nationalist trend which is affecting the current underwriting cycle.

However, if we look first at the demand side of the market, the economic depression of the mid-seventies has been particularly badly felt in world shipping. Though this is generally understood, it is easy to forget the severity of the shipping slump. To mention just a few indicative facts, the International Association of Independent Tanker Owners announced in October 1977 that tanker scrappings were exceeding deliveries for the first time in many years.[24] The Japan Line Company, one of Japan's largest shipping enterprises, revealed that it had taken a loss of $85.6 million in its last financial year and had sacked eight of the company's directors.[25] United States shipping companies were reported to be receiving $290 million in federal operating subsidies.[26] The Union of Greek Ship Owners called for voluntary laying up of 25 percent of

22. "IUMI: Marine Premiums Slide into the Abyss," *Review* 180/4521 (October 6, 1978): 45–51.

23. Statement by the Chairman, Mr. A. E. Mann, at the Annual General Meeting, Institute of London Underwriters, in *Policy* 77 (February 1978): 98–102.

24. *Financial Times,* Insurance Supplement (June 2, 1978).

25. *Business Week* (July 10, 1978).

26. Ibid. (July 17, 1978).

world capacity. By March 1978, one London shipbroking firm estimated that over one-third of world tanker tonnage was laid up.[27]

The position in shipbuilding was similar. *Lloyds Register of Shipping: Annual Report, 1977*, stated that the world's order books were lower than at any time for the past 10 yr and that all new shipbuilding orders that year amounted to less than one-sixth of their 1973 figure. Virtually every European country with a shipbuilding capacity was attempting to protect its own industry, to reduce capacity, and to find ways of subsidising orders from the developing and Communist countries—a process which went directly against the interests of their own shipping operators. In the United States, the industry was said to be operating at only 50 percent of capacity, and fierce recriminatory arguments continued between the shipbuilders and the U.S. navy whose orders alone were keeping many yards in business.[28] The OECD called on developing countries to show restraint in production—though it was likely that the call would go unheeded. Even the Japanese began to worry about South Korean competition.[29]

This considerable crisis may appear paradoxical in the light of UNCTAD figures showing a continually rising trend in tonnage of the world fleet and in the value of world seaborne trade. In shipbuilding, however, forward planning is based upon leadtimes of up to 10 yr. Some deliveries made in 1978 will have been the result of decisions taken in 1970 or earlier. As a senior executive of the unfortunate Japan Line Company put it: "We expected economic conditions to improve. . . . Well, they didn't. And the world tanker market is perhaps 20 percent of what it was in 1973."[30]

Thus, when the price of oil was raised in 1973 and the boom conditions of the late 1960s and early 1970s turned into recession, there was as much as 73.6 million g.r.t. of shipping on order.[31] The slump in tanker requirements had a ripple effect on dry-cargo and other types of ships as the growth of international trade slackened. Shipping companies were thus taking delivery of new ships while simultaneously laying them up. From the insurance point of view, this may have had the effect of lowering the average age of the world fleet, thus marginally reducing risks; but the result for most shipowners was a substantial rise in their debts to the banking system, a huge drop in the market value of their assets, and very fierce competition.

They consequently had every reason not only to seek out the lowest insurance premium rates, but also to insure at low market values and to insure as few ships as possible. The thought may even have crossed some unscrupulous minds that a few shipwrecks would miraculously ease their cash-flow problems and increase the liquidity of their assets.

27. *Financial Times* (March 25, 1978).
28. *Business Week* (May 1, 1978).
29. *Financial Times* (April 17, 1978).
30. *Business Week* (July 10, 1978).
31. *Lloyds Register of Shipping: Annual Report,* 1977.

Marine insurers have thus had fewer ships and cargoes to insure just when the shrinking market brought on a rate-cutting war. In normal circumstances, some insurers would have given up the struggle and left the market. Some undoubtedly did—though others joined in; the applications of new names, especially foreign ones, to Lloyds never ceased to rise. Many insurers, moreover, were kept in business by the protectionist policies of national governments, of which more in a moment.

According to one interesting study in *SIGMA*, the informative journal published by the Swiss Reinsurance Company in Britain based on information received from supervisory authorities and insurance associations in the United States, Canada, Japan, West Germany, France, Switzerland, Italy, and Australia, the bottom of the cycle for other branches of insurance came in 1974, followed by recovery in 1975/76 (see fig. 4). While motor insurance was the most volatile of all the branches, marine insurance was peculiar in that the recession came only after a time-lag in 1975, and the recovery in 1976 was dismally slow.[32] For the insurers, the effect on their profit-and-loss account of the decline in turnover was compounded by a simultaneous rise in underwriting losses and a reduced yield on their investments.

Moreover, insurance of all kinds has been particularly vulnerable to inflation. How is it possible to assess correctly the monetary value of a cargo 3 yr ahead and to set the appropriate permium if inflation is running at 10 percent or more and there is a highly competitive market? Hull claims may be subject to delays before settlement of up to 10 yr. As a result, a prudent underwriter is forced to retain approximately 37 percent of premium income long after the third and final accounting year should allow him to assess his profit or loss account for tax purposes.[33] With a premium calculated in 1968 as a fixed

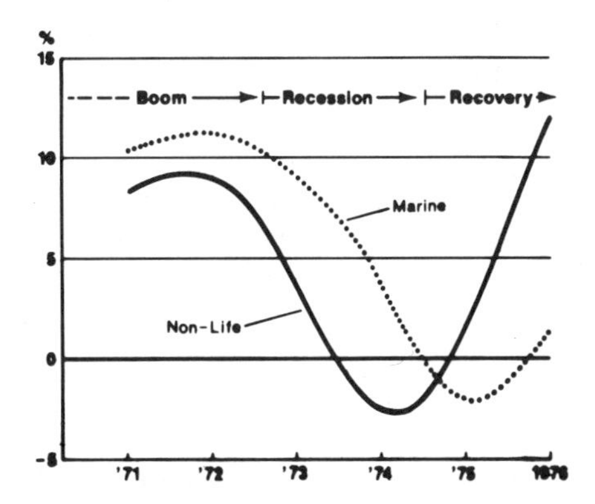

Fig. 4.—The 1970s slump in real premium growth in marine insurance. Prepared by Cartographic Services, University of Chicago, based on *SIGMA*, no. 6 (June 1978).

32. *SIGMA*, no. 6 (June 1978).

33. E. Rainbow, report to IUMI Conference, 1977, in *Policy* 76 (suppl.; November 1977): iii–x.

proportion of a hull value, an investment manager has to be very acute if he is to pay a 1978 claim without losing on it. Underwriters in countries with weak currencies in a floating exchange-rate system are particularly apt to suffer losses no matter how adequate the premium originally looked.

A final complication in the current picture has been the tremendous growth in recent years in numbers of "captive" insurance companies owned by large multinational corporations to do all their insurance business for them. A parallel development has been the growing practices of large companies in noninsurance of fairly standard risks (e.g., to company cars). The captive companies are very widely used now both as a means of avoiding tax and of adding to a company's financial resources. Millions of dollars that would otherwise go in premium payments to the insurance companies are in this way kept under the company's control. By locating the captive in Bermuda where at least 1,000 of them are now registered, or in some other tax haven, the multinational corporations have been able to significantly reduce the tax they pay the government. To hark back once again to the *Amoco Cadiz,* the oil which it spilled, valued at $23 million, belonged to Shell not Amoco and was insured with that company's captive, Petrol Insurance Limited registered in Bermuda. The parent companies owning captives have an incentive to keep premiums up, thus diverting profits by a special form of transfer pricing from the taxpaying parent's business to that of the tax-free captive—a factor which has largely insulated these insurance companies from the full effect of the underwriting cycle.

In the last 2 yr however, the United States tax authorities have begun to test in U.S. courts their recent claim that the captives are so closely identified with their parent that they are not proper insurance companies at all. The oil companies' joint subsidiary, OIL, or Oil Insurance Limited, recently formed to cover the pollution liabilities of 31 oil companies, was exempted on the grounds that spreading the risk among so many was a genuine insurance operation while a subsidiary taking all the risks of its parent and little or no other insurance business was not.

However this battle with the Internal Revenue Service (IRS) may end, the practice of self-insurance by multinationals is unlikely to be quickly reversed. Some are already taking on outside risks just in order to meet the tax authorities' charges. They are thus adding further to the competitive conditions in a depressed insurance market.

As if the difficulties of inflation and exchange loss in a purely accounting sense were not enough, they also affected replacement and repair costs. Indeed, one of the few bright spots in the marine underwriting world in 1978 was that shipyards, in their desperation to attract business, were at last cutting repair costs. According to the ILU, repair costs were increasing in 1978 at an annual rate of only 4.6 percent compared with 18 percent in 1976.[34]

34. "Chairman's Report," Annual General Meeting, Institute of London Underwriters, January 24, 1978. Reprinted in *Policy* 77 (February 1978): 98–102.

In such an unstable global economy, many insurers were themselves doubly at risk in facing a combination of market overcapacity, protectionism, and cut-throat competition.

TRENDS TO PROTECTIONISM

The trend toward protectionism in marine insurance is by now well established. Once a national merchant fleet is created, exporters are bound to make exclusive use of it, and both cargo and hull insurance is reserved to a national insurance enterprise. The Soviet Union set the example, adopting early on a practice in its trading relationships which others have tried to follow. When exporting goods, these were invoiced c.i.f. (i.e. including carriage, insurance and freight charges). But when importing, Soviet trade organizations insisted on buying f.o.b. (free on board), thus once more keeping the insurance charges to themselves. (A major bargaining issue in the arrangements for the import of U.S. grain was thus over what proportion of trade should be shared between United States and Soviet ships.) Indeed, so well insulated from the international market is the Soviet cargo-insurance enterprise, the Ingorstrakh Insurance Company, that it has been recently charging premiums as much as four times the current market rate.[35]

By now, not only the Soviet Union and the East Europeans follow this protectionist practice in insurance—so do the Brazilians, the Argentinians, and many developing countries. At first, the established insurance enterprises in Europe and the United States managed to cope with this protectionism. For, what they lost in direct insurance income they gained in increased profits from reinsurance. But now some secondary consequences are gradually becoming apparent.

First, the process of reinsurance, of whatever kind, places the reinsurer at least one step further away from the actual risk incurred, with the inevitable result that costs are increased. As with interbank lending in Eurocurrencies, a great many intermediaries may be involved in one transaction. A recent reinsurance wrangle concerning the reinsurance of New York property turned out to have involved no less than six intermediaries, ending with the Instituto de Resseguros do Brasil (IRB).[36] The reinsurer is taking a calculation on the reputation of intermediaries, not on the type of risk. Yet facultative reinsurance may be the result of a direct insurer's desire to pass on what he knows to be a particularly bad risk. Second, as noted above, the protected direct insurer may have the monopoly power to charge higher rates while the reinsurer still faces competitive rate cutting and has to take more risks for less premium. Finally, the effect of protection on the cycle is to reduce the number of insurers leaving

35. *Business Week* (June 12, 1978).
36. "The Sasse Reinsurance Wrangle," *Financial Times* (June 22, 1978).

the marine market. The necessary reduction in capacity is consequently concentrated in the residual, free-market sector. Ironically, protection pushes up the cost of insurance in the protected sectors and lowers it to unprofitable levels in the free market.

THE DEPENDENCY OF THE LESSER DEVELOPED COUNTRIES (LDCs)

Destabilizing to the insurance business as these protectionist practices have lately been, it must be conceded that they are the LDC's response to a perception—objectively undeniable—of over-whelming Western dominance in this service sector. The first Third World Insurance Congress held in the Philippines in October 1977 produced some interesting statistics. It was estimated that in 1975 only 6.3 percent of the total volume of world premium income originated from all the Third World countries put together. Only 16 percent of insurance companies operated in developing areas at all, and 61 percent of those operating in Asia were foreign owned. In the Philippines alone 80 percent of total insurance policies generated locally were reinsured abroad, and only 20 percent of the total premium income remained behind for domestic investment.[37] In short, if the local market for insurance—whether life, motor, or marine—is only a narrow one, and 50 percent of world sales of all insurance are made in the United States, then the local insurers are obliged in self-defense to buy reinsurance from larger foreign operators or to go without cover.

Since insurance profit is primarily derived from investment income, it follows that insurance or reinsurance with Western companies constitutes a considerable diversion of capital from national economic development—unless of course both investment and underwriting accounts happen to be taking a loss. Furthermore, since internal insurance capacity is small, in conditions of free trade the bulk of trade between a developing country and an industrialized one or between third countries will be insured on the Western insurance market. Often the same system is adopted by the West as by the Soviet Union, less for protectionist reasons than simply because developing countries have inadequate insurance capacity of their own so that the importer will be paying f.o.b. and the exporter will charge goods sent to developing countries c.i.f. Overall, although this foreign insurance may be cheaper, it constitutes a costly invisible import for the developing country; and the net balance-of-payments effects are a matter of dispute.[38]

As developing countries see it, the solution lies in several protectionist

37. "Report on First Third World Insurance Congress," *Policy* 76 (December 1977): 936–40.

38. G. M. Dickinson, "International Insurance Transactions and the Balance of Payments," *Geneva Paper on Risk and Insurance,* no. 6 (October 1977), pp. 17–35.

practices. Those countries, like Brazil, which maintain a merchant marine attempt to fix basic insurance and reinsurance tariffs at home and to place pressure on national carriers to insure through national insurers. In addition, reinsurance flows outward become the monopoly of state reinsurers like the IRB.[39] Western companies, if not excluded, are required to invest locally varying proportions of the income they earn, and restrictions are placed on repatriation of funds.

Those countries that do not have adequate shipping of their own have attempted through UNCTAD to enforce greater carrier liability on foreign shipping companies, thus reducing the burden of insurance on the exporter/importer. Since the fleets of developing countries amounted to only 6 percent of the world fleet in 1972 (though nearly double that now) such a shift is clearly in their interests.[40]

The result has been the acceptance by UNCTAD of the UNCITRAL Draft and the adoption of this Convention on the Carriage of Goods by Sea in Hamburg on March 31, 1978. These "Hamburg rules" replaced the two other Hague and Hague/Visby rules concerning cargo insurance. They increased the liability of carriers by 25 percent and put the burden of proof on the shipping companies.[41]

Whether or not enough countries ratify these revised rules to bring them into effect is not yet clear. But the proposed changes show that the dominance of Western marine insurance on international trade is under attack. In self-defense, the insurers argue that LDCs are acting emotionally more than rationally and that they will not necessarily be better off. On carrier liability, for example, the Western cargo insurers feel that the new rules will greatly over-burden the P&I clubs. These mutual-aid groups give carriers cover for risks not insured under normal hull and cargo accounts. Collision damage to other ships, damage to port facilities, accidents to crew and/or to passengers and third parties, and even losses of profit from strike action are all risks taken on by P&I clubs. They also insure against liabilities of shipowners as cargo carriers, and it is in this role that they are likely to be subject to claims under the new Hamburg rules, even though P&I liability is backed up through excess-of-loss cover by reinsurance. The point is that under the new rules the exporter or importer incurring a cargo loss will have to bring legal suit against the shipper—a far lengthier and less certain process than simply claiming from the insurer. In addition, the chances are that the extra cost imposed by the rules may be recouped by shipping operators through higher freight charges. And this may mean that exporters of low-value goods (e.g., the oil producers and exporters

39. See UNCTAD, Reinsurance Problems in Developing Countries (TD/B/C3/106/Rev 1), 1973, pp. 35–38.

40. UNCTAD, *Review of Maritime Transport,* 1972–1973 (TD/B/C4/1117), 1975, pp. 5–11.

41. *Review* 18/4515 (July 14, 1978): 11.

of minerals and raw materials from developing countries) will pay the increased cost of the insurance on high-value goods.[42]

With regard to the wider question of the protection of national insurance markets, Western insurers claim that costs can only be kept down if risks are widely spread internationally. Some developing countries are particularly prone to catastrophic earthquakes, floods, and other natural disasters. At any moment their national insurance systems could be faced with vast increases in claims far beyond their capacity to meet, even if they keep their reserves in a highly liquid form instead of investing them more profitably. As an instance of the need for a wide international spread, insurers quote the example of Spain which, in 1977, lost the equivalent of 25 percent of its yearly flow of marine premiums through the loss of one ship.[43]

Another point is that the educated manpower needed to administer insurance is by no means neglible. Some economists would argue that developing countries have more urgent priorities in industry, agriculture, or commerce on which to use their educated manpower.

With these considerations in mind, the solution for developing countries is sometimes thought to lie in regional insurance pools—like the one developing in Singapore or that set up by the oil-producing states in the Middle East—rather than in direct protection of national insurance. The advantages of regional pooling are particularly strong at present because the risks are growing. While technological developments may reduce the risks of seaborne trade, provided the total of world tonnage remains constant, disasters when they happen may be more damaging to the environment and ships lost may be more expensive. The first oil tanker carried a mere 2,000 tons; the *Amoco Cadiz,* by no means the largest at sea, carried 200,000 tons. Although there have as yet been no serious accidents to liquid natural gas (LNG) or liquid propane gas (LPG) ships, these have an explosive potential which can only be imagined.

The whole question of ecological risk and pollution liability raises difficult and fundamental problems. The first arises from the necessarily transnational character of any litigation to assign liability. Given the free flow of oil tankers through national and international waters, claims can be made on behalf of the victim in one country against the shipowner in another, who is, in turn, insured in a P&I club in a third. While U.S. courts have to some extent been successful in adjudicating on pollution liability cases, this has been yet another demonstration of the economic power of the United States in the world economy and over transnational factors rather than an indication of the strength of the legal case.

42. IUMI Carriers' Liability Committee, "The Essential Role of Marine Cargo Insurance in Foreign Trade," *IUMI* (October 1975).

43. "Ocean Hull Business, Spain," IUMI Conference, 1977, *Policy* 76 (suppl.; November 1977): iv.

A more efficient system would require international agreement and the institution of international arbitration processes, but this has not so far proved possible.

Second, there is the difficult question of setting a limit to liability. As the rather similar product liability cases have also shown, an insurance system cannot work well if there is virtually no upper limit on the extent of claims made through it. The insurers face open-ended multi-million-dollar claims but can collect only low premiums to cover the risks because the incidence of accident is still relatively low. Some attempt was made in the 1975 IMCO Convention on Civil Liability for Pollution Damage and the 1978 Convention for the Establishment of an International Fund for Compensation for Oil Pollution. The former set the limit of liability at $169 per g.r.t. or $16.8 million, whichever is less. This gave the liability insurers the basis for quoting a premium to cover smaller risks. A third convention allows compensation to be paid above this limit of liability up to a maximum of $36 million for each incident and supplements the compensation systems, Tovalop and Cristal, organized by the oil companies themselves. What is does not do, as the *Amoco Cadiz* case shows, is to arrange beforehand what happens and who has to pay up when the damage exceeds the limit. The result, in short, is a patchwork of uncertainties in which the situation is further confused by the fact that some countries are not signatories of the IMCO Conventions, some companies have not joined the voluntary oil company agreement, and the legal precedents are lacking to guide the courts. From the point of view of the insurance industry, any excess damage above the agreed level is the responsibility—as with most nuclear risks and war risks—of governments. But the question is which government? As the *Amoco Cadiz* case shows, this may be either the French or the American government depending on the decision of the U.S. courts. If they rejected outright all the French claims, the issue could rapidly become a highly political one. But other cases brought in U.S. courts, for example, against pharmaceutical companies by six developing country governments or against the McDonnel Douglas aircraft manufacturers by the relatives of passengers in the Turkish airline which crashed outside Paris some years ago, suggest that this is unlikely. *Faute de mieux,* it seems as if international law regarding pollution liability is being made in America.

In the meanwhile, it seems more than likely that the relative laxity of recent years in the enforcement of safety rules at sea, in the maintenance of standards of training, and of operational procedures in navigation will not be tolerated much longer. Though the responsibility for enforcement of regulations rests with governments, it has always been shared to some degree by the insurance industry. Insurers are, for example, the biggest single collectors of information on risk assessment and claims records. They have the ability to make judgments about the risks taken by persistent offenders against international regulations whether on pollution or navigation. Whether they make use of the information is a function of the pressures put upon them. To some extent this is reflected in the higher insurance rates applied to older ships.[44] And P&I clubs as mutual associations can and sometimes do exclude certain owners whose risks are known to originate more often in bad judgment than bad luck.

Unhappily, the fatal flaw in this notion of punitive insurance systems for constant offenders lies in the competitive conditions of the marketplace. Once the desire for premium income to invest exceeds the desire for underwriting profit and once insurance supply exceeds demand, underwriters are tempted to omit or skimp on the detailed scrutiny of a risk and to take it with their fingers crossed. A shipowner will find he can always get insurance from somewhere even if not from the old-established centers. And brokers can pressure underwriters to take good risks along with the bad. Thus, despite their knowledge of the risks involved and the constant warnings given by individual "leads," punitive action against individual owners is very difficult to take except in an undercapacity market.

In theory, it might be conceivable that this business, which is remarkable for evolving its own means of autonomous self-discipline as an adjunct to and partial substitute for government regulation, should be able to develop methods of exerting greater discipline. But once again the difficulty is that there are major centers in most of the industrialized countries, besides London and New York, and to get agreement between the operators in all of them would be every bit as difficult as getting international conventions signed and enforced by governments.

Currently one area which does allow insurers to pick and choose those they insure is the offshore oil-rig market. As of this writing, the value of a single offshore unit in the North Sea had reached approximately $600 million and was expected to double by 1980.[45] As the number of rigs increases, so does the casualty list. In 1976, for example, there were 16 major rig casualties. By June 1977 this total had risen to 25.[46] Given the huge investment required for undersea exploration and production, it is extremely difficult for oil companies both to operate rigs and to self-insure them. Such is the size of potential loss, through single and multiple blowouts, and so great the value of what is at stake, that the capacity of the marine insurance market is itself in doubt.[47]

Most insurers are still sanguine that capacity will be found, but the problem does reveal the new issues posed by higher technology. Even if individual companies balanced their portfolios by type of risk, reinsured, and spread the

44. F. M. Hunter, "Marine Insurance and the Problems of Substandard Vessels," *Review* 180/4506 (March 16, 1978): 19–20. Further details may be found in the OECD Annual Reports on Maritime Transport.

45. See also Edward Symonds, "Offshore Oil and Gas," in *Ocean Yearbook 1*, p. 128.

46. *Review* 18/4495 (September 23, 1977), p. 12.

47. Ibid., p. 11.

risks, the cost of one incident can have a ripple effect across the market. This may, in turn, produce a rapidly fluctuating underwriting curve instead of a smooth one and thus make future profitability even more unpredictable.

The insurance system is thus full of paradox, involving simultaneously the contradictory principles of cooperation in spreading the risks and competition in the processes of the market. When there is overcapacity, the relationship between the degree of risk and the cost of the premium is slowly weakened, and those who increase the danger by shortcuts in safety escape lightly. When there is undercapacity, the protection offered by the system is less than prudence might require.

CONCLUSIONS

Marine insurance, in short, is a tangle of paradox. Globally integrated so that few countries, outside the centrally planned economies of China and the Soviet bloc, are immune from market forces, it is yet subject to increasing economic nationalism. The whole system is an elaborate demonstration of complex cooperation in the division of economic functions, yet at each stage in the game competition is remorseless. The Ferris wheel of the underwriting cycle often rewards those who have deserved least of it and is more likely to punish the unwary or the unlucky than those who have added to the general level of risk by taking imprudent chances, whether at sea or in the market.

The function of states in relation to the market is equally contradictory. On the other hand, by making tighter safety rules they may lessen the burdens of risk on the insurance business. Yet by assigning liability and by granting legal claims for damage, they may increase the risks for shipping operators, their customers, or their insurers. When they act to protect the interests of national merchant fleets they may add to the problems of national insurers, while by protecting the insurers they may impose costly burdens on national exporters.

It is a business which operates best in periods of relative political calm and economic and monetary stability. If the volume of trade changes too fast, if the value of money depreciates too fast, or exchange rates are too unstable, insurance is more likely to add to the instability of the world system than to cushion it. Not only has the international political economy failed to provide insurance with the necessary stable environment, but its very instability has probably reduced the efficiency and increased the perceived inequity of the system. The big question is whether international agreement—or failing that, national regulation—can close the gaps opened by new technology.

Nonetheless, the marine insurance system does underpin the processes of international trade and will surely continue to do so. Its basic function is nonideological, and without insurance the benefits of trade would be smaller and the costs far greater. It makes the incentive to trade by sea much higher and the costs to the "fewe who do adventure" much smaller. To contemplate the potential disasters and see the compensatory system as a whole as inadequate is to forget that ancient civilizations could only regard natural calamity as the hand of fate and accept it. Not the least of the positive achievements of the system has been its continuing capacity, through ingenuity, to match risk with a mixture of regulation and insurance. The system may not be perfect but it is all there is. Noah's ark was not insured at all.

Soviet Shipowners and International Shipping

George A. Levikov
*United Nations Conference on Trade and Development**

The main process in current international relations is the deepening and broadening of detente, which has made the problems of international cooperation in the field of economics particularly important. This cooperation is the basis for the peaceful coexistence of states with different social systems. A great contribution toward working out the directions of political, economic, scientific, and technical cooperation among countries has been made by the Final Act of the Conference on Security and Cooperation in Europe. Considering, in particular, that improvement in transport conditions is one of the essential factors in the extension of cooperation, states party to the conference showed their readiness to promote both international transportation and adequate participation in trade on a mutually advantageous basis. This directly concerns merchant marines operating primarily on the international freight market.

The USSR's economic potential, the enlargement of its external economic activities, and its unique geographic position have resulted in an international Soviet merchant marine which became the main instrument of economic relations between the Soviet Union and a number of socialist, as well as developing and developed capitalist, countries. This intensifies the importance of maritime external transport (shipping) policy, which is an integral part of the foreign economic policy of the Soviet state and is expressed in the USSR's maritime legislation and in its practice of international law.

The shipping policy of the Soviet Union is determined by the special features and aims of its social system. Its foundations were first laid in the decrees signed by V. I. Lenin. It consists in securing and maintaining the independence of the USSR in its national maritime transport in cooperation with all interested states, based on the principles of equal rights, noninterference, and mutual advantage. Guided by the desire for a normal development of international merchant shipping, the Soviet Union has concluded bilateral agreements with many countries, including those in the West. These are being successfully implemented. Within the framework of these treaties and agreements, the problems of improving maritime trade were solved through the use of mixed commissions. Cooperation is also carried out through various

*Formerly of the Research Institute of Maritime Transport, Moscow.

© 1980 by The University of Chicago. 0-226-06603-7/80/0002-0018$01.00

international maritime organizations of the UN system and others. Soviet shipowners maintain joint lines with the companies of other countries, establish joint agency firms, and participate in liner conferences.

In the Soviet Union, great importance is attached to insuring the terms necessary for mutually advantageous development of international merchant shipping. These are the inadmissibility of discrimination and the broad extension of most-favored-nation or national treatment. This is in keeping with generally accepted principles of international trade relations and trade policy, under which economic relations between countries shall be based on sovereign equality, the self-determination of peoples, noninterference in the domestic affairs of others, and the elimination of discrimination based on differences in socioeconomic systems.

A serious obstacle to the extension of international economic relations is the policy of discrimination pursued by certain Western powers, particularly in the field of merchant shipping. In order to maintain their privileged positions in international shipping, these powers label as discrimination protective measures taken by other countries for the purpose of developing national marine transport or for the purpose of insuring fair participation in the transport of their country's foreign trade. At the same time, these Western powers themselves resort to measures hampering equal and mutually advantageous participation of all countries in merchant shipping. These measures include the exclusion of particular national ships from their ports, the prohibition of the carriage of the power's trade goods by such ships, nonrecognition of ship's documents, the selective restriction of the right of crew members to go ashore, restriction of the use of port facilities, etc.

On the other hand, the Soviet Union has been consistent in advocating the elimination of any form of discrimination in foreign economic relations. This position was confirmed in the Comprehensive Programme for the Further Extension and Improvement of Cooperation and the Development of Socialist Economic Integration of the CMEA (Council of Mutual Economic Assistance) Member Countries and in the Budapest Agreement on Co-operation in Merchant Shipping (1971), wherein the socialist countries stipulated the right of third parties' ships to participate in trade between the ports of one party to the agreement and ports of other contracting parties.

Despite increased attention to the problem of discrimination in merchant shipping, a common interpretation of the term itself does not exist. A solution to this difficulty has been advanced through the elaboration of the working document "Discrimination and Protection in International Shipping" (TD/B/C.4/L.112), which was submitted by seven socialist countries, including the USSR, to the seventh session of the UNCTAD Committee on Shipping in November 1975. In it "discrimination" is defined as "measures adopted by a country or group of countries in violation of the rules of international law which are designed primarily, owing to a difference in social and economic systems, to prevent participation on an equal footing of the fleet flying the flag of any country in international maritime shipping." Thus one should consider as discrimination those measures which infringe on generally recognized international law standards, lying, for example, in the fact that no state has the right to limit access to, or prevent from calling at any of its ports which are open to foreign ships, the ships of a particular state based on the display of a flag or other signs. Neither may it, as a general rule, prohibit the use of port facilities or similarly inhibit such a vessel. Such a definition increases precision and is not based on varying conceptions in some maritime countries or groups of countries which reflect only their own interests.

Soviet shipowners oppose the deliberate identification of the concept of protectionism with that of discrimination; they consider the former to be merely state assistance to national shipping, primarily in terms of the reservation of cargo. The formal position of more than 100 member countries of IMCO, including the Soviet Union, is reflected in the convention of this organization. It is emphasized that assistance and encouragement given by a government for the development of its national shipping and for purposes of security do not in themselves constitute discrimination, provided that such assistance and encouragement are not based on measures designed to restrict the freedom of shipping of all flags to take part in international trade. Therefore, each country's right to promote its own merchant marine is recognized. Every country also has the sovereign right to choose relations with other countries which it considers appropriate to the achievement of particular economic aims, providing it observes international agreements and allows the equitable participation of all countries in merchant shipping. In exercising its right to have a merchant marine a country may give it preference, otherwise development might prove impossible. However, in the allocation of goods among third-country ships no distinction may be drawn as to flag.

In the carriage of goods under a bilateral trade agreement, preference may be given to the respective merchant marines of the partners. The main purpose of such agreements is the implementation of trade between the two parties, and such agreements are realized only because they serve the national interests of each. It follows that a "third" carrier who is not a juristic person of the contracting states is not allowed to participate in the trade. In addition, bilateral lines set up for contractual trade between two countries are not shipping lines as the term is internationally understood, but regular transport services rendered under the agreement on mutual delivery of goods.

It goes without saying that in life very unusual situations may occur. For example, the partners to the line agreement may admit a third-country carrier to their trade. Similarly, when a given trip or carriage in one of the trade directions does not yield a full load, it may be permissible to carry, in part, the goods of other countries. Then the line becomes an outside participant with respect to conferences operating in this region.

Thus, protectionism may be characterized as measures taken by a state within its sovereign rights, in accordance with generally recognized principles and rules of international law, to create, develop, and encourage the national merchant marine.

In the opinion of the Soviet shipping interests, an important instrument of struggle against discrimination, for extension of economic cooperation and materialization of detente, is the concession of most-favored-nation treatment (MFNT) and national treatment (NT). At present, they are among the basic elements of almost all agreements on trade and shipping, and on regulating relations in the field of merchant shipping, concluded by the Soviet Union.

It is understood that under the MFNT one of the contracting parties commits itself in its territory to grant the other party the highest possible preferences and privileges in respect to a person or juridical body, goods, ships, and other objects that are given or will be provided for any third state. Hence, it appears that the MFNT does not extend special privileges; it is merely not less favorable than treatment to a "third" party. In other words, the MFNT is the "most" only for those states to which it does not apply, whereas the countries having an agreement on the MFNT among themselves are equal in their preferences and privileges. Side by side with the MFNT in a number of treaties and agreements between the governments of the USSR and other countries on merchant shipping, the NT is applied, signifying the concession, in certain respects, of privileges and preferences to another state which are at the disposal of persons and juridical bodies, ships, and goods of Soviet organizations. In certain cases the said treaties contain provisions on the MFNT and NT at the same time (e.g., the agreement between the governments of the USSR and Greece or that between the USSR and Finland). Thus, in their shipping policy the relevant Soviet organizations proceed from the interests of (1) fair and mutually advantageous cooperation, including cooperation with countries having different socioeconomic systems; (2) complete elimination of discrimination in international merchant shipping; and (3) observance of international law, including obligations arising from the MFNT.

The Soviet shipowners are extremely concerned by the absence of a genuine link between a vessel and a flag of registry (the problem of ships of "flag of convenience," or open registry) that, in their opinion, affects important aspects of international merchant shipping, such as the safety of navigation and marine environment protection, social and labor relations between seamen and shipowners, observance of the fair competition principle, etc. They think that the flag-of-convenience system is to blame for a lasting crisis in shipping, for it results in uncontrolled hunting for additional profits in shipping, surplus carrying capacity of merchant marines, and increasing competition on world sea routes. The Soviet shipowners are also greatly concerned by a rapid growth of multipurpose tonnage and container ships registered under flags of convenience, since this results in a surplus of cargo ship services in liner traffic. They are sure that the development of flag-of-convenience fleets seriously impedes the creation of the national marine transport of the developing countries.

Guided by their policy aimed at strengthening peace and normalizing the entire system of international economic relations, Soviet shipowners feel that the elimination of the flag-of-convenience ship registry practice would be a marked contribution to reorganizing these relations on a democratic and fair basis. In these circumstances they agree with the majority of the developing countries and are ready to support them in their struggle to make merchant shipping a genuine instrument of economic development for new states.

The Soviet shipowners also are confirmed opponents of "dumping," that is, the sale of marine transport services in the foreign market at prices considerably below world ones and often even lower than transportation costs. Dumping is one of the most important instruments of economic expansion for some Western shipping companies, who use it to suppress rivals and seize markets. Its characteristic feature is the dumping price system, but not separate operations. An example of dumping is the practice of closed liner conferences which deliberately incur losses during some periods, using "fighting" ships or analogous methods for ousting outsiders. In the long run such actions result in a general rise in tariff levels that hampers international trade and damages the interests of all countries, the developing ones in particular. Dumping is in fact an essential part of the shipping policy of some Western countries which grant subsidies to their shipowners. These subsidies enable shipowners to extract profits in spite of using tariff rates that do not cover their individual transportation costs. Such a practice leads to the ousting of fully competitive enterprises from international shipping and in the long run to the raising of marine transportation costs.

It should be mentioned that in a number of cases reduced rates reflect individual lower transportation costs, which cannot be defined as dumping. For example, it is known that Greek tonnage chartering costs are sometimes lower than those of the United States or of West European countries, but this fact is not grounds for charging Greek shipowners with dumping. The difference in the costs of production underlies the worldwide division of labor and international trade.

Marine transport is a profitable industry of the Soviet economy. On the world freight market the USSR seeks to obtain as high rates as possible under specific market conditions, and its competitive ability reflects the true level of transportation costs. Really, it cannot be otherwise because profit is among the enterprise plan indices, and the payments of enterprises and organizations made from their profits constitute a major part of the state budget revenues. This system is so constructed that the enterprises become interested in the best use of their production means and in increasing work effectiveness. In short, the greater an enterprise's profits, the greater the part of them remaining at its disposal. This creates favorable conditions for the development of production and material-incentive intensification in the workers in a given enterprise. At the same time, profits are naturally connected with budget expenditures as used by the state for common needs. That is why those who groundlessly charge the Soviet merchant marine with a lack of interest in gaining profits and dumping, thereby contradicting the real facts, are not actually concerned about the maintenance of the appropriate commercial practice, but only about securing a big shipowners' monopoly for a small number of the so-called traditional sea powers.

Obviously it is necessary to elaborate a scientific method for comparing the international prices of cargo carriage by sea as fixed by the shipowners of different countries. This problem has many objective complexities. The ship expenses of the Western companies may not include amortization, though it is taken into account in the freight rate level; rates of depreciation are different in Western countries; and because of subsidies and many other privileges the expenses denoted by foreign shipowners may not correspond to the real ones. Nevertheless, the elaboration of this methodology, whether it is conditional or not, may contribute to a clarification of this important problem and to the harmonization of the shipping policies of individual countries.

The Soviet shipowners proceed invariably from their interest in the general extension of international merchant shipping and are ready to cooperate in this field with all who follow the principle of freedom of shipping. This approach is based on the Leninist principles of Soviet foreign policy. It is determined in its character and not subjected to any modifying influences. The lengthy experience of the Soviet shipowners' organizations confirms their steady position on these questions and shows that unsolved problems do not arise in cases where their partners are interested in the development of cooperation. This is how cooperation with foreign countries in merchant shipping is being implemented.

It would be a mistake to think that the extension of this cooperation has reached a limit. On the contrary, the experience gained by the partners and the realistic approach to the cooperation in most cases convincingly confirm the favorable chances for its further extension and deepening. The economic relations in merchant shipping, tested by time and successful development, serve the radical interests of all peoples, as well as the strengthening of peace and the consolidation of detente.

International Cooperation in Marine Science

Warren S. Wooster
University of Washington

Cooperation implies a combining of effort for the sake of mutual benefit. It characterizes human activities in many spheres, one of which is science. The advance of human knowledge and understanding benefits particularly from the pooling of the experience and skills of scholars. Cooperation occurs among individuals, institutions, and governments. It has characterized marine science from its earliest days.

SPECIAL ATTRIBUTES OF THE OCEAN

The ocean has several attributes that make it particularly appropriate for cooperative research activities.

Size

The ocean is enormous, covering about 362 million km^2, or about 71 percent of the earth's surface. With a volume of about 1,350 million km^3, its mean depth is about 3.7 km.[1] Much of this vast region has yet to be described, and the detailed description of even the portion that is relatively time invariant (the seabed and the waters below a kilometer or so) requires resources far beyond those possessed by single institutions or even countries.

Variability

The ocean demonstrates a high degree of variability, in both space and time. Although the seabed away from the coastline changes only very slowly with time, it exhibits spatial variability similar to that of land. The surface water layer, to depths of a few hundred meters, displays marked changes horizontally, such as fronts and eddies, with scales of kilometers to hundreds of kilometers.

1. H. W. Menard and S. M. Smith, "Hypsometry of Ocean Basin Provinces," *Journal of Geophysical Research* 71 (1966): 4305.

© 1980 by The University of Chicago. 0-226-06603-7/80/0002-0003$01.00

Vertically, apart from the well-known thermocline discontinuity, the distributions of temperature and other variables show variations with scales of centimeters to tens of meters, the so-called fine structure. Marine organisms may have general distributions extending over hundreds or thousands of kilometers but in detail are often found in schools or patches of much smaller dimensions. And the whole hydrobiosphere, particularly in the upper layers, varies in time with periods of hours to centuries or longer. The study of processes in this highly variable regime, even within a single discipline, often requires the efforts of many scientists and several research vessels.

Interactions

The artificial nature of the distinctions among scientific disciplines is particularly evident in the ocean. The life histories and dynamics of populations of marine organisms are molded by their ever-changing physicochemical environment, which in turn receives the products of their metabolisms. The marine chemist investigates chemical species which are distributed by water motion and in many cases are transformed by passage through the biota. In his reconstruction of the past, the sedimentologist employs the remains of planktonic organisms and fish that have settled to the bottom. Water motion is driven by the atmosphere, and the properties of seawater derive in large part from ocean-atmosphere exchange. The comprehensive study of even simple ocean systems may call for the cooperation of scientists from several disciplines.

Jurisdiction

Artificial, man-made geographical boundaries are also ignored by ocean systems. Ocean currents and populations of marine organisms move past the coasts of many countries, and events in one region may be dominated as much by distant forces as by local processes. With the unity of natural phenomena being dissected by political boundaries and jurisdictions in the developing law of the sea, the study of such phenomena increasingly demands the joint efforts of scientists from both neighboring and distant countries.*

Most of these attributes occasion scientific cooperation within a country such as the United States as well as among scientists in different countries. It is the latter international cooperation on which the present paper focuses. Furthermore, although the personal collaboration of scientists has always been,

*Editors' Note: For additional discussions of international cooperation in the marine sciences, see Lord Ritchie-Calder, "Perspectives on the Sciences of the Sea," *Ocean Yearbook 1,* ed. Elisabeth Mann Borgese and Norton Ginsburg (Chicago: University of Chicago Press, 1978), pp. 286–88, 290–91; and G. L. Kesteven, "The Southern Ocean," ibid., pp. 488–89.

and continues to be, a major element, it is the institutionalized cooperation that concerns us here.

HISTORY OF FORMAL INTERNATIONAL COOPERATION

The history of formal international cooperation in marine science goes back to 1902 with the establishment in Copenhagen of the International Council for the Exploration of the Sea (ICES).[2] This stemmed from a proposal of Otto Pettersson, and negotiations during the intervening years involved such other well-known figures as Sir John Murray, Fridjtof Nansen, and Johan Hjort. Although from its beginning the council emphasized practical problems related to fisheries, other biological and hydrographical investigations were also incorporated. For example, an early project was to establish the quantity of halogen contained in seawater and its relation to the density of the water, an investigation in which Martin Knudsen played a central role.

The council, which continues its work to the present time, is undoubtedly a highly effective regional intergovernmental marine-science organization. It now has 18 members, 14 of them coastal states in the north Atlantic, seven members of the European Economic Community.[3] The ICES is concerned with promoting and encouraging research and investigations for the study of the sea, particularly those related to living resources. In addition, the council has advisory responsibilities to two fishery commissions and three pollution commissions.[4] The annual statutory meetings provide an important scientific forum for European marine scientists as well as an opportunity for conducting the council's business.

Although that business is formally the responsibility of national delegates meeting as the council, most of the annual meeting is devoted to scientific discussions within and among a number of standing committees.[5] Scientific proposals arise in these discussions, are accepted by the relevant standing

2. A. E. J. Went, "Seventy Years Agrowing: A History of the International Council for the Exploration of the Sea 1902–1972," *Rapports et procès-verbaux des réunions conseil permanent international pour l'exploration de la mer* 165 (1972): 1–252.

3. Present ICES members are Belgium, Canada, Denmark, Finland, France, German Democratic Republic, Federal Republic of Germany, Iceland, Ireland, Netherlands, Norway, Poland, Portugal, Spain, Sweden, the United Kingdom, the United States, and the Soviet Union.

4. Fishery commissions advised by ICES are the North East Atlantic Fishery Commission and the International Baltic Sea Fisheries Commission. Pollution commissions advised by ICES are the Oslo Commission and the Interim Paris and Helsinki Commissions.

5. As of 1978, ICES had the following standing committees: Fishing Technology, Hydrography, Statistics, Marine Environmental Quality, Mariculture, Demersal Fish, Pelagic Fish, Baltic Fish, Shellfish, Biological Oceanography, Anadromous and Catadromous Fish, and Marine Mammals.

committee, and passed to the consultative committee for approval. That committee consists of the chairmen of the various standing committees, and its decisions represent scientific, rather than political, judgments. Scientific recommendations of the consultative committee are usually accepted without change by the council.

This distinction between science and politics and the scientific origin of ICES programs are responsible for much of the success of the organization. There are other characteristics that should be noted. Although the statutory objectives are broad, the focus on fishery problems has been sharp. In recent years, increased attention has been paid to marine pollution; marine science per se tends to have a lower priority. Oceanographic programs are usually closely linked to applied interests. The council's contacts in its member countries reflect these emphases, being primarily with fishery laboratories rather than with universities or oceanographic institutions. Some fields of marine science—for example, geology and geophysics—are completely missing. Nonetheless, despite these limitations, ICES continues to be a highly effective mechanism for joint action among its members.

Another approach to international cooperation among marine scientists took place within the nongovernmental International Council of Scientific Unions (ICSU).[6] A predecessor organization, the International Research Council, first met in 1919 and was replaced by the present ICSU in 1931. The ICSU consists of both national members (national academies of science and equivalent bodies) and scientific unions, a number of which represent disciplines concerned in part with the ocean.

One of the earliest of these, the International Union of Geodesy and Geophysics (IUGG), was established in 1919. An original constituent of IUGG was the Section d'océanographie physique which first met in 1921 and concerned itself with the morphology of the sea bottom and of the surface of the oceans and seas, with the movements of water masses, and with physical and chemical studies of seawater.[7] The section was replaced with the Association d'océanographie physique in 1933; this was broadened to the International Association for the Physical Science of the Ocean (IAPSO) in 1967. Subsequently, two other scientific unions have established marine-science entities, the International Union of Biological Sciences and its International Association for Biological Oceanography (IABO) and the International Union of Geological Sciences and its Commission on Marine Geology (CMG).

Of these organizations, IAPSO and its predecessor bodies have by far the longest history. In 1933, the Committee on Mean Sea-Level and Its Variations was established and began compilation and publication of monthly and annual

6. *The Year Book of the International Council of Scientific Unions 1978* (Paris: ICSU Secretariat, 1978).

7. T. W. Vaughan et al., *International Aspects of Oceanography* (Washington, D.C.: National Academy of Sciences, 1937).

mean heights of sea level at a number of stations throughout the world.[8] In 1956, the Permanent Service for Mean Sea Level became part of ICSU's Federation of Astronomical and Geophysical Services. From 1948, the association assumed responsibility for the preparation and distribution of standard seawater.[9] The Standard Sea-Water Service was located at the Danish Hydrographical Laboratory until 1975 when it was transferred to the Institute of Oceanographic Sciences in England.

The principal activities of IAPSO and similar bodies have been the organization of scientific symposia and discussions at their periodic meetings. In more recent years, IAPSO, along with IABO and CMG, has worked through the Scientific Committee on Oceanic Research (SCOR).

Because of the interdisciplinary nature of oceanographic problems, the isolation of separate disciplines in their own associations makes it difficult both to exchange information across disciplinary boundaries and to organize investigations of the relevant interactions. These difficulties were recognized by ICSU which in 1957 established the Special (now Scientific) Committee on Oceanic Research for the purpose of "furthering international scientific activity in all branches of oceanic research."[10] To illustrate its interdisciplinary interests, SCOR at its first meeting identified three critical long-range problems: the use of the deep sea as a receptacle for the waste products of industrial civilization, the oceans as an important source of protein food, and the role of the oceans in climatic change.[11]

The activities of SCOR have been described elsewhere[12] and need only to be summarized here. They are pursued by national committees in 34 countries, by nearly 100 marine scientists who are nominated as members by those committees, and by many more scientists who participate in a wide variety of working groups on specific scientific problems. In addition, the officers of the specialized associations of scientific unions referred to earlier take part in the executive committee of SCOR. Besides its working groups, SCOR organizes scientific meetings and provides scientific advice to Unesco and to its Intergovernmental Oceanographic Commission (IOC), thereby linking the marine scientific community with the work of these organizations.

The officers of SCOR participated in the Unesco Intergovernmental Conference on Oceanographic Research which took place at Copenhagen on July

8. Association d'océanographie physique, "Monthly and Annual Mean Heights of Sea-Level 1937 to 1946," *Publication scientifique,* vol. 10 (1950).

9. Association d'océanographie physique, "General Assembly at Oslo August 1948," *Procès-verbaux,* vol. 4 (1949).

10. Warren S. Wooster, "The Scientific Committee on Oceanic Research," in *Ocean Yearbook 1,* ed. Elisabeth Mann Borgese and Norton Ginsburg (Chicago: University of Chicago Press, 1978), p. 563.

11. Special Committee on Oceanic Research, "Report on the 2nd Meeting at Paris, September 26–27, 1958," mimeographed (Seattle: Institute for Marine Studies, 1958).

12. Wooster, pp. 563–68.

11–18, 1960, and which led to creation of the Intergovernmental Oceanographic Commission of Unesco and to designation of SCOR as the scientific advisory body for the Unesco Office of Oceanography. The IOC was established by the eleventh General Conference of Unesco later in 1960 with the purpose "to promote scientific investigation with a view to learning more about the nature and resources of the oceans through the concerted action of its members."[13]

The subsequent histories of SCOR and IOC were closely intertwined for a number of years. At its first meeting, in 1957, SCOR had initiated planning for an International Indian Ocean Expedition, which commenced in 1959. Coordination of this large and complex operation, involving 23 participating countries, was difficult for a nongovernmental organization such as SCOR, and the coordinating responsibility was passed to IOC in 1962.

The IOC has evolved from a relatively small organization concerned with scientific questions to a large body (with nearly 100 members) with a much broader perspective.[14] From the early 1960s, the exchange of oceanographic data was given special attention, and a program for large-scale monitoring of the ocean, the Integrated Global Ocean Station System (IGOSS), was initiated. More recently, the importance of marine pollution has been recognized with the establishment of the Global Investigation of Pollution in the Marine Environment (GIPME).

To these elements of scientific investigation and science services was added that of Training, Education and Mutual Assistance (TEMA). Although the IOC had recognized from the beginning the desirability of assisting its members in developing their oceanographic capabilities, the developing countries did not play a major role in the early years of the organization. Their interest was aroused in the late 1960s by discussions in the United Nations General Assembly about the resources of the deep seabed and the need for a new legal regime for the ocean and its resources. As a consequence, many developing countries joined IOC, both to protect their jurisdictional interests and to participate in the possible benefits of the research activities promoted by the commission.

Since the developing countries gained a majority in IOC, the organization has paid more attention to their concerns and less to the promotion of scientific research. The political content of debates has overshadowed their scientific content, and the role of diplomats on national delegations has increased. Although TEMA has become a priority objective, it has been difficult for IOC

13. Intergovernmental Oceanographic Commission (IOC), "Intergovernmental Oceanographic Commission (Five Years of Work)," *IOC Technical Series,* vol. 2 (1966); see also IOC, "Report of the Intergovernmental Oceanographic Commission on Its Activities," in *Ocean Yearbook 1,* pp. 545–62.

14. At the time of the tenth session of the IOC Assembly, in late 1977, 95 countries were members of the commission. See Intergovernmental Oceanographic Commission, *Biennial Report for 1976–77* (Paris: Unesco, 1978).

to develop an effective program to deal with it. As the law of the sea negotiations draw to a close and as the scientific capabilities and interests of members increase, IOC is likely to revert to its original purposes.

Although IOC remains the only global intergovernmental organization with a primary responsibility for marine science, there are several other United Nations organizations that deal with related problems. These include the Food and Agriculture Organization (fisheries), the World Meteorological Organization (weather and climate), the Inter-governmental Maritime Consultative Organization (maritime safety, marine pollution), the United Nations Environment Programme (marine pollution), and the United Nations (law of the sea, minerals, oil, and gas).* These organizations cooperate in support of IOC and coordinate their programs through the Inter-Secretariat Committee on Scientific Programmes Relating to Oceanography (ICSPRO). The International Hydrographic Organization (maritime charting and navigation) is outside the United Nations system. There are also many regional and specialized fishery commissions which engage in some cooperative scientific activities.[15]

Before examining the roles played by international organizations in oceanographic cooperation, some of the characteristics of such cooperation will be discussed.

CHARACTERISTICS OF OCEANOGRAPHIC COOPERATION

As noted above, 20 years ago oceanographic cooperation was seen primarily as participation in large-scale cooperative investigations. This viewpoint is exemplified in the report of the SCOR Executive Committee meeting in April 1961 where it is stated "the IOC will probably be set up late this year by countries ready to engage in large joint oceanographic programmes."[16] Such programs were conceived to describe large ocean areas using common methods from a number of vessels. Noteworthy examples include the following.

1. NORPAC Expeditions: During the summer of 1955, 19 research vessels of 14 institutions from Canada, Japan, and the United States made physical, chemical, and biological observations at 1,002 locations across the North Pacific

15. These include the Northwest Atlantic Fisheries Organization (formerly the International Commission for Northwest Atlantic Fisheries), the International North Pacific Fisheries Commission, the International Pacific Halibut Commission, the Inter-American Tropical Tuna Commission, and the International Whaling Commission.

16. Special Committee on Oceanic Research, "Report on the Meeting of the Executive Committee of SCOR at Unesco with the Office of Oceanography and ICSU, Paris, April 10–12, 1961" (Seattle: Institute for Marine Studies, 1961).

*EDITORS' NOTE.—See "Reports from Organizations" (Appendix A) in this volume.

between the latitudes 20° and 60° N. The results of this quasi-synoptic look at the upper kilometer of the ocean were described in a comprehensive atlas.[17]

2. International Indian Ocean Expedition (IIOE): During the years 1959 to 1965, 40 research vessels from 14 countries conducted 180 cruises in the Indian Ocean, making observations in all the oceanographic disciplines.[18] The results have been described in a series of atlases.[19]

3. Equalant: Within the International Cooperative Investigations of the Tropical Atlantic of the IOC, three Equalant expeditions were conducted in 1963–64. The physical, chemical, and biological results of Equalant I and II, which involved 19 ships from eight countries, are described in two atlases.[20]

4. EASTROPAC: During the period January 1967–April 1968, 16 research vessels of five countries participated in seven survey and monitoring cruises in the eastern tropical Pacific east of 199° W between 20° N and 20° S. Observations were focused on the physics, chemistry, and biology of the upper kilometer, although some deeper observations were also made. The results are described in a series of atlases.[21]

This list is far from complete, but it illustrates some of the characteristics of such operations.[22] The IIOE was dedicated to exploration of a largely unknown region. The NORPAC and the Equalant expeditions attempted to obtain quasi-synoptic descriptions of large regions as a basis for planning more detailed oceanographic studies. Repeated observations at 2-month intervals were made by EASTROPAC to learn something about low-frequency variability in the tropical oceanic environment as it might affect populations of tropical tunas.

17. NORPAC Committee, *The NORPAC Atlas, Oceanic Observations of the Pacific: 1955* (Berkeley: University of California Press, 1960).

18. IOC, "Intergovernmental Oceanographic Commission (Five Years of Work)," p. 5.

19. These atlases include (1) K. Wyrtki, *Oceanographic Atlas of the International Indian Ocean Expedition* (Washington, D.C.: National Science Foundation, 1971); (2) *Meteorological Atlas of the International Indian Ocean Expedition* (Washington, D.C.: National Science Foundation, 1972), vol. 1, *The Surface Climate of 1963 and 1964,* ed. C. S. Ramage, F. R. Miller, and C. Jefferies (1972); vol. 2, *Upper Air,* ed. C. S. Ramage and C. V. R. Raman (1972); (3) G. B. Udintsev, ed., *Geological-Geophysical Atlas of the Indian Ocean* (Moscow: Academy of Sciences, 1975); (4) J. Krey and B. Babenard, *Phytoplankton Production Atlas of the International Indian Ocean Expedition* (Kiel: Institut für Meereskunde, 1976).

20. A. G. Kolesnikov, ed., *Equalant I and Equalant II Oceanographic Atlas* (Paris: Unesco) vol. 1, *Physical Oceanography* (1973); vol. 2, *Chemical and Biological Oceanography* (1976).

21. C. M. Love, ed., *EASTROPAC Atlas,* Circular 330, 11 vols. (Washington, D.C.: National Marine Fisheries Service, 1970–77).

22. For example, in addition to those studies already mentioned, IOC has organized the Cooperative Investigation in the Caribbean and Adjacent Regions (CICAR) and the Cooperative Investigation of the Mediterranean (CIM); it is now sponsoring the Regional Study of the Phenomenon Known as "El Niño" (ERFEN).

From an organizational point of view, IIOE was initiated by SCOR and later was coordinated by IOC; Equalant was organized by IOC. But NORPAC and EASTROPAC were developed and coordinated on a more informal and ad hoc basis, in both cases under the general aegis of the Eastern Pacific Oceanic Conference. Analogous cooperative studies have been organized by other regional bodies, most notably by ICES, ICNAF, and INPFC.

Although there remain large ocean regions which are nearly devoid of modern oceanographic observations—for example, much of the South Pacific—oceanographers show little interest in organizing further large-scale surveys. This lack of interest can be attributed in large part to the growing recognition of the highly variable nature of oceanic phenomena.[23] Such variability is poorly sampled by cruises such as those described above. Thus the more recent cooperative investigations have focused on studies of processes that lead to variability. Examples include studies of coastal upwelling processes, such as those of the Coastal Upwelling Ecosystem Analysis (CUEA) and the Cooperative Investigation of the North East Central Atlantic (CINECA)[24] and those of mesoscale eddies in the North Atlantic (POLYMODE).[25] Variability on the geological time scale is being investigated from *Glomar Challenger* in the International Program of Ocean Drilling (IPOD).[26]

While process-oriented studies can be organized and coordinated within established intergovernmental organizations, the trend appears to be toward more specialized arrangements. Because the operations are more complex than those of the large-scale surveys and require much more closely integrated use of instruments, the studies usually involve fewer countries and must be much more tightly coordinated. Thus coordination is a task for actively participating scientists rather than a secretariat function.

International cooperation in modern oceanographic research requires that the cooperating parties possess comparable skills and technology. Furthermore, the close integration of measurements requires that coordination be flexible as well as effective. These conditions tend to work against the development of such projects within intergovernmental bodies such as IOC.

23. A useful review of physical variability in the ocean can be found in A. S. Monin, V. M. Kamenkovich, and V. G. Kort, *Variability of the Oceans* (New York: Wiley, 1974).

24. Although CUEA is a project of the U.S. International Decade of Ocean Exploration, it has involved cooperation with Peru and Ecuador off the west coast of South America and with a number of countries participating in CINECA, an international program off northwest Africa organized by ICES on behalf of IOC.

25. POLYMODE is a cooperative investigation between Soviet and U.S. scientists organized within the framework of the US/USSR Agreement on Cooperation in Studies of the World Ocean.

26. The International Program of Ocean Drilling is a cooperative investigation involving the United States, United Kingdom, Federal Republic of Germany, France, Japan, and the Soviet Union. The program is coordinated by the Joint Oceanographic Institutions for Deep Earth Sampling (JOIDES), and its U.S. component is funded by the National Science Foundation.

IMPACT OF CHANGES IN MARITIME JURISDICTION

Recent changes in maritime jurisdiction, coupled with the rapid increase in the number of coastal states, have had a profound effect on international cooperation in marine science. Until 1945, the high seas began at a distance of three nautical miles from the coast, and nearly the entire oceanic region was freely open to all maritime activities, including research. In 1945 Truman proclaimed that "the Government of the United States regards the natural resources of the subsoil and sea bed of the continental shelf beneath the high seas but contiguous to the coasts of the United States as appertaining to the United States, subject to its jurisdiction and control."[27] In a parallel proclamation, the United States declared its right "to establish conservation zones in those areas of the high seas contiguous to the coasts of the United States where fishing activities have been or in the future may be developed and maintained on a substantial scale."[28] This proclamation did not claim exclusive jurisdiction but provided for agreements with other states where their nationals were involved; both proclamations made clear that the high-seas character of these zones was not affected.

Shortly thereafter, in 1947, Chile and Peru extended their jurisdictions over the continental shelves and adjacent seas and established protection and control over the seas out to 200 nautical miles. In 1952, Ecuador joined these countries in the Declaration of Santiago in which they each claimed sole sovereignty and jurisdiction over the adjacent sea, seabed, and subsoil "extending not less than 200 nautical miles from the coast."[29] A number of other states—including Argentina, Costa Rica, El Salvador, and Korea—made early claims to similarly extended jurisdiction.

In 1958, the United Nations adopted four treaties,[30] one of which dealt with the continental shelf, which it defined as "the seabed and subsoil of the submarine areas adjacent to the coast but outside the area of the territorial sea, to a depth of 200 metres or, beyond that limit, to where the depth of the superadjacent waters admits of the exploitation of the natural resources of the said areas."[31] The convention consolidated the hitherto unilateral action of a few states by stating that "The coastal State exercises over the continental shelf

27. "Presidential Proclamation Claiming Jurisdiction over Resources of the Continental Shelf," *Federal Register* 10, 1945, 12303.

28. "Presidential Proclamation with Respect to Coastal Fisheries in Certain Areas of the High Seas," *Federal Register* 10, 1945, 12304.

29. E. Ferrero, "The Latin American Position on Legal Aspects of Maritime Jurisdiction and Oceanic Research," in *Freedom of Oceanic Research,* ed. W. S. Wooster (New York: Crane, Russak, 1973), pp. 111–12.

30. These were (1) Convention on the Territorial Sea and the Contiguous Zone, (2) Convention on the High Seas, (3) Convention on Fishing and Conservation of the Living Resources of the High Seas, and (4) Convention on the Continental Shelf.

31. Convention on the Continental Shelf, art. 1.

sovereign rights for the purpose of exploring it and exploiting its natural resources."[32]

From the point of view of international cooperation in marine science, the convention made the first step in extending international approval to coastal state control over research by stating, "The consent of the coastal State shall be obtained in respect of any research concerning the continental shelf and conducted there. Nevertheless, the coastal State shall not normally withold its consent if the request is submitted by a qualified institution with a view to purely scientific research into the physical or biological characteristics of the continental shelf, subject to the proviso that the coastal State shall have the right, if it so desires, to participate or to be represented in the research, and that in any event the results shall be published."[33]

The Convention on the Continental Shelf went into effect in 1964. Meanwhile, additional countries unilaterally extended their jurisdiction over research in the water column offshore for distances up to 200 mi. A study completed in 1972 reviewed the consequences of coastal state control for the conduct of marine scientific research in the intervening years.[34] On numerous occasions, permission has been inordinately delayed, access has been denied, or work has been redirected to more hospitable regions. Although it appears that access to coastal regions has become more difficult since 1972, the experience has not yet been adequately documented.[35]

The United Nations Third Conference on the Law of the Sea (UNCLOS) has been underway since 1973. Although agreement has not yet been reached, an apparent consensus with regard to coastal state control over marine scientific research on the continental shelf and within a 200–nautical mile exclusive economic zone is reflected in the latest draft text.[36] The conditions to be imposed on research activities and the possible consequences on their conduct have been analyzed in a number of publications and need not be repeated here.[37] However, some observations on their impact on international cooperation in marine science are appropriate.

32. Ibid., art. 2(1).

33. Ibid., art. 5(8).

34. J. A. T. Kildow, "Nature of the Present Restrictions on Oceanic Research," in Wooster, *Freedom of Oceanic Research.*

35. The National Science Foundation has recently funded a project to document U.S. research vessel clearance experience since 1972; investigators are W. S. Wooster and M. L Healy, University of Washington.

36. "United Nations Third Conference on the Law of the Sea: Informal Composite Negotiating Text" (A/CONF.62/WP.10 and ADD.1), July 15, 1977, in *Ocean Yearbook 1,* pp. 655–58; see also Arvid Pardo, "The Evolving Law of the Sea," in ibid., pp. 19–20.

37. Ocean Policy Committee, *Procedures for Marine Scientific Activities in a Changing Environment* (Washington, D.C.: National Academy of Sciences, 1978), and "The Marine Scientific Research Issue in the Law of the Sea Negotiations," *Science* 197 (July 15, 1977): 230–33.

First, since all research by other countries in areas under the jurisdiction of the coastal state will require the permission of that state, all such research will require cooperation among the countries involved. Second, that cooperation may extend beyond the countries directly involved if the research is sponsored by an international organization. The relevant provision reads as follows:

> A coastal State which is a member of a regional or global organization or has a bilateral agreement with such an organization, and in whose exclusive economic zone or on whose continental shelf the organization wants to carry out a marine scientific research project, shall be deemed to have authorized the project to be carried out, upon notification to the duly authorized officials of the coastal State by the organization, if that State approved the project when the decision was made by the organization for the undertaking of the project or is willing to participate in it.[38]

EXTENDING MARINE SCIENTIFIC CAPABILITIES

The role of international organizations in supporting marine science must be considered in the light of the above considerations. Before doing so, one other contributing factor remains to be discussed, that of developing capabilities to conduct marine scientific research. As noted earlier, international cooperation requires that the cooperating parties possess comparable skills and technology. In the early stages of international cooperation which occurred primarily among countries in northern Europe and in North America, this was generally the case. Since the end of World War II, the number of independent states has more than doubled. The independence of most countries in Africa and of those surrounding the Indian Ocean dates from that postwar period. The countries in Central and South America are much older but, like the newer countries in the aforementioned regions, they have until recently had little indigenous marine scientific activity.

Interest in developing a marine scientific capability has grown in these countries for a variety of reasons. Their jurisdiction over offshore resources, both living and nonliving, is being confirmed in the Law of the Sea Conference. These resources have assumed increased perceived importance in the national economies. The necessity of a scientific basis for rational utilization of offshore resources is becoming recognized, as is the paucity of relevant scientific information and the necessity to be able to obtain and evaluate such information nationally.

As a consequence, the developing countries have assumed a dominant role in the law of the sea negotiations and have greatly increased their participation

38. "United Nations Third Conference on the Law of the Sea: Informal Composite Negotiating Text," art. 248, in *Ocean Yearbook 1,* p. 656.

in organizations such as the Food and Agriculture Organization and the Intergovernmental Oceanographic Commission. These organizations have, in turn, taken on major responsibilities in assisting the developing countries through programs of training, education, and mutual assistance.

THE ROLES OF INTERNATIONAL ORGANIZATIONS

What then should be the roles of international organizations in supporting marine science? The question can be dealt with at three levels: the global intergovernmental organization, the regional intergovernmental organization, and the nongovernmental organization.

The global intergovernmental organizations have the principal responsibility for assisting the developing countries in utilizing existing scientific organization and acquiring their own marine scientific capabilities. This responsibility is appropriate because of their broad, nearly universal, membership which includes both the countries that can provide technical assistance and those that require it. In carrying out this responsibility, the organizations need the assistance both of the industrialized and resource-rich countries who can provide the necessary funds and of the marine scientists who can provide the necessary skills. The participation of the latter can be arranged directly through member countries or with the aid of organizations at the other levels.

Once the marine science capabilities of developing countries have been strengthened, scientific cooperation between them and other countries should become more fruitful and problems of access to coastal waters should be eased. Both global and regional organizations may also assist in facilitating access through application of the proposed provision in the Law of the Sea Convention referred to earlier, should that eventually be approved.

Because of their global nature, these organizations have other important roles in supporting marine science. There is a set of ocean services which interests all countries and can best be promoted and coordinated by a global organization. For example, the World Meteorological Organization coordinates the global weather-observing network known as the World Weather Watch. In the marine field, the service of oceanographic data exchange is coordinated by the Intergovernmental Oceanographic Commission which is also developing an ocean monitoring network called the Integrated Global Ocean Station System.

Global intergovernmental organizations also can play an important part in supporting marine science directly. It is true, as noted earlier, that modern oceanographic experiments are often developed outside of formal organizations because of the special requirements for their coordination. Where there is no effective regional body, a global organization can provide strong support for a regional investigation; examples are the Estudio Regional del Fenómeno "El Niño" (ERFEN) in the eastern South Pacific and the Cooperative Investigation in the North and Central Western Indian Ocean (CINCWIO), both projects supported by the Intergovernmental Oceanographic Commission. Global organizations may also provide administrative and financial support to regional organizations, particularly when developing countries are involved, and may fund the work of nongovernmental organizations where this contributes to the advance of marine science and when the latter organizations provide scientific advice.

Regional intergovernmental organizations, such as the International Council for the Exploration of the Sea, provide a forum for the exchange of scientific information and a mechanism for organizing cooperative scientific work. Because of the relative homogeneity of interests and capabilities within such organizations, they are usually more effective in promoting joint scientific investigations than are the global organizations and thus can assist the latter in achieving their objectives. However, it should be noted that the common goals of countries in regional organizations tend to relate to the rational use of shared resources, and thus their activities are usually of an applied nature. Investigations that seek the advancement of fundamental knowledge may be supported by regional (or global) organizations but are often organized on an ad hoc basis among the countries directly concerned.

Among the relevant nongovernmental organizations, the Scientific Committee on Oceanic Research is the most broadly representative. While the activities promoted by SCOR may relate to applied problems (e.g., understanding the effects of marine pollutants or establishing the role of the ocean in determining climate), they are usually directed toward the solution of scientific problems, the development of improved scientific methods, or the exchange of scientific information. The nongovernmental organization provides the most effective international mechanism for bringing scientists from different countries together for such tasks and for mobilizing their support and advice on projects of intergovernmental interest.

In their several ways international organizations of various kinds can do much to promote the advance of marine scientific research and can facilitate the efforts of marine scientists to gain new understanding about the ocean. The different kinds of organizations each have their unique contributions to make, and their interaction can strengthen the effectiveness of the system as a whole.

Marine Archaeology: A Misunderstood Science

George F. Bass
Institute of Nautical Archaeology and Texas A&M University

Recent discoveries of Melian obsidian in the Franchthi Cave on the Greek mainland prove that Mesolithic man, and probably Paleolithic man, was able to reach the island of Melos 10,000–12,000 yr ago and return with precious volcanic glass for the manufacture of blades and scrapers.[1] Thus we can be sure that seafarers preceded farmers and shepherds in the Mediterranean area long before man built houses, fired pottery, and smelted metals.

At least twice as far back in time man had reached the continent of Australia by some type of watercraft. Even lowered sea levels would not have allowed a land crossing.[2] What were these most ancient watercraft like? Were they floats, rafts, dugout canoes, skin boats? We do not yet know. Nor can we yet estimate how much farther back in time the use of watercraft may go. Biruté Galdikas-Brindamour has reported that she and her colleagues studying orangutans on a river islet in Borneo had to guard their boats carefully lest the orangutans untie them, push off from shore, leap in, and cross to the opposite bank.[3] This was probably in imitation of man—although the orangutans made no attempts to paddle—but one must wonder if primates other than man have been observed voluntarily crossing wider expanses of water on logs or branches.

We must also wonder what first prompted man to risk his life on the open sea. Bones of large fish excavated in the Mesolithic levels of the Franchthi Cave suggest that fishermen more than 9,000 yr ago ventured far from shore in quest of deep-water catches.[4] Were fishermen carried to neighboring islands by winds and currents? Or did curiosity alone provoke early explorers to reach unknown islands, starting with the nearest, seen clearly from the mainland?

1. Thomas W. Jacobsen, "Evidence for Neolithic Trade in Greece" (paper delivered at a symposium on the Archaeology of Trade in the East Mediterranean, Duke University and the University of North Carolina at Chapel Hill, February 23, 1979, the proceedings to be published in *Archaeological News*).
2. D. J. Mulvaney, *The Prehistory of Australia* (New York: Praeger, 1969), p. 66; R. and M. E. Shutler, *Oceanic Prehistory* (Menlo Park, Calif.: Cummings, 1975), p. 31.
3. Lecture at Texas A&M University, April 24, 1978.
4. T. W. Jacobsen, "Excavation in the Franchthi Cave, 1969–1971, Part I," *Hesperia* 42 (1973): 59, 62.

© 1980 by The University of Chicago. 0-226-06603-7/80/0002-0005$01.00

Did economic necessity drive Mesolithic settlers to Aegean Skyros?[5] Whatever the underlying reasons, Crete and Cyprus had human populations more than 7,000 yr ago,[6] and by the third millennium B.C. a flourishing maritime traffic was active throughout the islands and along the coasts of the eastern Mediterranean.[7]

From these early times, ideas as well as goods seem to have been transported from east to west largely by watercraft, from concepts of agriculture and animal husbandry to those of coinage, monumental sculpture, and the alphabet.[8] Nevertheless, the study of the origins and development of seafaring has not yet been accepted as a valid anthropological concern.[9]

If the study of seafaring is of anthropological and historical importance, however, it is through marine, maritime, nautical, and underwater archaeology (terms having separate definitions, but used interchangeably here to denote generally the study of early shipwrecks) that we will learn about the ships that have, throughout history, provided the most economical means of transporting bulk cargoes[10] and the fleets on whose strengths empires rose and fell. Whether or not Helen's face "launched a thousand ships" toward Troy, we know that a nearly contemporaneous fleet of "Sea Peoples" devastated much of the eastern Mediterranean coastline before being defeated in a naval engagement with Egyptians under Pharaoh Ramses III (ca. 1190 B.C.). Yet there is no scholarly consensus even on who these Sea Peoples were, much less on their seafaring tradition.[11]

Naval battles at Salamis, Syracuse, and Actium surely affected the course of Western history, and neither Athens nor Rome would have prospered without

5. S. S. Weinberg, "The Stone Age in the Aegean," in *The Cambridge Ancient History*, ed. I. E. S. Edwards, C. J. Gadd, and N. G. L. Hammond, 3d ed. (Cambridge: Cambridge University Press, 1970), 1, pt. 1: 563–64.

6. J. D. Evans, "Excavations in the Neolithic Settlement of Knossos, 1957–60. Part I," *Annual of the British School at Athens* 59 (1964): 140; H. G. Buchholz and V. Karageorghis, *Prehistoric Greece and Cyprus*, trans. Francisca Garvie (London: Phaidon, 1973), p. 138.

7. C. Renfrew, "Cycladic Metallurgy and the Aegean Early Bronze Age," *American Journal of Archaeology* 71 (1967): 2–20.

8. Charles Seltman, *Greek Coins*, 2d ed. (London: Methuen, 1955), p. 33; Rhys Carpenter, *Greek Sculpture* (Chicago: University of Chicago Press, 1960), pp. 3–26, and "The Greek Alphabet Again," *American Journal of Archaeology* 42 (1938): 58–69.

9. Anonymous reviews, presumably by professors of anthropology, for proposals to the National Science Foundation for the study of ancient ships, 1977–78: "Underwater archaeology, no matter how competant [*sic*], does suffer an unusual divorce from normal anthropological concerns"; "fun . . ."; etc. The most recent textbook on archaeology to pass my desk had no mention of underwater archaeology.

10. A. H. M. Jones, *The Later Roman Empire 284–602* (Oxford: Blackwell, 1964), p. 842: "It was cheaper to ship grain from one end of the Mediterranean to the other than to cart it 75 miles."

11. N. K. Sandars, *The Sea Peoples* (London: Thames & Hudson, 1978); A. Nibbi, *The Sea Peoples and Egypt* (Park Ridge, N.J.: Noyes, 1975), reviewed by K. A. Kitchen, *Journal of Egyptian Archaeology* 64 (1978): 169–71.

maritime traffic. Yet we know more about almost every physical aspect of Greek and Roman life than we do of the ships with which they fought, and we have not a single example of the ships on which Rome was dependent for Egyptian grain.

For later periods we have only to think of medieval Arab trade with China, the Italian Maritime Republics, Vasco da Gama, Columbus, Lepanto, the Spanish Armada, the *Mayflower,* Yorktown, and Trafalgar to realize how different our world would be without ships. Nevertheless, diverse specialists within the field of archaeology have been slow to recognize the importance of marine archaeology,[12] as is evidenced by the following statement made for the Parliamentary Assembly of the Council of Europe in 1978:

> One of the most significant, but underlying, reasons for the difficulties in securing protection for the underwater cultural heritage and its proper excavation, is the lack of recognition in most academic circles of underwater archaeology as a valid scientific discipline. Very few countries in Europe have even one university institution prepared to support an official course in underwater archaeology: certain researchers in this field are tolerated on a temporary basis and a few more may be subsumed under the discipline most closely approaching their subject. . . . Discrimination against the underwater, as opposed to the land, archaeologist can in fact be found at all stages from courses, to funding excavations, to access to learned journals for publication, and inevitably to career possibilities. . . . Naturally the field has therefore been left exposed to others rather less meticulous in their methods and motives in exploring the sea bed.[13]

How has this general situation come about? If the importance of the evolution of seafaring is not recognized by many, the sheer abundance of artifactual materials lying in shipwrecks should be clear to all. If only one craft had sunk in each year since Paleolithic seafarers reached Melos, 12,000 wrecks would have taken place in the Mediterranean alone. Ships and boats have always sunk in alarming numbers, however. "Statistics for the eighteenth and nineteenth centuries indicate that approximately 40 per cent of all wooden sailing ships ended their careers by running onto reefs, rocks, or beaches made of rock, sand, or coral. This is a fantastic number of ships, perhaps as many as three hundred thousand per century for the world's oceans."[14] Certainly there were fewer ships and boats in earlier centuries, but fire, collision, rot, warfare, and piracy are additional possible causes for sinking.

Once on the seabed, or even covered by beach sands, countless wrecks are reasonably well preserved—protected against man, the most destructive agent

12. G. F. Bass, *Archaeology beneath the Sea* (New York: Walker, 1975), pp. 127–29.
13. John Roper in *The Underwater Cultural Heritage,* report of the Committee on Culture and Education, Council of Europe, doc. 4200-E (Strasbourg, 1978), p. 7.
14. Willard Bascom, *Deep Water, Ancient Ships* (New York: Doubleday, 1976), p. 72.

of all. Sculpture, pottery, weights, tools, glass, and weapons are found in wrecks in larger numbers, in a better state of preservation, and usually more precisely dated than on land sites. As early as the first half of the nineteenth century, geologist Charles Lyell could write that "it is probable that a greater number of monuments of the skill and industry of man will in the course of ages be collected together in the bed of the ocean, than will exist at any one time on the surface of the Continents."[15]

Although underwater archaeology is still in its infancy, the largest number of classical Greek bronze statues,[16] the largest and most firmly dated collections of Byzantine pottery,[17] the largest hoard of preclassical bronze and copper implements,[18] the largest collection of medieval Islamic glass,[19] the most complete set of Byzantine tools,[20] the most complete set of Byzantine weights,[21] and much more have come from the sea, including from later periods the best sets of navigational instruments used on particular ships.[22] One can with reason predict, therefore, that in the coming century the best preserved and dated collections of many types of antiquities will be recovered from shipwrecks, especially as all types of objects, from obsidian scrapers to prefabricated Byzantine churches,[23] were carried at sea.

Perhaps it is the history of marine archaeology that has led to its low standing among some archaeologists. Except in Scandinavia, where nineteenth-century and later excavations of burial deposits led to an early interest in ships themselves,[24] marine archaeology has been "object oriented" for a longer time than land-based archaeology, although their beginnings were similar; it later became "diver oriented" and then "technique oriented."

15. C. Lyell, *Principles of Geology*, 1st ed. (London, 1832), 1:258, as quoted by Keith Muckelroy, *Maritime Archaeology* (London: Cambridge University Press, 1978), p. 11.

16. G. F. Bass, *Archaeology under Water*, 2d ed. (Harmondsworth, Middlesex: Penguin, 1970), pp. 70–80; David I. Owen, "Excavating a Classical Shipwreck" *Archaeology* 24 (1971): 118–29.

17. G. F. Bass, "The Pottery," in *Yassi Ada I: A Seventh-Century Byzantine Shipwreck*, ed. G. F. Bass and F. H. van Doorninck, Jr. (College Station: Texas A&M University Press, in press), and "Pelagos, Northern Sporades," *International Journal of Nautical Archaeology* 1 (1972):192.

18. G. F. Bass, *Cape Gelidonya: A Bronze Age Shipwreck*, in *Transactions of the American Philosophical Society*, n.s., vol. 57, pt. 8 (Philadelphia, 1967).

19. G. F. Bass, "The Shipwreck at Serçe Liman, Turkey," *Archaeology* 32 (1979): 36–43, and "Glass Treasure from the Aegean," *National Geographic* 153 (June 1978): 786–93.

20. M. L. Katzev, "Iron Objects," in Bass and van Doorninck.

21. G. Kenneth Sams, "The Weighing Implements," in Bass and van Doorninck.

22. Muckelroy, pp. 118–25.

23. Gerhard Kapitän, "The Church Wreck off Marzamemi," *Archaeology* 22 (1969): 122–33.

24. Arne Emil Christensen, "Scandinavian Ships to the Vikings," in *A History of Seafaring Based on Underwater Archaeology*, ed. G. F. Bass (London: Thames & Hudson; New York: Walker, 1972), pp. 160–80; Arne Bang Andersen in Roper, p. 189.

There seem to be five distinct stages in the development of marine archaeology,[25] but due to the newness of the field and the time lag in scientific publications, only the earlier stages are known to most outside the field.

CIRCA 1800–1948, FISHERMEN AND SPONGE DIVERS

Chance finds by fishermen and sponge divers revealed to the world some of the finest examples of classical bronze sculpture, including the Piombino Apollo in the Louvre (netted in 1812 off the Italian coast),[26] the Antikythera Youth and head of a bearded "Philosopher" (raised by sponge divers in Greek waters at the turn of the century), a variety of bronzes from Mahdia (salvaged by Greek sponge divers between 1907 and 1913 off the coast of Tunisia), the Marathon Boy (netted in Greek waters in 1925), and the Zeus or Poseidon and Jockey and Horse from Artemision (again, raised by sponge divers in Greek waters).[27]

Such chance finds were sometimes accompanied or followed by crude attempts at excavation by helmeted sponge divers, with the result that glass, ceramics, jewelry, marble statues, furniture, and bits of wood were raised from sites like Antikythera, Mahdia, and Artemision. These objects were not excavated scientifically but were valued for themselves.

1948–60, PIONEERING EXCAVATIONS WITH SCUBA

Pioneering attempts at true excavation were made possible by the invention of the aqualung by Jacques-Yves Cousteau and Emile Gagnan in France during World War II. For the first time, divers could move freely over the seabed, performing delicate tasks among fragile remains. Some of the tools and techniques still standard in marine archaeology were introduced at this time, including the use of the air lift (a type of suction tube for removing sand and mud overburden), grids and photomosaics for mapping, and air-filled balloons for lifting heavy objects.[28]

None of the early aqualung excavations in the Mediterranean or Caribbean was conducted by archaeologists on the seabed, however, and in no case was a wreck excavated in its entirety. Consequently, no scientific publications of shipwrecks were produced. The book on the French Grand Congloué excavation is exemplary in its descriptions of artifacts, but since no plan of the site was

25. G. F. Bass, "Nautical Archaeology Comes of Age: The Vital Fifth Stage," *Sea History* 13 (Winter 1979): 27, summarizing stages first defined in a 1977 National Science Foundation proposal by the author.

26. Brunilde Sismondo Ridgway, "The Bronze Apollo from Piombino in the Louvre," *Antike Plastik* 7 (1967): 43–44.

27. Bass, *Archaeology under Water,* pp. 70–79.

28. Ibid., chaps. 6 and 7.

made or published, it remains uncertain if these artifacts came from a single wreck and, therefore, from a single period.[29]

Even so, because of their monopoly in the field, divers stressed the hardships of underwater work, stating that only professional divers, and those with many years of experience, could possibly excavate under water. Some even wrote that no archaeologist could possibly learn to dive well enough to perform satisfactorily on shipwreck excavations.[30]

During this phase the public, and archaeologists, perceived marine archaeology as the simple raising of countless amphoras and cannon, by professional and amateur divers, from the Mediterranean and Caribbean Seas. That such a perception remains today is indicated by the public television presentation of recent "excavations" at Antikythera which amounted to a frantic search for statuary and the careless destruction of wooden hull remains. The program made clear that neither the excavators nor responsible governmental officials had the slightest notion of developments in marine archaeology since 1960. Ironically, the program was recommended widely for viewing by schoolchildren so that they might learn about "underwater archaeology."

1960–70, ARCHAEOLOGISTS ON THE SEABED: THE DEVELOPMENT OF TECHNIQUES

At the beginning of the decade, the first ancient shipwreck was excavated in its entirety on the seabed. This excavation of a Bronze Age wreck at Cape Gelidonya, Turkey, was directed by a Bronze Age specialist who had learned to dive for the project, and was executed to the accepted standards of land archaeology.[31]

During the decade, "white-water" archaeology in the rapids of Minnesota and Canada revealed traces of the early North American fur trade;[32] the seventeenth-century Swedish warship *Wasa* was raised nearly intact from Stockholm harbor for continuing conservation and display;[33] a coffer dam built around five Viking ships in Roskilde Fjord, Denmark, allowed their excavation by archaeologists,[34] as did a coffer dam around a Roman-period vessel in the

29. F. Benoît, *L'Épave du Grand Congloué à Marseille; Gallia,* suppl. 14 (1961); review by G. F. Bass in *Antiquity* 37 (1963): 155–57.

30. Honor Frost, *Under the Mediterranean* (Englewood Cliffs, N.J.: Prentice-Hall, 1963), pp. 15–18, 171, 254; review by G. F. Bass in *Antiquity* 38 (1964): 70–72.

31. Bass, *Cape Gelidonya.*

32. R. C. Wheeler, W. A. Kenyon, A. R. Woolworth, and D. A. Birk, *Voices from the Rapids* (St. Paul: Minnesota Historical Society, 1973).

33. Anders Franzen, *The Warship Vasa* (Stockholm: Norstedts, 1961).

34. Ole Crumlin-Pederson, "Viking and Hanseatic Merchants," in Bass, *A History of Seafaring,* pp. 183–86; Olaf Olsen and O. Crumlin-Pedersen, "The Skuldelev Ships," *Acta Archaeologica* 38 (1967): 73–174.

River Thames in London;[35] the first excavation of a Classical Greek ship was begun off Kyrenia, Cyprus;[36] and still other wrecks were being located with newly developed sonar devices, metal detectors, and magnetometers. In many parts of the world, from Swiss lakes to the Yazoo River, Mississippi, there was an explosion of underwater archaeology.[37] At the same time, archaeologists at Yassi Ada, Turkey, and Kyrenia, Cyprus, who understood better than the professional divers who preceded them the questions that demanded answers on archaeological sites, developed many of the techniques still standard in underwater excavations.[38] The techniques were not an end to themselves, but only aimed for greater efficiency in underwater work.

Because the same archaeologists were still analyzing their sites through research and publishing only brief preliminary reports, it must have seemed to the public and to other archaeologists that underwater archaeology was mainly a technological field filled with diving bells, submarines, sonar, underwater decompression chambers, underwater television, stereophotogrammetric cameras, differential depth gauges, underwater telephone booths, magnetometers, polysulfide rubber casts of corroded iron, underwater dredges, and polyethylene glycol treatment of wood.[39] Archaeologists tended to emphasize their futuristic gadgetry, sometimes because funding for their work came from navies or foundations devoted to applied science rather than to the humanities, sometimes because historical conclusions were still years away, and sometimes from mere pride.

Progress toward the true ends of underwater excavation, however, was being made. For the first time the hull of an ancient ship, a seventh-century Byzantine merchantman, was reconstructed from its scanty seabed remains. However, as this still has appeared only in a University of Pennsylvania doctoral thesis,[40] only a few experts are aware of the methods or results of the work. At the end of the decade, only the Cape Gelidonya wreck had appeared in a full

35. Peter Marsden, *A Roman Ship from Blackfriars, London* (London: Guildhall Museum, 1966), and "Ships of the Roman Period and after in Britain," in Bass, *A History of Seafaring,* pp. 119–22.

36. Michael L. Katzev, "Resurrecting the Oldest Known Greek Ship," *National Geographic* 137 (June 1970): 840–57.

37. Bass, *Archaeology under Water,* passim.

38. G. F. Bass and M. L. Katzev, "New Tools for Underwater Archaeology," *Archaeology* 21 (1968): 164–73; E. J. Ryan and G. F. Bass, "Underwater Surveying and Draughting—a Technique," *Antiquity* 36 (1962): 252–61; M. L. Katzev and F. H. van Doorninck, Jr., "Replicas of Iron Tools from a Byzantine Shipwreck," *Studies in Conservation* 11 (1966): 133–42; Muckelroy, pp. 16–17.

39. G. F. Bass, "New Tools for Undersea Archaeology," *National Geographic* 134 (September 1968): 402–23.

40. F. H. van Doorninck, Jr., "The Seventh-Century Ship at Yassi Ada: Some Contributions to the History of Naval Architecture" (Ph.D. diss., University of Pennsylvania, 1967).

and detailed publication,[41] a startling fact that has held true until the most recent times.

The Cape Gelidonya wreck was, however, atypical in that the rocky, barren seabed onto which it sank did not hold sufficient sand to cover and protect the ship's hull from the ravages of shipworms. Thus, the excavation and following analyses were directed almost entirely at an understanding of cargo and personal possessions carried on board. Research on these remains, using comparable Bronze Age artifacts excavated on land, contemporaneous writings, and scenes painted and carved on Egyptian walls, along with a study of present and recent metalworking practices, provided a remarkably clear picture of a trading voyage of about 1200 B.C., with surprising historical conclusions.

The wreck was dated by a stylistic analysis of its pottery and an independent radiocarbon dating of brushwood dunnage found beneath its cargo. The dunnage may explain the brushwood Odysseus had put in the bottom of his boat, in a passage that had confused generations of translators, and a few remains of pegged planks at Cape Gelidonya compare favorably with those described by Homer in writing about boat construction in the same passage. Distribution of cargo on the seabed at Cape Gelidonya suggests a modest vessel about 30 ft long.

Personal possessions on board (a lamp, scarabs, a merchant's cylinder seal, tools) are Syro-Palestinian in origin, leading us to believe that the ship was Canaanite or early Phoenician (the terms are synonymous). Most of the cargo, however, comprising nearly a ton of four-handled copper ingots and baskets of scrap bronze, seems to have been taken on during a call at Cyprus. Metalworking tools (stone hammers, a swage, a whetstone, stone polishers, and a possible stone anvil), along with the scrap bronze, copper ingots, and tin ingots of unknown origin, suggest the presence of a tinker on board. That he could ply his skills throughout the eastern Mediterranean was shown by the presence of balance-pan weights based on standards of weight common in Egypt, Syria, Palestine, Cyprus, Troy, and Crete. His diet included fish that he netted on the voyage and olives.

Although specific artifacts on the wreck were of interest to Egyptologists, ceramicists, students of ancient metallurgy and metrology, and other specialists, the true value of the excavation at Cape Gelidonya was that it led to a reexamination of artifacts excavated previously on land, and this, in turn, indicated that Bronze Age Greeks, or Mycenaeans, did not hold a monopoly on maritime trade as had been supposed by most preclassical scholars; and that Homer's frequent mention of Phoenician sailors in the *Odyssey* was not as improbable for the Bronze Age as supposed by most classicists. The Cape Gelidonya publication was unique for its time, however, and both national and international conferences on "underwater archaeology" dwelt far more on techniques of underwater survey and excavation than on historical results.

41. Bass, *Cape Gelidonya.*

1970–76, EMPHASIS ON SHIPS AND SHIPPING

The restoration of the fourth-century B.C. Greek ship excavated off Kyrenia, Cyprus, under the direction of Michael Katzev, for the first time treated the ship itself as perhaps the major artifact of an underwater excavation in the Mediterranean.[42] The project brought together, also for the first time on a classical site, experts both in archaeology and ship construction, a combination found before only in northern Europe, especially Scandinavia, where excavations of ships buried on land had led originally to a different tradition of study.

"Underwater archaeology" was gaining more respectability as "nautical archaeology" or "maritime archaeology," especially with the publication of the *International Journal of Nautical Archaeology* in England and the *Cahiers d'archéologie subaquatique* in France, similar to scholarly journals devoted to Egyptian archaeology or medieval archaeology. No longer were all archaeologists and anthropologists who had learned to dive lumped together because of the method of reaching their sites, regardless of whether these were inundated Indian mounds or sunken Civil War blockade runners.

Various institutes for scientific marine archaeology were founded, including the Institute of Maritime Archaeology at the University of St. Andrews, Fife, Scotland; the University of Haifa Center for Maritime Studies; and the Institute of Nautical Archaeology (formerly the American Institute of Nautical Archaeology); joining the Institut d'Archéologie Méditerranéene at Aix-en-Provence. Just as American, British, French, German, Italian, Swedish, and other institutes and schools had long before been established in Athens, Rome, Cairo, Ankara, and elsewhere to serve as centers for full-time scholars and students to pursue archaeological research in specific areas and periods, these new institutes allowed for the first time at least a few marine archaeologists to conduct their research on more than a summer-vacation basis.

Similarly, a growing number of countries and states formed official agencies to protect and study their underwater cultural heritage. The Texas Antiquities Committee, for example, undertook the first scientific excavation and publication of a shipwreck in the United States—a wreck of the 1554 Spanish Fleet lost off Padre Island.[43]

The Institute of Nautical Archaeology, now based in Texas, conducted surveys and excavations in Turkey, Italy, Kenya, and the United States on sites ranging from the Middle Bronze Age (ca. 1600 B.C.) in the Mediterranean to those of a seventeenth-century Portuguese warship off Mombasa and an eighteenth-century American privateer scuttled in Maine during the War of

42. Susan and Michael Katzev, "Last Harbor for the Oldest Ship," *National Geographic* 146 (November 1974): 618–25.

43. J. Barto Arnold III and Robert S. Weddle, *The Nautical Archaeology of Padre Island* (New York: Academic Press, 1978); David McDonald and J. B. Arnold III, *Documentary Sources for the Wreck of the New Spain Fleet of 1554* (Austin: Texas Antiquities Committee No. 8, 1979).

Independence.[44] Most of its excavations have served simultaneously as field schools for archaeology students from more than half a dozen countries.

Because methods of excavation were fairly well standardized by this time, allowing for modifications necessitated by varying water conditions, books[45] and conferences[46] tended to be more concerned with the historical ends of marine excavation than its purely technical means.

Despite rapid maturation, marine archaeology remained generally misunderstood by the public. Since 1960, advances in almost all aspects of this new discipline, from the development of techniques to the interpretation of historical results, were due to professional archaeologists who had built upon the experience of pioneer divers. Yet, virtually anyone who raised objects from shipwrecks was recognized popularly as an "underwater archaeologist." Magazines, newspapers, and television glorified treasure hunters who often destroyed entire sites in their search for salable artifacts.

The cause of this misunderstanding is easily explained. Diving and spectacular antiquities are photogenic. Library and museum research are not. A number of popular articles appeared describing four summers of diving on a seventh-century Byzantine shipwreck off Yassi Ada, Turkey,[47] and thousands of people attended slide-illustrated lectures about the excavation. Only one popular article has been published on the years of research by a dozen archaeologists that followed those four summers.[48] Yet it was only through this research, leading to its forthcoming complete publication,[49] that the site became meaningful. This point is best illustrated by the following summary prepared for a symposium on the archaeology of ancient trade in the eastern Mediterranean.

> We had learned how a particular seventh-century ship was built and the significance of its design for the history of seafaring. This was not without economic interest, for we concluded that the trend away from older, mortise-and-tenon joined hull types of the Greco-Roman period was

44. Reports in the *International Journal of Nautical Archaeology* 5 (1976): 293–303; 6 (1977): 331–47; 7 (1978): 205–26, 301–19; and in the INA (formerly AINA) *Newsletter* (Drawer AU, College Station, Texas 77840).

45. Colin Martin, *Full Fathom Five, Wrecks of the Spanish Armada* (New York: Viking, 1975); P. Marsden, *The Wreck of the 'Amsterdam'* (London: Hutchinson, 1974); A. McKee, *King Henry VIII's 'Mary Rose'* (London: Souvenir, 1973); R. Stenuit, *Treasures of the Armada*, trans. Francine Barker (London: Sphere, 1974).

46. D. J. Blackman, ed., *Marine Archaeology, Proceedings of the Twentythird Symposium of the Colston Research Society Held in the University of Bristol, April 4–8, 1971* (London: Butterworths, 1973).

47. G. F. Bass, "Underwater Archaeology—Key to History's Warehouse," *National Geographic* 124 (July 1963): 138–56; Henry S. F. Cooper, Jr., "The Wreck at Yassi Ada," *Horizon* 10, no. 2 (Spring 1978): 82–90.

48. G. F. Bass, "A Byzantine Trading Venture," *Scientific American* 224 (August 1971): 22–33.

49. Bass and van Doorninck.

probably a labor-saving device initiated by modest Byzantine entrepreneurs who were coming into prominence at this time. We had learned the size of the ship and its estimated tonnage, and by using figures from the approximately contemporaneous Rhodian Sea Law[50] we were able to estimate the cost of its construction. From other prices preserved from antiquity we also could calculate the cost of the cargo of wine carried on board in nearly a thousand amphoras of two basic types.

A study of the Byzantine ship's eleven iron anchors suggested that statutes governing the sizes and numbers of anchors carried at sea in thirteenth-century Italy probably derived from this earlier time, for the seventh-century anchors and their stocks seemed to have been based on an arithmetic progression of 50-unit multiples of a pound of 316 grams: a hundred pounds for a stock, 250 pounds for bower anchors, 350 pounds for best bowers, and 450 pounds for sheet anchors. A pound of 315 or 316 grams was found also to have been that calibrated on the beam of the largest steelyard found on board (a shipwreck provides an almost unique opportunity to find a steelyard with its chains and counterpoise together, enabling one to determine its standards of weight). Balance-pan weights from the ship—possibly the most complete set known from Byzantine times—taught us, however, of a previously unknown pound of 284 grams divided into 14 ounces of 7 *nomismata* each, rather than the customary 12 ounces of 6 *nomismata.*

We even learned the name of the ship's captain, at the same time discovering that there were probably various grades of captain—or *naukleroi*—on shipboard; and we may have learned the name of the *emporos,* or merchant, on board, unless he was represented by an agent.

A study of long vanished iron tools, known to us only through rubber casts made inside the hollows of concretions that had formed over the now disintegrated metal, provided a unique set of Byzantine shipwright's tools—everything needed to construct or repair a ship—and these undoubtedly were carried by the ship's *naupegos,* or carpenter, in a storage area just forward of the ship's galley. Careful plans of the wreck revealed, however, that the boatswain, or *karabites,* had stored his tools, used in foraging for firewood and water ashore, inside a locker at the very stern of the ship, near a grapnel for the ship's boat that must have been in his charge.

We know, further, where the helmsman stood to man a pair of giant steering oars, and where the poorly paid *parascharetes*—literally the one beside the hearth—must have worked near a tile fire-box covered with metal grill set on the port side of the galley; a smoke-hole in the tiled galley roof allowed smoke to escape. Although pieces of fine ware—including the

50. For a detailed discussion of the Rhodian Sea Law, see John Wilkinson's chapter in this volume, pp. 89–93.

earliest examples of Byzantine glazed pottery—were limited and perhaps intended for officers, the large number of cooking pots and pitchers suggests the possibility of passengers, since such passengers were required to provide their own utensils. Water was provided for all from a common store, however, and the ship's large water jar stood in the galley.

The ship was lit at night by oil lamps, with a supply of unused, unblackened lamps kept in a cupboard with other valuables. These valuables included gold and copper coins whose purchasing power, we know from seventh-century accounts, could have fed fifteen people for a month.

We had learned, from the excavation, something about the storage, preparation, and cooking of food on board, and what some of this food was. We had learned of religious practices, of clothing worn, and of various types of fishing conducted from the ship.

A study of pottery, lamps, and about 70 coins suggests that the ship sailed from the Black Sea or Constantinople before she sank after striking a reef near Yassi Ada, not far from Halicarnassus (now Bodrum) in Asia Minor. The date of the wreck can be assigned with certainty to about 625/6 by the coins.[51]

For the first time an ancient ship's construction, lading, final voyage, and ultimate loss had been detailed from archaeological evidence, although only ten percent of its hull had survived on the seabed.

One voyage, however, does not provide sufficient data on which to base sound historical or anthropological theories. Dozens of wrecks from various past eras must be excavated, interpreted, and published before such data, used with conventional documentation, can form the basis for such theories.

1976–PRESENT, NAUTICAL ARCHAEOLOGY AS AN ACADEMIC DISCIPLINE

The fifth stage in the development of marine archaeology is still embryonic, but some of the directions it may take already are apparent.

Just as people in their middle years, after achieving certain successes in life, are thought by some to seek out younger people to continue their work,[52] the field of marine archaeology seems to be moving toward a generative stage.

Texas A&M University, for example, has since 1976 offered an M.A. with a

51. G. F. Bass, "A Medieval Islamic Merchant Venture" (paper delivered at a symposium on the Archaeology of Trade in the East Mediterranean, Duke University and the University of North Carolina at Chapel Hill, the proceedings to be published in *Archaeological News*).

52. Gail Sheehy, *Passages* (New York: Bantam, 1977), p. 405.

specialization in nautical archaeology, having brought together a number of the pioneers in underwater excavation, ship reconstruction, and the conservation of underwater finds.[53] Graduate students take seminars on preclassical, classical, medieval, colonial, and Far Eastern seafaring, learning in detail of the economics and patterns of early maritime trade, of the history of naval warfare, of the history and theory of wooden hull construction, of principles of harbor design, of the politics of sea power, and of the legal aspects of underwater archaeology. Most will develop a greater appreciation for the significance of the evolution of hull design than for the few spectacular objects that may be discovered within hulls.

Laboratory courses offer these students firsthand experience in reconstructing the lines of ancient ships from broken seabed remains, and in conserving finds of wood, metals, and other materials that have been raised from the sea. Most have the opportunity to receive practical experience on the excavations of ships from the areas and periods that interest them most.

This training has proved to be more important than technical advances for underwater archaeology in reasonable diving depths. For years it was thought by most that the solution to excavating in murky water would derive from experiments for clarifying water for photographic mapping. Archaeology students with a thorough understanding of wooden hulls, however, are now producing highly accurate drawings and plans in near-zero visibility without benefit of the sophisticated techniques thought essential only a few years ago.[54] In fact, because of these trained students, techniques used on the Institute of Nautical Archaeology's most recent projects are far simpler than those developed by its staff during the 1960s.[55]

Other technological advances, however, are required. Better methods of locating ancient wrecks must be developed, superceding existing sonar, magnetometers, submersibles, and underwater television, all of which have been used successfully in the past. Many or most of the best preserved wrecks will, in the future, be found either hidden completely under sand and mud or too deep for conventional diving.

Already the Institute of Nautical Archaeology, in conjunction with Sub Sea Oil Services, an Italian diving firm based in Milan, has begun the excavation of a wreck lying 200–300 ft deep off the island of Lipari, using saturation diving for

53. E. Walraven, "Deep Seas and Texas Caves Reveal Our Heritage," *Texas Aggie* 59, no. 6 (October 1978): 2–5.

54. P. F. Johnston, J. O. Sands, and J. R. Steffy, "The Cornwallis Cave Shipwreck, Yorktown, Virginia," *International Journal of Nautical Archaeology* 7 (1978): 205–26; Robert Adams, "Report on the Survey of the Steamboat *Black Cloud*" (report submitted to the Texas Antiquities Committee, April 1979).

55. G. F. Bass and F. H. van Doorninck, Jr., "The 11th Century Shipwreck at Serçe Liman, Turkey," *International Journal of Nautical Archaeology* 7 (1978): 119–21.

the first time in archaeology.[56] But this is only a beginning. Amphoras photographed more than a mile deep in the Mediterranean show them to lie remarkably clean and unburied on the seabed,[57] suggesting that much could be learned by photographing entire wrecks at that depth without actual excavation.

Faster, cheaper, and more effective methods of conservation are needed as well, especially for waterlogged wood which splits and warps out of recognition if allowed to dry without immediate chemical treatment.[58]

THE FUTURE

If the fifth stage of marine archaeology, just described, belongs to a new generation of students, the sixth and perhaps most vital stage will belong to the public at large.

The most pressing need is the protection of shipwrecks that have been preserved by the sea for centuries or even millennia. During a survey for ancient ships around western Sicily, I dived on half a dozen sites that had been destroyed by looters. Several of these sites might have told us how Carthaginians, the western Phoenicians, built their ships, but all were stripped clean for private gain. Most scholars in the field of nautical archaeology are aware of the clandestine sales of antiquities, including classical bronze statues, raised from Greek, Turkish, Italian, Israeli, and Lebanese waters in recent years, some publicized and others not. In the Caribbean and elsewhere, what passes in the press for "underwater archaeology" often is simple treasure hunting, without regard for scientific controls or publication. The claim that nothing new can be learned from the wood of Spanish galleons because detailed contemporary plans of them exist is incorrect.

Much of this wanton destruction is simply illegal, calling for better enforcement of existing laws, although such enforcement at sea is extremely

56. D. A. Frey, "The Capistello Project: Saturation Archeology for Deep Wrecks," *Sea Technology* 19, no. 12 (December 1978): 14–18, 49; D. A. Frey, F. D. Hentschel, and D. H. Keith, "Deepwater Archaeology: The Capistello Wreck Excavation, Lipari, Aeolian Islands," *International Journal of Nautical Archaeology* 7 (1978): 279–300.

57. F. N. Spiess and J. K. Orzech, "Location of Ancient Amphorae in the Deep Waters of the Eastern Mediterranean Sea by the Deep Tow Vehicle" (paper delivered at the Tenth Annual Conference on Underwater Archaeology, Nashville, Tenn., January 2–5, 1979).

58. W. A. Oddy, ed., *Problems in the Conservation of Waterlogged Wood,* Maritime Monographs and Reports, no. 16 (Greenwich: National Maritime Museum, 1975); Karl Borgin, "Wood in Archaeological Research," *AINA Newsletter* 2, no. 4 (Winter 1976): 1–4.

difficult. In other cases, a bewildering complex of laws calls for new legislation.[59] Antiquities laws in some countries cover underwater sites and in other countries do not. Some countries claim territorial waters to a distance of 3 mi, others to 12 mi, and still others to 200 mi. Even the definition of an antiquity varies from country to country. A wreck, whether of archaeological importance or not, does not necessarily belong to the country in whose waters it lies. Ownership of Dutch East India Company vessels by the Dutch government, which took over the company in 1789, has been recognized recently by the British Crown, even though the remains of the vessels lie in British waters.

Farther at sea, the law at present offers no protection at all. International agreements do not seem to cover ancient sites discovered on the outer continental shelf, for example, and wrecks found in truly international waters belong to their discoverers.

Suggestions that antiquities found in international waters should belong to the country of historical or cultural origin are meaningless.[60] If scholars cannot agree on the origins of the Cape Gelidonya wreck and its cargo, some holding that it is Syrian, others that it is Greek, and still others that it is either Cypriot or of mixed nationality, how could claims to modern ownership be argued in courts of law?[61] Similarly, it is known that classical Greek statues were cast not only in what is modern Greece, but also in Italy, Asia Minor, and elsewhere. Would a unique bronze from far at sea belong to Greece, Italy, or Turkey today? Ancient cargoes of raw metals found in the Mediterranean show that analyses of trace elements in the bronzes might eventually suggest only the sources of the ores mined, and not where the statues eventually were made.

A realization that many ships sank near lands far from their home ports would be beneficial for the formulation of principles governing ancient wrecks, for such wrecks represent the common cultural heritage of many nations. National pride today is often so great, however, that some countries which suffer extensive looting and which lack either the means or desire to excavate wrecks in their waters prohibit other nations from conducting scientific excavations there. The Institute of Nautical Archaeology, with staff, directors, and financial sponsors representing a variety of nations, strives to eliminate this narrow-mindedness and is already active in field research on four continents.

59. Richard Evans Hamilton, "A Proposal for Federal Legislation for the Protection and Preservation of Submerged Cultural Resources on the Outer Continental Shelf" (M.S. thesis, Texas A&M University, 1978); L. V. Prott and P. J. O'Keefe in *The Underwater Cultural Heritage,* pp. 45–135; Bascom, pp. 205–12.

60. See ICNT, Article 149, *Ocean Yearbook 1,* ed. Elisabeth Mann Borgese and Norton Ginsburg (Chicago: University of Chicago Press, 1978), p. 635.

61. G. F. Bass, "Cape Gelidonya and Bronze Age Maritime Trade," *Orient and Occident* (1973), pp. 29–38.

The solutions to problems in any country's waters will not, however, be simple. The United States, for example, offers a bizarre tangle of legal principles based on laws going back to the thirteenth century in England and confused by federal and state laws:

> It has been well established . . . that the United States cannot, except by positive legislation, claim prerogative rights at common law to abandoned and derelict property found at sea. In the absence of any such legislation and in the absence of any superior claim, the courts have had no choice but to apply the law of "finds" and award title to abandoned property to the finder. It seems equally clear, however, that the individual states can, if they so choose, establish a claim to such property, superior to that of the finder, based upon the common law principle of sovereign prerogative. It was to the individual states, and not the federal government, that this right passed in the adoption of the common law, by those states, as it existed in England at the time of the American Revolution.[62]

Will laws protecting our underwater cultural heritage be passed? The public, ultimately, will decide. At the moment, it appears, sympathy rests with the treasure hunters, representatives of "free enterprise" struggling against, as one nationally syndicated columnist put it, the "clammy hand" of governmental bureaucracy. There seems to be little understanding that archaeologists seek knowledge and not objects, that possession plays no part in their uncovering of the past.

The public would not allow the sale of Mt. Vernon, brick by brick, for private gain, nor would it tolerate the destruction of the Parthenon for the sale of souvenirs. It has yet to perceive antiquities beneath the sea as being similarly unique and precious to the people as a whole.

62. Hamilton, p. 39.

Approaches to Regional Marine Problems: A Progress Report on UNEP's Regional Seas Program

Peter S. Thacher and Nikki Meith
United Nations Environment Programme

INTRODUCTION

In the few years since the United Nations Environment Programme's (UNEP) Governing Council decided to adopt a regional approach in its oceans program rather than to impose a global plan from above to regions which clearly have different needs and problems, an encouraging consistency has been seen in the response of governments bordering each of the selected "Regional Seas" (fig. 1). Regardless of other immediate problems faced by these nations, most recognize the value of individual and collective action now to prevent future pollution of the marine environment, to achieve efficient management of their coastal resources on a *sustainable* basis. They admit the need for truly comprehensive plans reflecting the idea that for the effective protection and development of a marine region, all factors, maritime and land based, affecting their ecoregion should be taken into account when formulating development strategies for their individual nations. In other words, rather than addressing only the problems which appear to be the consequences of poor resource management and environmentally inappropriate development practices, the key to a successful protection of and development within an ecosystem lies in proper and sustainable resource management and careful application of development practices which are consistent with the health of the environment. It appears, too, that despite the continued inability of scientists and environmentalists to give a precise definition of a "healthy environment," people everywhere understand what is meant and have come to expect their governments to take the necessary steps "to protect and enhance the environment for the sake of present and future generations." Aiding this process is the declared objective of UNEP.

The purpose of this paper is to review and update the progress of the Regional Seas Programme in the last 2 yr (mid-1977 to mid-1979). Since the authors' initial report in *Ocean Yearbook 1*[1] was quite general, some additional

1. Peter S. Thacher and Nikki Meith-Avcin, "The Oceans: Health and Prognosis," in *Ocean Yearbook 1,* ed. Elisabeth Mann Borgese and Norton Ginsburg (Chicago: University of Chicago Press, 1978), pp. 293–339.

© 1980 by The University of Chicago. 0-226-06603-7/80/0002-0015$01.00

background information will be presented. But first, an attempt will be made to explain the theory and process by which the regional programs are initiated and guided by UNEP in cooperation with governments and specialized bodies of the United Nations family.[2]

UNEP'S REGIONAL STRATEGY

The Theory[3]

Not only does the nature of environmental problems differ greatly among the seas in UNEP's focus, but so do the cultural and socioeconomic realities of their bordering states and the status of national, bilateral, and multilateral environmental legislation in each region. For such reasons, it is thought that the initial success of the Regional Seas Programme is due largely to UNEP's regional strategy. Then there is the additional, perhaps obvious, factor that a sense of participation and control is essential for governments to consider that a program is truly *theirs* and not something imposed upon them which may be unrelated to their individual needs as perceived by themselves and their neighbors. It will, in the end, be their responsibility to ensure the wise management of their common resources which results in optimal socioeconomic growth without environmental degradation.

Any comprehensive regional action plan is felt to include necessarily the following functional tasks: environmental assessment, environmental management including legislation, and supporting measures.

Assessment

The provision of comparable environmental assessment data is usually the task of first priority in a region, since all other activities must be based on this information. These data should describe such characteristics of the regional sea

2. The following UN bodies have worked in close cooperation with UNEP on the Regional Seas Programme: United Nations (UN); Economic Commission for Europe (ECE); UN Industrial Development Organization (UNIDO); UN Development Programme (UNDP); UN Food and Agriculture Organization (FAO); UN Educational, Scientific, and Cultural Organization (Unesco); Intergovernmental Oceanographic Commission (IOC of Unesco); World Health Organization (WHO); World Meteorological Organization (WMO); Inter-Governmental Maritime Consultative Organization (IMCO); and International Atomic Energy Agency (IAEA). In addition, the International Union for Conservation of Nature and Natural Resources (IUCN), the World Tourist Organization (WTO), and the International Juridical Organization (IJO) have been involved in some aspects of the program.

3. The following discussion was adapted from S. Keckes, "Theory and Practice of the United Nations Environment Programme in Dealing with Regional Marine Problems," unpublished manuscript (September 1977). Stjepan Keckes is the director of the Regional Seas Programme Activity Centre in Geneva.

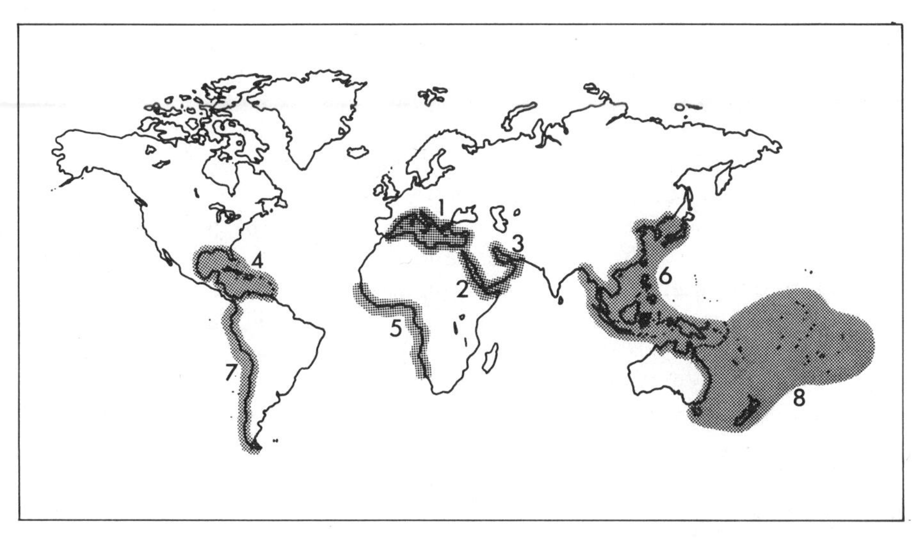

FIG. 1.—Areas covered by UNEP's Regional Seas Programme. The land and seaward limits to the regional seas indicated in this map are merely illustrative. The definition of the boundaries is the responsibility of the governments concerned. *Action plans in effect:* (1) Mediterranean, (2) Red Sea, (3) Kuwait Action Plan Region; *Action plans in preparation:* (4) Caribbean, (5) West Africa, (6) East Asian Seas, (7) Southeast Pacific, (8) Southwest Pacific.

as the sources, amounts, behavior, and effects of pollutants in seawater, sediments, and biota; effects of these pollutants on human health and on marine and coastal ecosystems; status and trends of exploitation of living and nonliving resources; ongoing socioeconomic development practices which have direct or indirect effect on the environment; and the status and proficiency of local institutions and experts available for participation in the action plan.

Management

The environmental management aspect of the regional plans has a dual goal: to assist the governments in making environmentally wise decisions about development, and to improve their ability to make such decisions on their own. But in order to make a choice among the options related to alternative patterns of development and resource allocation, the governments require information on the potential impact of the major development activities in a region, and on the various ways to reduce either the risk or severity of their effects. Within the UN system, efforts are under way to develop integrated plans for coastal area development which meet environmental quality requirements, and such plans will be available for application in the regions once data are made available identifying the specific nature of socioeconomic growth in each area and the environmental problems associated with it.

Legislation
International agreements, accepted globally or regionally and accompanied by national legislation, reflect the political commitment of states to maintain the environmental quality of their shared region, be it a small bay or an ocean.

In most regions, there is an initial need for some degree of support to be provided to governments, especially those of developing countries, in order to enable them to participate in all aspects of the action plan for their region. Such support may include technical assistance in the form of specialized training, provision of equipment for environmental assessment activities, visits by experts to national institutions, and supervision of intercalibration exercises to ensure comparable data from various institutions and regions, all carried out in an atmosphere of expectation that the participants will eventually assume full responsbility for even the most technically rigorous aspects of their program.

The Process

One reason for the relative speed with which most of the Regional Seas Programmes are becoming operational is UNEP's practice of taking full advantage of existing national and international structures. The role of UNEP is to mobilize these, and to "catalyze" a sequence of events leading to the development of a vigorous and smoothly functioning "action plan" for the particular region.

Experience thus far tells us that a good way to begin the first, or *preparatory*, phase of a regional program is within the UN system, especially the specialized agencies, through the mechanism of interagency consultation. Here, a general strategy for the region is developed, responsibilities are divided among participating agencies for specific aspects of the program, and a tentative schedule is prepared. Subsequent consultations may be held if needed to complete these preparations.

Next in the logical sequence of events, background documentation is assembled under agency supervision. This is based on the reports of fact-finding missions, reviews of activities already under way in the region, results of feasibility studies for aspects of the proposed program, and the like.

At this point it is necessary for this documentation to be reviewed and modified by experts from the region, each acting in his or her personal capacity but also reflecting in an informal way the viewpoint of their respective governments, at workshops, seminars, or expert consultations.

Next, official reactions of governments to the action plan, which is beginning by this time to take shape, are sought. One efficacious way of accomplishing this is to send an official of the coordinating organization (UNEP), perhaps accompanied by representatives of the other participating agencies, to visit some or all of the governments with the draft action plan in hand.

Then comes the critical, first intergovernmental meeting, by which time governments should have had enough time to study the proposals in detail and be prepared to discuss the plan, agree on final modifications, and finally adopt the action plan. By this stage, the draft action plan will have incorporated into it the reactions and suggestions made by the governments through earlier contacts, direct and indirect, and should be fit for approval with little delay. Additions may be made at this time, however, and any new recommendations approved by the meeting may require further preparatory work.

At this point in the process, the second, or *operational*, phase of the program begins. On behalf of the UN system, UNEP coordinates this phase, but the actual work—the implementation of the projects outlined in the action plan—is increasingly carried out by the governments of the region through their national institutions. The governments must first nominate the institutions which will participate in the program, and these are then organized into cooperative networks according to the nature of the project or projects in which they are involved. Each network is supported at the outset by UNEP together with the appropriate specialized UN agency or other component of the UN system.

We should note here that participation in the program is not limited to well-developed research centers already able to deal with the specified tasks at the highest level of technical competency. It is open to *all* institutions capable of contributing in even a limited manner. It is expected that through their participation these institutions will develop and improve their capabilities and the quality of their work. This strengthening of institutional capabilities at the national level is considered one very important indirect benefit of UNEP's regional strategy.

As results from the operational phase begin coming in, they may be reviewed by expert meetings or workshops before submission to the regular intergovernmental meetings which are periodically convened in the region. On the basis of these results the governments decide whether to continue or to redirect specific program activities or to terminate those which have served their purpose.

Once UNEP and the specialized agencies have assisted the governments of a region to initiate their regional action plan, UNEP's responsibility and financial support will gradually diminish, eventually leaving the program predominantly in the hands of the governments and national institutions. The assumption here is that ultimate responsibility for each regional program must be shared by those who benefit from it—the states bordering the particular regional sea. From the first, the plan is developed and carried out under the authority of governments and experts of the region, with assistance from UNEP and the UN agencies as they may require in their common endeavor. It is also necessary that UNEP gradually withdraw its support from these programs, due to its own limited financial and other resources.

REVIEW OF REGIONAL ACTION PLANS

In each of the regional seas, action plans are at one of the above stages of development. A brief examination of recent progress of each program will demonstrate the extent to which the theory described above is being put into practice.

Mediterranean Action Plan

The first Intergovernmental Review Meeting of Mediterranean Coastal States on the Mediterranean Action Plan[4] was held in Monaco from January 9 to 14, 1978, and attended by government representatives from 17 of the 18 Mediterranean countries. The meeting gave its full support to the various components of the action plan (legal, scientific, and socioeconomic) and made a number of specific recommendations which will be described according to these categories in the following paragraphs.[5]

The second Intergovernmental Review Meeting, which was also the first meeting of the contracting parties to the Barcelona Convention (see below), was held in Geneva (February 5–10, 1979) with the purpose of reviewing the progress made to date in this 4-yr-old action plan and deciding on the direction of the plan during the next 2 yr. From the institutional point of view the major accomplishment of the meeting was the establishment by the Mediterranean states of a $3.28 million trust fund by which they will provide the major financial support for their program. A total budget of $6.4 million for 1979 and 1980 was approved, of which half will come from the countries, 25 percent from UNEP, and 25 percent will be contributed in services, staff, and facilities by the participating international organizations.[6]

4. *Mediterranean Action Plan* and the *Final Act of the Conference of Plenipotentiaries of the Coastal States of the Mediterranean Region on the Protection of the Mediterranean Sea* (Nairobi: UNEP, 1978); the latter document is also found in *Ocean Yearbook 1,* pp. 702–33.

5. *Report of the Intergovernmental Review Meeting of Mediterranean Coastal States on the Mediterranean Action Plan* (held in Monaco, January 9–14, 1978) (UNEP/IG.11/4) (Nairobi: UNEP, 1978). Annex 4 of this report, "Recommendations for the Future Development of the Mediterranean Action Plan," will be found in this volume of *Ocean Yearbook,* pp. 547–54.

6. *Report of the Intergovernmental Review Meeting of Mediterranean Coastal States and First Meeting of the Contracting Parties to the Convention for the Protection of the Mediterranean Sea against Pollution* and its related protocols (Geneva, February 5–10, 1979) (UNEP/IG. 14/9) (Nairobi: UNEP, 1979).

Legal Aspects of the Plan

Following the intergovernmental conference in Barcelona in 1976, at which the Barcelona Convention and the two related protocols[7] were adopted, six ratifications were required in order that these legal agreements enter into force. This came about in only 2 yr—a very short time by international legal standards—and on February 12, 1978, the three instruments became law, having been ratified by Spain, Tunisia, Monaco, Lebanon, Malta, and Yugoslavia (in that order). As of April 1979, a total of 13 Mediterranean countries and the European Economic Community had ratified, and the first meeting of the contracting parties was held in Geneva coinciding with the second intergovernmental meeting.

A second meeting of the contracting parties to the Barcelona Convention and its protocols will be held in 1981, by which time it is hoped that all Mediterranean coastal states will have deposited their instruments of ratification. The present contracting parties, and their dates of ratification, are shown in table 1.

Environmental assessment data having confirmed the suspicion that most pollution in the Mediterranean comes from land-based sources (municipal sewage, industrial wastes, agricultural fertilizers, and pesticides), Mediterranean countries have pursued the development of a new protocol to control these sources. Following intensive preparatory work, including a survey of national legislation related to land-based pollution in the Mediterranean countries,[8] and a number of technical studies, an intergovernmental consultation was held in Athens, February 7–11, 1977, at which representatives revised the principles that had been suggested for inclusion in this protocol.[9] A second intergovernmental consultation was held in Venice, October 17–21, 1977, to incorporate these principles into the preliminary draft.[10] At present negotiations are continuing, and it is expected that the protocol will be ready for signature in Athens early in 1980.

7. "Protocol for the Prevention of Pollution of the Mediterranean Sea by Dumping from Ships and Aircraft" and "Protocol concerning Co-operation in Combating Pollution of the Mediterranean Sea by Oil and Other Harmful Substances in Cases of Emergency," in *Ocean Yearbook 1,* pp. 724–33.

8. *Protection of the Mediterranean Sea against Pollution from Land-based Sources: A WHO/UNEP Survey of National Legislation* (Nairobi: UNEP, 1976).

9. *Report of the Intergovernmental Consultation concerning a Draft Protocol for the Protection of the Mediterranean Sea against Pollution from Land-based Sources* (Athens, February 7–11, 1977) (UNEP/IG.6/6) (Nairobi: UNEP, 1977).

10. *Report of the Second Intergovernmental Consultation concerning a Draft Protocol for the Protection of the Mediterranean Sea against Pollution from Land-based Sources* (Venice, October 17–21, 1977) (UNEP/IG.19/5) (Nairobi: UNEP, 1977).

TABLE 1.—CONTRACTING PARTIES TO THE BARCELONA CONVENTION AND DATES OF RATIFICATION
(As of June 1, 1979)

Contracting Parties	Ratification
Egypt	August 24, 1978
European Economic Community	March 16, 1978
France	March 11, 1978
Greece	January 3, 1979
Israel	March 3, 1978
Italy	February 3, 1979
Lebanon	November 8, 1977
Libya	January 31, 1979
Malta	December 30, 1977
Monaco	September 20, 1977
Spain	December 17, 1976
Syria	December 26, 1978
Tunisia	July 30, 1977
Yugoslavia	January 13, 1978

Several other legal matters were decided at the February 1979 meeting. For one, UNEP was directed to cooperate with IUCN, FAO, and Unesco in preparing background documentation relating to development of a new protocol on specially protected marine and coastal areas prior to convening an intergovernmental meeting later in 1979 to review this material. The meeting also directed that steps be taken to develop an additional protocol related to pollution from the exploration and exploitation of the seabed, taking note of work already in progress within UNEP[11] and following a recent IJO meeting on the subject.[12] Also recommended were the study by a group of experts on the possibility of establishing an Inter-State Guarantee Fund and the formulation of procedures for determining liability and compensation for pollution caused by violations of the Barcelona agreements.

In connection with the Barcelona *Protocol concerning Co-operation in Combating Pollution of the Mediterranean Sea by Oil and Other Harmful Substances in Cases of Emergency,* the Regional Oil Combating Centre was established in Malta by UNEP with IMCO and the host government in December of 1976 for the purpose of facilitating cooperation among the Mediterranean states to combat

11. This refers to work being carried out by the UNEP Working Group of Experts on Environmental Law regarding corrective and preventive measures for pollution damage arising from offshore mining and drilling carried out in the areas within national jurisdiction.

12. IJO/UNEP Meeting of Experts on the Legal Aspects of Pollution Resulting from Exploration and Exploitation of the Continental Shelf and the Seabed and Its Subsoil in the Mediteranean (Rome, December 11–15, 1978).

massive pollution in the event of emergencies and to help them develop national contingency and antipollution capabilities. Since that time it has carried out the following activities: collection and dissemination of information relating to marine pollution in the region; establishment of communication systems appropriate to the needs of states being served by the center; promotion of technological cooperation and training programs for combating oil pollution; assistance to the Mediterranean states in the development of national, sectoral, and subregional contingency plans; and assistance in strengthening UNEP's INFOTERRA, a global referral system dealing with environmental problems.

Experts from 14 Mediterranean countries and the EEC met in Malta from September 4 to 7, 1978, to promote and exchange information on national contingency plans for oil pollution emergencies and to discuss possible regional contingency plans.[13] The meeting was organized by the Regional Oil Combating Centre.

Environmental Assessment

The seven pilot projects which were first identified at the 1975 Barcelona meeting have already provided a wealth of data from throughout the Mediterranean. Eighty-four institutions from 16 countries are collaborating with UNEP, FAO, IOC, WHO, WMO, and IAEA in carrying out these projects (table 2). Up to now, UNEP has provided $1,031,000 in equipment and $216,000 in training to participants, and intercalibration and maintenance services are continuing. These projects are expected to lead to a long-term pollution monitoring and research program to be drawn up in the next 2 yr. This long-term program should ensure systematic and regular information on the sources, amounts, levels, pathways, and effects of pollutants in the Mediterranean basin and will be based largely on the experience gained and expertise developed during the pilot phase.

The recommendations of the Monaco intergovernmental meeting called for the development of additional projects on the assessment of river-borne and airborne pollutants, the buildup of modeling capabilities of Mediterranean scientists, and creation of a central data bank for all information and statistics generated by the action plan. Action has been taken on these and other recommendations, and projects were begun on monitoring of the open waters of the Mediterranean (MEDPOL VIII), assessment of the role of sedimentation in Mediterranean pollution (MEDPOL IX), study of airborne pollutants (MEDPOL XII), and the development of conceptual and predictive models of biogeochemical cycles and water mass movement (MEDPOL XIII).

A project on land-based sources of pollution (MEDPOL X) has already been completed, resulting in a comprehensive survey covering the sources,

13. *Report of the Workshop on Oil Pollution Contingency Planning for the Mediterranean Sea* (Malta, September 4–7, 1978) (Nairobi: UNEP, 1978).

TABLE 2.—RESEARCH CENTERS PARTICIPATING IN MEDPOL (June 1979)

Country	MED I	MED II	MED III	MED IV	MED V	MED VI	MED VII	MED VIII	RAC	Total
Albania	...	...	...	...	...	...	...	...	...	...
Algeria	...	1	...	...	1	...	...	...	1	1
Cyprus	1	1	1	1	1	1	...	...	...	1
EEC	...	...	...	...	...	...	...	...	...	1
Egypt	2	2	2	2	2	2	1	...	1	2
France	5	3	2	1	1	3	2	1	1	12
Greece	3	5	5	3	3	3	3	1	...	13
Israel	1	1	1	3	1	1	4	1	...	7
Italy	3	5	2	3	1	3	7	...	1	18
Lebanon	1	1	1	1	1	...	1	...	...	1
Libya	...	...	...	...	...	...	...	...	...	...
Malta	1	1	1	1	...	1	1	1	1	2
Monaco	...	...	...	...	...	1	1	...	...	1
Morocco	1	3	3	3	2	1	2	...	...	3
Spain	4	3	3	2	2	3	3	1	...	8
Syria	1	1	1	...	...	...	...	...	...	1
Tunisia	1	1	1	1	1	1	1	...	...	3
Turkey	1	4	3	3	2	1	1	...	1	5
Yugoslavia	3	4	4	3	4	2	3	...	1	5
Total	28	36	30	27	22	23	30	5	7	84

NOTE.—Laboratories participating in more than one pilot project are indicated only once.

types, and amounts of various pollutants entering the Mediterranean.[14] The purpose of this project was to provide information needed by governments during their negotiations on the land-based sources protocol.

The 1979 Intergovernmental Review Meeting recommended that, due to limited financial resources, work should be postponed on certain of these projects (VIII, XII, and XIII), and they will be considered anew at the next review meeting.

In the few years since MEDPOL began, a great deal of monitoring and research has taken place as scientific laboratories throughout the region have begun to generate comparable data by standardized methods. Results from MEDPOL, and from other sources identified by the Mediterranean governments, were reviewed in mid-1977 at expert consultations in Dubrovnik,[15] Barcelona,[16] and Rome,[17] at the Mid-Term Review Meeting on MEDPOL in Monaco,[18] and at the Meeting of Experts on Pollutants from Land-based Sources in Geneva.[19] Then, at a joint workshop on Pollution of the Mediterranean, cosponsored by the International Commission for the Scientific Exploration of the Mediterranean (ICSEM) and UNEP, convened at Antalya, Turkey, November 24–27, 1978, 113 papers based on the initial results of MEDPOL were presented to the scientific community at large.[20] Participants then discussed elements for possible inclusion in the long-term pollution monitoring program being designed on the basis of the initial MEDPOL results. Presentation of the final report on the pilot phase of environmental assessment and the state of pollution in the Mediterranean is planned for 1980, with adoption by governments of the long-term monitoring and research program in 1981.

14. *Pollutants from Land-based Sources in the Mediterranean: Report Prepared in Collaboration with ECE, UNIDO, FAO, UNESCO, WHO, IAEA* (UNEP/IG.11/INF.5) (Nairobi: UNEP, 1977).

15. Co-ordinated Mediterranean Pollution Monitoring and Research Programme (MEDPOL), *Report of the Mid-Term Expert Consultation on the Joint FAO (GFCM)/UNEP Co-ordinated Project on Pollution in the Mediterranean (MED II, III, IV, and V)* (Dubrovnik, May 2–13, 1977) (Rome: FAO, 1977).

16. MEDPOL, *Summary Report of the Mid-Term Review Meeting on IOC/WMO/UNEP and IOC/UNEP Pilot Projects with Three Supplements* (Barcelona, May 23–27, 1977) (IOC-WMO-UNEP/MED-MRM/3) (Paris: Unesco, 1977).

17. *Mid-Term Review of the Joint WHO/UNEP Co-ordinated Pilot Project on Coastal Water Quality Control in the Mediterranean* (report of the meeting of principal investigators of collaborating laboratories, Rome, May 30–June 1, 1977) (Geneva: WHO, 1977); see also WHO report in this volume, pp. 437–38.

18. *Report on the Mid-Term Review Meeting on the Progress of the Co-ordinated Mediterranean Pollution Monitoring and Research Programme and Related Projects of the Mediterranean Action Plan* (Monaco, July 18–22, 1977) (UNEP/WG.11/5) (Nairobi: UNEP, 1977).

19. *Report on Meeting of Experts on Pollutants from Land-based Sources* (Geneva, September 19–24, 1977) (UNEP/WG.13/5) (Nairobi: UNEP, 1977).

20. *Proceedings of ICSEM/UNEP Joint Workshop on Pollution of the Mediterranean* (Antalya, November 24–27, 1978) (Monaco: ICSEM, 1979).

Two sets of documentation associated with the development of the land-based sources protocol have recently been issued. The UNEP's International Register of Potentially Toxic Chemicals (IRPTC) is responsible for preparing comprehensive data profiles for the use of experts faced with the task of evaluating the hazards associated with chemicals in the marine environment, and has now completed the first part.[21] Although this first installment is addressed to the Mediterranean, it should have relevance to all international legal instruments dealing with chemical pollution of the aquatic environment.

The second document recently issued in connection with the land-based sources protocol deals with principles and guidelines acceptable to the discharge of wastes into the marine environment, published under the join sponsorship of UNEP and WHO.[22] This manual constitutes the first of a six-part advisory code of practice on the control of the discharge of potentially harmful substances, and is comprised of two sections, one dealing with the general philosphy of the subject and the methodology and machinery of control, and the second with technical aspects of the effects of discharges and the formulation of remedial measures. The code of practice is the result of a recommendation of the 1979 Monaco meeting which urged the development of a scientific rationale for criteria applicable to the quality of recreational waters, shellfish-growing areas, waters used for aquaculture, and seafood.

Other activities under way as part of the environmental assessment chapter of the Mediterranean Action Plan include: preparation of a report on the state of pollution of the Mediterranean, based primarily on the results from MED-POL, to be issued in book form; preparation of a manual on reference methods for Mediterranean marine pollution studies by FAO, IOC, WHO, WMO, and UNEP. Specific parts of this manual are already available and the methodologies described are being tested by participants in MEDPOL;[23] preparation by the specialized agencies and UNEP of a selected bibliography on the

21. UNEP, *Data Profiles for Chemicals for the Evaluation of Their Hazards to the Environment of the Mediterranean Sea,* IRPTC Data Profile Series no. 1, vols. 1 and 2 (Nairobi: UNEP, 1978).

22. WHO, *Principles and Guidelines for the Discharge of Wastes into the Marine Environment* (Geneva: WHO, 1979).

23. *Manual for Monitoring of Oil and Petroleum Hydrocarbons in Marine Waters and on Beaches: Supplement to Manuals and Guides no. 7* (Paris: Unesco, 1977); *Manual of Methods in Aquatic Environment Research,* part 2, *Guidelines for the Use of Biological Accumulators in Marine Pollution Monitoring,* FAO Fisheries Technical Paper no. 150 (Rome: FAO, 1976); *Manual of Methods in Aquatic Environment Research,* part 3, *Sampling and Analysis of Biological Material,* FAO Fisheries Technical Paper no. 158 (Rome: FAO, 1976); *Manual of Methods in Aquatic Environment Research,* part 4, *Bases for Selecting Biological Tests to Evaluate Marine Pollution,* FAO Fisheries Technical Paper no. 164 (Rome: FAO, 1977); *Manual of Selected Bioassay Techniques* (Rome: FAO, in press); *Manual for Investigations of Pollution Induced Modifications of Marine Ecosystems* (Rome: FAO, in press); *Guidelines for Health Related Monitoring of Coastal Water Quality* (Geneva: WHO, 1977); *Manual of Methods in Aquatic Environment Research,* part 5, *Statistical Tests,* FAO Fisheries Technical Paper no. 182 (Rome: FAO, 1979).

pollution of the Mediterranean Sea, with the assistance of the MEDPOL Regional Activity Centres.

Two other documents relating to MEDPOL have already been completed. A second, updated version of the *Directory of Mediterranean Marine Research Centres,* describing detailed information (programs, staff, publications, facilities, etc.) on more than 140 institutions was issued in November 1977, and a report on the progress in MEDPOL, covering the activities of the 84 participating institutions in the period 1975–78, was prepared and submitted to the February 1979 intergovernmental meeting.[24]

The Blue Plan

Financial and administrative difficulties have slowed progress on the Blue Plan somewhat after its initial approval at a 1977 intergovernmental meeting.[25] In order to get the first phase of the Blue Plan under way, a meeting of the Blue Plan national focal points was held just prior to the 1979 Intergovernmental Review Meeting to discuss the provisional operational document of the plan, propose additional surveys that might be incorporated into it, and agree on means of coordinating Blue Plan activities. The subsequent intergovernmental meeting proposed that another meeting of focal points be held in the near future, to review, orient, and supervise implementation of the proposals contained in the operational documents. The meeting also allocated $1 million for the next 2 yr for a first-phase, systematic survey of major development and environment protection activities in the Mediterranean, coupled with a careful study of development trends as observed from data contributed by the Mediterranean states. This phase will also see the setting up of channels of information exchange crucial to the program.

The Centre d'Activités Environnement–Développement en Méditerranée (MEDEAS) in Cannes has been designated as the Regional Activity Centre for the Blue Plan.

PAP

The Priority Actions Programme (PAP), which also dates from the 1977 Split meeting, was designed to complement the Blue Plan be demonstrating that the theory of sustainable, environmentally sound development can work in practice, and indeed has an extremely pragmatic base. Certain of its six projects have gotten off to a good start in the short time since they were conceived. The Regional Activity Centre was recently established in Split, at the Town Planning Institute of Dalmatia.

24. MEDPOL, *Administrative Report: February 1975–September 1978* (UNEP/IG. 14/Inf.4) (Nairobi: UNEP, 1979).

25. *Report of the Intergovernmental Meeting of Mediterranean Coastal States on the Blue Plan* (Split, January 31–February 4, 1977) (UNEP/IG.5/7) (Nairobi: UNEP, 1977).

The UN Development Programme recently decided to increase its participation in UNEP-sponsored Mediterranean activities and identified the PAP as a special area of collaboration.

The cooperative regional project on mariculture is one of the fastest moving of the PAP projects. At a 5-day meeting in Athens in March of 1978, government-appointed experts decided on specific activities designed to fill the gaps in their knowledge and strengthen their capabilities in the field of mariculture by the exchange of scientific information, transfer of technology, and training of mariculture specialists. Thus, the program will rely heavily on the already existing activities in the Mediterranean countries before setting up new regional mariculture centers. The Athens meeting was hosted by the Greek government and jointly sponsored by FAO and UNEP.[26]

A mission jointly launched by UNDP, FAO, and UNEP has been visiting Mediterranean countries which have expressed an interest in participating in the project to help them to select suitable locations of mariculture production centers and for the pilot projects which will help determine the most appropriate technologies for particular areas and their economic feasibility. The mission will identify the monitoring, research, and training activities needed to develop these centers and will attempt to determine the specific needs of each center—equipment, personnel, technical assistance, etc. The report of the mission should be issued during the latter half of 1979, after which decisions on financing and managerial structure of the project can be taken.

The practical use of renewable sources of energy was treated by a joint UNDP/UNEP fact-finding mission which visited 16 Mediterranean countries, the EEC, and the relevant regional and international organizations from March to June of 1978 to explore the interest of these countries in such regional activities. Based on the results of the mission, a joint UNDP/UNEP Meeting of Government Experts was held in Malta, in October of 1978. Thirty-eight experts from 16 Mediterranean countries and the EEC attended the meeting to review the state of research on and utilization of renewable sources of energy in the Mediterranean region, to examine their potential use, and to propose specific activities to become part of the cooperative projects. The experts devoted most of their discussion to the use of solar energy for such purposes as household water and space heating, pumping and desalification of water for drinking and irrigation, drying and preserving of food, and greenhouse cultivation. Interest was also expressed in harnessing wind power for generating electricity and pumping water, and in the production of biogas from organic wastes for cooking and localized electricity supplies. Within each area it was agreed that the program should focus on the relative and potential importance of each energy resource in the energy balance of a given community, exchange of

26. *Report of the Expert Consultation on Aquaculture Development in the Mediterranean Region* (convened by the government of Greece in co-operation with FAO(GFCM) and UNEP, Athens, March 14–18, 1978) (UNEP/WG.15/5/Rev. 1) (Nairobi: UNEP, 1978).

information regarding present possibilities and technologies, training of installation and maintenance personnel, and comparative assessment of results.[27]

As a follow-up to the Malta meeting, a mission visited several Mediterranean countries during the first half of 1979 to determine how prepared the Mediterranean governments and national institutions are to participate in the program, and to hold discussions on the establishment of a network of institutions which would collaborate on technical matters.

Freshwater resources in the Mediterranean was the topic of a seminar convened by the International Training Centre for Water Resources Management (ITCWRM) in cooperation with UNEP in April 1978. The meeting recommended the methodology to be used for a survey of freshwater resources to be carried out as part of the Blue Plan and drew up a cooperative program to be implemented on regional, subregional, and bilateral levels. The proposed program includes exchange of technology and methodology for efficient use of water resources, establishment of an information system concerning water resources in the Mediterranean basin, recycling of used waters, studies on chronic and periodic water shortages in the region, and remote sensing of the evolution of water resources and their interaction with the sea.[28] The next step is for the Mediterranean governments and the EEC to decide how to begin developing one or several cooperative projects in the field following the recommendation of the February intergovernmental meeting to accelerate their progress.

Three other current projects (protection of soil, human settlements, and tourism) of PAP are at present the objects of feasibility studies, and activity should increase on these with the coming into operation of the PAP Regional Activity Centre. It is expected that the operative phase of the project on protection of soil (especially from the effects of erosion and desertification) may begin in late 1980 or early 1981.

Other Activities

Other activities related to environmental management include the establishment of specially protected areas, first discussed at a consultation held in Tunis, January 1977;[29] consideration of waste disposal and management practices in the Mediterranean, partially treated within the project on pollutants from land-based sources (MEDPOL X) and analyzed by the expert group meeting in

27. *Report of the UNDP/UNEP Meeting of Government Experts for Developing a Co-operative Programme on the Practical Applications of Renewable Sources of Energy in the Mediterranean Region* (Malta, October 9–13, 1978) (UNEP/WG.20/5) (Nairobi: UNEP, 1978).

28. *Report of the UNEP/CEFIGRE Meeting of Experts on Fresh Water Resources Management in the Mediterranean Region* (Cannes, April 25–29, 1978) (UNEP/WG.16/5) (Nairobi: UNEP, 1978).

29. *Report of Expert Consultation on Mediterranean Marine Parks and Wetlands* (Tunis, January 12–14, 1977) (UNEP/WG.6/5) (Nairobi: UNEP, 1977).

September 1977;[30] and training in environmental management, exemplified by a week-long international training program for environmental managers held in Urbino, September 1978, and jointly sponsored by the government of Italy and UNEP.

Outlook for the Mediterranean Action Plan

At the February meeting, the contracting parties to the Barcelona Convention approved a work plan for the next 2 yr. The general outline of this plan is described in table 3.

Kuwait Action Plan Region

The eight countries from this, the world's richest oil-producing region, reached agreement on two antipollution treaties and an action plan for environmentally sound development at the Kuwait Regional Conference of Plenipotentiaries on the Protection and Development of the Marine Environment and the Coastal Areas, held in Kuwait, April 15–24, 1978.[31] The 10-day conference was convened by UNEP and hosted by the government of Kuwait, with the support of various UN organizations.

Close to 80 high officials, experts, and legal and scientific advisers from Bahrain, Iran, Iraq, Kuwait, Oman, Qatar, Saudi Arabia, and the United Arab Emirates participated in the meeting and unanimously approved several measures, including: (1) the establishment of a marine emergency mutual aid center to coordinate action against oil spills in the region; (2) the creation of a regional trust fund; and (3) the setting up in Kuwait of a Regional Organization for the Protection of the Marine Environment to manage the action plan.

The Framework Convention[32] states in its preamble that eight countries "realize that pollution of the marine environment in the region shared (by them) by oil and other harmful or noxious materials arising from human activities on land or at sea, especially through indiscriminate and uncontrolled discharge of these substances, presents a growing threat to marine life, fisheries, human health, the recreational uses of beaches and other amenities." More specifically, the signatory states pledge to "prevent, abate and combat pollution . . . caused by intentional or accidental discharges from ships and aircraft . . . caused by discharges from land reaching the sea whether water-borne, air-borne or directly from the coast, including outfalls and pipelines . . . and resulting from land reclamation and associated suction dredging and coastal dredging."

30. UNEP, *Report on Pollutants from Land-based Sources* (see n. 19 above).

31. *Final Act of the Kuwait Regional Conference of Plenipotentiaries on the Protection and Development of the Marine Environment and the Coastal Areas* (Kuwait, April 15–23, 1978) (Nairobi: UNEP, 1978); will also be found in this vol., pp. 516–46.

32. Ibid., this vol., p. 527.

TABLE 3.—WORK PLAN FOR MEDITERRANEAN, 1979–82

Objective	Cooperating Agency	Target Dates (1979–82)	Impact
Promotion of regional conventions	WHO, UNEP	1979	Adoption of protocol concerning land-based sources of pollution
	UN, ILO, IMCO, UNESCO, IUCN, UNEP	1979–82	Development of additional protocols in areas such as specially protected areas and exploration and exploitation of the seabed for consideration by contracting parties
Assessment of the states, sources, and trends of marine pollution	ECE, UNIDO, UNESCO, IOC, FAO, WHO, WMO IMCO, IAEA, UNEP	1980	Final report on pilot phase of environmental assessment and state of pollution in Mediterranean Sea
	ECE, UNIDO, UNESCO, IOC, FAO, WHO, WMO, IMCO, IAEA, UNEP	1981	Formulation and adoption by governments of long-term program for continuing monitoring system
Coordination of environmental management efforts	ECE, ECA, ECWA, UNIDO, FAO, UNESCO, WHO, WMO, IMCO, WTO, IUCN, ALECSO, UNEP	1980	Report on sectoral studies carried out during first phase of Blue Plan
	ECE, ECA, ECWA, UNIDO, FAO, UNESCO, WHO, WMO, IMCO, WTO, IUCN, ALECSO, UNEP	1980	Report on second phase of Blue Plan
	UNDP, UNIDO, FAO, UNESCO, WHO, WMO, WTO, UNEP	1981	Report on implementation of activities in selected areas of priority actions program
Exchange of information on protection of marine and coastal resources; coordination of environmental management efforts	IUCN, UNESCO, FAO, UNEP	1980	Establishment of association of protected areas
Review of all activities and review of work program	All agencies listed above	1981	Second meeting of contracting parties to Barcelona Convention

The second legal instrument, a protocol,[33] calls upon signatories to "co-operate in taking the necessary and effective measures to protect the coastline . . . from the threat and effects of pollution due to the presence of oil or other harmful substances . . . resulting from marine emergencies." Such emergencies include collisions of ships, stranding of ships (principally tankers), blow-outs caused by petroleum drilling, and failure of coastal industrial installations.

The importance of these instruments is highlighted by the fact that about 60 percent of all the oil carried by ships throughout the world—around a billion tons per year—is exported from this region, most of it transported in tankers as large as or larger than the *Amoco Cadiz*. They are also significant in view of the region's extraordinarily rapid industrial development. There are 20 existing or planned major industrial centers along the coast, and investment figures vary from $20 million to $40 million per kilometer of coastline.

By April 1, 1979, five states (Bahrain, Iraq, Kuwait, Oman, and Qatar) had ratified the Kuwait Convention. Ninety days after this date, or on June 30, the convention and protocol entered into force—only a little more than 1 yr after the Kuwait Conference.

The action plan[34] emphasizes the need for applied research and development rather than academic research. For example, the plan deals with such practical matters as the origin and magnitude of oil pollution, an inventory of industrial waste and municipal sewage, stock assessment of commercially important species of fish and shellfish, and the ecological effects of coastal engineering and mining. The action plan also gives consideration to contingency planning for accidents arising from oil exploration, exploitation and transport, environmental engineering, public health problems, aquaculture, marine parks, port pollution, and freshwater management. It foresees intensive training programs and public awareness campaigns.

Based on the decisions of the Kuwait Conference, representatives of the Regional Seas Programme Activity Centre and 11 other UN organizations and IUCN met in Geneva for 2 days (July 24–25, 1978) to prepare a program document for the action plan. The document describes the objectives, background, work plan, timetable, outputs, financial implications, and institutional framework of 29 cooperative projects designed to implement the action plan.

The overall goals of these projects are: (1) assessment of the state of the environment, including socioeconomic development activities related to environmental quality and the needs of the region in order to assist governments to cope properly with environmental problems, particularly those concerning the marine environment; (2) development of guidelines for the management of those activities which have an impact on environmental quality or on the protection and use of renewable marine resources on a sustainable basis; (3) provision of supporting measures, including national and regional institutional mechanisms and the structure needed for the successful implementation of the

33. Ibid., this vol., pp. 537–43.
34. Ibid., this vol., pp. 521–27.

action plan. The program document has been sent to the eight governments of the region. After its thorough study, a meeting of government-nominated experts will be held, possibly in late 1979, to review the program and its 2½-yr budget of $5,800,000 and to assign priorities to the proposed projects. It is expected that work on some of the projects will begin in 1979.

Caribbean

The problems faced by the Caribbean countries in selecting appropriate development strategies are as distinctive as their tropical ecosystems. Patterns of development pursued by countries in temperate climates and designed for continental land masses can be applied to this tropical region, which includes 19 island nations, only with great care. Most of the Caribbean states are developing countries whose major problem is to fulfill the basic needs of their people. Fortunately there is widespread conviction in the region that environmental management is a necessary feature of development.

As in most other areas of the globe, there is a paucity of information on the natural systems and resources of the area. Training of professionals and technicians in environmental management skills is urgently required. Existing institutions in the region, such as monitoring stations and research institutes, need strengthening and support. And, perhaps most urgent, there is a need to create awareness of the nature of environmental problems and solutions through public education.

The first steps toward developing a comprehensive action plan for the wider Caribbean region were to identify the area's environmental problems and the manpower and institutions available to deal with them, to look at the environmental activities already under way, and to fix objectives for future action. Toward these ends, an international workshop was convened in Trinidad, in December 1976, by IOC, FAO, and UNEP. The workshop reviewed the marine pollution problems of the region and recommended a set of projects with the view of generating information needed for a proper understanding of the causes and consequences of marine pollution.[35]

Preparatory activities for the Caribbean Environment Project (CEP) began in 1976, coordinated by a small project team established in Trinidad in April 1977 by the Economic Commission for Latin America (ECLA) and UNEP. The final version of the project will take the form of a draft action plan, which has already begun to take shape as the result of extensive consultations among governments of the region, the UN and specialized agencies, interested regional and nongovernmental organizations, ECLA and UNEP.

35. *Report of the IOC/FAO/UNEP International Workshop on Marine Pollution in the Caribbean and Adjacent Regions,* IOC Workshop Report no. 11 (Port of Spain, December 13–17, 1976) (Paris: Unesco, 1977).

In its present form the draft action plan is a reflection of the complexity of the Caribbean's environmental problems. It is action-oriented and contains specific recommendations concerning the incorporation of environmental criteria into preferred modes of development. Its environmental assessment component calls for a survey of regional ecosystems, environmental conditions, and factors which influence human health and the quality of the environment. Specific activities will include: (1) assessment of the origin and magnitude of oil pollution in the region, including reports on the frequency of oil spills, the effects of oil pollution on mangroves, coral reefs, beaches, and coastal fisheries, and the ability of existing methods to handle oil spills; (2) an assessment of the effects of coastal and land-based activities on marine resources, with special attention to the effects of industrial, agricultural, and domestic waste discharges on natural communities and coastal fisheries; (3) a survey and evaluation of nonconventional energy sources; and (4) a survey of existing environmental health problems with special reference to water-borne diseases, malnutrition, and substances concentrated in the food chain.

Furthermore, there are plans for identifying major economic growth activities, their trends and influence on environmental quality. The plan also provides for assistance to countries in such matters as environmental impact assessments for major development projects, the formulation of coastal zone management schemes, and the promotion of public awareness of environmental issues. The plan also calls for governments to ratify existing international treaties and offers to provide assistance and advice for drafting national legislation.

An advisory panel was appointed to help ECLA and UNEP in the preparation of CEP. The first meeting of the panel, held April 5–7, 1978, contributed several improvements to the draft action plan. A second meeting was scheduled for July 1979.

An interagency meeting of 16 international organizations working in the region was convened in Mexico City, August 23–25, 1978, to review the draft plan for CEP and the proposals of the advisory panel and to discuss their involvement in the preparation of the project.[36] As a result of one of the meeting's recommendations, several international organizations active in the Caribbean region have collaborated in the preparation of a set of overviews intended to give a general picture of the state of the environment in the Caribbean and problems arising from a variety of human activities. The overviews highlight the problems of human settlements, energy, agriculture and fisheries, human health, natural disasters, and marine pollution. They have been used as the basis for a document which synthesizes the major issues linking development and environment and constitute an important contribution to preparation of the draft action plan. These documents will be reviewed by

36. *Report of the Interagency Meeting of the Joint UNEP/ECLA Carribbean Environment Project* (Mexico City, August 23–25, 1978) (Mexico City: ECLA, 1978).

government experts and the action plan considered for adoption at the intergovernmental meeting now scheduled for 1980.

A directory of marine research centers in the wider Caribbean region is being prepared by UNEP in cooperation with the Secretariat of the IOC Association for the Caribbean and Adjacent Regions (IOCARIBE). The directory will describe the staff, achievements, work plans, and facilities of the various research institutions in the region. Its prototype, the *Directory of Mediterranean Marine Research Centres,* has proved an invaluable source of information and guidance to environmental scientists and planners in that region, and UNEP foresees publication of similar directories for each of its regional seas.

Several other events of the past 2 yr have contributed to development of the action plan. The second session of IOCARIBE was held in San José, Costa Rica, August 7–11, 1978. It recommended that IOCARIBE implement a regional oil pollution monitoring project in cooperation with the Regional Seas Programme Activity Centre and other regional action plans.[37]

Then, from October 23 to 27, 1978, 42 experts from 17 independent Caribbean states and dependent territories met in Cartagena with the aims of identifying pollution problems in the region caused by ships and considering possible ways to deal with them. The discussion focused on oil pollution, considered the most prevalent problem associated with maritime transport. Participants examined such matters as the effects of oil spills on the open ocean, harbors, estuaries, and rivers; the likelihood of a spill in any particular country and its probable location; problems of cleanup and containment after a spill; the behavior of oil in the environment; appropriate disposal methods; and possible preventive measures.[38]

A large part of the Cartagena workshop's time was also devoted to a review of the status, implications, and means of enforcing existing international law. It also discussed possible national arrangements, their technical problems and financial implications, and the setting up of contingency arrangements in the case of massive oil spills. The results of these two meetings concerning oil pollution in the wider Caribbean region are expected to have significant effect on the corresponding chapters of the Caribbean Action Plan. The UNEP, UNDP, and the Cuban government have put finishing touches onto a 3-yr Cuban project entitled "Investigation and Control of Marine Pollution." The project document was signed and approved in March 1979. This national project is expected to become a model, or "pilot project," for subsequent pollution monitoring and control activities within CEP.

37. *Summary Report of the IOC Association for the Caribbean and Adjacent Regions* (IOCARIBE), 2d sess. (San José, Costa Rica, August 7–11, 1978, and Paris, December 22, 1978) (Paris: IOC, 1978).

38. *Final Report of the IMCO/UNEP International Workshop on the Prevention, Abatement and Combating of Pollution from Ships in the Caribbean* (Cartagena, October 23–27, 1978) (London: IMCO, 1978).

West Africa

As a follow-up to the UNEP survey mission to West Africa in 1976 two meetings were held in May 1978 in Abidjan to complete preparatory work for the draft action plan. The first meeting was a scientific workshop which reviewed the major problems related to marine pollution by petroleum hydrocarbons, by industrial and agricultural wastes, and by sewage areas on which the research and monitoring activities associated with the action plan will be focused.[39] The workshop singled out oil pollution of coastal waters, mainly from tankers transporting crude around the Cape of Good Hope to Europe, as a problem for the whole region. They noted also that chemical residues from industry and agriculture seem to create considerable problems locally. Untreated or inadequately treated sewage is probably the third major source of pollution and may seriously affect human health directly or through consumption of contaminated seafood.

The workshop recommended that the UN assist in the training of local scientists and technicians in: (1) analytical techniques for measuring pollutant concentration; (2) techniques used to measure effects of pollutants on human health, fishery resources, and marine coastal ecosystems; and (3) methods for establishing water-quality criteria and effluent standards. The UNEP was requested to help equip the national institutions so they could effectively participate in the regional program.

Immediately after the workshop, UNDP Resident Representatives from 13 West African states and representatives from 11 international organizations met for 2 days in Abidjan to discuss their role in development of the plan, prepare a joint strategy for preparatory activity, and pinpoint the elements they thought could be effectively incorporated into the regional program.

On the basis of that meeting, the first draft of the action plan for the West African region was prepared and circulated to the UN organizations and appropriate UNDP Resident Representatives for their comments. In the meantime, FAO, IMCO, and the UN Department of International Economic and Social Affairs (UN/DIESA) prepared the necessary supporting documentation. In October 1978, the draft action plan was sent to the West African governments.

The action plan follows the basic four-part scheme seen in other plans sponsored by UNEP. Assessment will underlie all the other activities, providing information on the particular pollution problems of the region. The management component emphasizes training of managers and policymakers to make management decisions based on the priorities of the region and on principles which governments will have defined as acceptable and applicable, as well as on

39. *Report of the IOC/FAO/WHO/UNEP International Workshop on Marine Pollution in the Gulf of Guinea and Adjacent Areas,* IOC Workshop Report no. 14 (Abidjan, May 2–9, 1978) (Paris: Unesco, 1978).

environmental considerations. The envisaged regional convention may provide the legal backbone for cooperation among the countries of West Africa to protect the long-term interests of the region. It may include protocols incorporating specific obligations, such as measures to be taken in cases of pollution emergencies or measures to control pollution from specified sources.

In order to get the reactions and comments of West African governments to the draft action plan, UNEP appointed a mission to visit the region in early 1979. The mission had visited Benin, Cameroon, Gabon, Gambia, Ghana, Ivory Coast, Liberia, Nigeria, Senegal, Sierra Leone, and Togo and reported that authorities in those countries were anxious to see a regional action plan implemented in West Africa and that reactions to the initial draft had been quite positive.

A meeting of government-nominated experts was tentatively scheduled for November 1979 to review and further revise the action plan. This should be followed by yet another expert meeting in early 1980, and then, if all goes according to schedule, an intergovernmental meeting or conference of plenipotentiaries may be held at which the action plan and convention will undergo a final review leading to their adoption.

East Asian Seas[40]

"East Asia is probably one of the most complicated areas of the world, with its great diversity of geography, races, societies, economies, and environments. It contains the largest part of the world's population. Environmental problems have received inadequate attention. The marine environment is one area in which international co-operation is essential if pollution problems are to be solved, since marine pollutants recognize no national boundaries." This statement introduces the summary report of the IOC/FAO(IPFC)/UNEP International Workshop on Marine Pollution in East Asian Waters, which was held in Penang in April of 1976.[41] Several activities relating directly or indirectly to the envisaged regional action plan for East Asian seas have grown out of the recommendations of this meeting. It will be the function of the Regional Seas Programme to integrate and harmonize these varied activities under the umbrella of the regional action plan once it has been approved by the East Asian governments.

Recently, environmental experts from the five member nations of the Association of South East Asian Nations (ASEAN) met in Jakarta (December

40. The term "East Asian Seas" refers to the oceanic embayments, gulfs, and semienclosed seas of both East and Southeast Asia.

41. *Report of the IOC/FAO(IPFC)/UNEP International Workshop on Marine Pollution in East Asian Waters,* IOC Workshop Report no. 8 (held at University Sains Malaysia, Penang, April 7–13, 1976) (Paris: Unesco, 1976).

18–20, 1978) to consider a proposed ASEAN Sub-regional Environment Programme (ASEP), which is a comprehensive regional program with a strong marine component. At the meeting, experts from Indonesia, Malaysia, the Philippines, Singapore, and Thailand identified specific projects as subjects for regional collaboration. Working groups were formed to discuss the subjects of marine environment, rural and urban development, environmental education, environmental planning and development, nature conservation, and the social aspects of environmental development. Four areas were finally identified as having top priority: (1) protection and management of the marine environment; (2) environmental development, especially the development of methodologies for environmental impact assessment; (3) water and urban air quality monitoring; and (4) pollution control technologies.

The meeting report[42] clearly delineates those problems related to the marine environment which the experts considered most pressing—siltation and sedimentation, organic and nutrient loading, pollution by petroleum hydrocarbons, heavy metals, organochlorine pesticides and PCBs, and thermal pollution. It also emphasizes the need for water quality standards for effluent discharge and urges that special consideration be given to the protection of mangrove and coral ecosystems and areas where seabed oil exploration and exploitation, submarine pipelines, and nuclear power stations are found. Several recommendations deal with specific actions to be taken, such as the establishment of training programs, workshops, and seminars for scientists and technicians who will be involved in the research and monitoring projects. A call is made for laws and regulations to protect the marine environment, and special mention is made of the necessity of intercalibration exercises for heavy metal, organochlorine, and petroleum hydrocarbon analyses. The experts were determined to give ASEP a firm scientific base for the policy decisions and legal instruments to come.

Another important recommendation of the expert meeting dealt with the creation of a subcommittee on environment under the Committee on Science and Technology of the ASEAN Secretariat, which will be responsible for making sure that environmental considerations are incorporated into the policies and programs initiated by ASEAN, and for coordinating ASEAN cooperative environmental projects.

The experts requested assistance from UNEP in the formulation of an action plan for the protection and development of the marine environment and coastal areas of the Southeast Asian region. The immediate response of UNEP was to organize a mission to the region to draw up an initial draft of the plan according to the advice and guidance of regional experts and reflecting the recommendations of the ASEAN meeting. The mission was completed in

42. *Report of the ASEAN Experts Meeting on the Environment* (Jakarta, December 18–20, 1978) (Kuala Lumpur: ASEAN, 1978).

March of 1979 and the draft action plan drawn up. It consists of environmental assessment, environmental management, and legal components, with a fourth section on institutional and financial arrangements. The legal aspects of the plan include a Southeast Asian Regional Convention, proposals for supplementary protocols detailing obligations for the control of pollution from specific sources, such as land-based pollution and pollution from exploration and exploitation of the seabed, for the establishment of specially protected areas, and for combating pollution in cases of emergency. The draft action plan is now ready for review by a meeting of government-nominated experts early in 1980 and possible adoption toward the end of that year.

Oil pollution was identified by the Penang workshop as a concern of first priority in the area of the Straits of Malacca/Singapore, which is the most heavily used shipping lane between the Indian Ocean and the South China Sea. One effort to alleviate the threat of pollution from collisions or groundings of tankers in the area resulted in the passage of a resolution by IMCO on November 14, 1977, adopting a new routing system for the Straits which include traffic separation schemes, deep-water routes, and rules of navigation. The resolution states that a series of navigational aids (light beacons), some equipped with radar reflectors, are to be installed to facilitate compliance with the routing system.

A proposal has been made by IMCO and UNEP that an International Workshop on the Prevention, Abatement, and Combating of Pollution from Ships in East Asian Waters be held in November 1979. The purpose of this project will be to provide participating countries with a nucleus of personnel trained in methods of abatement of massive pollution of the marine environment by oil. The workshop will provide instruction in mechanical and chemical methods of dealing with spills and on ways of selecting the best method in a particular circumstance.

On the basis of a proposal made by the Penang workshop, a draft program has been prepared as part of the FAO/IOC/UNEP Pollution Monitoring and Research Programme on the assessment of oil pollution and its impact on living aquatic resources. Its long-term objectives are to establish an oil-pollution monitoring network, to assess the chronic effects of oil on littoral communities and fish stocks, to investigate water circulation and exchange as they affect transport of pollutants, and to examine the processes of biodegradation and weathering of oil in tropical waters. It is expected that this program will become one of the major elements of the action plan once it is approved.

In response to the perceived threat to the widespread but fragile mangrove ecosystems of the South China Sea and elsewhere, UNEP, FAO, Unesco, WHO, and IUCN have proposed a project on the impact of pollution on the system and its productivity. The initial phase of the project will involve convening an expert consultation to prepare a summary of present knowledge of the mangrove system, with consideration of the results of previous meetings—notably the UNESCO Regional Seminar on Human Uses of the Mangrove

Environment and Management Implications, held in Bangladesh, November 1978,[43] and the work of the SCOR (Scientific Committee on Oceanic Research of the International Council of Scientific Unions)–UNESCO Working Group on Mangrove Ecology.

The Economic and Social Commission for Asia and the Pacific (ESCAP) and the Swedish Environmental Protection Service (SEPS) are collaborating on plans for a series of seminars on protection of the marine environment and related ecosystems in Asia and the Pacific. The Swedish participation will provide information on the technology and practical solutions (such as residue recycling) adopted in Sweden to combat marine pollution. Plans include a series of national and regional seminar/workshops, a study tour to Sweden, and follow-up activities emphasizing the development of legislative guidelines.

As part of a series of short courses on coastal water pollution control sponsored by WHO and the Danish International Development Agency (DANIDA), a 4-week course is scheduled for the East Asian seas with the WHO Centre for the Promotion of Environmental Planning and Applied Science (PEPAS). The course will be held in the summer of 1980 at the Universiti Pertanian in Kuala Lumpur and will deal with general aspects of pollution control as well as the specific topics of (1) effects of pollution on coastal water ecology with special reference to coral reefs, and (2) problems of sewage disposal into mangrove swamps.

A survey of pollutants from land-based sources in metropolitan agglomerations has been initiated by WHO and should continue through 1980. The study has focused initially on metropolitan Jakarta and Bangkok. It will be carried out along the lines of a similar survey already completed for the Mediterranean and contain an inventory of all major pollution sources and an assessment of the total pollution load. Such data provide the necessary basis for pollution control measures and enforcement of pertinent legislation and should have great influence on the management and legislative components of the regional action plan.

The UN Department of International Economic and Social Affairs (UN/DIESA) is collaborating with the Committee for Co-ordination of Joint Prospecting for Mineral Resources in Asian Offshore Areas (UNDP/CCOP) and UNEP is planning a regional workshop on coastal areas development and management in Southeast Asia. The project aims at assisting governments to incorporate environmental considerations into development planning and to identify conflicts between activities on land and protection of marine resources.

Several other activities in recent years have contributed to the preparatory work for the action plan for East Asian seas, especially: Indo-Pacific Fisheries

43. "Report on the UNESCO Regional Seminar on Human Uses of the Mangrove Environment and Management Implications" (Bansoc, Dacca, Bangladesh, December 4–8, 1978), unpublished document (Paris: Unesco, 1978).

Council (IPFC), Third Session of the IPFC Working Party on Aquaculture and Environment, Bangkok, August 31–September 3, 1976; IMCO/UNEP Oil Pollution Contingency Planning for the Straits of Malacca and Singapore Region, preparatory stage, November 1976; ASEAN Council on Petroleum (ASCOPE), First Petroleum Conference and Exhibition, Jakarta, October 11–13, 1977; ESCAP/UNEP Intergovernmental Meeting on Environmental Protection Legislation, Bangkok, July 4–8, 1978; Committee for Co-ordination of Joint Prospecting for Mineral Resources in Asian Offshore Areas (CCOP), fifteenth session, Singapore, October 24–November 6, 1978; Fifth Meeting of the ASEAN Experts Group on Marine Pollution, Manila, February 7–9, 1979; UNESCO/IOC Workshop on the Western Pacific (WESTPAC), Tokyo, February 19–20, 1979; preparation of Directory of Indian Ocean Marine Research Centres, National Institute of Oceanography, India, and UNEP.

Red Sea

The Red Sea and Gulf of Aden region was one of the first of the Regional Seas to adopt an action plan. That was in Jeddah in 1975, and since that time the Arab League Educational, Cultural, and Scientific Organization (ALECSO) has been coordinating the program.

Work has begun on a Regional Convention for the Protection of the Red Sea and Gulf of Aden Environment, and a revised draft will be presented to a meeting of plenipotentiaries to be held at a yet unspecified date. Within the environmental assessment of the action plan, activities thus far have included training of scientists and technicians, the setting up of new marine science institutions and evaluation of the needs of existing ones, and reviewing of past scientific work in the region. These activities are aimed at establishing a solid base for a pollution monitoring and research program.

In 1977 a team of scientists and a specialized architect visited the Arab Republic (AR) of Yemen, the People's Democratic Republic (PDR) of Yemen, Somalia, and Ethiopia to select sites for future marine research stations, while Unesco prepared a project for the establishment of a marine sciences and resources institute in PDR Yemen.

A marine scientist was commissioned by ALECSO to visit the University of Jordan in Amman to discuss the organization of teaching and research. Institutes in Jordan, AR Yemen, Port Sudan, and Ghardaqa prepared reports and equipment lists, and negotiations began for the acquisition of research vessels.

A project proposal for a survey of marine turtles and dugongs has been drafted, and a bibliography of Red Sea marine research was prepared for the use of local institutions. A provisional report was made concerning the establishment of a marine park off the Jordanian coast of the Gulf of Aqaba, and with technical assistance from IUCN, a project on the ecology and conservation of coral reefs was initiated by the Institute of Oceanography at Port Sudan.

Other proposals involve a pilot monitoring program for evaluation of oil pollution levels at selected sites, monitoring of phosphate pollution in coastal waters off Egypt and Jordan, a survey of mangrove systems off the Saudi coast and elsewhere, and preparation of a taxonomic atlas of fishes of the northern Red Sea and Gulf of Aden.

In December 1977 ALECSO organized a seminar on oil pollution as a preparatory step for a future regional meeting on the subject. Also, IMCO issued a report prepared under contract with the Saudi Arabian government on the state of coastal pollution off the Saudi Arabian coast. Seminars and symposia have been held on such subjects as the effects of coastal development on the marine environment and the status of existing laws and regulations relating to protection of the marine environment. A training course on Red Sea ecology was held in Ghardaqa, attended by 16 trainees from seven countries. And several fellowships or study grants in marine sciences and oceanography have been awarded to area university students.

A symposium on the coastal and marine environment of the Red Sea, Gulf of Aden and tropical western Indian Ocean is under preparation by ALECSO and Unesco and is tentatively scheduled for 1980 at the University of Khartoum.

Pacific

Not even the most ambitious of programs could attempt to deal with the vast area of the Pacific Ocean as a whole, so UNEP has identified two subregions for separate attention in its initial efforts to deal with the special problems of the Pacific.

Southeast Pacific

In order to begin developing a regional action plan for the Southeast Pacific, scientists, managers, and legal experts attended an international workshop held in Santiago, Chile, November 6–10, 1978. Organized by the Permanent Commission for the South Pacific (CPPS), in cooperation with FAO, IOC, WHO, and with the support of UNEP, 17 experts from Chile, Colombia, Ecuador, Panama, and Peru gathered to discuss the area's environmental problems in general, the state of marine pollution in particular, and the legislation relevant to the region's environmental protection.[44]

Oil and land-based sources of pollution, particularly domestic and agricultural wastes, are seen to be the major problems in this area, which refers to the coast of South America, stretching from Colombia in the north to the southern

44. *Report of the CPPS/FAO/IOC/UNEP International Workshop on Marine Pollution in the Southeast Pacific,* IOC Workshop Report no. 15 (Santiago de Chile, November 6–10, 1978) (Paris: Unesco, 1978).

tip of Chile, and its coastal waters. The meeting emphasized the importance of controlling local sources of pollution, especially domestic, industrial, and mining wastes. Oil was singled out as a special problem in the northern part of the area.

Technical and legal aspects of the envisaged action plan were discussed by the experts, who approved elements of a pilot project on research and monitoring of marine pollution, guidelines for the formulation of a regional convention on the protection of the marine environment from pollution, and a draft agreement on regional cooperation in combating pollution by petroleum hydrocarbons and other toxic substances in cases of emergency. The recommended studies should reveal the causes and trends of marine pollution and should provide a better basis for administrative and technical measures the governments may wish to take. The draft action plan will undergo review by experts in mid-1980, after which it will be submitted to an intergovernmental meeting for approval.

Southwest Pacific

The major activity currently under way for the Southwest Pacific region is the South Pacific Conference on the Human Environment, planned for 1980. The conference will review the major environmental problems of the area, identify priority actions necessary to reverse the present trend of environmental deterioration, formulate environmentally sound guidelines and management policies, including legislation, and adopt an action plan for the protection and development of the region.

Preparatory activities for the conference consist of extensive consultation with the governments concerned, presentation of country reports, meetings of regional experts and international organizations supporting the conference, field missions, and study tours. With UNEP's support, the Economic and Social Commission for Asia and the Pacific (ESCAP), the South Pacific Bureau of Economic Co-operation (SPEC), and the South Pacific Commission (SPC) will play the leading roles in preparation of the conference.

Conclusions

The above review of recent activities related to UNEP's Regional Seas Programme serves to demonstrate that the imagination and determination needed to develop and launch these complex and comprehensive regional environmental programs are abundant in the countries bordering on the regional seas. In every case, action plans have been or are being developed with the cooperation of individuals at all levels of responsibility—government authorities and advisers, scientists, lawyers, technicians, and supporting staff.

One especially interesting aspect of the various programs, as they begin to emerge from their preparatory stages and to take on shape and substance, is the similarity in the design of programs from seemingly disparate areas of the globe. In every case, the same tripartite structure tends to develop, consisting of scientific assessment, environmental management, and legal components, accompanied by the necessary support measures. This is the result of independent processes in various regions and is an encouraging indication that, despite the variability of specific environmental problems, there exists a common ground for their solution which can help governments to cooperate in practical steps.

The purpose of the United Nations, as set forth in article 1 of the UN Charter, is to be a "centre for harmonizing the actions of nations" in the attainment of certain common ends: peace and security, friendly relations, and international cooperation in solving international problems of economic, social, cultural, and humanitarian characters. In all areas where UNEP has developed a comprehensive plan for a regional sea, the governments in those areas have come together and, with advice and assistance from virtually all agencies and components of the UN system, have overcome obstacles that divide them and now cooperate to protect their future well-being. So long as this cooperation is maintained, one can be confident that the common problem of marine pollution can be surmounted by the logic of the process now under way; and in the long run, the experience of cooperation in preventing future problems may help governments to resolve other problems that continue to divide them.

The Role of Recreation in the Marine Environment

Susan H. Anderson
California Marine Parks and Harbors Association

THE IMPORTANCE OF MARINE RECREATION

Advanced technologies, increases in disposable income, mobility, education, overall standard of living, and, in some countries, labor laws all have contributed to increasing time available for leisure pursuits. Recreation—"any experience voluntarily engaged in largely during leisure . . . from which the individual derives satisfaction"[1]—has thus become an integral part of twentieth-century life-styles in the developed countries.

Even among those people who do not have blocks of time for recreational activities or who cannot afford to take vacations away from home, recreational activities are important. Because about 70 percent of the world's people live within 50 mi of the coast, recreation on the coast and in the nearshore waters is in demand by both long- and short-term users.

Certainly in the United States, water-oriented recreation ranks among the top recreation priorities. According to the 1962 report of the U.S. Department of Interior's Outdoor Recreation Resources Review Commission: "Most people seeking outdoor recreation want water to sit by, to swim and fish in, to ski across, to dive under and to run their boats over. Swimming is now one of the most popular outdoor activities and is likely to be the most popular of all by the turn of the century. Boating and fishing are among the top 10 activities. Camping, picnicking, and hiking, also high on the list, are more attractive near water sites."[2]

The coastal environment provides especially rich resources for recreation pursuits. Wide, fine-grained sand beaches are especially attractive to sun lovers for sunbathing, people watching, volleyball, walking, and jogging. Depending on the natural shelter of the beach from offshore waves, the depth profile of the bottom, and the prevalent longshore currents, beaches also attract swimmers and surfers. If the area is rich in nearshore fish, these same sandy beaches lend themselves to surf fishing.

1. Robert B. Ditton and Mark Stephens, *Coastal Recreation: A Handbook for Planners and Managers* (Washington, D.C.: Office of Coastal Zone Management, National Oceanic and Atmospheric Administration, Department of Commerce, 1976), p. 1-1.

2. Outdoor Recreation Resources Review Commission, *Outdoor Recreation for America* (Washington, D.C.: Department of the Interior, Bureau of Outdoor Recreation, 1962), p. 4.

© 1980 by The University of Chicago. 0-226-06603-7/80/0002-0011$01.00

Rocky coastlines, made rugged by wind or wave erosion, often feature rich biological communities. The rugged nearshore terrain provides many surfaces to support bottom-dependent plants and mollusks. The diversity of underwater terrain in these areas provides potential prime sites for skin diving and scuba. Nearshore kelp beds, providing protected habitats for many midwater and bottom fish, also lend appeal to these sites for divers.

Within the area of tidal variation, there are many marine organisms that are dependent on exposure to both air and water. These tidepool areas support intriguing biological communities. In many areas, however, it has been necessary to effect regulations to stop the excessive collection of samples of the many species found in the tidepools.

Where rocky headlands feature pounding waves, they offer particular visual pleasures for photographers, painters, and nature observers. High bluffs provide ocean vistas for picnicking, camping, and sight-seeing by people in all walks of life.

Sheltered inlets found along a sinuous coastline form natural harbors for recreational boats as well as beaches with little or no wave action. These harbors provide protected waters for small-boat sailing, waterskiing, and fishing from rowboats, as well as necessary protection for boat berthing and launching.

Where inlets are fed by freshwater sources, nutrient-rich wetlands are often prevalent, providing a nursery and breeding ground for many fish species as well as a buffer zone for storm waves. These wetlands areas support a wide variety of birds and other wildlife and are a haven for bird-watchers and nature observers.

Each of these aspects of the ocean environment offers a variety of recreational opportunities. For many, the motion of the sea and its changing moods provide special solace. For some, the smell of the ocean and its movement provide a delightful setting for relaxing and observing the activity around them. For others the unique interface between land and sea offers an invitation to explore.

No matter how passive or active a person's recreation is, the recreational opportunity is valuable. In his book, *Shoreline for the Public,* Dennis Ducsik suggests that arguments concerning the extent to which outdoor recreation may be a cure-all for mental illness in our society may be overstated, but that recreation does appear to enhance the participant's feeling of well-being and to refresh his or her outlook.[3] Quoting from Herbert Gans, Ducsik indicates that although it may do little to cure mental illness, ". . . recreation, especially outdoor recreation, provides one of the most promising approaches to the elusive goal of mental health as a form of 'primary prevention' of mental ill health."[4]

3. Dennis W. Ducsik, *Shoreline for the Public: A Handbook of Social, Economic, and Legal Considerations Regarding Public Recreational Use of the Nation's Coastal Shoreline* (Cambridge, Mass.: M.I.T. Sea Grant Project Office, 1974), p. 19.

4. Ibid., p. 23.

The increased emphasis on recreation seems to be not only a result of the availability of time and disposable income but also a catalyst for the continued improvement of an overall standard of living. In fact, in recent years of inflation and recessions, recreation expenditures have remained at a high level. Recreation demands continue to grow despite the economy, although some reluctance to spend so much money on recreation may be evident.

Thus, we see increasing demands on shoreline resources even as population increases are slowing. This is not only a U.S. phenomenon. A recent study in Ireland showed that over 70 percent of holiday activity there—both by nationals and foreigners—was accommodated in the coastal area.[5] Coastal tourism in northwestern Belgium has shown a remarkable increase in overnight stays from 67,444,000 in 1957 to over 165,971,000 in 1973.[6] At the same time, the pursuit of marine recreation has brought economic prosperity to coastal towns, particularly where facilities and setting encourage the influx of overnight visitors. This is especially true for those towns on the periphery of large metropolitan areas that are accessible by air transportation.

Although recreation pressures on coastal resources are increasing, recreational use of the oceans is nothing new. Summer and vacation homes were prevalent in the United States even at the turn of the century. Along the northwestern coast of Belgium the end of the eighteenth century signaled the onset of tourism.[7] Mediterranean coasts have long attracted vacationers from Europe and, more recently, from the United States. Today, while many coastal vacation homes are still privately owned and used, some have been converted into inns to provide lodging for the many more who come to the coast for only a few weeks each summer. Where feasible, whole communities have been built around clusters of older vacation homes to accommodate the needs of the increasing numbers of people who want to recreate on the coasts. Stretches of some coastal roads have become a string of commercial services to respond to the influx of visitors. For example, many areas along the Mediterranean coasts are stacked with high-rise hotels and apartments.[8]

In the United States, marine recreation has become a multibillion dollar industry, with much of the dollar value exchanged between many small businesses rather than multimillion-dollar corporations.

At the 1975 National Conference on Marine Recreation, Patrick Doyle, manager of Environmental Communications of the Outboard Marine Corporation in Milwaukee, presented an overview of the importance of recreation

5. G. Bagnall, "The Irish Coastline," *Ocean Development Policies: Proceedings of the Second International Colloquium on the Exploitation of the Oceans, Bordeaux* (Paris: Association pour l'organisation du colloque océanologique, 1974), 1:5.

6. A. Haulot, N. Vanhove, L. R. A. Verheyden, and R. H. Charlier, "Coastal Belt Tourism, Economic Development and Environmental Impact," *International Journal of Environmental Studies* 10 (1977): 161–72.

7. Ibid.

8. W. Hipple, "Almeria, Orphan of Costa del Sol, Adopted by the World's Visitors," *Los Angeles Times* (March 19, 1978).

and leisure pursuits for American society: "Recreation spending . . . which includes recreational products, equipment, vacation spending, recreational trips, second homes, and the like . . . contributed more than $105 billion to the economy last year, and employed over 4 million people . . . or about one in 20 jobs in this country."[9]

The boating industry component of U.S. recreation spending includes retail sales exceeding $6.6 billion.[10] A look at the import and export of boats around the world will begin to show the extent of involvement in boating and the recreational boat industry in a variety of countries. Australia's pleasure-boat business is one of the country's fastest growing recreational industries, with boat imports in 1976 totaling $11,841,000.[11] In France, imports of recreational boats almost equaled exports in 1976, with imports valued at $43,950,800 and exports valued at $48,580,400.[12] In Sweden, dollar value of exports in the first 3 mo of 1978 was $12,466,520 (up $3,368,640 from the same period the previous year), although imports dropped from $12,271,820 in the first 3 mo of 1977 to $8,244,040 for the same period in 1978 because of an economic slump.[13]

The United States extensively exports boats to Canada, West Germany, the Netherlands, Venezuela, the United Kingdom, the Bahamas, Saudi Arabia, Switzerland, the Arab Emirates, Sweden, and France—a total of $74,836,609 in 1977.[14] Countries producing recreational boats imported into the United States include Taiwan, Hong Kong, Singapore, Italy, Japan, and Finland as well as Canada, the United Kingdom, the Netherlands, Sweden, and France (valued in 1977 at $79,965,379).[15] Boat sales alone are only a portion of the total revenue generated by recreational boating. Wet storage provides over $0.6 billion in revenues annually in the United States. Charter cruising, both sailing and power, further increases the revenues associated with recreational boating.

A survey of marine recreational fishing by the U.S. National Marine Fisheries Service "revealed an estimated 10.8 million marine recreational fishermen . . . during 1973–1974. These people included recreational shell-fishermen as well as finfishermen, without respect to their ages, frequency of

9. Patrick Doyle, "Impact of the Recession on the Recreational Marine Industry," in *Recreation: Marine Promise*, ed. Susan H. Anderson (Los Angeles: University of Southern California Sea Grant Program, 1976), p. 107.

10. MAREX and National Association of Engine and Boat Manufacturers, *Boating '78: A Statistical Report on America's Top Family Sport* (Chicago: MAREX, 1978), available from the Marketing Department of MAREX, 401 North Michigan Avenue, Chicago, Illinois 60611.

11. National Association of Engine and Boat Manufacturers, *Intercom* (September 22, 1977).

12. Ibid. (August 4, 1977).

13. Ibid. (July 20, 1978).

14. Ibid. (April 6, 1978).

15. Ibid. (March 2, 1978).

participation, or level of expenditure."[16] While many of these anglers are pier and jetty fishermen who may spend relatively little on the sport, a substantial number of saltwater anglers spent $1.25 billion on fishing excursions in 1970.[17] Where the catch of trophy fish is involved, fishermen frequently spend considerable money for travel and charter boat arrangements.

Scuba diving is an equipment-intensive recreational activity, requiring an initial average expenditure of $500 per diver if equipment is purchased rather than rented. Equipment usually added if the diver remains active costs an additional $300–$400. According to a recent government report on civil diving in the United States, the estimated annual expenditures for recreational diving total $241.3 million, including basic training equipment, advanced equipment and replacements, and personnel expenditures (instruction).[18] Additional expenditures not calculated here include expenditures for charter boat excursions, compressed air, and travel.

Worldwide there are an estimated 3 million surfers, annually supporting a $50 million industry.[19] A small tourist industry is also attributed to surfing with an estimated average travel distance of 500 miles per surfer, based on a survey of over 1 million readers of *Surfer Magazine*.[20]

Other marine recreational activities are, as a rule, less equipment intensive. (Although a coastal photographer could certainly take exception to this comment, photographic equipment is not usually purchased solely to pursue marine recreation.) Incidentals purchased for swimming and picnicking, plus parking fees, add nearly $3 billion to U.S. marine recreation expenditures. The expenditures for sightseeing are more difficult to obtain. However, in southern California alone, more than ½ million people pay boat fares annually to view ocean birds, whales, sea otters, sea lions, and other marine wildlife.[21]

There is no question that the hotel, condominium and cottage rentals, and restaurant businesses associated with coastal recreation are by themselves multibillion-dollar businesses throughout the world. Second homes and recreational vehicles also contribute to the total economic value attributed to marine recreation.

A small but significant portion of marine recreational activities involve extensive foreign travel. For general beach activities, any place that offers wide expanses of sandy beach is susceptible to tourism if service amenities are

16. Richard H. Stroud, "Competition for Recreation Resources—Introductory Remarks," in Anderson, ed., p. 74.

17. John Harville, "The Ocean's Living Resources for Marine Recreation," in Anderson, ed., p. 47.

18. Manned Undersea Science and Technology Office (MUS&T), *An Analysis of the Civil Diving Population of the United States* (Rockville, Md.: National Oceanic and Atmospheric Administration, U.S. Department of Commerce, 1975), p. 8.

19. Steve Pezman (ed., *Surfer Magazine*), telephone interview, April 11, 1977.

20. Ibid.

21. Stroud, p. 75.

available. The setting of a quaint coastal village may be particularly appealing in contrast to the high-rise or condominium complexes that are prevalent near a number of major resorts in the western hemisphere. For example, Almeria on the coast of Spain has recently received considerable tourist attention because it retains much of its Spanish charm, having been bypassed during the earlier tremendous growth in tourism along the Costa del Sol.[22] The Caribbean and the Bahamas boast a certain intimacy that attracts many return visitors. Many of these islands offer "great gleaming expanses of uncrowded beach," unobstructed by development. The resorts there offer warm hospitality and a variety of ways to immerse oneself in the beauty of the marine environment, from lazily soaking up the sun on a sandy beach to exploring the offshore reefs with mask and snorkel or scuba, to sailing in clear coastal water.[23]

Charter sailing has been especially popular in the Caribbean and might lend itself to other seas of the world that offer multiple ports within 1–2 days' sailing distance. Charter fishing is popular and easily arranged in a number of countries. Mexico, both on the Caribbean (for example, Cozumel)[24] and in Baja California, offers excellent charter opportunities, from 1-day trips to more extensive excursions. The waters near Cairns, Australia, are reported to be among the best big-game-fishing waters in the world,[25] providing the Australians with a good basis for a thriving charter fishing industry.

Although scuba divers are found in the most uninviting, cold, and murky waters, good diving locations are ideal for building a successful tourist resort. Waters like those of the Caribbean that are clear, warm, and abundant with diverse colorful fish attract both the experienced and the novice (or even uninitiated) diver. A resort operator with a reputation for honesty and dependability in providing multiple diving opportunities most successfully capitalizes on the scuba tourist industry. In the Caribbean, there are numerous dive resorts and a number of package tours that can be arranged through the airlines. The quality of the diving packages varies considerably, with the best ones tending to have fewer divers per guide, multiple dives feasible on each day, and methods of operation based on principles of diver safety.[26] It is also important that the areas have not been "fished out" by unrestricted spearfishing in the past. Diving tourism is growing in the Red Sea, Truk and Palau, the Galapagos, and the Philippines through coordination with American travel programs.

22. Hipple.

23. I. Keown, "Intimate Islands," *Travel and Leisure* 6 (October 1976): 35–39, 78, 80, 82.

24. D. Gold, "Report from Cozumel," *Travel and Leisure* 6 (November 1976): 32.

25. C. Garrison, "Along the Waterfront," *Western Outdoor News* (March 3, 1978), p. 19.

26. J. Olivo, "Shopping for a Dive Vacation," *Skin Diver* 27 (February 1978): 73, 84, and 86.

ENVIRONMENTAL QUALITY AND RECREATION

The pursuit of recreation and the continued growth of economic prosperity in relation to marine recreation are dependent on a high-quality environment. It is the unique environment which attracts people to the coast. Thus, the features that make that environment unique must be maintained and, where possible, enhanced.

Various forms of industrial pollution have had significant impacts on marine recreation. Waters have been closed to swimmers because of health hazards posed by pollution. Shellfish beds have also been closed because shellfish often not only filter pollutants from the water column but also concentrate them, intensifying their impact on consumers. Filth and scum on the water make it unacceptable for recreation even if the actual health hazard is tested as negligible.

The offshore oil industry and ocean transport of oil pose increasing threats to marine recreation. Dead fish resulting from spills may wash up on beaches, creating an unacceptable stench and mess. The oil itself carrying oiled debris may wash ashore and mix with the sand. Coastal wildlife is often destroyed or debilitated by the spreading oil.

The impact of oil spills is not limited to coastal cities near offshore oil-drilling operations or to those that are the sites of oil importation. Since the closure of the Suez Canal, enormous stretches of beaches along the west coast of Africa are being polluted because of the petroleum discharge from oil tankers en route to ports in the Indian Ocean.[27]

Noise levels can also destroy the environmental ambience for recreation. In order to meet community standards of noise abatement, airports often restrict takeoff to runways leading out over the water. Beaches near or under the flight path are less desirable for recreational use.

Whereas recreation itself usually contributes only low-level pollution to the environment, the need for support facilities may have major impacts on environmental quality. Little more than a decade ago, an innovative recreation developer could buy an area of what was then called "swamp," fill it, and dredge channels to make a new marina or even a whole resort built around a marina. Because it was believed that reclaiming useless land from the sea was a virtue, it also was assumed that the coastal resource, even under increasing pressures for development, could be expanded to meet the increasing demand.

It is now recognized, however, that wetlands, formerly referred to as swamps, have a high value in the life cycles of many fish species. These areas are often essential for reproduction and as juvenile habitats. To dredge and fill

27. Unesco, *Marine Questions–Coastal Area Management and Development: Report of the Secretary General,* U.N. (E/5648), May 8, 1975, annex p. 4.

them for development depletes fish stocks and leads to extinction of some coastal wildlife.

The demand for increased coastal homes, resorts, marinas, and even parking lots to meet the needs of recreation, has diminished many of the wide open spaces associated with the land/sea interface. Furthermore, the heavy influx of visitors to the coast has frequently burdened sewage systems beyond their capacities and overcrowded and overtaxed the very environment the visitors come to enjoy.

The more pristine the environment chosen for recreation, the more danger there is of negative impacts. Opening up marsh areas and tidepools for photography and educational exploration is desirable in the enhancement of recreational opportunities, but care must be taken to limit the numbers of people in these areas in an effort to protect the environment.

We do not have adequate information on the number of people who can walk over a tidepool area every day and still allow a vital biological community to maintain itself within the tidepool. And we do not know how often a fragile environment, such as a tidepool, should be totally closed just to reestablish the health of the biological community. However, with controlled use many of these areas can survive the impacts of recreation.

Where natural processes have depleted sand beaches or have in some way impaired pursuit of a desired recreational activity, engineering technology has been employed to alter these processes. For a number of years great pride has been taken in the ability to stabilize beach erosion with use of groins and to provide breakwater protection for areas without sufficient natural shelter for commercial and recreational harbors.

More recently, however, we have discovered that these alterations frequently create new problems. While making one beach better than before, we may be depleting the sand from another beach downcurrent from the improved area. Groins have had to be modified to lessen this problem. "The building of jetties or other types of projecting walls to develop harbors along straight coasts has had disastrous effects on many beaches."[28] Typically, the sand is trapped next to the jetty on the side of the prevailing current, greatly widening the beach. On the downcurrent side, the beach may gradually diminish to the extent that coastal buildings may be undermined. Pumping sand from the upcurrent side of the jetty to the downcurrent side is usually necessary to alleviate the problem.

Through engineering it has been possible to increase the usable coastal interface by using landfill to create islands and to extend existing shorelines. We

28. Francis P. Shepherd, *Submarine Geology,* 2d ed. (New York: Harper & Row, 1963), p. 196. For a general discussion of both natural and man-induced coastal erosion, see James K. Mitchell, *Community Response to Coastal Erosion,* University of Chicago Department of Geography, Research Paper no. 156 (Chicago: University of Chicago, Department of Geography, 1974), chap. 1.

are not yet certain to what extent these may disrupt the environment. Particularly in urban areas, they may prove to be the only hope for increasing the usable recreation coastline.

ACCESS

The question of access—that is, who is going to be able to use this recreational resource and how they are going to get there—is a critical component of the provision of recreational opportunities along coastlines today. Along many sections of the world's coasts, only those who are fortunate enough to own coastal property already (and who can afford the soaring taxes on that property) or who can afford the increasingly expensive coastal resorts are able to enjoy marine recreation. Certainly much coastal property is privately owned and is not available for the use of the general public.

Many of the coastal states in the United States have declared by constitution or by law that the public has the right to use the shoreline to the mean high tide level or higher. In some states, the entire land/sea interface is considered to be a common property resource. In some countries—for example, the Dominican Republic—the right of the public to use coastal beaches seems to be undisputed.[29] The use of these areas by all people is accepted and respected. In Rio de Janeiro there is no question—from the extensive and varied use made of the beaches—that these beaches are a public resource to be used by all classes of people.[30]

Many nations have, without pause for thought, left the seaward side of coastal roads open to public use. Harbor areas in many countries outside the United States not only support an active commerce but also provide walkways and promenades for members of the community. In contrast, this kind of access to the waterfront has only been realized in heavily developed countries such as the United States through restoration of the urban waterfronts to resemble their historical character. In Boston, Haymarket Square until recently had been almost abandoned as newer and bigger buildings expanded the city inland. Today Haymarket Square has been restored as an eighteenth-century marketplace. The New England Aquarium has been added to the waterfront along with restaurants and small shops. Restoration of the area and better lighting have brought people back to enjoy the waterfront as a kind of passive recreation.

In New York City, South Street Museum is another renovation of an urban waterfront making the shoreline again accessible to the public. Here donations of old sailing ships have assisted in the restoration of the area as an eighteenth-century port. Other examples are the Cannery in San Francisco, Cannery Row

29. L. G. Beatty, personal communication (June 1, 1978).
30. J. Reedy, "Rio!" *Odyssey* (July–August 1975), pp. 6–11.

in Monterey, and a similar development in Vancouver. These restorations each have built shops and restaurants in the interior of old buildings to re-create a historical atmosphere. By combining tourism with access, they seem to be successful ventures. The benefits of these restorations can provide a valuable lesson to those countries that continue to have active accessible urban harbor areas.

Through coastal zone management laws around the United States, increasing attention has been focused on the question of access. The California Coastal Law[31] provides several approaches to alleviating the problems of access. The basic policy relating to access states that a major long-term goal of coastal conservation and development is the provision of maximum amounts of oceanfront area for public use and enjoyment. Although responsive to the need to protect coastal areas from destructive overuse and to protect both public rights and the rights of property owners, the policy directs action to provide areas large enough to permit significant opportunities for public use and enjoyment of the land/sea interface. In urban areas an active program of public acquisition was recommended by the Coastal Commission to preserve remaining open oceanfront areas for public use, especially where development precludes effective access. Acquisition was also outlined for remaining fragile coastal areas away from urban centers.[32]

Increasing numbers of public utilities and public works construction projects in the United States have incorporated public recreation access into development plans to mitigate possible negative effects of the developments. Many of these projects require proximity to the ocean environment or are, in fact, designed to provide sheltered harbor areas or beach protection for recreational use. With a little extra planning, these facilities have provided access for inexpensive public use, such as fishing or painting from walkways along the breakwaters and jetties or picnicking adjacent to coastal facilities. Even the tops of constructed buildings have been suggested for park areas providing coastal views. Inland dams built for power and flood control set the precedent for this kind of cooperation in providing lakes and ramps for recreational boating access.

The issue of access, however, is more complicated than the legal right to use the land/sea interface or even the provision of recreational opportunities. There may not be sufficient corridors perpendicular to the shoreline, or lack of parking facilities nearby may prevent use of the shoreline by those who do not reside near it. Public transportation to the shoreline may be inadequate, preventing use by those without cars.

Shoreline accessibility depends on the ability of the public to get there. Along many coastlines, private development has created an effective wall between the nearest public road and the shoreline. One can drive along a

31. State of California, *California Coastal Act,* Stat. 1330, §§30210–13.
32. Ibid., §31352.

coastal road with little or no awareness that the ocean is just behind the houses or apartments built to take maximum advantage of the coastal view—for the residents only. Along many urban waterfronts, outdated warehouses and shipping terminals have been abandoned, leaving visual blights on the shoreline and obstacles to access.

In reality, the legal right to use the shoreline is of little value to the public unless there is specific action providing for corridors or accessways from the nearest public land (usually road) to the shoreline, action to improve transportation and parking, and action to clean up port areas and to restore unused buildings to make the urban shoreline again accessible to the public.

While the United States probably has the most extensive coastal zone laws dealing with access, several European countries are also beginning to address this issue. The tendency, however, is to deal with the coast in a piecemeal fashion instead of with comprehensive planning. The focus of management plans in the United Kingdom, for instance, is on "wildlife conservation with subsidiary aims such as education, research, recreation, and amenity."[33] Specific natural areas may be identified as needing a management plan, which may incidently include requirements for access. Management plans are often devised for areas perceived as recreation areas or industrial areas, but in many areas no overall plan seems to be in effect to determine which areas shall be designated for what purpose—such as natural reserves, recreation areas. A further concern is that some of the regulatory agencies setting policy on access and other aspects of coastal management do not have financial responsibility for their policies. Thus the policies providing for access may not be implemented because of lack of funds for maintenance of access ways or for compensation to owners of land condemned for access.

ALLOCATION

There is tremendous competition concerning the use of the land/sea interface, especially near developed cities. Cities were begun on the coast because of access to commerical shipping routes. Many industries either prefer being adjacent to water or are water dependent. Private homeowners prefer coastal locations for both climate and regular enjoyment of the coastal environment. Private developers capitalize on the coastal environment with apartment complexes and tourist accommodations.

Thus, to take maximum advantage of both the beauty and practicality of the land/sea interface, development has placed increasing pressures on the coast.

33. M. B. Usher, "Coastline Management: Some General Comments on Management Plans and Visitor Surveys," in *The Coastline,* ed. R. S. K. Barnes (London: Wiley, 1977), p. 293.

Because coastal land is in such tremendous demand, in the United States land values have soared, and taxation, usually based on potential use for high-density, high-profit development, has increased. Many private properties once surrounded by open space have been subdivided and sold in smaller lots so that the taxes are more affordable. The smaller lots are, in turn, filled with homes creating an effective barrier to public access. And even these smaller lots are too expensive for middle-class purchase.

As long as the decision concerning who will use coastal lands remains in the economic marketplace, those who can afford to buy coastal property or afford to vacation in the luxury coastal resorts will have the greater opportunity to enjoy the full variety of coastal recreations. If marine recreational opportunities are to be available for free or low-cost public use, it is often necessary for government to intervene in the marketplace allocation of coastal resources, and thus for the taxpayers to subsidize the recreational experience.

Government intervention may take the form of zoning regulations. These can serve to lower taxation on coastal lands, but, because down-zoning lowers the tax base for the responsible jurisdiction, there is little incentive to down-zone valuable coastal property. Even when coast related businesses have been forced to move inland because of continued high taxation, the responsible jurisdictions do not always consider it beneficial to rezone the lands adjacent to the land/sea interface.

Public beaches and coastal parks are usually made available through public acquisition. Once they are in the public domain, these properties are no longer taxed. Thus, while increasing the public recreational opportunity, a coastal community increases its maintenance and security loads and decreases its effective budget for implementing the necessary management.

Because the beaches frequently serve a larger constituency than the population of the coastal community adjacent to the beach, the question arises of who should bear the cost burden. If there is no beach user fee, the local community bears the cost through taxes. But the availability of the beach may be the attraction that brings many tourists to the community.The economic disadvantage of publicly owned beaches can be effectively offset by the substantial increase of tourism to the area generated by the availability of coastal resources. This is particularly true in areas that attract visitors from considerable distances so that once visitors are in the coastal settings they usually spend one or more nights in local hotels, hostels, or campgrounds. Successful marine recreation–oriented tourism relies on this availability of coastal resources.

If private entrepreneurs are expected to leave land as open space for picnicking, hiking, bicycling, and horseback riding to enjoy the coastal scenic environment, there must be an incentive to make the venture viable. Private ownership and management of coastal recreation property rely heavily on the economic marketplace for determination of needed support facilities and services. In order for a private developer to supply recreational opportunities, such as a campground or coastal park, there must be a market for those opportunities—people who are willing to pay for access to this coastal open space.

For the most part, the availability of open space around a tourist-serving facility increases the attractiveness of the facility and encourages repeat visitations. While many people thrive on the interactions made possible by a crowded sandy beach, the availability of other open space lending itself to more secluded recreational experiences will frequently increase the appeal of the location because of the diversity offered.

One recommendation coming from the 1975 National Conference on Marine Recreation was that tax incentives be used aggressively in coastal-zone management.[34] The concept of tax incentives to encourage and help existing landowners to maintain the open-space quality of land is not widely used in the United States.

English national parks set a sophisticated precedent for public/private parks carried out, in part, through tax incentives.[35] English parks are almost entirely in private ownership, but through land-use regulations their scenic beauty and open space for recreation are maintained. Landowner opposition to regulation is reduced through "real estate taxation at existing use value, modification in tort liability, and the negotiation of access agreements to compensate private landowners for public use of private land."[36] Another option for encouraging the maintenance of open space around coastal development is to acquire parkland in public ownership while leasing service opportunities to the private entrepreneur.

Many developing nations, in emphasizing the conservation of unique historical, archaeological, and scenic sites, have planned to establish national coastal parks, which might also use the public/private park concept. Barbados has established a Parks and Beaches Commission with special interest in preserving areas for underwater parks. Kuwait is proposing that parts of the North Coast and the Failaka Islands be made into national parks with accommodations and interpark transportation. The United Republic of Tanzania has proposed a marine reserve and park near Dar es Salaam to protect the coral reefs near the islands of Mbudya and Sinda.[37]

Several public/private parks do exist in the United States for national coastal recreation areas such as Cape Cod National Seashore, Indiana Dunes, and Fire Island. Martha's Vineyard and Nantucket are examples of state public/private parks. The County of Los Angeles is now planning a conservation management plan for much of privately owned Catalina Island, which has been entrusted to the county for 50 years for public recreational use. In this

34. Anderson, ed., p. 7 (see n. 9 above).

35. Jon A. Kusler, "Techniques for Minimizing Coastal Zone Recreation Conflicts: An Overview," in ibid., p. 93.

36. Ibid.

37. Unesco, p. 14.

case the private landowner wanted to maintain the open-space character of the island but taxation based on "highest and best use" was making it increasingly impractical. By opening the island to public recreation through a public/private park concept, increased access will be provided but at a level that is consistent with the character of the fragile open-space environment. The landowner in turn receives a significant tax reduction.

Where this kind of public/private partnership exists, care must be taken that park information includes descriptive material about the total facility. Clare Gunn noted about Mission Bay, San Diego, that whereas over 75 percent of the area was devoted to public access for picnicking, swimming, and beach use, the information literature available referred only to the commercial facilities within the area.[38] This wall between recreation and tourism is an unnecessary impedance to access.

Governmental intervention, preferably in the form of incentives, may also be necessary to encourage diverse recreational opportunities. For instance, marinas could be more than simply a dockage area for those with boats. The upland area can be used to include a total marine park in the environment of the marina, with areas for pier fishing and picnicking. Nonboaters, perhaps handicapped, can thus also enjoy the boats from a "landlubber's" perspective.

In both public and private development of coastal recreation opportunities, it is important to maintain perspective concerning the resource that is the attraction. Excessive scenic depreciation of the major attraction—the coast—can occur if too many people and facilities that are not directly involved in the appreciation of the attraction are clustered around the focal point.

CONCLUSION

Marine recreation is an integral part of lives worldwide. It supports a substantial industry, not only in the United States, but also in coastal states around the world. It is a growing activity utilizing a resource constantly diminished by demands for development. The marine environment is the feature attraction for tourism—for events ranging from the Dubrovnik Summer Festival, to the Laguna Arts Festival, to diving excursions in the Truk Lagoon. It offers recreational opportunities to both the young and the old, to those who are handicapped as well as to those who are physically able.

38. Clare A Gunn, "Toward Joint Tourism-Recreation-Conservation Policy," in Anderson, ed., p. 201.

ADDITIONAL REFERENCES

Arthur Young & Co. *Boating Resources Development Planning Study.* Sacramento, Calif.: Department of Navigation and Ocean Development, 1973.

Bongartz, Roy. "Freedom of the Beach." *New York Times Magazine* (July 13, 1975), pp. 12, 13, 27, 30, and 31.

Burke, James E. "Recreation Planning in the Coastal Zone: Analytical Techniques, Information and Policies." Working Paper no. 278. Berkeley: University of California, Berkeley, Institute of Urban and Regional Development, 1977.

California Coastal Zone Conservation Commission. *California Coastal Plan.* Sacramento, Calif.: Documents and Publications Branch, 1975.

California Department of Parks and Recreation. *California Coastline Preservation and Recreation Plan.* Sacramento, Calif.: Department of Parks and Recreation, 1971.

California Department of Parks and Recreation. *California Outdoor Recreation Resources Plan.* Sacramento, Calif.: Department of Parks and Recreation, 1974.

Clepper, Henry, ed. *Marine Recreational Fisheries.* Washington, D.C.: Sport Fishing Institute, 1976.

Commission on Marine Science, Engineering and Resources. "Recreation." In *Our Nation and the Sea,* pt. 7, pp. 235–52. Washington, D.C.: Government Printing Office, 1969.

Conservation Foundation. *Barrier Islands and Beaches: Technical Proceedings of the 1976 Barrier Islands Workshop.* Washington, D.C.: Conservation Foundation, 1976.

Ditton, Robert B. "The Social and Economic Significance of Recreation Activities in the Marine Environment" (WIS-SG-72-211). Technical Report no. 11. Green Bay: University of Wisconsin Sea Grant Program, 1972.

Ditton, Robert B.; Strang, W. A.; and Dittrich, M. T. *Wisconsin's Lake Michigan Charter Fishing Industry* (WIS-SG-75-411). Madison: University of Wisconsin Sea Grant College Program, 1975.

Fitzsimmons, Allan K. "National Parks: The Dilemma of Development." *Science* 191 (February 6, 1976): 440–44.

Gonen, Amiram. "Mediterranean Tourism: Some Geographic Perspectives." In *Tides of Change: Peace, Pollution, and Potential of the Oceans,* edited by Elisabeth Mann Borgese and David Krieger. New York: Mason/Charter, 1975.

Idaho Cooperative Fishery Unit. *Sportfishing Economics: A Report to the National Marine Fisheries Service.* Moscow: University of Idaho, 1973.

Kusler, Jon A. "Public/Private Parks and Management of Private Lands for Park Protection." Report no. 16. Madison: University of Wisconsin, Institute for Environmental Studies, 1974.

McConnell, K. E. "Some Problems in Estimating Demand for Outdoor Recreation." *American Journal of Agricultural Economics* 57, no. 2 (May 1975): 330–34.

Neuman, Michael T. *Public Access to the Great Lakes.* Madison: Wisconsin Department of Natural Resources, 1976.

Pacific Southwest Airlines Magazine (January 1978).

Roy Mann Associates, Inc. *Aesthetic Resources of the Coastal Zone.* Cambridge, Mass.: Roy Mann Associates, 1975.

Roy Mann Associates, Inc. *Recreational Boating on the Tidal Waters of Maryland.* Cambridge, Mass.: Roy Mann Associates, 1976.

Urban Design Group, Inc., and Economics Research Associates. *Marinas and Pleasure Boating Facilities Study.* Newport, R.I.: Urban Design Group, 1975.

U.S. National Park Service, Gateway National Recreation Area Project. *Gateway: Basic Information.* New York: Gateway National Recreation Area Project, 1975.

Walgenbach, Frederick E., ed. "Resources Related Recreation Evaluation: A Symposium to Investigate." Sacramento: California Department of Fish and Game, 1975.

Williams-Kuebelbeck and Associates, Inc.; Moffatt and Nichol, Engineers; and Environmental Assessment and Resource Planning. *Development Feasibility Analysis for Stacked Boat Dry Storage.* Redwood City, Calif.: Williams-Kuebelbeck & Associates, 1975.

Wulfsberg, R. M., and Lang, D. A. *Recreational Boating in the Continental United States in 1973.* Washington, D.C.: Coast Guard, 1974.

Naval Forces[1]

Andrzej Karkoszka
Stockholm International Peace Research Institute

Trends, both in numbers of warships procured and in the pace of their technological advancement, indicate that the global naval arms race is becoming more intensive than ever before. The political and military importance of states' naval power is determined by all or some of the following factors: (1) The role of submarine-launched ballistic missiles (SLBM) in strategic deterrence is constantly growing. (2) The potential of conventional naval forces, especially those equipped with various types of ship-launched missiles and possessing large fleet air-arms, may seriously influence the outcome of an eventual war, either directly (projection of force from sea to land) or indirectly (interdiction of vital supplies). (3) The mobility and flexibility of naval forces support political coercion or the use of force in a limited conflict. (4) The economic importance of sea resources is now realized more than ever before, intensifying competition for their distribution, especially since the number of developing states actively securing their national interests at sea is constantly on the increase. (5) Technological progress in both ships and their weapons greatly enhances the effectiveness of naval forces in every aspect of their action—in speed, range, and accuracy—or in lethality against a large range of targets.

CENTRAL NAVAL BALANCE

The largest fleets of warships are operated by states belonging to the two major military alliances—NATO and the Warsaw Treaty Organization (WTO). These states, of which 18 deploy naval forces, operate (see Appendix tables 1H-6H, pp. 674–677): (*a*) all 11 nuclear-powered surface warships in the world; (*b*) 18 of 21 attack aircraft carriers and another 18 of 20 antisubmarine and assault aircraft carriers in the world; (*c*) all 167 nuclear-powered attack submarines; (*d*) all 121 nuclear-powered, and all 23 conventionally powered, ballistic missile submarines; (*e*) 288 out of 500 conventional patrol submarines; (*f*) 66 out of 76

1. Editors' Note: Portions of this piece are adapted from the author's chapter in the *SIPRI Yearbook of World Armaments and Disarmament, 1979* (London: Taylor & Francis, 1979).

© 1980 by The University of Chicago. 0-226-06603-7/80/0002-0011$01.00

coastal submarines; (*g*) 390 out of 512 missile-armed major warships; and (*h*) 290 out of 553 conventionally armed major warships.

The United States and the Soviet Union dominate the two respective alliances; their positions are particularly crucial with regard to strategic naval forces, since these two powers deploy 135 of the world's 154 submarines. The seaborne element of strategic arsenals tends to acquire more and more importance. There are 121 nuclear-propelled ballistic missile submarines, exactly twice as many as there were a decade ago. More important, however, is the fact that the numbers of independently targetable nuclear warheads carried by U.S. and Soviet submarines increased sixfold during the same period, that is, from about 800 to 5,200 for the United States and from 250 to nearly 1,500 for the USSR. Whereas in 1970 the submarine-borne, independently targetable warheads represented only 18 percent of the total number of such warheads in the arsenals of both superpowers (i.e., 1,028 out of 5,800), in 1979 they represented 48 percent of the total number of nuclear warheads of the two strategic arsenals (i.e., 6,700 out of 14,000). The accuracy of these weapons has been rapidly improved, thus increasing their potential to attack well-protected military targets. In this way the SLBMs (so far believed to be a purely second-strike strategic weapon), being the most stabilizing element of nuclear arsenals, are increasingly becoming an organic part of the future war-fighting machinery. It is therefore logical that both the great nuclear powers and their allies alike are intensifying their efforts in antisubmarine warfare (ASW) and designating large air, surface, and submarine forces for this purpose.[2]

The numerical strengths of navies composing the "central" balance are given in tables 1 and 2. These tables take into consideration not only the numbers of ships possessed by the United States and the USSR but also the total numbers of warships possessed by the respective alliances. West European fleets (France included) represent nearly 55 percent of all NATO's major surface ships (frigates and above) with different shares in various classes of ships. Thus, omitting this force, as is often done, in the central balance is a distortion of reality. However, the capabilities of the East European members of the WTO, as far as the major naval combatants are concerned, are very small. In the case of the United States and the Soviet Union, all their forces are taken into account in tables 1 and 2, although large parts of these forces are deployed in the Pacific and Indian Oceans. In the Pacific, the Soviet Navy might also be confronted by the navies of other U.S. allies, such as Japan and South Korea. Tables 1 and 2, showing the numbers of surface ships, indicate that in all categories of major combatants—including those armed with surface-to-surface missiles (SSMs)–NATO forces together with France enjoy a great numerical advantage over WTO naval forces. This superiority is greatest in the

2. See Owen Wilkes, "Ocean-based Nuclear Deterrent Forces and Antisubmarine Warfare," pp. 226–49 in this volume.

category of aircraft and helicopter carriers (32 to 2) and in the category of frigates (228 to 44). The only category in which WTO naval forces possess numerical superiority is the category of light naval forces (corvettes and fast patrol boats [FPBs]). These vessels are suitable for operations in coastal and closed-sea waters but are of no or only marginal utility in open-sea operations. Mine warfare forces are not taken into consideration here; the WTO is believed to be preponderant in this category of ships. All aforementioned proportions would differ in various regional waters. To a large degree, the proportions of ship inventories correspond to the length of coastlines and to the dependence of some states on their sea lines of communication and supply.

One of the more controversial figures in table 1 is the number of U.S. helicopter carriers. This figure represents all U.S. amphibious assault ships of LHA- and LPH-class, of large displacement (three Tarawa-class of 39,300-ton and seven Iwo Jima-class of 17,000-ton displacement), able to carry from 30 to 24 helicopters, respectively, and said to be able to operate vertical or short take-off and landing (V/STOL) aircraft. (The same ships are also included in table 7 on amphibious forces.)

The numbers of warships in both alliances have been gradually declining over the past decade, however, with a concomitant transition from old gun-armed ships to modern missile-armed ones. On the Soviet side, the fastest growing categories are helicopter carriers (five expected by 1982) and guided-missile cruisers. The new production of other classes is offset by the need to replace older ships. The NATO forces, especially the U.S. and British navies, cut their inventories nearly by half during the first 2 decades after World War II. This reduction has been the main factor behind the relative decline of numerical strengths of NATO navies in comparison with the Soviet one. This process of reduction came to a stop in the early and mid-1970s and has now been reversed. The NATO plans include: in the U.S. Navy to build some 30 improved Spruance-class guided-missile destroyers and some 50 new SSM FFG-7 frigates; in the British Navy three new antisubmarine helicopter cruisers are to be constructed; the *Bundesmarine* to acquire up to 12 F.122 frigates; Italy to get one helicopter cruiser and six frigates of Maestre-class; the Netherlands to procure 11 Kortenaer-class SSM frigates; and France to go ahead with its "Blue Plan" for substantial modernization of the entire navy, in part by including a new nuclear-powered aircraft carrier and over 20 destroyers. Similarly, an extensive expansion is planned by several NATO states in their light naval forces.

General numerical comparisons alone say little about military effectiveness, since this also depends on the actual concentration and configuration of warships in a given sea area, on the timing of an engagement, and on several other technological and organizational conditions. For example, a small number of dispersed ships can threaten important lines of communication and supplies or may carry out a successful operation against a potentially stronger navy. All such gains are, however, only tactical or temporary; in a protracted

TABLE 1.—SURFACE NAVAL STRENGTH OF NATO COUNTRIES (Including France), 1978

	United States	Belgium	Canada	Denmark	Germany	Greece	Italy	Netherlands
Aircraft carriers, nuclear	3	...	...	...	...	...	...	...
Aircraft carriers, conventional	12	...	...	...	...	...	...	...
Helicopter carriers	10	...	...	...	...	...	...	...
ASW carriers	4	...	...	...	...	...	...	...
Total aircraft carriers	29	...	...	...	...	...	...	...
ASW helicopter cruisers	...	...	...	...	...	...	1	...
SSM cruisers, nuclear	7	...	...	...	...	...	...	...
SSM cruisers, conventional	22	...	...	...	...	...	2	...
SAM/gun cruisers	5	...	...	...	...	...	...	...
Total cruisers	34	...	...	...	...	...	3	...
SSM destroyers	39	...	...	...	7	1	4	2
SAM/gun destroyers	54	...	4	...	4	11	3	9
Total destroyers	93	...	4	...	11	12	7	11
SSM frigates	7	4	...	5	...	...	4	2
SAM/gun frigates	58	...	19	5	6	4	10	5
Total frigates	65	4	19	10	6	4	14	7
SSM major combatants	75	4	...	5	7	1	10	4
SAM/gun major combatants	146	...	23	5	10	15	14	14
Total major combatants	221	4	23	10	17	16	24	18
SSM corvettes	...	...	...	...	...	...	...	...
Gun corvettes	...	...	...	3	5	...	9	6
Total corvettes	...	...	...	3	5	...	9	6
SSM FPBs	1	...	...	10	30	12	1	...
Gun/torpedo FPBs	2	...	1	6	10	19	9	...
Total FPBs	3	...	1	16	40	31	10	...
Total warships	224	4	24	29	62	47	43	24

	Norway	Portugal	Turkey	United Kingdom	France	Total (NATO+France)
Aircraft carriers, nuclear	...	...	...	...	...	3
Aircraft carriers, conventional	...	...	...	1	2	15
Helicopter carriers	...	...	...	...	...	10
ASW carriers	...	...	...	...	...	4
Total aircraft carriers	...	...	...	1	2	32
ASW helicopter cruisers	...	...	...	2	1	4
SSM cruisers, nuclear	...	...	...	...	...	7
SSM cruisers, conventional	...	...	...	8	1	33
SAM/gun cruisers	...	...	...	...	...	5
Total cruisers	...	...	...	10	2	49
SSM destroyers	...	...	...	5	12	70
SAM/gun destroyers	...	...	12	...	8	105
Total destroyers	...	...	12	5	20	175
SSM frigates	5	...	...	17	18	62
SAM/gun frigates	...	7	2	39	11	166
Total frigates	5	7	2	56	29	228
SSM major combatants	5	...	...	30	31	172
SAM/gun major combatants	...	7	14	42	22	312
Total major combatants	5	7	14	72	53	484
SSM corvettes	...	...	...	...	...	...
Gun corvettes	2	10	...	...	...	35
Total corvettes	2	10	...	...	...	35
SSM FPBs	27	...	7	...	5	93
Gun/torpedo FPBs	19	...	13	...	...	79
Total FPBs	46	...	20	...	5	172
Total warships	53	17	34	72	58	691

SOURCES.—J. L. Couhat, ed., *Combat Fleets of the World 1976–1977: Their Ships, Aircraft, and Armament* (London: Arms & Armout, 1976), p. 449; J. E. Moore, ed., *Jane's Fighting Ships 1978–1979* (London: Macdonald & Jane's, 1978); *NATO's Fifteen Nations*, special ed., vol. 23 (1978); and G. Albrecht, comp., *Weyers Flotten Taschenbuch 1977–1978, Warships of the World* (Munich: Bernard & Graefe Verlag fuer Wehrwessen, 1978).

TABLE 2.—SURFACE NAVAL STRENGTH OF WTO COUNTRIES, 1978

	USSR	Bulgaria	German Democratic Republic	Poland	Romania	Total (WTO)
ASW SSM carriers	2	...	...	...	...	2
ASW helicopter cruisers	2	...	...	...	...	2
SSM cruisers	23	...	...	...	...	23
SAM/gun cruisers	13	...	...	...	...	13
Total cruisers	36	...	...	...	...	36
SSM destroyers	39	...	...	...	...	39
SAM/gun destroyers	71	...	...	1	...	72
Total destroyers	110	...	...	1	...	111
SSM frigates	...	...	...	...	...	...
SAM/gun frigates	41	2	1	...	...	44
Total frigates	41	2	1	...	...	44
SSM major combatants	64	...	...	...	...	64
SAM/gun major combatants	127	2	1	1	...	131
Total major combatants	191	2	1	1	...	195
SSM corvettes	17	...	...	...	...	17
Gun corvettes	180	...	...	...	...	180
Total corvettes	197	...	...	...	...	197
SSM FPBs	123	5	15	12	5	160
Gun/torpedo FPBs	252	8	70	21	21	372
Total FPBs	375	13	85	33	26	532
Total warships	763	15	86	34	26	924

SOURCES.—See table 1.

naval war, the more numerous navy would take over especially if technologically equal or superior.

A different and more professional method of naval comparisons was used in *The Military Balance: 1978–1979*. Here, comparison was based on the mission to be fulfilled by a specific task group of warships (though the Soviet Navy would not, most probably, operate in any "task group," in the Western sense of the term). According to this publication, NATO forces—especially the so-called sea-control forces—are "considerably greater by any assessment than their Soviet counterparts."[3]

Table 3 gives numerical inventories of all types of submarines operated by the two alliances. Here the balance in numbers is tipped in favor of the WTO, which possesses some 100 submarines more than NATO. A particularly large difference in numbers is visible in the categories of diesel and nuclear submarines equipped with guided missiles—72 to 1 in favor of the WTO—and in diesel ballistic missile submarines (obsolescent in view of modern ASW). A rough balance exists in smaller, conventionally powered patrol and attack submarines. The general trend on both sides is to replace older diesel submarines by new nuclear-propelled ones. A substantial number of the Soviet diesel patrol submarines are old-fashioned, being withdrawn from use quicker than the rate of replacement; thus, the overall number of Soviet submarines is declining. It is interesting that only a very small portion of the Soviet nuclear submarine fleet is out at sea. On a day-to-day basis, only 15 percent of the Soviet ballistic missile-carrying nuclear-powered submarine (SSBN) force is out of port in comparison with 55 percent for the U.S. SSBNs.[4]

Aircraft

Naval forces include large numbers of aircraft and helicopters. More than 30 different types (several of them in a number of versions) of fixed-wing aircraft and about 20 types of helicopter are in use by the Eastern and Western blocs. Numerical balances between the NATO and WTO naval air forces are summarized in table 4. The table shows that NATO naval air power leads in practically every category of aircraft. This superiority is especially great in sea-based aircraft, giving NATO a complete air command over the NATO operation areas as well as providing it with a powerful means of attacking land targets from the sea. The sea-based NATO aircraft are to some extent matched in number by the Soviet land-based long- and medium-range bombers carrying various antiship weapons. The biggest role in this connection can be played by the Tu-26 Backfire, here considered as a long-range attack aircraft—the only supersonic Soviet naval aircraft. This aircraft, being introduced into the inventory at a rate of some 12 per year, is gradually replacing other aging Soviet long- and medium-range maritime bombers. Although the Soviet bombers armed with long-range air-to-surface missiles (ASMs) constitute a powerful force, they have to operate from very distant bases; and when flying to open-sea stations, they are exposed to strict surveillance and would be attacked by NATO's land air forces in war (first Norwegian, and later British, U.S., and other forces). No land-based Soviet fighters have the range required for escorting bombers to

3. International Institute for Strategic Studies (IISS), *The Military Balance: 1978–1979* (Boulder, Colo.: Westview, 1978), p. 117.

4. *Interavia Air Letter* no. 9184, February 1, 1979.

TABLE 3.—SUBMARINE INVENTORIES OF NATO (Including France) AND WTO COUNTRIES, 1978

	United States	Canada	Denmark	Federal Republic of Germany	Greece	Italy	Netherlands	Norway
SSBN	41	...	...	...	...	...	...	...
SSB	...	...	...	...	...	...	...	...
SSGN	...	...	...	...	...	...	...	...
SSG	1	...	...	...	...	...	...	...
SSN	70	...	...	...	...	...	...	...
SS	8	3	6	24	7	8	6	15
Total submarines	120	3	6	24	7	8	6	15

	Portugal	Turkey	United Kingdom	France	Total (NATO+ France)	USSR	Bulgaria	Poland	Total (WTO)
SSBN	...	...	4	4	49	72*	...	...	72
SSB	...	...	...	1	1	22	...	...	22
SSGN	...	...	...	...	...	46	...	...	46
SSG	...	...	...	...	1	26	...	...	26
SSN	...	...	10	...	80	41	...	...	41
SS	3	12	17	23	132	170	2	4	176
Total submarines	3	12	31	28	263	377	2	4	386

SOURCES.—See table 1.

*Two of these submarines are not operational, and eight of older design are not counted under the SALT I provisions.

their distant station areas. Western airborne early-warning (AEW) capabilities are extensive and numerous, now being updated with the new Orion (EP-3E) and airborne warning and control system (AWACS) aircraft entering service. Although NATO does not deploy any naval long-range antiship aircraft, there are a large number (75) of U.S. SAC B-52D aircraft technically prepared and trained for maritime operations, such as mine laying, sea surveillance, and possibly bombing of surface vessels. The NATO maritime reconnaissance and ASW air forces are twice as large as their WTO counterparts. In helicopter inventories the picture is similar as far as forces afloat are considered, NATO possessing substantially more of them. A more balanced situation exists in land-based helicopter forces. On the NATO side there is a greater reliance on helicopters for other than ASW duties, especially search and rescue and assault.

The figures given in table 4 are all approximations. As in any other comparisons here, the qualities of the weapon systems and their mode of operation may have a great influence on any conclusions on real balances. Also, land-based air forces, not generally considered for naval operations, may, in reality, be used in battle over the sea if their range permits it. These important factors are not considered further here.

Developments in naval air forces on either side indicate that increasing importance is being attached to naval air power. The Soviet Union is enlarging its fleet of aircraft and helicopter carriers. The comparatively small size of carriers and the small number of aircraft they carry indicate that their chief role is in ASW and reconnaissance, and that they have a secondary role in providing air defense. Despite these constraints, the deployment of these vessels has, for the first time, provided Soviet ships with an organic air defense in remote seas. On the NATO side, nearly all member states are planning fast modernization and expansion of their naval air arm. The United States is procuring large numbers of F-14 Tomcat fighters and is developing another air superiority F-18 fighter and V/STOL aircraft. The U.S. ASW capabilities will also be increased by the introduction of LAMPS III helicopters. The United Kingdom plans to procure two or three through-deck cruisers and has already ordered 35Harrier V/STOL aircraft, 165 MRCA Tornado aircraft, and over 80 Sea King and Lynx helicopters. The Federal Republic of Germany plans to acquire 110 Tornado aircraft to replace its F-104Gs. Denmark, Norway, and Turkey have ordered several Lynx helicopters, and France has ordered about 30 Super Etendard fighters and a number of Lynxes. These plans prove that NATO intends to enhance its indisputable naval air superiority.

Guided Missiles

The guided missile is the main offensive weapon of today's ship. Thanks to the deployment of extremely able and compact fire-control systems (FCSs), missiles and guns play a major role in the defense against ASMs. In order to ascertain

TABLE 4.—SEA- AND LAND-BASED NAVAL AIRCRAFT IN NATO AND WTO COUNTRIES, 1978

	United States	Belgium	Canada	Denmark	Federal Republic of Germany	Greece	Italy	Netherlands
Afloat aircraft:								
Fixed-wing aircraft:								
Air combat	~750	...	...	...	...	...	...	...
ASW	130	...	...	...	...	...	...	...
Total, all types*	~1,100	...	...	...	...	...	...	...
Helicopters†:								
ASW	~200	...	16	5	...	...	~30	9
Other types‡	~320	...	9	...	...	...	...	...
Total helicopters	~520	...	25	5	...	...	~30	9
Ashore aircraft:								
Fixed-wing aircraft:								
Long-range§ recce/ASW‖	~450#	...	26	...	19	...	18	8
Long-range attack**	...††	...	...	...	...	...	...	...
Medium-range recce/ASW	...	...	30	...	...	8	8	15
Medium-range attack	...	...	...	...	...	...	...	...
Air combat##	...	...	...	...	112	...	...	...
Total, all types*	~450***	...	65	...	154	8	26	23
Helicopters:								
ASW	...	8	32	15	...	4	~50	36
Other types	...	...	...	...	21	...	~20	...
Total helicopters	...	8	32	15	21	4	~70	36

TABLE 4.—SEA- AND LAND-BASED NAVAL AIRCRAFT IN NATO AND WTO COUNTRIES, 1978 *(Continued)*

	Norway	Portugal	Turkey	United Kingdom	France	Total (NATO+ France)
Afloat aircraft:						
Fixed-wing aircraft:						
Air combat	...	...	...	32	40	~822
ASW	...	...	...	...	40	170
Total, all types*	...	...	...	37	80	~1,217
Helicopters†:						
ASW	...	...	...	~90	~20	~370
Other types‡	...	...	...	~55	~20	~404
Total helicopters	...	...	...	~145	~40	~774
Ashore aircraft:						
Fixed-wing aircraft:						
Long-range§ recce/ASW‖	...	6	...	35	35	~597
Long-range attack**	...	...	...	...	...	...
Medium-range recce/ASW	5	...	22	11	20	119
Medium-range attack	...	...	...	14	24	38
Air combat##	...	...	...	14	20	~145
Total, all types*	5	6	22	86	~200	~1,045
Helicopters:						
ASW	10	...	9	~80	60	~304
Other types	...	...	...	~100	...	~141
Total helicopters	10	...	9	180	60	~445

TABLE 4.—SEA- AND LAND-BASED NAVAL AIRCRAFT IN NATO AND WTO COUNTRIES, 1978 (*Continued*)

	USSR	Bulgaria	German Democratic Republic	Poland	Romania	Total (WTO)
Afloat aircraft:						
Fixed-wing aircraft:						
Air combat	24	...	...	...	...	24
ASW	...	...	...	...	...	...
Total, all types*	24	...	...	...	...	24
Helicopters†:						
ASW	101	...	...	...	...	101
Other types‡	...	...	...	...	...	...
Total helicopters	101	...	...	...	...	101
Ashore aircraft:						
Fixed-wing aircraft:						
Long-range§ recce/ASW‖	135‡‡	...	...	...	...	135
Long-range attack**	50§§	...	...	...	...	50
Medium-range recce/ASW	150‖‖	...	...	...	...	150
Medium-range attack	~300	...	...	...	...	~300
Air combat##	60	...	...	60	...	120
Total, all types*	695†††	...	...	60	...	755
Helicopters:						
ASW	~200	6	13	25	4	248
Other types	...	...	...	...	...	...
Total helicopters	~200	6	13	25	4	248

Services, *Department of Defense Authorization for Appropriations for FY 1979: Hearings on S. 2571*, 95th Cong., 2d sess., 1978; J. E. Moore, ed., *Jane's Fighting Ships 1978–1979* (London: Macdonald & Jane's, 1978); International Institute for Strategic Studies, *The Military Balance: 1978–1979* (Boulder, Colo.: Westview, 1978); *NATO's Fifteen Nations*, special ed., vol. 23 (1978); N. Polmar, "Soviet Naval Aviation," *Air Force Magazine* 61 (March 1978): 67; P. H. Rasmussen, "The Soviet Naval Air Force: Development, Organization and Capabilities," *International Defense Review* 11, no. 5 (1978): 689–95; and U.S. Congress, Congressional Budget Office, *U.S. Sea Control Mission: Forces, Capabilities, and Requirements, Background Paper, June 1977* (Washington, D.C.: Government Printing Office, 1977), p. 9.

*Includes also tankers, AEW, and EW aircraft.

†Helicopters on all naval ships, including amphibious helicopter assault.

‡Includes commando assault, command, utility, etc.

§"Long-range" meaning about 2,500 nautical miles (4,000 km) range or more.

||Including maritime patrol and in some cases AEW aircraft.

#This figure does not include 10 EC-121 AEW aircraft having maritime role, removed from active service; ~200 Orion P-3Cs in reserve being included in the figure.

**"Attack" meaning antiship capability (ASMs, bombs, torpedoes); "air combat" planes with such capability not taken into account.

††75 B-52Ds may be used in this role.

‡‡4 Tu-95 Bears of long-range aviation not included, although may have naval utility.

§§100 Tu-95 Bears of long-range aviation not included, although may have naval utility.

|| ||115 Tu-16 Badgers of long-range aviation used for ELINT and reconnaissance not included.

##These are exclusively used for naval missions.

***117 C-130 Hercules transport and tankers not included.

†††230 transports and tankers not included.

the balance in numbers and qualities of naval weapon systems possessed by the NATO and WTO states, all available weapons would have to be taken into account, including those on aircraft and submarines. This is, however, next to impossible, since both aircraft and submarines can carry mixes of weapons depending on mission. Thus, only SSMs and surface-to-air missiles (SAMs) will be considered here. There are three possible ways of indicating the number of naval missiles: (*a*) number of launchers (from single, to twin, quadruple, and octuple); (*b*) number of tubes or rails from which a missile is launched—in other words, number of missiles on launchers theoretically "ready" for firing; and (*c*) number of missiles in magazines, or reloads. The two latter seem to be more instructive than the first, and tables 5 and 6 give information on numbers of rails/tubes, equal to the number of missiles ready for fire. The quantity of missiles in ships' magazines is very difficult to obtain from open sources. It is, however, known that NATO ships are more spacious and carry larger amounts of reloads than do Soviet ships. Also, the number of SAMs for reload on warships is much greater than the number of SSMs. Thus, for example, the U.S. cruisers of the Leahy class, fitted with two twin launchers for Standard ER missiles, have 80 reloads, whereas the *Providence* cruiser has 120 reloads of the Terrier missile for its single twin launcher, and the *Oklahoma* cruiser has 46 Talos reloads for a similar launcher. The majority of U.S. destroyers carry about 40 Standard ER missiles for reloading. The Netherlands' De Ruyter-class destroyer has one complete Harpoon reload for its single octuple launcher, 40 Tartar reloads for one single launcher, and 60 Sea Sparrow missiles for reloading a single octuple launcher. In comparison, the Soviet Moskva-class helicopter cruiser, armed with four twin SAN-3 launchers, is said to carry 180 reloads, whereas Kresta-class cruisers, fitted with two twin SSN-3 launchers and one twin SAN-1 launcher, have no reloads for the first type and an unknown number of reloads for the second type of missile. A Kynda-class cruiser, equipped with two quadruple SSN-3 launchers and one twin SAN-1 launcher, are supposed to carry only one SSN-3 reload per tube and about 30 SAN-1 reloads.

Over a period extending from the 1960s to the early 1970s, the Soviet Union was considered to be superior in numbers and even in technical advances in antiship missiles. However, during the past decade, NATO countries have undertaken intensive programs for developing several types of such missiles and deploying them in large numbers. The most recent to join this effort was the United States, but now, with the Harpoon SSM program completed and with the Tomahawk cruise missile in development, that country plans rapid and extensive armament in antiship long-distance guided missiles. It has been decided that by 1985 the U.S. Navy is going to acquire over 2,100 Harpoons and about 1,100 Tomahawks.

The WTO forces deploy over 600 SSMs more than NATO. However, nearly half of the total WTO numbers—that is, 610 missiles—are on FPBs. More than half of the total number of SSMs are short-range SSN-2 missiles. The number of SSN-3 missiles is made up of those carried by both surface ships

TABLE 5.—NUMBER OF NATO SHIP-TO-SHIP (SSM) AND SHIP-TO-AIR (SAM) MISSILES ON LAUNCHERS, BY TYPE AND COUNTRY, 1978

Type of Missile*	United States	Belgium	Canada	Denmark	Federal Republic of Germany	Greece	Italy	Netherlands	Norway	Portugal	Turkey	United Kingdom	France	Total (NATO + France)
SSMs:														
SS11	...	...	...	...	...	...	...	...	...	...	...	...	4	4
SS12	...	...	...	...	...	...	...	...	...	...	...	...	24	24
Exocet MM 38	...	16	...	...	136	32	...	...	...	12	...	84	78	258
Penguin	...	...	...	...	...	...	...	...	150	...	...	...	...	150
Otomat	...	...	...	...	...	...	34	...	...	...	...	...	...	34
Harpoon	n.a.†	...	...	88	12	...	...	32	...	...	64	...	...	196
Total SSMs	n.a.	16	...	88	148	32	34	32	150	12	64	84	106	666
SAMs:														
Crotale	...	...	...	...	...	...	...	...	...	...	...	...	32	32
Masurea	...	...	...	...	...	...	...	...	...	...	...	...	6	6
Sea Cat‡	...	...	...	...	...	...	...	48	...	...	...	444	...	492
Sea Dart‡	...	...	...	...	...	...	...	...	...	...	...	12	...	12
Sea Slug‡	...	...	...	...	...	...	...	...	...	...	...	14	...	14
Sea Sparrow‡	540	32	48	40	...	8	32	24	...	...	...	...	...	724
Sea Wolf	...	...	...	12	...	...	...	...	...	...	...	12	...	24
Standard MR‡	12	...	...	...	...	...	4	...	...	...	...	...	4	20
Standard ER‡	60	...	...	...	...	...	...	...	...	...	...	...	...	60
Talos	10	...	...	...	...	...	...	...	...	...	...	...	...	10
Tartar§	56	...	...	...	3	...	...	2	...	...	...	...	...	61
Terrier§	26	...	...	...	...	...	6	...	...	...	...	...	...	32
Total SAMs	704	32	48	52	3	8	42	74	...	...	...	482	42	1,487

SOURCES.—J. L. Couhat, ed., *Combat Fleets of the World 1976–1977: Their Ships, Aircraft, and Armament* (London: Arms & Armour, 1976), p. 449; J. E. Moore, ed., *Jane's Fighting Ships 1978–1979* (London: Macdonald & Jane's, 1978); R.T. Pretty, ed., *Jane's Weapon Systems 1978–1979* (London: Macdonald & Jane's, 1978); *NATO's Fifteen Nations*, vol. 22, no. 6 (December 1977–January 1978); and G. Albrecht, comp., *Weyers Flotten Taschenbuch 1977–1978, Warships of the World* (Munich: Bernard & Graefe Verlag fuer Wehrwessen, 1978).

* Antisubmarine missiles (such as SUBROC and Malafon) are not taken into account.

† The exact number is not available, since they are installed in large numbers on several surface ships and submarines.

‡ SSM capability.

§ Difficult to ascertain in many cases which of the two missiles is actually deployed on a given ship.

TABLE 6.—NUMBERS OF WTO SHIP-TO-SHIP (SSM) AND SHIP-TO-AIR (SAM) MISSILES ON LAUNCHERS, BY TYPE AND COUNTRY, 1978

Type of Missile*	USSR	Bulgaria	German Democratic Republic	Poland	Romania	Total (WTO)
SSMs:						
SSN-1	4	...	...	...	...	4
SSN-2	572	20	48	48	20	708
SSN-3	345	...	...	...	...	345
SSN-7	140	...	...	...	...	140
SSN-9	114	...	...	...	...	114
SSN-12	16	...	...	...	...	16
Total SSMs	1,191	20	48	48	20	1,327
SAMs:						
SAN-1	132	...	...	4	...	136
SAN-2	2	...	...	...	...	2
SAN-3	80	...	...	...	...	80
SAN-4	184	...	...	...	...	184
Total SAMs	398	...	...	4	...	402

SOURCES.—See table 5.

*Antisubmarine missiles, such as SSN-14 (over 100 deployed), are not taken into account.

(48) and by submarines (297). In addition to 1,327 SSMs possessed by the WTO, there are well over 300 ASMs on long-distance Soviet bombers and several hundred torpedoes on submarines, all of them representing a threat to surface ships.

NATO's inventory of SSMs on launchers, indicated in table 5, is half that of the WTO. However, the figures given in the table do not include those types of SAMs which also possess antiship capability, such as Sea Dart, Sea Cat, Sea Slug, Standard, and some versions of Sea Sparrow missiles. The number of these SAMs with antiship capability is about 600–800. In addition a growing number of Harpoon missiles are now being installed in several U.S. ships, submarines, and naval aircraft, partly in replacement for older types of missiles. Altogether, the SSM inventory on the NATO side is rapidly closing the gap with WTO SSM forces, especially after the deployment of air-launched antiship missiles, such as the West German Kormoran. The pace of arming the Western warships with SSMs is such that in the near future some 75 percent of the NATO fleet will be thus equipped. When U.S. naval and air forces acquire the Tomahawk cruise missile, the balances in antiship missiles will undoubtedly be reversed, giving the numerical and technological advantage to NATO forces.

Both sides are deploying large quantities of antisubmarine missiles and rockets, such as ASROC, SUBROC, and Malafon on the NATO side, and

SSN-14 on the WTO side. These weapons have not been included in tables 5 and 6 although some, like SUBROC, have antisurface-ship capability.

As far as SAMs are concerned, NATO forces, possessing nearly four times more such missiles on launchers, are much better prepared for defense against attacking SSMs and aircraft than WTO forces. This is a logical outcome of NATO's awareness of the early development of Soviet antiship capabilities.

In assessing the actual balance between the opposing weapon systems, numbers naturally play a secondary role. The effectiveness of a missile attack on ships depends on the ability to saturate the ship's defense in a massed and concentrated attack. Therefore, the problem of target acquisition, surveillance and warning, control and coordination, as well as communication, is of decisive importance for the outcome of any engagement.

It has been said that the effectiveness of any weapon system depends as much on the characteristics of the weapon itself as on the whole chain of technical and organizational elements allowing this weapon to traverse long distances and to hit the target accurately. A missile flying over the horizon of a ship's radars must be able to recognize a target ship among several other ships in the given area. It must be resistant to electronic countermeasures (ECMs), be able to distinguish between decoys and target, and should be able to take evasive action against the target's defenses. Hence, for the effective use of long-distance missiles, surveillance and targeting technologies and real-time relay of data are crucial. The existence of long-range guided missiles makes the side using them dependent on external targeting sensors which are located on different platforms and all exposed to growing threats. In all these areas of military technology the Western states claim to be in a more advanced stage of readiness than the Soviet Union is believed to be.

It is sometimes presumed that the only chance for a successful Soviet attack with antiship missiles would lie in sending a large number of missiles at a group of ships in a short span of time. However, such a saturation attack from several platforms placed at different distances from the target requires that the exact positions of both firing ships and targets be simultaneously known. Hence, the proficiency of C3 systems is a first priority. Some Western observers, on evidence from large WTO maneuvers, such as the "Okean" exercises, believe that Soviet warships have rather good communication with their command centers but rather poor ability to communicate between themselves as far as combat data exchange is considered. However, this area of Soviet technical capability is a matter of guesswork. Thus, the previous statement about the good communication links of the Soviet Navy is questionable to some extent on the grounds of the frequent disturbances to which high and very high frequencies, used by Soviet ships, are exposed in time of solar activity, especially in the Northern regions. At the same time, there are indications that Kiev-class ASW cruisers and Kara-class cruisers may have a first generation of naval tactical data systems (NTDS)-like capability. The problems of communication between the command center and individual ships could be alleviated by the development of a satellite network. This fact seems to be recognized by all states able to launch satellites and possessing large ocean-going fleets. However, this trend would make all the states concerned more and more satellite dependent, offering a further justification for preparing for antisatellite warfare.

Most of the Soviet antiship missiles are deployed on submarines. However, a single Soviet submarine has rather a limited number of them, from 3 to 10. To obtain the required saturation of attack, a large number of these vessels would have to approach rather close to the target. The exposure of the submarines to ASW sensors and weapons would then be certain. The combat suitability of cruise missiles depends on the scenario of their use; they can be considered dangerous in a salvo attack against a group of ships. However, such a mass attack would be very unprofitable in a prolonged war at sea.

Several of the Soviet-produced missiles were designed some 10–20 years ago. They are large and have unsophisticated homing systems, as the naval engagements in the 1973 Middle East War showed. In comparison, the majority of NATO missiles have a capability for various modes of attack, from sea skimming to underkeel and steep diving on the target. Thus, defense against them is more difficult. Moreover, their guidance and homing systems are designed to be highly resistant to ECMs. The importance of "software," composed of radars and computers, is even more decisive in FCSs for air defenses. The Western systems are said to have the capability for multiple target handling, probably unmatched by the present Soviet designs.

Many authors writing about naval balances stress the preponderance of the Soviet fleet in its number of submarines. This statement should, however, be seen against the background of the existence of extensive and advanced antisubmarine capabilities of NATO navies and the geographic constraints under which the Soviet Navy has to operate.

As U.S. Navy Admiral T. B. Hayward states: "Air superiority is the *sine qua non* of successful surface ship operations in modern war. . . . "[5] The figures given in table 4 prove that such a superiority is on the NATO side. With aircraft like the F-14 and F-18 armed with modern air-to-air missiles (AAMs) (e.g., Phoenix) coming into inventories in large numbers, this superiority will be further strengthened.

Amphibious Forces

Amphibious forces constitute a highly mobile, powerful, and flexible armed service, effective in military action or in political coercion in a crisis situation. They are especially useful in the case when a country considers it vital to deploy

5. *NATO's Fifteen Nations,* special ed., 23 (1978): 107.

its armed strength in any crisis involving its interests. This sort of "crisis management" is especially in line with U.S. doctrine, as is indicated by the pronouncements of prominent U.S. officers and by the strength of the U.S. marine forces. Amphibious operations are nothing new, but, for example, today's well-armed and heavy-lift helicopters and hovercraft permit an assault to be made with a speed of over 20 knots, that is, twice as fast as in the Second World War. The emerging precision-guided munitions (PGMs) (such as semiactive laser-guided projectiles) and helicopter "gunships" provide obvious advantages in the projection of forces ashore, especially against an unprepared or less modern defense.

Amphibious forces are being improved in both NATO and the WTO. Partly going against the general trend in naval shipbuilding toward smaller hulls, both the U.S. and the Soviet navies are in the process of procuring large amphibious helicopter-equipped ships. In 1978 the Soviet Navy acquired a new 14,000-ton *Ivan Rogov* amphibious assault ship, the first of its size in this fleet, with the capability for long-range ocean operations, carrying a force of one naval infantry battalion with landing craft, air-cushion vehicles (ACVs), helicopters, and troop transports. It is well armed with SAN-4 missiles, several dual-purpose guns, and a large rocket launcher, presumably for giving fire support to the landing force.

The U.S. Navy has received its third of five ordered LHA Tarawa-class amphibious assault ships, each costing about $230 million. These carrier-like ships mark a further expansion of the U.S. amphibious forces and are an important step in designing multipurpose warships. The ship has a 39,300-ton displacement and combines within one hull the operational capabilities of six different types of amphibious vessel: assault helicopter carrier, transport ship dock, landing ship dock, tank-landing ship, troop transport, and attack transport ship. Thus, it will carry about 2,000 marines with all their artillery, tanks, and vehicles, disembarking them in some 40 landing craft with a speed exceeding 20 knots on to the shore. The hangars on the ship can accommodate 20–30 helicopters (depending on type), and when required, the ship's deck is able to accommodate V/STOL aircraft, such as the AV-8A Harrier or even the older AV-10 Bronco types. The ship is armed with two eight-cell launchers of Sea Sparrow SAMs, and three 54-mm and six 30-mm guns.

The strengths of NATO and WTO amphibious forces—both ships and manpower—are shown in table 7. The figures indicated here prove that there exists a strong numerical superiority of NATO amphibious forces over the WTO. The latter's forces are mainly composed of small landing craft, useful only in close-water operations along the shorelines close to the borders of WTO countries. On the NATO side, the bulk of amphibious forces belongs to the United States, supporting the point that this country is the only one able to carry our mass amphibious operations around the world. Such a capability is all the more desirable to the United States since its reliance on overseas bases is shrinking all the time.

TABLE 7.—AMPHIBIOUS FORCES, NATO AND WTO COUNTRIES, 1978

	Approximate Full Load Displacement (Tons)	United States	Federal Republic of Germany	Greece	Italy	Netherlands	Norway	Portugal	Turkey
Ships and landing craft:									
Helicopter commando cruiser	12,000–20,000	...	...	...	...	...	...	...	...
Amphibious command ship (LCC)	19,000	2	...	...	...	...	...	...	...
Amphibious assault helicopter ship (LHA, LPH)	18,000–40,000	10	...	...	...	...	...	...	...
Landing dock, cargo, transport (LPD, LPA, LKA)	10,000–20,000	36	...	1	...	...	...	...	...
Landing ship	8,000	23	...	...	2	...	...	...	...
Landing ship	4,000– 6,000	...	...	10	...	...	...	...	2
Landing ship	1,000– 2,000	...	...	5	...	...	...	...	...
Landing ship	300– 1,000	...	20	6	19	7	...	1	35
Landing craft	20– 200	108	28	47	61	...	12	9	43
Specialized marine forces	...	192,000	...	2,000	1,700	2,000	...	...	...

	United Kingdom	France	Total (NATO + France)	USSR	Bulgaria	German Democratic Republic	Poland	Romania	Total (WTO)
Ships and landing craft:									
Helicopter commando cruiser	2	1	3	...	...	...	...	...	...
Amphibious command ship (LCC)	...	...	2	...	...	...	...	...	...
Amphibious assault helicopter ship (LHA, LPH)	...	...	10	...	...	...	...	...	...
Landing dock, cargo, transport (LPD, LPA, LKA)	2	...	39	...	...	...	...	...	...
Landing ship	...	2	27	...	...	...	...	...	...
Landing ship	7	5	24	20	...	...	...	...	20
Landing ship	2	2	9	74	...	4	23	...	101
Landing ship	3	11	102	99	18	8	...	...	125
Landing craft	82	36	426	135	...	12	15	42	204
Specialized marine forces	7,750	4,500	208,950	14,500	...	2,100	7,000	...	23,600

SOURCES.—See table 1.

Approximate Equilibrium

The emerging picture of the naval balance between the NATO and WTO forces does not corroborate with the doomsday-like descriptions of the rapid decline of Western naval capabilities. It is more appropriate to say that an approximate equilibrium—to use an expression from the SALT lexicon—exists as far as the numbers of ships are concerned and when the reliance on, and importance of, the sea lines of communication of the two alliances are taken into account. The numerical preponderance of NATO over the WTO in warships persists if amphibious ships are also taken into account but vanishes if coastal defense ships are additionally included. A rough balance also exists in the offensive and defensive capabilities of ships, to the accompaniment of a steady expansion of NATO ASW, SSM, and SAM capabilities. The total number of ships largely remains stable, since the more sophisticated and capable ships tend to be restrictively costly, smaller numbers thus being procured by either side for the same amount of money. Therefore, the mainstream of the contemporary naval arms race is the technical sophistication of ships and their weapon suites, with a strong reliance on the software side of weapon technology.

Developments in naval weapon systems provide a good example of the evolutionary character of the present qualitative arms race. New types of naval weapons do not differ much in their external characteristics from their predecessors. Where they do differ is in the greater efficiency of action of their elementary components: more energy-rich composition of fuel; greater engine efficiency; higher accuracy of guidance packages; greater resistance to electronic, optical, and other countermeasures; greater adaptability to various firing platforms; and smaller size and lower weight. In addition to these features of weapons per se, several technological improvements have taken place in entire weapon systems in which the weapons are only links in a whole chain—from remote surveillance, discovery, and location of a target, through quick and accurate delivery of the weapon in order to execute the attack and destruction of the target, to the assessment of the postattack damage, and possible reattack. The effective integration of all these stages and their technical perfection is undertaken by all the parties concerned, with the highly industrialized countries setting the pace. The technological improvements are based on a wide military research and development (R & D) basis. Though costly, military R & D allows great economies to be made by decreasing the numbers of weapons without any corresponding loss in their efficiency. Savings made in this way are, however, of short duration. Once a weapon has been procured, efforts begin to look for a counterweapon. Moreover, once a state obtains possession of the weapon in question, other states feel obliged to follow suit, thus starting the numbers game. As a result, countries end up with comparable arsenals of highly sophisticated—and hence costly—weapons. Thus, the U.S. tenet of matching quantity with quality yields only a short-term advantage.

It is a dubious and even dangerous policy to believe, according to the following line of reasoning, that a dominant position can be attained by pursuing intensive technological programs: "Recognizing that modern technology offers us and our allies the opportunity to dominate the oceans without necessarily building vast fleets of ships is the key to making favorable changes in our naval forces."[6] Naval balances seem to be based on far too many components for any side to be able to achieve clear superiority across the board in all technological aspects of naval warfare.

WORLDWIDE TRENDS IN THE NAVAL ARMS RACE

The relative position of the nonaligned European and other developed countries has not changed during recent years in the global naval context. The major naval possessions of these countries are indicated in table 1H (p. 676) concerning the stock of aircraft carriers, table 4H (p. 678) on patrol submarines, table 5H (p. 679) on coastal submarines, and table 6H (p. 679) on world stock of major surface warships.

What is, however, worth noting in the general worldwide naval trends is that, first, the small-displacement, missile-armed fast patrol boat (FPB) rises quickly to the position of most fashionable naval vessel; and second, variously armed FPBs and other modern coastal patrol craft proliferate rapidly throughout the world and especially among the fleets of the developing nations.

The prohibitive costs of modern large-displacement warships impose on ship designers and users the necessity of seeking more economical solutions—generally in the form of reduced displacement—yet high capability. The FPB is the best example of how technology can offset the effects of current economic trends on warships, which force designers to produce ships of lower tonnage. A FPB costs about one-fifth of a missile frigate. Despite the smaller size of FPBs, their performance and strike power has been brought up to the former level of much larger ships such as frigates or destroyers. According to some, the advent and development of missile-armed FPBs had as great a consequence for naval warfare as did the application of nuclear propulsion to submarines.

The FPBs are evolving from small-size and single-purpose vessels to larger, heavily armed, multipurpose warships. This trend is exemplified in the French Combattante III, West German Type 143, and Israeli Reshef-class FPBs. New FPBs have improved combat capability owing to the provision of better protection against nuclear radiation and chemical weapons, as in the Swedish Spica II and the two previously mentioned West German and French designs. Moreover, advances in gun automation and light modular designs of SAMs

6. S. J. Deitchman, "Designing the Fleet and Its Air Arm," *Astronautics and Aeronautics* 16 (November 1978): 29.

have, for the first time, given FPBs a good self-defense capability against air threats. The attainment of this capability by FPBs is remarkable. Thus, FPBs can now operate all types of weapon systems: torpedoes, guns, and various missiles—the make-up of the weapon suite being determined by the customer's requirements.

Available figures clearly show that in the last decade there has occurred a steep rise in the numerical and qualitative characteristics of the worldwide light naval forces (table 7H, p. 679). It is best visualized in table 8. The term "light naval forces" as it is used here comprises several different vessels: corvettes, coastal escorts, gunboats, large and coastal patrol craft (PC), coast guard and maritime safety agencies' armed vessels, and last but not least, all types of FPB. The common denominator of this class is a displacement of less than 1,000 tons and close-to-shore operation. From the purely military point of view, these vessels represent different types of weapon systems of largely varying capabilities—from missile and torpedo FPBs to slow gun- or machine-gun-armed patrol boats. However, being so different, all of them can be used equally well in quasimilitary or police-like patrols, or in purely military duties on vastly different operational scales.

There are several reasons for the expansion of the light naval forces. Those which can be specified by an outside observer can be listed as follow:

1. Since the early 1960s, the number of states has grown steadily with the decolonization process. Many of these states have long coastlines, usually unprotected by naval forces. It is therefore natural, although costly, for these states to acquire such forces. Expansion of these forces was at first confined to slow patrol craft, the numbers of which have been growing throughout the last decade at about 10 percent yearly. Among these states, slowly developing their economies and armed forces, a strong demand for the further growth of coastal naval forces will exist in the future.

2. A number of the Third World countries find themselves embroiled in conflicts, some of them open, others only potential, driving them into extensive arms purchases including the acquisition of naval vessels. This is best exemplified by countries in the Middle East, in South Asia, and in the Far East.[7] It is indicative that these regions were the first among Third World countries to acquire large numbers of modern FPBs. Even after they had acquired large numbers of these vessels in the first half of the last decade, the rate of growth of their light forces is still one of the fastest in the world. The Middle Eastern countries more than doubled their large patrol craft forces (excluding FPBs), and expanded their missile-armed FPB inventory by nearly two-and-a-half times (including outstanding orders) and all types of FPBs by half during the short span of the past 4 years. A similar high rate is exhibited by countries in

7. Middle East—Cyprus, Egypt, Iran, Iraq, Israel, Kuwait, Lebanon, Saudi Arabia, and Syria; South Asia—India, Pakistan, and Sri Lanka; Far East (excluding Japan)—Burma, China, Indonesia, Khmer Republic, Malaysia, Singapore, North Korea, South Korea, Taiwan, Thailand, the Philippines, and Vietnam.

TABLE 8.—NUMBER OF COUNTRIES DEPLOYING MISSILE-ARMED FPBs, AND NUMBER OF MISSILE-ARMED FPBs DEPLOYED: 1960–78 AND 1978 INCLUDING ORDERS

	1960	1965	1970	1975	1978	1978+
Number of countries	1	7	17	33	45	53
Number of FPBs	5	141	282	398	691	811

SOURCES.—See table 1.

South Asia and the Far East (including China, North Korea, and South Korea). The common feature of these acquisitions is that they are oriented to craft most suitable for purely warfare operations, possessing high firepower, great maneuverability, and quick reaction time. Unless the regional tensions and open conflicts are resolved, the acquisitions of this type of naval force will certainly continue.

3. A striking lesson of the military effectiveness of the missile-armed FPBs was provided by their successful use in the 1967 Middle East War and in the 1971 Indo-Pakistani War. With the delayed realization of the potential of fast, small yet large firepower vessels, and pressed by the skyrocketing costs of large modern warships, West European countries undertook extensive FPB production programs. Strangely enough, countries which have the biggest shipbuilding potential and which produce modern FPBs are not necessarily identical with those which procure them for their own navies. This has partly to do with their strong feelings about their insular position—as in the case of the United Kingdom and the United States—in relation to naval threats. Such beliefs tend to emphasize the importance of larger vessels as a means of defense against sea-borne attack. Even more important, these states still enjoy naval supremacy and are not therefore worried about their coasts—on the contrary, historically speaking they were used to projecting their military power far away from their shores. It seems, however, that even those countries which have not yet begun to procure FPBs for their own navies, will soon do so, given the growing effectiveness of these vessels. This is true of the United States, the United Kingdom, and France, all possessing great capacity and technical expertise in constructing modern FPBs and—especially in the case of the United Kingdom and France—in selling them in large quantities abroad.

Among the West European countries which have procured large FPB forces, either indigenously built or bought, are the Federal Republic of Germany, Norway, Denmark, Sweden, Greece, and Yugoslavia. The WTO member countries, excluding the USSR, developed their light naval forces in the 1960s, and after 1970 procured rather small quantities. Europe as a whole was the fastest growing market for FPBs, and this tendency still holds.

4. So far the major reason for the acquisition of lightweight naval forces has been to fulfill the military interests of states. At present, an additional stimulus for enlarging these forces is provided by the growing exploitation of

the seabed and by the extension of states' exclusive fishery zones. An increasing number of countries justify their new purchases of patrol vessels of various kinds on the grounds of the necessity for protecting their extended interests at sea. Further extensions of territorial waters by certain countries are expected and the establishment of so-called exclusive economic zones (EEZs) at sea is certain, despite the fact that United Nations Conference on the Law of the Sea (UNCLOS) has not yet clarified the legal regime of the seas. Because of the fact that their actions are not as yet regulated by international law, states are anxious to secure their interests by means of naval force. Modern FPBs are the best answer to these needs, since they provide a strike potential deterring much larger naval forces than the less up-to-date coastal military vessels. Thus, a further rapid growth of their number must be expected.

5. Disregarding regions of open hostilities, there exist several other regions where a dynamic naval arms race is taking place, connected with the phenomenon of the rise of new "regional superpowers." The most striking example was provided by Iran, with its ambitious plans for arming. The official reason given was the need for counteracting the growth of leftist states in the region and for filling the alleged "strategic vacuum" left by the withdrawal of the British forces guarding Western interests in the Persian Gulf. These extensive programs, if they had been carried out, would without doubt have made Iran the most powerful state in the region. Saudi Arabia attempted to follow suit by ordering from the United States in 1977 nine corvettes armed with Harpoon missiles, guns, and torpedoes, and several minesweepers; from France four missile-armed FPBs; and from the United Kingdom a large number of gun-armed coastal patrol craft and several dozen other patrol craft.

Another example of similar development is Nigeria which recently ordered six missile-armed FPBs from the Federal Republic of Germany and France. These boats, of Type 143 and La Combattante III-class, are armed with Exocet antiship missiles, wire-guided torpedoes, two automatic 76-mm guns, and two twin 30-mm guns. These craft, together with the corvettes and patrol craft already in possession of the Nigerian naval forces, will make the country an indisputably superior force in the region.

Yet another development of this kind is taking place in the Indian Ocean where India is also building up new powerful naval forces, apart from expanding its ocean-capable fleet by procuring a number of new frigates and destroyers. In the last 2 years or so, India has acquired eight new Osa-class missile-armed FPBs and as many as six Nanuchka-class corvettes armed with SS-N-11 missiles. India also plans to strengthen its local naval shipbuilding capabilities, both in surface warships and submarines—all this in order "to meet any contingency by creating a naval force equal in size and competence to the naval forces of any one of the superpowers normally operating in the area."[8]

8. A. K. Chatterji, "Indian Navy in the Eighties," in *The Chanakya Defence Annual 1978*, ed. R. Kaul (Allahabad: Chanakya, 1978), p. 89.

6. It is sometimes difficult to find a logical explanation for the creation by some states of an ultramodern if sparse naval force consisting of a few FPBs, when in the surrounding region no other state possesses such craft. This applies particularly to a number of African countries. It is probable that the best explanation for their early acquisitions of FPBs is the prestige attached to these modern, electronically sophisticated vessels. Now, when new reasons are being put forward for the expansion of coastal patrol forces, several other states will follow suit, partly justified in their acquisitions by the actions of their neighbors.

7. Several navies operate numerous but dated patrol forces which are not suitable for extended patrols further offshore. Since the supply potential is vast, it is reasonable to expect vigorous modernization and replacement, most probably connected with the desire to increase the technical qualities and hence the military potential of these forces. This may be particularly true in the case of the Latin American and East Asian countries.

8. Among the reasons for the rapid expansion of light naval forces are the political, military, or economic interests of several industrialized states in supplying these types of ships to other countries. In the case of the Soviet Union and China, these interests are usually connected with the political or ideological affiliation of the recipient country. Apart from such political motivations, it seems that the upsurge of popularity of FPBs may partly be a response to the marketing activities of the Western and other industrialized countries (e.g. France, the Federal Republic of Germany, the United Kingdom, and Israel), which through this new fashion in armaments expect to gain greatly in orders for their otherwise not-too-busy shipyards. By competing in the market, the potential suppliers accelerate the technological naval arms race among the Third World countries.

All of these above-mentioned factors act more or less simultaneously. As table 8 indicates, the cumulative effects of these factors are staggering. Whereas in 1960 only one state possessed FPBs, in 1970 17 countries, and in 1978 more than one-third of the 122 countries having direct access to the sea—45 to be precise—were already operating FPB forces.

Available figures indicate that in the naval arms race, light vessels have now gathered momentum. The several factors indicated above will continue to urge states throughout the world to acquire new modern patrol vessels. This development cannot fail to aggravate regional conflicts and tensions. It will also deprive several countries of much needed resources, despite the fact that the FPBs and other small patrol vessels are cheaper to buy and to maintain in comparison with larger vessels. The poorer states must measure the costs according to their resources and not to the theoretical cost of large modern warships. In addition, the highly sophisticated equipment and weapons on several vessels will increase the dependence of less advanced states on spare parts and repair services from the supplying states.

Ocean-based Nuclear Deterrent Forces and Antisubmarine Warfare

Owen Wilkes
Stockholm International Peace Research Institute

INTRODUCTION

Nuclear deterrence is based on the assumption that no nation will contemplate nuclear attack on another if the nation to be attacked has an assured capability to respond with prompt and massive destruction of the attacking nation's cities.

A capability for mutually assured retaliation requires that at least some nuclear weaponry (the second-strike forces) of each side be deployed in such a way as to be immune to a first strike by the other side. For this reason most land-based ICBMs are hidden in underground silos which can withstand the sort of overpressures generated by nearby explosions. However, the most important way of guaranteeing invulnerability has been for each side to keep a proportion of its deterrent forces hidden within the oceans on board nuclear-powered submarines.

The invulnerability of the submarine-launched ballistic missile (SLBM) deterrent rests on two important assumptions. These are (*a*) that communication systems can reliably convey missile-launch orders to submarines even in the chaos accompanying the outbreak of general nuclear war, and (*b*) that the nuclear submarines are essentially undetectable and, hence, untargetable.

The first assumption is open to doubt. Despite the enormous research effort that has gone into seeking more secure communications modes, there are serious doubts as to whether communications really are reliable enough to guarantee that during a nuclear exchange the orders for a retaliatory attack will actually get from command authorities to individual submarines.[1]

The second assumption is concerned with the question of just how effective antisubmarine warfare (ASW) detection systems are becoming. Although there is no breakthrough imminent which is suddenly likely to render the oceans militarily transparent, there is evidence of a number of lines of technological progress currently coming to fruition which will, taken together, offer a high probability of detecting missile submarines anywhere in the ocean.

1. Stockholm International Peace Research Institute, *SIPRI Yearbook of World Armaments and Disarmament, 1979* (London: Taylor & Francis, 1979), chap. 7.

© 1980 by The University of Chicago. 0-226-06603-7/80/0002-0012$01.00

Once detection and location are achieved, destruction is easy, and ASW weapon systems are already adequate for the task. Even if they are not already operational, these combined developments may soon make it possible to detect, locate, and destroy all adversary missile submarines within a time period so short as to effectively eliminate the adversary's sea-based retaliatory capability.

If invulnerability of both the submarines and their communications systems cannot be guaranteed, then there will be an increased temptation to adapt the SLBM system as a whole for use in a first-strike counterforce role. In other words, the SLBMs may be targeted against the missile silos and other strategic weapons of the other side instead of against cities (or "countervalue" targets). A counterforce doctrine is inherently destabilizing in that it creates pressures for both sides to launch preemptive attacks, while the second-strike doctrine, despite all its faults, does have defensive connotations and does seem to have had a stabilizing effect.

U.S. AND SOVIET MISSILE SUBMARINE FLEETS

The United States has 41 operational ballistic missile–carrying nuclear-powered submarines (SSBNs), about half of which are on station and ready to fire their missiles at any particular time. Each submarine carries 16 missiles. The original Polaris missile, first deployed in 1960, had a range of 2,000 km and a single reentry vehicle (RV). The still-current Polaris A3 has a range of up to 4,600 km, with each missile generally carrying three RVs of about 200 kt each.[2] Only 10 submarines still carry Polaris missiles; the remainder have been converted to Poseidon. Poseidon has about the same range as the Polaris A3 but is much more accurate and carries multiple independently targetable reentry vehicles (MIRVs) each of about 50-kt energy yield. The use of MIRVs allows one missile to be used against 10–14 separate targets, and the United States can now land warheads on 5,440 separate targets using missile submarines alone. Each of these targets would be subjected to a blast about three times as powerful as that delivered by the 1945 Hiroshima bomb.

The Soviet Union has about 80 ballistic missile submarines (of at least four different types) carrying at least five different types of missile. There are still perhaps as many as 19 of the 1960 vintage Golf-class diesel-powered submarines carrying SS-N-4 or SS-N-5 1,300-km range missiles with 1–2-Mt warheads.[3] There are seven Hotel-class submarines which are similar to Golf but nuclear powered. The Yankee-class, first deployed in 1968 and with about 30 now operational, is a large submarine comparable to the U.S. Polaris boats, each carrying 16 SS-N-6 (3,000-km range) missiles with 1–2-Mt warheads. A

2. kt = kiloton, i.e., the amount of high explosive measured in thousands of tons which yields an equivalent amount of energy.

3. Mt = megatons, i.e., millions of tons (see n. 2 above).

later version of the SS-N-6 carries three RVs on each missile. The Delta-class submarine first went on patrol in 1974. The earlier version, of which there are 14, carries 12 missiles. A "stretched" version, the Delta II, carrying 16 missiles is now in production, and about 12 of these are now in the water. The Delta submarine carries SS-N-8 missiles, of about 8,000-km range. This range allows submarines to be within range of targets in both the United States and China while cruising in the Barents Sea.

The oldest U.S. submarines are now reaching retirement age and are due to be replaced by a new class of submarine carrying the Trident missile. The new submarine will carry 24 missiles and will cruise at greater depths, at higher speeds, and with less emission of noise than present submarines. The Trident I missile, currently under test, will be notable for its 7,000-km range, comparable to that carried by the Soviet Delta submarines, and each missile will carry up to 17 MIRVs. Trident I will be backfitted to at least 10 of the existing Poseidon boats beginning in 1980. Development of a Trident II missile is also under way. Trident II will be a bigger missile which will fit only Trident submarines, and its most important feature will be the high accuracy of its mid-course and/or terminally guided RVs.

The Soviet Union is also testing new missiles. The SS-N-X17, a follow-on to the SS-N-6, of about 5,000-km range, will probably carry a payload of three 1–2-Mt MIRVs. The SS-N-18 is a similarly MIRVed follow-on to the SS-N-8. While all earlier Soviet SLBMs had storable liquid propellants, the SS-N-6 and SS-N-18 will have solid propellants as have all U.S. SLBMs from the beginning.

If only the figures for numbers of submarines are examined (table 1), the Soviet Union appears to have a distinct advantage—of about 80 to 40. Even allowing for the geographic disadvantages of the Soviet Union, these figures are significant—the large number of Soviet submarines could make U.S. preemptive destruction more difficult. If numbers of missiles are compared, the Soviet Union again appears to have an advantage—972 to 656. In terms of RVs, however, this advantage is reversed—the United States has about 5,000 as compared with about 1,500 for the Soviet Union. This ratio gives a reasonable indication of the relative "countervalue" capabilities of the two great powers—the United States can inflict damage greater than that inflicted on Hiroshima on about 5,000 targets, while the Soviet Union can inflict much greater damage on about 1,500 targets (a 1-Mt explosion will destroy all housing in an area of about 50 km^2). The greater energy yield of the Soviet warheads is more or less balanced by the greater accuracy of the U.S. warheads.

If counterforce capabilities are to be compared, then somewhat different considerations apply. Here the weapon must deliver a large amount of explosive energy to a very small target, and therefore accuracy becomes much more important than for countervalue targets. The counterforce effectiveness of a warhead can be summarized in the concept of "lethality" or K, derived from the equation

$$K = \frac{Y^{2/3}}{CEP^2},$$

TABLE 1.—U.S. AND SOVIET BALLISTIC MISSILE SUBMARINES 1978

Submarine Type	No. Submarines	Total No. of Missiles	Explosive Yield and No. of RVs per Missile	Total No. of RVs (ca.)
United States:				
Polaris	10	160	3 × 200 kt or 1 × 1 Mt	320
Poseidon	31	496	10 × 50 kt (ca.)	4,960 (ca.)
Total	41	656	...	5,280
Soviet:				
Golf (diesel)	19	57	1 × 1–2 Mt	57
Hotel	7	21	1 × 1–2 Mt	21
Yankee	34	544	3 × (kt range) or 1 × 1–2 Mt	1,080
Delta	26	350	1 × 1–2 Mt	350
Total	86	972	...	1,508

SOURCES.—International Institute for Strategic Studies, *The Military Balance 1978–79* (London: International Institute for Strategic Studies, 1978), pp. 80–81; R. T. Pretty, ed., *Jane's Weapon Systems* (London: Jane's Yearbooks, 1977).

where Y = yield in Mt, and CEP = accuracy (circular error probable) in nautical miles.[4] Table 2 gives approximate figures for the lethalities of the SLBM forces of the two great powers which indicate the vastly greater effectiveness of the U.S. SLBM arsenal.

A further factor that should be taken into account, especially when considering the possibility of a first strike, is *readiness,* that is, the number of submarines, or missiles, which are ready for use at any particular time. The United States has on average over 50 percent of its submarines at sea and will have 65 percent at sea when Trident is deployed. The Soviet Union, however, according to U.S. figures, has only 15 percent of its submarines at sea on average, and never demonstrates a "surge capacity" of putting more to sea at any particular time. Assuming that the numbers of submarines of different classes at sea are in proportion to the numbers of submarines in each class, this would seem to indicate that if general nuclear war were to break out suddenly and without warning, the United States might have about 2,600 SLBM RVs safe from a first-strike attack, while the Soviet Union might have as few as 166.

The question of readiness also has important implications in assessing for each side the magnitude of the problem of preemptively destroying the other side's missile submarines. If there were no advance warning, a preemptive attack by the United States would only have to locate and destroy about 10 submarines, while a similar attack by the Soviet Union would have to locate and destroy about 20.

DEVELOPMENTS IN DETECTION TECHNOLOGY

Detection of submarines is primarily achieved by the use of sonar, that is, by sensing underwater sound waves reflected from or generated by submarines.[5] Sound waves travel very long distances underwater. However, sonar detection suffers from three limitations. First, the velocity of sound in sea water is extremely low (compared with light waves) and highly variable because it is influenced by salinity, temperature, currents, and other factors. Second, the paths followed by sound waves in sea water are difficult to predict because they are subject not only to multiple reflections at the surface and on the sea floor but also to refraction between layers of different density and absorption by chemical, biological, and gaseous constituents of the sea water. Third, the ocean is a very noisy place due to abundant marine life, crashing waves, rumbling underwater volcanoes, increasing volumes of merchant shipping, and, more recently,

4. Stockholm International Peace Research Institute, *Offensive Missiles*, Stockholm Paper no. 5 (Stockholm: Almqvist & Wiksell, 1974).

5. For an additional discussion of antisubmarine warfare, see "The ASW Problem: ASW Detection and Weapons Systems," *Ocean Yearbook 1,* ed. Elisabeth Mann Borgese and Norton Ginsburg (Chicago: University of Chicago Press, 1978), pp. 380–85.

TABLE 2.—HARD TARGET LETHALITY OF SLBMs (1978)

Missile	Accuracy of RV (naut mi) (*CEP*)	Explosive Yield of Warhead (Mt) (*Y*)	Lethality per RV (*K*)	No. of RV per Missile (*n*)	Total No. of Missiles (*m*)	Total Lethality of Missiles Force (*K*)×(*n*)×(*m*)
United States:						
Polaris A3	.5	.2	1.37	3	160	658
Poseidon C3	.3	.04	1.30	10	496	6,448
Total	...	...	...	...	...	7,106
Soviet:						
SS-N-4/5	2.0	Assume 1	.25	1	78	20
SS-N-6	1.5	1.0	.45	1	544	245
SS-N-8	.8	1.0	1.56	1	350	546
Total	...	...	...	...	...	811

SOURCES.—Stockholm International Peace Research Institute, *Offensive Missiles*, Stockholm Paper no. 5 (Stockholm: Almqvist & Wiksell, 1974); R. C. Aldridge, *The Counterforce Syndrome: A Guide to U.S. Nuclear Weapons and Strategic Doctrine*, Pamphlet Series no. 7 (Washington, D.C.: Transnational Institute, 1978).

increasing seismic exploration and offshore oil-drilling activities on the sea floor. These noises tend to mask those of the submarines or may be misinterpreted as noises made by submarines. Advances in sonar detection of submarines have been primarily achieved by overcoming these limitations. The following sections describe the progress made in the United States.

Surveillance Sonar Arrays

Long-range detection of submarines is mainly carried out by means of large fixed sea-bottom arrays of hydrophones that passively listen for sounds generated by submarines. These arrays are individually and collectively known as a sound surveillance system (SOSUS). There were stated to be 22 SOSUS installations around the world in 1974.[6] Both coasts of the United States are supposed to be covered by SOSUS, and they are deployed in other areas in which Soviet submarine movements are particularly intense or restricted. The global distribution of SOSUS and related arrays is shown in figure 1.

Each SOSUS installation consists of an array of hundreds of hydrophones laid out on the sea floor, or moored at depths most conducive to sound propagation, and connected by submarine cables for transmission of telemetry. In such an array a sound wave arriving from a distant submarine will be successively detected by different hydrophones according to their geometric relationship to the direction from which the wave arrives. This direction can be determined by noting the order in which the wave is detected at the different hydrophones. In practice, the sensitivity of the array is enhanced many times by adding the signals from several individual hydrophones after introducing appropriate time delays between them. The result is a listening "beam" that can be "steered" in various directions toward various sectors of the ocean by varying the pattern of time delays. The distance from the array to the sound source can be calculated by triangulation.

The SOSUS had its beginnings in 1952 when attempts were first made to exploit the sound duct that lies between the sun-heated surface layers of the ocean and the deeper, permanently frigid water below. The first arrays, called Caesar, had no beam-forming capability and functioned simply as detection barriers with several hydrophones per kilometer of barrier. In the 1960s, additional barriers were built along the Pacific coast, in much deeper water, and in several overseas locations under various names such as Trident, Artemis, Barrier, and Bronco. There was a continuous improvement program. Early Caesar had poor range-estimation capability and had to be supplemented either by explosive sound sources for calibration or by sonar picket ships. In

6. U.S. Congress, House, Committee on Appropriations, *Hearings on Department of Defense Appropriations, Fiscal Year 1975*, 93d Cong., 2d sess., 1974, pt. 3, p. 444.

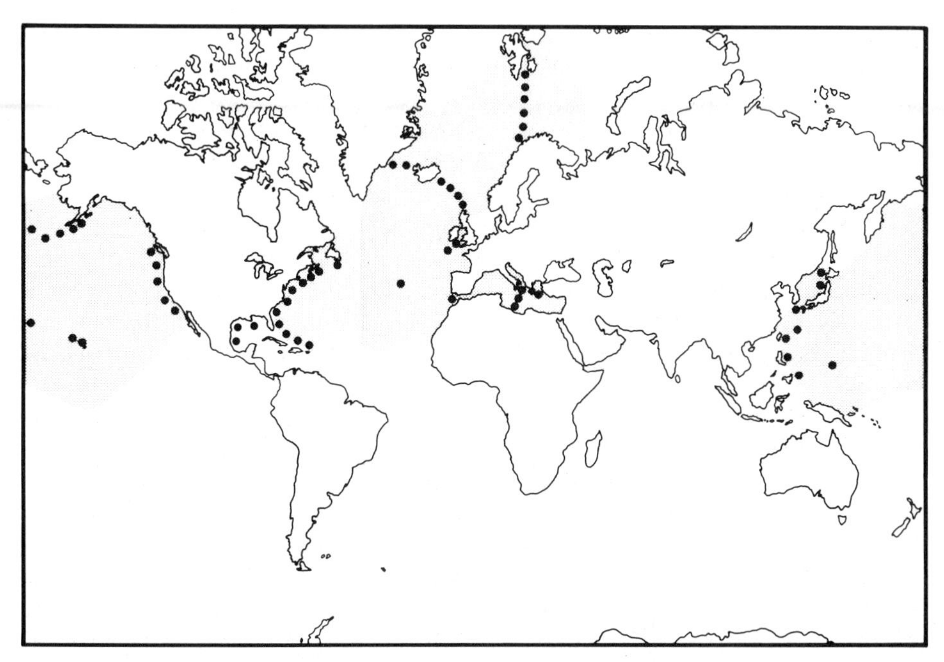

Fig. 1.—Known and probable locations of U.S. and allied sea-bottom sonar arrays

general, the early systems were able to locate submarines at distances of up to about 150 km.[7]

In the early 1960s, techniques for array signal processing began to be introduced at the shore monitoring stations and the detection range was extended to several hundred kilometers.[8] Such increased sensitivity, however, led to problems in identifying all the sound sources detected. The solution was to develop an integrated surveillance system combining sea-bottom arrays, surface ships, and aircraft.

Currently the detection and identification range of sea-bottom arrays is being further extended under a major Defense Advanced Research Projects Agency (DARPA) program called Project SEAGUARD, which is focused on three areas: (1) large acoustic array technology—improvements in hydrophones, telemetry, and mooring techniques; (2) signal processing—new array processing techniques, automatic search, detection, and recognition; and (3) ocean hearing—establishing further the spatial and temporal variation of all

7. "Undersea 'R & D' Covers Up for Caesar's Ghost," *Electronics* 42, no. 21 (October 13, 1969): 66; "Navy Testing Antisub Alarms," *Electronics* 34, no. 2 (January 13, 1961): 26–27, 29.

8. L. Booda, "ASW: A Holding Action?" *Sea Technology* 19, no. 11 (November 1978): 12.

the factors that influence sound propagation.[9] Overall, Project SEAGUARD is intended to determine the fundamental physical and technical limitations of acoustic surveillance.

In the field of array technology, DARPA has developed more sensitive hydrophones which are cheaper to deploy and which counter the effects of submarine silencing techniques. They are optimized to detect noise resulting from the submarine's passage through the sea rather than from engine noise. The signal processing study has involved feeding enormous amounts of acoustic data in real time from widely scattered sea-bottom arrays into what is probably the world's largest computer. This computer, called Illiac 4, has applied the signal processing techniques originally developed for seismic arrays to sonar array data.

It has been found that, unlike seismic propagation, acoustic propagation within the oceans is far more coherent than was formerly suspected. In other words, signal processing can filter out all other noise while amplifying the signal from a submarine, and submarines become potentially detectable from thousands of kilometers away. The quantities of data required are enormous, and assembling the data is in itself a large-scale process. The navy and DARPA have, however, been acquiring experience in this using ARPANET, a network of large computers of diverse types linked by high speed, high volume data links. Its operation requires millions of messages a minute in a complex pattern from sensors to computers, between computers, and from computers back to sensors. The seemingly insoluble "switchboard" problem was solved by the development of a technique called packet switching.

Airborne Surveillance

The P3 Orion provides the basis of U.S. airborne antisubmarine capability. The Orion is a four-engined aircraft capable of flying 2,500 km, patrolling for 4 hr in search of a submarine, and returning to base. It is equipped with a variety of submarine detection systems but relies mainly on sonobuoys.

The P3 Orion has undergone continuous improvements in the 18 years it has been operational, and the current P3-C version carries over 300 "black boxes," that is, 300 discrete electronic systems performing various detection and navigation functions. The Orion is now equipped to use active sonobuoys which determine the azimuth as well as the range of echoes from submarines, and current passive sonobuoys which monitor 10 times the frequency spectrum of earlier models. In an updating program which began in 1974, the Omega

9. U.S. Congress, Senate, Committee on Armed Services, *Authorization for Military Procurement, Fiscal Year 1978: Hearings on S. 1210,* 95th Cong., 1st sess., 1977, pp. 6179, 6188–89, 6224, 6292.

navigation system was fitted to the Orions. This resulted in considerable augmentation of Orion capabilities by limiting to within 3 km the formerly quite large navigational errors that built up during 10-hr missions while relying on inertial and Doppler navigation aids which are subject to cumulative errors over time.

In a current updating program, aircraft are being fitted with improved electronics for fixing the positions of sonobuoys relative to the aircraft and for recording data from them.

The next cycle of improvements to the Orion, called Update III, will, it is claimed, provide the best ASW aircraft in the world. This will be largely due to the provision of an advanced signal processor called Proteus, which will process data from the sonobuoys and other sensors to fix positions of target submarines and determine the best offensive measures. A new, fully integrated and programmable communications suite will provide simultaneous links via satellite and other radio nets. An improved magnetic anomaly detection (MAD) unit, using cryogenic SQUID magnetometers, will double the range at which submarines can be detected by their magnetic signature.

The United States has 400 P3 Orions, of which over 22 are in active service. They are based worldwide (fig. 2). Assuming a usual operating range of 2,300 km, the map shows the area of ocean accessible to U.S. Orions, the areas

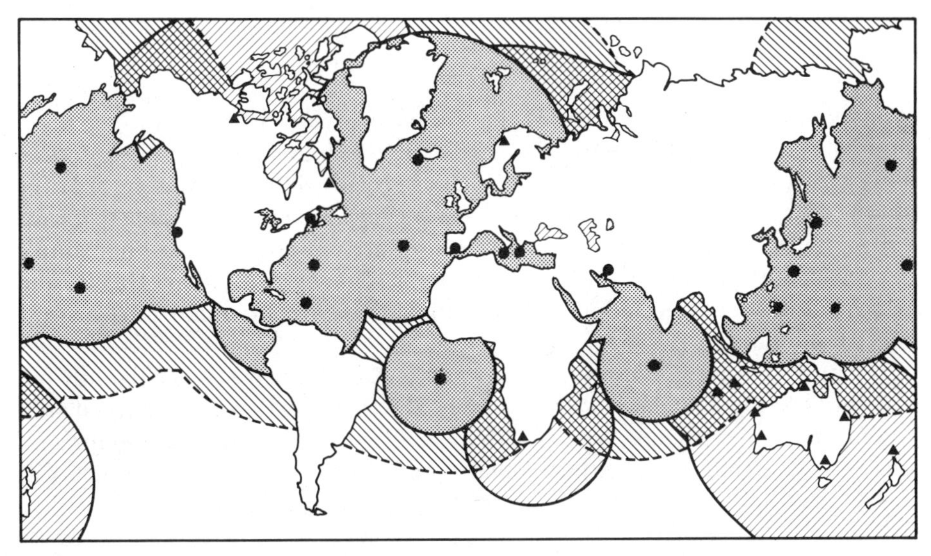

FIG. 2.—U.S. and allied ASW airborne reconnaissance. Dots indicate airfields from which U.S. Orions operate. Triangles indicate airfields of U.S. allies. Dark, shaded areas indicate Orion ASW coverage assuming 4 hr on station. Areas bounded by dashed line indicate additional coverage if no time is spent on station or if aerial refueling were to become available. Areas bounded by solid line indicate additional coverage provided by U.S. allies.

covered by the ASW aircraft of U.S. allies, and the extra area that could be covered if shorter patrol times were acceptable or if aerial refueling capability were added (something which is under consideration). It can be seen that, without assistance from allied or aerial refueling, an area of about 51.5 million km^2—including all ocean areas in which Soviet missile submarines are likely to be found—is covered. A single Orion can search 725,000 km^2 in the course of one mission. The capabilities of the Orion, however, are best used not in wide area searches but in narrowing down the location of a hostile submarine already detected by SOSUS-type fixed arrays. The SOSUS arrays are said to be able to locate a submarine within a circle of about 50-km diameter, and it is widely believed that they can locate to within 15 km or so. Thus an Orion or other ASW plane need only search an area of a few hundred square kilometers which SOSUS intercepts have indicated contain a submarine. The aircraft are sent out to such areas on what are known as vectored intercepts. Current improvements in SOSUS accuracy and reliability will allow ASW aircraft to be more accurately vectored to their targets, and fewer air missions would need to be flown.[10]

There are many other aircraft besides the U.S. Orions involved in hunting Soviet and Warsaw Treaty Organization (WTO) submarines. There are something like 600 Orion-type aircraft in NATO and to this must be added the aircraft of countries such as Australia, New Zealand, and Japan which share the task of patrolling the open ocean and monitoring ocean choke points.

Other Developments

The sonobuoys sown from aircraft have limited lifetimes, are very much more limited in range, and have less directional discrimination than the sea-bottom sonars. Sea-bottom sonars can only be laid in relatively secure waters, near the shores of friendly states, and in time of war they are very vulnerable to a wide range of countermeasures. In particular they are exposed to the rather simple process of cutting marine cables, which can be done by an ordinary trawler and which often occurs accidentally. Two innovations are intended to overcome these limitations.

The first is the use of towed sonar arrays. The United States currently has six towed array surveillance systems (TASS) in operation. The surveillance towed array surveillance system (SURTASS), currently under development, incorporates technology with capabilities an order of magnitude better than TASS and comparable with those of the sea-bottom arrays. The SURTASS data will be relayed to shore processing sites by satellite. The U.S. Navy is currently procuring three SURTASS units, each valued at about $33 million. In order to

10. U.S. Congress, Senate, Committee on Armed Services, *Authorization for Military Procurement, Research and Development, Fiscal Year 1976: Hearings on S. 920,* 94th Cong., 1st sess., 1975, p. 2976.

operate in the low frequency part of the acoustic spectrum, SURTASS requires a very long array which can be towed only at very slow speeds.

The second innovation is the rapidly deployed surveillance system (RDSS). This consists basically of a very large sonobuoy of the same dimensions as a Mark 46 torpedo and capable of being delivered by any vehicle that can carry a torpedo. The RDSS can thus be dropped from Orions, B-52 bombers and various carrier-based aircraft, or it can be more secretively positioned by release through the torpedo tubes of a submarine. The RDSS has obvious advantages for surveillance operations in such high interest areas as ocean choke points and the approaches to Soviet SSBN bases. It is described as intended for use in crisis areas requiring quick reaction surveillance.

Once delivered, the RDSS deploys a string of hydrophones down toward the bottom and anchors itself. The moored sonobuoy system (MSS) will be relatively unaffected by future advances in submarine quieting, since it is optimized for the detection of noise resulting from water flow around submarine hulls and cavitation noise associated with submarine propellors rather than engine-room noise. A buoy at the surface periodically transmits compressed bursts of data which are so short that hostile direction finding on them is difficult, and these data are picked up by aircraft in the vicinity or, in the future, by satellite. Each RDSS buoy can remain active for up to a month or so, and it is responsive to radio command so that it can be turned on only when needed. Both SURTASS and RDSS will function as part of the integrated surveillance system of which SOSUS is the principal component. According to navy testimony to Congress, "The integration of SURTASS detection information with that obtained from other undersea-surveillance system elements will enable the dissemination of highly accurate, near real-time, evaluated target data for follow-up action by tactical ASW forces."[11]

The Proteus advanced signal processor being designed for the Orion is capable of handling RDSS data. Proteus will also be provided for other ASW platforms including surface ships and hunter-killer submarines. By using the same equipment in all platforms, the interoperability of and coordination between various ASW platforms are made possible. Coordination is further enhanced by providing Omega navigation aids to all ASW platforms, including hunter-killer submarines, which can receive very low frequency (VLF) signals at antenna depths of 15 m. This allows all platforms to navigate on a common navigational grid, in which relative errors are limited to about 200 m.

Global Oceanographic Monitoring

One of the key problems in the analysis of all kinds of sonar data, whether collected by sea-bottom, submarine-mounted, or sonobuoy sonar, is the determination of propagation parameters—how fast the sound waves are traveling

11. Ibid., p. 4469.

in the ocean, the extent to which they are being bent as they pass through various layers, the number of bounces between the surface and the floor of the ocean, and so on. Acoustic propagation is strongly influenced by the physical and chemical variables of sea water, and considerable effort has gone into the investigation of these variables and into devising systems for the global collection of synoptic information on them.

This experimentation began in the 1960s, at the same time as the potential long-range capabilities of sea-bottom sonar began to be recognized. In 1964, the U.S. Naval Ordnance Laboratory carried out Project Neptune in which 120 depth charges were dropped by ships and aircraft over wide ranges of ocean between Bermuda in the Caribbean and Perth in Australia. This experiment conclusively demonstrated long-range propagation—detonations near Capetown were heard 10,000 km away near New Zealand—with a primitive sonar array.[12] A major experiment was carried out in 1972 and aimed at achieving a near-synoptic calibration of the Pacific Ocean. In this experiment, called Kiwi One, 300 small depth charges were dropped in the course of a flight from Panama to New Zealand and back, with other aircraft dropping bathythermograph buoys so that oceanwide thermal profiles could be charted.[13]

There are numerous satellites which contribute to global monitoring of the oceans.[14] The list includes the Defense Meteorological Satellite Program (DMSP), the Tiros operational weather satellite, the Stationary Meteorological Satellite (SMS/GOES), the Landsat series, the Applications Technology Satellites (ATS), the Nimbus series, the Skylab manned observatories, Tiros-N, and Seasat. Still more are under development including Stormsat, the Applications Explorer Mission (AEM-4), the Remote Ocean Measurement System (ROMS), and the Synchronous Earth Observatory Satellite (SEOS).

The DMSP is a weather satellite originally launched to meet Air Force requirements in the Southeast Asian War. The current satellites have sensors of 0.6-km resolution and have proved excellent for mapping sea ice and determining sea-surface temperatures. Knowledge of sea-ice distribution is important not only for surface ASW operations but also for determining sonar-propagation conditions—some sea-ice undersurfaces tend to absorb rather than to reflect sound waves. The Tiros-N satellite is a civilian version of DMSP with military participation in its operation and in the use of the data. The navy operates its own ground terminal for the Geostationary Operational Environmental Statellite (GOES).

12. A. C. Kibblewhite, R. N. Denham, and P. H. Barker, "Long Range Sound Propagation Study in the Southern Ocean and Project Neptune," *Journal of the Acoustical Society of America* 38, no. 4 (October 1965): 629–43.

13. W. H. Thorp and W. R. Schaumacher, "Project Kiwi One Cruise Report," Naval Underwater Systems Center Report no. TD-4455 (Washington, D.C., February 1973).

14. See Bhupendra Jasani, "Ocean Surveillance by Earth Satellites," in this volume, pp. 250–67.

The ATS, Nimbus, and, perhaps, other satellites are important because of the equipment they carry for relaying data from remote, unattended buoys to land stations. Such buoys can simply send data on currents as indicated by daily changes in the position of the buoy, or they can send back temperature profiles or even direct sound-velocity measurements made by a series of sensors suspended at various depths beneath the buoy.

Seasat is a militarily important satellite flown by the U.S. National Aeronautics and Space Administration (NASA) to test several concepts in ocean sensing. One Seasat can monitor the oceans of the world once every 36 hours, and the combination of sensors carried can measure sea state (wave height), wind speed and direction, wave direction, and ocean temperature. The original Seasat only lasted 100 days, but there are plans to eventually have six Seasat-type satellites in orbit continuously so that all oceans will be monitored every 6 hr. Seasat has an important advantage over the weather satellites because its microwave sensors not only see through cloud but also work as well in the dark as in the sunlight. It is thus better equipped to map Arctic ice during winter. Seasat is a civilian program, but its operation is controlled by a committee that includes Defense Department representation, and it is partly financed out of the defense budget. Data were supplied by NASA from the Seasat terminal in Alaska, via a communications satellite, to the Navy's Fleet Numerical Weather Central in California.

An important innovation with Seasat was the incorporation of equipment to enable the satellite itself to determine its own position to within 10 m by receiving signals from the Defense Department Navstar satellites. The synthetic aperture radar aboard Seasat is capable of resolving targets on the sea surface as small as 25 m. This means that even quite small fishing vessels can be accurately located on the ocean. The value of this for ASW is that it enables identification of some of the noises that enter the SOSUS arrays: once identified, they can be filtered out. Identification and location of sound sources also aid in real-time calibration of the range-estimating function of the SOSUS arrays.

Another powerful technique for ocean monitoring over large areas is Over the Horizon Back-scatter radar (OTH-B).[15] Back-scatter radar uses high frequency radio waves which bounce off the ionosphere and return to the earth's surface 1,500 km and more away. Here they are scattered and reflected by land and sea surfaces and by targets of military interest such as ships and aircraft. Some of the scattered radio energy returns by the same path to sensitive receivers located near the transmitters. The average wavelength, the direction

15. S. R. Curley, "Measurements of Mid-Ocean Conditions by Over-the-Horizon Radar," (U. S.) *Naval Research Reviews*, vol. 26, no. 11, p. 1; E. E. Barrick, J. M. Headrick, R. W. Bogle, and D. D. Crombie, "Sea Backscatter at HF: Interpretation and Utilization of the Echo," *Proceedings of the Institute of Electrical and Electronics Engineers (IEEE)* 62, no. 6 (June 1974): 673.

of the waves, and the velocity of the wind that drives them can all be determined by spectral analysis of the clutter signal returned to the receiver from the sea surface. The U.S. Navy has for some years been operating a trial OTH-B radar in Maine, which was able to determine these parameters over an area of 16 million km^2, including most of the likely North Atlantic patrol area for Soviet missile submarines. A second trial radar was more recently built on San Clemente Island on the Pacific coast of the the United States. An operational OTH-B radar is now under construction to cover the North Atlantic and another one may be built in the Pacific Northwest.

Satellite-borne lasers offer another possiblity for global sounding of the oceans. Blue green lasers have exceptional penetration capabilities—to depths of 100 m or more in clear weather—and are under investigation as possible communication modes with submerged submarines and for detection of submerged submarines. A spin-off is that coupling of the laser beam with the molecular resonance of the water molecules can generate Raman frequency shifts and polarization shifts: analysis of the signal return as a function of depth can generate temperature and salinity profiles.

Another aspect of global ocean monitoring is the mapping of all the sound sources that are being registered by the SOSUS and other sonars, so that surface ships of no interest can be distinguished from hostile submarines. The director of DARPA has pointed out that "at any particular time, there are several thousands of maritime merchant ships crossing the world's oceans. Each of these ships constitutes a potential source of acoustic interference to [the U.S.] undersea surveillance system. Merchant ships are not designed for quietness, they are designed for economical transport and their high powered propulsion systems generate a great deal of noise which is well coupled to the ocean's acoustic propagation path."[16]

As already noted, Seasat has demonstrated a potential for contributing data that help filter out these noise sources. The OTH-B is also, at least potentially, useful in this role and does not suffer from the 6-hr delay involved in use of Seasat data. Orion aircraft, including ferret EP-3A aircraft that detect ships by their own radar emissions and ferret satellites that perform the same function from space, probably also help. More recently, the "White Cloud" ocean-surveillance satellites have become available to passively track ships by their own radar emissions, and soon "Clipper Bow" radar satellites may be actively tracking all vessels from space—including those maintaining radio-frequency silence. Finally, there is the AMVER system, in which the merchant ships of many nations report their own positions to the U.S. Coast Guard for search and rescue purposes. The U.S. Coast Guard relays this information to the U.S. Navy's ocean surveillance information system (OSIS).

16. U.S. Congress, Senate, Committee on Appropriations, *Department of Defense Appropriations, Fiscal Year 1978: Hearings on H. R. 7933,* 95th Cong., 1st sess., 1977, pt. 5, pp. 83–84.

Exotic Detection Techniques

In anticipation of Soviet progress in submarine-quieting, and in recognition of the fact that acoustic detection systems can be flooded with noise either unintentionally, as with merchant ships and sea-bottom mineral exploitation, or intentionally, by screening submarines with surface vessels or using noise generators to jam sonars, there has been a continuous effort to expand the range of nonacoustic detection techniques.

Magnetic detection is one of the most promising lines of inquiry. It has already been mentioned that cryogenic SQUID magnetometers have doubled or tripled the range of magnetic airborne detection. This detection range now makes it feasible to install magnetic detectors on the sea bottom, and it is reported that this technique is being investigated for the Greenland-Iceland-United Kingdom (GIUK) gap.

The U.S. Navy maintains a satellite series called Solrad-Hi which measures variations in solar activity, and the resulting variations in the earth's magnetic field. Knowledge of these events is necessary for MAD to distinguish between magnetic anomalies caused by submarines and those caused by geophysical factors.

Submarines also produce perturbations of the electric field of the ocean, which are potentially detectable. One approach uses large electric coils laid out on the sea bottom. The U.S. Navy is also sponsoring research into the methods by which some marine organisms can detect extremely small perturbations in electric fields. One species of ray, for example, can detect field changes of .01 μv. Submarine-induced perturbations are considerably larger.

Submerged submarines also generate thermal anomalies—which are potentially detectable at the surface—either from the heat released from the reactor through the condensor heat-exchangers or as a result of wake turbulence mixing cold deep water into the warmer surface layers. Present infrared sensors on board weather and oceanographic satellites have a very high degree of thermal resolution but lack the spatial resolution to detect the relatively small upwellings from submarines. This situation may well change in the near future, however, with the continuing rapid development of new kinds of mosaic sensors composed of very small detector elements.

Blue green lasers are another possibility for submarine detection from space. Laser detection of submarines from the air has already been successfully demonstrated with the optical ranging identification and communication system (ORICS).[17] Lasers, however, suffer rather fundamental depth limitations and limited swath widths because the laser beam must be at near-perpendicular incidence to the sea surface. A great many satellites would therefore be needed to successfully detect submarines with lasers.

17. "Advanced Sensors Major Anti-Submarine Warfare Goal," *Aviation Week and Space Technology* 106, no. 5 (January 31, 1977): 134–45.

Investigations of surface wake effects of submerged submarines are more promising. Just as a surface ship produces a V-shape wake which may persist for several kilometers behind the vessel, so a submerged submarine produces a conical wake which intersects the surface at some distance behind the submarine. The turbulence caused by the passage of a submarine is also expressed at the surface in various ways which may be summarized under the term hydrodynamic signature.

Hydrodynamic signatures are potentially detectable with Over-the-Horizon (OTH) radar. An upwelling of water alters the morphology of surface waves—something that is detectable with OTH-B. The wake itself may be expressed as a much longer wavelength undulation, potentially measurable with OTH-B thanks to a higher-order interaction between the ocean waves and the electromagnetic waves giving rise to second-order harmonically generated Doppler lines in the radar return. In May 1974, the director of DARPA noted that DARPA was investigating the application of OTH radar "to the detection of ships, *submarines,* SLBMs and cruise missiles" (emphasis added).[18]

Satellites of the Seasat type also have the potential of detecting hydrodynamic signatures. Upwellings resulting from the passage of a submarine can bring water of different temperature, salinity, and biological content to the surface; these can all result in a water mass of different dielectric constant, which is detectable by microwave radiometry from satellite altitudes.[19] Also the long wavelength undulation is potentially detectable by the radio altimeter carried by Seasat, which has 10-cm vertical resolution.

The infrared techniques already described are another way of detecting hydrodynamic signature.

Antisubmarine Weaponry

Having examined the antisubmarine detection systems, it remains to take a brief look at the weapons that can be used against submarines once they are located. The United States has three weapon systems of importance for strategic ASW—the Orion, the hunter-killer submarine, and the Captor mine.

Perhaps the most important ASW weapon platform in the U.S. forces is the P3 Orion, already described as one of the detection platforms. The Orion can carry Mark 46 acoustic torpedoes and nuclear depth charges. The Mark 46 torpedo has a range of about 10 km, a running speed of 45 knots, and homes in on its target acoustically. If it misses or overshoots a target on its first attempt, it is capable of turning and making further attempts. It has a potential vulnerability to acoustic and other countermeasures.

18. R. T. Pretty, ed., *Jane's Weapon Systems* (London: Jane's Yearbooks, 1978), p. 613.

19. C. T. Swift, "Preface," *IEEE Transactions on Antennas and Propagation* AP725, no. 1 (January 1977), p. 1.

Nuclear depth charges are capable of creating overpressures within the ocean capable of imploding any submarine within MAD range.

It can be assumed that these weapons, the Mark 46 torpedo, and the Mark 57 and Mark 101 nuclear depth charges, have almost complete certainty of destroying any submarine already detected and located by the aircraft's sensors. To guard against possible future developments in quieting and acoustic countermeasures, the U.S. Navy has embarked on the Neartip improvement program for the Mark 46 and an advanced lighweight torpedo is being developed to replace the Mark 46. The new torpedo will go faster, dive deeper, reach targets at greater range, and have a higher probability of success than the Mark 46.

The other important U.S. ASW platform is the hunter-killer or nuclear-powered attack submarine (SSN)—considered to be the best ASW platform in existence today. It has the advantage of operating in the same medium as the target. Sonar detection is enhanced by the hunter-killer submarine's ability to operate at depths optimal for propagation. It can use more sophisticated hydrophones than can be built into an Orion sonobuoy, and it creates lower self-noise levels than surface vessels.

The U.S. Navy is currently involved with the construction of the SSN-688 or Los Angeles-class of hunter-killer submarine.[20] Four have already been built, 32 are already authorized by Congress, and a total of 42 is planned for delivery during the 1980s. Including other hunter-killer submarine classes, the United States will have 90 killer-submarines by 1983. The SSN-688 has been described as the most combat-capable submarine in the world, with higher speed, lower noise levels, and more advanced sensors and countermeasures than any other U.S. hunter-killer submarine.

The heart of the SSN-688 detection capability is its 15-ton AN/BQQ-5 sonar system which includes a digital signal processor of unprecedented complexity and performance. The signal processing routines are said to be effective against such countermeasures as torpedoes equipped with recorded submarine sounds intended to make the sonar lose track.

Hunter-killer submarines carry Mark 48 torpedoes, which are larger, and more capable, than the Mark 46 already described. They have a higher speed, longer range, and are either acoustic-homing or wire-guided. Hunter-killer submarines also carry the nuclear warheaded SUBROC ASW missile. This is launched from an ordinary torpedo tube, rises to the surface, flies as a missile for up to 50 km, reenters the ocean in the vicinity of the target and explodes as a nuclear depth charge. The Tarpon, a crossbreed of the Mark 46 torpedo with a

20. U.S. President, *Fiscal Year 1979 Arms Control Impact Statements,* statements submitted to the Congress by the president to the Congressional Committee on International Relations and the Senate Committee on Foreign Affairs (Washington, D.C.: Government Printing Office, 1978), pp. 164–68.

Harpoon cruise missile, which will have a range of about 100 km, will replace SUBROC.

The Captor ASW mine[21] basically consists of an encapsulated Mark 46 torpedo with equipment to enable anchoring to the sea floor. This mine is equipped with acoustic sensors which can distinguish between surface vessels and submarines. When a submarine is detected, the torpedo is released and homes in on the submarine at distances of up to 10 km.

According to the U.S. undersecretary of defense for research and development: "Analyses show that, within the limits in which it can be employed, CAPTOR will kill more submarines per dollar than any other ASW system."[22]

In time of war, Captor minefields can be rapidly sown across all the choke points through which Soviet submarines must pass. Although it can be delivered by ship or submarine, Captor is most likely to be delivered by air. The B-52s can carry 18 Captors, while Orions and carrier-based A-6s or A-7s can each carry six. Only some 500 Captors would be needed to seal off the GIUK gap. It would take only 28 B-52 sorties to deliver this quantity, and the United States has 60 B-52s prepared for minelaying.[23] Thus the GIUK gap could be sealed off in a few hours at most. Any breaches opened up on this barrier as a result of mines being activated against penetrating submarines could be readily filled by later B-52 sorties.

The United States has many other ASW systems and platforms, including various classes of ASW destroyer, light airborne multipurpose system (LAMPS) and other ASW helicopters, and carrier-based S-3A Viking aircraft. The latter carry basically the same sensors and weapons as the Orion. However, these are generally thought of as tactical ASW systems to be used for defending aircraft-carrier task forces, convoys, the North Atlantic sea lanes, and so on. They make an indirect contribution to U.S. strategic ASW capability, nonetheless, insofar as their existence frees the Orions and SSNs to concentrate on strategic roles.

TREND TOWARD FIRST-STRIKE CAPABILITY AGAINST MISSILE SUBMARINES

A 1974 Stockholm International Peace Research Institute (SIPRI) study of ASW dismissed the possibility of a first-strike attack against missile submarines by noting that it would require continuous trailing of every missile submarine

21. Ibid., pp. 175–80.

22. U.S. Congress, Senate, Committee on Armed Services, *Department of Defense Authorization for Appropriations for Fiscal Year 1979: Hearings on S. 2571*, 95th Cong., 2d sess., 1978, pt. 8, p. 5820.

23. B. Day, "The B-52: Growing More Vital with Age," *Air Force Magazine* 62, no. 2 (February 1979): 32–37.

from the moment it left port until the moment it returned.[24] Five years later this no longer seems to be the case as regards U.S. ASW against Soviet missile submarines. A first strike against Soviet submarines might even be considered a more attractive alternative than a strike launched during the course of a nuclear war in an attempt to limit the destruction caused by retaliatory submarine-launched missiles.

The feasibility of such a first strike can be illustrated in terms of a purely speculative scenario. (This scenario is not in any way intended to suggest that the United States is preparing for such a contingency.)

In the scenario it is supposed that the USSR has only 10 SSBNs at sea, perhaps two in the Pacific, four in the North Atlantic, and four in the Barents Sea. Submarines departing from Murmansk to the Atlantic Ocean and the Barents Sea will have been monitored by photographic and, perhaps, by electronic reconnaissance satellites, and the progress of the Atlantic submarines will have been successively monitored by Norwegian Orions, sea-bottom sonar between Norway and Bear Island, the well-documented SOSUS barrier across the GIUK gap, and by British Nimrod ASW aircraft. Once the submarines have passed into the Atlantic proper, they are monitored by SOSUS installations, SURTASS, and RDSS.

The SOSUS fixes will be calibrated with oceanographic data obtained in real time by various satellites, mid-ocean buoys, and Orion-dropped acoustic-velocity measuring buoys. From time to time, the Orions will make vectored intercepts of the submarines to verify the SOSUS recordings. The SSBNs in the Pacific will be subject to a similar pattern of suveillance. They will be monitored by fixed sonars as they pass through the various Japanese straits, and then monitored by SOSUS, Japanese aircraft, and U.S. Orions.

If the United States were to decide to launch a first strike, still assuming surprise, it would have between 200 and 400 Orions available to hunt down those 10 submarines. Such a ratio obviously provides plenty of surplus capacity to check out with sonobuoy and MAD the more dubious SOSUS signals and plenty of overkill to ensure that all hostile SSBNs are hit within the very limited time span within which a first strike must be carried out. This is without taking into consideration more tactically oriented ASW forces, such as carrier-based S-3As and ASW destroyers, which could also be directed against SSBNs as opportunity permitted.

The Delta-class submarines remaining in the Barents Sea pose more of a problem. The Barents Sea is shallow, however, and partly covered by sea ice; two factors which pose problems for ASW. In particular, shallow seas provide unfavorable conditions for long-range sonar propagation and the sea ice prevents sonobuoys from being sown from aircraft. The proximity of the Barents

24. Stockholm International Peace Research Institute (SIPRI), *Tactical and Strategic Antisubmarine Warfare* (Stockholm: Almqvist & Wiksell, 1974), pp. 39–42.

Sea to the USSR makes Orion aircraft, surface vessels, and submarines vulnerable to attack.

It may be that, for the time being, the USSR has established a successful sanctuary for some of its SSBNs, thanks to the long-range missiles carried by the Delta-class submarines. There are however indications that the United States does not intend to let the Barents Sea remain a SLBM sanctuary much longer.

The U.S. secretary of defense has virtually served notice that the United States does not intend to let the USSR keep the Barents Sea as a sanctuary. His fiscal report for 1979, which has been widely interpreted as proclaiming a further shift toward counterforce doctrines noted that the United States has a "strategic stake in such distant places as the Sea of Japan . . . and the Barents Sea" and that Soviet naval forces have to "invest in the defense of the Barents Sea and the Sea of Japan."[25]

If the United States were to attempt to eliminate SSBNs from the Barents Sea, it would have at its disposal, among other weapon systems, a considerable number of hunter-killer submarines, including Los Angeles-class submarines. The very features of the Barents Sea that make Soviet submarines hard to find would also make it difficult for the USSR to oserve or hit U.S. hunter-killers.

It remains to examine the fate of the four-fifths of the Soviet SLBM force still in port. These submarines are all highly vulnerable to U.S. intercontinental ballistic missile (ICBM) and SLBM attack, and possibly to bomber attack. The U.S. chief of naval operations has described attacks on ". . . the submarine bases from which the nuclear powered submarines operate" as being part of the United States' "first line of defense."[26]

If complete surprise were not achieved, a proportion of the remaining fleet might be able to put to sea. Yankee-class and older submarines would have to transit the GIUK gap in order to be within firing range of the United States: this could be prevented by a barrier of Captor mines laid in a matter of hours across this gap and either further north between the northwest corner of Norway and Greenland or across the Norway-Spitzbergen gap. Delta-class submarines might be subjected to a "rolling barrage" of megaton-size nuclear explosions before they fully dispersed over all of the Barents Sea. Each such explosion can create overpressures sufficient to crush a submarine over an area of about 350 km^2.

25. U.S. Department of Defense, *Annual Report for Fiscal Year 1980* (Washington, D.C.: Government Printing Office, 1979), pp. 31, 36.

26. U.S. Congress, Senate, Committee on Armed Services, *Authorization for Military Procurement, Fiscal Year 1979: Hearings on S. 1210,* 95th Cong., 1st sess., 1978, pt. 5, p. 4321.

Survivability of U.S. ASW Forces

An important consideration when attempting to assess the extent to which any weapon system is intended for first- or second-strike roles is its ability to survive the opening stages of a nuclear war. Credible retaliatory or damage-limiting forces must be survivable; first-strike systems need not be. By this criterion some aspects of ASW would seem more suited to a first-strike role.

The ASW surveillance systems are particularly vulnerable. The sea-bottom sonar arrays are highly vulnerable to jamming and spoofing, and the cables linking the hydrophones are easily cut even by ordinary fishing trawlers. This happens continuously, by accident or design, in peacetime. The SOSUS shore stations are also very vulnerable. A U.S. Navy spokesman has noted that "SOSUS stations are vulnerable to a wide range of physical threats" but are provided merely with the nominal protection of chain-link fences and small arms.[27]

SURTASS is even more vulnerable. These mobile sonar arrays will be towed by vessels which "are to be built as non-combatants,"[28] which will be unarmed, have civilian crews, and will be capable of only 11 knots without array and 4 knots with array extended.

Other aspects of ASW are, however, much better adapted to survival. In particular the hunter-killer submarines are well suited to a damage-limiting role during nuclear war. But, as Senator McIntyre of the U.S. Senate Armed Services asked recently: "It seems that the starting point and the foundation for ASW is the SOSUS system. Yet the SOSUS system appears to be very vulnerable, particularly the land based terminals. . . . Are we basing our ASW on a capability that could easily be wiped out in the opening days of a war?"[29]

Soviet ASW

It is generally agreed that Soviet ASW forces do not constitute a significant threat to U.S. SSBNs.

In part this is due to the more sophisticated nature of U.S. missile submarines and the electronic systems which support them. The U.S. nuclear submarines dive more deeply and are quieter than the Soviet equivalents. The U.S. submarines have access to global VLF communication systems that can be received 15 m under water, and to an operational ELF (extremely low frequency) system that, in the future, may be receivable at 400-m antenna depth.

27. Ibid., pt. 8, p. 6350.
28. Ibid., p. 6345.
29. Ibid., p. 6350.

Over most of their patrol area U.S. submarines can navigate by Loran-C, which is accurate enough for missile targeting and which is receivable at 4 m depth.[30]

The Soviet ASW capability suffers from geographic factors —in particular, the lack of friendly coastline bordering U.S. SSBN deployment areas so that sea-bottom sonar arrays cannot be used. The Soviet Union also lacks airfields in the vicinity of many of the U.S. SSBN deployment areas.

Soviet ASW detection technology is also much less advanced than that of the U.S., particularly in the field of signal processing and in sensor design.

Most of the effort that the Soviet Union has put into ASW has gone into tactical ASW which is intended to disrupt merchant shipping and into defensive strategic ASW which is intended to protect Soviet missile submarines in such areas as the Barents Sea. A U.S. Congressional Research Service report, which criticized U.S. ASW advances as having a potential for first-strike doctrines, stated that the "Soviets apparently have no effective capability for open-ocean ASW, regardless of the scenario envisaged."[31]

A detailed evaluation of the situation was made by U.S. Defense Secretary Brown in November 1977. He said that there is "no definitive Soviet threat today to Polaris Poseidon SSBNs" and that "there is no evidence that any Soviet weapon or equipment have been developed solely to meet the Polaris/Poseidon threat" although Soviet ASW in general could be so used. The United States had assurances from particular tests carried out that the Soviet Navy was not trailing U.S. SSBNs. Brown noted that the 135 "principal combatants" as well as all patrol combatants in the Soviet Navy have ASW roles, as do 250 submarines and 400 aircraft. However, only about 50 aircraft, the Ilyushin 38s, have ASW as a prime mission.[32]

He further noted that the USSR was "conducting applied research in a variety of ASW related acoustic and non-acoustic areas." Yet, in his fiscal 1980 report, Brown was still able to state that neither the hunter-killer submarine, which "constitutes the most capable Soviet ASW platform . . . nor any other currently deployable Soviet ASW systems represent a serious threat to [U.S.] ballistic missile submarines."[33]

30. Stockholm International Peace Research Institute (SIPRI), *SIPRI Yearbook of World Armaments and Disarmament 1979* (London: Taylor & Francis, 1979), chap. 7.

31. *Evaluation of Fiscal Year 1979 Arms Control Impact Statements: Toward More Informed Congressional Participation in National Security Policymaking,* report prepared for the House Subcommittee on International Security and Scientific Affairs of the Committee on International Relations (Washington, D.C.: Government Printing Office, 1978), chap. 5, pt. 2, pp. 103–17.

32. U.S. Congress, House, Committee on Armed Services, *Supplemental Authorization for Appropriations for Fiscal Year 1978 and Review of the State of U.S. Strategic Forces: Hearings on H.R. 8390,* 95th Cong., 1st sess., 1977, pp. 203–9.

33. U.S. Department of Defense, *Department of Defense Annual Report for Fiscal Year 1980* (Washington, D.C.: Government Printing Office, 1979), p. 74.

CONCLUSION

It is still the declared policy of the United States not to acquire the capability to eliminate the other side's deterrent; hence it is never officially admitted that the ASW forces are directed against missile submarines. However, both the nature of these forces and their global distribution suggest that the distinction between tactical ASW against attack submarines and strategic ASW against missile submarines is already becoming indistinct.

The dilemma of the present situation is that deployment of constantly improved ASW systems, even if they are only intended for tactical roles, inevitably lends toward a strategic first-strike capability. When this capability exists, there will be temptations to use it; and there will be a perception by the other side that it can be used.

Expressions of U.S. confidence in ASW are abundant. Admiral Kauffman testified that the United States had a predominant lead because of advances in sensors, weapons, and submarine quietening.[34] Secretary of the Navy W. G. Claytor is reported to have said that the "qualitative edge that we hold over the Soviets is awesome . . . and our ability to orchestrate the many components of the United States' anti-submarine warfare team into an effective submarine killer force has enormously improved in recent years."[35]

Concern about the implications are well summarized in a U.S. Congressional Research Service report prepared for the House Committee on International Relations.[36] This report notes the ambiguity between strategic ASW and tactical ASW, particularly in areas close to the USSR, and quotes official sources to the effect that in time of war the United States would not discriminate between missile submarines and other types. It accepts the fact that the United States does not, at this moment, have the capability to eliminate Soviet sea-based missiles, but points out that research and development is leading in this direction and that Soviet decision makers have grounds for fearing such a capability.

The situation as a whole demands urgent attention. If the United States achieves a first-strike capability against Soviet ICBMs, as appears to be one of the objectives of the M-X program, and if this is coupled with maintenance of the present lead in ASW, there are serious grounds to fear that the concept of mutual assured destruction, with all its faults, will have been abandoned in favor of a war-fighting and war-winning strategy.

34. U.S. Congress, Senate, Committee on Armed Services, *Authorization for Military Procurement, Fiscal Year 1978: Hearings on S. 1210,* 95th Cong., 1st sess., 1977, pt. 10, p. 6756.

35. B. Weintraub, "Navy Chief Says U. S. Has Best Submarines," *New York Times* (May 25, 1978), p. 6.

36. See n. 31 above; also summarized in W. Pincus, "Study Warns of Effect on SALT. U.S. and Anti-Sub Systems Seen 'Far Ahead,'" *Washington Post* (January 10, 1979).

Ocean Surveillance by Earth Satellites

Bhupendra Jasani
Stockholm International Peace Research Institute

The first man-made earth satellite was successfully orbited in 1957 when the Soviet Union launched its Sputnik 1 satellite. Since then over 2,200 artificial earth satellites have been launched; of these, over 1,600 were launched for military purposes. Of the latter, over 50 percent have been for military reconnaissance purposes.

Reconnaissance satellites can be divided into five groups on the basis of their mission: photographic reconnaissance, electronic reconnaissance, early warning, nuclear explosion detection, and ocean surveillance. A number of ocean-surveillance satellites have been launched by the United States and the USSR to monitor the location of naval fleet and shore facilities. Like most military activities in outer space, the development and the existence of such satellites have gone relatively unnoticed. However, attention was once more focused on the use of such satellites when, at the beginning of 1978, control was lost over a Soviet ocean-surveillance satellite carrying a nuclear reactor. Although several U.S. satellites carrying nuclear power sources have also crash-landed,[1] it was Cosmos 954 which received most publicity. Some of the U.S. satellites as well as the Soviet Cosmos 954 contaminated part of the earth's surface and the atmosphere with radioactive materials from the spacecraft's nuclear power sources. The United States has also launched ocean-surveillance satellites, but they probably do not carry nuclear power sources.

Whereas ocean-surveillance satellites are used to detect and track military surface ships, oceanographic satellites are used to determine various ocean properties. These are, for example, departures of the ocean surface from the geoid, wave heights, ocean currents, surface temperatures, salinity, wind speeds, and coastal features. The latter are measured to improve maps and charts for navigation. All of these tasks are performed by sensors such as long-range radars, microwave and infrared radiometers, radar altimeters, photographic and television imaging sensors, and microwave scatterometers aboard satellites.

*Editors' Note: Portions of this piece are adapted from the author's chapter in the *SIPRI Yearbook 1979,* pp. 267–302.

1. Arthur H. Westing, "Military Impact on Ocean Ecology," in *Ocean Yearbook 1,* ed. Elisabeth Mann Borgese and Norton Ginsburg (Chicago: University of Chicago Press, 1978), pp. 453–54.

© 1980 by The University of Chicago. 0-226-06603-7/80/0002-0007$01.00

The transmission of sound by submarine, for example, is partly dependent upon temperature and salinity. To a varying extent waves on the surface of the sea add to the general background noise of the ocean. A knowledge of these factors is essential for the better design and efficient deployment of sensors used on and below the surface of the ocean for the detection of submarines.[2]

In the following sections, therefore, the capabilities of some of the above-mentioned sensors and the ocean-surveillance satellite programs of the USSR and the United States are briefly reviewed.

SENSOR TECHNOLOGY

Two types of radiometer are used for measuring temperatures and ocean images—infrared radiometers and microwave radiometers.

The Infrared Radiometer

An infrared radiometer is basically a very sensitive thermometer designed to respond to electromagnetic radiation of wavelengths between 8 and 13 μm emitted from the surface of the sea. At these wavelengths, a considerable amount of radiation is transmitted by the earth's atmosphere. Moreover, there is very little interference from the scattered radiation. The sensor can either be fixed or be made to scan at right angles to the orbital path of the satellite. The output from the sensor can be displayed either in digital form or in the scanning mode so that the result resembles a photograph. The scan rate is synchronized with the speed of the satellite. Such infrared images can show the temperature gradients along the surface of the ocean together with ocean currents and eddies.

Infrared radiometers with high resolutions have been used, for example, aboard the U.S. weather satellites Nimbus 1 and 2. The field of vision of these instruments was about 0.5° with a resolution of about 9 km at an altitude of some 1,000 km. The recent U.S. ocean-surveillance satellite, Seasat A, launched on June 27, 1978 (table 1), has an infrared radiometer to provide ocean surface images and coastal features from an altitude of about 800 km with a resolution of 4 km over a 1,500-km swath.[3] However, the spacecraft experienced some problems and ceased to transmit data on October 9, 1978.

2. See Owen Wilkes, "Ocean-based Nuclear Deterrent Forces and Antisubmarine Warfare," in this volume, pp. 226–49, and "The ASW Problem: ASW Detection and Weapons Systems," *Ocean Yearbook 1,* pp. 381–83.

3. "Seasat, an Ocean Dedicated Satellite," *Sea Technology* 19 (May 1978): 33.

TABLE 1.—POSSIBLE U.S. OCEAN-SURVEILLANCE SATELLITES

Satellite Name and Designation	Launch Date and Time (GMT)	Orbital Inclination (km)	Perigee Height (km)	Apogee Height (km)	Comments
1971:					
USAF/USN (1971-110A) ..	Dec. 14; 1214	70	983	999	Navy ocean-surveillance satellites; quadruple launch
USAF/USA (1971-110C) ..	Dec. 14; 1214	70	983	999	
USAF/USN (1971-110D) ..	Dec. 14; 1214	70	982	997	
USAF/USN (1971-110E) ..	Dec. 14; 1214	70	981	997	
1975:					
NASA/GEOS C (1975-27A) ...	Apr. 10; 0000	115	839	853	Geodynamic Experimental Ocean Satellite; the satellite was used to calibrate and to determine positions of NASA and other agency C-band radars, and to perform a satellite-to-satellite tracking experiment with ATS-6 spacecraft using an S-band transponder system
1976:					
USN/NOSS-1 (1976-38A) ...	Apr. 30; 1912	63	1,092	1,128	Navy ocean-surveillance satellites; quadruple launch
USN/SSU-1 (1976-38C) ...	Apr. 30; 1912	63	1,093	1,129	
USN/SSU-2 (1976-38D) ...	Apr. 30; 1912	63	1,093	1,130	
USN/SSU-3 (1976-38J) ...	Apr. 30; 1912	64	1,083	1,139	

TABLE 1.—POSSIBLE U.S. OCEAN-SURVEILLANCE SATELLITES (*Continued*)

Satellite Name and Designation	Launch Date and Time (GMT)	Orbital Inclination (km)	Perigee Height (km)	Apogee Height (km)	Comments
1977:					
USN/NOSS 2 (1977-112A) . .	Dec. 8; 1746	63	1,054	1,169	Navy ocean-surveillance satellites; quadruple launch
USA/NOSS2 (1977/112C) . .	Dec. 8; 1746	63	1,054	1,169	
USN/NOSS2 (1977-112D) . .	Dec. 8; 1746	63	1,054	1,169	
USN/NOSS2 (1977-112E) . .	Dec. 8; 1746	63	1,055	1,168	
1978:					
NASA/SeasatA	June 27	108	776	800	

NOTE.—GEOS = Geodynamic Experiment Ocean Satellite; NASA = National Aeronautics and Space Administration; NOSS = Navy Ocean Surveillance Satellite; USAr = U.S. Army; USAF = U.S. Air Force; USN = U.S. Navy.

The Microwave Radiometer

A microwave radiometer is used to measure the sea surface temperature under all weather conditions. The sensor aboard Seasat A (table 1), for example, measured ocean surface temperature with an accuracy of 1° C and, by determining the brightness of the foam, ocean surface wind speed of up to 50 m per second (m/s) could be determined.[4] The sensor also provided data for atmospheric corrections to the satellite's active radars by measuring liquid and water-vapor content in the atmosphere. The observations were made over a surface area 650 km wide beneath the satellite.

The Radar Altimeter

The unevenness of the surface of the sea owing to the presence of waves is well known. Less well-known variations are the bumps and dips created by gravity and the configuration of the ocean floor. These fluctuations in sea level are measured by radar altimeters aboard ocean-surveillance satellites. Such an

4. Ibid.

instrument measures the distance between the satellite and the sea surface immediately beneath it by determining the time required for the radar pulse generated by the altimeter to travel from the antenna to the sea surface and back. With this technique, subtle variations caused by wind stress, ocean depth, and atmospheric pressure gradients can be measured. However, from the military point of view, the important measurements are those of the large variations in the elevation of the sea surface caused by local gravity anomalies. From these measurements local changes in the strength of the gravitational field can be determined. Knowledge of variations in the gravitational field is essential for submarine activities.

Modern radar altimeters can measure the height of ocean surface waves to within 0.5–1 m. Differences between the ocean surface and the geoid as small as 10 cm can be measured.[5]

The Scanning Radar Scatterometer

If a beam of radar energy is directed at the surface of the sea, the reflected energy is strong near the vertical if the surface of the sea is smooth and weak at large oblique angles. However, if the sea surface is rough, then the reflected radar signal weakens near the vertical and increases at large angles. This phenomenon has been used to determine the state of the ocean surface using a so-called radar scatterometer.

Knowledge of the state of the sea could, therefore, be improved in order to allow small vessels to be detected by radar. Since small-scale irregularities on the wave surface are produced by local winds, wind speed and direction could also be determined by the use of a radar scatterometer. Wind speeds between 3 and 25 m/s could be measured and wind direction could be determined to within 20°.[6] The area covered by a scatterometer aboard the Seasat A satellite was a swath more than 1,000 km wide, covering 95 percent of the surface of the earth every 36 hr.

The Synthetic Aperture Radar

One of the most important instruments aboard the Seasat A satellite is a synthetic aperture radar. In a side-looking radar, the resolution of the radar deteriorates as the distance between the antenna and the objects increases. This is because the radar beam fans out so that the beam is wider at greater distances (see fig. 1). Moreover, resolution is proportional to beam width. Thus two objects at the same range separated by a distance less than the beam width will

5. Ibid.
6. Ibid.

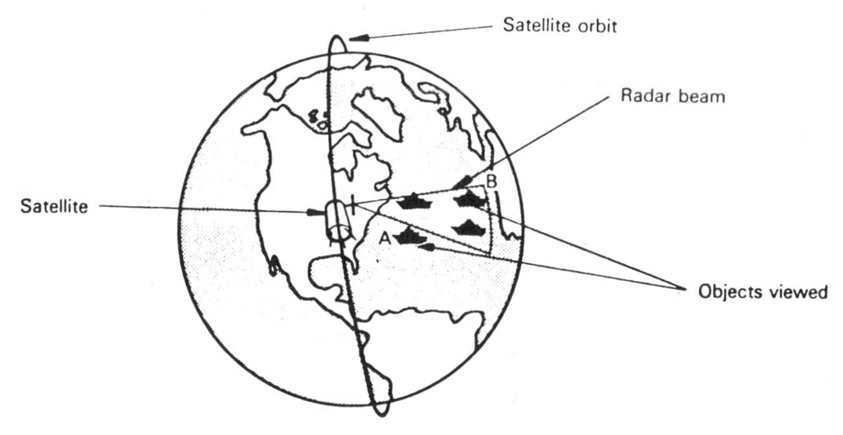

FIG. 1.—Effect on resolution as the distance between the objects viewed and the radar increases. Objects at A will be resolved as two objects but those at B will be seen as one.

not be resolved. The angular width of the beam generated by the radar antenna is inversely proportional to the length of the antenna. The resolution could therefore be improved by using a longer antenna. However, the size of the antenna which can be carried by a satellite is limited, thus limiting the resolution of a side-looking radar.

This problem is overcome by a synthetic aperture radar. This is a side-looking radar with a relatively short antenna which is made to behave like a very long antenna with a narrow beam. A long antenna can be synthesized by taking advantage of the motion of the satellite in its orbit. As the satellite progresses along its orbit, the short antenna of its radar transmits pulses of radiation at regular intervals toward the earth. As the satellite approaches an object on the earth, for example, the beam of the antenna falls upon, moves across, and finally leaves the object. During this time it reflects the microwave pulses received from the radar antenna back to the antenna. As seen from figure 1, the greater the distance between the object and the antenna, the longer the object remains in the beam. Seen from the object, therefore, the radar antenna appears much larger than it is, and this apparent length will depend on the distance between the object and the real antenna.

The effective length of the antenna is, therefore, proportional to the range of the object. Theoretically, the resolution for an unfocused synthetic aperture radar is given by $r = \frac{1}{2}\sqrt{\lambda R}$, where R is the slant range and λ is the wavelength. For a focused synthetic aperture radar with an antenna of length l, however, the theoretical resolution is given by $r = l/2$. Thus, in the latter system, the resolution of the image remains almost the same at all ranges. High-resolution images of the earth's surface can thus be obtained from great distances.

Such a synthetic aperture radar aboard the Seasat A satellite was capable of covering a 100-km swath. The radar provided all-weather photographs of ocean waves and ice fields. It detected icebergs, ice leads (openings in sea ice), and ships and other objects larger than 25 m. The radar employed a 2.1×10.7 m deployable planar antenna which provided data in real time when the satellite was within the line of sight of a receiving station on earth. The images could be transmitted at a rate of 110 megabits per second.[7]

NUCLEAR POWER SOURCES

On January 24, 1978, the Soviet ocean-surveillance satellite, Cosmos 954, entered the earth's atmosphere and partially burned up. The remaining debris landed in northern Canada. The satellite was carrying a nuclear reactor to provide power for the radar and probably for other equipment. Considerable interest in nuclear power sources for use in satellites has recently been generated, since Cosmos 954 was the second satellite which had contaminated the atmosphere and the earth's surface with radioactive materials. It is, therefore, useful to consider briefly the extent to which such power sources are being used, particularly aboard military satellites. In a radar system, power is needed to operate the transmitter, the receiver and associated circuits, and the data-handling circuits. The amount of power required for a transmitter, for example, depends on such parameters as orbital altitude, swath width, resolution, and antenna size. A high-resolution, fully focused synthetic aperture radar system will consume more than twice as much power as required for an unfocused and moderate- to low-resolution system. The amount of power needed in various types of equipment aboard ocean-surveillance satellites is shown in table 2. In most satellites the power is generated by solar cells. However, many such cells have to be used, and concern has recently been expressed because such a power source becomes vulnerable to nuclear or beam weapon attack. In order to make military satellites capable of surviving nuclear attack and possible attack from hunter-killer satellites, therefore, considerable impetus was given to the development of nuclear power generators.

The two most commonly used nuclear energy sources are the energy released when a radionuclide decays and the energy released when a fissile atom fissions. In the first instance the energy source contains a highly radioactive substance, whereas the second generates a number of highly radioactive substances during operation.

In the United States, such compact nuclear power generators are described under the general title of Systems for Nuclear Auxiliary Power (SNAP). The SNAP devices using radionuclides were assigned odd numbers, and those employing nuclear reactors were given even numbers. However, SNAP number designations are no longer given to new devices.

7. Ibid.

TABLE 2.—POWER REQUIREMENT FOR VARIOUS SENSORS AND EQUIPMENT ABOARD OCEAN-SURVEILLANCE SATELLITES

Sensor or Equipment	Power (Watts)
Microwave scatterometer	30
IR radiometer	7
Microwave radiometer	20
Synthetic aperture radar	3,400
TV sensor	150
Multispectral camera	160
Recording equipment on board	115
UHF communications link	5.4
S-band communications link	1.2
Total	3,888.6

SOURCE.—"The Potential of Observation of the Oceans from Space," report prepared for the National Council on Marine Resources and Engineering Development, Executive Office of the President, December 1967.

The heat produced by decaying radionuclides can be converted into electricity in two ways: (*a*) by dynamic conversion using a turbogenerator, or (*b*) by static conversion which uses mainly thermoelectric devices.

Of more than 1,300 available radionuclides, only eight have characteristics suitable for use as power device fuels. The important characteristics are half-life, power density, gamma-ray emission, physical and manufacturing properties, and cost. The most commonly used radionuclide is plutonium-238, an emitter with a half-life of 87.8 yr. It is produced in a reactor either by neutron irradiation of neptunium-237 or by producing curium-242 which decays into plutonium-238 by emitting α rays. In the center of the typical radionuclide thermoelectric generator (RTG), there is a thick cylindrical fuel capsule which serves as the heat source. Surrounding the fuel capsule are thermoelectric energy converters. Such power sources have been used on several satellites and space probes by the United States (table 3). The power output has varied from 2 w to about 65 w.

A U.S. Navy satellite launched on April 21, 1964, carried a SNAP-9A RTG. It failed to orbit, however, and the payload reentered the earth's atmosphere in the Southern Hemisphere. The Pu-238 content was 17 kCi or about 1 kg. The RTG was completely burned up during reentry, and the resulting radioactive particles were distributed at about 50 km above the earth's surface. The radioactivity from this source was measurable until the end of 1970, and some 95 percent of the Pu-238 from SNAP-9A was deposited on the surface of the earth.[8]

8. E. P. Hardy, P. W. Krey, and H. L. Volchok, "Global Inventory and Distribution of Fallout Plutonium," *Nature* 214 (February 16, 1973): 444–45.

As can be seen from table 3, RTGs do not produce high-level power. Equipment aboard modern satellites needs power in the region of a few kilowatts (table 2). Compact nuclear reactors have been developed to generate such high levels of power. The U.S. SNAP-10A, launched into orbit on April 3, 1965, used a reactor consisting of uranium-zirconium fuel elements surrounded by a beryllium reflector. The heat from the reactor was removed by a liquid sodium-potassium coolant circulating within the reactor core. The uranium used was enriched to about 93 percent uranium-235, and the electrical power generated was about 600 w. The reactor produced 42 kw of thermal power and it operated for 43 days. However, after this time a spurious command resulting from a number of failures aboard the satellite shut the reactor down on the 555th orbit.[9]

In the USSR, the U.S. SNAP-1 type of RTGs have been used on two communications satellites.[10] The SNAP-1 used cerium 144, a B-emitter with a half-life of 290 days. The heat produced was used to drive a small turboelectric generator to produce electricity. However, this project was abandoned in favor of a thermoelectric conversion system called SNAP-1A. Two groups of five satellites were launched by the USSR on September 3, 1965, and September 18, 1965. One satellite in each of these groups carried a SNAP-1 type of nuclear power source. The Lunokhod Moon spacecraft also carried radionuclide power generators.

As for reactor-type nuclear power sources, it had been speculated for some time that Soviet ocean-surveillance satellites carry such power generators. The proof of this was provided by the Cosmos 954 accident, and it has been reported that the reactor of the satellite was fuelled with about 50 kg of highly enriched uranium.[11]

THE U.S. PROGRAM

The efficient use of naval weapons, either on the surface of the ocean or below it, is dependent upon detection and location of targets. The systems used for surveillance of targets below the surface of the ocean depend on a knowledge of the physical and chemical state of the ocean. Information about the effects of ocean salinity and the ocean current on the transmission of sound over long ranges, for example, is essential in the design of long-range sonar systems for the detection of submarines. The design and use of over-the-horizon radar systems require accurate knowledge of the atmospheric conditions above the

9. "US Admits to Nuclear Generator Accidents," *Flight International* 113 (April 8, 1978): 997.

10. P. J. Klass, "Russians Believed Deploying Comsat Net," *Aviation Week and Space Technology* 83 (October 11, 1965): 29.

11. "Cosmos Re-entry Spurs Nuclear Waste Debate," *Aviation Week and Space Technology* 108 (January 30, 1978): 33.

TABLE 3.—NUCLEAR POWER GENERATORS ON SATELLITES AND SPACE PROBES

Satellite	Date of Launch	SNAP No.	Power [*W(e)*]	Comments
USN Transit-4A (1961-01)	June 29, 1961	3	2.7	Test for developing integrated navigation system; first nuclear power supply; Pu-238 fuel
USN Transit-4B (1961-AH1)	Nov. 15, 1961	3	2.7	Similar to Transit-4A, SNAP-3, lifetime 8 mo
USAF/USN (1963-38B)	Sept. 28, 1963	9	...	Navigation satellite
USAF/USN (1963-49B)	Dec. 5, 1963	9	...	Navigation satellite
USN navigation satellite	Apr. 21, 1964	9	2.5	Satellite failed to orbit; about 17 kCi of Pu-238 were distributed at about 50 km altitude; by 1970 about 95% of this was deposited on earth's surface; 1 kg of PU-238 fuel
USAF Snapshot (1965-27A)	Apr. 3, 1965	10A	580	First nuclear reactor launched into space; fuel was 93% U-235; thermal power output 33.5 kw
Cosmos 80-84 (1965-70A-E)	Sept. 18, 1965	1 type	...	Communications satellites; power source in one of the five satellites; probably used cerium-144 as fuel
Cosmos 86-90 (1965-73A-E)	Sept. 18, 1965	1-type	...	Communications satellites; power source in one of the five satellites; probably used cerium-144 as fuel
NASA Nimbus 2 weather satellite	May 18, 1968	19	25	Two power units were carried by the satellite but guidance malfunctioned and the satellite was exploded; power units recovered; Pu-238 fuel in each
NASA Nimbus 3 (1969-37A)	Apr. 14, 1969	19	30	Two power units were carried by the satellite; Pu-238 fuel

TABLE 3.—NUCLEAR POWER GENERATORS ON SATELLITES AND SPACE PROBES *(Continued)*

Satellite	Date of Launch	SNAP No.	Power [*W(e)*]	Comments
NASA Apollo 11 lunar module (1969-59C)	July 16, 1969	SNAP–	15 *W(th)*	Early Apollo Scientific Experiment Package was kept warm during lunar night by two Pu-238 power sources
NASA Apollo 12 lunar module (1969-99C)	Nov. 14, 1969	27	63	Apollo Lunar Surface Experiment Package
Apollo 13 lunar module (1970-29C)	Apr. 11, 1970	27	63.5	The power source from the Lunar module was jettisoned in the South Pacific Ocean; no contamination was found; 3.8 kg of the P-238 fuel (44.5 kCi)
Luna 17/Lunokhod 1 (1970-95A)	Nov. 10, 1970	–	...	RTG power generator
NASA Apollo 14 lunar module (1971-8C)	Feb. 1, 1971	27	30	Third lunar module landed on Feb. 5, 1971; strontium-90 used as a fuel
NASA Apollo 15 lunar module (1971-63C)	July 26, 1971	27	...	Lunar module landed on the Moon on July 30, 1971
NASA Pioneer-10 (1972-12A)	Mar. 3, 1972	...	30	RTG, umanned spacecraft flew by Jupiter in December 1973
NASA Apollo 16 lunar module (1972-31C)	Apr. 16, 1972	27	...	...
USAF Triad-01-1X transit navigation (1972-69A) ...	Sept. 2, 1972	...	30	RTG power generator
NASA Apollo 17 lunar module (1972-96C)	Dec. 7, 1972	27	...	...
Luna 21/Lunokhod 2 (1973-1A)	Jan. 8, 1973	–	...	RTG power generator
NASA Pioneer-11 (1973-19A)	Apr. 6, 1973	–	30	Spacecraft flew by Jupiter in December 1974 and will encounter Saturn in September 1979

TABLE 3.—NUCLEAR POWER GENERATORS ON SATELLITES AND SPACE PROBES *(Continued)*

Satellite	Date of Launch	SNAP No.	Power [*W(e)*]	Comments
NASA Viking-1 lander (1975-75G)	Aug. 20, 1975	–	35	RTG; lander landed on Mars on July 20,1976
NASA Viking-2 lander (1975-83C)	Sept. 9, 1975	–	34	Lander landed on Mars on Sept. 3, 1976
USAF Les-8 (1976-23A)	Mar. 15, 1976	–	145	RTG power generator
USAF Les-9 (1976-23B)	Mar. 15, 1976	–	145	RTG power generator
Soviet Cosmos 954 (1977-90A)	Sept. 18, 1977	–	...	Satellite entered earth's atmosphere on January 24, 1978; it mainly burnt up but some pieces were recovered which were radioactive; 17 such ocean-surveillance satellites have been launched; most of these have carried nuclear power reactors on board fueled with highly enriched uranium-235

ocean surface and the state of the ocean surface. In the United States such data have, among other means, been provided by various satellites.

The U.S. Navy's interest in the problems of ocean surveillance dates from 1965. The air force has since joined in the program. Surveillance from space by the U.S. Navy is carried out mostly by means of an ocean-surveillance satellite system using available technology and some advanced sensor systems developed for use aboard satellites. Data gathered from space can then be relayed via communications satellites to land bases. The main aim is to provide almost real-time ocean surveillance.

In the initial stages, the data collected by the U.S. Air Force reconnaissance satellites have been used for comparison with those obtained from naval satellites. Moreover, some of the data collected by civilian satellites have also been used by the navy. Sea surface temperatures, for example, have been measured using the Nimbus weather satellites, and the state of the sea has been determined by the Tiros weather satellites.[12] The manned flights have also contributed considerably to the field of oceanography. On April 9, 1975, a Geodynamic Experiment Ocean Satellite (GEOS-C) was launched to an altitude of about 850 km. A radar altimeter was placed aboard the satellite to measure wave heights in the sea and the state of major ocean current systems. The instrument measured ocean surface heights with an accuracy of 5 m.[13]

A newer and more accurate radar altimeter was installed aboard the recent Seasat A ocean-monitoring satellite, launched on June 27, 1978 in near-circular orbit at an altitude of about 800 km. The satellite was equipped with a multifrequency microwave radiometer to measure, among other parameters, sea surface temperature and wind speed; a radar scatterometer to measure sea surface effects (which can be used to determine wind speeds and direction); a radar altimeter to monitor wave heights; and a synthetic aperture radar to provide all-weather high-resolution photographs of ocean waves, ice fields, icebergs, coastal features, and, of course, ships on the ocean surface. The satellite, costing about $95 million, failed on October 9, 1978, after operating normally for 105 days. There was a loss of power on board the satellite due to an electrical short circuit.[14]

The U.S. Navy's first ocean-surveillance satellite, designed to monitor locations of surface ships, was built under the code name White Cloud and launched on April 30, 1976. It was designed to monitor surface ships and carried three small subsatellites which were placed in near-circular orbits similar to that of the parent satellite. The basic technology of using several satellites to monitor electronic signals generated by naval vessels, and to determine the direction of ships, was demonstrated in 1971 by the launching of multiple satellites (see table 1). The main satellite contains a number of sensors, including passive infrared and microwave radiometers and radio-frequency antennas for detecting shipborne radars and communications signals. Each of the three subsatellites is believed to carry an infrared and a microwave sensor so that, together with the main satellite, it is possible to cover a large part of the ocean surface.

The navy launched a second ocean-surveillance satellite on December 8, 1977. As in the case of the previous White Cloud satellite, the main satellite is accompanied by a constellation of three subsatellites. It is suggested that both

12. Missile and Space Division, General Electric Co., "The Potential of Observation of the Oceans from Space," report PB177726 prepared for the National Council on Marine Resources and Engineering Development, Executive Office of the President, December 1967.

13. "GEOS-C Flight Exceeds Expectations of Planners," *Aviation Week and Space Technology* 102 (April 28, 1975): 97.

14. "Communication Problems Cited in Failure of Seasat Spacecraft," *Aviation Week and Space Technology* 110 (February 5, 1979): 14.

these groups of satellites are being used to determine the precise location of surface ships.[15] The telemetry indicates the transmission of large amounts of information such as radar pulses. These White Cloud satellites monitor communications and radar transmission from surface ships and submarines. The satellites can detect signals from a range of some 3,000 km. They are positioned 3,000 km apart to allow continuous monitoring of naval vessels.[16] Under the Clipper Bow program, advanced ocean-surveillance systems, consisting of radars, will be built for the accurate determination of the positions of surface naval vessels. The full-scale engineering development was expected to begin in 1979. The first launching with such sensors will be in about 1983.[17] With the electronic signals monitored by the White Cloud satellites and the information from the Clipper Bow satellites, a ship could then be identified. The U.S. ocean-surveillance satellites are listed in table 1.

THE SOVIET PROGRAM

The USSR probably began its ocean-surveillance program from space in 1967 when the first research and development satellite, Cosmos 195, was launched from Tyuratam on December 27. After only 21 orbits, the satellite was moved from its low altitude of 250 km to a higher perigee of about 900 km.[18] This type of maneuver has characterized most of the Soviet ocean-surveillance satellite operations. The satellites are launched using type F-1-m launchers. Such launchers are also used to orbit intercepter/destructor satellites. The next test satellite, Cosmos 209, remained in low orbit for 6 days before maneuvering to high altitude. In subsequent operational satellites, the period for which such satellites remained in low-altitude orbit increased to a maximum of 74 days for Cosmos 654, which was launched on May 17, 1974.

The true nature of these satellites was discovered in 1974, and the reasons for maneuvers such as those described above are now clear. These ocean-surveillance satellites are equipped with radar systems. Radar systems require large power sources. The Cosmos 954 accident in January 1978 has now shown such a power source to be a nuclear reactor. This would then explain, to some extent, the reason for changing the orbits of the satellites from low to high altitudes. At high altitudes the satellites will remain in orbit for some 500 years,

15. "Expanded Ocean Surveillance Effort Set," *Aviation Week and Space Technology* 109 (July 10, 1978): 22–23.

16. "US Navy Plans Surveillance Network," *Flight International* 113 (October 14, 1978): 1427.

17. "'DoD' Major New Satellite Programs," *Armed Forces Journal International* (February 1978), p. 25.

18. G. E. Perry, "Russian Ocean Surveillance Satellites," *Royal Air Force Quarterly* 18 (Spring 1978): 60–67.

a sufficient time for the short-lived radioactive fission products generated within the reactor to decay. When the satellites return to earth, however, there will still be some long-lived radioactive materials in the reactor of the satellite, including some of the unburned, highly enriched uranium-235 reactor fuel.

The launches of the Cosmos 198 and 209 satellites might be considered as the first phase of the Soviet ocean-surveillance satellites. Cosmos 367, 402, 516, and 626 belong to the second phase, during which the interval between the launch of each satellite is nearly constant. The exception is the period of 495 days between Cosmos 516 and 626. It is possible that a satellite was launched some 250 days after Cosmos 516 but that it was unsuccessful.[19]

Since 1974 two satellites have been launched each year, with the exception of 1975 when one extra satellite, Cosmos 785, was launched. This latter satellite may have been a failure. Satellites in each pair have been launched at a maximum of 5-day intervals. Moreover, Cosmos 651 and 654 were in the same orbital plane, about 25 min apart. This suggests that the satellites operate in pairs.[20]

The time difference between Cosmos 723 and 724 in their orbits (about 27 min) was similar to that between Cosmos 651 and 654, but the orbital planes differed by about 23°. By about the middle of May 1975, the time difference between Cosmos 723 and 724 was reduced to zero and, at about that time, Cosmos 723 was placed in its higher parking orbit. This suggests that this time difference is important to the operation of the satellites in pairs.[21] It is interesting to note that the United States uses ocean-surveillance satellites in a somewhat similar manner—four satellites are used at a time. Satellites in such a group are separated from each other in time and distance along their orbital paths, their orbits being in the same plane. The use of such pairs (90° apart) indicates that they are probably used to determine the position and velocity of the naval vessels being surveyed.

Satellites launched in 1976, Cosmos 860 and 861, were orbited in the same orbital plane. The orbits of Cosmos 952 and 954, launched on September 16 and 18, 1977, respectively, were also coplanar, and their time difference in the orbit was about 27 min. On October 8 the orbit of Cosmos 952 was raised to some 900 km.

From figure 2 it can be seen that from about October 26, Cosmos 954 had stopped making any maneuvers in order to remain in orbit. In fact it had begun the course of natural decay. However, this pattern suddenly changed on January 6, 1978, when the satellite began to come rapidly down to earth and, as can be seen, it landed on the earth's surface on January 24, bringing with it the

19. Ibid.
20. Ibid.
21. "'DoD' Major New Satellite Programs."

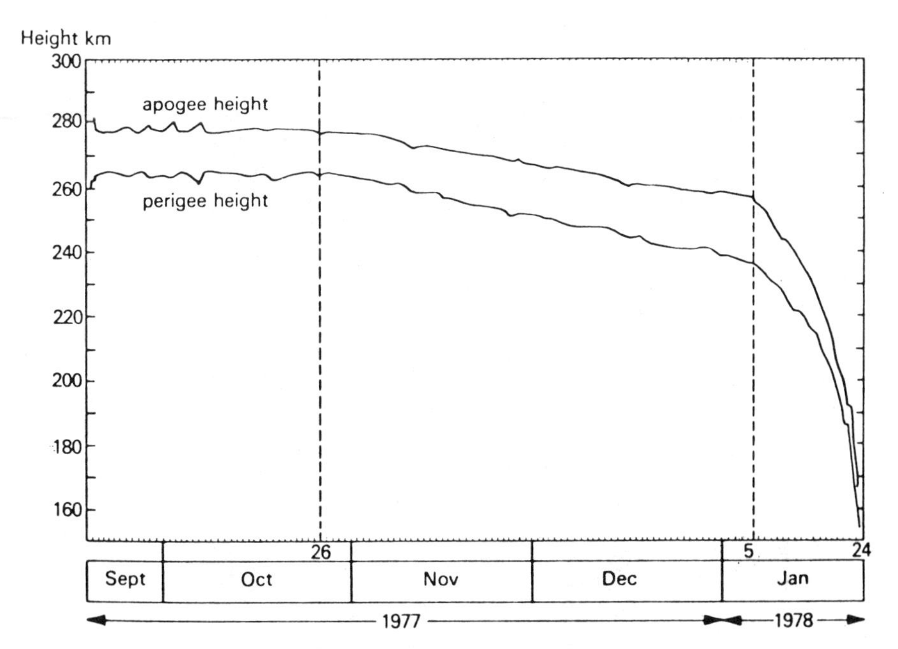

FIG. 2.—The perigee and the apogee heights of Cosmos 954 between September 17, 1977, and January 24, 1978, when the remains of the satellite landed in northern Canada.

radioactive parts of the satellite's reactor. In figure 3 the last few ground tracks of the failed Cosmos 954 satellite are shown. During track 2060 the satellite's unburned parts fell on Canadian soil (figure 4).

It has been suggested that, in addition to using satellites carrying radars, the USSR also uses spacecraft with television or elctronic signal monitoring sensors.[22] The first of this type of satellite, Cosmos 699, appears to have been orbited on December 12, 1974. One of the reasons for considering that this type of satellite performs ocean-surveillance tasks is that Cosmos 699 and 777 were observed to be testing microthrusters, which maneuvered the satellites for station keeping in a manner similar to Cosmos 723 and 724.[23] The latter satellites carried radars for ocean-surveillance purposes. Moreover, the satellites in this new group also transmitted on 166 MHz, a frequency used by the radar-carrying ocean-surveillance satellites. The Soviet ocean-surveillance satellites are listed in table 4.

22. "Soviets Seen Operating Two Types of Ocean Surveillance Satellite," *Aerospace Daily* 79 (June 2, 1976): 169–70.

23. Ibid., and Perry.

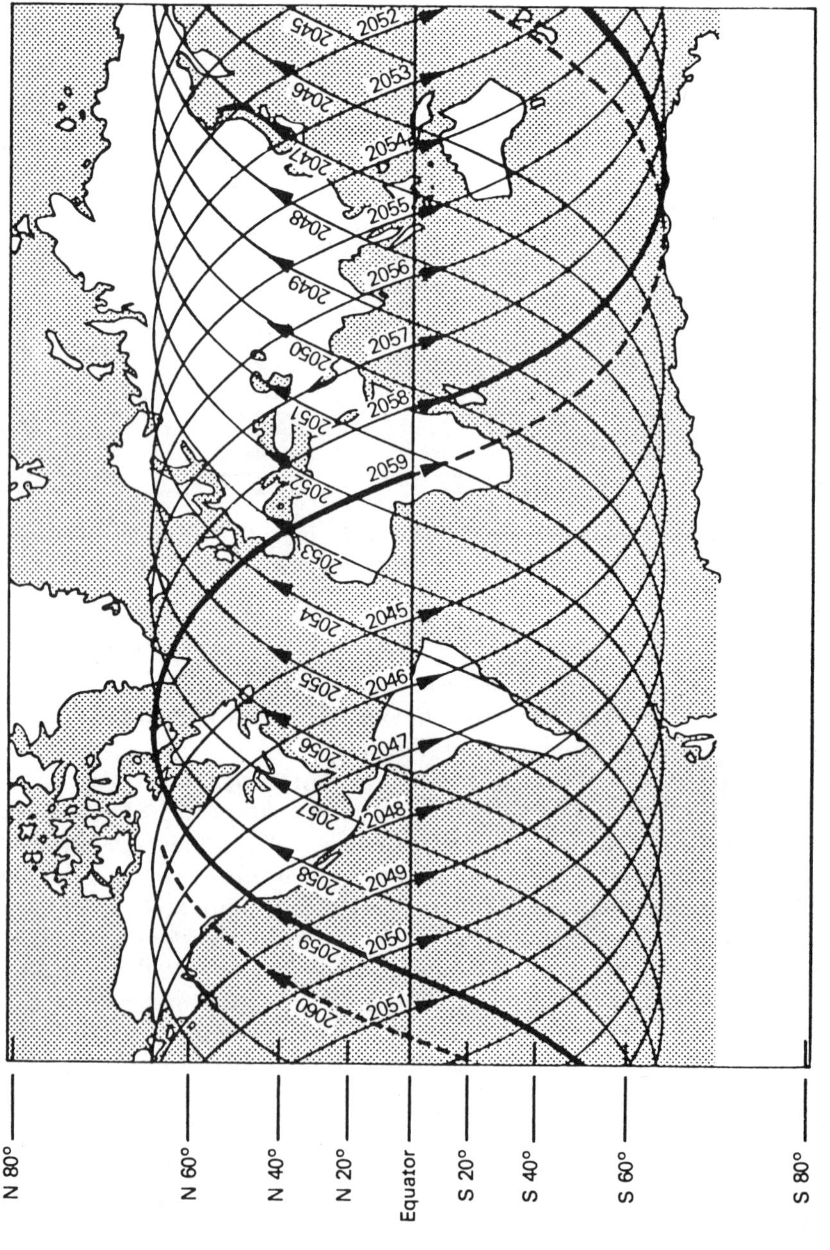

FIG. 3.—Ground tracks corresponding to the last 15 orbits of Cosmos 954. The number on the ground track is the orbit number.

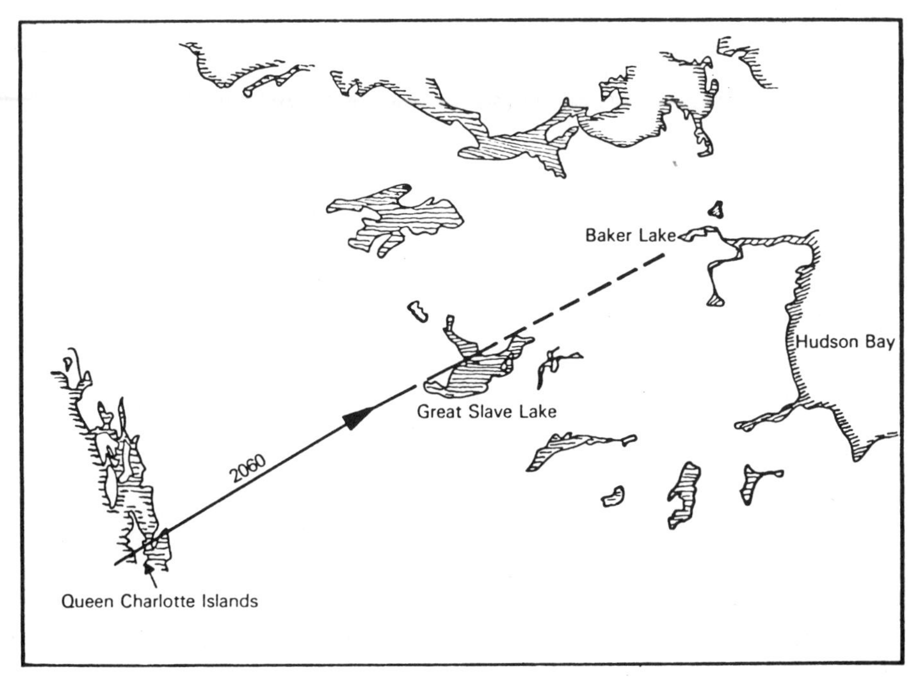

FIG. 4.—The reentry ground track of Cosmos 954 before its debris landed in northern Canada. The number on the ground track is the orbit number.

CONCLUSIONS

It can be seen that both the United States and the USSR have several satellites concentrating on ocean-surveillance tasks. It is also apparent that a considerable amount of ocean-surveillance data is generated by other types of satellites, such as weather and photographic reconnaisance satellites. Recently the U.S. Navy has completed a study on assessments of how Soviet spacecraft affect U.S. naval operations.[24] A method of avoiding detection of naval ships by satellites has been discussed in an earlier study. Knowing the orbital characteristics of a satellite, it is possible to determine the time and location of detection areas along a ship's route. By varying the ship's speed, such detection areas could be avoided. However, this becomes more complicated when more than one surveillance satellite is deployed. In this case, probability of detection by a given

24. "Navy Assesses Use of Space Systems," *Aviation Week and Space Technology* 109 (November 13, 1978): 67.

TABLE 4.—POSSIBLE SOVIET OCEAN-SURVEILLANCE SATELLITES

Satellite Name and Designation	Launch Date and Time (GMT)	Orbital Inclination (°)	Perigree Height (km)	Apogee Height (km)	Comments
1967:					
Cosmos 198 (1967-127A) . .	Dec. 27; 1131	65, 65	249, 894	270, 952	. . .
1968:					
Cosmos 209 (1968-23A) . . .	Mar. 22; 0936	65, 65	183, 871	343, 944	. . .
1970:					
Cosmos 367 (1970-79A) . . .	Oct. 3; 1033	65, 65	250, 922	280, 1,024	Moved to its higher orbit rapidly so that the orbital parameters of this orbit were announced
1971:					
Cosmos 402 (1971-25A) . . .	Apr. 1; 1131	65, 65	247, 948	274, 1,036	. . .
Cosmos 469 (1971-117A) . .	Dec. 25; 1131	65, 65	249, 941	262, 1,023	. . .
1972:					
Cosmos 516 (1972-66A) . . .	Aug. 21; 1033	65, 65	251, 920	263, 1,030	. . .
1973:					
Cosmos 626 (1973-108A) . .	Dec. 27; 2024	65, 65	257, 910	259, 990	. . .
1974:					
Cosmos 651 (1974-29A) . . .	May 15; 0726	65, 65	250, 892	264, 954	. . .
Cosmos 654 (1974-32A) . . .	May 17; 0658	65, 65	248, 913	265, 1,024	. . .
1975:					
Cosmos 723 (1975-24A) . . .	Apr. 2; 1102	65, 65	249, 916	266, 951	. . .
Cosmos 724 (1975-25A) . . .	Apr. 7; 1102	65, 65	248, 870	266, 934	. . .
Cosmos 785 (1975-116A) . .	Dec. 12; 1300	65, 65	251, 898	261, 1,023	. . .
1976:					
Cosmos 860 (1976-103A) . .	Oct. 17; 1814	65, 65	252, 919	265, 1,008	. . .
Cosmos 861 (1976-104A) . .	Oct. 21; 1702	65, 65	251, 919	265, 1,005	. . .
1977:					
Cosmos 937 (1977-77A) . . .	Aug. 24; 0712	65	424	444	. . .

TABLE 4.—POSSIBLE SOVIET OCEAN-SURVEILLANCE SATELLITES (*Continued*)

Satellite Name and Designation	Launch Date and Time (GMT)	Orbital Inclination (°)	Perigree Height (km)	Apogee Height (km)	Comments
Cosmos 952 (1977-88A) . . .	Sept. 16; 1424	65, 65	251, 910	265, 998	. . .
Cosmos 954 (1977-90A) . . .	Sept. 18; 1355	65	251	265	The satellite crash-landed in Canada with parts of its nuclear reactor

NOTE.—The second figure in the orbital inclination, perigee height, and apogee height columns is that of the final orbit.

satellite can be computed, particularly when weather conditions have to be taken into account.[25]

As a result of the Cosmos 954 accident, the United Nations has agreed to begin a technical study on the future of nuclear power systems aboard satellites. The results were to have been examined by the UN Space Committee in June 1979 and the recommendations discussed in the UN General Assembly. Both the United States and the USSR have agreed to participate in the technical study. There may be some difficulties raised by this study since, ideally, all types of nuclear power source have to be discussed and their use controlled. The United States, however, differentiates between a nuclear reactor and an RTG. The science and technology subcommittee met in February 1979 to formulate the study group which will analyze the questions of nuclear power systems aboard satellites.

The type of questions to be discussed would concern the altitude at which satellites should be allowed to use nuclear power sources aboard; when notification, if any, should be given of the launch and malfunction in orbit of such a satellite; and whether RTGs and reactors should be treated differently.

In 1978, President Carter stated the United States' willingness to ban the use of nuclear power sources aboard satellites. Such a ban would, however, hamper the use of certain military satellites only.

25. S. Barclay and R. Rancall, "Satellite Surveillance Avoidance Optimization Aid," Department of Defense, Defense Advanced Research Project Agency Report PR 77-14-36, December 1977.

Review of the Seabed Treaty

Jozef Goldblat
Stockholm International Peace Research Institute

The Treaty on the Prohibition of the Emplacement of Nuclear Weapons and Other Weapons of Mass Destruction on the Seabed and the Ocean Floor and in the Subsoil Thereof (the seabed treaty), which entered into force on May 18, 1972, provides for a review of its operation with a view to assuring that its purposes and provisions are being realized.[1] Such a review was carried out at a conference held in Geneva, Switzerland, from June 20 to July 1, 1977, with the participation of 42 states party to the seabed treaty.[2] Three signatories—Argentina, Brazil, and Greece—which had not ratified the treaty attended the conference without taking part in the decisions, while one state, Nigeria, which had neither signed nor ratified the treaty was accorded observer status at its own request. Representatives of the United Nations and the International Atomic Energy Agency (IAEA), as well as certain nongovernmental organizations, were also present. The conference adopted by consensus—that is, without a vote being taken—a final declaration assessing, article by article, the working of the treaty.[3]

SCOPE AND GEOGRAPHICAL EXTENT OF THE SEABED PROHIBITIONS

Under article 1, paragraph 1, the parties to the seabed treaty undertook not to emplace on the seabed and the ocean floor and in the subsoil thereof beyond a given seabed zone any nuclear weapons or other types of weapons of mass destruction as well as structures, launching installations, or any other facilities

1. For the text of the treaty, see Jozef Goldblat, "The Seabed Treaty," in *Ocean Yearbook 1*, ed. Elisabeth Mann Borgese and Norton Ginsburg (Chicago: University of Chicago Press, 1978), pp. 402–11.
2. Australia, Austria, Belgium, Bulgaria, Byelorussia, Canada, Cyprus, Czechoslovakia, Denmark, Finland, the German Democratic Republic, the Federal Republic of Germany, Ghana, Hungary, Iceland, India, Iran, Iraq, Ireland, Italy, Japan, Jordan, Malaysia, Mauritius, Mongolia, Morocco, Nicaragua, the Netherlands, New Zealand, Norway, Poland, Portugal, Romania, Sweden, Switzerland, Tunisia, Turkey, Ukraine, the United Kingdom, the United States, the USSR, and Yugoslavia.
3. For the test of the final declaration, see this vol., pp. 555–58.

© 1980 by The University of Chicago. 0-226-06603-7/80/0002-0010$01.00

designed for storing, testing, or using such weapons. The seabed zone was defined in article 2 as a zone the outer limit of which is coterminous with the 12-mi outer limit of the zone referred to in the Convention on the Territorial Sea and the Contiguous Zone of April 29, 1958, and measured in accordance with the provisions of that convention and international law.

No complaints with regard to compliance with article 1 were made, and the review conference stated that the obligations assumed by the parties had been "faithfully" observed. In fact, the activities prohibited by the treaty are militarily so unattractive that they are unlikely to occur even in the absence of internationally binding commitments not to engage in them.

It will be recalled, however, that according to article 1, paragraph 2, the nuclear weapon powers are not prevented from installing nuclear weapons (or other weapons of mass destruction) within the seabed zone adjacent to their coasts, that is, in that portion of the seabed which is more suitable for this purpose than are outlying areas, or even beneath the territorial waters of other states (if these states are willing to authorize such installation and if the operation is carried out within the 12-mi seabed zone). To fill this rather significant gap in the treaty, Japan appealed to the parties voluntarily to refrain from emplacing nuclear weapons within territorial waters.

Another problem raised in connection with articles 1 and 2 of the seabed treaty concerned the legal regime which may emerge from the Law of the Sea Conference and the implications of such a regime for the geographical extent of the area covered by the treaty.[4] In particular, the Convention on the Territorial Sea and the Contiguous Zone, referred to in article 2, is expected to be replaced by new rules. The overwhelming opinion of the participants in the seabed treaty review conference was that the outcome of the Law of the Sea Conference should have no effect on the rights and obligations deriving from the seabed treaty nor reduce the extent of the denuclearized area by altering the delimitation of that portion of the seabed to which the undertakings set out in article 1 of the treaty apply. This opinion was reflected in the final declaration, in which the conference reaffirmed its support for the provisions defining the zone covered by the treaty.

VERIFICATION OF THE PROHIBITIONS

Parties to the seabed treaty have the right to verify through observation the activities of other states on the seabed and the ocean floor and in the subsoil thereof beyond the seabed zone (article 3, par. 1). The possibility of "appropriate" inspection of objects, structures, installations, or other facilities that reasonably may be expected to be of the kind prohibited is also envisaged (article 3, par. 2).

4. Goldblat, pp. 393–95.

Since no charges of violation had been made, the effectiveness and practicability of the verification procedure set out in the treaty remained untested. Nevertheless, several nations found the control provisions insufficient and pointed out that most countries have no technical means to follow the developments on the seabed and cannot, therefore, acquire information which would justify setting in motion the consultative and investigative machinery described in paragraphs 2 and 3 of article 3.

As far as verification operations are concerned, the treaty envisages the possibility of carrying these out with the full or partial assistance of other states. However, many nations, especially those which are nonaligned or neutral, would be reluctant to resort to the aid of the technologically advanced powers and to rely for their security on the goodwill of these powers. It was therefore proposed that the modalities for appropriate, but unspecified, international procedures within the framework of the United Nations, as stipulated in article 3, paragraph 5, should be so interpreted as to include the good offices of the UN Secretary-General in providing assistance to states not possessing satisfactory verification means of their own and desiring such assistance. This interpretation proved unacceptable to the USSR.

Another proposal was to establish a consultative, fact-finding committee to clarify problems relating to the application of the treaty before complaints concerning compliance are referred to the UN Security Council in accordance with paragraph 4 of article 3. (Such a committee is provided for in the 1977 convention prohibiting the use of environmental modification techniques for hostile purposes.) This proposal was strongly supported by several states, but instead of amending the treaty, which would require a special procedure under article 6, the review conference participants agreed to include the following formulation in the final declaration: "The Conference considers that the provisions for consultation and co-operation contained in paragraphs 2, 3 and 5 include the right of interested States Parties to agree to resort to various international consultative procedures, such as *ad hoc* consultative groups of experts and other procedures."

India attempted to obtain recognition for a principle that it had proclaimed whereby no state could, without the consent of the coastal state, take measures with a view to implementing verification provisions which might restrict the sovereign rights of a coastal state in the "exclusive economic zone." In other words, the coastal state would have to be consulted and its permission obtained before any activities pursuant to article 3 could be undertaken in the waters extending up to 200 mi from its shores. Also, Yugoslavia interpreted article 3 to mean that states exercising the right stipulated therein should give the coastal state concerned prior notification of their verification activities beyond the seabed zone referred to in article 1. (A similar view expressed by Yugoslavia upon ratification of the seabed treaty was objected to by the United States and the United Kingdom.)

However, the tendency to extend the coastal states' national jurisdiction over large areas of the high seas met with opposition. The majority of conference participants were of the opinion that parties to the seabed treaty should not have the right to restrict the verification operations provided therein and that paragraph 6 of article 3 gave sufficient assurance that coastal states' rights under international law would not be encroached upon by the activities of other states. In this context, Switzerland regretted the absence of a provision for impartial procedures to settle disputes that might arise over verification of activities on the seabed. In any event, the above discussion had little practical relevance, because the seabed treaty's verification machinery is not likely to be put to a test.

FURTHER SEABED DISARMAMENT MEASURES

Article 5 calls upon the parties to the seabed treaty to continue negotiations concerning further measures in the field of disarmament for the prevention of an arms race on the seabed, the ocean floor, and the subsoil thereof. Such negotiations had not taken place by the time the review conference was convened.

The United States stated that no arms race on the seabed had occurred and that there was little prospect of it occurring in the future. It was convinced that negotiations on further multilateral arms-control measures focusing exclusively on the seabed were not necessary to satisfy the purpose of article 5. The USSR, however, reiterated its view that the treaty was only a first step toward the complete demilitarization of the seabed. Yugoslavia said that because of the increased militarization of the seas it was impossible to consider the seabed and the ocean floor in isolation from the ocean space as a whole and suggested negotiations aimed at a gradual reduction and withdrawal of fleets from sensitive regions of the world.

On the other hand, Japan found it unrealistic, under the present circumstances, to aim at a comprehensive prohibition of military activities that would ban the use of the seabed even for purely defensive purposes. In the same vein, Canada pointed out that certain military uses of the seabed must be regarded as an important part of national defense and should be viewed as compatible with the principle of the peacful uses of the deep ocean floor. Similarly, Iran expressed the opinion that if further steps were taken to prevent an arms race on the seabed exception would have to be made for acoustic submarine detection systems.[5] As possible further restrictive measures concerning the seabed, Iran mentioned a prohibition on the military use of the continental shelf by foreign

5. For a discussion of such systems, see Owen Wilkes, "Ocean-based Nuclear Deterrent Forces and Antisubmarine Warfare," this vol., pp. 226–49.

states without the consent of the coastal states concerned, as well as an extension of the proscription under the treaty to reduce or prevent military support activities on the seabed.

These and other states drew attention to a possible overlap between military and civilian uses of new techniques in emplacing sophisticated structures under the sea and considered it necessary that a future legal regime for the peaceful exploitation of the ocean floor resources should be complemented by a ban on the military use of the ocean floor.

Finally, the review conference agreed that the Committee on Disarmament should be requested, in consultation with the parties to the treaty and taking into account the proposals made during the conference, as well as any relevant technological developments, to proceed "promptly" with consideration of further measures in the field of disarmament for the prevention of an arms race on the seabed, the ocean floor, and the subsoil thereof. No such consideration took place during the 2 yr which followed the review conference.

AMENDMENTS TO THE TREATY

According to article 6 of the seabed treaty, any party may propose amendments. Amendments would enter into force for each party accepting them upon their acceptance by a majority of the parties to the treaty and, thereafter, for each remaining party on the date of acceptance by it.

The conference noted that during the preceding 5 yr of the operation of the treaty no state party proposed amendments to the treaty according to the procedure described above. During the conference itself, proposals were made to modify the control provisions set out in article 3 in order to insure that all parties have access to the means of verification necessary to determine whether or not a breach of the treaty has occurred, but the need for introducing a formal amendment was avoided due to an agreed interpretation of article 3, as discussed above.

PERIODIC REVIEW OF THE TREATY

Article 7 of the seabed treaty, which established a legal basis for the review conference, stipulates that "relevant technological developments" shall be taken into account in the review of the operation of the treaty. It is obvious that only the great powers which possess sophisticated underwater technologies, as well as military resources, are in a position to identify developments that might affect the purposes and the provisions of the treaty. However, the United Kingdom, the United States, and the USSR did not provide any such information. Finland saw the possibility of support equipment being emplaced on the seabed, which, without being designed for launching, storing, testing, or using nuclear weapons, could indirectly facilitate their deployment. In its view, the seabed could also be used for the deployment of controllable mines or mine-torpedo systems which might even be equipped with nuclear warheads.[6]

The Netherlands and Sweden found it hard to believe that no relevant military or peaceful technological developments had taken place since the seabed treaty entered into force. In this connection, Sweden suggested an examination of the following questions: (*a*) Have technological developments facilitated the emplacement of weapons of mass destruction on the seabed, the ocean floor, and in the subsoil thereof? (*b*) Has it become technologically and economically easier to emplace weapons other than weapons of mass destruction on the seabed? (*c*) Has technology created new motives for the emplacement of weapons on the seabed? (*d*) Do present trends imply a greater danger of an arms race on the seabed, and, therefore, should further measures in the disarmament field be the subject of studies and negotiations? (*e*) How should new technological developments be assessed as regards installations for peaceful uses which might also be used for military purposes? and (*f*) How are new technologies for controlling activities on the seabed, which might be relevant to the application of article 3, to be evaluated? Thus, consideration of article 7 concerning the review of the operation of the treaty was linked with the substance of article 5 providing for further measures in the field of disarmament for the prevention of an arms race on the seabed.

The review conference recognized the need to keep under continuing review the technological developments relating both to the treaty's applicability and to possible further seabed disarmament measures and invited the Committee on Disarmament to establish an ad hoc expert group for these purposes. It also invited the UN Secretary-General to collect information relevant to the treaty from "officially" available sources and to publish it in the *UN Yearbook on Disarmament.* (A proposal that experts should be able to consult all sources, whether official or unofficial, was not accepted.) By 1979, no expert group had been established to review the technological developments on the seabed, and no information about such developments had been provided to the UN Secretary-General.

After a prolonged debate the review conference decided that a further review of the seabed treaty, in accordance with its article 7, should be carried out at a conference convened in 1982, unless a majority of the parties indicated to the depositaries that they wished it to be postponed. In any case, a further review conference would take place not later than 1984. It would determine, in conformity with the views of a majority of the parties attending, whether and when an additional conference should be held.

6. Ibid., p. 244.

WITHDRAWAL FROM THE TREATY

Under article 8 of the treaty, each party has the right to withdraw from it if it decides that extraordinary events related to the subject matter of the treaty have jeopardized the supreme interests of its country. The right of withdrawal had not been invoked by any party, and this was noted by the conference with satisfaction. However, Jordan expressed the view that article 8 made withdrawal from the treaty too easy and that the shortness of the notice period (3 mo) might tempt the parties to take action which would render the treaty ineffective. This point of view was disputed. Several countries indicated that withdrawal would no doubt be exercised only if another party violated the treaty's provisions or if a nonparty performed acts which endangered the interests of the parties.

DISCLAIMERS

Article 9 stipulates that the provisions of the treaty should in no way affect the obligations assumed by the parties under international instruments establishing zones free from nuclear weapons. The validity of this clause was reaffirmed by the review conference. Its importance is particularly obvious in the case of the treaty of Tlatelolco prohibiting nuclear weapons in Latin America, according to which the whole area of the seabed in that region of the world has been denuclearized.

The question of the application of the seabed treaty to the seabed adjacent to demilitarized territory was also raised. Turkey stated that, in its view, the seabed treaty did not grant the right to install arms in the demilitarized zones coming within the definition contained in article 1, paragraph 2 of the treaty. This was a reference to the demilitarized regime of certain Greek islands, but the same could apply to other demilitarized areas, such as the Antarctic or Spitsbergen (Svalbard) islands. In view of the politically sensitive nature of the issue, no clear-cut position could be agreed upon. The review conference simply noted that obligations assumed by parties to the seabed treaty arising from other international instruments "continue to apply."

ADHERENCE TO THE TREATY

The review conference called upon states that had not yet become parties to the seabed treaty, particularly those possessing nuclear weapons or any other types of weapons of mass destruction, to do so at the earliest possible date. Only a handful of states have so far responded to this appeal. By December 1978, the total number of parties was no more than 66 (see the Appendix to this article). China and France remained outside the treaty.

CONCLUSIONS

The main purpose of the seabed treaty review conference was to examine the effects of developments in underwater and weapons technology on the military uses of the ocean floor and the implications of such developments for efforts aimed at the demilitarization of the seabed. But such examination proved impossible because relevant information was not made available to the conference participants. Consequently, no impulse was provided for consideration of further measures. A potentially major issue—the competing peaceful and military uses of the seabed—was mentioned but not discussed. The conference did not arouse interest for the seabed treaty among states which had not adhered to it. Evidently, arms-control agreements relating to the seas lack attractiveness as long as they are restricted to the seabed and to weapons of mass destruction alone. It is possible that the expected convention on the law of the sea may require for its effectiveness the adoption of more comprehensive arms-control measures.

APPENDIX

COUNTRIES WHICH HAVE SIGNED, RATIFIED, OR ACCEDED TO THE SEABED TREATY AS OF DECEMBER 31, 1978

Country	Signed	Ratified*
Afghanistan	Feb. 11, 1971 (L,M,W)	Apr. 22, 1971 (M) Apr. 23, 1971 (L) May 21, 1971 (W)
Argentina†	Sept. 3, 1971 (L,M,W)	...
Australia	Feb. 11, 1971 (L,M,W)	Jan. 23, 1973 (L,M,W)
Austria	Feb. 11, 1971 (L,M,W)	Aug. 10, 1972 (L,M,W)
Belgium	Feb. 11, 1971 (L,M,W)	Nov. 20, 1972 (L,M,W)
Benin (Dahomey)	Mar. 18, 1971 (W)	...
Bolivia	Feb. 11, 1971 (L,M,W)	...
Botswana	Feb. 11, 1971 (W)	Nov. 10, 1972 (W)
Brazil‡	Sept. 3, 1971 (L,M,W)	...
Bulgaria	Feb. 11, 1971 (L,M,W)	Apr. 16, 1971 (M) May 7, 1971 (W) May 26, 1971 (L)
Burma	Feb. 11, 1971 (L,M,W)	...
Burundi	Feb. 11, 1971 (M,W)	...
Byelorussia	Mar. 3, 1971 (M)	Sept. 14, 1971 (M)
Cambodia (see Democratic Kampuchea)	...	...
Cameroon (see United Republic of Cameroon)	...	...
Canada§	Feb. 11, 1971 (L,M,W)	May 17, 1972 (L,M,W)
Central African Empire	Feb. 11, 1971 (W)	...
Colombia	Feb. 11, 1971 (W)	...

APPENDIX (*Continued*)

Country	Signed	Ratified*
Congo	...	Oct. 23, 1978 (W)
Costa Rica	Feb. 11, 1971 (W)	...
Cuba¶	...	June 3, 1977 (M)
Cyprus	Feb. 11, 1971 (L,M,W)	Nov. 17, 1971 (L,M)
		Dec. 30, 1971 (W)
Czechoslovakia	Feb. 11, 1971 (L,M,W)	Jan. 11, 1972 (L,M,W)
Dahomey (see Benin)	...	...
Democratic Kampuchea (Cambodia)	Feb. 11, 1971 (W)	...
Democratic Yemen (South)	Feb. 23, 1971 (M)	...
Denmark	Feb. 11, 1971 (L,M,W)	June 15, 1971 (L,M,W)
Dominican Republic	Feb. 11, 1971 (W)	Feb. 11, 1972 (W)
Equatorial Guinea	June 4, 1971 (W)	...
Ethiopia	Feb. 11, 1971 (L,M,W)	July 12, 1977 (L)
		July 14, 1977 (M,W)
Finland	Feb. 11, 1971 (L,M,W)	June 8, 1971 (L,M,W)
Gambia	May 18, 1971 (L)	...
	May 21, 1971 (M)	...
	Oct. 29, 1971 (W)	...
German Democratic Republic	Feb. 11, 1971 (M)	...
	July 27, 1971 (M)	...
Germany, Federal Republic of#	June 8, 1971 (L,M,W)	Nov. 18, 1975 (L,W)
Ghana	Feb. 11, 1971 (L,M,W)	Aug. 9, 1972 (W)
Greece	Feb. 11, 1971 (M)	...
	Feb. 12, 1971 (W)	...
Guatemala	Feb. 11, 1971 (W)	...
Guinea	Feb. 11, 1971 (M,W)	...
Guinea-Bissau	...	Aug. 20, 1976 (M)
Honduras	Feb. 11, 1971 (W)	...
Hungary	Feb. 11, 1971 (L,M,W)	Aug. 13, 1971 (L,M,W)
Iceland	Feb. 11, 1971 (L,M,W)	May 30, 1972 (L,M,W)
India**	...	July 20, 1973 (L,M,W)
Iran	Feb. 11, 1971 (L,M,W)	Aug. 26, 1971 (L,W)
		Sept. 6, 1972 (M)
Iraq¶	Feb. 22, 1971 (M)	Sept. 13, 1972 (M)
Ireland	Feb. 11, 1971 (L,W)	Aug. 19, 1971 (L,W)
Italy††	Feb. 11, 1971 (L,M,W)	Sept. 3, 1974 (L,M,W)
Ivory Coast	...	Jan. 14, 1972 (W)
Jamaica	Oct. 11, 1971 (L,W)	...
	Oct. 14, 1971 (M)	...
Japan	Feb. 11, 1971 (L,M,W)	June 21, 1971 (L,M,W)
Jordan	Feb. 11, 1971 (L,M,W)	Aug. 17, 1971 (W)
		Aug. 30, 1971 (M)
		Nov. 1, 1971 (L)
Kampuchea (see Democratic Kampuchea)	...	...
Korea, South¶	Feb. 11, 1971 (L,W)	...

APPENDIX (*Continued*)

Country	Signed	Ratified*
Lao People's Democratic Republic	Feb. 11, 1971 (L,W)	Oct. 19, 1971 (L)
	Feb. 15, 1971 (M)	Oct. 22, 1971 (M)
		Nov. 3, 1971 (W)
Lebanon	Feb. 11, 1971 (L,M,W)	...
Lesotho	Sept. 8, 1971 (W)	Apr. 3, 1973 (W)
Liberia	Feb. 11, 1971 (W)	...
Luxembourg	Feb. 11, 1971 (L,M,W)	...
Madagascar	Sept. 14, 1971 (W)	...
Malaysia	May 20, 1971 (L,M,W)	June 21, 1972 (L,M,W)
Mali	Feb. 11, 1971 (W)	...
	Feb. 15, 1971 (M)	...
Malta	Feb. 11, 1971 (L,W)	May 4, 1971 (W)
Mauritius	Feb. 11, 1971 (W)	Apr. 23, 1971 (W)
		May 3, 1971 (L)
		May 18, 1971 (M)
Mongolia	Feb. 11, 1971 (L,M)	Oct. 8, 1971 (M)
		Nov. 15, 1971 (L)
Morocco	Feb. 11, 1971 (M,W)	July 26, 1971 (L)
	Feb. 18, 1971 (L)	Aug. 5, 1971 (W)
		Jan. 18, 1972 (M)
Nepal	Feb. 11, 1971 (M,W)	July 6, 1971 (L)
	Feb. 24, 1971 (L)	July 29, 1971 (M)
		Aug. 9, 1971 (W)
Netherlands	Feb. 11, 1971 (L,M,W)	Jan. 14, 1976 (L,M,W)
New Zealand	Feb. 11, 1971 (L,M,W)	Feb. 24, 1972 (L,M,W)
Nicaragua	Feb. 11, 1971 (W)	Feb. 7, 1973 (W)
Niger	Feb. 11, 1971 (W)	Aug. 9, 1971 (W)
Norway	Feb. 11, 1971 (L,M,W)	June 28, 1971 (L,M)
		June 29, 1971 (W)
Panama	Feb. 11, 1971 (W)	Mar. 20, 1974 (W)
Paraguay	Feb. 23, 1971 (W)	...
Poland	Feb. 11, 1971 (L,M,W)	Nov. 15, 1971 (L,M,W)
Portugal	...	June 24, 1975 (L,M,W)
Qatar	...	Nov. 12, 1974 (L)
Romania‡‡	Feb. 11, 1971 (L,M,W)	July 10, 1972 (L,M,W)
Rwanda	Feb. 11, 1971 (W)	May 20, 1975 (L,M,W)
Saudi Arabia	Jan. 7, 1972 (W)	June 23, 1972 (W)
Senegal	Mar. 17, 1971 (W)	...
Seychelles	...	June 29, 1976 (W)
Sierra Leone	Feb. 11, 1971 (L)	...
	Feb. 12, 1971 (M)	...
	Feb. 24, 1971 (W)	...
Singapore	May 5, 1971 (L,M,W)	Sept. 10, 1976 (L,M,W)
South Africa	Feb. 11, 1971 (W)	Nov. 14, 1973 (W)
		Nov. 26, 1973 (L)
Sudan	Feb. 11, 1971 (L)	...
	Feb. 12, 1971 (M)	...
Swaziland	Feb. 11, 1971 (W)	Aug. 9, 1971 (W)

APPENDIX (*Continued*)

Country	Signed	Ratified*
Sweden	Feb. 11, 1971 (L,M,W)	Apr. 28, 1972 (L,M,W)
Switzerland	Feb. 11, 1971 (L,M,W)	May 4, 1976 (L,M,W)
Taiwan	Feb. 11, 1971 (W)	Feb. 22, 1972 (W)
Tanzania (see United Republic of Tanzania)	...	...
Togo	Apr. 2, 1971 (W)	June 28, 1971 (W)
Tunisia	Feb. 11, 1971 (L,M,W)	Oct. 22, 1971 (M)
		Oct. 28, 1971 (L)
		Oct. 29, 1971 (W)
Turkey	Feb. 25, 1971 (L,M,W)	Oct. 19, 1972 (W)
		Oct. 25, 1972 (L)
		Oct. 30, 1972 (M)
Ukrainian SSR	Mar. 3, 1971 (M)	Sept. 3, 1971 (M)
Union of Soviet Socialist Republics	Feb. 11, 1971 (L,M,W)	May 18, 1972 (L,M,W)
United Kingdom§§	Feb. 11, 1971 (L,M,W)	May 18, 1972 (L,M,W)
United Republic of Cameroon	Nov. 11, 1971 (M)	...
United Republic of Tanzania	Feb. 11, 1971 (W)	...
United States	Feb. 11, 1971 (L,M,W)	May 18, 1972 (L,M,W)
Uruguay	Feb. 11, 1971 (W)	...
Viet Nam¶¶	...	...
Yemen (North)	Feb. 23, 1971 (M)	...
Yugoslavia##	Mar. 2, 1971 (L,M,W)	Oct. 25, 1973 (L,M,W)
Zambia	...	Oct. 9, 1972 (L)
		Nov. 1, 1972 (W)
		Nov. 2, 1972 (M)

NOTE.—L = London, M = Moscow, W = Washington (place of signature and/or deposit of the instruments of ratification or accession).

* Deposit of instruments of ratification or accession.

† On signing the treaty, Argentina stated that it interprets the references to the freedom of the high seas as in no way implying a pronouncement of judgment on the different positions relating to questions connected with international maritime law. It understands that the reference to the rights of exploration and exploitation by coastal states over their continental shelves was included solely because those could be the rights most frequently affected by verification procedures. Argentina precludes any possibility of strengthening, through this treaty, certain positions concerning continental shelves to the detriment of others based on different criteria.

‡ On signing the treaty, Brazil stated that nothing in the treaty shall be interpreted as prejudicing in any way the sovereign rights of Brazil in the area of the sea, the seabed, and the subsoil thereof adjacent to its coasts. It is the understanding of the Brazilian government that the word "observation," as it appears in paragraph 1 of article 3 of the treaty, refers only to observation that is incidental to the normal course of navigation in accordance with international law.

§ In depositing the instrument of ratification, Canada declared that article 1, paragraph 1 cannot be interpreted as indicating that any state has a right to implant or emplace any weapons not prohibited under article 1, paragraph 1 on the seabed and ocean floor, and in the subsoil thereof, beyond the limits of national jurisdiction, or as constituting any limitation on the principle that this area of the seabed and ocean floor and the subsoil thereof shall be reserved for exclusively peaceful purposes. Articles 1, 2, and 3 cannot be interpreted as indicating that any state but the coastal state has any right to implant or emplace any weapon not prohibited under article 1, paragraph 1 on the continental shelf, or the subsoil thereof, appertaining to that coastal state, beyond the outer limit of the seabed zone referred to in article 1 and defined in article 2. Article 3 cannot be interpreted as indicating any restrictions or limitation upon the rights of the coastal state, consistent with its exclusive sovereign rights with respect to the continental shelf, to verify, inspect, or effect the removal of any weapon, structure, installation, facility, or device implanted or emplaced on the continental shelf, or the subsoil thereof, appertaining to that coastal state, beyond the outer limit of the seabed zone referred to in article 1 and defined in article 2. On April 12, 1976, the Federal Republic of Germany stated that the declaration by Canada is not of a nature to confer on the government of this country more far-reaching rights than those to which it is entitled under current international law, and that all rights existing under current international law which are not covered by the prohibitions are left intact by the treaty.

APPENDIX (*Continued*)

¶ A statement was made containing a disclaimer regarding recognition of states party to the treaty.

On ratifying the treaty, the Federal Republic of Germany declared that the treaty will apply to Berlin (West).

** On the occasion of its accession to the treaty, the government of India stated that as a coastal state India has, and always has had, full and exclusive sovereign rights over the continental shelf adjoining its territory and beyond its territorial waters and the subsoil thereof. It is the considered view of India that other countries cannot use its continental shelf for military purposes. There cannot, therefore, be any restriction on, or limitation of, the sovereign right of India as a coastal state to verify, inspect, remove, or destroy any weapon, device, structure, installation, or facility which might be implanted or emplaced on or beneath its continental shelf by any other country, or to take such other steps as may be considered necessary to safeguard its security. The accession by the government of India to the seabed treaty is based on this position. In response to the Indian statement, the U.S. government expressed the view that, under existing international law, the rights of coastal states over their continental shelves are exclusive only for purposes of exploration and exploitation of natural resources, and are otherwise limited by the 1958 Convention on the Continental Shelf and other principles of international law. On April 12, 1976, the Federal Republic of Germany stated that the declaration by India is not of a nature to confer on the government of this country more far-reaching rights than those to which it is entitled under current international law, and that all rights existing under current international law which are not covered by the prohibitions are left intact by the treaty.

†† On signing the treaty, Italy stated, inter alia, that in the case of agreements on further measures in the field of disarmament to prevent an arms race on the seabed and ocean floor and in their subsoil, the question of the delimitation of the area within which these measures would find application shall have to be examined and solved in each instance in accordance with the nature of the measures to be adopted. The statement was repeated at the time of ratification.

‡‡ Romania stated that it considered null and void the ratification of the treaty by the Taiwan authorities.

§§ The United Kingdom recalled its view that if a regime is not recognized as the government of a state neither signature nor the deposit of any instrument by it, nor notification of any of those acts, will bring about recognition of that regime by any other state.

¶¶ South Viet Nam signed the seabed treaty on February 11, 1971. On April 30, 1975, the Republic of South Viet Nam ceased to exist as a separate political entity. As from July 2, 1976, North and South Viet Nam constitute a single state under the official name of the Socialist Republic of Viet Nam. The government of the unified state may decide whether it will adhere to international commitments undertaken by the former administration.

On February 25, 1974, the ambassador of Yugoslavia transmitted to the U.S. secretary of state a note stating that in the view of the Yugoslav government, article 3, paragraph 1 of the treaty should be interpreted in such a way that a state exercising its right under this article shall be obliged to notify in advance the coastal state, insofar as its observations are to be carried out "within the stretch of the sea extending above the continental shelf of the said state." On January 16, 1975, the U.S. secretary of state presented the view of the United States concerning the Yugoslav note, as follows: "Insofar as the note is intended to be interpretative of the Treaty, the United States cannot accept it as a valid interpretation. In addition, the United States does not consider that it can have any effect on the existing law of the sea." Insofar as the note was intended to be a reservation to the treaty, the United States placed on record its formal objection to it on the grounds that it was incompatible with the object and purpose of the treaty. The United States also drew attention to the fact that the note was submitted too late to be legally effective as a reservation. A similar exchange of notes took place between Yugoslavia and the United Kingdom. On April 12, 1976, the Federal Republic of Germany stated that the declaration by Yugoslavia is not of a nature to confer on the government of this country more far-reaching rights than those to which it is entitled under current international law, and that all rights existing under current international law which are not covered by the prohibitions are left intact by the treaty.

East Asian Ocean Security Zones

Bruce D. Larkin
University of California at Santa Cruz

One object of sea law is to define rights of access to designated ocean regions. Access regulation is a common theme shared by several key issues at the Third United Nations Conference on the Law of the Sea. What access shall be permitted in the territorial sea? in the proposed economic zone? through the waters of an archipelagic state? More is at stake than the right to be physically present: presence and purpose, the activity for which a right to presence is claimed, are not separable.

The conference deliberates; and as it proceeds, states declare unilateral claims and joint undertakings. The focus of this paper is upon claims by the People's Republic of China and the Democratic People's Republic of Korea that a coastal state may properly regulate foreign naval vessels in zones beyond the outer limit of the territorial sea. Chinese and North Korean claims are distinct; there is no public record of action in concert. Nonetheless, the effect of their independent claims is to pose more insistently an issue of principle and profound practical consequence. It is important and appropriate to consider the lines along which such claims might develop and the arguments which might be adduced to justify them. It is useful to examine the record with care and wise to consider the form such claims might take in the future.

The controversy is most simply one between naval powers seeking broad rights of access and coastal states seeking to keep foreign naval forces at a distance.

If the strategic consequences were not compelling in themselves, we should explore this issue to clarify the distinctions among transit, use, occupation, and patrol of a region hitherto designated common. Other law of the sea problems raise questions concerning the global commons, but the issue posed by security zones is distinctive. For example, it is not that of the "tragedy of the commons" (destruction of a common zone because those with a right of access fail to exercise restraint) nor is it that of enclosure (removal of a zone from common use and its dedication instead to the private gain of a sole owner) with which the doctrine of exclusive economic zones is in some respects congruent. Here the issue is whether the right to walk through the commons carries with it an unlimited right to bear arms as well, or to be accompanied by a guard bearing arms, even in sections of the commons which have been dedicated to the

© 1980 by The University of Chicago. 0-226-06603-7/80/0002-0002$01.00

exclusive economic use of another party (which has made a solemn commitment to keep the peace in its zone).

Does the right to walk across another's field imply the right to go fully armed? Posing the problem in this way emphasizes the difference between sovereign states and private persons within the state: There is in principle no reason to carry arms because the state keeps the peace; much less is there reason to send private armed patrols across the right of way. Moreover, the case made for a right to send armed patrols through another state's exclusive economic zone is that it is required for self-defense and the protection of interests outside that zone: even if a coastal state is conscientious and effective in keeping the peace within its zone, since it cannot guarantee the peace elsewhere, it must not limit the freedom of movement of others' naval vessels through its zone according to this argument.

SECURITY ZONES

Security zones are an issue because (i) states are declaring such zones although (ii) they contradict prior understandings in international law. Three changes in the last 30 years have placed security zones on the agenda. It is increasingly possible for a coastal state, even with only a modest naval capability, to enforce exclusion from a security zone. Several ad hoc precedents of exclusion and regulation have entered the record. And there is broad consensus that claims to exclusive economic zones extending up to 200 nautical miles from the baseline of the territorial sea will be acknowledged. These are the trends of possibility, precedent, and creeping sovereignty. Ironically, it is the high-seas states which have pioneered changes in weapons technology, set some of the precedents for exclusion or regulation of foreign craft, and set in motion successively expansive claims on ocean space beyond the limit of the territorial sea.

Possibility

Two changes have made possible declaration of zones which 50 years ago would have been foolhardy. Missile technology has created the possibility that coastal craft and land-based aircraft can seriously damage a major warship. Granted that shipborne defensive systems are now routinely installed by the principal navies, such systems remain untested in practice against land-based batteries and an array of missile-armed attackers. Challenging a security zone risks severe political costs and—even more sobering—steps into the manifold uncertainties of quasi war. Constraint falls more heavily on the high-seas state which would provocatively enter a prohibited security zone than on the coastal state, which is simply declaring that foreign naval vessels must remain at some distance from her shores. Political and technological possibility coincide.

Precedent

The 1950s and 1960s produced several unilateral claims by states to sovereignty over portions of the high seas. Each is important; each will be examined more closely below. These cases include designation of rocket test and nuclear test areas and of air defense identification zones (ADIZs). The most interesting are the Chinese military zones and the North Korean military areas.

Security zones may be temporary or continuing; prohibitive or inhibitive or regulatory; blanket or selective (as to type of ship and states affected). Zones differ in purpose; some service comprehensive security needs, but others are limited special-use zones. Whether stated or not, the declaring state is always free to grant permission to foreign naval vessels to enter the zone, but it may explicitly insist upon applications to enter. Recent examples illustrate the range of forms which security zones may take: Chinese zones (1955): comprehensive, continuing, prohibitive; U.S. ADIZs: special-use, continuing, regulatory; nuclear test zones: special-use, temporary, inhibitive; and North Korean zone (1977): comprehensive, continuing, prohibitive zone.

Examination of declared security zones shows that they fall into two groups; though the distinction is not absolute, it is helpful to an understanding of the issues they pose. Some are slack, others are taut. Table 1 makes the difference clear.

A security zone would not mean much if it were not in some respect "taut": even the temporary, inhibitive zones threaten trespassers with nuclear radiation or a descending rocket and, to that extent, probably have the effect of closing merchant marine entry to the declared zone. The likelihood that a falling rocket will actually strike one small ship is near nil, but insurance requirements work against entry. On the other hand, foreign naval vessels are

TABLE 1.—EXAMPLES OF TAUT AND PARTIAL ZONES

	Partial and Qualified Zonal Characteristics			Taut Zonal Characteristics			
Zone	Temporary	Regulatory	Special Use	Continuing	Inhibitive	Prohibitive	Comprehensive
U.S. air defense identification zone (ADIZ)	...	X	X	X	...	...	...
U.S. nuclear test	X	...	X	...	X	...	...
USSR rocket test	X	...	X	...	X	...	...
U.K. nuclear test (Australian site)	X	...	X	...	...	X	...
French nuclear test (Pacific site)	X	...	X	...	X	...	...
Chinese military security zone	...	...	...	X	...	X	X
North Korean military zone	...	...	...	X	...	X	X

not prohibited from entering an inhibitive security zone, where they may attempt to monitor nuclear or rocket tests in progress.

Among these zones, some are small, others large. Some overlap the territorial sea and portions of a fishery conservation zone, but others are located only on the high seas, beyond any other claims of the declaring state. Commentators have distinguished security zones located in much-used seas from those in areas of infrequent use. These are significant distinctions, and security zones superimposed upon other jurisdictional regions suggest the issue of creeping sovereignty.

Creeping Sovereignty

On September 28, 1945, President Harry S. Truman issued a proclamation on the continental shelf which declared, inter alia, that the United States considered "the natural resources of the subsoil and sea bed of the continental shelf beneath the high seas but contiguous to the coasts of the United States as appertaining to the United States, subject to its jurisdiction and control."[1]

On the same day, the United States declared the right to establish fishery conservation zones in areas of the high seas along its coasts. These steps appear to have been the impetus to claims by (i) Mexico (1945)—possession of the continental shelf; (ii) Argentina (1946)—continental shelf and adjacent seas; (iii) Chile, Peru, and Costa Rica—sovereignty over the continental shelf and surrounding seas; (iv) El Salvador (1950)—territory included seas to 200 nautical miles and superadjacent airspace; and (v) Chile, Peru, and Ecuador (1952)—jointly declared sovereignty over seas to not less than 200 nautical miles from their coasts.[2] In the early 1950s, it was unclear whether the traditional 3-nautical-mile territorial sea would stand or, instead, be replaced by an alternatively agreed-upon territorial sea or a welter of uncoordinated national claims.

The U.S. assertions also prompted a distinction between resource and nonresource claims. Washington gambled that it could stake a claim to resources on the continental shelf and impose fisheries conservation zones without cost to the U.S. Navy's mobility. The United States could bar foreign states from exploiting oil on the U.S. continental shelf, but foreign states would not interfere with U.S. naval maneuvers just beyond 3 nautical miles from their shores.

In fairness, it is not clear that a well-considered decision was made, in 1945, to chance economic advantage at the risk of naval mobility. We can guess that U.S. officials, flush with World War II victory and enjoying unparalleled naval

1. Cited in Lawrence Juda, *Ocean Space Rights: Developing U.S. Policy* (New York: Praeger, 1974), p. 22.
2. Ibid., pp. 24–27.

supremacy, foresaw no serious challenge to the fleet's moving where it chose. But by its action, the United States set in motion the process which has led to current calls for exclusive economic zones and—from a few states—security zones in areas previously designated as high seas.

CHINESE POLICY

Since the founding of the People's Republic of China in 1949, she has taken several measures which limit access by foreign naval vessels to waters along her coast. She has designated some areas as internal waters, refused unpermitted access to her territorial sea, and declared military security zones embracing both territorial sea and areas hitherto considered part of the high seas.

Internal Waters

China has declared as internal waters (*a*) the Bohai Gulf, (*b*) the Qiongzhou Straits, and (*c*) those coastal waters landward of the baseline of the territorial sea. The Bohai is a large, shallow sea east of Beijing, which can be reached from the Pacific only by passing through Chinese territorial waters (assuming a 12-nautical-mile territorial sea); its status as internal waters can be argued by several routes and is not likely to be challenged.[3] The Qiongzhou Strait divides Hainan Island from mainland Guangdong Province. In her 1958 declaration defining territorial waters,[4] China defined both the Bohai and the Qiongzhou Strait as within her "straight baseline" and therefore as inland waters. Similarly, any waters between the straight baseline and the coast also were defined as internal waters: "China's territorial sea along the mainland and its coastal islands takes as its baseline the line comprising straight lines connecting base-points on the mainland coast and those on the coastal islands on the outer fringe, and the water area extending twelve nautical miles outward from the baseline is China's territorial sea. The water areas inside the baseline, including the Pohai Bay, and the Chiungchow Strait, are Chinese inland waters."[5] The claim that these are internal waters implies unqualified jurisdiction. China may set whatever rules she pleases and exclude whomever she pleases. It is as if the waters were Chinese soil.

3. China has claimed that any of three tests qualify the Bohai as internal waters: as a bay, as waters within "straight baselines," or by claim to historic waters (Jerome Alan Cohen and Hungdah Chiu, *People's China and International Law* [Princeton, N.J.: Princeton University Press, 1974], pp. 482–83.

4. English text in *International and Comparative Law Quarterly* 8 (1959): 182n.

5. Ibid. Pohai and Chiungchow are spellings used prior to general adoption of pinyin romanization.

Twenty years' usage, the claims' limits, and the political consequences of challenge suggest that the probability of a practical challenge to China's internal waters is vanishingly small. There are two grounds on which such a challenge could be made. Customary international law among the naval powers has held that extending claims across avenues through straits which link seas cannot undo the transit right of foreign states. Moreover, the straight baseline method of establishing the baseline of the territorial sea is not regarded as an option available for election by all but as a special method to be employed in extraordinary geographic cases, as along the deeply indented Norwegian coast. These legal cavils, however, are insignificant by comparison with the broader issues of economic and security zones. China's claimed internal waters are not likely to be challenged.

Territorial Sea

China's 1958 declaration on the territorial sea specifies a "strong" Chinese authority in the coastal band: "No foreign vessels for military use and no foreign aircraft may enter the Chinese territorial sea and air space above it without the permission of the Government of the People's Republic of China. Any foreign vessel while navigating the Chinese territorial sea must observe the relevant laws and regulations laid down by the Government of the People's Republic of China."[6]

Chinese comment during the Third United Nations Conference on the Law of the Sea has confirmed that she maintains the distinction between naval and merchant vessels and the claim to "strong" authority. A coastal state has the right, in China's view, to require foreign warships to notify it or acquire its permission beforehand "in accordance with its laws and regulations" before passing through its territorial sea.[7] Moreover, the extent of a coastal state's regulations governing movement of foreign warships through the territorial sea is a "matter of national sovereignty and no restrictions whatever should be applied."[8]

This position challenges the "innocent passage" doctrine of the naval powers. According to that tradition, naval vessels may transit the territorial sea of a foreign state if their passage is "innocent." China denies that innocence is sufficient to create a right of transit; instead, the coastal state may grant or withhold permission and may impose regulations.

6. Ibid.

7. U.S. Department of Commerce, National Technical Information Service, Foreign Broadcast Information Service, *People's Republic of China: Daily Reports,* no. 69 (1976), pp. A1–A2 (original dated April 8, 1976) (hereafter cited as FBIS).

8. Ibid., summarizing approvingly the position of "dozens of countries."

Security Zones

In 1955 a Chinese fisheries organization, in accordance with Chinese government instructions, defined two security zones and a fisheries conservation zone from which Japanese fishing vessels could be barred. Each reaches well beyond the Chinese territorial sea.[9] The existence of these zones set the precedent for broader claims to regulate foreign naval vessels in the exclusive economic zone, or to exclude them from an economic zone:

i) A military alert zone [Junshi jingjie qu] lying west of a line from the Yalu River, at China's border with Korea, to a point off the Shandong Peninsula.[10] The exchange of letters in 1955 accorded entry to this zone only with express Chinese permission. Although the zone includes the Bohai, claimed as internal waters, and areas of claimed territorial sea, it also includes waters east of the Bohai Strait which have been considered high seas.

ii) A military exclusion zone (literally, a military zone in which navigation is forbidden) [Junshi jinhang qu].This zone surrounds Shanghai and a number of islands to the east of Hangzhou Bay. Since China has not drawn a chart to show where the baseline of the territorial sea would run, it is possible that this zone does not extend beyond the territorial sea. Small islands dot the region, and a baseline hung on the outer fringe would reach well offshore.[11] Japanese fishing boats are forbidden to enter this zone.

iii) A fisheries conservation zone, from which Japanese fishing boats were excluded, almost certainly includes water areas more than 12 nautical miles beyond the baseline of the territorial sea.[12]

In addition, China cautioned that the region south of 29° N in the neighborhood of Taiwan and the Taiwan Strait was an area in which military activity could occur and warned Japanese fishing craft against entry.

9. People's Republic of China, Ministry of Foreign Affairs, *Zhonghua Renmin Gongheguo tiaoyueji* [Treaties of the People's Republic of China], Treaty Series (Beijing: Renmin Chubanshe, 1955), pp. 280–83. For identification of agreements to 1966, see Douglas M. Johnston and Hungdah Chiu, *Agreements of the People's Republic of China, 1949–1967: A Calendar,* Harvard Studies in East Asian Law, no. 3 (Cambridge, Mass.: Harvard University Press, 1968), p. 251.

10. The geographic coordinates are from 39°46′48″N, 124°10′00″E to 37°20′00″N, 123°3′00″E. In the 1963 agreements, the line was shifted slightly to a new starting point at the Korean end: from 39°45′00″N, 124°9′12″E to 37°20′00″N, 123°3′00″E (*Renmin ribao* [People's daily] [Beijing], November 10, 1963). The revised line is reflected in the 1975 agreement, exchanged on August 15, 1975, and ratified December 22, 1975.

11. The coordinates are from 31°00′00″N, 122°00′00″E to 30°55′00″N, 123°00′00″E to 30°00′00″N, 123°00′00″E to 29°30′00″N, 122°30′00″E to 29°30′00″N, 122°00′00″E. The coordinates transmitted to the Japanese side during a 1963 exchange are identical.

12. The coordinates are from 37°20′00″N, 123°03′00″E to 36°48′10″N, 122°43′00″E* to 35°11′00″N, 120°38′00″E to 30°44′00″N, 123°23′00″E† to 29°00′00″N, 122°45′00″E. In the 1963 version, the second coordinate (*) is given as 122°44′30″E and the fourth coordinate (†) as 123°25′00″E. In addition, the 1963 version added some additional points to the south: to 27°30′00″N, 121°30′00″E to 27°00′00″N, 121°10′00″E. The 1963 changes are also preserved in the 1975 agreements.

CHINESE COMMENT ON SECURITY PROVISIONS

Recent Chinese references to the exclusive economic zone confirm that Beijing envisions strong coastal state claims. China has argued that coastal states should exercise sovereignty over their exclusive economic zones[13] and has also argued that "the new law of the sea should clearly stipulate that the exclusive economic zone is a sea area under the jurisdiction of a coastal country and not part of the high sea, and that the coastal country should enjoy exclusive jurisdiction over this sea area."[14]

The same text admits the reasonable inference that a coastal state could bar foreign military installations and military activities in its exclusive economic zone, if it were clearly understood that the zone was not a part of the high seas.[15] In effect, the economic zone would serve as a broad defensive buffer from which unwanted foreign naval vessels could be barred.

NORTH KOREAN ZONES DECLARED AUGUST 1, 1977

The Democratic People's Republic of Korea had given notice that it would institute and enforce an exclusive economic zone in waters off its coast, effective August 1, 1977. When the economic zone became effective, North Korea also instituted a security zone: "The military boundary is up to 50 miles from the starting line of the territorial waters in the East Sea [Sea of Japan] and to the boundary line of the economic sea zone in the West Sea [Yellow Sea]." Foreign ships, both civilian and military, were banned from the zone "on the seas, in the seas, and in the air" unless they had obtained "appropriate prior agreement or approval," but foreign fishing boats were exempted from the prohibition.[16]

The maritime states were prompt to respond verbally to North Korean claims, rejecting the zones as contrary to international law. The effect of their statements was to ensure that they were not later charged with tacit assent and

13. According to an unofficial report of remarks by the Chinese delegate to a plenary meeting of the Sixth Session of the Third United Nations Conference on the Law of the Sea, June 28, 1977, reported in United Nations Press Release SEA/267, June 28, 1977, p. 6.

14. FBIS no. 78 (1976), pp. A4–A7, A5 (original dated April 20, 1976).

15. Ibid.

16. *New York Times* (August 2, 1977), quoting an announcement of the official North Korean press agency. The version reported by the *Asahi Evening News* (August 1, 1977), p. 1, has it that "foreign warships and military planes will be barred from within the military demarcation lines" but that "civilian ships and planes with advance permission from North Korean authorities can sail through or fly over these areas"; *Asahi* stated that fishing boats were exempt from the rule. Note that the respected *Far Eastern Economic Review* (Hong Kong) (August 12, 1977), p. 28, reported the zone's extent erroneously as "covering 200 miles to the west and 50 miles to the east of its shorelines," which would have had it overlap China's internal waters of the Bohai Gulf.

to deter other states from imitating Pyongyang's action. The U.S. position was set forth by a Department of State spokesperson, who said " . . . we and many other nations exercise a freedom of navigation and overflight beyond the 12-mile territorial sea and we do not recognize the right of any nation unilaterally to regulate navigation and overflight on the high seas."[17]

However directly Japan put its official position on record, the comment of Prime Minister Fukuda Takeo was more revealing. For the moment, he said, nothing could be done except to try to prevent trouble.[18] In fact, Japan's interests lay far more heavily in prospective access to marine resources. During the following year, Japanese fisheries access was negotiated between Japan and North Korea, with the Japan-Korea Fisheries Consultative Council representing Japanese requirements. The agreement finally reached in mid-1978 allowed Japanese fishing boats to operate in the Korean economic zone, but only in those portions outside the security zone.[19]

Several distinct issues raised by the North Korean security zone are illustrated by figure 1. The zone is measured from a straight baseline, which would place two rather wide-mouthed sea areas, not lying behind an island array, in the category of internal waters. The passage of foreign naval vessels is prohibited, and foreign merchant ships and fishing craft must obtain permission. Finally, wedges at the junctions with the USSR and South Korean borders extend into seas beyond the "hypothetical equidistant lines" which would divide the 200-mile exclusive economic zones, were such zones to be agreed upon. Figure 1 does not attempt to define Korea's western security zone in the Yellow Sea.[20]

17. *New York Times* (August 3, 1977).

18. *Asahi Evening News* (August 1, 1977), p. 1.

19. *Japan Times* (Tokyo) (July 2, 1978), reporting a Radio Pyongyang broadcast on conclusions of the June 26–29, 1978, talks between a delegation of Japanese dietmen and North Korean officials. In December 1977, North Korean officials told a Japanese dietman they would "consider" permitting fishing in the security zone (*Japan Times* [Tokyo] [December 5, 1977]). On April 18, 1978, a Japanese fishing boat was seized by North Korea for alleged violation of the military zone; it was released after 5 days (*Japan Times* [Tokyo] [April 27, 1978]). In May 1978, North Korean President Kim Il Sung told Japan Socialist party chairman Asukata Ichio that the military zone was necessary to protect North Korea in its confrontation with U.S. "puppets" (*Japan Times* [Tokyo] [May 14, 1978]). At that time Japanese fishermen still wished to obtain North Korean assent to their fishing in the zone (ibid.). Soon thereafter a Japanese fishing boat, in the economic zone and near the military zone, was ordered to leave, apparently by a North Korean warship which may have mistaken it for a South Korean craft (*Japan Times* [Tokyo] [June 1, 1978]). Unofficial negotiation of agreements is the consequence of absence of diplomatic relations between Japan and North Korea.

20. A map roughly similar to that of fig. 1 appears in the *Yomiuri shimbun* (Tokyo) (March 12, 1978). A less-detailed map in the *Yomiuri shimbun* on May 16, 1978, depicting both the Japan Sea and Yellow Sea zones, departs from fig. 1 in two respects: in the west it extends far southwest from the north-south border, to perhaps 36°N, and in the east the military zone appears to assume not a straight baseline but a baseline following the coast.

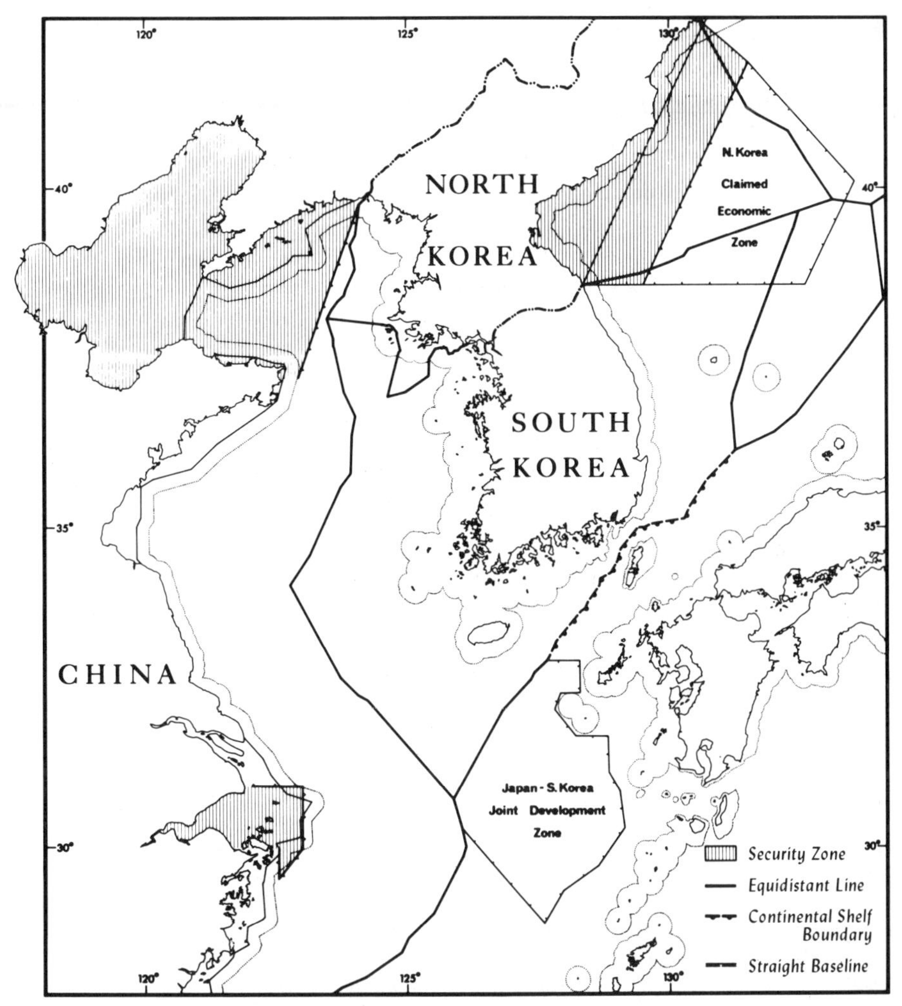

FIG. 1.—Adapted from the map "Potential Maritime Zones of Northern East Asia," Office of the Geographer, U.S. Department of State, December 1977. The map contains this important disclaimer: "Other than the Japan–Republic of Korea agreement, this map is intended only to depict hypothetical maritime zones as calculated by the Office of the Geographer, U.S. Department of State. These lines do not necessarily reflect United States Government position nor necessarily those of the states involved. The map should serve only as a general reference document." It is cited here in that spirit.

The North Korean economic zone in the Yellow Sea—and therefore its security zone—reaches to the yet undefined Chinese line. We assume that China will in the future declare an economic zone and, further, that it will be the subject of discussion between China and North Korea. Until then, however, claims at the center of the Yellow Sea are not firm. The 1955 Chinese line from the Yalu to a point east of the Shandong Peninsula, behind which lies her security zone, does not reach to the equidistant line separating China and North Korea.

A reasonable conjecture is that China and North Korea envision the closure of the Yellow Sea, or the northern half of the Yellow Sea, against unpermitted entry. Their security zones would abut. At some time in the future, South Korea, or a united Korea, might be drawn into the arrangement and the entire Yellow Sea declared a security zone.[21]

If the trend to stronger national claims in the territorial sea and offshore economic zones continues, the region between Korea and Taiwan may be the site of further security zones. China, Japan, South Korea, and Taiwan all have claims in the area, and there are grounds for dispute. For example, China claims rights in accordance with the doctrine of the continental shelf, which could carry Chinese jurisdiction into areas Korea has designated for drilling. In June 1977, 8 months after Hua Guofeng became chairman, China responded in strong terms to attempts by Tokyo and Seoul to establish a joint development zone in the waters bordering Korea and Kyushu. China makes a broad claim to sovereignty but tempers it with a call for consultations: "The East China Sea continental shelf is the natural extension of the Chinese continental territory. The People's Republic of China has inviolable sovereignty over the East China Sea continental shelf. It stands to reason that the question of how to divide those parts of the East China Sea continental shelf which involve other countries should be decided by China and the countries concerned through consultations."[22]

21. The Republic of Korea (Seoul) has not hastened to declare a 200-nautical-mile economic zone, because of its active far-ocean fishing industry. In December 1977, the South Korean National Assembly authorized a 12-nautical-mile territorial sea. Naval vessels would be required to obtain permission to enter the territorial sea, and "innocent passage" provisions for merchant vessels could be suspended "temporarily" for reasons of national security (*Japan Times* [Tokyo] [December 17, 1977]). The delicate problem of passage through the strait between Korea and Japan was solved by fixing a 3-nautical-mile territorial sea there. As finally fixed in the enforcement decree, naval vessels and foreign noncommercial government vessels were required to give 72-hour advance notice of intention to enter the South Korean territorial sea. Procedures are set out for approval of foreign fleet exercises, submarine passage, and surveys by foreign vessels and for temporary suspension of innocent passage by foreign ships (*Japan Times* [Tokyo] [February 10, 1978, July 13, 1978, August 11, 1978]).

22. The Statement of the Ministry of Foreign Affairs of the People's Republic of China, June 13, 1977 (*New China News Agency* [English] [June 13, 1977]), in FBIS, June 13, 1977, pp. A9–A10. On this area, see fig. 1, "Potential Maritime Zones of Northern East Asia," Office of the Geographer, U.S. Department of State, December 1977. A detailed view from the vantage of petroleum issues is rendered in Selig S. Harrison, *China, Oil and Asia: Conflict Ahead?* (New York: Columbia University Press for the Carnegie Endowment for International Peace, 1977); the maps, prepared by the Ford Studios, Arlington, Va., are superb examples of clear and informative figures (11 maps following p. 46). Japan is concerned both about her southwest frontier and her seas neighboring the Soviet Union (for a useful discussion, see Frank Langdon, "Japan-Soviet 200-Mile Zone Confrontation," in *Pacific Community* 9, no. 1 [October 1977]: 46–58).

POLITICS AND PRECEDENT

Claim and response form a duality. A state's claim and the response of foreign states are each meaningless without the other. The object of a claim is to set boundaries around the actions of foreign states; the object of response is to reinforce, question, or challenge the boundaries which have been claimed, and similar boundaries which might be claimed elsewhere. There is always a problematic quality to boundaries among sovereign states: they are sustained because of the profound value of order—provided the order is in some sense reasonable—but are in jeopardy because of the real temptation to profit from an alternative order. This is apparent in land borders, but it is no less true with respect to ocean space, contract rights, and access to global commons.

The force of precedent is that it defines an order which is familiar and apparent and which has for some time been acceptable. Whether sketched in practice or etched in law, precedent raises one possibility among all others to a prominence, a distinct visibility. In the shouldering and testing among nation-states, the line drawn by precedent becomes the "obvious" line of reference. Schelling effectively describes the general case: tacit bargainers focus on a salient outcome.[23]

Because boundary claims are inherently relational, they entail bargaining with others: they imply politics. It may be the politics of complementary interest in maintaining inviolable high seas . . . or the politics of challenge to the prior high-seas order. What is ironic about growing claims to security zones is that they were first used by naval powers who saw them as a natural extension of their control and use of the oceans. Bounds on others were assumed, and it was not clearly perceived that U.S. and U.K. security zones would in turn give rise to Chinese and North Korean zones through the exercise of similar logic. The others were written out of politics. A similar lapse concerning economic zones was promptly exposed, as we have noted above, by the reactive claims of coastal Latin America.

Thus, politics and precedent move in intimacy. International law, in one view, is the depository of precedent, and it is indeed not law—not positive law—but the articulation of precedent and agreement by self-appointed guardians. It is a norm, but a norm which some subset of nation-states has taken it on

23. Thomas Schelling, *The Strategy of Conflict* (New York: Oxford University Press, 1963), pp. 89 ff., 111–12. As an illustration of tacit bargaining, consider the "coordination game" among friends who agree to meet in New York on a given day but fail to specify a time or place. Until the Biltmore closed, there were perhaps two solutions; now there is one. Nation-states acting unilaterally are in an analogous position: they could set the boundary of the territorial sea anywhere and make any regulations they choose, but instead they choose (with certain exceptions) from a small number of salient solutions now endowed with the power of precedence. Schelling terms these "focal-point solutions."

itself to declare. Arguments about right international law are therefore arguments about who may fix norms and who must follow them. They are political arguments. And both the efficacy of norms and sanctions against violations must remain the result of intensely political relations.

In the dialectic between politics and precedent, however, politics shapes the precedents which come to bear consequence, and precedent defines some portion of the politics through which the sovereign states enact their disagreements. Moreover, both the guardians of the status quo and the advocates of a new order rely on precedent, at the minimum as a guide to the provocative, and commonly as an authority cited to justify an action.

North Korea is on familiar ground when it cites the need to defend itself as reason for its security zone: the "inherent right of self-defense." What is startling in the North Korean claim is that it so exceeds precedent: on its face it appears to be a provocative act. On the other hand, in whose interest would it be to challenge the North Korean zone? Only the Soviet Union would normally run naval vessels southward along the Korean coast: and Pyongyang appears unlikely to wish a confrontation with Moscow on this point.[24] Only as precedent does the Korean zone threaten the established routes of the great navies. Does the *possibility* of future losses in strategic maneuverability, were another state to imitate North Korea, justify the *near certainty* that North Korea would attempt to enforce its zone against naval trespass? Of course, the zone could be a grand bluff: perhaps there are sealed instructions that, if challenged, North Korean coastal craft are to retreat into the 12-nautical-mile territorial sea. But the spectacle of a trespassing superpower destroying North Korean naval craft suggests that there are no political gains to be won by such a test. Japanese Prime Minister Fukuda's hope to avoid trouble is further guarantee to North Korea that the United States would not force an encounter. The factors which made for the Tonkin Gulf and *Mayaguez* affairs are absent; the controlling geopolitics are more nearly those which dictated U.S. caution following seizure of the *Pueblo*.

PRIOR ZONES

The inseparability of politics and precedent suggests that a state weighing whether to declare a security zone like North Korea's will be largely governed by its actual, mundane political relations with foreign states which will care: those whom the zone is intended to bound out and those who deem extended zones bad precedents. But zonal claims can be sustained better in the absence of

24. It remains odd that the declared language of the prohibition extends to all foreign naval vessels and that the zone appears to overlap waters of the Soviet economic zone. Its declaration coincides with a trough in Pyongyang-Moscow relations, marked by some modest Soviet overtures to South Korea.

a collective effort to delegitimate them. Among the strategies which high-seas states could attempt are (i) a multinational test of the zone, (ii) collective economic or administrative retaliation, and (iii) treaty law rejection. Draft provisions of a new treaty being considered in the Third United Nations Conference on the Law of the Sea leave no room for a North Korean military zone. But if the legitimacy of broad security zones were debated there, if collective sanctions were proposed, or—in the most extreme case—if a collective flotilla sought to enforce high-seas law by an actual transit, many states would turn to precedent as a guide to correct policy. Proponents of the right of a coastal state to set military security zones would no doubt adduce historic cases; if high-seas states could be shown to have accepted "national security" exceptions to the high-seas principle, so much the better. Within high-seas states, domestic opinion—even among political leaders—would turn in part on the clarity with which it could be shown that coastal state acquisition of hitherto high seas was a reprehensible, in some sense criminal, act. Of course, the issue is already confused by the proposed distinction between sovereign rights over resources and territorial sovereignty.

The cases most likely to be adduced follow.

Air Defense Identification Zones

In late 1950, President Harry S. Truman issued an executive order establishing air defense identification zones (ADIZs), including coastal ADIZs in waters off the coast of the United States. These are expressly described as security zones: " . . . it is necessary in the interest of the national security to establish security provisions for the use of aircraft in designated areas in the airspace above the United States, its territories, and possessions (including areas of land or water administered by the United States under international agreement)."[25]

The order extends to the coastal ADIZs. Rules are set, aircraft "shall not be operated into or within an Air Defense Identification Zone . . . in violation of" those rules, and penalties are defined for failure to comply. Foreign aircraft entering the United States through a coastal ADIZ must either make position reports as required of U.S. aircraft or " . . . in lieu thereof, the pilot . . . shall report to an appropriate aeronautical facility when the aircraft is not less than one hour and not more than two hours average cruising distance via the most direct route, from the United States. Thereafter, reports shall be made as instructed by the facility receiving the original report."[26]

25. U.S. Civil Aeronautics Administration, "Security Control of Air Traffic," pt. 620 of Title 14, *Federal Register* 15, no. 250, December 27, 1950, 9319; see also U.S. Presidential Executive Order 10197, "Directing the Secretary of Commerce to Exercise Security Control over Aircraft in Flight," *Federal Register* 15, no. 248, December 22, 1950, 9180.

26. Ibid., 9320.

The executive order provides further authorization to make whatever provision the administrator of the Civil Aeronautics Board might choose to make governing aircraft in the coastal ADIZ: "*Air defense security instructions.* Under emergency defense conditions which may involve the national security, aircraft shall be operated into or within an ADIZ in accordance with such additional special security instructions issued by the Aministrator as may be deemed necessary for the identification, location and control of the particular flight."[27]

How far from the coast of the United States do the coastal ADIZs reach? The order establishes both an Atlantic coastal ADIZ and a Pacific coastal ADIZ. The original 1950 zone extended more than 150 nautical miles from the U.S. coast: it is a broad, far-reaching body. From time to time, the boundaries have been altered, but zones remain in effect.[28]

A glance at aeronautical charts demonstrates the continuing role of the ADIZs as components of the U.S. air security system.[29]

Although the waters over which the U.S. ADIZs extend are waters for which the United States has asserted economic and fisheries control rights, it has not claimed that it may regulate naval vessels in that zone. It has not claimed, for example, that it may require submarines to run on the surface beyond the limit of the territorial sea, despite the fact that missile-armed submarines may fire nuclear warheads from waters close to the United States. The 1950 provisions requiring flight information at a given distance from the United States reflect the technology of that time; and in a real sense the coastal ADIZs are technologically obsolescent. Nonetheless, for almost 30 years the

27. Ibid. The executive order continues with a note: "Such instructions shall be designated as 'Air Defense Security Instructions.' Emergencies involving the national security may require the exercise of special control for identification purposes. In such cases instructions may be issued requiring identification turns, flight in a specified direction or location, or such other instructions as may be necessary."

28. The 1950 coastal ADIZs are bounded as follows. Atlantic coastal ADIZ: (*a*) 44°30′N, 66°45′W; (*b*) 43°10′N, 70°00′W; (*c*) 42°40′N, 70°10′W; (*d*) 42°00′N, 69°30′W; (*e*) 41°15′N, 69°30′W; (*f*) 46°15′N, 73°30′W; (*g*) 39°30′N, 73°45′W; (*h*) 37°00′N, 75°30′W; (*i*) 36°10′N, 75°10′W; (*j*) 35°10′N, 75°10′W; (*k*) 33°30′N, 78°00′W; (*l*) 32°00′N, 74°00′W; (*m*) 40°00′N, 64°00′W; and (*n*) 44°30′N, 66°45′W; Pacific coastal ADIZ: (*a*) 48°30′N, 125°00′W; (*b*) 48°00′N, 125°15′W; (*c*) 46°15′N, 124°30′W; (*d*) 43°00′N, 124°40′W; (*e*) 40°00′N, 124°45′W; (*f*) 38°50′N, 124°00′W; (*g*) 34°00′N, 120°30′W; (*h*) 33°15′N, 118°30′W; (*i*) 32°30′N, 117°45′W; (*j*) 32°30′N, 117°15′W; (*k*) along a line parallel to and approximately 7 miles from the Mexican coast to the points (*l*) 29°00′N, 114°40′W; (*m*) 27°00′N, 121°30′W; (*n*) 38°00′N, 129°00′W; (*o*) 50°00′N, 132°00′W; (*p*) 51°00′N, 130°00′W; and (*q*) 48°30′N, 125°00′W.

29. E.g., those of the U.S. Air Force, Aeronautical Chart and Information Center, St. Louis, 1:1,000,000 Operational Navigation Charts (ONC). Chart ONC H-25 displays the southern end of the Atlantic coastal ADIZ and portions of a Gulf of Mexico coastal ADIZ. Charts ONC F-19 and ONC G-21 show more successive northerly sections along the Atlantic coast. The ADIZs are supremely prominent features on these charts. The most recent ONC consulted contained air information current through January 19, 1970, but I believe the ADIZs remain valid.

United States has claimed a right to regulate foreign traffic in airspace beyond the limit of the territorial sea. Its claim has explicitly rested on national security needs. It has not distinguished civil from military craft.[30] It has even claimed—in a section quoted above—that it may require foreign aircraft in this airspace to maneuver in accordance with its instructions. Here there is no innocent passage. Despite its name, the ADIZ requires far more of foreign aircraft than "identification."

But the force of the ADIZ as precedent for Chinese, North Korean, or any other's action is that the ADIZ is not confined to North America. The extent of the ADIZs declared by America's Asian allies is apparent from the U.S. Air Force Global Navigation and Planning chart for Southeast Asia, showing the Asian Pacific littoral from Korea and Japan to Sri Lanka (Ceylon).[31] A continuous chain of ADIZs reaches from Japan to Korea, Okinawa, and Taiwan to the Philippines. In 1963 the Korea ADIZ reached north to Pyongyang itself and northwest into the waters which now constitute the North Korean economic zone, and even into waters which then lay in the Chinese military security area. The 1963 Taiwan ADIZ extended beyond coastal Fujian and Zhejian to include portions of inland Jiangxi Province. In retrospect, the right of control which these zones implied appears brazen and provocative. At the minimum, they attest to the principle that states may claim to control traffic at some distance beyond their borders by citing needs of national security.

United States Nuclear Tests

Designation of an area including high seas as a nuclear test site is well exemplified by the 1958 HARDTACK nuclear test series. The national security justification for the zone is explicit: "The efficient and early completion of this

30. No doubt in part because of the ambiguity between civil and military aircraft, each of which could carry nuclear weapons. But the more settled nature of sea law, and the U.S. operational imperative to maintain untrammeled access to landward seas for U.S. naval forces, has prevented the United States from making the same argument with respect to foreign merchant vessels, which may just as easily be outfitted to carry nuclear weapons.

31. Published by the U.S. Air Force, Aeronautical Chart and Information Center, St. Louis, Global Navigation and Planning Chart GC 13-N. One chart examined here was revised June 1963, with air information current through August 8, 1963. Revision and replacement of these series is continuous. The zones appear on other Aeronautical Chart and Information Center charts, including chart ONC J-12 (scale 1:1,000,000), showing the southern part of Taiwan (revised December 1974 with air information current through January 22, 1975); on Jet Navigation Chart JNC-25N (scale 1:2,000,000), showing the Korean ADIZ (revised August 1974 with air information current through December 18, 1974); and on chart JNC-38 (scale 1:2,000,000), showing the Taiwan ADIZ reaching into China (revised October 1973 with air information current through November 27, 1973). It is my understanding that the ADIZs remain current.

test series . . . is of major importance to the defense of the United States and to the free world."[32]

The initial test region encompassed Bikini and Eniwetok Atolls, ranging more than 8° of latitude and 14° of longitude.[33] An extension of the zone was made effective July 25, 1958, and maintained for 1 month. The addition, designated as a "danger area," was "that area established, effective July 25, 1958, encompassing Johnston Island and which is a circle of 400 nautical miles radius centered at the following geographic coordinates: longitude 169°31′00″ West and latitude 16°45′00″ North."[34] The United States did not forbid foreigners from entering the zone, although it did prohibit those over whom it could claim jurisdiction.[35] It then successfully prosecuted a pacifist group which sought to sail from Hawaii into the zone to protest the tests.

The issue here is whether the United States might appropriate a vast area of the high seas, even temporarily, for a purpose which *effectively* closed it to use by others except at great risk. The traditional defense rests on national security requirements, and mitigations include the remoteness of the site and temporary nature of the encumbrance. Criticisms have been based not only on high-seas interference but also on the environmental consequences of the test.

USSR Rocket Test Site

Although the Soviet Union had protested designation of high-seas nuclear test areas, on January 8, 1960, she declared an intention to conduct ballistic missile tests during which missiles were "expected" to fall in a portion of the Pacific.[36]

32. U.S. Atomic Energy Commission, "HARDTACK Nuclear Test Series, 1958, Miscellaneous Amendments," F.R. Doc. 58-4983, *Federal Register* 23, no. 127, June 28, 1958, 4782 (hereafter cited by F.R. Doc. no.). Notification of the first part of the series is in "Eniwetok Nuclear Test Series, 1958," F.R. Doc. 58-2716 (filed April 11, 1958), *Federal Register* 23, no. 73, April 12, 1958, 2401. See also "HARDTACK Nuclear Test Series, 1958, Definitions; Prohibition," F.R. Doc. 58-6966 (filed August 25, 1958), *Federal Register* 23, no. 168, August 27, 1958, 6617; and "HARDTACK Nuclear Test Series, 1958, Revocation," F.R. Doc. 58-7392 (filed September 8, 1958), *Federal Register* 23, no. 177, September 9, 1958, 6974.

33. F.R. Doc. 58-4983. The area bounded by a line connecting these points: (*a*) 18°30′N, 156°00′E; (*b*) 18°30′N, 170°00′E; (*c*) 11°30′N, 170°00′E; (*d*) 10°15′N, 166°16′E; and (*e*) 10°15′N, 156°00′E. Presumably, the line returns to (*a*) 18°30′N, 156°00′E, although the notice does not state this.

34. F.R. Doc. 58-4983.

35. "This part applies to all United States citizens and to all other persons subject to the jurisdiction of the United States, its Territories and possessions. . . . No United States citizen or other person who is within the scope of this part shall enter, attempt to enter or conspire to enter the danger area during the continuation of the HARDTACK test series, except with the express approval of appropriate officials of the Atomic Energy Commission or the Department of Defense" (F.R. Doc. 58-2716).

36. The zone was defined as follows: (*a*) 9.6N, 170.47W; (*b*) 10.22N, 168.22W; (*c*) 6.16N, 166.16W; and (*d*) 5.3N, 168.40W. The original Tass announcement, printed in the *New York Times* (January 8, 1960), presents portions of degrees in decimal notation.

Like the United States in nuclear test areas, the Soviet Union did not undertake to forbid foreign nationals outside its jurisdiction from entering the area. Rather, it invited the assistance of other states: " . . . the Government of the Soviet Union asks the governments of the nations whose ships or aircraft may find themselves during this period in the vicinity of the area where the rockets might fall to see that the authorities concerned instruct the ship masters and aircraft captains to refrain from entering the aquatorium [water] area and airspace of the Pacific designated by the above mentioned coordinates."[37]

British Nuclear Tests in Australia

The Australian Parliament adopted a law on June 10, 1952, to make provision for British nuclear tests. It designated as a prohibited area a circle 45 miles in radius centered on Flag Island of the Monte Bello group, 20°27.5′S, 115°35′E. The text further read that "a person shall not be in, enter or fly over a prohibited area unless he is the holder of a permit. . . . "[38]

CONCLUSION

The object of this paper is to suggest that the prospect of future encumbrances upon high-seas prerogatives be assessed realistically.

In the absense of very broad agreement that security zones beyond the limits of the territorial sea are illegitimate and unacceptable, the weight of interest favors the likelihood that more coastal states will declare security zones. Although many will be deterred by the risks and costs, and others will favor casting their choice on behalf of high-seas freedom and the protection of commerce by the global naval powers which that implies, the power of local arguments and the desire to achieve a perceived increment of security and control can prove as persuasive for other states as it has for China and North Korea.

Elsewhere I have argued that it is in China's interest, both in the short term and from a longer perspective, to achieve support for the norm that the seas open to naval vessels should be restricted.[39] It is not in China's interest that the Soviet Union and United States be able to impress, threaten, or protect smaller states simply by reason of their capacity to effect naval presence far from home.

37. *New York Times* (January 8, 1960).

38. Reprinted in *White Paper on the French Nuclear Tests,* Ministere des Affaires Etrangeres, Paris, 1973.

39. "China: Law and Naval Strategy," in the *Proceedings of the Fifth Leverhulme Conference,* Centre of Asian Studies, University of Hong Kong, v. 4 (Hong Kong: Centre of Asian Studies, University of Hong Kong, in press).

Of course, in the absence of restrictions on the Soviet Union, China could genuinely favor U.S. deployment in the western Pacific, and numerous visitors to China have reported comments of members of the Chinese leadership in just that vein. But it is difficult to read in any other way Beijing's insistence that an exclusive economic zone should imply "jurisdiction" and that the waters of the economic zone must not be designated as high seas. Some of this mystery will disappear when China establishes an exclusive economic zone: at this writing she has not given any indication when that might be.

In a logical sense, though not necessarily in practice, the high-seas states have disarmed themselves by their own casual appropriation of the high seas. The security zones which they have declared do not take the precise form of the zone declared by North Korea, but the analogy to ADIZs is close and the argument from security requirements no less compelling. Test sites are a less apt parallel. The distinction suggested above between "qualified" and "taut" security zones is useful. There is a difference between a duration of 1 month and an indefinite term. Both nuclear testing and coastal buffering, however, say that the high-seas right is thought by states to be subject to unilateral limitations they may choose to impose.

Thus we return to politics. In international law the only mechanism of enforcement is self-help. Though we talk of sanctions and collective measures, they are in each case a political matter undertaken among a set of states in a deliberate concert of action. High-seas freedom as a lubricant of commerce and therefore of human wealth is subject to an asymmetry quite analogous to that of free passage across the commons: as long as merchant commerce is not in fact subject to interference, the case for the social utility of naval force is not immediate, nor is there present reason to doubt the good will of nation-states. Our experience with the harassment and interdiction of commerce in time of war is now ample; provisions for high-seas freedom will not guarantee that goods can flow. But what of the case in times short of war, or short of general war? And what of outcast and unpopular states? Is it poisonous to distinguish security zones which block access to a foreign port from such a zone which merely forces a longer journey?

In short, there appear to be two principles at issue. One is the principle that any departure from high-seas freedom threatens the expectations of untrammeled commercial movement: since there exist plausible, but merely plausible, arguments for specific exceptions to high-seas freedom, the principle can be maintained only by resisting those specific claims which appear reasonable or of little danger in themselves. The second principle is that foreign naval force is threatening and that, in the longer term, means should be found to ensure commerce against piracy and interruption which do not justify maintenance of large national navies, now equipped with unprecedented nuclear destructive power, and their cruising the global sea.

The question to which this inquiry leads is what form such assurance of commerce might take. Implicit in the declaration of security zones by coastal states is an undertaking by that coastal state to ensure the movement of permitted vessels through the zone. But will all commercial vessels be permitted? Will political favor be bestowed on some while others are denied? Issues reminiscent of those concerning the movement of neutral vessels at the time of the First World War would certainly be posed.

Patrol and protection of a substantial area of the sea is a costly matter, now borne by the governments of naval powers and by those governments which volunteer to be helpful beyond the limits of their territorial sea. Would a coastal state argue that the practice of regulating passage through offshore security zones required it to impose reasonable fees? The practice is now general that aircraft passing through a state's territorial air space may be required to pay a fee for the privilege of passage.

One objective of the proposed Law of the Sea language is to forestall such nuisance. The problem is raised here to show how intimately economic zone and military zone issues can join in the more comprehensive question of jurisdiction. Local interest favors more expansive jurisdiction. It is not clear that more expansive local claims will guarantee adequate levels of order and security. Nevertheless, insistent unilateral pressures may force us to conceive of political and multilateral avenues to commercial security and to accept important restrictions on the movement of global navies, with important strategic consequences.

Offshore Oil Development in the China Seas: Some Legal and Territorial Issues

Choon-ho Park
East-West Center, Hawaii

Offshore oil development in the China seas[1] has to be seen from the perspective of the overall oil situation in the coastal states involved, namely, China, Japan, North and South Korea, the Philippines, Taiwan, and Vietnam. With the exception of China, which produces oil in great quantities and exports a fair amount of it, these countries are heavily or totally dependent on foreign sources to meet their demand for oil. For instance, Japan—one of the world's largest oil consumers—imports nearly 100 percent of its crude oil, its own domestic supply scarcely reaching half a percent of the total demand which exceeds 250 million tons a year;[2] in the other coastal states, the situation is similar to or even worse than Japan's. In these oil-poor countries, therefore, crude oil imports weigh very heavily on their balance of payments. Since China as an "oil power" and a major coastal state in the region is in a decisively important position with respect to the future of offshore oil development in the China seas, it is first necessary to take a brief and general look at China's oil situation.

According to some Chinese records, use of mineral oil in China dates back to centuries before Christ. The "burning stuff" was called by different names up until the eleventh century, when the word *shiyou*, meaning petroleum, was first used by Shen Kuo (1031–95), a Sung dynasty scholar. China also claims proudly that the world's first oil well was drilled in the present Sichuan Province in 1521, preceding the first one drilled in the United States in 1859 (the Drake well of Titusville, Pennsylvania) by 339 years.[3] For nearly 4 centuries following the first drilling, however, there was no significant development of oil in China until the early part of the present century, when old wells were improved and

1. As used in this study, "the China seas" refers to the Yellow Sea, East China Sea, South China Sea, the Gulf of Tonkin, and part of the Sunda Shelf, severally or collectively, as the case may be.

2. Sekyu Tsushinsha, *Sekyu shiryo* [Oil data] (Tokyo: Sekyu Tsushinsha, 1977), pp. 2–3. For crude oil statistics, Japan uses "kiloliter," not "ton." A ton amounts to 7.3 bbl, and a barrel to 0.159 kl. Therefore, a kiloliter of crude oil is 0.8615 ton.

3. Zhang Ming-nan, Chen Ru-xi, Xu Wen-jun, and Yang Zhen-yu, *Zhongguo shiyou qingliufendi zucheng* [The fractional distillation of Chinese petroleum] (Beijing: Kexue Chubanshe, 1962), p. 6.

© 1980 by The University of Chicago. 0-226-06603-7/80/0002-0013$01.00

new ones were developed. But it was the People's Republic, founded in 1949, that launched an extraordinary drive to build an oil industry in the modern sense of the term. Nevertheless, the production of crude oil was so inadequate in meeting the demand that, up until the mid-1960s, China had to rely on foreign sources for the major part of its oil supply.

In the traditional view of Western oil geologists, China was not richly endowed with oil resources, its sedimentary basins being continental in origin, with the exception of a few that are of marine origin. This view was not taken seriously by certain Chinese oil specialists, among them a Western-trained geologist by the name of Li Siguang (1899–1971). Li was convinced that oil could be present in continental sedimentation as distinct from marine sedimentation, which comprises major oil centers elsewhere. Li successfully proved that China was not poor in oil, and, in recognition of this monumental contribution, he is sometimes called the father of oil exploration in China.

Currently, China produces over 100 million tons of crude oil a year from its onshore oilfields alone and has emerged as the largest oil producer throughout East Asia, with Indonesia running close behind. China has also begun to export its crude oil, first to North Korea in 1964 and to North Vietnam in 1965; well over 10 percent of its annual production is now exported to a number of countries in East Asia, with Japan as the major customer. Since the last decade, outside observers have been engaged in an endless numbers game to determine the volume of oil deposits in the Chinese continent and to assess China's potential to become a major exporter as well as a major producer in the future. One analysis, done by the U.S. Central Intelligence Agency (CIA) and believed to be possibly conservative, may be noted with interest:

> Analysis of the limited body of information available on onshore liquid reserves . . . has yielded broad agreement on a range centering on about 40 BB [billion barrels] of ultimately recoverable reserves, with the possibility that there may be as much as 100 BB. In comparison, as of mid-1976, remaining proved plus probable reserves were estimated to be 390 BB in the Middle East, 64 BB in Africa, 47 BB in North America, and 42 BB in Latin America.
>
> China's onland reserves, though considerable, cannot support the predictions of China becoming a world oil power. Moreover, a large and growing domestic demand for oil, the quality of many of the reserves, technological problems in extracting oil, and geopolitical considerations argue against continuous increases in exports.
>
> There are kerogen or oil shale deposits in China said to be comparable to the vast oil shale deposits in the United States. Soviet geologists in China through 1960 reported 153.3 BB of shale reserves.[4]

4. U.S. Central Intelligence Agency, *China: Oil Production Prospects,* ER 77-10030U (Washington, D.C.: Central Intelligence Agency, 1977), p. 7.

As in the case of the inland areas, the offshore waters of the China seas were also thought by marine geologists of the West to be generally unlikely to contain oil and remained unexplored by them up until the late 1960s. In the late 1950s, however, China's own marine geologists published reports of geophysical surveys they had conducted in the Yellow and East China Seas as well as in the Po Hai.[5] China had thus already been putting some effort into the search for offshore oil before other coastal states of the region began to turn their attention to the sea for oil. In fact, by the mid-1960s, China's interest in its seas as another (potential) source of oil was so serious that offshore oil development, in its third 5-year economic plan (1966–70), represented one of the major points of emphasis.[6] It was also during this period that informal cooperation began between Chinese and Japanese specialists.[7]

In the other coastal states that border on the China seas, however, serious efforts toward searching for oil from the sea were not made until October 1968, when, inspired by the preliminary findings of eminent marine geologists such as Kenneth Emery of the United States and Hiroshi Niino of Japan, a joint geophysical survey was conducted in the Yellow and East China Seas by a team of scientists from Japan, South Korea, Taiwan, and the United States. The survey was sponsored by the Committee for the Coordination of Joint Prospecting for Mineral Resources in Asian Offshore Areas (CCOP) of the United Nations Economic Commission for Asia and the Far East (ECAFE).[8] The report of the survey published in 1969 said: "The shallow sea floor between Japan and Taiwan appears to have great promise as a future oil province of the world, but detailed seismic studies are now required."[9] With respect to China's offshore oil potential, however, the U.S. CIA analysis cited above says: "Offshore reserves, although possibly very large, are as yet the subject of conjecture only. Even if very large, they may prove difficult and expensive to locate and extract. Neither the Chinese nor foreigners have yet acquired enough data on offshore sedimentary deposits to make valid estimates. Predictions about China's future as an oil power based on exploitation of offshore deposits are premature."[10]

5. Fan Shih-ching and Chin Yun-shan, "Zhongguo Donghai he Huanghai nanbu dizhidi chubu yenjiu" [Preliminary study of submarine geology of China's East Sea and the southern Yellow Sea], *Haiyang yu huzhao* [Oceanologia et limnologia] 2 (April 1959): 82–85; English trans., *Translations on Communist China,* no. 97, Joint Publication Research Service (JPRS) 50252 (April 7, 1970), pp. 12–36. For further details, see, e.g., S. Harrison, *China, Oil, and Asia: Conflict Ahead?* (New York: Columbia University Press, 1977), p. 58.

6. Jingjibu [Ministry of economic affairs], *Dalu shiyou gongye gailan* [Summary of the petroleum industry of the mainland] (Taipei: Ministry of Economic Affairs, June 1967), pp. 6–7.

7. Ibid., p. 7.

8. K.O. Emery et al., "Geological Structure and Some Water Characteristics of the East China Sea and the Yellow Sea," *Technical Bulletin* (ECAFE) 2 (1969): 3–43.

9. Ibid., p. 4.

10. U.S. CIA, p. 8.

Against the foregoing background, the present study attempts to analyze the problems by which the efforts of the coastal states to develop oil from some (supposedly) promising areas of the China seas have been stalled continually since 1969. Particular reference is made to the law-of-the-sea and territorial issues which, apparently, some of the coastal states involved are not yet prepared or willing to settle in the immediate future.

THE LAW-OF-THE-SEA ISSUES

In the Yellow and East China Seas

In terms of consequent results, the ECAFE report above of 1969 can be said to have been the origin of the "seabed oil war" among the coastal states of Northeast Asia. Partly due to exaggeration in the report,[11] the oil-poor coastal states were instantly overtaken with excitement and premature expectations which are still unrealized. In the countries with debilitating demands for oil, this initial reaction was perhaps natural in light of statements made by the report in its conclusion: "A high probability exists that the continental shelf between Taiwan and Japan may be one of the most prolific oil reservoirs in the world." "A second favorable area for oil and gas is beneath the Yellow Sea where three broad basins are present."[12] In fairness to the report, however, it has to be pointed out that the conclusion also recommended "Further seismic studies . . . in order to adequately portray the shapes and extents of these small structures."[13]

But the actual reaction of Japan, South Korea, and Taiwan was to attempt to grab as much of the seabed area as possible by extending their respective jurisdictions over the area that each of them regarded as its own continental shelf. In so doing, they interpreted the law of the sea to the advantage of their individual national positions. In specific terms, at issue was the applicability of the relevant provisions of the 1958 Geneva Convention on the Continental Shelf, particularly the median-line principle as given in article 6 (1), which was basically undermined by a new criterion invented by the International Court of Justice (ICJ) in its judgment of the North Sea Continental Shelf Cases of 1969, namely, the natural prolongation of land territory principle.[14] As a result, Japan insisted on the median-line principle, Taiwan on the natural prolongation of land territory principle, and South Korea on a combination of both, that

11. "The project leader was removed following a barrage of criticism. It is true that the ship's equipment was unable to penetrate all of the formations most likely to contain oil . . . " (ibid., p. 6).

12. Emery et al., p. 41.

13. Ibid.

14. International Court of Justice, *The North Sea Continental Shelf Cases* (1969), p. 53.

is, the median-line principle toward China in the Yellow Sea and the natural prolongation of land territory principle toward Japan in the East China Sea.

When, based on the delimitation principle(s) of its own choice, each of them staked out claims over the continental shelf, most of the Yellow and East China Seas was divided into 17 seabed mining zones of Japan, Taiwan, and South Korea.[15] Their claims heavily overlapped, especially in the areas which the ECAFE report held to be promising for oil, leaving only four of the 17 zones uncontested. Furthermore, each claimant hastened to involve Western oil companies in order to enhance its claims; concession arrangments with them for most of the unilaterally claimed zones had been completed by September 1970.[16]

Each of three coastal states was thus concentrating on the consolidation of its unilateral claims but ignoring the question of mutual boundaries. It was in this state of deadlock that the idea of joint development was conceived as a possible breakthrough in a situation that could otherwise remain an endless legal controversy. Japan, South Korea, and Taiwan were to proceed with oil development, leaving the surface boundaries to future negotiations. In mid-November, 1970, they agreed to form a three-party oil development consortium. At this time, however, Peking hastened to intervene, lodging a strong protest in early December of the same year. Despite its apparent practicality, the first attempt at joint development had to end abortively even before its merits could be tested.[17]

In subsequent years, Japan and South Korea made overtures to China to seek negotiated agreement on their seabed boundaries, but they were invariably greeted with silence. Impatient at China's indifference, they made a second attempt at joint development, this time without Taiwan, by signing an agreement in January 1974.[18] As expected, China resumed its protest against the alleged infringement of its sovereignity. South Korea ratified the pact in December 1974, but it was not until June 1978 that Japan did the same, following repeated procedural manipulations in its parliament.[19] The reluctance of the Japanese parliament to approve an agreement that the government had willingly signed is related to the emergence of a new regime in the law of the sea, namely, the exclusive 200-mile economic zone (EEZ).

15. Choon-ho Park, "Oil under Troubled Waters: The Northeast Asia Sea-Bed Controversy," *Harvard International Law Journal* 14 (1973): 226 and map at p. 219.

16. Ibid., pp. 223–24.

17. Ibid., pp. 227–29.

18. Choon-ho Park, "The Sino-Japanese-Korean Sea Resources Controversy and the Hypothesis of a 200-Mile Economic Zone," *Harvard International Law Journal* 16 (1975): 42–45. For the English text of the agreement and a map, see *Continental Shelf Boundary and Joint Development: Japan-Republic of Korea,* Limits in the Seas, no. 75, (Washington, D.C.: Department of State, Office of the Geographer, 1977).

19. For details, see, e.g., Choon-ho Park, "China and Maritime Jurisdiction: Some Boundary Issues," *German Yearbook of International Law* (Kiel: Institut für Internationales Recht, 1980), vol. 22, chap. 2, and sources cited therein.

The 200-mile EEZ was first proposed by Kenya to the Geneva session of the United Nations Seabed Committee in August 1972.[20] This new form of maritime jurisdiction has since been adopted for fishing and other economic purposes, by 79 states to date (as of September 1979), including Japan, the Soviet Union, the United States, and other major maritime powers of the world.[21] To the extent that the 200-mile EEZ makes depth and bottom topography of the sea irrelevant in delimination of a coastal state's economic jursdiction, the natural prolongation of land territory principle is basically undermined in cases where opposite coastal states are less than 400 miles apart. However, the applicability of the median-line principle is, as a consequence, enhanced, apparently, to the advantage of Japan. This assumption is of great importance to Japan, as the joint development area is situated almost entirely on its side of the median line toward China and South Korea. In Japan, therefore, opponents to the pact may have counted on the potentiality that, on the strength of the 200-mile EEZ, the seabed area in question would eventually fall under Japanese jurisdiction, hence destroying the need for parliamentary approval to ratify it.

As yet the 200-mile EEZ has had no dramatic impact on the Yellow and East China Seas because North Korea has been the only contiguous country to declare such a zone (in August 1977).[22] To settle fishing-rights problems with the Soviet Union, Japan adopted a provisional 200-mile fishing zone in May 1977[23] on the west; therefore, it does not apply beyond Japanese territorial waters toward China and Korea. China (and South Korea also) has been reluctant to declare a 200-mile EEZ, obviously because of the difficulty of delimiting maritime boundaries with its adjacent and opposite neighbors. In the northern part of the Yellow Sea, no boundary delimitation or dispute is reported to have taken place, to date, between China and North Korea.[24]

In the Yellow and East China Seas, the major law-of-the-sea dispute over offshore oil development began with Japan, South Korea, and Taiwan as the

20. United Nations, General Assembly, Official Records, *Report of the Committee on the Peaceful Uses of the Sea-Bed and the Ocean Floor beyond National Jurisdiction,* suppl. 21 (4/9021), 6 vols. (1973), 3:72 (hereafter cited as *Report*).

21. Robert D. Hodgson, "The Delimitation of Maritime Boundaries between Opposite and Adjacent States through the Economic Zone and the Continental Shelf," *Proceedings* (13th Annual Conference of the Law of the Sea Institute, University of Hawaii, forthcoming), n. 6 (hereafter cited as "Delimitation of Maritime Boundaries").

22. *Pyongyang Times* (July 9, 1977), p. 3; or *Foreign Broadcast Information Service* (FBIS) (Asia and Pacific) 4 (July 1, 1977): D2. For a commentary, see *People's Korea* (July 13, 1977), p. 2.

23. United Nations, Doc. ST/LEG/SER.B/19, preliminary issue, June 13, 1978, pp. 226–40. For details with a map, see Seiichi Yoshida, "200-kairi jitai: wagakunino gyogyo suiekinimo" [The 200-mile era: in our seas as well] *Tokino horei* [Current laws], no. 974 (1977), pp. 11–19.

24. Park, "China and Maritime Jurisdiction," and "The 50-Mile Military Boundary Zone of North Korea," *American Journal of International Law* 72 (1978): 866–75.

principal disputants. Peking has since emerged, in place of Taiwan, to argue against the other two parties, just as North Korea has occasionally protested against Japan and South Korea for their alleged attempts to carve up the sea resources to which it insists it is also entitled. In relation to Japan, however, the interests of Peking and Taiwan would coincide, as would those of North and South Korea. Because of this, the adjacent Yellow and East China Seas would, in the first instance, have to be delimited into three sectors (by a trijunction in the East China Sea), leaving the Chinese and the Korean sectors to be shared, where appropriate, between the two sides of each divided coastal state.[25] But it should be pointed out that final delimitation of boundaries among the parties involved is not possible unless two basic problems are overcome by them. One is the ambiguity in the law of the sea regarding boundary delimitation between adjacent and opposite coastal states (see Current State Practices below), and the other is the territorial dispute over the ownership of offshore islands (see Territorial Issues below).

In the South China Sea

The main basin of the South China Sea is contiguous to the Gulf of Tonkin in the northeast and, through the Sunda Shelf, to the Gulf of Thailand in the west and is also surrounded by Brunei (due to become independent from Britain in 1983), China, Indonesia, Malaysia, the Philippines, Taiwan, and Vietnam.[26] On account of such geographical circumstances, the law-of-the-sea and territorial issues are even more complicated here than in the Yellow and East China Seas. With respect to seabed oil development, there are two current law-of-the-sea issues: one is between China and Vietnam in the Gulf of Tonkin and the other between Indonesia and Vietnam in the Sunda Shelf.

1. In the Gulf of Tonkin, China and Vietnam are reported to have been anxious to develop oil in recent years, although to date they have no agreement on their seabed boundary.[27] At present the political relations between the two socialist neighbors have so deteriorated that the delimitation of their boundary even in this small and semi-enclosed gulf would not be an easy undertaking. Their political relations aside, however, the dispute represents one of a series of

25. For a hypothetical illustration, see U.S. Department of State, Office of the Geographer, *Potential Maritime Zones of Northern East Asia,* 503591 12-77 (Washington, D.C.: Department of State, 1977).

26. For the geophysical circumstances of the South China Sea, see M. L. Parke, Jr., K. O. Emery, R. Szymankiewicz, and L. M. Reynolds, "Structural Framework of Continental Margin in South China Sea, *American Association of Petroleum Geologists Bulletin* 55 (May 1971): 723–51.

27. Reportedly China found oil in the Gulf of Tonkin (the *Beibu* Gulf in Chinese and the *Vinh Bac Phan* in Vietnamese); see *Dagongbao* (October 12, 1979), as quoted in *Sankei shinbun* (October 13, 1979), p. 5.

highly complicated and interrelated offshore problems for them, problems which neither coastal state would find practical to settle as self-contained issues. With respect to the general legal principle to be applied in delimitation of the continental shelf, both China and Vietnam officially support the natural prolongation of land territory principle.[28] Their respective overtures and unofficial practices, however, do not necessarily appear to substantiate this, as may be seen from the instances below.

In April 1973, Vietnam was reported to have signed an agreement with Ente Nationale Idrocarburi (ENI), the Italian state oil company, for drilling in the Gulf of Tonkin. The deal is suspected to have since fallen though. It may be noted, however, that the outer edge of the ENI zone roughly coincided with the median line toward China,[29] although this fact alone is not sufficient to represent Vietnam's official position on the extent of its offshore claims in the gulf. Now Vietnam is concerned about the three contracts that China has signed with Western oil consortia in the spring and summer of 1979 to explore oil in the disputed waters around Hainan Island, because the outer edge of the Chinese zones also approximately coincides with the median line toward Vietnam (fig. 1). Specifically, Vietnam would fear that the exploration arrangements might eventually result in de facto recognition of the Chinese claims.

Owing to a series of offshore boundary problems with its adjacent and opposite neighbors, China as well as Vietnam would feel it inadvisable to adhere rigidly to, or to depart conspicuously from, any particular principle of delimitation. For instance, China has persistently pressed for the natural prolongation of land territory principle with Japan and South Korea (and presumably North Korea as well) with respect to the continental shelf dispute in the Yellow and East China Seas; for geomorphological reasons, most of the disputed waters would then fall under Chinese jurisdiction. In contrast, the same principle—if rigidly applied to the Gulf of Tonkin—would be more advantageous to Vietnam than to China because of the seabed topography.

2. In the Sunda Shelf, the continental shelf is adjacent to the territories of Brunei, Indonesia, Malaysia and Vietnam, but is delimitation would not necessarily involve all these coastal states.[30] Brunei's sector would probably be delimited only with Malaysia, and China's only on the strength of its claims to the

28. Working paper presented to the 1972 Geneva session of the UN Seabed Committee by China in *Report,* p. 74; and Vietnamese declaration of 200-mile economic zone, *FBIS* (Asia and Pacific) (May 24, 1977), p. K6.

29. *Petroleum News: Southeast Asia* 7 (January 1977): 46. A Chinese proposal on the delimitation of boundaries in the Gulf of Tonkin reads: "[Paragraph] 4. [T]he two sides shall demarcate their respective economic zones and continental shelves in the Beibu Gulf and other sea areas in a fair and reasonable way in accordance with the relevant principles of present-day international law of the sea" *(Beijing Review* [May 4, 1974], p. 17).

30. For the geophysical circumstances of the Sunda Shelf, see Zvi Ben-Avraham and K. O. Emery, "Structural Framework of Sunda Shelf," *American Association of Petroleum Geographers Bulletin* 57 (1973): 2323–66.

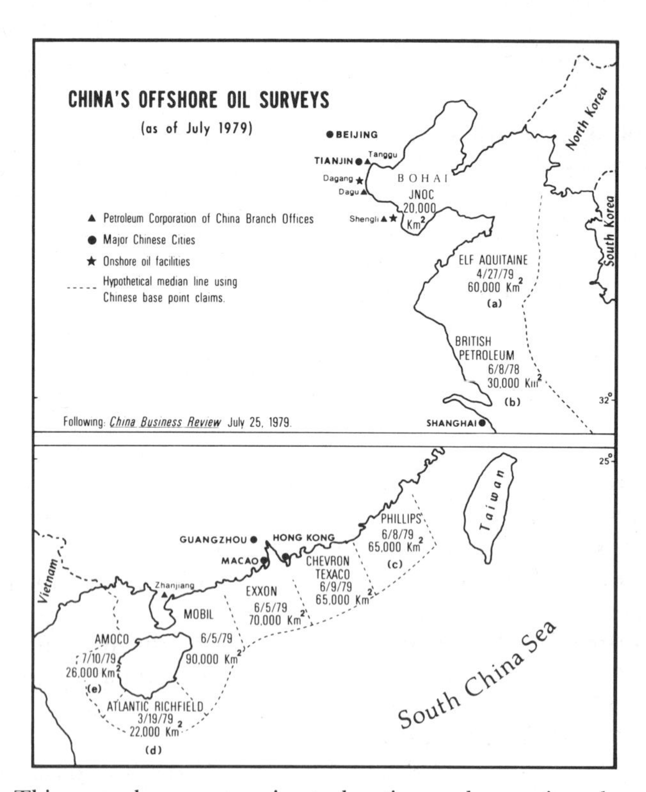

FIG. 1.—This map shows approximate location and area size where foreign oil companies have signed contracts with the Chinese to conduct seismic surveys in offshore areas. Exact locations have not been announced, at the request of the Chinese. The date listed is that of the contract signing. Companies are committed to completing the surveys and submitting the analyzed results within 12 months of starting. The Chinese have said they will try to conduct the first round of bidding for actual exploration contracts within 12 months of receiving the data. To quote one authority, the contracts will be "Chinese in concept, service in name, risk in nature." Companies now undertaking seismic surveys were told they will be given the final exploration contract if their bid matches the best one submitted. Japan National Oil Corp. (JNOC) reached an agreement in principle with the Chinese last summer on exploration and development of oil in south Bohai Gulf, but has not yet signed a firm contract. Developments as of July 25, 1979, include: (1) Unified Pearl River Mouth Area Basin: On July 4 the Chinese notified 55 U.S. and foreign companies, who had already shown some interest, that they had 60 days to inform the Chinese of their serious intent in becoming early participants in this area composed of the Mobil, Exxon, Chevron, and Phillips blocks. After the Chinese receive this response, they will give the companies another 30 days to accept formally. Members of the first round of early participants are expected to sign with the original four operators on August 7 in Dallas. (2) Amoco block: In mid-July the Chinese sent telexes to at least 20 U.S. and foreign companies with an offer to contact Amoco if interested in early participation in its block. (3) Yellow Sea blocks: A set of letters was also extended in June to a number of companies inviting them to become early participants in these two blocks. *a* = Total, British Petroleum (BP), Petro Canada, JNOC, and Shell had been asked to participate but had not decided as of July 1, 1979; *b* = Eff Aquitaine, Exxon, Phillips, Shell, and Union Oil had been asked to participate but had not decided as of July 1, 1979; *c* = Original participants are AGIP, Eff Aquitaine, Allied Chemical's Union Texas, and Total; *d* = Santa Fe International is an original participant; *e* = Original participants are Cities Service, Pennzoil, Union Oil of California, AGIP, and BP.

(Legend adapted from *China Business Review,* July/August 1979, p. 62. © National Council for US China Trade, 1979.)

ownership of the Spratly Islands. This would leave the major part of the Sunda Shelf to be shared among Indonesia, Malaysia, and Vietnam, by a pair of trijunctions (because Malaysia would have two separate—the Peninsular and the Eastern—sectors).

Following the downfall of Saigon to Hanoi in April 1975, the unified Vietnam has been anxious to resume oil drilling in the waters off the Mekong Delta. Oil had been found here previously by Western oil companies operating under contracts with the defunct South Vietnamese regime.[31] To consolidate its claims in this promising area, Vietnam sought to settle its seabed boundary first with Indonesia, but the four rounds of talks, held alternately in Hanoi and Jakarta between June 1978 and January 1979, were reported to have ended without agreement.[32] The two parties apparently disagreed on the principle to be applied, Indonesia insisting on the median-line principle as against the natural prolongation of land territory principle preferred by Vietnam for the same seabed topography reasons as in the Gulf of Tonkin. In an area where the former South Vietnamese claims were contested by both Indonesia and Malaysia, however, final boundary agreement would not have been possible without the participation of Malaysia, even if Indonesia and Vietnam had reached agreement on their segment of the boundary. Since, among the coastal states of the Sunda Shelf, only Vietnam has declared a 200-mile EEZ, the delimitation of a multilateral seabed boundary will probably be deferred, pending declaration of similar zones by Indonesia and Malaysia. In passing, a similar situation is to be found over the final delimitation of a multipronged boundary in the greater South China Sea region, because Brunei, China, and Thailand have also not declared their 200-mile EEZs.[33]

Current State Practices

When its 6-week resumed eighth session adjourned in New York in August 1979, the Third United Nations Law of the Sea Conference (UNCLOS III) had been in session for a total of 63 weeks since 1973. Nevertheless, the delimitation of the continental shelf and the exclusive economic zone between adjacent and opposite states still remains one of the hard-core issues (at the conference) on

31. *Petroleum Economist* (June 1975), p. 206; *Oil and Gas Journal,* (July 16, 1973), p. 80. For further details, see Katsutoshi Murakami, "Tonan ajiani okeru sekyu keizaino doko [Trends of oil development in Southeast Asia] (1)," *Refaransu* [Reference], no. 297 (1975), pp. 38–39.

32. British Broadcasting Corporation (BBC), *Summary of World Broadcast (SWB)*, pt. 3, Far East, 2d ser., FE/5835/A3, June 10, 1978, p. 8; FE/5894/A3, August 18, 1978, p. 5; FE/5966/A3, November 11, 1978, pp. 8–9; and FE/6024/A3, January 24, 1979, p. 3.

33. Among the coastal states of the China seas, the following maintain 200-mile limits: Japan, for fishing only, May 1977; North Korea, EEZ, August 1977; the Philippines, EEZ, May 1979; Taiwan, EEZ, September 1979; and Vietnam, EEZ, May 1977.

which no consensus has been reached. The relevant provisions of the revised Informal Composite Negotiating Text (ICNT, rev. 1) of UNCLOS III, simply read: "The delimitation . . . between adjacent and opposite states shall be effected by agreement in accordance with equitable principles, employing, where appropriate, the median or equidistant line, and taking account of all the relevant circumstances."[34] The ninth session, scheduled to meet for 5 weeks each in New York and Geneva in the spring and summer of 1980, respectively, is expected to refine the ICNT, revision 1, text before incorporating it into the future Caracas Convention on the Law of the Sea which the conference plans to conclude in 1980.

The core of the articles above is "agreement," which is wrapped in a shroud of ambiguities such as "equitable principles," "where appropriate," and "all the relevant circumstances." The geographical circumstances of the world are so complicated that it is simply not possible for a universal convention to accommodate them all in specific terms. Deliberations at Negotiating Group 7 of UNCLOS III, to which this subject was assigned, have resulted in futile exchanges between the advocates of equidistance and equitable principles, whereas, in the view of an adept observer, the former is a methodology and the latter an aim.[35]

Pending conclusion of UNCLOS III, many coastal states have either declared their maritime economic jurisdictions in unspecific terms or deferred negotiation of seabed boundaries with adjacent or opposite neighbors. For example, China, as well as all the "200 milers" in the China seas, has been invariably saying that, in waters with overlapping claims, the boundary should be determined "through consultations." Another point of ambiguity is found in declarations which specify the spatial extents of claims but not the baselines from which the extents are measured; in fact, all the "200 milers" in the China seas have followed this "evasive" practice. It is noteworthy that, according to one estimate, "If all existing coastal states and territories declared 200-mile economic zones, there would result approximately 331 maritime boundaries, [of which] fewer than 25 percent . . . have been negotiated."[36]

THE TERRITORIAL ISSUES

In the China seas, there are two territorial disputes that have obstructed the development of seabed oil in the region. One is the Senkaku-Tiaoyutai dispute in the East China Sea among China, Japan, and Taiwan, and the other is the

34. UNCLOS III, *Informal Composite Negotiating Text,* rev. 1 (ICNT, rev. 1) (A/CONF. 62/WP. 10/Rev. 1) April 28, 1979, articles 74(1), 83(1) (hereafter cited as ICNT, rev.1).

35. Robert D. Hodgson, to the author, September 21, 1979; see also *U.S. Delegation Report,* 8th sess., UNCLOS III, Geneva, March 19–April 27, 1979, pp. 3, 33.

36. "Delimitation of Maritime Boundaries," p. 2.

Paracel-Spratly dispute in the South China Sea among China, the Philippines, Taiwan, and Vietnam. Both cases concern the ownership of obscure offshore islands which are valuable to the owners because of their location rather than their physical usefulness. Most of them are uninhabited and so tiny that they are hardly visible on an ordinary map, but the claimants sometimes exaggeratedly call them "the sacred territory of the fatherland."

The Senkaku-Tiaoyutai Dispute[37]

In July 1970, Taiwan and Gulf Oil signed an oil concession contract over an offshore area northeast of Taiwan. Eight uninhabited islands, situated northeast of Taiwan and west of Okinawa (the *Senkaku* in Japanese and the *Tiaoyutai* in Chinese) were included in the Taiwan-Gulf Oil concession area.[38] Almost instantly, Japan protested against Taiwan, alleging that the islands belonged to Japan. The territorial controversy thus begun in 1970 has been a major political issue—initially between Japan and Taiwan, but now Peking has assumed the burden of the argument against Japan.

Each side bases its contention on history, geography, international law, and even geology, and invariably insists that the territory in question has always and indisputably been its own. As the dispute involves not only potential oil but also ownership of territory, it becomes even more difficult for either side to compromise its claims simply in the interest of negotiated settlement, because, for historical reasons, national feelings run high over territorial issues in East Asian countries, and such issues are often regarded as too important to be negotiated.

Currently, the islands are not in the full control of either claimant, each side cautiously trying to avoid grating on the raw nerves of the other, while trying to change the status quo in its favor. In consideration of such political sensitivity, the United States government has reportedly decided to discourage American oil companies from operating in disputed waters. As a result, the Taiwan-Gulf Oil contract of 1970 has been suspended indefinitely, as have many other such contracts Taiwan signed with various American operators.[39]

37. For the contentions of China and Japan, see Park, "Oil under Troubled Waters," pp. 253–58.

38. Ibid., maps at pp. 219 and 240.

39. *Petroleum News: Southeast Asia* (January 1977, 1978, 1979 [feature issues entitled, "Exploration Annual"]).

The Paracel-Spratly Dispute[40]

In the South China Sea there are approximately 200 islands, many of them coral outcrops without vegetation or otherwise incapable of "supporting human habitation."[41] They are grouped into four archipelagoes, namely, the Pratas Reef, the Macclesfield Bank, and the Paracel and the Spratly Islands. Chinese ownership of the Pratas Reef (the *Dongsha* in Chinese) has not been contested. The Macclesfield Bank (the *Zhongsha* in Chinese) consists of some 30 underwater elevations, but Chinese ownership of these submerged "islands" does not seem to have been contested to date. However, it remains to be seen whether contemporary international law would recognize claims placed on submerged mid-ocean islands. According to some Chinese observers, the highest island in the Macclesfield Bank is approximately 10 meters (32 feet) below sea level and, being a coral island, grows up at a rate of 10 centimeters (4 inches) a year![42]

The Paracel Islands (the *Xisha* in Chinese and the *Hoangsa* in Vietnamese) have been under Chinese control since January 1974, when, in a two-day *blitzkrieg*, the former South Vietnamese armed forces were wiped off the islands by Chinese forces. Vietnam nevertheless insists that the change of hands was merely a temporary relinquishment of its sovereignity, whereas China seems to regard it as a resumption of control of its own territory. The Spratly Islands (the *Nansha* in Chinese and the *Truongsa* in Vietnamese) are claimed by China, the Philippines, Taiwan, and Vietnam. With the exception of China, each claimant is in control of some of the islands. Furthermore, the Philippines is reported to have confirmed the presence of oil in the Reed Bank area.[43]

The nature of the Paracel-Spratly dispute is basically similar to that of the Senkaku-Tiaoyutai dispute, although, with more parties and a larger number

40. Choon-ho Park and Hungdah Chiu, "Legal Status of the Paracel and Spratly Islands," *Ocean Development and International Law Journal* 3 (1975): 1–28; Choon-ho Park, "The South China Sea Disputes: Who Owns the Islands and the Natural Resources," ibid., 5 (1978): 27–59; *Beijing Review* (May 4, 1979), p. 17; *Viet Nam's Sovereignty over the Hoang Sa and Truong Sa Archipelagoes,* Vietnamese *White Book* (September 27, 1979), abridged in *FBIS* (Asia and Pacific) (October 1, 1979), pp. K1–K13, and reproduced in full in UN Doc. A/34/541:S/13565, October 19, 1979; and the Chinese rebuttal, *Some Documentary Evidence Showing That the Vietnamese Government Recognized the Xisha and Nansha Islands as Chinese Territory,* reproduced in UN Doc. A/34/712:S/13640, November 23, 1979.

41. According to ICNT, rev. 1, "Rocks which cannot sustain human habitation or economic life of their own shall have no exclusive economic zone or continental shelf" (pt. 8, "Regime of Islands," article 121, par. 3).

42. Chen Dong-kang, *Wo guo de nan hai zhu dao,* 2d ed. (Peking: Zhongguo Qingnian Chubanshe, 1964), pp. 31–32: English trans. of the 1st ed. in JPRS (n. 5 above), no. 18–424 (1963), pp. 24–25; and Huang Jiu-shun, *Zhongguo dili gailun* (The natural geography of China) (Hong Kong: Shanghai Shuju), p. 85.

43. *Nihon keizai shinbun,* Tokyo (August 3, 1976), p. 4; *Petroleum News: Southeast Asia* (January 1977), p. 37.

of islands involved, the former is more complicated by far. In strategic terms, the Paracel-Spratly dispute is also deeply couched in another dimension of great importance to major maritime powers as well as to other users of the sea for communications, because the South China Sea provides a predominantly important maritime route that links East Asia with the rest of the world, with the exception of the Americas. Furthermore, one of the world's most important and densely used straits for international navigation, namely, the Malacca Strait, is situated at the threshold of the South China Sea. For this geographical reason, it would be correct to say that the importance of the straits is contingent on the safety of passage through the South China Sea, and vice versa.

The Impact on Seabed Oil Development

Most of the small offshore islands in the China seas whose ownership is currently under dispute were previously held to be almost negligible in terms of physical value. The emergence of the 200-mile EEZ at the beginning of the present decade has dramatically increased their value, because of the mounting expectation that their owners may eventually be entitled to the resources underlying the surrounding waters. For instance, the ownership of a tiny mid-ocean island (depending on its location) can be the basis on which the owner can claim jurisdiction to approximately 130,000 square nautical miles of adjacent sea. For these reasons, UNCLOS III has experienced considerable difficulty in the course of its deliberations to define an island and its legal status.

As they can affect the division of sea resources between adjacent and opposite states, offshore islands situated at critical locations have become extremely important, and the China seas are studded with numerous small islands which, seldom heard of otherwise, have become prominent in their obstruction of seabed oil development in the region. Aside from their traditional attitude toward territorial integrity, the intransigence of the parties with regard to the two territorial disputes in the China seas should be looked at in this context. In brief, they realize that in order to own the sea resources, it is necessary for them to own the islands. Consequently, until territorial issues are settled, it is impossible to delimit seabed boundaries in disputed waters. This, in turn, blocks the development of resources, as has happened, for example, in the East China Sea.

OBSERVATIONS

In the late 1960s and early 1970s, the coastal states of the semi-enclosed China seas, particularly those with no domestic supply of crude oil, were overjoyed at the "sniff" of oil in their own offshore waters and plunged into a sea-grab race with one another. In the flurry, however, they stumbled on a series of problems, including delimitation of the continental shelf with adjacent and opposite states. The law of the sea, which was just beginning to be reviewed under United Nations sponsorship, was not specific enough to be readily applied to their highly complicated geographical circumstances. Although boundary negotiations have been taking place sporadically in less controversial areas, most coastal states are not likely to undertake serious negotiations, pending conclusion of UNCLOS III in 1980 (as scheduled).

Problems of boundary delimitation were further exacerbated by two territorial disputes, which will continue to be obstacles to oil development in the disputed waters. Ironically, one of them, namely, the Senkaku-Tiaoyutai dispute between China and Japan, originated in one claimant's endeavor to search for oil in the area. Since, in East Asia, territorial issues involve national sentiments more seriously than in other regions, however, the settlement of the two cases is not going to be easy, unless all parties to each dispute come to an agreement whereby no one loses face.

The coastal states also failed to appreciate the procedures involved in moving oil from underneath the seabed to the filling station, a process none of them was economically or technologically capable of undertaking without the participation of extraregional oil interests. In a region chronically plagued by ideological feuds—up until the demise of South Vietnam in April 1975, East Asia had three of the world's four divided nations and still has two at present—the arrival of foreign oil companies enhanced political sensitivity to seabed oil development. Under such circumstances, some of the coastal states simply deferred the delimitation of sea boundaries, while others tried joint development. In a deadlocked situation, the joint endeavor of Japan and South Korea was an alternative signifying a breakthrough. But this new approach has been found to be fraught with problems unforeseen at the beginning, making its future still quite uncertain. On balance, any attempt to settle an issue without the participation of all the parties involved has merit as a provisonal measure, but as a partial settlement its inherent weakness always threatens to cancel out its strength.

Inevitably, the alternative approach open to the coastal states of the China seas, as elsewhere, has been to proceed with seabed oil development in undisputed near-shore waters and to gradually expand their operation seaward. The security of crude oil supply having become a global concern, obstacles to seabed oil development, such as problems of boundary delimitation, will increasingly come under the scrutiny of the vigilant coastal states concerned. The expected conclusion of UNCLOS III will also expedite settlement of boundary issues in the China seas that have been shelved to await codification of universal criteria. In the final analysis, however, the question still remains whether the coastal states are arguing impatiently over actual or mythical oil.

Existing and Potential Maritime Claims in the Southwest Pacific Ocean[1]

J. R. V. Prescott
University of Melbourne

For the purposes of this analysis the Southwest Pacific Ocean is considered to lie south of the equator between meridians 120° east and 155° west. This definition includes a small sector of the Indian Ocean and that part of the Southern Ocean lying between Australia and Antarctica. It seemed important to include Australia and its adjoining seas for three reasons. First, Australia is a prominent member of the South Pacific Forum, an important regional political organization which has addressed itself to some maritime questions. Second, there are some difficult problems, involving important questions of principle, involved in drawing sea boundaries between Australia and its northern neighbors. Third, Australia is one of the countries which claims sovereignty over a sector of Antarctica, part of which marks the southern coast of the Pacific Ocean.

THE RELEVANT PHYSICAL CHARACTERISTICS OF THE SOUTHWEST PACIFIC OCEAN

There are five major physical characteristics which bear upon any analysis of maritime boundaries in this region. First, most political organizations consist of archipelagos, which frequently include very small islands. In no other part of the world is there such a concentration of archipelagic states, and the continuing debate over the definition and nature of archipelagic waters, together with the importance of islands as special or relevant circumstances in the construction of common sea boundaries,[2] emphasizes the importance of this area in the

1. The author gratefully acknowledges the assistance provided by the Department of Foreign Affairs, Canberra, in obtaining copies of proclamations of national claims; by Professor H. C. Brookfield in obtaining references on fishing in the Southwest Pacific; and by Dr. J. M. Massey who made available his computer program for measuring areas.

2. United Nations, *Informal Composite Negotiating Text,* Third Conference on the Law of the Sea (New York: United Nations, 1977), pp. 24–25, 50, and 55 (also in *Ocean Yearbook 1,* ed. Elisabeth Mann Borgese and Norton Ginsburg [Chicago: University of Chicago Press, 1978], pp. 619–21, and 631); and Arvid Pardo, "The Evolving Law of the Sea: A Critique of the Informal Composite Negotiating Text (1977)," in *Ocean Yearbook 1,* pp. 12–13.

© 1980 by The University of Chicago. 0-226-06603-7/80/0002-0013$01.00

evolution of international maritime law. There is likely to be a reciprocal process by which proposed laws are tested in their application to the Southwest Pacific Ocean, and by which solutions, adopted by countries in this area, are recommended for general consideration. The proximity of these islands ensures that every state or territory must agree on a common boundary with at least one neighbor. All the outer sea limits claimed by Western Samoa and American Samoa must be agreed with neighboring states, and this condition also applies to more than half the length of the outer limits of Niue, New Caledonia, New Hebrides, Wallis and Futuna, Fiji, Tonga, Tokelau, the Solomon Islands, and Papua New Guinea.

The second relevant characteristic is that the pattern of islands ensures that there will be some enclaves of unclaimed seas surrounded by the exclusive economic zones of adjoining countries. There will also be some long *culs-de-sac* of unclaimed seas separating some national claims. This irregular pattern of sovereignty will create problems both for navigators of vessels fishing in this region and for authorities charged with the enforcement of regulations in claimed seas. The problems associated with enclaves of high seas have not been considered by the UN Conferences on the Law of the Sea, although this was a subject raised at the international conference at The Hague in 1930.[3]

The volcanic origin of most of the islands in the eastern part of this region is the third important physical feature. It ensures that if any continental shelf exists, it is very narrow, and that there is little chance of mining useful resources under the sea. The composition of the rock, the steepness of the continental slope, and the powerful surges of the surf on many coasts conspire to make undersea mining unattractive. The rapid descent of the continental slope to the abyssal plain which surrounds the islands guarantees that their exclusive economic zones will enclose considerable areas of the deep seabed. Although such areas contain manganese nodules, their collection in commercial quantities would bring individual countries into conflict with the Authority, which will be established by the UN Conference on the Law of the Sea to supervise mining of the deep seabed. Plainly, Australia, New Zealand, and Papua New Guinea are more fortunate than other countries in the region because they are surrounded by continental margins which have produced mineral wealth in the case of Australia and which could yield valuable resources for the other two countries.

There is generally a more even distribution of fish stocks throughout the region than there is of submarine mineral wealth. However, it is important to record as the fourth relevant characteristic that the potential for fisheries on the high seas diminishes toward the east and north of the Southwest Pacific. The variations in the size of fish stocks result from the operation of many factors which influence the distribution of phytoplankton, which is the basic food source for fish. The Geographer of the U.S. Department of State has produced

3. S. W. Boggs, "Delimitation of the Territorial Sea," *American Journal of International Law* 24 (1930): 541–55.

an excellent map[4] showing the average daily production of phytoplankton in the world's oceans north of parallel 60° south. The map reveals the high level of production around the world's coasts which contrasts with the phytoplankton deserts in the central areas of the oceans. Phytoplankton need a supply of mineral salts and sunlight for growth. Because sunlight and salts such as sodium chloride and calcium carbonate are available over wide areas of the oceans, the most critical factor is the availability of phosphates and nitrates, which are unevenly distributed. These salts exist in the highest concentrations in deep water, where they have accumulated through the decay of unused phytoplankton. Phosphates and nitrates become available in the surface layers of the sea as a result of strong vertical movements of water. These movements are less common around tropical archipelagos than higher-latitude continents, because the islands do not form such a definite barrier to currents and because the climate encourages the formation of a deep thermocline. The steepest temperature gradients occur in tropical waters, where the surface layers might be 10° C warmer than deep waters. Such a condition discourages the vertical movement of waters. Of course, those islands with coral reefs, in the eastern part of the region, possess rich inshore fisheries associated with the reefs and lagoons, but these have been more important in the subsistence economy than for commercial production for international trade. The major commercial fish stocks are located along sections of the coasts of Australia, New Zealand, and Papua New Guinea, and around Antarctica.

The last physical feature which must be mentioned is the icebound Antarctic coast. No rules have yet been proposed for establishing baselines along such coasts. Because of the Antarctic Treaty, questions of national sovereignty in Antarctica have been in abeyance since 1959, and for this reason it may be thought that the baseline issue is unimportant. Such a view may be complacent as several developed countries begin to express an interest in harvesting krill, a shrimplike crustacean about 4–5 cm long. This creature swarms in surface waters south of parallel 50° south and forms the basis of most of the food chains in the Antarctic.[5]

THE POLITICAL ORGANIZATION OF THE SOUTHWEST PACIFIC OCEAN

The territory of this extensive region is organized among nine independent countries and 10 dependent authorities. The countries show wide variations in the extent of territory controlled, the number of citizens, the scale and variety

4. Office of the Geographer, *Phytoplankton Production* (Washington, D.C.: Department of State, 1972).

5. G. L. Kesteven, "The Southern Ocean," in *Ocean Yearbook 1,* pp. 477–80, 491–92, and 494.

of available resources, and the length of the period of independent existence. Australia stands at one end of the spectrum. It has an area of 7.6 million km^2, a population of more than 13 million, rich and diverse mineral and farming resources, and has been independent since 1901. At the other end of the scale stands Nauru, with an area of 21 km^2, a population numbering about 7,000, and an economy totally dependent on the mining and export of phosphate. It has been independent since January 1968. The most recent country to achieve independence is Kiribati, formerly the Gilbert Islands, which achieved independence on July 12, 1979.

The 10 dependent territories are supervised by five countries. American Samoa, an unorganized, unincorporated territory of the United States of America, is administered by the U.S. Department of the Interior. The Cook Islands, Niue, and Tokelau are overseas territories of New Zealand; the Cook Islands have internal self-government. New Caledonia and Wallis and Futuna are overseas territories of France, which also shares a condominium over the New Hebrides with the United Kingdom. Australia, France, and New Zealand also claim sectors of Antarctica.

Australia and New Zealand are both fortunate in having island outposts which increase the maritime claims they can make. In addition to the overseas territories already mentioned, New Zealand also owns the Kermadec Islands, which lie between New Zealand and Tonga. Norfolk and Macquarie Islands allow Australia to increase its claims to areas of the Southwest Pacific Ocean.

The usual sharp distinction between independent countries and dependent territories has been blurred in the Southwest Pacific region. Some dependent territories have attended international gatherings as separate members; for example, the Cook Islands and Niue and Tuvalu and Kiribati before they became independent have all attended meetings of the South Pacific Forum as full members. The South Pacific Forum is an important association which enables governments in the region to discuss a wide variety of common problems and issues. It was formed in 1971, and all the independent countries of the region are members, together with those dependent territories mentioned earlier (see also George Kent's chapter in this volume).

EXISTING MARITIME CLAIMS IN THE SOUTHWEST PACIFIC OCEAN

The South Pacific Forum has been responsible for a higher degree of uniformity in existing maritime claims than might have been expected in view of the diversity of imperial powers which previously controlled this entire area. At the eighth meeting of the South Pacific Forum, in Port Moresby in August 1977, the members made a declaration on the law of the sea and the need for a regional fisheries agency.[6] In separate paragraphs all the members recognized

6. "South Pacific Forum Communiqué," *Australian Foreign Affairs Record* 48 (September 1977): 466–70.

the dangers of fish stocks being depleted and agreed to extend their fishing zones to 200 nautical mi:

> The members of the South Pacific Forum meeting at Port Moresby . . .
> 3. Recognize that in the continued absence of a comprehensive international convention on the Law of the Sea and in view of the action taken by a large number of countries, including distant-water fishing countries exploiting the valuable highly migratory species in the region, the countries in the region should move quickly to establish fishing or exclusive economic zones and should take steps to coordinate their policies and activities if they are to secure more than a very small part of the benefits from their resources for their peoples;
> 4. Undertake to complete as early as practicable and, if possible, by 31 March 1978, the legislative and administrative actions necessary to establish extended fisheries jurisdiction to the fullest extent permissible under international law, and to apply within their zones principles and measures for the exploration, exploitation, management and conservation of the living resources; . . .[7]

With the exception of Tuvalu, all the independent countries have claimed or indicated their intention to claim an exclusive economic zone stretching 200 nautical mi from the baseline used to fix the inner edge of the territorial waters. The same claim or intention of claim has also been made for all the dependent territories with the exceptions of the Australian Antarctic Territory and New Zealand's Ross Dependency.

Territorial seas are claimed around all territories with the exception of the Antarctic territories of Australia and New Zealand. In five cases the territorial seas measure 3 nautical mi; in all other cases the zone of territorial waters is 12 nautical mi wide. Narrow seas are claimed by Australia, the Solomon Islands, and Tuvalu, and Britain and France and the United States claim only 3 nautical mi around the New Hebrides and American Samoa, respectively.

By October 1978 four countries had proclaimed sections of straight baselines. Australia and New Zealand restricted their claims to closing lines across bays which meet the specifications of the *Informal Composite Negotiating Text.*[8] The Australian authorities proclaimed baselines along two sections of the coast on October 31, 1974.[9] The first section concerns 250 nautical mi of the east and south coast of Tasmania between parallel 41°45′ south on the east coast and parallel 42°45′ south on the west coast. Along this section closing lines are drawn across 24 bays and three river mouths. Great Oyster Bay and Storm Bay

7. Ibid., pp. 468–69.
8. United Nations, pp. 21–25 (also in *Ocean Yearbook 1,* p. 613).
9. J. R. V. Prescott, "Drawing Australia's Marine Boundaries," in *Australia's Offshore Resources,* ed. G. W. P. George (Canberra: Australian Academy of Science, 1978), pp. 28–29.

are the largest features enclosed; the former has an entrance 19 nautical mi wide, and the latter, which is linked to the D'Entrecasteaux Channel, has a mouth measuring 23.5 nautical mi wide. The second section occupies the southern section of the coast of New South Wales between parallels 36° south and 37°30′ south, and consists of very short closing lines across bays and river mouths. The Australian authorities have taken a very conservative attitude to the proclamation of baselines; indeed, they have followed the spirit of the rules to such an extent that they have apparently declined the opportunity to increase Australia's legitimate claim off southern Tasmania.

The New Zealand administration simply recites the rules which will govern the construction of baselines in its legislation proclaiming the territorial sea and exclusive economic zone.[10] The rules would enable the Bay of Islands and Hauraki Gulf to be closed on the North Island and Golden Bay to be closed on the South Island.

Fiji and Papua New Guinea have proclaimed straight baselines enclosing their archipelagos. According to the *Informal Composite Negotiating Text* archipelagic baselines must satisfy four conditions: they must conform closely to the outline of the archipelago; no segment may be longer than 125 nautical mi; not more than 3 percent of the baseline segments may exceed 100 nautical mi; and the ratio of sea to land within the baselines must not be less than 1:1 nor exceed 9:1.[11] Although it has not been possible to inspect charts showing the location of Fiji's baselines, the declaration on April 12, 1978, specified that the system excluded the Rotuma Islands, which are a detached northern group, and the isolated Ceva-i-Ra Island, which lies 240 nautical mi southwest of Kadavu Island. It is perfectly feasible to draw a baseline around the remaining islands which meets all the requirements outlined above. The longest segment would measure 120 nautical mi between Vuata Ono and Matuku Islands, and there would be only one other segment measuring 100 nautical mi. The ratio of water to land within the archipelagic baseline would be 8:1.

Papua New Guinea proclaimed its archipelagic baseline and other maritime limits in legislation published on March 30, 1978.[12] The baseline's longest segment stretches for 120 nautical mi between the northern terminus of the land boundary between Papua New Guinea and Indonesia and Wuvulu Island; less than 3 percent of the segments would exceed 100 nautical mi, and the ratio of water to land within the baseline is 1.3:1. The only exception which can be taken to this baseline concerns the manner in which is diverges from the general configuration of the archipelago. The baseline includes the tiny Malum and Southern Nguria Islands, when it appears from the charts of the region

10. New Zealand Government, *Territorial Sea and Exclusive Economic Zone* (Wellington: Government Printer, 1977), p. 5.

11. United Nations, pp. 37–38 (also in *Ocean Yearbook 1*, pp. 619–20).

12. Papua New Guinea Government, *Offshore Seas Proclamation 1978*, no. 7, schedule 2 (March 30, 1978), pp. 4–6.

that the main outline of the archipelago would be better represented by a line connecting Tanga, Feni, Green, and Buka Islands.

Papua New Guinea has also proclaimed archipelagic baselines around the Tauu and Nukumanu Islands. These groups consist, respectively, of 23 and 40 low coral islets on atoll reefs. Since the *Informal Composite Negotiating Text* specifically allows "the seaward low-water line of the reef" to be used in such situations, it seems unnecessary to use archipelagic baselines in this case.[13] However, the proclamation refers to the lines enclosing all islands, shoals, rocks, and reefs whether or not submerged within two areas measuring 30′ of latitude and 1° of longitude. It is not clear why such large areas were defined when the atolls appear on current large-scale charts as single, compact features. Perhaps it is just an illustration of a public servant's caution.

Papua New Guinea also proclaimed the line which marks the outer edge of its maritime claims. Although the northernmost parts of this line are defined as arcs drawn from the baseline with a radius of 200 nautical mi, the remainder, and longest part, separates the claims of Papua New Guinea from the maritime zones of Indonesia, the American Trust Territory of the Pacific Islands, the Solomon Islands, and Australia.[14] The boundary with Indonesia south of the island of New Guinea follows a number of points agreed upon by Australia and Indonesia in May 1971 and January 1973.[15] The agreement in May 1971 defined only a solitary point north of the island of New Guinea; Papua New Guinea has defined another four points on the line of equidistance, and the last is located 200 nautical mi from the baselines of both countries. Whereas Papua New Guinea has defined its common boundary with the American Trust Territory of the Pacific as a median line, it has not been equally consistent in its definition of its common boundaries with the Solomon Islands and Australia. This matter is considered later when potential boundary problems in the Southwest Pacific Ocean are explored.

Australia and Indonesia have negotiated a common marine boundary. The present seabed boundary consists of two segments separated by a gap of 140 nautical mi south of the former territory of Portuguese Timor. The first segment was drawn between points A12 and B1 through the Arafura Sea in May 1971 (fig. 1); points B1 to A3 now form part of the boundary between Papua New Guinea and Indonesia. Point A12 is located at meridian 133° 23′ east on the line of equidistance between the two countries, where the water attains a depth of 200 m. This was not a difficult line to draw because it traversed a shallow continuous continental shelf and the baseline points for each country were undisputed. The continuation of the boundary westward from point A12 presented some difficulties. Those arose because Australia and Indonesia

13. Ibid., p. 6.
14. Ibid., pp. 3–7.
15. J. R. V. Prescott, H. J. Collier, and D. F. Prescott, *Frontiers of Asia and Southeast Asia* (Melbourne: Melbourne University Press, 1977), pp. 78–79.

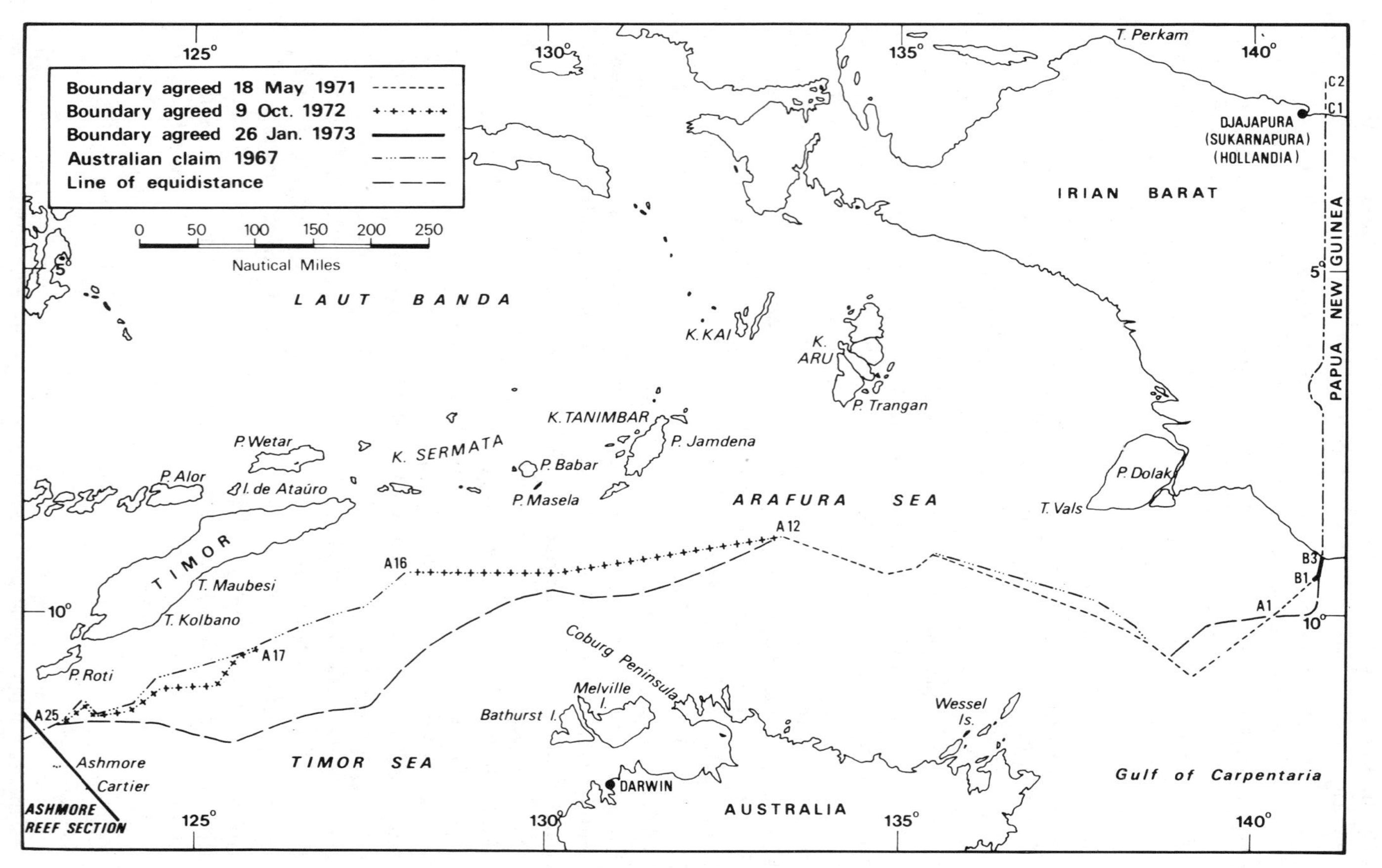

FIG. 1.—The Australian-Indonesian seabed boundaries

placed different interpretations on the significance of the morphology of the seabed between the two countries. The main feature of the seabed is the Timor Trough which descends to a maximum depth of about 3,100 m, along an axis following the alignment of the south coast of Timor, at a distance of 30–60 nautical mi from the coast. The views of the Australian authorities were revealed by the proclamation of Petroleum Adjacent Area boundaries in 1967, and by the issuing of petroleum exploration permits at various dates. The Australian government considered that there were two distinct continental shelves; their own stretched for 200 nautical mi or more to the southern flank of the Timor Trough and was distinguished by this depression from the very narrow shelf bordering Timor (fig. 2). Australian administrators would have been happy either with a boundary drawn along the axis of the Timor Trough or with two boundaries drawn along the northern and southern margins of the Trough. The Indonesian authorities took the view that the Timor Trough was merely an incidental depression in a single continental margin joining Australia and Timor. Accordingly, Indonesia pressed for a line which lay well south of the axis of the Trough. In view of the existence of Portuguese Timor at that time, the disputed area between Australia and Indonesia consisted of two parts; the part lying west of Portuguese Timor measured 8,100 nautical mi^2, while the part to the east measured 12,800 nautical mi^2.

This serious disagreement in principle could have strained relations between countries not well disposed toward each other. Fortunately, the two countries were able to reach rapid agreement, and on October 9, 1972 two new segments of boundary were drawn. East of Portuguese Timor the existing line was continued from point A12 to point A16; this latter point was located on the line of equidistance between Portuguese and Indonesian waters. West of Portuguese Timor a boundary was drawn for about 195 nautical mi from point A17 to point A25; point 17 was also located on a line of equidistance separating the claims of Portuguese Timor and Indonesia. The total area gained by Indonesia through this agreement measured about 3,000 nautical mi^2. The annexation of Portuguese Timor by Indonesia in July 1976 means that eventually Australia and Indonesia will have to negotiate a common boundary connecting points A16 and A17. This question is considered later in the section dealing with potential problems connected with maritime boundaries.

POTENTIAL MARITIME CLAIMS IN THE SOUTHWEST PACIFIC OCEAN

It seems unlikely that those countries and dependent territories which only claim territorial waters of 3 nautical mi will continue to do so for much longer. The majority of coastal states around the world now claim 12 nautical mi, and it is expected that the Solomon Islands and Tuvalu will soon proclaim wider territorial seas and that Australia and the United States will eventually follow

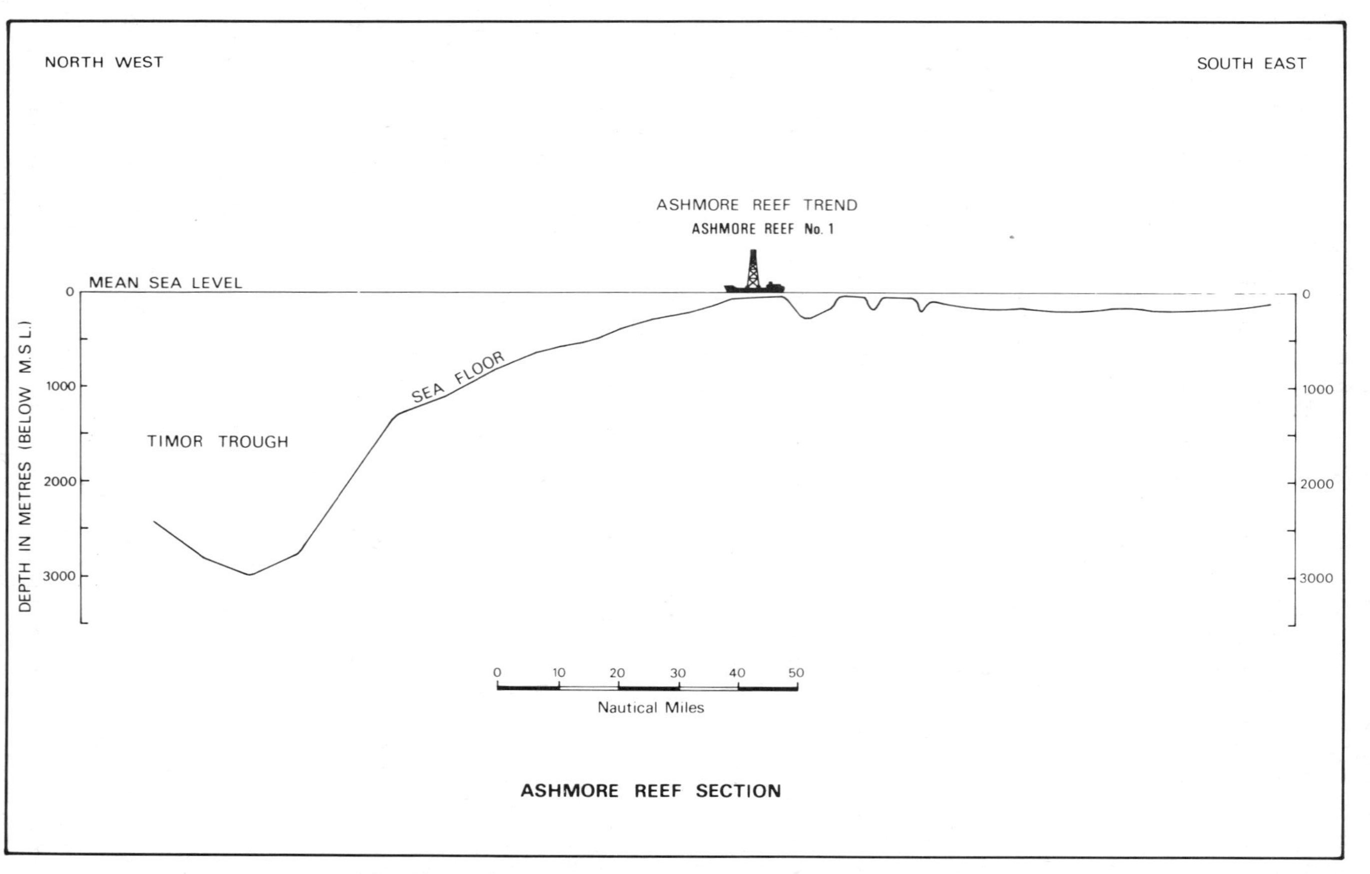

Fig. 2.—Section through Australia's continental shelf south of the Timor Trough

suit, thus establishing a uniform pattern of claims throughout the Southwest Pacific Ocean.

In the same fashion it is probable that Tuvalu and Tonga will also claim an exclusive economic zone of 200 nautical mi to bring them into line with all other territories throughout the area.

It is a much more open question whether Australia and New Zealand will follow France's lead in Adélie Land and make claims to territorial waters and exclusive economic zones around their Antarctic regions. It is unlikely that either country would take any steps which might jeopardize the conclusion of a convention for the conservation of Antarctic marine living resources. Meetings to draft such a convention were held by members of the Antarctic Treaty in Canberra and Buenos Aires in March and July 1978, respectively. However, if it proved impossible to agree on the terms of a convention, claimant states in Antarctica might be tempted to proclaim maritime zones so that they could enforce conservation measures. Fortunately, there is no evidence that such a drastic course is necessary at present.

It is now necessary to consider which territories might join Fiji and Papua New Guinea in proclaiming archipelagic baselines. In addition to setting out the four conditions which archipelagic baselines must meet, and which have already been considered, the *Informal Composite Negotiating Text* defined the terms "archipelagic state" and "archipelago" in the following manner: "(*a*) 'Archipelagic State' means a State constituted wholly by one or more archipelagos and may include other islands; (*b*) 'Archipelago' means a group of islands, including parts of islands, interconnecting waters and other natural features which are so closely interrelated that such islands, waters and other natural features form an intrinsic geographical, economic and political entity, or which historically have been regarded as such."[16] The Antarctic dependencies can immediately be discounted from this discussion. Inspection of charts quickly rules another seven territories out of consideration. Nauru and Niue consist only of single islands, and the islands which constitute the Cook Islands, Tuvalu, Tokelau, Wallis and Futuna, and American Samoa are too small and too scattered to satisfy the ratio test. Even if the stipulation that the ratio of water to land must not exceed 9:1 were waived, there would be little advantage in Tuvalu, Tokelau, and Wallis and Futuna claiming archipelagic waters, because their constituent islands are arranged in straight linear patterns; baselines drawn around these groups would enclose only small areas of water.

There are thus seven territories where archipelagic baselines have not been delimited, but where it would be reasonable to assume that the authorities have considered, or are considering, this option. They are Australia, New Caledonia, New Hebrides, New Zealand, Western Samoa, the Solomon Islands, and Tonga. These seven territories can be conveniently considered in three

16. United Nations, p. 37 (also in *Ocean Yearbook 1,* p. 619).

groups. The first group consists of the Solomon Islands, New Caledonia, and New Hebrides, which are comparatively compact archipelagos with many similarities to Fiji and Papua New Guinea which have proclaimed archipelagic baselines. Australia, New Zealand, and Samoa would enclose relatively small areas if they drew archipelagic baselines and the water to land ratio in each case would be less than 1:1. Tonga stands by itself because the distribution of its islands makes it impossible to satisfy the ratio test, even though it is a much more compact archipelago than Tuvalu or the Cook Islands, which have already been discounted.

There have been reports that the administration in the Solomon Islands has prepared archipelagic baselines, but no details have been published of their location. The Solomon Islands consist of two separate archipelagos; the main group extends from Choiseul Island in the northwest to San Cristobal Island in the southeast, while the smaller group is formed by the Santa Cruz Islands. The longest baseline in the main archipelago would measure 96 nautical mi between San Cristobal and Rennell Islands, and the ratio of water to land within the baseline would be 3.6:1. The longest segment of the baseline drawn around the Santa Cruz Islands would measure 72 nautical mi, and the ratio of water to land would be 7.6:1. Since the nearest points of the two archipelagos are 190 nautical mi apart, it would not be possible to connect them in a single system without breaching the rules proposed in the *Informal Composite Negotiating Text.* It would be possible for the Solomon Islands' administration to increase its claim to archipelagic waters by extending the baselines from Rennell Island to the islets on the Indispensable Reefs, which lie 40 nautical mi to the south. Such an extension would add another 970 nautical mi^2 to the country's archipelagic waters and would increase the water to land ratio marginally to 3.7:1. However, it is much harder to make a case for regarding this emerging atoll as part of the main archipelago, and an extension to enclose the Indispensable Reef might be considered by some to be a breach of the spirit, though not the rules, of the *Informal Composite Negotiating Text.*

All the islands of the New Hebrides could be enclosed by a baseline of which the longest segment would measure 90 nautical mi. This elongated archipelago, which extends from the Torres Islands in the north to Aneityum Island in the south, would have a water to land ratio of 4.4:1. It would not be possible in similar fashion to draw archipelagic baselines around all the islands which make up New Caledonia. This territory consists of the main archipelago stretching from Huon Island in the north to the Island of Pines in the south; Walpole, Matthew, and Hunter Islands are isolated and lie southeast of the main group; while the islets associated with the Chesterfield and Bellona Reefs are located away to the west. If a baseline is drawn only around the main group of islands, the longest segment would measure 70 nautical mi and would stretch from Maré Island to the Island of Pines; the water to land ratio would be 3.1:1. The system could be extended to include Walpole Island, if it could be established that the island and interconnecting waters were "closely interrelated."

The longest segment of baseline would then measure 75 nautical miles between Maré and Walpole Islands, and the water to land ratio would increase to 3.6:1. It would not be possible to link Matthew Island to the archipelagic baseline and satisfy the condition dealing with the longest segment, because Matthew and Maré Islands are 185 nautical mi apart. The islands and cays in the Chesterfield Group only measure an area of 101 ha, and any attempt to draw an archipelagic baseline around them would breach the ratio rule.

New Zealand and Western Samoa both consist of two main islands separated by a narrow strait. Both countries could draw baselines linking their islands and enclosing the intervening straits as archipelagic waters. The longest segment of baseline for New Zealand would measure 75 nautical mi across the western mouth of Cook Strait; the corresponding segment for Western Samoa would measure 16 nautical mi across the northern entrance of Apolima Strait. The water to land ratios for New Zealand and Western Samoa would be 0.26:1 and 0.9:1, respectively, and therefore fail to satisfy the proposed minimum value. It should be noted that the definition of archipelagic waters in the case of Western Samoa would not affect any waters which did not already form part of that country's territorial waters, and that the measurement of New Zealand's ratio of water to land assumes that Stewart Island, off the southern tip of the South Island, is included in the archipelagic system. Further, it is important to record that the use of archipelagic baselines by these two countries would not increase the area of the exclusive economic zone which each country claims.

It is very unlikely that if individuals were asked to nominate some of the archipelagos of the Pacific and Indian Oceans very many would mention Australia. However, there can be no question that Australia consists of a group of islands and that most of those islands are closely interrelated with each other and the waters separating them. If Australia decided to proclaim archipelagic baselines, the principal effect would be to erect segments across the entrances to Bass Strait. The longest segment in the eastern mouth would measure 23 nautical mi between South East Point on Wilsons Promontory and Curtis Island; the longest segment in the western mouth would lie between Franklin Point, just east of Cape Otway, and King Island and would measure 47 nautical mi. The other large islands which would be included within the Australian archipelago would be Kanagaroo Island, off the coast of South Australia; Melville Island and Groote Eylandt, off the coast of the Northern Territory; and Fraser and Wellesley Islands off the east and north coasts of Queensland, respectively. The longest segment linking any of these islands to the archipelagic system would measure 25 nautical mi and link Tasman Point of Groote Eylandt with Rantyirrity Point on Arnhem Land. Minor groups of islands included within the system would be the Archipelago of the Recherche off the south coast of Western Australia, the Wessel Islands off the coast of the Northern Territory, and the southern Torres Strait Islands off Cape York. It would be much harder to justify the inclusion of islands lying further from the coast such as Barrow Island, Browse Island, and Cartier Island.

Because the water to land ratio would be much less than 1:1, it is very unlikely that the present Australian authorities would consider seriously the establishment of archipelagic baselines around the main islands, although all the baselines described above could be drawn without recourse to the archipelagic concept. Australia has shown some reluctance to establish straight baselines, even though there are many sections of coast where classical conditions exist for the use of straight baselines according to the existing and proposed rules. It seems probable that Australia's reticence in this matter stemmed from a desire to set a good example to its neighbors and to discourage the proclamation of extravagant baselines, such as that established by Burma, which closed the Gulf of Martaban with a baseline measuring 222 nautical mi. The Australian administration hopes to proclaim its baselines in 1979. Delay has been caused by disagreement between the federal authorities and some state departments over the scale at which the baselines should be plotted. States such as Western Australia prefer the baseline to be shown on maps or charts with a scale of at least 1:100,000; it is believed federal authorities were proposing to map the baselines at a scale of about 1:250,000. The precise definition of baselines in Australia will be more significant in the area of federal-state relations regarding authority in coastal areas than in the exact definition of the outer edge of Australia's various maritime claims.

Western Australia has another reason for being concerned with a very precise definition of the baseline. North of Dampier there is a section of coast which experiences very large mean spring tide ranges. Between Dampier and Broome, tidal ranges of 3.6–8.5 m expose wide intertidal zones and make the selection of a particular line critical.[17]

Australia is probably the only country in the Southwest Pacific area which could claim historic bays, but it has not done so.[18] It would not be difficult to make out a case for claiming Spencer Gulf and the Gulf of Saint Vincent in South Australia and Shark Bay in Western Australia as historic bays. It seems likely that such claims have been avoided for two reasons. First, as already mentioned, Australia has adopted a consistently modest approach to maritime claims. Second, the federal authorities might be trying to ensure that they do not provide some of the states with grounds for claiming authority over parts of the adjoining continental shelf. According to section 14 of the *Seas and Submerged Lands Act (1973)*, the Australian states retain rights over those parts of the coastal waters and continental shelf which were within the limits of the state on January 1, 1901.[19] The existence of these pre-federal waters, as they are sometimes called, must be established by the states, and attempts to do this should prove an interesting exercise for lawyers, historians, and geographers.

17. E. C. F. Bird, "Metric Tide Ranges on the Australian Coast," *Australian Marine Science Bulletin,* no. 37–40 (1972), pp. 14–15.
18. United Nations (in *Ocean Yearbook 1,* p. 614).
19. Prescott, p. 36.

Their task would be simplified if Australia proclaimed title to historic bays on the basis of control and use during the previous century.

Tonga consists of 169 islands with a total area of 700 km^2; they are aligned along a north-south axis between parallels 15° south and 23° south. It would not be possible for Tonga to draw archipelagic baselines enclosing the majority of its islands without exceeding the proposed water to land ratio of 9:1. A series of lines enclosing the islands south of Fonualei would produce a water to land ratio of 45:1; if the system was extended to enclose all the islands the ratio would be increased to 65:1.

Tonga does have a claim to historic waters which can be traced back at least to the royal proclamation of June 11, 1887.[20] That proclamation defined the limits of Tonga as parallels 15° south and 23° south and meridians 173° west and 177° west. There is no specific provision in the *Informal Composite Negotiating Text* for claims to historic waters. However, the section dealing with the delimitation of territorial seas between adjacent or opposite states does permit countries to take into account historic title as a special circumstance.[21] It is not clear whether the provision refers to title to land only or to sea and land. India and Sri Lanka drew a common boundary through historic waters between the two countries in June 1974.[22] The waters in question were located in Palk Bay and Palk Strait and were deemed to be historic waters by a judgment in the Appellate Criminal Division of the Indian High Court in Madras in 1904.[23] The proper basis of claims to historic bays and historic waters seems to be a matter which the international community will eventually have to establish.

In order to obtain an impression of the potential claim which each country and territory could make in the Southwest Pacific Ocean, theoretical boundaries were constructed on British Admiralty Charts 780 and 788 according to the following principles. First, it was assumed that every country would eventually claim an exclusive economic zone of 200 nautical mi. Second, where adjoining territories were separated by less than 400 nautical mi, lines of equidistance were drawn. Because these charts are drawn on the Mercator projection the scale increases southward. This means that the lines of equidistance which trend east-west will be situated slightly north of their true position. Fortunately most of these lines are found north of parallel 20° south where the distortion is only very slight. Third, the proposal of the *Informal Composite Negotiating Text,* that rocks which cannot sustain human habitation or economic life of their own should not generate claims to either the continental shelf or exclusive economic zone, was followed.[24] Any features which were labeled

20. *Tonga Government Gazette,* vol. 11, no. 55, August 24, 1887.

21. United Nations, pp. 24–25 (also in *Ocean Yearbook 1,* p. 614).

22. Government of India, *Notice to Mariners,* 9th ed. (New Delhi: Government Printer, April 15, 1975), notices 133–56.

23. Office of the Geographer, "Historic Water Boundary: India–Sri Lanka," *Limits in the Sea,* no. 66 (Washington, D.C.: Department of State, December 1975).

24. United Nations, p. 69 (also in *Ocean Yearbook 1,* p. 631).

rocks on the charts and which lay outside the territorial waters generated by an island were discounted during the construction of lines of equidistance. The most important application of this principle involved Havre and Esperance Rocks south of the Kermadec Islands. If these features had been used as base points for the measurement of the exclusive economic zone, the maritime claims around New Zealand's North Island and the Kermadec Islands would have been continuous. When they are ignored, a corridor of high seas remains between New Zealand and the Kermadec Islands.

The results of this exercise are shown in the accompanying map (fig. 3), and they suggest four main conclusions. First, a very considerable area of the Southwest Pacific Ocean is involved in the definition of exclusive economic zones. The total area of sea contained within all the claimed maritime zones is 6,407,300 nautical mi^2. This extensive area is divided unequally between the countries and dependent territories of the Southwest Pacific (table 1). There is no point in describing in detail the area which each country can claim by the application of principles of equidistance, because in any bilateral negotiations this principle might not be paramount. Further, the area of sea which can be claimed is not an accurate reflection of the wealth which might be generated from the resources contained in the waters or on the seabed of that particular area. However, the areas available for each country do give an impression of the extent to which countries have been fortunate or unfortunate in their geographical location. This is an important matter because recent meetings of the Third UN Conference on the Law of the Sea have witnessed some sharp divisions between the 53 landlocked and geographically disadvantaged countries and coastal countries which are not geographically disadvantaged.

By comparing the area of seas which can be claimed with the area and population of the claimant country or territory it is evident that Western Samoa is geographically disadvantaged and that the Cook Islands and Tuvalu are geographically fortunate. Western Samoa could only claim 38,100 nautical mi^2 if principles of equidistance are applied. Western Samoa is confined on the east and north by American Samoa, and it is the northern boundary which probably seems most inequitable to Western Samoa's administration, because it is entirely based on Swain's Island which was annexed by the United States in 1925. Swain's Island lies 190 nautical mi from the main group in American Samoa; it is an atoll with an area of about 1 nautical mi^2 and a maximum elevation of 6 m. On the south, Western Samoa's claim is restricted by Tafahi and Niuatobutabu, which are the northern islands of Tonga lying 127 nautical mi north of the main group. On the west, Wallis Island restricts Western Samoa's potential claims.

The Cook Islands and Tuvalu can claim 566,100 and 211,500 nautical mi^2, respectively, because they consist of small islands which are fairly widely scattered and comparatively remote in some directions from neighboring territories.

The second important point revealed by the maps is that the countries and territories of the Southwest Pacific vary considerably in the extent to which they

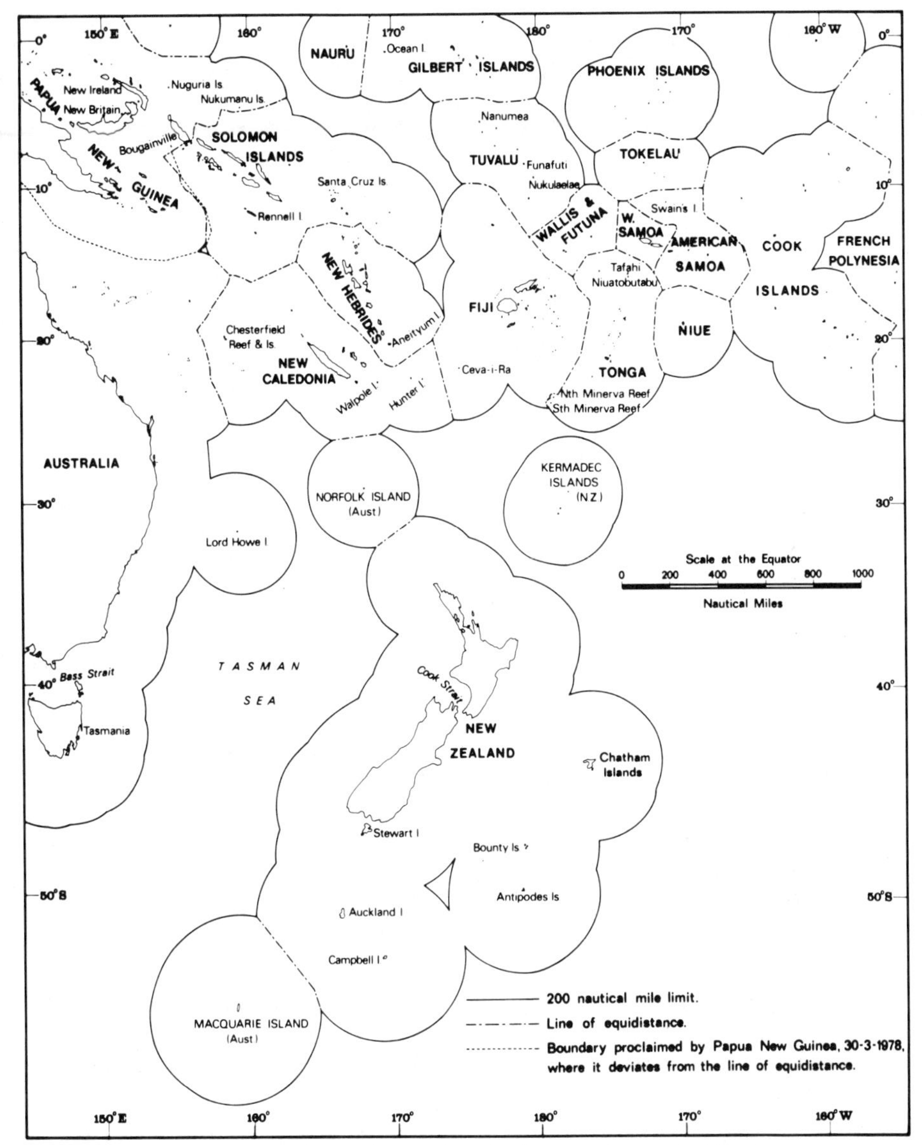

FIG. 3.—Potential maritime claims in the Southwest Pacific Ocean

will have to negotiate common boundaries with neighbors. Although all territories will have to negotiate with at least one neighbor, it is possible to distinguish five which will face a larger negotiating burden. Western Samoa and American Samoa are both entirely surrounded by neighbors less than 400 nautical mi distant. This means that the entire limit of their exclusive economic zone will have to be negotiated with neighbors if the maximum claim of 200 nautical mi is made. The New Hebrides, Papua New Guinea, Wallis and

TABLE 1.—POTENTIAL MARITIME CLAIMS, LAND AREA, AND POPULATION OF TERRITORIES IN THE SOUTHWEST PACIFIC OCEAN

Country	Potential Maritime Claim (Nautical mi²)	Area (Nautical mi²)	Population (Thousands)
Australia	1,854,000	2,240,771	13,548
New Zealand	1,058,100	78,320	3,129
Papua New Guinea	684,200	134,586	2,490
Cook Islands	556,100	70	18
Solomon Islands	458,400	8,682	197
New Caledonia	382,400	5,567	132
Fiji	386,900	5,326	588
Tuvalu	211,500	7	6
New Hebrides	179,900	4,303	97
Tonga	158,400	203	90
American Samoa	114,000	57	31
Nauru	92,800	6	8
Tokelau	91,300	3	2
Niue	87,300	75	4
Wallis and Futuna	71,900	80	10
Western Samoa	38,100	828	151

SOURCE.—The potential maritime claim is based on the author's calculations; the data on area and population are taken from *The Statesman's Yearbook 1978/1979.*

Futuna, and Tokelau will have to negotiate about three-quarters of the outer limit of their exclusive economic zones with neighbors. Papua New Guinea has already made a good start on this task through its negotiations with Australia, and because it inherited certain maritime limits which Australia had negotiated with Indonesia before 1975. New Zealand is the most fortunate country in terms of bilateral negotiations; no negotiations will be necessary in respect of the Kermadec Islands, and only short, easily constructed boundaries will be necessary between claims from the North Island and Norfolk Island belonging to Australia, and between Aukland Island and Macquarie Island, which also belongs to Australia.

The third main conclusion is that a number of countries have been fortunate in owning detached islands which significantly increase the areas of sea which they can claim. Australia seems to be most fortunate because it owns Lord Howe, Norfolk, and Macquarie Islands in the Pacific, all of which allow claims close to the maximum possible, and because its ownership of Ashmore and Cartier Islands in the Timor Sea ensure that the seabed boundary agreed with Indonesia lies distant from the Australian mainland in the western part of that sea. It might be noted in passing that Australia also has other possessions in the Indian Ocean, including Heard Island and the Cocos and Keeling Islands, which also generate claims to extended maritime zones. New Zealand's ownership of the Kermadec Islands has already been mentioned. Other fortunate territories include the Solomon Islands, which benefit from the easterly Cherry and Mitre Islands; New Caledonia, which increases its maritime claims through the possession of Hunter Island; and Fiji, which is able to extend its claim southwestward from the remote Ceva-i-Ra Island.

The final main point revealed by the maps is that the delimitation of national claims in the Southwest Pacific will create some enclaves of high seas and some deep corridors penetrating between exclusive economic zones. The largest enclave separates the claims of the Solomon Islands from those of Nauru and Tuvalu; although not shown on this map, this area of high seas is bounded on the north by the claims based on the Marshall and Caroline Islands. Smaller enclaves of high seas lie between Fiji and the New Hebrides; near the tri-junction of the claims of the Solomon Islands, Australia, and Papua New Guinea; and south of New Zealand. The principal corridors of high seas lie east of Tuvalu, in the vicinity of the Kermadec Islands, and between Lord Howe and Norfolk islands. The present regulations on the delimitation of maritime claims fail to deal with the problems presented by these irregular patches of high seas. There is an obvious case for eliminating the smaller enclaves by incorporating them in adjoining national claims.

EXISTING PROBLEMS CONNECTED WITH MARITIME CLAIMS IN THE SOUTHWEST PACIFIC

There appear to be only two current problems, and both of these are likely to be solved fairly quickly. The first involves the construction of a maritime boundary through Torres Strait; the second concerns disagreements about the membership of the fisheries organization proposed by the South Pacific Forum.

The dispute in Torres Strait has a long and interesting history which had its origins in 1859. Since the late 1960s, successive Australian governments have worked to reach agreement with the authorities in Papua New Guinea. No foreign minister has worked harder to achieve this goal than Andrew Peacock.

When the colony of Queensland was created in 1859, it included "all and every adjacent islands, their members and appurtenances, in the Pacific Ocean."[25] The governor was permitted to extend his jurisdiction over all islands within 96 km of the coast in 1872; this effectively placed the limit of Australian authority about parallel 10° south, in the middle of Torres Strait, and ensured that the only navigable course through the Strait for large vessels was controlled. Then, as now, the Prince of Wales Channel provided the route which had to be followed by merchant vessels. The first regular steamship mail service was initiated through the Strait in 1873, and it served to focus attention on a region which was attracting some concern in Queensland. The administration in Brisbane tried to persuade the British government to annex the southern coast of Papua so that the entire Strait would be controlled by British forces.

25. Prescott et al., p. 39.

This policy was advocated for a number of reasons: British commerce through the Strait would be protected from foreign interference; Papua might provide gold; foreign powers, particularly France and Germany, would be prevented from securing a foothold near Australia; and the regulation of pearling, fishing, and the recruiting of indigenous labor would be simplified. In 1878 Queensland was authorized to annex certain islands in Torres Strait; the islands were not named, but as a means of geographical shorthand a line enclosing the islands was defined, and the definition was repeated in the parliamentary act on June 24, 1879:

> Certain islands in Torres Straits and lying between the Continent of Australia and the Island of New Guinea that is to say all islands included within the line drawn from Sandy Cape northward to the south-eastern limit of Great Barrier Reefs to their north-eastern extremity near the latitude of nine and a half degrees south thence in a north-westerly direction embracing East, Anchor and Bramble Cays thence from Bramble Cays in a line west by south (south seventy-nine degrees west) true embracing Warrior Reef, Saibai, and Tuan Islands thence diverging in a northwesterly direction so as to embrace the group known as the Talbot Islands thence to and embracing the Deliverance Islands and onwards in a west by south direction (true) to the meridian of one hundred and thirty-eight degrees of east longitude.[26]

Five years later, when Britain overcame its reluctance to annex the Papuan coast, the *raison d'être* for Queensland's ownership of the islands disappeared and moves were immediately started to transfer some of the islands to British control in Papua. These protracted negotiations reached fruition just as Australia became a federation, and this matter of foreign relations passed from the Queensland to the Australian government. The new Australian constitution eventually thwarted implementation of the agreement, and all the islands in Torres Strait remained Australian.

The latest phase of this dispute began as Papua New Guinea approached independence and its political leaders started to draw attention to the inequitable distribution of sovereignty in Torres Strait. At that time Papua New Guinea's demands for a share of the waters and seabed of Torres Strait was supported by two other arguments besides the claim that the historic arrangement was unfair. First, it was maintained that the 1879 line was a colonial boundary which was not binding on Papua New Guinea. That argument may have a certain chauvinist appeal, but the fact remains that almost all the boundaries in Southeast Asia are colonial boundaries, and it is such a colonial

26. Commonwealth of Australia, Joint Committee on Foreign Affairs and Defence, Subcommittee on Territorial Boundaries, *The Torres Strait Boundary,* Parliamentary Paper no. 416/1976 (Canberra: Commonwealth Government Printer, 1977), p. 185.

boundary that excised Bougainville Island from the Solomon Islands and attached it to Papua New Guinea. Second, it was argued that this was a unique situation which would not be tolerated by any self-respecting country. Anyone with a good mental map of the world will be able to provide examples of similar situations. The Philippine islands close to Sabah; the Greek islands close to Turkey; the French islands close to Canada; and the British islands close to France are only the most obvious examples.

Nevertheless, the transparent weakness of both these arguments did not detract from the force of the argument that it was unconscionable that an accident of colonial division affecting a few small islands a century ago should serve to deny a newly independent country rights over land, sea, and seabed to which it could otherwise, in its view, lawfully lay claim. Australian governments were properly nervous of the damage which might be caused to Australia's reputation if it was dragged before the International Court of Justice by a former dependency, and of the possibility that the Court might make a judgment less favorable than could be secured by negotiations.

Only about 17 of the 124 Torres Strait islands are occupied, but all the unoccupied islands are used in various ways by the Islanders; they use some for growing vegetables, on others they collect wood, and still others provide shelter on long fishing trips and places where birds' eggs and turtles can be collected. Papua New Guinea wanted Australia to cede some of the uninhabited islands, and to leave the three inhabited islands close to the coast of Papua as an Australian enclave in waters owned by Papua New Guinea; these three islands are called Saibai, Boigu, and Dauan. Australia could not comply with the first part of this request because of the relevant provisions of the constitution. This known fact was responsible for the general surprise exhibited in Australia on March 31, 1978, when the Foreign Minister announced that three islands called Kawa, Mata Kawa, and Kussa, which had previously been considered part of Australia, would in future be part of Papua New Guinea.

This remarkable development was possible because the Foreign Minister had been satisfied by historic research that the three islands had never belonged to Australia. In the definition of the boundary line provided above there is reference to a northward detour to enclose the Talbot Islands; the Australian government decided that Kawa, Mata Kawa, and Kussa were not part of the Talbot Islands. This view was defended by the publication of the evidence in a short monograph.[27] The main piece of evidence was a chart which accompanied the draft Letters Patent sent to Australia in October 1878. This chart does not in fact show the three islands in question, but it does show the line hugging the northern coast of Boigu and then trending south. If the islands were marked in their correct position on the chart they would lie north of the line. It is doubtful that an academic conducting research on this question would

27. Foreign Affairs Department, *Status of the Islands of Kawa, Mata Kawa, and Kussa* (Canberra: Commonwealth Government Printer, 1978).

base such a definite opinion on this piece of evidence; indeed there is plenty of compelling evidence to show that the authorities in Queensland and Papua regarded at least Kawa and Mata Kawa as part of the Talbot Islands in the 1880s. However, this is not an academic question but a political one, and the evidence justifies the political action which has been taken.

This effective change in the ownership of three small, uninhabited islands was a major advance in the resolution of the dispute. The next major advance occurred on May 25, 1978, when the principal basic elements of the proposed boundary treaty in Torres Strait were announced. The main features of this treaty are shown in the accompanying maps (figs. 4, 5).

The agreement can be considered in two main parts. First, there will be a seabed boundary stretching for 1,200 nautical mi from point *A3* on the Australian-Indonesian boundary agreed in 1971, to a point near the tri-junction of the potential claims of Australia, Papua New Guinea, and the Solomon Islands. Apart from 47 nautical mi, the seabed boundary coincides with a line separating the swimming fisheries zones of the two countries. If the seabed boundary is compared with the line of equidistance based on Australia's ownership of most of the islands in the Strait, it is evident that Australia has conceded 8,800 nautical mi^2 of seabed. However, Papua New Guinea has not obtained any seabed rights within the territorial waters of Australian islands located north of the seabed line.[28]

The second important feature of the agreement is the creation of a Protected Zone within which the Torres Strait Islanders and Papua New Guineans who live in the adjacent coastal area will be able to pursue their traditional activities without hindrance. The Protected Zone has an area of 12,478 nautical mi^2 and extends between meridians 141°20′ east and 144°28′ east north of parallel 10°28′ south. The seabed boundary divides the Protected Zone in such a way that one-third of the area falls to Papua New Guinea north of the line. One of the sections of the treaty places an embargo on mining and oil drilling of the seabed of the Protected Zone for 10 yr after the treaty comes into force. The same section will also provide for holders of Australian petroleum exploration permits for areas which will pass to Papua New Guinea to negotiate equitable arrangements with the government of Papua New Guinea. Finally on this point, the treaty will ensure that if any hydrocarbon deposit is found to straddle the seabed boundary the two countries will consult on the best method of exploiting the deposit and sharing the profits equitably. Clauses of this nature are common in recent seabed agreements around the world.

The arrangements for control of the commercial fishing industry within the Protected Zone is the most complicated part of the treaty. The line separating fishing zones in the Protected Zone deviates from the seabed boundary between meridians 142°03′30″ east and 142°51′ east, and passes north of the

28. Letter from the Foreign Minister of Australia, September 25, 1978.

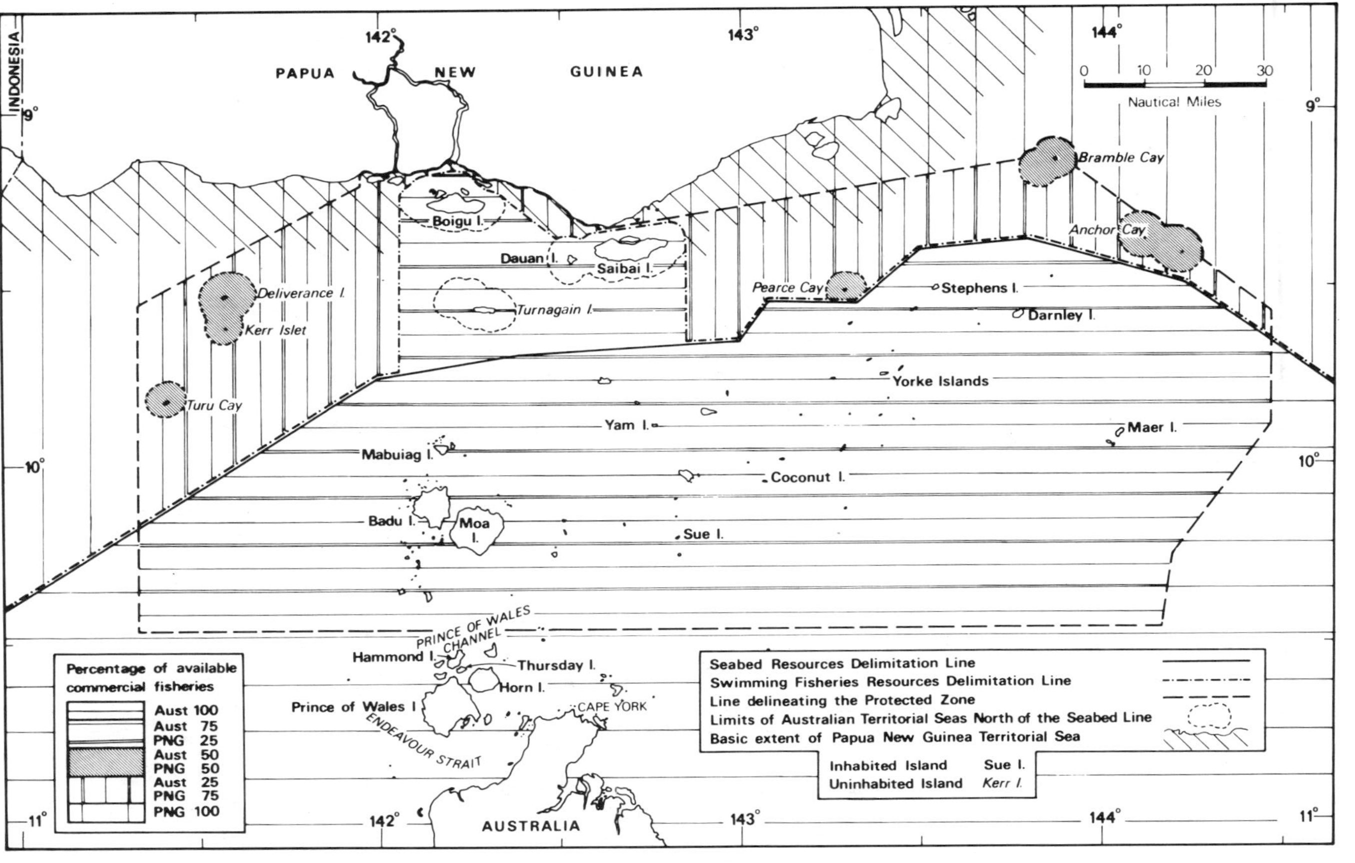

FIG. 4.—The maritime boundaries between Australia and Papua New Guinea

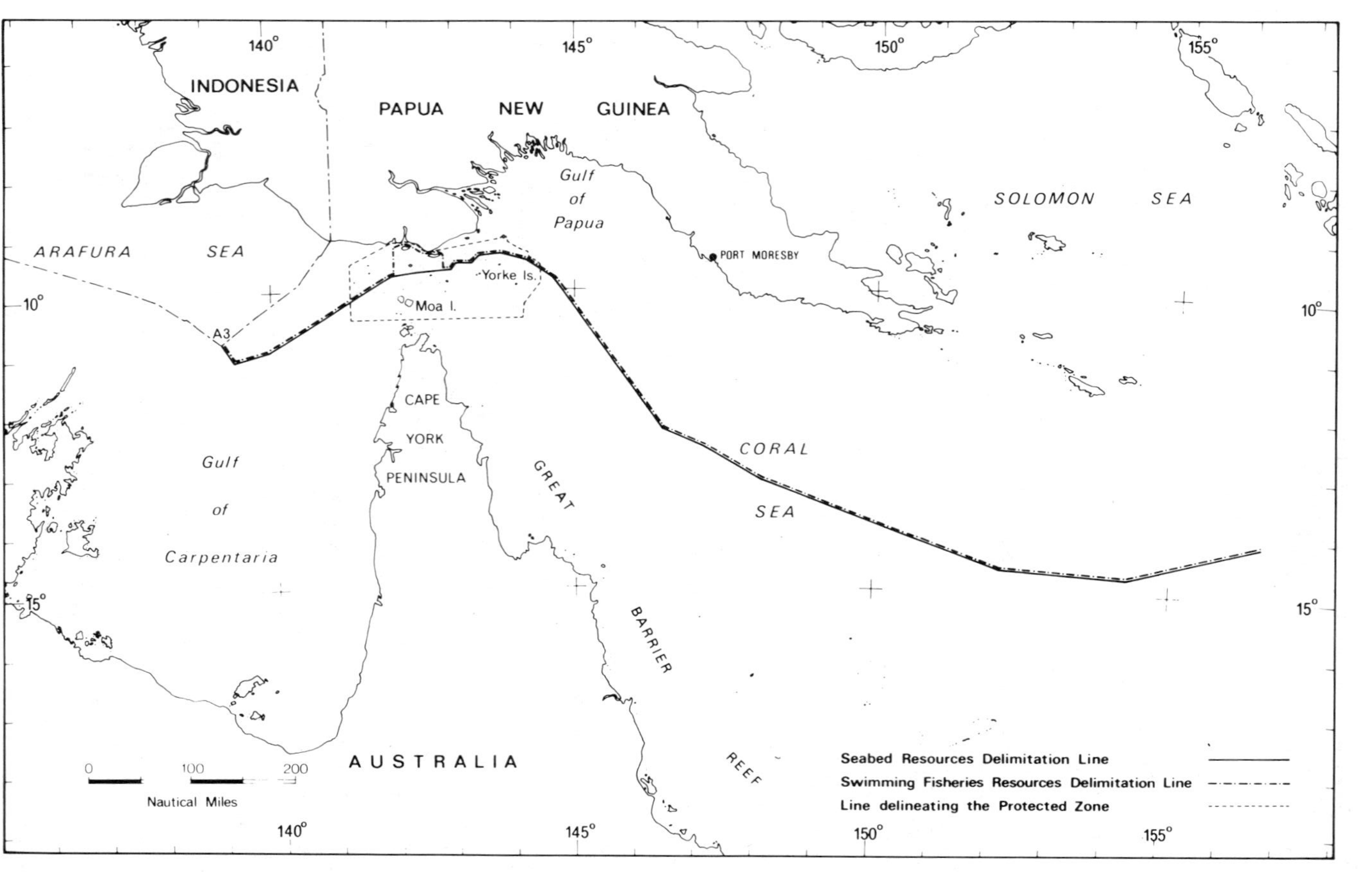

Fig. 5.—The Protected Zone and fisheries boundaries in Torres Strait

inhabited Australian islands called Boigu, Dauan, and Saibai. In Australian waters south of the fisheries boundary, Australia will have authority to issue licenses to commercial fishermen who may take 75 percent of the available commercial catch. The total area through which that proportion applies amounts to 9,504 nautical mi^2, of which 1,143 nautical mi^2 lie north of the seabed boundary. This rule means that Papua New Guinea will be able to issue licenses for the remaining quarter of the commercial catch in this area, which includes some Australian territorial waters. In the territorial waters of Australian uninhabited islands north of the fisheries boundary the two countries will share the commercial catch equally; the total area where this rule applies is 219 nautical mi^2. Further, Australia has undertaken that it will not increase the width of territorial seas around islands north of the fisheries boundary. In the remaining 2,755 nautical mi^2 of the Protected Zone which lies north of the fisheries boundary, and outside Australian territorial waters, Papua New Guinea will be able to issue licenses for 75 percent of the available commercial catch. Australia will issue licenses for the other 25 percent, except that Papua New Guinea will have sole rights to the barramundi fishery near the coast of Papua New Guinea. The arrangements for commercial sedentary fisheries will be the same as for swimming fisheries, but separate arrangements will be made for the other seabed resources, including coral, seaweed, and pearl shell. If any country is not able to take up its full entitlement to commercial fisheries, then the other country will be offered first choice against any third country, and in addition vessels of third countries will not normally be licensed to exploit commercial fisheries unless there is agreement by both Australia and Papua New Guinea.

Plainly the mechanism for measuring the available commercial fishery in the Protected Zone and surveying the activities of operators will have to be very carefully devised. However, it is unlikely that technical problems will prove insuperable given the cooperative manner in which much more intractable political problems have been solved. The Australian authorities still face some difficulties in securing the enthusiastic cooperation of the Queensland government and the Torres Strait Islanders, but it will be surprising if these domestic political problems cause much delay in the production and implementation of the treaty. [EDITOR'S NOTE.—The treaty in the terms described was signed on December 18, 1978, and can be found in this volume, pp. 559–74.]

The second problem connected with maritime claims concerns the attempt by members of the South Pacific Forum to establish a South Pacific Regional Fisheries Organisation. A declaration was adopted at the meeting in Port Moresby in August 1977. It drew attention to the absense of a comprehensive convention on the law of the sea and to the actions of a number of countries whose distant fishing fleets were exploiting the valuable highly migratory species in the South Pacific, and decided that the members would agree on a common basis for negotiating with distant-water fishing interests and establish a South Pacific Regional Fisheries Agency. A meeting was held in Suva in

November 1977 to draft the articles of a convention to establish the agency, and these draft proposals were presented to a full meeting of the South Pacific Forum at Niue in September 1978. The draft convention envisaged that the agency would study and collect information on highly migratory species; prepare proposals for regional cooperation for the effective use and conservation of living resources; facilitate a regional approach to management, licensing, and enforcement; and collect technical fisheries information. The September meeting found the proposed aims of the agency unexceptionable, but there was sharp disagreement about the membership of extraregional countries. Papua New Guinea, the Solomon Islands, and Fiji opposed the membership of the United States, Britain, and France, while Western Samoa and the Cook Islands favored their membership. The smaller territories supported the inclusion of the major powers because it would afford opportunities of securing technology and assistance in supervising extended fishing zones. This disagreement prevented any final decision being taken, and it was decided to set up an agency comprising only Forum countries and examine further the question of a more broadly based organization. There seems little doubt that the most effective organization will include the distant fishing interests. [EDITOR'S NOTE: The convention establishing this fisheries agency can be found in this volume, pp. 575–78; for an extensive discussion of the convention see George Kent's chapter in this volume.]

POTENTIAL PROBLEMS CONNECTED WITH MARITIME CLAIMS IN THE SOUTHWEST PACIFIC

It is not proposed to examine the potential problems in detail because the extent to which they prove hard to solve depends more on the nature of bilateral relations between the countries concerned than on the characteristics of the potential problems.

There seem to be four cases where the settlement of a common marine boundary might be attended with some difficulties. It was noted earlier that there is a gap of 140 nautical mi in the seabed boundary between Australia and Indonesia; it exists because eastern Timor was a Portuguese colony at the time the two governments drew that line, and the Portuguese refused to negotiate with Australia until the final Convention on the Law of the Sea had been adopted. Since 1975 Indonesia has controlled the former Portuguese area, and so the boundary segment completing the line will have to be settled with that country. Indonesia's annexation of eastern Timor was not recognized by the Australian government until January 1978 when the situation was given de facto recognition. The problem along this section of boundary is identical with that already solved in the agreed sections and concerns the existence of the Timor Trough. There seems to be no practical reason why rapid agreement would be difficult once the political and legal obstacles connected with de jure recognition have been solved. However, there is always the possibility that Indonesia will try to strike a harder bargain this time because of Australia's reluctance to give early recognition to the incorporation of eastern Timor and because of Australia's generosity in the settlement it reached with Papua New Guinea in Torres Strait. In December 1978 the Indonesian Foreign Minister observed that his country had been "taken to the cleaners" during previous seabed negotiations with Australia.

A more serious problem arises from Indonesia's decision to distinguish between claims to the exclusive economic zone and claims to the continental margin. The effect of this distinction is to allow Indonesia to claim an exclusive fishing zone south of the agreed seabed boundary, where that boundary is not a line of equidistance. This is a new development caused by the failure of the Law of the Sea Conference to resolve this anomaly, which allows states to rely either on the concept of the exclusive economic zone or on the natural prolongation of their territory. Each state will make the choice which suits it best, and reliance on different concepts will definitely complicate the delimitation of marine boundaries between adjacent states and between states separated by less than 400 nautical mi of sea.

The line delimiting the outer edge of Papua New Guinea's offshore seas does not follow the line of equidistance through the Solomon Sea. Between points 23 and 29 the proclaimed line is consistently east of the line of equidistance and encloses an area of 6,500 nautical mi^2 which could be claimed by the Solomon Islands. However, the proclamation notes that this outer line is "without prejudice to the ultimate location of appropriate boundary lines."[29] Presumably therefore Papua New Guinea would be prepared to withdraw its claim to the line of equidistance if it was unable to persuade the Solomon Islands administration that there were special circumstances justifying the proclaimed boundary.

The proclamation defining Tonga's waters in 1887 fixes the western boundary as meridian 177° west. This overlaps the line of equidistance between Fiji and Tonga, and it will be necessary for the two countries to come to some agreement about the allocation of the area concerned. The same two countries may also face a problem over how to deal with the question of the significance to be attached to Minerva Reefs which were annexed by Tonga on June 15, 1972. The proclamation refers to the islands of Teleki Tokelau and Teleki Tonga, but all charts and gazetteers consulted describe them as reefs. The distinction between these two forms may be hard to draw given the dynamic processes associated with island construction in the Southwest Pacific, but the matter will have to be resolved because if they are used as part of the baseline from which the line of equidistance is constructed, the increase in Tonga's area will be about 56,500 nautical mi^2. Fiji's area would be reduced by 18,500 nautical mi^2, and

29. Papua New Guinea Government, p. 2.

that attributable to the Kermadec Islands would be reduced by 5,000 nautical mi^2.

It was noted earlier that Western Samoa appears to have a definite claim to be considered as a geographically disadvantaged state. At present two different factions at the Law of the Sea Conference are stressing the equitable and equidistant principles in the delimitation of maritime boundaries between states. If equitable principles became paramount, Western Samoa could probably make out a case for concessions by the neighboring countries, especially American Samoa, which benefits to such a marked extent from the detached Swain's Island.

The last potential problem concerns the baselines from which marine claims around Antarctica should be measured. This is unlikely to be an urgent problem because there is no evidence that any of the countries claiming sovereignty in Antarctica intend to enforce their authority in the adjoining seas. The question of defining baselines along icebound coasts has not been discussed at the Law of the Sea Conference. Article 6 of the Antarctic Treaty could be interpreted to imply a parity between the landmass and the ice shelves of Antarctica, suggesting that the baseline would at least follow the outer edge of the ice shelves. It seems worthwhile to promote discussion of this subject during the lifetime of the Antarctic Treaty, so that if territorial claims in Antarctica become an established fact, conflict over the extent of maritime claims will be avoided.[30]

CONCLUSION

The Southwest Pacific Ocean may be likened to a giant laboratory in which nearly all the interesting questions related to the Law of the Sea Conference can be studied. In respect of maritime zones and international maritime boundaries the region provides a very large area subject to national claims, as internal waters, archipelagic waters, exclusive economic zones, and continental shelves, and many overlapping areas where common maritime boundaries will have to be negotiated between neighbors. There is a multitude of examples of islands as special circumstances and a greater concentration of archipelagic countries than in any other sea. The issue of equitable versus equidistant boundaries is raised by Western Samoa's geographical disadvantage of near neighbors on all sides, and Australia and Indonesia have faced and solved the problem of deciding what significance should be given to deep depressions in the seabed when negotiating boundaries on the continental margin.

30. For a further discussion of these claims, see Kesteven, pp. 491–92; and F. M. Auburn, "Legal Implications of Petroleum Resources of the Antarctic Continental Shelf," *Ocean Yearbook 1,* pp. 505, 514.

When the economic use of the sea is considered, the region is equally obliging. There is a conflict of interest between distant and domestic fishing fleets, and Australia has a continental margin which extends beyond 200 nautical mi. In addition, the narrow margins of most of the countries in the Southwest Pacific raise the question, which has not yet been tackled, of the possible conflict between mining the deep seabed outside national jurisdiction and mining those areas of the deep seabed which are part of national claims.

Two other problems not yet faced by the Conference on the Law of the Sea are evident in this region. The first concerns inconvenient pockets of high seas as enclaves in national areas or as corridors between national claims. The second relates to the problem of deciding where the baseline should be on icebound coasts.

It can be asserted with confidence that existing problems in the area appear to be capable of early solution, and that those potential problems which have been identified will cause little serious friction if the present, admirable spirit of the South Pacific Forum is maintained.

Fisheries Politics in the South Pacific[1]

George Kent
University Of Hawaii

In late 1976 the South Pacific Forum began work toward the creation of a new South Pacific regional fisheries organization. As in the case of the global Third United Nations Conference on the Law of the Sea, the hopes held by the separate parties at the outset have proven incompatible, and visions of the possible have been greatly narrowed. Aspirations for a strong regional fisheries management organization have succumbed to political realities.

Part I of this essay presents a descriptive account of the situation in fisheries in the South Pacific. Part II raises the question of what in the situation ought to be regarded as significantly problematic, and thus ought to be understood as being in need of deliberate international management. These judgments of course depend on the value perspectives brought to the analysis. In Part III, the global and regional responses to the problems of fisheries management are critically assessed.

I. THE SITUATION

Basic Data

The islands of the Pacific are divided into three major groups. The Polynesian islands are contained within the triangle bounded by Easter Island to the east, Hawaii to the north, and New Zealand to the south. The western Pacific is divided at the equator, with Micronesia to the north and Melanesia to the south. Although north of the equator, Micronesia is commonly recognized as part of the South Pacific. The region is also sometimes described as the tropical Pacific or as Oceania.

1. This essay is based on a study on South Pacific Fisheries Management made possible by the assistance of the Food and Agriculture Organization of the United Nations. The views expressed are those of the author and are not necessarily those of FAO. A more extensive version is to be published by the Westview Press of Boulder, Colo., under the title, *The Politics of Pacific Islands Fisheries.* (EDITORS' NOTE: For a general discussion of developments in marine fisheries, see S. Holt's chapter in this volume.)

© 1980 by The University of Chicago. 0-226-06603-7/80/0002-0006$01.00

Many of the islands share similar environments and similar political histories. All except Tonga have been colonies of major metropolitan powers. The islands are among the last to gain formal independence. They are also among the last to feel the impact of Western technology and culture.

The islands of the South Pacific generally share the same cluster of problems, although in different degrees. They have low per capita incomes, increasing imports in relation to exports, high unemployment, high rates of emigration of their youth, high birthrates, rapid urbanization, exhaustion of their agricultural lands, and high reliance on external aid. Their social systems are stressed by the conflict between the desire to preserve traditional values and the desire to participate in the Western world's twentieth century.

Many of the islands also suffer from common remedies to their problems. They are heavily subsidized by their former mother countries and other sources of foreign aid, with a resulting high level of dependence, subordination, and lack of initiative. Where the remedies place great emphasis on large-scale, high-technology enterprises geared for the export market, the islanders suffer from the alienation of their traditional ways. The dilemmas of modernization are becoming increasingly clear to the Pacific islanders. Like many of the remedies that are introduced, this new consciousness is a mixed blessing.[2]

Basic descriptive data on the territories are presented in table 1, and a map is provided in figure 1. Table 2 provides a rough picture of the pattern of fisheries production for the countries listed by the FAO as included in Oceania. Tables 3 and 4 provide data on the values of fisheries exports and imports.

The overall pattern of fisheries production may be described very simply. On a per capita basis, subsistence production appears to be relatively high, but it is almost certainly declining. Given the small populations, the total quantities of fish produced on a subsistence basis are small. A much larger share of the fish taken from the region had, at least until 1978, been taken by outsiders either using distant-water fleets operating from bases outside the region or using bases established through contractual arrangements in the territories of the region.

International Organizations

Apart from the territories themselves, there are a number of other factors involved in fishing in the South Pacific, including, in particular, international organizations. These are of three major types: the international governmental

2. A good recent review of conditions in the Pacific is provided in a series of three monographs by Donald M. Topping on *The Pacific Islands,* published in the American Universities Field Staff's Southeast Asia series as vol. 25 (January, February, and March 1977). A great deal of valuable background information may also be found in Stuart Inder, ed., *Pacific Islands Year Book,* 13th ed. (Sydney: Pacific Publications, 1978). Current information may be obtained from the magazine, *Pacific Islands Monthly.*

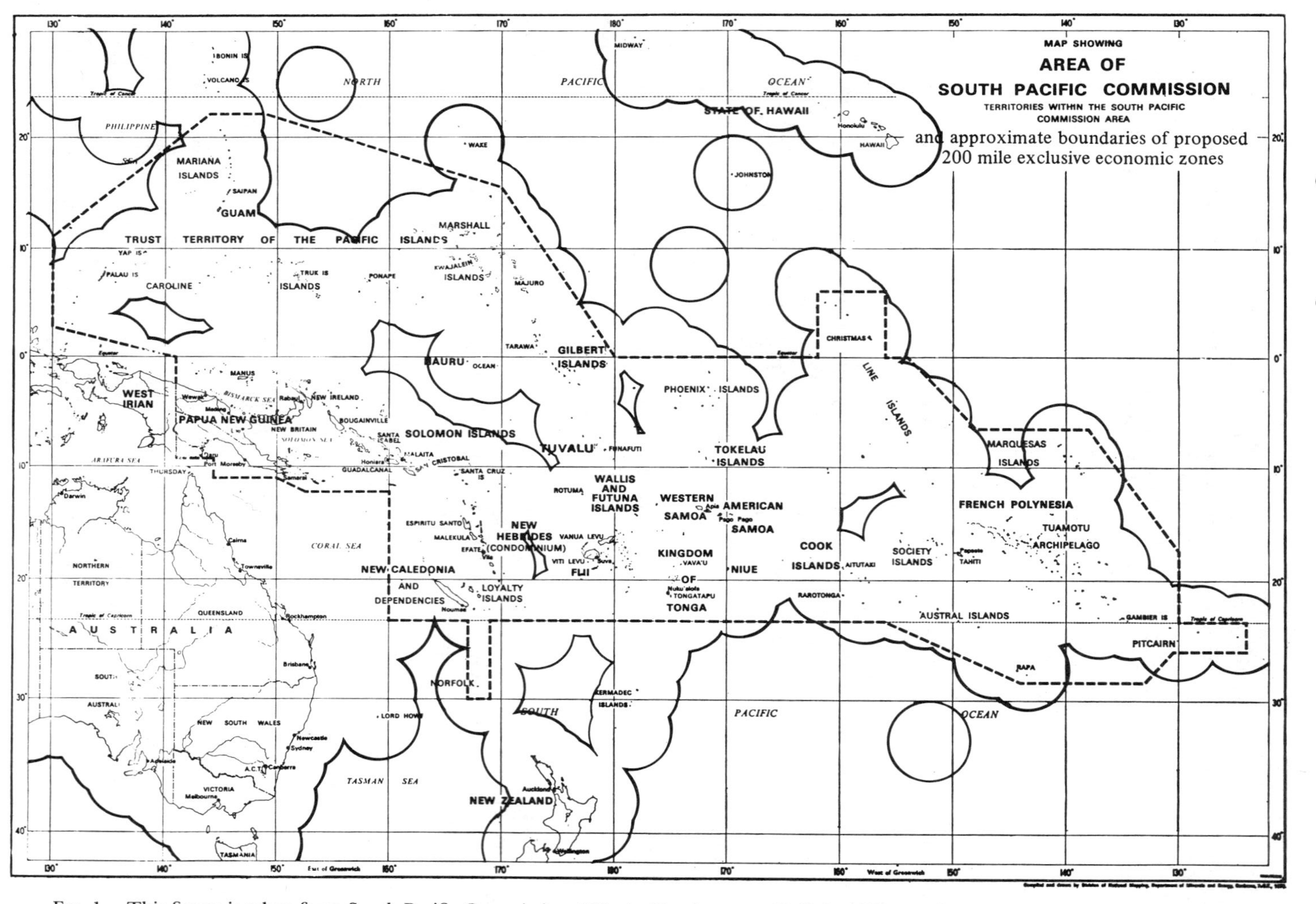

FIG. 1.—This figure is taken from South Pacific Commission, *Fisheries Newsletter*, no. 14 (July 1977), p. 37; it is originally from the Centre National pour l'Exploitation des Océans (CNEXO), Tahiti.

TABLE 1.—BASIC DATA ON TERRITORIES OF THE SOUTH PACIFIC

Territory	Capital	Population (1977)	GNP (1974 Millions $US)	GNP/ Capita (1974, $US)	Coast-line Length (km)	Land Area (km²)	200-mi Economic Zone Area (Thousands of km²)
American Samoa	Pago Pago	30,500	30	1,100	...	197	410
Cook Islands	Avarua	18,500	...	...	120	240	2,200
Fiji	Suva	592,000	470	840	1,129	18,200	1,370
French Polynesia	Papeete	137,000	340	2,530	2,252	4,000	5,380
Gilbert Islands ...	Bairiki	53,500	40	730	1,143	684	4,430
Guam	Agana	102,000	460	4,420	...	541	...
Nauru	...	7,300	120+	6,500+	24	30	290
New Caledonia	Noumea	134,000	550	4,170	2,254	19,100	1,540
New Hebrides	Vila	99,500	40	480	2,528	11,880	670
Niue	Alofi	3,800	...	...	...	260	350
Norfolk Island	Kingston	1,900	...	...	...	35	570
Papua New Guinea ...	Port Moresby	2,908,000	1,250	470	5,152	461,693	2,300
Pitcairn Island	...	65	...	...	...	4	970
Solomon Islands ...	Honiara	206,000	60	310	5,313	28,500	1,520
Tokelau	...	1,600	...	...	...	10	330
Tonga	Nuku'alofa	90,000	30	300	419	697	720
Trust Territory	Saipan	120,000	50	500	...	1,800	7,460*
Tuvalu	Funafuti	7,500	...	...	24	25	760
Wallis and Futuna ...	Mata Utu	9,700	...	...	129	16	280
Western Samoa	Apia	152,000	50	300	403	2,934	160

SOURCES.—Donald M. Topping, *The Pacific Islands*, American Universities Field Staff Southeast Asia Series, vol. 25, nos. 2–4 (January, February, and March 1977); Stuart Inder, ed., *Pacific Islands Year Book*, 12th ed. (Syndey: Pacific Publications, 1977); *Pacific Islands Monthly*, various issues; *National Basic Intelligence Factbook* (Washington, D.C.: Government Printing Office, 1977); *Population 1974: Statistical Bulletin of the South Pacific No. 7* (Noumea, New Caledonia: South Pacific Commission, 1975); South Pacific Commission, *Fisheries Newsletter*, no. 14 (July 1977), p. 37.

*200-mi zone area for Trust Territory includes that for Guam.

TABLE 2.—CATCHES BY TERRITORIES OF OCEANIA
(in Metric Tons)

	1970	1971	1972	1973	1974	1975	1976
Less Developed:							
American Samoa ...	0	0	0	100	82	136	113
Canton Island	0	0	0	0	0	0	0
Christmas Island	0	0	0	0	0	0	0
Cocos Islands ...	0	0	0	0	0	0	0
Cook Islands ...	1,000F	1,000F	1,000F	1,000F	1,000F	1,000F	1,000F
Fiji	4,000F	4,000F	4,700F	4,700	4,805	5,001	5,456
French Polynesia	2,300	2,200	2,400	2,600	2,386	2,169	2,826
Gilbert Islands ...	500F	500F	500F	500F	750F	750	750F
Guam	100	100	100	100	92	122	122F
Johnston Island	0	0	0	0	0	0	0
Midway Island	0	0	0	0	0	0	0
Nauru	0	0	0	0	0	0	0
New Caledonia	500F	500F	500F	800F	868	900	1,000
New Hebrides	8,000F	8,000F	8,000F	8,000F	8,000F	8,000F	8,000F
Niue Island	0	0	0	0	0	0	0
Norfolk Island	0	0	0	0	0	0	0
Pacific Island Trust Territory	1,000F	1,000F	400	6,300	3,360	7,795	6,053
Papua New Guinea ...	23,400F	38,000F	38,000F	57,800F	60,010	45,731	63,029
Pitcairn Island	0	0	0	0	0	0	0
Samoa (Western)	900F	900F	900F	900	900F	1,000	1,100
Solomon Islands ...	1,000F	5,700F	8,800F	7,300F	11,585	8,711	18,600
Tokelau	0	0	0	0	0	0	0
Tonga	400F	500F	500F	600	726	901	1,019
Tuvalu	0	0	0	0	0	0	0

TABLE 2.—*(Continued)*

	1970	1971	1972	1973	1974	1975	1976
U.S. misc. Pacific Islands . . .	. . .	. . .	. . .	. . .	. . .	. . .	. . .
Wake Island	0	0	0	0	0	0	0
Wallis, etc. . .	0	0	0	0	0	0	0
Less developed subtotal . . .	43,100	62,400	66,300	90,700	94,564	82,216	109,068
Developed:							
Australia . . .	101,400	111,400	124,300	130,000	137,519	108,564	113,961
New Zealand . .	59,300	66,000	58,300	66,100	68,732	63,262	70,449
Developed subtotal . . .	160,700	177,400	182,600	196,100	206,251	171,826	184,410
Oceania total .	203,800	239,800	248,900	286,800	300,815	254,042	293,478

SOURCE.—FAO, *Yearbook of Fishery Statistics, Catches and Landings, 1975,* vol. 40, table A-2, p. 9
NOTE.—F = FAO estimate.

organizations, the international nongovernmental organizations, and the multinational corporations.

The major international governmental organizations with an interest in fishing in the South Pacific are the Food and Agriculture Organization of the United Nations (FAO) and the United Nations Development Programme. The FAO has regional subsidiaries devoted to fishing, including the Indo-Pacific Fisheries Council. There are several other international bodies, not affiliated with the FAO, which are concerned with fishing in the Pacific, but all of them are focused on the northern or the eastern Pacific, rather than the southwest Pacific, the area of primary interest here.

Several regional organizations include fisheries among their many concerns. The South Pacific Commission is a consultative and advisory body devoted to promoting the economic and social welfare of the people of the South Pacific. It has undertaken numerous fisheries development projects in the region, and it publishes a regular *Fisheries Newsletter.* The annual South Pacific Conference provides policy guidance for the South Pacific Commission.

TABLE 3.—VALUES OF FISHERIES IMPORTS
(Thousands of U.S. Dollars)

	1970	1971	1972	1973	1974	1975	1976	1977
American Samoa	450	412	731	...	1,157	1,353	570	897
Cook Islands	139	...	...	62	...	...	164	296
Fiji	4,901	6,008	8,357	7,433	9,835	5,354	8,054	...
French Polynesia	1,233	1,167	1,638	1,762	2,801	2,631	2,875	...
Gilbert and Ellice Islands	...	52	78	91	215	98	85	83
Guam	...	...	1,468	...	...	...	...	...
Nauru	...	...	...	...	75	161	133	118
New Caledonia	953	1,257	1,563	1,688	1,006	2,370	1,268	2,547
New Hebrides	642	908	1,112	1,390	2,309	505	...	...
Niue	21	27	39	41	80	71	44	61
Norfolk Island	...	...	...	...	...	...	...	...
Papua New Guinea	4,562	5,451	6,026	6,609	14,781	6,971	9,389	...
Solomon Islands	138	196	193	211	467	205	...	...
Tonga	83	169	110	215	483	220	130	...
Tuvalu	...	...	...	...	...	...	6	28
Western Samoa	504	638	624	813	1,672	1,153	...	925

SOURCE.—South Pacific Commission, *Overseas Trade: Statistical Bulletin of the South Pacific,* nos. 8, 11, and 13 (1975, 1977, 1978).

Another regional organization, the South Pacific Forum, similarly provides policy guidance for its operating arm, the South Pacific Bureau for Economic Co-operation, known as SPEC. The forum and SPEC have taken the major initiatives toward the creation of a new regional fisheries organization.

The major international nongovernmental organizations concerned with fishing in the South Pacific are the International Center for Living Aquatic Resources Management (ICLARM), the Foundation for the Peoples of the South Pacific, and the University of the South Pacific. These are research, educational, and granting agencies which provide support to the territories of the region in the development of their fisheries.

The international organizations with the greatest impact on fishing in the South Pacific are the multinational corporations. These are commercial enterprises having operations based in more than one nation. One common type of

TABLE 4.—VALUES OF FISHERIES EXPORTS
(Thousands of U.S. Dollars)

	1970	1971	1972	1973	1974	1975	1976	1977
American Samoa:								
Tuna	33,018	38,248	47,270	55,017	73,600	48,664	58,224	73,098
Total	34,492	39,284	49,571	56,342	74,105	48,787	58,783	73,587
Cook Islands .	...	...	...	...	...	...	...	...
Fiji:								
Tuna	4,811	5,801	7,951	254	303	252	1,201	...
Total	5,086	5,997	7,996	279	316	258	1,231	...
French Polynesia:								
Tuna	...	...	...	...	...	...	...	...
Total	...	...	4	6	15	4	...	...
Gilbert and Ellice Islands	...	...	...	...	...	...	...	...
New Caledonia	...	...	...	...	...	...	...	141
New Hebrides:								
Tuna	5,226	7,293	10,423	11,154	8,363	3,657	6,674	13,813
Total	5,226	7,293	20,423	11,154	8,363	3,657	6,674	13,813
Nuie	...	...	...	...	...	...	...	...
Norfolk Island:								
Tuna	...	...	...	...	...	...	...	...
Total	11	10	2	2	0	...	...	...
Papua New Guinea:								
Tuna	175	1,475	3,255	3,872	15,338	11,398	7,169	17,141
Total	1,010	2,567	6,002	5,966	20,840	13,385	11,117	24,056
Solomon Islands:								
Tuna	...	1,387	4,157	1,970	5,293	3,172	7,995	...
Total	...	1,387	4,210	2,063	5,492	3,597	8,269	...
Tonga:								
Tuna	...	...	...	...	...	...	...	...
Total	7	12	2	3	12	...	...	...
Trust Territory Pacific:								
Tuna	989	878	886	309	...	...	...	3,546
Total	991	878	889	340	...	...	...	3,755
Western Samoa:								
Tuna	...	...	...	...	...	...	...	...
Total	2	...	...	...	...	...	...	...

SOURCE.—South Pacific Commission, *Overseas Trade: Statistical Bulletin of the South Pacific,* nos. 8, 11, and 13 (1975, 1977, 1978).

operation, perhaps just barely fitting the meaning of multinational corporation, is that in which bases are established in host countries to service the owning country's fishing fleet. The Republic of Korea, for example, having started with just a few vessels operating out of American Samoa in 1961, has operated a fleet of as many as 500 distant-water fishing vessels working from bases all over the world. At the base in American Samoa, the South Koreans, along with the Taiwanese, supply the canneries in Pago Pago. The Korean longline fleet was expanded not only with the assistance of the Korean government and a variety of international lending agencies, but also with the cooperation of U.S. tuna processors.

Taiwan, too, began to concentrate on long-distance fishing in the 1960s. In 1970 Taiwan's worldwide fleet included 116 vessels operating out of bases in the Pacific Ocean, and in 1976 it had 80 vessels in the Pacific, operating out of bases at American Samoa, Fiji, the New Hebrides, and Papua New Guinea.

The Japanese have large multinational corporations in fishing, most with a very high degree of vertical integration, concerns such as Taiyo, Nippon Suisan, Mitsui, Marubeni, and Nichiro. The Taiyo Fishery Company, Ltd. (Taiyo Gyogyo Kabushiki Kaisha), as one example, has offices, subsidiaries, and affiliates throughout the world, including, in the South Pacific, American Samoa, Australia, and New Zealand. It is a widely diversified conglomerate, engaged in food processing and preservation, cold storage, warehousing, shipbuilding, and trading, all augmenting its operations in fishing.

There is a deep historical basis for Japan's extensive involvement in fisheries joint ventures. In the 1920s and 1930s, the Japanese-owned South Seas Development Corporation controlled and operated the fishing industry in the Marianas, Caroline, and Marshall islands of Micronesia. The first of Japan's modern foreign-based skipjack fisheries was established in northern Borneo in 1960. After a lapse of a few years many more fisheries ventures were initiated in the late 1960s and early 1970s.

In Fiji, the Pacific Fishing Company, Ltd. (PAFCO), was formed in 1963, with five-sixths owned by three different Japanese companies and one-sixth owned by private interests in Fiji. As a result of new negotiations in 1974, 70 percent is now owned by Japanese interests, 25 percent is owned by the Fiji government, and 5 percent is owned privately by Fijians. PAFCO is engaged in the freezing and transshipment and, since 1976, the canning of tuna.

In the Solomon Islands, Solomon-Taiyo, Ltd., was established in 1972. The government holds 25 percent of the authorized shares on a "free ride." That is, rather than having to pay cash, the government obtained its share for in-kind considerations. The government was also assured the option of purchasing an additional 24 percent of the paid-up shares, and it may press to become the majority shareholder when the Joint Venture Agreement is subject to renegotiation in 1982. Like PAFCO, Solomon-Taiyo is wholly devoted to the tuna business, including freezing, transshipment, and canning. Unlike PAFCO, however, Solomon-Taiyo has its own tuna-catching operations.

As a large and resource-rich country, especially with respect to marine resources, Papua New Guinea hosts a broad variety of fisheries joint ventures, primarily in tuna but also in prawn and lobsters. Most of the enterprises involve Japanese interests, but there is also extensive American and Australian involvement.

The vigor of the Japanese in pursuing joint ventures is demonstrated by the fact that in 1972 alone 27 new Japanese-American fisheries joint ventures were established. On a worldwide basis, as of March 1976, the Japanese were involved in 173 fisheries joint ventures in 51 countries.

The Japanese are negotiating many new joint ventures as a result of the displacement of their distant-water fleet by the new 200-mi fishing zones. Almost all of the South Pacific territories within reach of substantial marine resources are engaged in negotiations for new joint ventures, most of them with the Japanese.

The United States has vast corporate conglomerates which include within them corporations involved in fishing in many nations. For example, the H. J. Heinz Company, a major American food conglomerate based in Pittsburgh, acquired Star-Kist Foods, Inc., in 1963. It has collection and cold-storage stations at Rabaul in Papua New Guinea, Papeete in Tahiti, and Greymouth in New Zealand, and another outside the Pacific region, at Dakar in Senegal. Star-Kist Foods, Inc., a wholly owned subsidiary of the H. J. Heinz Company incorporated in California, is in turn the parent corporation wholly owning Star-Kist Caribe, Inc., Star-Kist Samoa, Inc., and Star-Kist Papua New Guinea Pty., Ltd. Star-Kist Caribe, Inc., incorporated in Delaware, has a cannery in Mayaguez, Puerto Rico. Star-Kist Samoa, Inc., incorporated in California, has a cannery in Pago Pago. Star-Kist International, S.A., operates in Panama City, with a factory and collection station in Tema, Ghana. Star-Kist is building a new tuna-processing and canning operation at Manus, in Papua New Guinea's Admiralty Islands. Apart from the different Star-Kist operations, the H. J. Heinz Company of Australia operates the Greenseas fish packing plant at Eden, Australia.

The Ralston Purina Company of St. Louis has canneries at San Diego and at Terminal Island in California, Manta in Ecuador, Ponce in Puerto Rico, and Pago Pago in American Samoa. There are seafood-receiving stations at each of the canneries, and also at Agana in Guam and Koror in Palau. A buying office is maintained at Tema in Ghana. The company's involvement with seafood was established with the acquisition of the Van Camp Sea Food Company in 1963. Van Camp was merged into Ralston Purina and thus is not a subsidiary or a separate legal entity. Ralston Purina does own Estrella Blanca, S.A., of Panama as a wholly owned subsidiary. Estrella Blanca is the legal entity which operates the Van Camp facility in Palau. Estrella Blanca also owns, as a subsidiary, INECAPA, which runs the cannery at Manta in Ecuador. Its status in the country is under negotiation.

Apart from the fixed, land-based facilities, Ralston Purina/Van Camp also has 10 small fishing vessels. In addition, as of 1977 it had an equity interest in 19 tuna seiners. Of these, the company had a 50 percent or more interest in 14, and had a less than 50 percent interest in five.

The company purchases most of its raw tuna from independent fishing enterprises, many of which operate their tuna-fishing vessels under United States flags. Some tuna vessels are leased by the company, renting for about $5.7 million annually through 1986. Toward the end of 1976, the company had also guaranteed about $10.8 million in bank loans and other obligations incurred by some of the tuna-boat operators and suppliers.

The Castle & Cooke Company, headquartered in Hawaii, is involved in real estate, manufacturing, and services, but its major business is in foods, in operations spread all over the world. In seafood, the company's principal products are canned tuna and canned salmon, and it also is a major producer of processed and frozen crab, shrimp, and other specialties. The company has no fixed facilities in the South Pacific apart from a fish-receiving and cold-storage station in Tahiti shared with Star-Kist. Nevertheless, it is a major force in the larger Pacific region, especially in relation to tuna. Castle & Cooke first entered the seafood business in its home state of Hawaii in 1948 when it acquired a 41 percent interest in the Hawaiian Tuna Packers cannery. In 1951 the ownership was increased to 99.9 percent. Castle & Cooke now also has tuna canneries in Astoria, Oregon, and in Mayaguez, Puerto Rico.

Until 1969 raw tuna for the Castle & Cooke–owned canneries was purchased entirely from independent fishermen, both in the United States and abroad. In that year, the company put into operation its first long-range tuna purse-seining vessel. A second was added in 1972. In 1975 the company purchased a modern fleet of tuna vessels, at a cost of about $30 million. The fleet of 14 vessels has a combined carrying capacity of about 13,000 tons of fish, making Castle & Cooke one of the three largest operators in the tuna-catching business in the United States. One of the vessels was wrecked, but counting the purse seiner acquired at Mayaguez, in 1978 the fleet still numbered 14 vessels. For a time, the company continued to operate the vessels out of their San Diego base, using many of the same personnel who operated the vessels before the purchase. Many of the vessels have now been shifted from the Pacific to the Atlantic Ocean to supply the Mayaguez cannery.

The policies of the American tuna industry are very vigorously represented in both domestic and international politics by the companies' own lobbyists and by the American Tunaboat Association. The American tuna industry has a very strong influence on American foreign policy on fisheries. One clear indication of this strength is the exclusion of tuna from the U.S. 1976 Fisheries Conservation and Management Act, an exclusion which the industry hoped would assure continued access to tuna within the 200-mi zones of other nations. Another indication of the tuna industry's strength is the regular presence of its lobbyists on the U.S. delegation to the Law of the Sea negotiations at the United Nations and at the negotiations leading to the creation of the new South Pacific Regional Fisheries Organization. No other nation in these negotiations has had such a strong representation of private fishery interests in what is in essence supposed to be a governmental delegation.

The economic and political capacities of the American tuna companies are also suggested by the fact that the U.S. Department of Justice initiated an inquiry into the industry in January 1975. A consent order against industry members was directed against tuna canners, tuna-boat operators, tuna cooperative marketing associations, fishermen's unions, and individuals in the tuna industry. The order prohibits their acting in ways designed to fix prices or conditions of sale in the purchase or sale of raw, frozen, or canned tuna at processing, wholesale, or retail levels of distribution. Included within the order are prohibitions against joint or collective negotiations in the purchase or sale of tuna, activities designed to curtail the importation of tuna into the United States, and practices designed to otherwise limit competition in the purchase or sale of tuna.

The Department of Justice action was followed by the launching in October 1975 of a Federal Trade Commission investigation, which is still continuing in 1979. The investigation is looking into the importation, purchasing, processing, sale, and distribution of tuna and tuna products within and outside the United States. The commission's investigation is particularly directed to determining whether members of the tuna industry have engaged in unfair methods of competition, or whether there may have been acquisitions, mergers, joint ventures, or consolidations in violation of federal antitrust laws.

II. PROBLEMS

Problems of Production, Conservation, and Information

The traditional problems of fisheries management have certain distinctive qualities in the South Pacific. With respect to problems of production, even allowing for enormous rates of underestimation, it is certain that the catch rates of the Pacific islands are miniscule. The total recorded catch for all the less developed territories of Oceania combined has been running at around 0.1 percent of the world's total fish catch. This corresponds with the region's share of the world's population, but it is far lower than would be expected on the basis of the region's access to major fish stocks.

The catch rates *of* the islands are very different from the catch rates *around* the islands. The production figures of table 2 greatly underrepresent the quantities taken from the waters of the South Pacific because of the entry of long-distance fishing fleets of other nations, chiefly nations of the northern Pacific such as Japan, Korea, Taiwan, and the Soviet Union.

The importance of long-distance fishing is demonstrated by table 5, which shows production in the region designated by the FAO as the Southwest Pacific. The developed nations of the region, Australia and New Zealand, have taken substantial quantities, but Japan, Korea, and the Soviet Union have taken the lion's share. As the table shows, from 1972 onward, these three nations have taken more than half of the total production from the region.

If we consider tuna alone, an even higher share of the production is taken by foreign fleets. In 1974 the countries of the South Pacific Commission took 54,876 tons of skipjack, while the total taken by foreign fleets in the area was

TABLE 5.—CATCHES IN THE SOUTHWEST PACIFIC
(in Metric Tons)

	1970	1971	1972	1973	1974	1975	1976
Australia	29,600	30,100	31,400	33,700	32,134	32,907	31,050
Japan	56,400	65,100	69,600	74,200	90,526	83,157	133,402
Korea, Republic of	...	30,300	40,500	43,600	44,261	38,474	25,165
New Zealand	58,300	64,500	56,100	64,700	67,803	61,918	69,049
Norfolk Island ..	0	0	0	0	0	0	0
Pitcairn Island ..	0	0	0	0	0	0	0
USSR	...	10,400	53,700	74,300	88,800	44,767	78,020
Others	21,000F	23,300F	23,900F	25,200F	40,000F	44,347F	43,446F
Total	165,300F	223,700	275,200	315,700	363,524	305,570	380,132

SOURCE.—FAO, *Yearbook of Fishery Statistics, Catches and Landings, 1976,* vol. 42, table C-81, p. 218.
NOTE.—F = FAO estimate.

estimated at 145,000 metric tons (MT). Thus, foreign fleets took about 72 percent of the skipjack in the area of the South Pacific Commission. Indeed, the foreign catch of tuna was almost double the 78,268 MT reported caught by SPC countries of all species, including tuna.

Apart from the fishing fleets sailing from home ports outside the region, the fleets operating from ports within the region are to a large extent controlled by foreign interests. Where these islands are hosts to overseas bases for foreign fishing companies, or are parties to joint ventures, the benefits to the host island are likely to be quite different from what they would have been in the case of independent production and trade. The shipment of fishery products from a Japanese fishing base in the Solomon Islands to Tokyo, for example, is essentially a transfer within a single corporation, even though it may be recorded as a Solomon Islands export and a Japanese import.

Where large proportions of capital, technology, and possibly even labor originate in countries of the northern Pacific, but the fishing operations work out of a base in a small South Pacific island, it becomes difficult to distinguish between local and long-distance fishing. Sometimes, as in the case of Taiwanese operations from overseas bases, the products are not shipped to the home country but are instead sold to some third country. Although the home country reaps major benefits, the data may suggest that it was not even involved, leaving the impression that the production and sale of the fish was accomplished wholly by the country hosting the base.

The problem of production, then, should not be conceived so narrowly as to call only for an enlargement of gross production figures. It seems evident that the catch rates of the territories of the South Pacific ought to be higher, but the imperative is not really very clear. It is widely agreed that the islanders ought to be gaining a greater share of the benefits to be obtained from fishing in South Pacific waters. That is not exactly the same thing as having the territories show higher catch rates, however. Through the licensing of foreign fishermen, for example, it is possible to benefit from fishing without being the one to do the fishing. But would that be enough? To what degree is it important that the islanders themselves become increasingly engaged in the production, and also the processing, of fish?

Benefits are to go to the islanders, not to the islands, which suggests the question, which islanders? The benefits from fishing enterprises may be distributed very unevenly within territories, and the character of that distribution will be different for different kinds of enterprises. In some cases the greatest share of the benefits might go to the local entrepreneur who facilitates the investment of foreign capital, while the wage earners get very little. For any given quantity of production, it makes a big difference whether that tonnage is consumed locally or is caught to be exported for processing and consumption elsewhere.

Like problems of production, environmental problems too have been conceived narrowly. The environmental problem which has dominated the concerns of traditional fisheries managers is that of conservation, in the sense of limiting exploitation to a level which would permit sustained production through time. There has been little concern for conservation beyond that instrumental purpose, no sense of conservation for its own sake. Apart from the destruction of whale stocks, there has not yet been any seriously excessive and widespread depletion of stocks in the South Pacific. There have been many localized cases of overexploitation.

There is little appreciation of the extent to which the market system is itself a major factor leading to overexploitation. As Johannes explains:

> Prior to western contact it was customary in most of Oceania to share one's catch with one's fellow villagers and give some of the products of their labor in return. It is difficult to convey the fundamental importance of this custom to westerners whose most basic assumptions about the distribution of goods and services are firmly rooted in a money economy. The introduction by westerners of this money economy, the development of distant markets (i.e., district centers, foreign countries) and the consequent growth of the profit motive started the process of environmental decay around many Pacific islands.
>
> Under this funadmentally new economic order goods are bought and sold, not shared, and the fisherman finds himself competing for money, and therefore for fish. In order to compete effectively he has to buy better equipment and fish harder. This process is self-reinforcing. The need to spend more money to get more efficient gear to harvest more intensively increases as number of fish decrease.[3]

3. Robert E. Johannes, "Traditional Marine Conservation Methods in Oceania and Their Demise," *Annual Review of Ecology and Systematics* 9 (1978): 349–64.

Apart from conservation, the other major category of environmental issues is that of pollution control. In some cases the fishing industry itself has been the source, rather than the victim, of pollution. At times the discarding of by-catches has been a source of pollution. At the processing level, in American Samoa the canneries have dumped primary treated sewage into Pago Pago harbor. In addition to the effects on the esthetic quality of the harbor, the practice has contributed to the deterioration of the subsistence fishery.

The lack of information on the underlying fish stocks is particularly acute in the Pacific. Almost all systematic efforts at assessing coastal stocks have focused on the waters of Australia and New Zealand, with very little knowledge having been generated for the less developed territories. For highly migratory species, however, a major project to assess the skipjack tuna resource in the Pacific has been launched by the South Pacific Commission.

Information on production levels is also incomplete, especially in relation to subsistence production. It can be assumed that where production figures are available, they tend to systematically underestimate the true quantities of fish caught in the islands.

Most fisheries research has focused on things like patterns of underlying stocks and on gross production figures, but a greater share of research resources ought to be devoted to other kinds of questions. Perhaps issues such as economic returns to fishermen and trade patterns should be investigated more systematically. With catch and trade data so difficult to interpret because of the roles of foreign interests, perhaps joint ventures ought to be analyzed more carefully.

Problems of Trade

All nations of the South Pacific, with the exception of New Zealand, are major importers of fishery products. The largest importer by far, however, not only in the Pacific region but in the world as a whole, has consistently been the United States, importing over 1 million MT of fish every year, about twice as much as all poor countries combined. In 1977, however, partly as a result of the rising relative value of the yen, and partly as a result of the displacement of Japan's distant-water fishing fleet from the 200-mi zones of many countries, Japan replaced the United States as the world's largest importer of fishery products, in terms of value.

A major characteristic of the global trade pattern is that, like other primary products, fish in the international market tends to flow from the less developed to the more highly developed countries. This trade pattern is demonstrated by the fact that the typical trading country purchases its fish imports from countries of lower levels of development than those to which it sends its fish exports. Table 6 shows the average gross domestic product per capita for the leading suppliers and the leading buyers of each of several fish-trading nations of the

TABLE 6.—AVERAGE GROSS DOMESTIC PRODUCT PER CAPITA FOR FISH-TRADING PARTNERS

	Average GDP/Capita of Leading Suppliers of Imports (U.S. Dollars)	Average GDP/Capita of Leading Buyers of Exports (U.S. Dollars)
Fiji	2,189	3,027
Indonesia	1,629	1,737
Japan	434	2,522
New Hebrides	2,406	2,886
Papua New Guinea	2,054	3,210
Philippines	1,507	2,887
Solomon Islands	1,981	2,508
Thailand	533	1,737
United States	2,344	2,765

SOURCE.—Various national fisheries data reports.

larger Pacific region for which detailed trade data are available. The table shows that imports consistently are drawn from countries of lower levels of development than those to which exports are sent. In other words, most countries buy fish from countries which are poorer than the countries to which they sell fish. This results in a trade pattern in which the fish tend to move from the poorer to the richer countries.

Another indication of this pattern is that, in the trade in simply preserved (e.g., frozen) fish among the market economies of the world, developed countries export around 70 percent by value but import 90 percent by value. Developed countries account for over 80 percent of both imports and exports of canned and otherwise prepared fish in the total value traded among market economies. This dominance of trade shares by developed countries is far greater than their dominance of production shares. As shown later, in table 8, 39 developed market economies together have been producing almost 40 percent of the world's total fish catch, while 175 developing market economies have been producing a smaller share, a little over 30 percent of the world's total.

Thus the general pattern in the world fish market is that (1) most of the trade is among developed countries; (2) there is very little trade among the less developed countries; and (3) where there is trade between the two groups, the fish tends to flow from the less developed to the more highly developed. There are exceptions to the global pattern in that there is a large flow of canned fish from developed to less developed countries. There also is some trade between less developed countries, as in the case of the large exports from Thailand to Malaysia. Fiji does some exporting to other less developed South Pacific territories.

Sometimes, however, the exceptions are only apparent. Often imports by less developed countries are destined not for local consumption but for processing and reexport to developed countries. The large quantities listed as imports into American Samoa, for example, are processed in the Van Camp and Star-Kist canneries and then shipped to the United States. The Soviets supply fishery products to Singapore only to be processed and exported to Japan. Similarly, Korea's plan to import some 1,000 tons of squid from the United States and Canada was not for consumption in Korea, but for processing and export to Japan and to other countries.

The less developed islands of the South Pacific import and consume large quantities of canned fish, mostly cheap canned mackerel from Japan. In Fiji in 1974, for example, canned-fish imports accounted for fully 75 percent of the local fish consumption. The new cannery is supposed to provide an increasingly large share of the supply of canned fish over the coming years. In 1975 the imports of canned fish were reduced to 2,677 MT, but in 1976 the imports rose once again to 3,595 MT. In 1973–74, Papua New Guinea imported over 22,100 MT of canned fishery products. Over 20,860 MT of that was canned mackerel, mostly from Japan. In 1973, Tonga spent 146,000 pa'anga, or about $186,880, on imported canned fish, almost all of it from Japan. In 1971, Western Samoa imported 2,510 tons of fish, more than 90 percent of the total consumed.

In fiscal year 1974, the Trust Territory of the Pacific Islands imported $944,747 worth of canned fish, 28 percent of which went to the district of Palau. Local production of tunas and reef fish in Palau, not including the Van Camp landings, were valued at about $60,000 in 1974. This figure is greatly overshadowed by the $268,287 worth of canned fish imported into Palau in the same period. The story was the same in other districts. Truk produced about $66,500 worth of fish, but imported $385,290 worth of canned fish. Ponape produced about $71,000 worth of fish, and imported about $124,700 worth of canned fish. Yap was the only district to show signs of self-sufficiency, catching about $2,100 worth of fish, while importing only $550 worth of canned fish in 1974.

The high dependence on imported canned fish is part of the larger trend of increasing urbanization and increasing dependence on a cash economy, tied in with an increasing export orientation, and increasing imports of all kinds of products, but particularly food. Among the more immediate reasons are the factors of convenience, the lack of available refrigeration, the desirability of the oil in which the fish is packed, and the fact that canned fish offers especially strong flavoring so that, eaten with rice, smaller quantities may be used. One of the major factors is that, for a given weight, imported canned fish is often cheaper than fresh fish, even with duties imposed on the imports.

Locally produced foods are relatively expensive partly because local productive resources and development efforts are directed to the export sector. The deterioration of the capacity to produce food for local consumption and the dependence on imports creates great vulnerabilities. The danger is clearly signaled by the rising prices of the Japanese canned mackerel. In one observer's view the mackerel "will not only become more costly soon, it may also become far more scarce and eventually disappear as anything but a luxury item."[4]

Given the dominance of the developed countries over the international trade in fisheries products, it might be argued that the less developed countries ought to try to capture a larger share of that lucrative trade. In particular, the territories of the South Pacific which have the potential ought to increase their involvement in the tuna business. That view certainly has a great deal of merit, but there are other considerations to be taken into account.

It is an illusion to believe that the tuna business can be as rewarding to the less developed countries as it has been to the developed countries. The poorer countries may gain control over the waters, and they may even become engaged in the fishing and in some processing, but if they are to operate on the world market, they will inescapably be dependent on the richer countries. The richer countries would be capable of squeezing the poorer countries in specific negotiations, and the general trend of inflation would tend to make the earnings less and less valuable. The history of export-oriented agriculture in less developed countries is relevant here. One characteristic of less developed countries is that a very small number of commodities account for a very large share of their exports. Countries exporting only a few commodities are subject to considerable damage as their markets swing up and down. The recent weakening of the sugar market, for example, has hurt many of the territories of the South Pacific.

One response, of course, is to export a larger variety of items. Perhaps the territories of the South Pacific ought to enter into the tuna business in order to diversify their exports. Conceivably, however, this could result only in a change from concentration on some items to concentration on other items. Moving into the tuna business could provide diversification of exports, but at the same time it could create new vulnerabilities. There is the problem of subordination to the capital, processing, and marketing facilities controlled by outsiders, and there is the very real danger that the returns may have been overestimated. There are now so many nations in the Pacific region planning to enter or enlarge their participation in the skipjack business that the industry could well become overcapitalized. This may be foreshadowed by the fact that in summer of 1978 Japanese fishermen actually voted to suspend operations for 2 mo because of an oversupply of skipjack. What if, perhaps because of rising energy costs, new regulations, or a new fish disease, the bottom drops out of the tuna market? What if salmon somehow regains its once strong standing in the canned-fish market, displacing tuna from its niche in the supermarket environment? Countries investing their resources into the tuna business need to consider these possibilities.

4. John E. Bardach, *The Fisheries of Papua New Guinea: Promises and Problems* (Honolulu: Hawaii Institute of Marine Biology, 1977), p. 19.

It appears that the South Pacific territories are more in need of basic foods than they are in need of foreign exchange. One trade problem common to less developed nations is that far too much of their food is imported. The use of scarce foreign exchange may be a roundabout and inefficient way for a poor country to get food, and reliance on the importation of food can create enormous dependency.

Some countries, including many in Latin America, now rely so much on food and other imports that they have built up enormous external debts. They are now in a kind of debt slavery, apparently compelled to enlarge their exports in a hopeless race to try to meet their obligations. Their gains tend to be dissipated by inflation and by disadvantageous terms of trade. The debt can be built up specifically in the fish trade, as it has been in the Philippines: "The Philippine government is currently pursuing a policy of production for export. With reference to Japan, this policy is particularly urgent in order to offset the record $355 million total trade deficit the Philippines incurred in 1976. In the fishing industry, such a policy is questionable as local demand for fish has yet to be met, and most Filipinos have been made to do with a diet of low quality fish as the better kinds are now beyond their means."[5]

With care, the countries of the South Pacific might avoid the debt trap. Whereas one way to compensate for overdependence on a few export commodities is to diversify exports, another way is to cut back on the need for foreign exchange by reducing imports. Reducing food imports is especially advantageous because it can sharply reduce vulnerability, increase self-reliance, and thus increase bargaining power.

More attention needs to be given to the development of local fisheries for local consumption. This does not necessarily mean forgoing the bountiful tuna resources. In Tonga, for example, a good share of the skipjack that is caught is brought back home for consumption, despite the seductively higher prices offered for the fish at foreign ports. Ultimately, the issue is one of balance:

> So far, the development of high-seas fisheries has claimed the major share of national and international attention. Small-scale fishermen in inland and coastal areas, who now contribute about 50 per cent of the production of fish used for direct human consumption, have yet to benefit significantly from existing development policies and production programs. It must be recognized that high-seas fisheries development is strongly skewed toward capital- and energy-intensive technology and production schemes—technologies and schemes that have an inherent tendency to magnify existing social and economic inequities. Therefore, development strategies are needed which will promote a more balanced

5. Third World Studies, University of the Philippines System, "Japanese Interests in the Philippines Fishing Industry," *AMPO: Japan-Asia Quarterly Review* 10, nos. 1–2 (1978): 52–60.

pattern of development and which address not only food production but employment and the overall economic status of the rural poor.[6]

Each of the nations of the South Pacific will have to determine for itself how it will deal with these dilemmas to find a suitable balance between production for domestic use and production for export.

Problems of Employment

Since local ownership in the fisheries joint ventures is commonly limited to minority holdings, much of the return to capital is exported. Similarly, despite the hopes that such enterprises would provide new employment opportunities for local people, much of the return to labor also goes to outsiders. Most of the fishermen working on the vessels are Okinawans, Koreans, and Taiwanese, and much of the management in the shore-based processing facilities is American, Japanese, and Australian. The importation of foreign labor, bypassing local people, dates back to the 1920s and 1930s when the Japanese South Seas Development Corporation relied almost exclusively on Japanese and Okinawan labor for the sugar and fishing industries.

The failure of the local people to seek employment or to demonstrate high productivity in the fishing industries may be attributed to a broad mixture of factors. The islands are generally well endowed by nature so that, with the possibilities for subsistence fishing and gardening and the modest needs for clothing and shelter, there is relatively little need to work. Also, most of the islands are so heavily subsidized in one form or another that adequate living conditions can be maintained with little or no work.

There are also some important cultural factors limiting the motivation to work. For example, there is a tendency to work only as long as it takes to earn "enough" money, and "enough" is often exceedingly modest. In the traditional economies there has been little concern for accumulation of personal wealth or for planning and investing for the future. Thus, people often just do not show up for work. Traditional living and working patterns simply do not mesh well with the clockwork patterns of industrial operations. Another problem is the traditional obligations to share one's income with family and friends. When workers find that they regularly retain only a small fraction of their take-home pay, while the rest must be given away, the motivation to work long and hard for the money is dissipated.

While taking these considerations into account, the failure to motivate local people is surely also related to the prevailing wage rates. At the Van Camp operation in Palau, most fishermen earn less than the lowest-paid government

6. International Center for Living Aquatic Resources Management, *Program Development Statement* (Quezon City, Philippines: ICLARM, 1976), p. 24.

employees. In the Trust Territory as a whole, in 1976 the average wage for government workers was $4,010, and the average wage in the private sector, included the Van Camp operation, was $1,869—annually. In American Samoa the minimum wage is maintained at less than half of the minimum wage in the United States, and Van Camp and Star-Kist take full advantage, keeping a large proportion of the cannery workers near to the minimum. In 1978 workers at the PAFCO cannery in Fiji earned about a dollar an hour. In 1976 the average wage paid to Solomon Islanders employed by Solomon-Taiyo was $562 Australian or $686 American—annually. In 1978 the minimum wage was raised to $31 a month.

Low wage rates prevail throughout the South Pacific and are not unique to the fishing sector. Many people participate in both the cash and the subsistence economies, gardening and fishing and bartering on a communal basis after their wage-earning work is completed. Thus, a small annual income goes a long way. Moreover, wage rates in the fishing industry are generally favorable when compared with other forms of employment available in the islands. Seen in this light, the wages paid to workers in the fishing industry might be accepted as reasonable.

However, the export-oriented fishing industry, strongly tied to the outside, ought to be assessed not only by local standards but also by the standards prevailing in that outside world. The conclusion seems inescapable that, from a global perspective, the wage rates in the fishing industries in the Pacific Islands are extremely low. Workers for these companies who are so fortunate as to be located in developed countries not only earn more but also improve their lot much faster than their counterparts in the South Pacific. In not taking a fuller share of the benefits of the enterprise, each low-income worker in the islands who helps to furnish canned tuna to distant Americans or Japanese or Europeans in effect subsidizes them. It should be possible for the global economy to move toward equal pay for equal work, across nations as well as within nations.

Problems of Nutrition

The distribution of marine fishery products between developed and developing countries is as shown in table 7. Supplies for any given country are comprised of its production, plus its imports, minus its exports. Supplies include fish for both direct and indirect use. Consumption refers to direct use alone, and thus excludes fish used to feed livestock. About 70 percent of the world catch is used for direct human consumption.

Table 7 shows that the developed market economies together use about two and a half times as much fish as the developing market economies (56 percent vs. 22 percent). The skew is especially clear when per capita measures are used. As table 7 shows, on a per capita basis people in developed market economies use five and a half times as much fish as people in developing market economies

TABLE 7.—DISTRIBUTION OF FOOD ORIGINATING FROM MARINE RESOURCES IN 1970
(Estimates in Parentheses)

	Production	Consumption	Supply	Consumption	Supply
	% of Total			kg per Capita	
Developed market economies	52	49	56	25	33
Developing market economies	26	26	22	6	6
Centrally planned economies	(22)	(25)	(22)	(9)	(10)

SOURCE.—Sidney Holt, "Marine Fisheries," in *Ocean Yearbook 1*, ed. Elisabeth Mann Borgese and Norton Ginsburg (Chicago: University of Chicago Press, 1978), table 21, p. 75.

(33 kg vs. 6 kg). With the likely underreporting of domestic production for local use, the disparity is probably really not so great, but it is surely substantial.

The disparity between production shares and consumption shares is the result of the pattern of trade described earlier whereby fishery products in international trade tend to flow from less developed to more highly developed countries. People in developed countries consume more fish, but they consume more of everything, so they cannot be said to rely on that fish. People of the less developed countries, however, tend to be far more dependent on fish because it accounts for a far higher proportion of their animal protein intake. Historically, fish has been of major importance in providing protein for the people of the South Pacific.

Australians and New Zealanders are well nourished, even overnourished, but the situation is quite different in the less developed territories of the South Pacific. On the basis of infant mortality data, health surveys, and other more fragmentary evidence, it is clear that there is significant malnutrition in the islands. The effect of the growing dependence on imported foods is sketched out in *Fiji's Seventh Development Plan:* "Nutrition surveys undertaken in Fiji show that imported cereals are replacing the traditional staple foods in one or two meals daily while canned meat and canned fish are used in place of fresh vegetables. . . . Increasing dependence on sugar, cereals, canned meat and canned fish have all contributed to a deterioration in the total nutritive value of meals. . . . Increasing numbers of malnutrition cases have been reported at health centres and hospitals in recent years as an effect of the changing food habits."[7]

The concern for nutritional value does not necessarily imply a return to inefficient techniques and organization. It does mean that any plan for fisheries development, with whatever mix of old and new methods, should be designed with a full sensitivity to likely effects on the nutritional and other concerns of the local population. The strengthening of local fisheries has the potential for significantly improving the nutritional status of local people, a potential which implies a major responsibility for management.

Problems of Comparative Disadvantage

The system of essentially free trade among nations is commonly advocated as a highly efficient means of distributing tasks: nations which are better endowed with certain resources and which can produce certain products more efficiently should trade their products for other items produced more efficiently by other people—to the benefit of all. In this context, the pattern of trade in fishery products may be considered reasonable because, although the fishery products

7. *Fiji's Seventh Development Plan, 1976–1980* (Suva: Central Planning Office, 1975), p. 61.

tend to flow from poor to rich, there is a flow of payments from rich to poor. The question arises, however, whether the compensation is sufficient. Do the poorer nations obtain a fair share of the total benefits produced by the fishing enterprises with which they are involved?

Evidently, the economic advantages which accrue to some nations are only marginally due to their special endowments of natural resources. The Trust Territory, for example, with direct access to millions of square miles of richly stocked Pacific Ocean, produces far less from those waters than many nations which are much further removed. The really significant advantages which certain nations enjoy are their advantages in capital and technology or, more generally, their wealth. Richer nations generally produce more fish, as table 8 shows very clearly. From 1972 onward, the 39 developed market economies have been producing more fish than the 175 developing market economies.

Table 8 presents data on production shares for the world total catch, including both marine and inland production. Table 9 presents similar data on production shares, but for marine resources alone. Comparison of the two tables shows that the discrepancies in favor of the developed countries are far larger if we concentrate on marine resources alone.

The advantages of capital and technology are not distributed haphazardly or in accordance with natural endowments, but are themselves the effects of earlier patterns of advantage and disadvantage. In the South Pacific, surely a major factor accounting for the disadvantaged condition of the islands is their colonial history. Like colonies elsewhere, the islands became suppliers of raw materials to the metropolitan countries, and they became increasingly dependent on imports to fulfill their own basic needs. The islands' economies and social structures became even more distorted than those of other colonies such as those in Latin America and Africa. Despite their meager supplies of land and their abundant access to the sea, the islands were induced to undertake large-scale agricultural operations based on continental development models and continental ecologies. The myopic colonists failed to take account of the natural "comparative advantage" of the islands, the potential for development of their marine resources. The transplantation of inappropriate development models, without regard for indigenous potentials, helps to account for the low level of development of fishing in the islands. Capacities for fishing which were once considerable were displaced by the new demands imposed by the outsiders.

With some important exceptions, the fish caught by less developed countries tends to be of lower value per ton than that caught by the developed countries. The less developed countries tend to catch species of lower market value partly because their inferior technology and their lack of capital limit their capacity to pursue the more valuable species. Another factor accounting for the lower unit values is that the less developed countries simply have less bargaining power. Even when they catch the same species as the developed countries, they tend to get paid less for the product. Fishermen bringing skipjack tuna into Palau, for example, regularly get paid less than half the rate

TABLE 8.—SHARES OF THE WORLD TOTAL CATCH (%)

	1970	1971	1972	1973	1974	1975	1976
Developed market economies (39 countries and territories)	36.21	36.72	39.74	40.48	38.24	37.53	37.76
Developing market economies (175 countries and territories)	39.66	38.04	32.32	30.31	33.09	32.16	33.41
Centrally planned economies (13 countries and territories)	22.97	24.11	26.65	27.83	27.53	28.95	27.62

SOURCE.—FAO, *Yearbook of Fishery Statistics, Catches and Landings, 1976,* vol. 42, table A-2, pp. 6–10.

TABLE 9.—SHARES OF THE WORLD TOTAL MARINE CATCH
(%; Estimates in Parentheses)

	1964	1970	1971	1974	1975
Developed market economies	56.9	53.3	53.6	48.8	48.0
Developing market economies	23.2	24.7	23.4	26.4	25.6
Centrally planned economies	19.9	22.0	(23.0)	(24.8)	(26.4)

SOURCE.—Sidney Holt, "Marine Resources," in *Ocean Yearbook 1*, ed. Elisabeth Mann Borgese and Norton Ginsburg (Chicago: University of Chicago Press, 1978), table 17, p. 70.

that skipjack draws in California ports. With only one buyer in Palau, and lacking the capacity to transport the product elsewhere, the fishermen have no alternatives. Thus they lack the bargaining power to demand higher prices.

It might be argued that the tuna industry in Palau nevertheless contributes to the development of the local economy. Moreover, the price paid for skipjack in Palau has been rising steadily. However, if we compare the rates paid over time, as shown in table 10, we see that the prices paid have been rising even more rapidly in California ports. Indeed, the gap between Palau prices and California prices, shown in the third column, has been increasing steadily. Of course the growth of this gap is in large measure due to inflation.

As a side-note, in 1972, when Van Camp was paying $150 per metric ton for skipjack of 5 lb or more landed at the dock, it paid $75 per ton for fish under 5 lb. It then allowed the fishermen to buy back the undersized fish at 10¢ per pound, and local residents could buy them at 12¢ per lb. One metric ton equals 2,204.6 lb. Thus these fish, not as suitable for canning as the larger fish, for

TABLE 10.—PRICES PAID TO FISHERMEN FOR SKIPJACK
(U.S. Dollars per Metric Ton)

	A. Prices Paid at West Coast Canneries	B. Prices Paid at Palau (at Midyear)	C. Price Gap (A−B) (Based on Midrange West Coast Price)
1964	220	75	145
1965	220–315	75	192.5
1966	287–432	75	284.5
1967	226–323	105	169.5
1968	283–295	105	184
1969	299–313	105	201
1970	326–383	110	244.5
1971	398–414	140	266
1972	446–450	150	303
1973	473–519	230	266
1974	590–601	275	320.5
1975	595–596	250	345.5

SOURCE.—Paul Callaghan, "Employment and Factor Productivity in the Palau Skipjack Fishery: A Production Function Analysis" (Ph.D. diss., University of Hawaii, 1976), pp. 129, 130).

which the company paid $75 per metric ton, were sold back to the local people by Van Camp at the rate of $220.46 or $264.55 per metric ton. There were no processing or transportations costs involved in these dockside transactions.

It seems reasonable to take economic development to mean not simply gaining something, but catching up. That is, the criterion of fairness suggested here is that the distribution of benefits should be such as to help narrow the gap between rich and poor. These data show that the skipjack industry, as it has been operating in Palau, contributes to the widening of that gap.

The process has been similar elsewhere. For example, Doumenge has estimated that in 1962 the fishing and transshipment operations in the New Hebrides yielded an income to the Japanese economy of about $700,000, while their contribution to the New Hebridean economy amounted to only about $200,000. "The main share of the profits go to Japan in the form of payment for fish or are distributed outside the territory in various forms." Similarly, on estimating gross incomes for the 1961-62 fiscal year for the Van Camp cannery in American Samoa, Doumenge found that: "it is therefore impressive to see the biggest effort toward the economic development of American Samoa work out at an increase in revenue of 1 million dollars for the territory, 2 million dollars for Japan and South Korea, and no less than 9,200,000 for the United States."[8] Thus, while the cannery may help in the development of the American Samoan economy, it contributes much more to the economic growth of the United States.

In Palau, Van Camp enjoys great advantages as a strategically placed middleman. The company could make substantial profits simply by shipping the fish it purchases in Palau and selling it in California, with no further processing. The company could make at least the difference in the two prices, less shipping and related costs. Of course the value of the tuna to the company as a whole is considerably higher than it would be for an independent transshipment operator in Palau since it is engaged in processing as well. It has been estimated that canning increases the value of the fish landed at the dock by about two and a half times. In the tuna industry, as in many others, higher benefits tend to go to those operating at higher levels of processing. Several South Pacific territories which have been serving as transshipment stations or as suppliers of raw fish have contemplated moving into the highly profitable canning phase as well. To prevent this, the United States, controlling one of the major markets, has imposed heavy tariffs on imports of canned tuna.

The distribution of benefts is skewed on the consumption side as well as the production side. It is sometimes suggested that while there may be inequities in the short run, the growth of an enterprise and the increased production which results will lead to a broader distribution of benefits. In the case of the tuna

8. F. Doumenge, *The Social and Economic Effects of Tuna Fishing in the South Pacific*, Technical Paper no. 149 (Noumea, New Caledonia: South Pacific Commission, 1966), pp. 1, 15, 25, 26.

industry, however, the rapid growth of production has been accompanied by a corresponding rise of consumption in the United States, Japan, and western Europe. The rise in American consumption has slightly outpaced the rapid growth of world production, so that as a result, the United States has been consuming an increasingly large portion of an increasingly large total. Describing tuna as a luxury product may help to explain this, but it does not justify it.

The gap-widening process takes place not only in the large-scale international fisheries but also in the smaller fisheries functioning within the territories:

> As equipment becomes more sophisticated its price ultimately rises byond the means of the average fisherman. A new profession, money-lending, materializes to enable him to finance his purchases and he often falls into debt. Employment opportunities diminish as more efficient modern boats drive out native craft. The fisherman becomes further impoverished, and profits, such as they are, end up largely in the pockets of a few entrepreneurs. This pattern is all too familiar in tropical artisanal fisheries. It is part of the oft-repeated sequence of events whereby self-sufficient, internally regulated subsistence economies are converted to money-based economies, governed ultimately by decisions made in market centers thousands of miles away.[9]

The process of concentration of holdings, and thus of benefits, in fisheries has followed the same dynamics as it has in agriculture where small family farms have been replaced by large agribusiness operations, and as it has in industry where large conglomerates have bought up and consolidated previously independent smaller corporations.

Thus, while there may be comparative advantages for some, there are comparative disadvantages for others, and it has little to do with natural endowments. By all indications, the dynamics of the market system lead not to a convergence, an evening out, but to a steadily widening gap between the advantaged and the disadvantaged.

III. REMEDIES

The Emerging International Law of Fisheries

The international law of fisheries in effect up to the late 1970s had been shaped by the four conventions codifying ocean law signed in Geneva in 1958, particularly the Convention on Fishing and Conservation of the Living Resources of

9. Johannes; he cites five different studies to support this observation.

the High Seas, by a large number of bilateral and multilateral agreements, and by formal and informal unilateral actions and the accumulation of customary practices. This law had been undergoing a major transformation through the 1970s as a result of the deliberations of the Third United Nations Conference on the Law of the Sea and as a result of a series of unilateral declarations. Virtually all major coastal nations have unilaterally extended their jurisdictions over resources out to 200 mi from their baselines, anticipating one of the major points of likely agreement at the United Nations conference.

The basic law relating to fishing proposed in the draft convention, the Informal Composite Negotiating Text (ICNT), is that coastal states would have jurisdiction over the fisheries resources in their exclusive economic zones, with other nations having a right of access to these stocks only under special conditions. These exclusive economic zones would extend up to 200 nautical mi from the coastal state's baselines—and these baselines could be some distance out from shore.

The management of tuna, the resource of major interest in the South Pacific, is handled quite vaguely in the ICNT. According to article 64, for tunas and other highly migratory species, the concerned states "shall cooperate directly or through appropriate international organizations with a view to ensuring conservation and promoting the objectives of optimum utilization of such species throughout the region, both within and beyond the exclusive economic zone." Moreover, "in regions where no appropriate international organization exists, the coastal State and other States whose nationals harvest these species in the region shall cooperate to establish such an organization and participate in its works." The ICNT simply calls for international cooperation for conservation of the species, and deliberately leaves open the question of whether they are to be included within the 200-mi state jurisdictions.

The major public argument for excluding the highly migratory species from the jurisdiction of coastal states within the exclusive economic zones is that, because of their widespread patterns of migration, they cannot sensibly be managed by individual nations but must be managed on an international basis. The major underlying motivation, however, is that major distant-water fishing nations such as the United States want their tuna fleets to continue to enjoy free access to tuna stocks near the coasts of other nations.

Apart from the fact that important issues have been left to future bilateral and multilateral negotiations, one of the major weaknesses of the United Nations Law of the Sea negotiations on fisheries has been that the terms of reference have been overly narrow, so that problems are artificially truncated. Fishing needs to be recognized as an integrated system, involving not only production but also scientific research, the development of technology, transport on sea and on land, processing, preservation, trade, and marketing. Effective management needs to span the entire system. Otherwise, improvements in one sector, such as the distribution of production rights, can be vitiated by countervailing patterns in other sectors, such as patterns of trade. The

Informal Composite Negotiating Text focuses on production rights and ignores other elements of the system.

Joint ventures are increasingly being used to blunt the impact of the 200-mi zones. The motivation of the most active joint venturers is quite plain: "The Japanese Government is encouraging fishing and also food companies to invest in processing plants and in other fishing companies in the US and Canada in order to circumvent the 200-mile zone quotas by importing fish from Japanese companies established within them."[10]

The most important failing of the ICNT with regard to fisheries, however, is the wholly misplaced faith in the extension of jurisdictions. It has now become clear that the greatest gains in areas of jurisdiction go to developed countries, with the largest gains of all going to the United States. Australia and New Zealand together gain more area from the 200-mi extensions than all of the less developed territories of the South Pacific combined. Moreover, given their greater capacities, the developed countries will be able to draw far more benefit from each square mile of extended jurisdiction than the less developed countries.

The problem of continuing inequities in the world cannot be met by rearranging the geography of jurisdictions. That can only be a temporary corrective, in the same way that a one-time gift of resources from the rich to the poor would only temporarily alter the balance. Direct control over natural resources is certainly *a* cause of inequities, but it is greatly overshadowed by the role of social structures, and particularly the structure of commercial relationships.[11]

Extending jurisdictions also fails to meet the problem of threats to the integrity of the environment, whether through pollution or through depletion of resources. Despite the fact that the problems of production, conservation, and allocation of fisheries resources are so much more intense now than they have ever been in the past, the proposals embodied in the ICNT do not begin to meet these urgent problems of fisheries management on a global scale.

The New South Pacific Regional Fisheries Organization

The draft convention emerging from the Third United Nations Conference on the Law of the Sea does not adequately address the problems of fisheries management, but it does provide a context and some space for more satisfactory arrangements on a regional basis. The nations of the South Pacific have

10. "Joint Deals Are Japan's Answer to 200 Miles," *World Fishing* 26, no. 2 (February 1977): 30.

11. For a fuller treatment of this argument, see my essay, "Equity in Global Fisheries Management," *Oceans* 10, no. 5 (September 1977): 60–64, or my article, "Fisheries and the Law of the Sea: A Common Heritage Approach," *Ocean Management* 4 (1978): 1–20.

been moving to respond to the need. Early considerations of the problems of fisheries management in the South Pacific were fragmented and uncoordinated, but in late 1976 the South Pacific Forum began to exercise leadership toward the creation of a South Pacific Regional Fisheries Organization. The term was not used, but there seemed to be some possibility that at least the highly migratory species, particularly tuna, might somehow be jointly recognized and managed as the common heritage of all of the peoples of the South Pacific.

The idea of the new organization first took clear shape at a meeting of the South Pacific Forum held in Suva, Fiji, on October 13–14, 1976, to discuss Law of the Sea issues. At that meeting the members decided in principle to establish a South Pacific fisheries agency to promote the conservation and rational utilization of the fish stocks of the region. They also declared their intention to establish 200-mi economic zones and to generally coordinate their fisheries policies.

The sixteenth South Pacific Conference of the South Pacific Commission, meeting in Noumea on October 20–29, 1976, fully endorsed the forum's decisions. Plans became much firmer at the Eighth South Pacific Forum meeting held in Port Moresby, Papua New Guinea, on August 29–31, 1977. In its *Declaration on Law of the Sea and a Regional Fisheries Agency,* the members of the forum did, among other things, "*Decide* to establish a South Pacific Regional Fisheries Agency open to all Forum countries in the South Pacific with coastal state interests in the region who support the sovereign rights of the coastal state to conserve and manage living resources, including highly migratory species, in its 200 mile zone." A meeting to begin negotiation of a convention establishing the new organization was held in Suva on November 18–25, 1977, at the headquarters of the South Pacific Bureau for Economic Co-operation, the operating arm of the South Pacific Forum.

One of the major issues in the negotiations arose from the lack of consensus concerning the character of the organization being planned. In a preparatory report, the director of SPEC pointed out that two rather different types of organization were contemplated:

> One would aim primarily at ensuring conservation and promoting optimum utilization of the living resources throughout the sea in which they occur. . . . The other would aim primarily at ensuring maximum benefits for the peoples of the coastal countries in the region and for the region as a whole.
>
> To be fully effective, the first type of organization would need participation by all countries in whose waters the resources occur at various stages of their life cycle as well as by all the countries that exploit them. The second type of organization, on the contrary, would comprise only those countries in the South Pacific with a common interest as coastal states.

The first of these two types of organization would be based not so much on common interests as on complementary interests, with distant-water fishing nations from outside the region participating as members. Such a broad-membership organization, devoted primarily to conservation rather than to full management, would fulfill the mandate of article 64 of the ICNT, the article calling upon nations of a region to create an appropriate international organization to provide for the conservation of highly migratory species. In the second type of organization the members, essentially members of the South Pacific Forum, would join together out of their common interest in having coordinated policies with which to face the distant-water fishing nations. In this approach, the organization would amount to a kind of cartel or union, drawing its power from its control over resources which outsiders wanted.

At the signing of the Port Moresby Declaration in August, 1977, the general understanding was that it was the second of these two types that was being planned. The director of SPEC and many (but not all) of the delegates of the nations of the South Pacific approached the November 1977 meeting in Suva with the expectation that an organization of this second type would be created. As events unfolded, however, this determination became muddled. This occurred largely because the United States was admitted to full participation in the meeting. Chile, France, the United Kingdom, and the United States participated as voting members because they represented nonsovereign territories within the region. In the case of the United States, however, the representation of American Samoa, Guam, and the Trust Territory was only nominal. In fact, the United States spoke for its interests as a major distant-water fishing nation and as a major industrial fish-processing and -marketing nation.

The major divisive issue which had arisen earlier at Port Moresby and remained unresolved at the conclusion of the November 1977 meeting in Suva was the question of whether highly migratory species should be recognized as being included within the 200-mi zones of national jurisdiction over fishing. Fearing that the access of distant-water fishing fleets might be limited, the United States has consistently refused to recognize the inclusion of those species within national jurisdictions. The South Pacific nations, however, argue that the highly migratory species must be included within their 200-mi jurisdictions, especially since they are the only resource of significant commercial value within their zones.

Although the United States and the South Pacific nations have all consistently advocated the creation of a regional fisheries organization, there are very important differences between their approaches. If a regional organization were to be established on the basis of national rights in 200-mi zones (whether for highly migratory species or for fisheries generally), the mandate for the organization would derive from powers delegated by the separate nations. The organization would act as agent for the member nations by their consent. And it would be the delegation of national rights which would provide the basis for national participation in the decision making of the organization. Thus, national jurisdiction would be a prerequisite for management through a regional organization. According to the U.S. position, however, the separate nations would not be the source of those powers at the regional level, so far as highly migratory species are concerned, since they would never have those powers at the national level. Their standing would remain uncertain.

The United States indicated that it would recognize jurisdiction over tuna to the extent that it was exercised through the regional organization. The argument was that if this right was asserted in an international treaty establishing the organization, and the treaty was ratified by the U.S. Congress, that treaty would supersede U.S. domestic law. To many observers, however, this formulation would amount to nonrecognition of national jurisdictions over tuna in the 200-mi zones. This view is reinforced by the U.S. Fisheries Conservation and Management Act of 1976, which asserts, in Section 202e, that "the United States shall not recognize the claim of any foreign nation to a fishery conservation zone . . . if such nation . . . fails to recognize and accept that highly migratory species are to be managed by applicable international fishery agreements, whether or not such nation is a party to any such agreement. . . ." Moreover, Section 205 provides for the prohibition of imports of fisheries products from nations "not allowing fishing vessels of the United States to engage in fishing for highly migratory species in accordance with an applicable international fishery agreement. . . ." It was under this law that the United States cut off tuna imports from Canada in September 1979.

In effect, the United States said that it would recognize other nations' claims to jurisdiction over highly migratory species if such claims were made on terms acceptable to the United States. It is understandable that, to many nations, this condition substantively denies recognition of their claims to jurisdiction.

The U.S. proposal of a form of regional management for highly migratory species which did not derive from separate national powers would have one significant advantage: geographically, the jurisdiction would be continuous. With a jurisdictional area composed of the separate 200-mi zones, in contrast, there would be gaps and holes in the space covered, as shown by the map in figure 1. With discontinuous coverage, fleets would be able to station themselves in gaps like that north of the Solomon Islands, in the path of migration of some of the major stocks of tuna, but out of reach of the regional agency.

The question of merging economic zones is moot, however, since none of the participants in the Suva negotiations advocated such a proposal. All of the South Pacific nations stood fast behind the idea of maintaining the sanctity of national jurisdictions.

Provisional agreement was reached at the November 1977 Suva meeting on the draft text of most of the major articles of a convention which would create the new fisheries organization. These, and the outstanding issues, were carried over to a resumed session of the meeting held June 5–10, 1978, again at Suva. This session hammered out a full draft convention.

In that draft, it was agreed that the new organization would have only limited advisory functions (e.g., to conduct studies, to provide assistance to member nations on their request) and would not have any positive management powers. A weak organization would have the advantage of making it unnecessary to face up to some of the knottier issues. The problem of jurisdiction over highly migratory species, for example, was legalistically accommodated by draft article II, which simply called upon the member nations to state their positions on the issue. The incompatibilities certainly were not resolved. Similarly, if the organization were to be limited to advisory functions, gaps in the geographical coverage would not have any serious consequences.

The draft convention prepared at Suva was presented to the leaders of the nations of the South Pacific at a meeting of the South Pacific Forum held in Niue in September 1978. It generated a great deal of acrimonious debate. Peter Kenilorea, prime minister of the Solomon Islands, reportedly protested, "We do not interfere in the coal mines of America—why should America be able to interfere in the fisheries of the independent Pacific Forum countries?"[12] Fiji, Papua New Guinea, Nauru, Tonga, and Kiribati sided with the Solomons, while Australia, New Zealand, Western Samoa, the Cooks, and Niue took the opposite side. The core disagreement at Niue was over whether there should be an article 64 type of organization, as proposed in the Suva draft, to include the United States and other outside fishing nations, or whether membership should be limited to establish a common front of Pacific nations. The disagreement in part reflected the differences in interests between the larger nations of the region, which hope to exploit the fisheries resources themselves, and the smaller nations, which expect to benefit primarily from royalties and license fees obtained from outside fishing nations.

Thus, the tentative agreement which had been reached by lower-level negotiators at Suva came apart completely. There was no consensus on an article 64 type of organization, and there was still a strong tendency to favor a Pacific Forum–based organization.

The Niue meeting concluded with the acceptance of an interim proposal by which it was decided to establish a fisheries organization with the Port Moresby declaration of 1977 as a guide, and with membership restricted to the Pacific Forum nations. The more broadly based organization proposed in the draft convention was simply to be subject to "further examination." The new South Pacific Forum Fisheries Agency, as it was called, had its first meeting in Honiara in May 1979, under its new director, Wilfred Razzell. In July 1979, a

12. Ross Stevens, "Solomons Opposed to U.S. in Fisheries Agency," *New Pacific* 4, no. 2 (March–April 1979): 9. Also see Bob Hawkins, "The Niue Forum: Regionalism to Sub-Regionalism," *Pacific Islands Monthly* (November 1978), p. 17.

full meeting of the South Pacific Forum held in Honiara formally approved the convention creating the agency. The text of that convention is included in the Selected Documents and Proceedings Appendix to this volume (pp. 575–78).

The major functions of the new Forum Fisheries Agency, according to article VII of the convention, are: collecting, analyzing, evaluating, and disseminating statistical and biological information on the fisheries resources of the region; collecting and disseminating information on management procedures, legislation, and agreements; collecting and disseminating information on prices, shipping, processing, and marketing of fish and fishery products; providing technical assistance, on request, on fisheries development policies, negotiations, issuing of licenses, collection of fees, surveillance, and enforcement; and establishing working arrangements with regional and international organizations such as the South Pacific Commission.

Thus, the agency has been established as a rather weak service agency rather than as anything approaching a cartel or even a management agency. The assertion in article X that "this Convention is not subject to ratification" confirms the understanding that the agency is not being delegated any positive powers by the participating nations.

As these negotiations proceeded it became increasingly clear that the less developed nations of the South Pacific have enormous differences among themselves. Their interests are not so common as some observers may have wished them to be. Some less developed nations are much more highly developed than others, and therefore they have more options which they can put to good advantage. These special advantages have played an important role in the bilateral negotiations which many of the South Pacific nations have been undertaking on their own with major outside powers, especially Japan. They have been doing this without systematically informing one another of their activities, and certainly without coordinating their policies. Thus, with the many bilateral negotiations already underway, and the apparent eagerness of some nations to offer their resources to outsiders, the declared intention of the South Pacific nations to coordinate their negotiations with outsiders is open to question.

The South Pacific is simply reproducing the basic dynamics of world politics. Relationships typically found between developed and less developed nations are being reenacted within the region in the relationships between the more and less developed of these less developed nations. The seemingly inexorable movement toward the widening of the gap between the rich and poor appears to be continuing here as elsewhere. Nations which have more are able to strike harder bargains, whether with fellow islanders or with outsiders, and thus gain larger shares of the benefits. Just as at the Third United Nations Conference on the Law of the Sea, the nations most interested in sharing are those which have nothing to be shared. Policies are determined by rather pinched visions of short-term material interests, and cooperation is undertaken only incidentally, when it is seen as serving those interests.

Nevertheless, it is possible that the nations of the region will find specific bases for cooperation in, rather than in spite of, their own interests. At the opening of the November 1977 meeting in Suva, the Director of SPEC said that "some of the resources of the ocean belong to the region as a whole. They must be managed by the region as a whole, and the benefits should be shared by the region as a whole." Through progressive cooperation, it could be that the nations of the South Pacific will slowly move toward the recognition that at least some of the resources of the region should be regarded as the common heritage of all of the peoples of the region.

Appendix A

Reports from Organizations

The reports and abridged reports contained in this Appendix represent reviews of only some of the activities by organizations which deal with ocean-related matters. They provide the reader with basic coverage of important programs and directions being pursued by major international organizations. In addition, the chapters in this volume by Peter S. Thacher and Nikki Meith and by S. J. Holt and C. Vanderbilt also review activities and programs of certain international organizations. Future editions of the *Ocean Yearbook* will include reports from other organizations as well as subsequent accounts from the bodies reported on in this edition.

THE EDITORS

ECOR: Activities for 1977 and 1978[1]

1. An international workshop on ocean instrumentation was sponsored by ECOR in Washington, D.C., on May 1–3. The purpose of the workshop was to (*a*) support the development of improved standards and instrumentation for the international ocean engineering community, (*b*) encourage international calibration, and (*c*) enhance international information exchange. Paper presentations and workshop discussions were concentrated on the following major engineering activities: coastal and offshore structures, ocean floor, urban and industrial waste disposal, and mining and dredging.

2. ECOR's Third General Assembly, held in Washington, D.C., May 3–6, 1978, had as its technical theme, "Critical Elements in the Exchange of Ocean Engineering Technology." Papers were presented on three topics: the differing needs for the exchange of ocean engineering technology; training and education aspects of the exchange of ocean engineering technology; and national and international capabilities, limitations, and problems relating to the exchange of ocean engineering technology.

3. New Officers were elected by the delegates to the Third General Assembly. They will serve until the Fourth General Assembly in 1981. They are: president, Elmer Wheaton (United States); vice-president, Ron Goodfellow (United Kingdom); past president, Kenji Okamura (Japan); secretary, Donald L. Keach (United States); and treasurer, J. G. Th. Linssen (Netherlands). The new Council members, also elected by the delegates, are: I. Engelsen, Norway; G. A. Heyning, Netherlands; Neil Hogben, United Kingdom; Massashi Homma, Japan; Per Laheld, Norway; Ascensio Lara, Argentina; Siegfried Schuster, Federal Republic of Germany; Robert L. Wiegel, United States; Fernando Vasco Costa, Portugal; and J. A. Zwamborn, South Africa.

4. Future plans include an ECOR Council meeting in the Netherlands in September 1979, an ECOR supporting International Workshop on Coastal Engineering and Council meeting in Mexico City in October 1980, and the Fourth General Assembly on ECOR in London in April 1981.

5. An associate membership category is to be established based on the recommendation of the delegates to the Assembly. The category is designed to encourage participation by qualified individuals in countries which do not have adequate resources to form a national committee.

6. ECOR is compiling a bibliography of ocean engineering books and journals for use by developing national ocean engineering organizations which have limited budgets. Ron Goodfellow, vice-president, ECOR, will spearhead the effort on behalf of ECOR. William Gaither, United States, and I. Engelsen, Norway, will assist in the effort. The listing will be submitted to the IOC for possible use.

1. Prepared by Donald L. Keach, secretary of ECOR, with the addition of material from *ECOR Newsletter* (Fall 1978).

ILO Report of the Committee on Conditions of Work in the Fishing Industry[1]

1. The Committee on Conditions of Work in the Fishing Industry met at the International Labour Office in Geneva from 21 to 30 November 1978 in accordance with decisions of the Governing Body of the International Labour Office taken at its 201st Session (November 1976) and 205th Session (February–March 1978).

2. The Committee was attended by 21 members appointed by the Governing Body and accompanied by seven advisers on the Government side, two advisers on the Employers' side and two advisers on the Workers' side. Representatives of several international governmental and non-governmental organisations were also present as observers. A list of participants is given in Annex I to this report.*

3. The Officers selected by the Committee were: Chairman, Mr. J. H. Bertnes (Government member, Norway); Vice-Chairmen, Mr. T. Dahl (Employer member, Norway), Mr. H. Rake (Worker member, Federal Republic of Germany); Reporter, Mr. F. Davila (Government member, Argentina). The Secretary-General and Assistant Secretary-General were Mr. P. Astapenko (Assistant Director-General of the ILO) and Mr. E. Argiroffo (Chief of the Maritime Branch of the ILO) respectively.

Agenda

4. The agenda of the Committee, as fixed by the Governing Body, comprised the following six items:

(1) Working hours and manning.
(2) Stabilisation of employment and earnings.
(3) Medical care on board.
(4) Pensions and sickness insurance.
(5) Holidays with pay.
(6) Repatriation.

5. The International Labour Office had prepared reports dealing with the items on the agenda on the basis of data received from governments in reply to a questionnaire despatched by the Office on 16 June 1977.

1. Geneva, 21–30 November 1978 (CCF/3/6). Editors' Note: See "Maritime Activities of the ILO," *Ocean Yearbook 1*, ed. Elisabeth Mann Borgese and Norton Ginsburg (Chicago: University of Chicago Press, 1978), pp. 520–30, for a summary of other ocean-related activities of the ILO.

* Editors' Note: Due to limitations of space, Annex I has not been included here. The reader is referred to the original report, available through the ILO, Geneva.

Opening of the Committee

6. At the opening sitting Mr. Astapenko welcomed all participants on behalf of the Director-General of the International Labour Office. Mr. Astapenko reviewed the continuing attention devoted by the ILO to fishermen's questions since the Committee last met in 1962. Resolutions and conclusions which the Committee had put forward led to the adoption by the International Labour Conference in 1966 of two Conventions and one Recommendation. The Conventions, both of which had entered into force, concerned fishermen's competency certificates and crew accommodation in fishing vessels. The Recommendation dealt with fishermen's vocational training. Of the resolutions adopted by the Committee, that on safety on board fishing vessels resulted, as it requested, in the preparation and publishing, through co-operation with FAO and IMCO, of a comprehensive international code of practice on that subject. The Committee's resolution on employment injury benefits had been covered by a Convention on that subject pertaining to industry in general and applying equally to fishermen which had been adopted by the Conference in 1964.

7. The agenda of the present session of the Committee was based on a third resolution on the programme of future work, adopted by the Committee in 1962, as well as a resolution in similar terms adopted by the Conference in 1966. The work of the Committee would constitute advice to the Governing Body concerning possible further international action in respect of each item.

8. Mr. Astapenko concluded by emphasising the opportunity presented by the Committee to help further the progress already made in the field of improving fishermen's conditions throughout the world and said that it had an important role to play in setting the ILO on the right course during the years which promised substantial transition for the fishing industry.

Item 1—Working Hours and Manning

9. The Committee had before it for the discussion of item 1 on the agenda a report prepared by the International Labour Office. The report reviewed previous international action regarding the working hours of fishermen and manning standards for fishing vessels, presented a comprehensive survey and analysis of national law and practice in respect of these two questions and contained concluding remarks designed to assist the Committee in its considerations of this item.

10. The Worker members felt that the two questions—hours of work and manning—while complex as a result of the great variations between fishing industries in respect of types of vessels and methods of operations, were nevertheless extremely important to fishermen as there was a definite link between the very high accident rate in the fishing industry and excessive hours of work and insufficient manning. In particular, hours of work, minimum rest periods on the fishing grounds, shift systems for crews, as well as compensation for overtime, were significant issues in collective bargaining. They argued that fishermen needed standards established at the international level in addition to the Hours of Work (Fishing) Recommendation, 1920 (No. 7) to provide them with adequate protection and to improve their conditions of work and life. Various existing and relevant ILO instruments pertaining to merchant seafarers, as well as certain national studies concerning fishermen's working conditions and safety could in the view of the Workers serve as a basis for formulating new instruments suitable for fishermen.

In spite of the complications involved, the Workers held the view that the Committee could usefully consider hours of work and manning in sufficient depth to recommend appropriate international guidance be established which, they felt, ought to take the form of a Convention supplemented by a Recommendation. The consideration of such instruments, the Workers felt, could perhaps for simplicity initially focus on maritime fishing vessels of at least 24 metres in length or 100 gross registered tons or, alternatively, an equivalent size as determined by ship measurement systems based on the length, depth and beam of a vessel.

11. The Employer members fully recognised the intricate nature of hours of work and manning in fishing vessels as being a consequence of the diversity internationally not only of the factors cited by the Workers, but also such conditions as the operational needs and status of vessels which also varied considerably owing to different circumstances in different countries. They drew special attention to practical difficulties and inadequacies in use of a fishing vessel's size as a basis for applying international standards. The Employers therefore had reservations about the ability of the Committee to consider hours of work and manning in a meaningful manner, and hence of the appropriateness of it proposing new international standards on the two subjects which they felt were not feasible at the present time.

12. The Government members also acknowledged that the dissimilarities and composite nature of fishing activities and the fishermen's occupation tended to present difficulties in identifying, internationally, widespread and consistent patterns of working hours and manning aboard fishing vessels as a basis for the Committee to propose international standards on these subjects. They felt nevertheless that some form of ILO action with a view to establishing guidelines or even more formal standards further to the Recommendation of 1920 might be feasible, and therefore suggested that the Committee explore the possibility in the light of the proposed points for discussion outlined in the Office's report.

13. The Committee decided to establish a working group composed of four members from each of the three sides which considered the questions of hours of work and manning on the basis of the discussions which had transpired and the contents of the Office's report. The working group agreed on the draft resolution which was considered by the Committee and, following a minor amendment, was unanimously adopted. The text of the resolution as adopted is reproduced as Annex II.

Item 2—Stabilisation of Employment and Earnings

14. For its consideration of item 2 of the agenda, the Committee had before it a report prepared by the International Labour Office which summarised the replies received from governments concerning measures taken nationally through law or practice for stabilising the employment and earnings of maritime fishermen. Proposed points for discussion also contained in the report were intended as guidance to the Committee in considering this question.

15. The Worker members emphasised the importance of stabilising employment and earnings as a means of improving fishermen's conditions and said that little action had been taken nationally in this regard. They felt that the current economic problems of the fishing industry would eventually be overcome, that responsibility for maintaining economic stability in the fishing industry rested with the employers and that the Committee should take the opportunity now available to it to promote the establishment

of international standards on stabilisation of employment and earnings. The Workers considered that these standards should take the form of a Convention.

16. The Employer members stressed the general unstable and adverse economic situation of the fishing industry world-wide caused by such factors as changes in national fishing limits, seasonal variations in fishing, the operations of joint ventures in fishing and the expanding role of developing countries in fishing. In their view the viability of the fishing industry, and therefore the employment and earnings of fishermen depended on the prior solution of factors exogenous to the industry—including the state of entire national economies.

17. The Employers therefore questioned the advisability at the present time of proposing the adoption of any instrument. They felt that existing general employment schemes in various countries should also provide for the needs of fishermen, and that in any case, any international instrument should take account of the capabilities of different countries to provide for the stabilisation of fishermen's conditions. The Employers considered that action of the Committee should be limited to recommending international guidelines for solving problems of fishermen's employment and earnings based on national experiences.

18. The Government members also referred to the relation of economic conditions in the fishing industry to fishermen's employment and earnings. They suggested that either a Recommendation or less formal international guidelines should be the preferred means of action for the Committee owing to the flexibility such instruments offered countries in accommodating different national circumstances.

19. The Committee decided to set up a working group composed of several members from the Government, Employers' and Workers' sides, which considered the points for discussion set out in the Office report and formulated a draft resolution.

20. This draft resolution was considered by the Committee during which the Government member of Spain drew attention to the fact that governments should be conscious of the serious employment problems created for other States by the establishment of economic fishing zones and by excluding or restricting access to the waters of such zones by vessels of other countries which traditionally fished in them. Following the introduction of minor amendments, the resolution was unanimously adopted. The text is set out as Annex III.

Item 3—Medical Care

21. A report prepared by the International Labour Office largely based on information provided by member States was before the Committee for its consideration of item 3 of the agenda. The report discussed practices, issues and problems in regard to medical care on board fishing vessels, and also contained concluding remarks intended to assist the Committee in its discussions.

22. The Worker members submitted a draft resolution on substantive elements of medical care for fishermen while at sea which invited the Governing Body of the ILO to consider the possibility of initiating action with a view to the adoption of an international instrument on the topic. Conscious of the inherent constraints of medical care on fishing vessels at sea, the Workers felt that the proposed instrument ought to cover, and at the same time harmonise at the international level, such elements and activities as medically oriented training for crew members, medical guides, medicine chests, medical advice by radio, evacuation in the event of medical emergencies if necessary by helicopter, and the

optimum use of physicians in fishing and auxiliary vessels at sea where such personnel exist. They expressed strong views that consideration ought to be given to the development of some practical criteria for determining a minimum doctor/paramedic–fishing crews ratio. They also recalled the need for continued co-operation between the ILO and the World Health Organization in the field of medical care at sea, and suggested that all future work in this direction should take into account existing relevant standards, resolutions, guides and codes.

23. The Employer members associated themselves with the Workers on the need of extending the optimum possible level of medical care at sea and on the desirability of seeking the maximum possible regional and international harmonisation in its delivery. They felt, however, that the topic required further study. The Employers also questioned the indicative comparative rates in the Office's report on injury and illness frequencies between fishermen, miners and industrial workers, following which a further review of the source by the Office revealed that two studies in one country had concluded that comparative rates of the order indicated in the report applied in respect of mortality from all causes and fatality due to accidents.

24. The Employers considered that rather than establishing a new instrument, additional to the ILO ships' medicine chests, Recommendation, 1958 (No. 105) and Medical Advice at Sea Recommendation, 1958 (No. 106), it was more opportune for the ILO to convene a tripartite meeting for the purpose of establishing technical guidelines on medical care at sea. They concurred that there should be collaboration between the ILO and the WHO.

25. The Government members associated themselves with the special circumstances of medical care at sea which called for international solidarity and mutual assistance. They suggested that countries might make available special medical facilities for foreign fishermen.

26. After some discussion, during which certain doubt was expressed as to the direct practical applicability of Recommendations Nos. 105 and 106 to the fishing industry, the Committee agreed that instead of proposing the convening of a tripartite meeting as suggested by the Employers, it should invite the Governing Body of the ILO to include the question of medical care for fishermen in the agenda of a future session of the International Labour Conference with a view to the adoption of an international instrument.

27. There was also a consensus that the Workers' draft resolution should be considered by a working group composed of several members of the Committee from the three sides which formulated an agreed text. Subsequently, the Committee adopted the Workers' resolution as amended. The text as adopted appears as Annex IV.

Item 4—Pensions and Sickness Insurance

28. To assist the Committee in its consideration of item 4 on the agenda, the International Labour Office had prepared a report analysing national law and practice regarding pensions and sickness insurance for fishermen, discussing special problems of fishermen's social protection and containing suggested points for discussion based on conclusions drawn from the report.

29. The Committee noted that the Invalidity, Old-Age and Survivors' Benefits Convention (No. 128) and the Medical Care and Sickness Benefits Convention (No. 130) applied to fishermen, and that they permitted governments to exclude their application

to seafarers, including sea fishermen, only when such workers were covered by special schemes providing an equivalent level of protection. The Committee also noted that existing ILO instruments concerning social security protection for seafarers contained provisions permitting the exclusion of fishermen without any specified condition. As regards the rather limited social security coverage of fishermen in general, the Committee further noted the technical, administrative and financial difficulties involved in the extension of the scope of protection in the case of sickness, invalidity, old age and death, and emphasised that the ultimate goal was the full coverage of all categories of persons working in the fishing industry.

30. The Worker members felt that existing instruments concerning seafarers' social security should be revised so as to take account of the position of fishermen. They emphasised the need for providing effective social security protection to fishermen operating as a family unit and expressed concern whether the inclusion of some categories of self-employed fishermen whose activities were usually viable, might not give rise to an increased financial burden on the majority of protected persons who belonged to lower income groups.

31. The Employer members considered that before the adoption of any new international standards each member State should elaborate its own system of social security for the protection of fishermen. They noted that as regards the rather small number of ratifications of existing instruments concerning social security for seafarers, these instruments were formulated at a time when experience in that particular field was limited; they contained technical details which gave rise to a lack of flexibility in enabling many member States to accept the obligations. The Employers were of the view that the social security coverage of self-employed persons, including fishermen, was a matter for the decision of governments who played an important role in ensuring the necessary financial resources for such persons in respect of whom no employer's contribution was payable.

32. The Government members said that if covered by a contributory social security scheme, self-employed fishermen would be required to pay higher rates of contributions than those payable in respect of workers in a wage-earning employment, as long as the same level of protection was provided to them irrespective of their employment status. They also pointed out that incomes of self-employed fishermen were extremely variable. This demanded special measures to ensure financial resources for their social security protection, including the levy of tax on the sale of the catch.

33. The Committee considered a number of conclusions prepared by the Office in the light of the discussions which took place, and after having introduced several amendments to the text, adopted unanimously these conclusions. The text as adopted is given in Annex V.

Item 5—Holidays with Pay

34. The Committee had before it for discussion of item 5 on the agenda a report prepared by the International Labour Office which summarised national law and practice relating to holidays with pay for fishermen, and also contained concluding remarks designed to assist the Committee in its discussion of this item.

35. The Worker members pointed out that although in some countries the provisions of the Holidays with Pay Convention (Revised), 1970 (No. 132) were applied to

fishermen, those contained in the Seafarers' Annual Leave with Pay Convention, 1976 (No. 146) were more appropriate. They drew attention to a provision of Convention No. 146 which concerned extension of its application by ratifying member States to fishermen after consultation with the employers and workers concerned and suitable modification. Their views were embodied in a draft resolution which urged the Governing Body of the ILO to appeal to governments to ratify Convention No. 146 and to extend its provisions to persons employed in fishing operations.

36. The Employer members considered that Convention No. 132 gave sufficient margin for national legislation to extend its application to fishermen, a practice which was already being done in certain countries. For this reason, as well as the relative recent adoption of Convention No. 146, and its ratification so far by only a few countries, they felt that Convention No. 132 was the most appropriate instrument for the coverage of fishermen. They further emphasised that the Workers' draft resolution should request governments to ratify Convention No. 132 in addition to Convention No. 146.

137. Following some discussion, the draft resolution was amended, mainly to improve the linkage between the preambular and operative parts, and unanimously adopted by the Committee. The text of the resolution as adopted appears as Annex VI.

Item 6—Repatriation

38. Item 6 of the agenda was considered by the Committee in the light of a report prepared by the International Labour Office which summarised the situation of repatriation for fishermen in various countries, and closed with concluding remarks and five points for discussion formulated to assist the Committee in considering this question. These points for discussion essentially recalled a resolution adopted by the ILO in 1926 inviting governments to ensure the repatriation of fishermen by applying to them the provisions of the Repatriation of Seamen Convention, 1926 (No. 23); noted that many governments had already taken such measures; pointed out that in certain countries the question of repatriation of crews was neither inapplicable or limited; and suggested that governments which had not given effect to the afore-mentioned resolution should do so.

39. The Worker members accepted in principle the points for discussion in the Office report but said that Convention No. 23 was adopted as long ago as 1926 and did not specifically apply to fishermen. They also emphasised the inadequacy of Article 6 of the Convention in not assigning to public authorities the ultimate responsibility of repatriating fishermen in circumstances when this was not done by employers. Consequently, the Workers' members felt that the Committee should adopt a resolution on fishermen's repatriation, embodying the first four of the Office points and a revised text, which they submitted, of the fifth point so as to specify the responsibility of public authorities.

40. The Employer members pointed out that many countries were applying Convention No. 23 to fishermen and they emphasised that this instrument, together with the resolution adopted in 1926 were still relevant and constituted all that was required as regards international standards concerning repatriation of fishermen. They concurred with the five points in the Office report and felt that the action of the Committee should be limited to encouraging further implementation by countries of the existing resolution and Convention and should not recommend the adoption of new standards deviating from, or in addition to those two instruments.

41. The Government members agreed with the five points in the Office report as amended by the Worker members, as well as the view that it would be inadvisable for the Committee to propose the adoption of a new Convention or Recommendation concerning the repatriation of fishermen specifically.

42. The Committee accepted the proposal of the Worker members for the formulation of a resolution and agreed on an amended version of the fifth point of the Office report. The text of the resolution, as adopted, is set out in Annex VII.

Resolution on Technical Co-operation in Vocational Training for the Fishing Industry

43. The Employer members submitted a draft resolution requesting the International Labour Office to co-operate with other international organisations in developing training programmes for fishermen—particularly in small-scale fisheries—which would include safety, health and first-aid questions, and providing technical assistance to developing countries in this regard. The draft resolution also urged governments, and employers' and workers' organisations to co-operate with the ILO with a view to achieving the objectives of the ILO's World Employment Programme.

44. The Employers introduced a minor amendment to their resolution concerning the source of funds for implementing technical co-operation by the ILO and explained that the intent of the resolution was essentially to promote employment and acquisition of sea-food resources in developing countries through assistance in training their nationals. They referred to the constitutional obligation of the ILO in vocational training for all workers and the need for the ILO to expand such activities in respect of the fishing industries of developing countries. In their view, preliminary preparations for such assistance could be financed from the limited funds of the ILO's regular budget, while the subsequent implementation of activities should be funded by the United Nations Development Programme, and other extra-budgetary resources.

45. The Worker members endorsed the resolution and views of the Employers on condition that the technical co-operation activities mentioned supported the genuine needs of developing countries in improving their own fishing industries. The Workers were however opposed to such activities being used to the benefit of foreign shipowners who registered fishing vessels in developing countries for purposes of reducing labour costs.

46. The Government members also referred to the limited funds available for technical co-operation activities from the ILO's regular budget, and they therefore recommended that some attention should be directed to the ways and means of financing assistance to developing countries in fishermen's training through the UNDP.

47. The representative of the Food and Agriculture Organisation pointed out that the lack of trained manpower was the largest constraint to fisheries development throughout the world, and he therefore welcomed any international effort in improving the situation. He referred to the growing need of developing countries for trained manpower resulting from an expansion of their fishing activities. He also stated that the FAO had long-standing and wide experience in planning and implementing training programmes for all fisheries sectors in the developing countries which, when appropriate, were carried out in co-operation with other international organisations. FAO activities in this field already received the support both of UNDP and other various development and funding agencies and he noted that the present proposals could lead to duplication of effort and wastage of limited development funds.

48. The draft resolution as amended by the Employers was unanimously adopted by the Committee. The text appears as Annex VIII.

Resolution on Programme of Future Work

49. The Worker members submitted a draft resolution which referred to needed improvements in systems of remuneration and welfare afloat and ashore for fishermen, and which also called for the convening, at an early date, of a future session of the Committee to study these questions, with a view to establishing international standards.

50. The Workers said that the prevailing system of remunerating fishermen on the basis of a wage payment combined with a share of the catch was antiquated as well as inequitable since it forced fishermen to participate in the risk of the fishing operation which often entailed fluctuations in earnings. This was particularly unacceptable when failure or poor results of a particular expedition rested with the employers. No other group of workers carried such a burden; fishermen should therefore ideally be entitled to a system of remuneration similar to that applicable to merchant seafarers—a basic wage supplemented by overtime payments.

51. The Workers introduced an amendment to their draft resolution outlining their views on systems of remunerating fishermen.

52. The Government members also indicated that many fishermen were remunerated by sharing in the proceeds from the sale of the catch, and they acknowledged that this system was not always acceptable to fishermen because of fluctuations in incomes which often resulted. In one country legislation proposed the establishment of a system of guaranteed income, and in another national social security benefits already served to stabilise fishermen's income. Nevertheless, any action which could be suggested by the Committee in this regard would be welcomed.

53. The Employers had difficulty accepting a paragraph of the resolution which criticised existing practices of remunerating fishermen; the additional elaboration contained in the Workers' amendment made the resolution entirely unacceptable. The Employers referred to special problems existing in developing countries which precluded their ability to implement the standards of fishermen's conditions reached in highly industrialised economies, and they also considered that to a large extent existing welfare facilities for merchant seafarers could serve fishermen as well, thereby avoiding unnecessary duplication of effort and facilities in this regard.

54. Following the withdrawal by the Workers of their amendment to the resolution, the Employers submitted several proposed amendments reflecting their own views. In the light of these amendments, and subsequent to some discussion and further proposals of the Workers, the text was amended by the Committee and unanimously adopted. The text of the resolution is set out as Annex IX.

Resolution on Unemployment Problems Resulting from the Production of Fish Meal

55. The Worker members submitted a draft resolution which drew attention to the diminishing of employment in the fishing industry owing to a depletion of fish stocks caused partly by operations geared to fish meal production which were of low labour intensity and in addition utilised fish suitable for human consumption. The draft resolution invited the ILO to request governments to restrict the production of fish meal

to non-edible fish and offal resources so as to conserve fish reserves that were important to human nutrition.

56. The Workers stressed that exploitation of certain fish species for fish meal led to a disappearance of other dependent species, a reduction of fish stocks and decreased activity and employment in the fishing industry generally.

57. The Government members stated that in some countries efforts were being made to reduce the production of fish meal from edible fish and to conserve fisheries' resources generally, a practice which they felt deserved encouragement. Although noting the Workers' views on the effect on employment of fish meal production, some Government members questioned whether the over-all subject-matter of the resolution was appropriate for consideration by the Committee.

58. The Assistant Secretary-General of the Committee also pointed out that the issues raised by the Workers' resolution might not be fully within the Committee's terms of reference.

59. The Employer members, in recognition of the general concern about conservation and environmental questions and their relation to various labour matters, were sympathetic with the broad intent of the Workers' resolution. They, too, felt it raised problems outside the competence of the Committee, and therefore that the resolution would best be withdrawn. Nevertheless, in order to make the resolution receivable and within the terms of reference of the Committee, the Employers proposed a substantial revision of the text which noted with concern the loss of jobs in the fishing industry caused by depletion of fish stocks and invited the ILO to urge governments to examine methods of conserving such stocks through measures aimed at preventing indiscriminate fishing operations which might result in loss of jobs in the fishing industry.

60. The Workers were unable to accept the drastically revised text submitted by the Employers and the draft resolution was therefore withdrawn on the understanding that the essence of their views, as well as those of the Employers and Government members would be reflected in the Committee's report.

Resolution on a Model Employment Scheme for Wage-earning Fishermen

61. The Worker members submitted a draft resolution which, stated [*sic*] that instability of employment in the fishing industry and a trend toward general expansion of the fishing activities of various countries, and suggested that regularity and security of employment of fishermen would be facilitated through creation of national labour boards for fishing. The draft resolution proposed that such labour boards should be established in all maritime countries concerned.

62. The Workers elaborated on the suggested functions of the proposed national labour boards as listed in their resolution, many of which they felt had been endorsed by the Employers' members in the context of other issues previously discussed by the Committee. They emphasised that the proposed board could, where appropriate, delegate to local boards as many of its functions as possible, and should in any case pursue a policy of full consultation vis-à-vis local boards.

63. The Government members drew attention to different systems of labour bodies which existed in various countries for maritime and other workers, all of which operated satisfactorily in accordance with national conditions. They noted that the Workers' resolution proposed a particular system which might not work in some countries, and they therefore proposed an amendment to the resolution to satisfy that concern.

64. The Employers sympathised with the concept embodied in the Workers' resolution, but questioned not only the appropriateness of its title for the issue covered, but the proposed functions of the national labour boards, many of which they felt would be outside its competence and/or infringe upon matters normally dealt with through collective bargaining procedures by employers and workers. In their view the labour boards could be advisory bodies only.

65. The Workers concurred that it was not the intention that the suggested scheme should encroach on the freedom of the collective bargaining units, and that the function of governments would generally be to provide the necessary administrative backup.

66. In the light of the reservations expressed by the Government and Employer members, and following some discussion which included certain suggestions by the Workers, the resolution was amended by the Committee and then unanimously adopted. The text as adopted is set out as Annex X to the present report.

Discussion and Adoption of the Report

67. The Committee considered the draft report on its proceedings at its twelfth sitting and unanimously adopted the text thereof in its present form, including the resolutions and conclusions set out in Annexes II, III, IV, V, VI, VII, VIII, IX and X.

Geneva, 30 November 1978.

(signed) F. DAVILA,
Reporter.

(signed) J. H. BERTNES,
Chairman.

ANNEX I

List of Members

[EDITORS' NOTE: Due to limitations of space, Annex I, which includes government members, employer members, and worker members with their advisers, as well as various observers (58 people in all), has not been included here. The reader is referred to the original report (CCF/3/6) available from the International Labour Office, Geneva.]

ANNEX II

Conclusions concerning Fishermen's Hours of Working and Manning

The Committee on Conditions of Work in the Fishing Industry;

Considering that great changes have occurred in some aspects of the fishing industry over the past decades;

Noting that the establishment of hours of work for fishermen is a very complex problem in view of the great variety of fishing boats and types of fishing, the diversified nature of fishermen's employment and the manning standards of fishing vessels;

Recognising the direct link between hours of work and manning and the safety involved;

Noting the continuous work done by the ILO generally in regard to hours of work;

Noting the increasing international attention paid to the social and technical aspects of safety manning, particularly by the ILO and IMCO;

Noting the Hours of Work (Fishing) Recommendation, 1920 (No. 7);

Noting further that several other ILO instruments relating to hours of work and manning have been adopted in relation to the shipping industry which may have a bearing on fishermen's conditions of work, such as the:

Wages, Hours of Work and Manning (Sea) Convention (Revised), 1958 (No. 109);
Fishermen's Competency Certificates Convention, 1966 (No. 125);
Wages, Hours of Work and Manning (Sea) Recommendation, 1958 (No. 109); and
Protection of Young Seafarers Recommendation, 1976 (No. 153).

Requests the Governing Body of the International Labour Office:

(1) To instruct the Director-General:
- (a) to continue the studies relating to the hours of work and manning for the fishing industry, taking into account the existing instruments for the fishing industry and the provisions of the shipping industry's instruments which provide a suitable framework for fishermen to the extent that the conditions in the fishing industry are comparable with those of seafarers; and
- (b) to include fishermen's hours of work and manning in the possible items for future sessions of the International Labour Conference which he submits to the Governing Body for its consideration.

(2) To urge governments and employers' and workers' organisations concerned:
- (a) to pursue their efforts in improving the conditions of work of fishermen through legislation or other appropriate measures and collective bargaining, particularly in relation to hours of work and manning; and
- (b) to establish appropriate machinery at the national level to examine these questions where such machinery does not already exist.

ANNEX III

Resolution on Stabilisation of Employment and Earnings

The Committee on Conditions of Work in the Fishing Industry;

Noting the unstable and difficult economic situation of the fishing industry worldwide, caused by such complex factors as:

—the present economic recession prevailing in several countries;
—changes in territorial water limits and the resulting rearrangements within the national fishing fleets;
—seasonal fluctuations in fishing;
—differing types of fishing operations;

Considering also

—the expanding role of developing countries in this field and the development of joint ventures;

Recognising the urgent need for international action to stabilise employment and earnings in the fishing industry;

Recalling the provisions of Convention (No. 145) and Recommendation (No. 154) on Continuity of Employment of Seafarers, adopted at the 62nd (Maritime) Session of the International Labour Conference in 1976;

Fully agreeing that the principles expressed in Article 2 of Convention No. 145 and Paragraph 2 of Recommendation No. 154 should also be applied to fishermen to the extent that conditions are comparable in full consultation with representative organisations of fishing vessel owners and fishermen;

Recognising in this connection the need for the existence of strong and independent employers' and workers' organisations as provided for in the Freedom of Association and Protection of the Right to Organise Convention, 1948 (No. 87);

Urges the Governing Body of the International Labour Office to instruct the Director-General, in the light of the foregoing considerations, to carry out a study to determine to what extent the provisions of Convention No. 145 and Recommendation No. 154 might be applied to fishermen (as defined in Article 2 of the Fishermen's Articles of Agreement Convention, 1959 (No. 114);

Further requests the Governing Body, in the light of the results of the above-mentioned study, to include in the agenda of a session of the International Labour Conference an item on stabilisation of employment and earnings of fishermen, for such action as the Conference may deem appropriate.

ANNEX IV

Resolution on Medical Care for Fishermen While at Sea

The Committee on Conditions of Work in the Fishing Industry;

Considering that fishermen are prone to illness and injury of an occupational and non-occupational nature while at sea and conscious of the essential disparities in medical care arrangements for fishermen at work as compared with the medical facilities available to shore-based workers;

Noting the limited number of medical practitioners and other trained medical personnel on fishing vessels, leading to the use of non-medical crew members for the delivery of medical care;

Recalling that the question of medical care for fishermen at sea has not been covered by an international labour instrument except indirectly through some standards primarily related to the shipping industry;

Noting that medical care for fishermen at sea transcends national boundaries through the involvement of various international medical consultations and coordinated activities;

Convinced that medical care for fishermen at sea could be significantly improved for the health benefit of patients through measures aimed at the international harmonisation of medical training programmes of fishing personnel with close coordination of these programmes with international standards on related matters as medical guides, medicine chests, medical advice by radio and, where medically necessary and feasible, the evacuation of the patient;

1. Invites the Governing Body of the International Labour Office to include the question of medical care for fishermen at sea in the agenda of a future session of the

International Labour Conference with a view to the adoption of an international instrument.

2. In the course of the preparatory work for the proposed international instrument, the Committee invites the ILO to take due account of the following principles:

—in connection with medical care, crew members of fishing vessels require different levels of training related to the circumstances in which they will be required to operate and the functions expected of them. It is important that the question of retraining is also given detailed consideration;

—medical care training should be so designed as to correspond to the medical and surgical problems to which fishermen are exposed; it should be based on the medical guide; it should be related to the contents of the medicine chests and it should include instructions on the proper use of the radio medical advice system;

—the presence of a standard scale of medicine chests on board fishing vessels is strongly recommended. A basic minimum chest should be designed commensurate with the needs of the industry and added to in modular form according to requirements, i.e. greater distances, larger ships, larger crews and longer stays at sea. Medical guides and medicine chests should be regularly revised. Medicine chests should be regularly checked;

—medical advice by radio could be rendered more effective through measures to harmonise the orderly assembly and transmission of medical information from the ship to the medical consultant. In turn, this will facilitate the proper assessment of the patient's clinical condition, the making of a rapid diagnosis and the establishment of the correct treatment;

—prompt evacuation of the severely ill or injured patient should be arranged whenever this is medically indicated;

—to make optimum use of the limited number of fishing industry vessels carrying a physician, it is opportune to encourage whenever feasible further regional and international collaboration in terms of information and operating waters of such vessels so as to extend their availability to the greatest number of fishermen far out at sea.

3. In further studying this question, the Committee invites the ILO to take into account, as appropriate, the Ships' Medicine Chests Recommendation, 1958; the Medical Advice at Sea Recommendation, 1958; the resolutions adopted by the Fifth Session of the Joint ILO/WHO Committee on the Health of Seafarers in 1973, the International Medical Guide for Ships together with its current revision; the International Code of Signals; and Recommendation No. 8 of the IMCO International Conference on Safety of Fishing Vessels, 1977.

4. In view of the solidarity required in connection with medical care at sea, it is requested that an appeal be made to member States to provide or facilitate the delivery of the best possible level of medical care to foreign fishermen operating in waters close to the territories of these member States.

5. The Committee invites the ILO to continue to co-operate with WHO and IMCO on matters concerning medical care on board fishing vessels.

ANNEX V

Conclusions on Social Security Protection of Fishermen in the Case of Sickness, Invalidity, Old Age and Death

The Committee on Conditions of Work in the Fishing Industry;

Having examined the report prepared by the International Labour Office on questions concerning pensions and sickness insurance; adopts the following conclusions:

1. In view of the fact that the existing international labour Conventions concerning social security for seafarers allow the exception of workers on board fishing vessels, it is highly advisable to examine the possibility of including them in the scope of application, when the International Labour Office carries out the in-depth study on such standards in accordance with paragraph 3 of the Resolution concerning the revision of Conventions and promotion of maritime social legislation, adopted by the International Labour Conference at its 62nd (Maritime) Session.

2. In regard to national law and practice concerning social security protection of fishermen and their dependants in the case of sickness, invalidity, old age and death;

(a) it is necessary to extend as far as possible the range of persons protected by the national social security scheme so as to cover all fishermen, including self-employed and their dependants, with a view to ensuring greater social justice which should be expressed in the form of equal conditions for all;

(b) continuous efforts should be made to improve both quantity and quality of benefits to be provided to workers in the fishing industry, which should be supported by sound financial arrangements relative to the level of development of each country;

(c) where employment of fishermen is intermittent, or seasonal, and where entitlement to social security benefits is related to the length of employment, it is advisable to adapt the qualifying conditions to the particular circumstances in which fishermen are employed;

(d) where fishermen are remunerated by a share of profit or are self-employed, due account should be taken of the fluctuation in the levels and regularity of their income in the computation of contributions and the calculation of benefits under contributory social security schemes;

(e) for self-employed fishermen operating as family unit or on an extremely small scale, efforts should be made to improve the existing benefit structures so as to ensure comprehensive medical care, to provide suitable compensation in the case of incapacity for work due to sickness, involving suspension or substantial reduction of income, to guarantee adequate level of invalidity, old-age and survivors' pensions under conditions for entitlement which are compatible with those required for fishermen working for an employer, and to extend effective protection against invalidity through the provision of rehabilitation measures;

(f) in view of the hazardous nature of work and exceptional stress involved in the fishing industry, due consideration should be given to the possibility of lowering the age at which fishermen who have been engaged in the industry for a considerable number of years are entitled to old-age or retirement pensions.

ANNEX VI

Resolution on Holidays with Pay for Fishermen

The Committee on Conditions of Work in the Fishing Industry;

Noting that the Seafarers' Annual Leave with Pay Convention, 1976 (No. 146), provides for optional application to fishermen of the standards laid down therein;

Noting further that in several countries fishermen benefit from provisions governing holidays with pay under legislation relating to workers in general or under collective agreement; but that this is not the case in all countries and that where there is leave entitlement nevertheless wide variations may be found in regard to such specific provisions as length of holidays, or to the categories of fishermen enjoying such entitlement;

Considering it essential that fishermen should not be excluded from protection in respect to holidays with pay;

Strongly urges the Governing Body of the International Labour Office to appeal to governments—

(a) to ratify the Seafarers' Annual Leave with Pay Convention, 1976 (No. 146); and
(b) to extend the provisions of this Convention, in accordance with its Article 2, paragraph 4, to persons employed on board ships engaged in fishing or in operations directly connected therewith.

ANNEX VII

Resolution on the Repatriation of Fishermen

The Committee on Conditions of Work in the Fishing Industry;

Noting the substantial number of ratifications of the Repatriation of Seamen Convention, 1926 (No. 23) that have been communicated to the International Labour Office;

Noting also that national legislation governing conditions of employment of seafarers applies, to a very wide extent, to fishermen as well as to merchant seamen; and therefore that it would be inadvisable to adopt new international standards concerning the repatriation of fishermen specifically;

Recognising, however, that fishermen have no formal coverage as regards repatriation under internationally agreed standards, and where they do benefit from such standards it is by reason of national legislation;

Considering it essential that fishermen should not be excluded from provisions in regard to repatriation;

Requests the Governing Body of the International Labour Office to:

(1) call to the attention of member States the resolution adopted by the International Labour Conference at its Ninth Session in 1926, on the occasion of the adoption of Convention No. 23, which invites the governments of all maritime countries which have not already done so to take the measures required to ensure the repatriation of fishermen left in a foreign port;
(2) express its satisfaction that a large number of governments have already taken such measures, and that they either extend to fishermen the provisions of national legislation governing the repatriation of seafarers or include such provisions in fishermen's codes;
(3) note that in certain countries the particular nature of fishing operations makes inapplicable the question of repatriation of crews, but that in others fishermen's entitlement to repatriation is limited to certain circumstances and is not guaranteed on all occasions when a fisherman is landed in a foreign port for reasons not of his own fault;

(4) reaffirm the views expressed by the International Labour Conference in 1926 on this question; and to

(5) appeal to those governments which have not yet given effect to the resolution to now take measures to ensure that all fishermen who are landed in foreign ports for any reason other than their own fault will be repatriated to the port of engagement or other mutually agreed destination at no cost to themselves and that in this respect the ultimate responsibility of the State of registry of the fishing vessel is recognised.

ANNEX VIII

Resolution on ILO Technical Co-operation in the Field of Vocational Training within the Fishing Industry

The Committee on Conditions of Work in the Fishing Industry;

Considering that the fishing industry has a good potential for the creation of employment;

Noting that the new techniques developed among traditional industrialised maritime nations can be of great assistance to developing countries in meeting the basic needs of its population and in improving the nutritional value of their food;

Noting further that access to such techniques is not within the reach of developing nations mainly because of the lack of training facilities for fishermen;

Considering that the ILO has a clear mandate to assist developing countries in their training needs through its technical co-operation programme;

Recognising that the ILO's efforts have so far mainly been concentrated on assistance in merchant maritime matters and that projects for the fishing industry have been somewhat marginal;

(i) *Requests* the Governing Body of the ILO to instruct the Director-General:

(a) to establish a suitable programme of training for fishermen, particularly in small-scale fisheries, which would include safety and health aspects as well as first aid on board fishing vessels;

(b) to provide on request assistance to developing countries related to their training needs through its regular budget and through funds from UNDP and other extra-budgetary sources;

(c) to co-operate fully with other UN agencies particularly UNIDO, FAO, WHO, IMCO, etc.;

(ii) *Urges* governments, employers' and workers' organisations to co-operate fully with the ILO with a view to achieving the main goals spelled out in the World Employment Programme.

ANNEX IX

Resolution on Programme of Future Work

The Committee on Conditions of Work in the Fishing Industry;

Welcoming the resumption of work on fishermen's questions within the ILO;

Expressing the hope that its present recommendations will receive favourable consideration by the Governing Body;

Noting that there are nevertheless further areas of the fishing industry which require attention at national and international level, e.g. the system of remuneration and welfare facilities;

Considering that the remuneration of fishermen is based wholly or largely on a share of catch and that fixed payments play a relatively minor role which creates difficulties in some countries in the establishment of a scheme of social security benefits;

Considering further that provision should be made for adequate recreation facilities on board vessels and for welfare arrangements ashore in accordance with the guidance given by the Seafarers' Welfare Recommendation, 1970 (No. 138) and the Seamen's Welfare in Ports Recommendation, 1936 (No. 48);

Invites the Governing Body of the International Labour Office to give favourable consideration to the possibility of convening, at an early date, a further session of this Committee to study systems of remuneration and welfare facilities in the fishing industry with a view to establishing international instruments on these subjects.

ANNEX X

Resolution concerning the Establishment of a National Tripartite Labour Board for the Fishing Industry

The Committee on Conditions of Work in the Fishing Industry;

Considering that stability of employment is generally lacking in the fishing industry due to its fragmented structure;

Noting that, with the continuing trend towards the extension of national fishing limits, an increasing number of developed and developing nations are expanding their fishing industry, and that as a consequence overdevelopment of the industry may ensue which may ultimately have a detrimental effect on the fishermen concerned;

Convinced that a large measure of regularity and security of employment, and resulting therefrom greater efficiency, could be achieved in the fishing industry if the concept of a national labour board for fishing was introduced;

Recommends therefore that bearing in mind provisions of the ILO Freedom of Association and Protection of the Right to Organise Convention, 1948 (No. 87), and Right to Organise and Collective Bargaining Convention, 1949 (No. 98), in maritime States where appropriate and in accordance with national practice, a tripartite national labour board for fishing be set up for the purpose of advising on, and where operable, implementing advice regarding:

(1) establishing and revising as appropriate a list of registered ports;
(2) maintaining and adjusting as required in each registered port a register of trained and qualified fishermen as well as of registered employers;
(3) regulating the recruitment and training of fishermen;
(4) allocating jobs, and in this connection maintaining appropriate records;
(5) establishing criteria for the standardisation of employment conditions and fair practices of employment, including the establishment of an appropriate disciplinary and grievance procedure;
(6) reviewing and updating training methods and qualifications;

(7) introducing effective port medical services within the national medical system;
(8) introducing an appropriate redundancy payments scheme;
(9) developing a "medical severance" payments scheme;
(10) assisting registered fishermen wishing to transfer to shore-based employment with job placement;
(11) facilitating the establishment of sound industrial relations.

IOC: Biennial Report for 1976–1977[1]

INTRODUCTION

During the biennium the Intergovernmental Oceanographic Commission strengthened its activities under its mandate "to promote scientific investigations with a view to learning more about the nature and resources of the oceans through the concerted action of its members."

The Commission continued to implement projects of the Long-Term and Expanded Programme of Oceanic Exploration and Research, and of its acceleration phase, the International Decade of Ocean Exploration. Particular emphasis was given to ocean science projects in East Asia and the South Pacific, off the west coast of South America (investigations of "El Niño"), and off the east coast of Africa (investigations in the North and Central Western Indian Ocean). The field work of the Co-operative Study of the Kuroshio and adjacent regions was terminated but activities in the region will be continued under a new working group for the Western Pacific. Work has been started on an important programme for the Biological Investigation of Marine Antartic Systems and Stocks, the research programme for which has been developed by four non-governmental organizations.

The Commission also developed new activities in the field of "Ocean and Climate," particularly in support of the Global Atmospheric Research Programme (GARP) and its components, the GARP Atlantic Tropical Experiment, the First GARP Global Experiment, and the Polar Sub-Programme.

Two projects devoted to the morphological charting of the ocean floor—the General Bathymetric Chart of the Oceans and the International Bathymetric Chart of the Mediterranean—have been developing most successfully along planned lines.

The Working Committee for the Global Investigation of Pollution in the Marine Environment implemented a number of important projects, in close co-operation with the United Nations Environment Programme (UNEP).

The Commission continued to provide world-wide services under such programmes as its Integrated Global Ocean Station System and International Oceanographic Data Exchange. The Tsunami Warning System in the Pacific has been developing satisfactorily.

Under the Commission's programme in Training, Education, and Mutual Assistance in the Marine Sciences, the initiation of an IOC Voluntary Assistance Programme is expected to improve substantially the support which donor Member States can make available in response to requests from other members addressed through the IOC. Special attention has been given to the assessment of the needs of developing countries through a series of regional meetings which have now virtually covered the globe.

1. August 30, 1978 (IOC/INF-376). [Editors' Note: Due to limitations of space, figures and photographs accompanying the original text, as well as Annex II, have been omitted.)

Recommended actions stemming from these meetings are now being integrated into all major programmes and projects sponsored by the Commission.

Co-operation has been sought from other intergovernmental and non-governmental global and regional organizations in the marine sciences, such as the Committee for Co-ordination of Joint Prospecting for Mineral Resources in Asian Offshore Areas and the Committee for Co-ordination of Joint Prospecting for Mineral Resources in South Pacific Offshore Areas of the Economic and Social Commission for Asia and the Pacific, the Commission for Marine Geology of the International Union of Geological Sciences, the International Council for the Exploration of the Sea, the International Commission for the Scientific Exploration of the Mediterranean Sea, the Tsunami Committee of the International Union of Geodesy and Geophysics, etc.

The IOC Assembly met for its tenth session (IOC-X), in Paris, from October 27 to November 10, 1977; in addition the Commission's Executive Council met in three sessions (EC-VII, Bergen, June 21–26, 1976; EC-VIII, Paris, April 4–8, 1977; and EC-IX, Paris, October 24–26, 1977).

The Summary Report of the tenth session of the IOC Assembly is available separately as document SC/MD/60; this document also contains the resolutions adopted at the seventh and eighth sessions of the Executive Council.

OCEAN SCIENCE PROGRAMMES UNDER THE LONG-TERM AND EXPANDED PROGRAMME OF OCEANIC EXPLORATION AND RESEARCH (LEPOR)

Background

The International Decade of Ocean Exploration (IDOE), 1971–1980, designated as the acceleration phase of LEPOR, was intended to stimulate the acquisition of scientific knowledge of the oceans and to improve the capability of all Member States to participate in oceanographic research activities. Two years from now this initial decade will come to an end and planning for the 1980s is already well in hand. A major result of the IDOE is that marine scientific exploration and research activities are now widely acknowledged as important both for the economic exploitation of marine resources and for understanding man's impact on the environment.

To review the scientific validity of ongoing and proposed work in LEPOR research programmes, an IOC Scientific Advisory Board was established. The Board submitted a summary report and the recommendations of its two sessions held in 1976 and 1977 to the tenth session of the IOC Assembly. At that time it was decided to continue the Board's work until 1979 when the eleventh session of the Assembly is expected to make a final decision on the structure and terms of reference for an overall scientific subsidiary body for the Commission.

Projects worthy of special mention amongst the marine science and technology programmes of the Commission are described below.

El Niño

The phenomenon known as El Niño is the primary cause of large anomalies in the physical, chemical, and biological conditions in the sea off the west coast of South America.

El Niño, essentially a cessation of the normal wind conditions that effect marine upwelling along the coastal regions, causes higher than normal rainfall in northern Peru and Ecuador. More importantly, the reduction in upwelling brought on by El Niño deprives the normally abundant anchoveta of the main food source. The economic impact is drastic. The sudden decline of anchoveta stock influences the world fish and soya bean market and, indirectly, the production of fertilizer which is strongly dependent on the guano produced by the birds feeding on the anchoveta.

It has not yet been possible to predict accurately or even to describe precisely why the phenomenon occurs. Scientists now believe that El Niño is a local response to meteorological and oceanographic processes occurring over at least the whole of the eastern tropical Pacific, and possibly over much larger areas of the globe.

In view of the nature and magnitude of the research effort required, the IOC Assembly established a working group on the investigations of "El Niño," to be co-sponsored with the World Meteorological Organization and the Comisión Permanente del Pacífico Sur.

North and Central Western Indian Ocean

There is increasing evidence linking sea-surface temperature anomalies in these regions with rainfall over India. The Asiatic monsoon is the strongest, large-scale, long-period variable force that the atmosphere exerts on the Indian Ocean and its major surface current, the Somali Current. Better understanding of this current, its variability, and the related variability of its physical, chemical, and biological properties, should contribute to the development of the fishery industry and to the management of coastal resources of the region.

In 1977 a joint marine science mission studied the state of marine and fisheries sciences in three countries of the region as a basis for proposing a programme for strengthening existing, or for establishing new, research institutes. The IOC Assembly at its tenth session decided to convene a meeting of the countries of the region in 1978 to plan scientific investigations in this part of the Indian Ocean and to identify requirements for intergovernmental or international co-operation.

Northern Part of the Eastern Central Atlantic

Canary Current: upwelling and living resources. —A symposium on the upwelling and living resources of the Canary Current is scheduled to be held in Las Palmas, Gran Canaria, Spain, from April 11 to 14, 1978. This symposium, jointly organized with the International Council for the Exploration of the Sea and the FAO Fishery Committee for the Eastern Central Atlantic, will review results from the Co-operative Investigations of the Northern part of the Eastern Central Atlantic. The investigations consisted of a series of internationally co-ordinated expeditions during which research vessels from various countries studied the waters off north-west Africa. Its scientific aim was to develop a general oceanographic description of the Canary Current region between Gibraltar and 10°N and to undertake an analysis of the upwelling processes in the area and their biological consequences.

The Commission will terminate eight years of activity in this region with a comparison of results achieved here and in other regions of the world's oceans having similar oceanographic structures.

Caribbean and Adjacent Regions

Following its first session in Caracas in July 1976, the IOC Association for the Caribbean and adjacent regions (IOCARIBE) launched a six-year pilot programme to prove the viability of locally organized regional activities. A regional secretariat was set up in

Port-of-Spain, Trinidad and Tobago, making IOCARIBE IOC's first semi-autonomous regional programme.

In co-operation with FAO's Western Central Atlantic Fishery Commission, an interdisciplinary workshop on "Scientific Programmes in Support of Fisheries" was held in Martinique, November/December 1977. Complementary oceanographic research programmes in support of two important practical fishery problems in the Antilles and along the Caribbean coasts of Central America were identified. Marine pollution and environmental geology projects are also being developed as international collaborative exercises.

Western Pacific

At it eleventh and final session in Nouméa, New Caledonia (July 1977), the International Co-ordination Group for the Co-operative Study of the Kuroshio and adjacent regions decided to terminate the field phase of the study at the end of 1977. Over a period of nearly twelve years, the entire Kuroshio current system, its seasonal, annual, and multiannual variability, and its whole physical, chemical, and biological structure have been studied. The environmental conditions and productivity of the region, including the South China Sea, were also investigated.

The results achieved during this co-operative investigation, the longest in duration since the establishment of the Commission, will be presented at the Fourth CSK Symposium (to be convened in Japan early in 1979). This Symposium will be followed by a scientific workshop to define the priorities for continued marine research in the western Pacific. The workshop will be held in conjunction with the first session of the new IOC Working Group for the Western Pacific.

The programme of research developed by the IDOE Workshop on Metallogenesis, Hydrocarbons and Tectonic Patterns in Eastern Asia (Bangkok, September 1973) has been taken several stages further by a Joint Working Group on IDOE Studies on East Asia Tectonics and Resources.

The core of the research programme is a study along six transects in the following regions: Burma–North Thailand; Andaman Sea–Malay Peninsula–Sumatra–Malay Peninsula–Sunda Shelf; Timor–Banda Arc; across Northern Philippines; and SW Japan–Korean Peninsula. An important part of the work has now been implemented by the joint efforts of the CCOP countries and contributing countries from outside the region. As the study is scheduled to end in 1980, the Group proposed, at the Manila meeting, that a workshop should be convened in 1978 to assess and evaluate achievements under the project, and to define further work requirements not only for the time remaining but also for new studies in the 1980s.

South Pacific

The Commission provided considerable assistance for activities carried out under the programme of research for the South Pacific region, developed during the IDOE International Workshop on the Geology, Mineral Resources, and Geophysics of the South Pacific (Suva, Fiji, September 1975).

Countries of the region have made great efforts to carry out the 55 national and 20 regional projects outlined under this programme, despite the fact that most of the countries involved are island States with little available manpower. Among the co-operating countries are the Cook Islands, Fiji, Gilbert Islands, New Zealand, Papua New Guinea, Solomon Islands, Tonga, and Western Samoa. Assistance from outside is received from Australia, France, Japan, and the United States.

Southern Oceans

The Southern Oceans region has been the object of national research for a long number of years during which scientists from many countries have delved into the mysteries of the Antarctic and adjacent waters. It is thought that this area may not only be the key to understanding world climate, but will also one day offer new possibilities as a source of food and other natural resources.

The third session of the International Co-ordination Group for the Southern Oceans, London (September 1977), welcomed as an international co-operative study the Biological Investigations of Marine Antarctic Systems and Stocks (BIOMASS) programme. Member States encouraged SCAR and SCOR and their collaborators to continue the scientific planning of the various activities envisaged, paying specific attention to the standardization of methods involved. The Commission will assume responsibility for exchange of data from such scientific activities through the World Data Centre system, as well as the collection and dissemination to interested countries of information on scientific programmes and ship schedules.

WMO-ICSU Global Atmospheric Research Programme (GARP)

The First GARP Global Experiment, scheduled for the period December 1, 1978 to November 30, 1979, is the largest concerted international scientific effort ever contemplated. Although founded on atmospheric science objectives, it has many aspects which promise to be of great value to oceanographic science.

A publication entitled "Oceanographic Components of the Global Atmospheric Research Programme (GARP)" (IOC Technical Series No. 17), describes existing international oceanographic programme elements which will support, or receive support from, the WMO/ICSU Global Atmospheric Research Programme.

Primarily, the benefits to be derived from GARP should be the improvement of atmospheric and oceanic forecasting and climatological models; such improvements are expected to result from better understanding of oceanographic processes within these models. In particular, the First GARP Global Experiment, in which work will be concentrated on the Equatorial, Indian, and Southern Ocean areas, will support many IOC ocean science activities by providing unprecedented data for studying the many ocean processes linked to atmospheric forcing and feedback.

Because of the potential importance of FGGE to the world oceanographic community, IOC has greatly increased its support to this experiment since October 1976. In close collaboration with WMO, IOC is now playing a major role in oceanographic data management and in documenting, co-ordinating, and planning of all oceanographic aspects of the experiment. These activities involve many of the Commission's science and services programmes including the Voluntary Assistance Programme. In addition the IOC Secretariat is acting as an active link between the scientists taking part in the experiment and the governments of coastal States, in order to facilitate granting of permission for foreign research ships to work in waters under the national jurisdiction of coastal States.

Support for the Global Experiment is being concentrated on promoting an increase in the collection of oceanographic data, and co-ordinating, together with WMO, the subsequent data flow. Specific areas of promotion are the tropical Atlantic and the western Indian Ocean.

Under its data management programmes, IOC has promoted data exchange activities of significant potential benefit to the global oceanographic community. Through its promotional activities, IOC has encouraged Member States to support these activities; one result has been the establishment of a Responsible National Oceanographic Data Centre for oceanography during the Global Experiment. This data centre will produce a Global Ocean Climate Data Base which will not only enable oceanographers to obtain copies of data from a central depository, but should also pave the way for longer-term climate data bases in the 1980s and beyond.

It is clear that the success of the Global Experiment requires significant participation by both the developing and the developed nations of the world. Participation by developing countries will be facilitated by the active programmes of training, education, assistance, and technology transfer offered by WMO and IOC.

Global Investigation of Pollution in the Marine Environment (GIPME)

During the biennium the Commission published a Comprehensive Plan for the Global Investigation of Pollution in the Marine Environment together with Baseline Study Guidelines (IOC Technical Series No. 14). This Comprehensive Plan, available in all the working languages of the Commission, contains a scientific guide usable by Member States in the organization and implementation of their own marine pollution research and monitoring programmes. The Baseline Study Guidelines provide an approach to the study of present marine pollutant levels in the marine environment.

The Working Committee for GIPME is served by a Task Team on Marine Pollution Monitoring and a Group of Experts on Methods, Standards, and Intercalibration. The latter was established to consider all problems related to analytical procedures for estimating various kinds of pollutants in the marine environment.

A Programme for Monitoring Background Levels of Selected Pollutants in Open-Ocean Waters, proposed by UNEP as a component of its Global Environmental Monitoring System, was approved by the IOC. The programme, which includes an intercalibration exercise, training, and preliminary monitoring, will start in the Atlantic Ocean in 1978 and may be continued in other oceans. Such open-ocean studies should facilitate our understanding of the role of the oceans in pollutant transfer.

A book, "The Health of the Oceans" by E. D. Goldberg, was published in 1976. French and Spanish versions are in preparation.

Regional Marine Pollution Research and Monitoring

Several international workshops on marine pollution have been held in recent years to define the major pollution problems of particular importance to a given region [Fig. 1]; and to suggest solutions. During the biennium, one was held on the Caribbean and adjacent regions (Port-of-Spain, Trinidad, December 13–17, 1976). The workshop proposed seven projects (on petroleum, health aspects, coastal hydrology, coastal oceanography, effects of pollutants on ecosystems, monitoring of persistent chemicals, and controlled environment experiments). These recommendations will serve as the basis of a marine research programme in the region, under a UNEP Regional Action Plan.

Within the UNEP Co-ordinated Mediterranean Pollution Monitoring and Research Programme, the IOC shares responsibility for co-ordinating Pilot Projects on Baseline

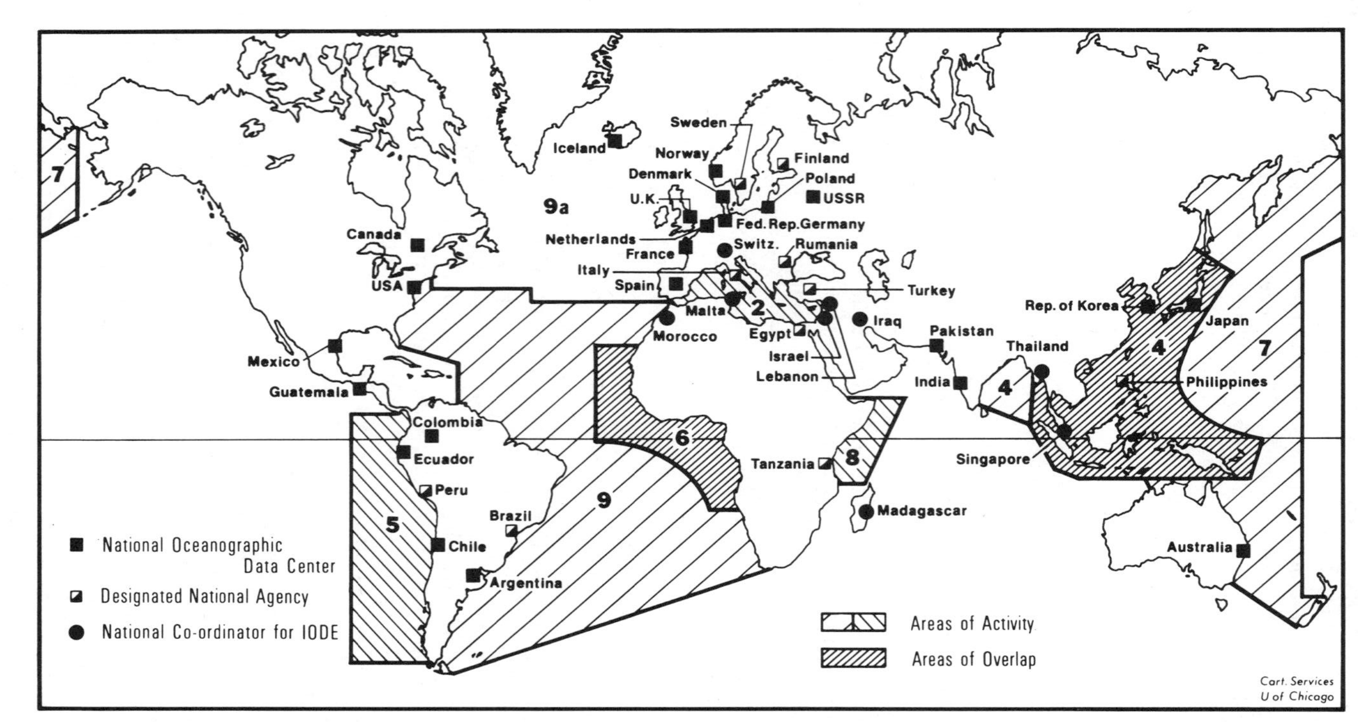

Fig. 1.—Areas in which activities are being carried out within the framework of the Global Investigation of Pollution in the Marine Environment (GIPME) and members of the International Oceanographic Data Exchange System (IODE). (Editors' Note.—Figure and legend adapted from figs. 1 and 6 in the original IOC Report [IOC/INF-376]).

Explanatory notes.—Global Investigation of Pollution in the Marine Environment (GIPME).

1. IGOSS Pilot Project on Marine Pollution (Petroleum) Monitoring. Area: all seas and oceans. Activity: project in progress; end of project mid-1979; Second Workshop on Marine Pollution Monitoring held in Monaco, June 14–18, 1976. Participating agencies: IOC, WMO, UNEP.

2. Mediterranean Sea. Area: Mediterranean Sea. Activity: two projects (MED-I and MED-VI) within the UNEP Co-ordinated Mediterranean Pollution Monitoring and Research Programme in progress; projects end 1978; Mid-Term Review Meeting held in Barcelona, May 23–27, 1977. Participating agencies: IOC, WMO, UNEP.

3. Caribbean and adjacent regions. Area: Caribbean Sea, Gulf of Mexico, and adjacent regions. Activity: International Workshop on Marine Pollution in the Caribbean and Adjacent Regions held in Port-of-Spain (Trinidad), December 13–17, 1976; seven projects proposed. Participating agencies: IOC, FAO, UNEP. Participating regional bodies: IOC Regional Association for the Caribbean and adjacent regions (IOCARIBE); Western Central Atlantic Fishery Commission (WECAFC) of FAO, Economic Commission for Latin America (ECLA) of UN.

4. East Asian Waters. Area: Bay of Bengal, Strait of Malacca, Gulf of Thailand, South China Sea, Sea of Japan, Yellow Sea, East China Sea, and Seas of Eastern Archipelago. Activity: International Workshop on Marine Pollution in East Asian Waters held in Penang (Malaysia), April 7–13, 1976, four regional and twenty subregional projects proposed. Participating agencies: IOC, FAO, UNEP. Participating regional bodies: Indo-Pacific Fisheries Council (IPFC) of FAO, Association of South East Asian Nations (ASEAN).

5. South-east Pacific Region. Area: Western coast of South America from Colombia to Chile. Activity: International Workshop on Marine Pollution in South-east Pacific Region planned for 1978. Participating agencies: IOC, FAO, UNEP. Participating regional body: Comisión Permanente del Pacifico Sur (CPPS).

6. South-east Atlantic/Gulf of Guinea Region. Area: along western coast of Africa from Mauritania to Angola. Activity: International Workshop on Marine Pollution in the Gulf of Guinea and adjacent Regions planned for May 1978. Participating agencies: IOC, FAO, WHO, UNEP. Participating regional body: Fishery Committee for the Eastern Central Atlantic (CECAF) of FAO.

7. Western Pacific. Area: within a line Bering Strait–New Zealand–Burma. Activity: The eleventh and final Session of the International Co-ordination Group for the Co-operative Study of the Kuroshio and adjacent regions held in Noumea (New Caledonia), June 30–July 4, 1977; meeting of an ad hoc Task Team for the Western Pacific held in Noumea, June 27–29, 1977. Participating agency: IOC. Participating regional body: IOC Working Group for the Western Pacific (WESTPAC).

8. North and Central Western Indian Ocean. Area: along the eastern coast of Africa from Somalia to Tanzania. Activity: Scientific Workshop to initiate planning for a co-operative investigation in the North and Central Western Indian Ocean held in Nairobi (Kenya), March 25–April 2, 1976. Participating agencies: IOC, Unesco, FAO. Participating regional body: Indian Ocean Fishery Commission (IOFC) of FAO.

9 and 9a. Programs for Monitoring Background Levels of Selected Pollutants in Open-Ocean Waters. Area: Central and southern Atlantic (9) and, in co-operation with ICES, northern Atlantic (9a); possibility of extension to other oceans. Activity: project document in preparation; the project to have started in 1978. Participating agencies: IOC, WHO, UNEP, ICES.

International Oceanographic Data Exchange System (IODE). World Data Centres (WDC): Washington, Moscow. Regional Data Centres (RDC): CIM—NODC, USSR; IOCARIBE—NODC, USA; ICES—Service Hydrographique ICES, Denmark; CSK (WESTPAC)—Japan. Responsible National Oceanographic Data Centres (RNODC): IGOSS—NODC, USSR; NODC, U.S.A.; NODC, France; NODC, Japan. MAPMOPP—U.S.A., Japan; Wave Data—U.K.

Studies and Monitoring of Oil and Petroleum Hydrocarbons in Marine Waters and on Problems of Coastal Transport of Pollutants.

Preparations for an International Workshop on Marine Pollution in the Gulf of Guinea and Adjacent Areas have started. The Workshop is expected to take place in May 1978 in the Ivory Coast.

GLOBAL AND REGIONAL OCEAN SERVICES ACTIVITIES

Integrated Global Ocean Station System

The Integrated Global Ocean Station System (IGOSS) is a joint IOC/WMO operational service programme for providing information on the state of the oceans required by various marine users.[2] These users deal with exploration and exploitation of biological and mineral resources of the ocean, shipping, recreation, search and rescue operations, ocean and offshore engineering, harbour control, and pollution abatement and control. The information is also used in support of meteorological and oceanographic research.

After several years of preparation by respresentatives of member States and IGOSS experts, the IGOSS General Plan and Implementation Programme for 1977–1982 (IOC Technical Series No. 16, WMO No. 466) was approved by the seventh session of the IOC Executive Council (June 1976) and the XXVIIIth session of the WMO Executive Committee (May–June 1976).

This plan is a guide for further development of IGOSS in the manner desired by participating nations. It contains guidelines for the development of the following basic components: the IGOSS Observing System, the IGOSS Data Processing and Services System; IGOSS telecommunication arrangements; research in support of IGOSS; and training and education related to IGOSS.

In 1976–1977, major efforts were concentrated on implementing the following aspects:

Operational Programme for the Collection and Exchange of Oceanographic Data
This programme [fig. 1] is considered as the initial operational phase of IGOSS aimed at establishing an internationally co-ordinated scheme for collection and exchange of oceanographic data (temperature, salinity, and currents) obtained mainly from ships. These data are used by national centres for the preparation of oceanographic and marine meteorological products required to support activities of various marine users, as well as for oceanographic and meteorological research.

The number of reports exchanged over the WMO Global Telecommunications System has increased considerably, i.e., 16,500 reports in 1975, 33,500 reports in 1976, and 38,200 in 1977. During this period Argentina, Australia, Iceland, and India joined the programme. An increasing amount of information from ocean data buoys is now also being reported.

Operational instructions for this programme are described in IOC Manuals and Guides Nos. 1, 3, and 4 and are kept under permanent review by IOC and WMO bodies.

2. Editors' Note: See also "WMO: Meteorology and Ocean Affairs," in this volume.

To facilitate the identification and archiving of meteorological and oceanographic data obtained from environmental data buoys, an international identifier system for buoy stations has been adopted for use by Member States of IOC and WMO.

In 1977, a regular information service on ocean data buoys was started by the IOC and WMO Secretariats; the first bulletin was issued in August 1977. The *IGOSS Data Processing and Services System (IDPSS)* is an integral element of IGOSS consisting of national, specialized, and world oceanographic centres for processing of observational data, preparation of products (oceanographic analyses and forecasts), and provision of services to various marine users.

The following steps towards developing this system were made in 1976–1977:

the IOC publication "Oceanographic Products and Methods of Analysis and Prediction" (IOC Technical Series No. 12) was published—this contains information on the activities of national oceanographic centres;

regular dissemination of information on oceanographic products issues by National Centres, was initiated in 1977;

special attention was given to system developments at a Joint IOC/WMO meeting of governmental experts held in Ottawa in August 1977.

A Seminar Workshop on Ocean Products and the IDPSS will be held in 1979; following the recommendations of IGOSS experts, specific projects pertaining to IGOSS products and services in support of fisheries and in support of the Global Atmospheric Research Programme, are being planned.

The *IGOSS Pilot Project on Marine Pollution (Petroleum) Monitoring*, launched in 1975, was extended until the end of 1978 after its evaluation at a workshop held in Monaco, in June 1976. During 1976–1977, special attention was given to the collection and exchange of data from this experiment. Two Responsible National Oceanographic Data Centres were designated to store and process these data—the United States NODC and the NODC of Japan.

Development of this project has been carried out with the financial support of the United Nations Environment Programme. Procedures to be followed by Member States participating in MAPMOPP have been published.

Within the framework of the Pilot Project, a training course was organized by UNESCO in November 1976 at Duke University, U.S.A.; it was attended by twenty trainees from developing countries.

In 1976, the Joint IOC/WMO Planning Group for IGOSS, together with UNEP, undertook the design of a programme for monitoring background levels of selected pollutants in open-ocean waters. Planning of the first phase—intercalibration, preliminary surveys, and training—is in hand.

Radio Frequencies for Ocean Data Transmission Allocated by the World Administrative Radio Conference (WARC)

Co-ordination of requirements and requests for use of frequencies allocated by WARC-67 (and reallocated by WARC-74) for ocean data transmission, is provided jointly by the IOC and WMO Secretariats. The initial result of this co-ordination was an Agreed Interim Frequency Utilization Plan, first published in September 1970. The latest edition of this Plan was distributed in October 1976 by Joint IOC/WMO Circular Letter.

International Oceanographic Data Exchange (IODE)

IOC activities in IODE are aimed at facilitating the exchange and handling of, and access and referral to, oceanographic data and information resulting from or relating to marine programmes.

The Working Committee on IODE was established to oversee the entire international oceanographic data exchange system and to standardize forms and formats for reporting and coding data. At present this IODE network consists of World Data Centres A and B (Oceanography), three Regional Data Centres (for the Caribbean, Mediterranean, and Western Pacific), CSK (WESTPAC), 24 National Oceanographic Data Centres, 7 Designated National Agencies, and 5 Responsible National Oceanographic Data Centres dealing with specific projects or data. In addition 41 countries have designated National IODE co-ordinators.

A number of subgroups have been established to develop formats and procedures for international data exchange in the fields of marine pollution, geology-geophysics, air-sea interaction, waves and physical oceanography, and also data obtained by remote sensing. Other expert groups are responsible for advising on the development of the Marine Environmental Data Information Referral System (MEDI), the Aquatic Sciences and Fisheries Information System (ASFIS), the Responsible National Oceanographic Data Centre scheme, and Marine Information Management.

Thanks to support of the IODE system by IOC Member States, the number of oceanographic data stations from which data is received by WDCs has quintupled since 1967. Also, a large volume of marine data related to marine geology and geophysics, current measurements, and marine biology have been received; all data are available to participating Member States. To date, 75 countries have provided input to the world oceanographic data banks.

Amongst the activities currently under implementation are the following:

Development of the Pilot Programme for RNODCs as new IODE system elements. RNODCs are designed to support specific international and regional projects.

A revised version of a general format for exchange of oceanographic and marine chemical data has been prepared and is expected to be approved in 1978.

The *ad hoc* group on Marine Pollution Data has accepted a format for the exchange, on magnetic tape, of data resulting from the IGOSS Marine Pollution (Petroleum) Monitoring Pilot Project.

With the assistance of the NODC of the Federal Republic of Germany, data forms for the reporting of marine geological data were printed and widely distributed for experimental use and comment.

Further actions have been undertaken by the IOC Secretariat, in co-operation with UNESCO, toward developing the Marine Environmental Data Information Referral System (MEDI) as a sectoral focal point for the International Referral System (IRS) of UNEP. The purpose of this system is to provide information on the availability, location, and characteristics of marine environmental data. It is operated by the IOC Secretariat in co-operation with FAO, WMO, IHO, IAEA, ICES, UNEP, and IMCO. UNEP has provided major financial support for the development of MEDI.

UNEP has also provided substantial support for further development of the FAO-IOC-UN Aquatic Sciences and Fisheries Information System (ASFIS) to broaden its scope and coverage. Here, progress has been considerably accelerated by the addition of the United Nations Ocean Economics and Technology Office (OETO) as an ASFIS partner. Through their MACTIS (Marine and Coastal Technology Information Service) programme a number of subject areas related to non-living resources and coastal technology will be given in-depth coverage within ASFIS. Recently, Mexico and Portugal have become new ASFIS centres and interest has been expressed by several other countries.

Acting on recommendations of the joint FAO-IOC Panel of Experts on ASFIS, the FAO and IOC Secretariats have initiated a number of projects. Some deal directly with thesaurus development and expansion of the disciplinary areas covered by the system's abstract journal, ASFA *(Aquatic Sciences and Fisheries Abstracts)*. Beginning in 1978, ASFA will be produced by, and available on, magnetic tape. Other projects are oriented towards making the system more widely known and in expanding information services to a larger segment of the marine science community.

A number of publications have been issued, including a popular brochure on ASFIS, the FAO-IOC *International Directory of Marine Scientists,* and the *World List of Aquatic Sciences and Fisheries Serial Titles*. Plans are well under way for the compilation of regional institutional directories and, for a later date, computerized registers of scheduled meetings and international research projects. During 1978, a pilot study will be carried out in the Caribbean to assess the information needs of the region and to formulate plans by which ASFIS can meet these needs to ensure expansion of the system into this and other regions when interest and funding are available.

In 1977, the fourth edition of the "Manual on International Oceanographic Data Exchange" was prepared and issued.

Tsunami Warning System in the Pacific (ITSU)

The IOC Assembly, at its tenth session, adopted the new Mandate and Functions of the International Tsunami Information Center (ITIC), in Honolulu. The Center is manned by a full-time Director seconded by the host government, and a full-time Associate Director seconded from one of the other Member States of the International Coordination Group for the Tsunami Warning System in the Pacific (ICG/ITSU), for a one- or two-year period. The present Associate Director comes from New Zealand.

The ICG/ITSU held its sixth session in Manila (February 20–25, 1978) to review the present status of the programme and to recommend further action. Plans were adopted for the expansion of the warning system including technical improvements in communication. Survey procedures to obtain accurate run-up data in the event of Tsunami inundation were also discussed. The group emphasized the need to increase public awareness of the dangers of Tsunamis through a broad educational programme.

Four scientists visited ITIC, each for a six-week period, for training and to establish co-operation between their agencies and the Center. The ITIC has become the focal point for advising Member Countries on the establishment of national warning systems within the framework of a regional automated network.

TRAINING, EDUCATION, AND MUTUAL ASSISTANCE IN THE MARINE SCIENCES (TEMA)

The IOC is the primary body within the United Nations system responsible for the evaluation of national needs and resources in Training, Education and Mutual Assistance in the marine sciences (TEMA). As such, it is responsible for developing, recommending, and co-ordinating programmes to satisfy those needs through the concerted efforts of its Member States and interested international organizations. In carrying out its functions, the Commission seeks to provide, selectively, small amounts of assistance, mainly to stimulate existing national mechanisms to develop or strengthen marine science and technology organizational structures, and to establish comprehensive national ocean policies.

IOC's mandate in the TEMA field is as an advisory catalytic body, rather than as an operational body, with IOC identifying the TEMA needs of the Member States and the ICSPRO agencies (UN, FAO, Unesco, WMO, IMCO) acting as operational arms for the Commission within their own fields of competence. It has been recognized that the IOC and the ICSPRO agencies undertake many joint activities in marine science, though seldom on TEMA matters, and that the mechanisms for interaction between them need to be improved to maximize international assistance to developing countries without duplicating effort.

During the biennium, there has been a clear trend towards regionalization. This is based on the concept that regionalization should evolve out of regional needs. It has also been suggested that greater use should be made of the facilities of the regional offices of United Nations agencies for information, advice, and assistance. The policy of the Commission is to encourage Member States to co-operate actively in regional research projects, training courses, and similar activities which, otherwise, would be beyond their individual capabilities.

An analysis has been made of the difficulties faced by the Commission in financing its TEMA activities. At the present time, the only funds available to the IOC for TEMA operational activities come from the IOC Trust Fund. It has been clearly recognized that if TEMA is to gain in depth and scope, more funding and additional sources of support are needed. As a result the Commission has established a Voluntary Assistance Programme (IOC-VAP) which is expected to become a most important source of funding and support for TEMA. The Assembly, at its tenth session (October–November 1977), approved rules for the utilization of the IOC-VAP.

IOC Voluntary Assistance Programme (IOC-VAP)

The IOC Voluntary Assistance Programme (IOC-VAP) is entirely dependent on contributions made by Member States in response to requests for assistance expressed to the IOC by other Member States. Contributions may be in the form of funds, equipment, or the offer of services. Funds are administered through the IOC Trust Fund; equipment and services are furnished under agreements arranged by the IOC. The IOC-VAP is similar to the WMO-VAP which has operated successfully for many years.

Under the IOC-VAP, assistance may take the form of: training of personnel; assistance in the development of marine science organizational structures, research laboratories, or educational facilities; information on technology for the exploitation of marine resources; and the advice of visiting experts in training, education, and technology related to the marine sciences.

Regional Meetings

In response to the resolution on Assessment of Training Needs adopted by the Assembly at its eighth session, regional *ad hoc* TEMA meetings for assessing the training needs at the regional level and for studying ways and means by which such needs could be satisfied were continued.

The fifth regional *ad hoc* TEMA meeting for South American countries was held in Montevideo, Uruguay, November 15–19, 1976. Member States in attendance from the region were Argentina, Brazil, Ecuador, France, and Uruguay. Member States in attendance from outside the region were the Federal Republic of Germany and the United States of America. FAO, OAS, WMO, and Unesco also sent representatives. Recommendations were adopted concerning Education and Training of Marine Scientists in the Region, Marine Science Administrators, Training of Technicians, Participation in Oceanographic Cruises, Oceanographic Instrumentation, and Scientific Programmes of the Commission.

In preparation for the sixth regional *ad hoc* TEMA meeting for the countries bordering the Northern Indian Ocean (to be held in Karachi, March 11–16, 1978) a consultant visited Bangladesh, India, Pakistan, Sri Lanka, and Thailand, and obtained information on the degree of interest of these countries in the marine sciences, the facilities available, future plans, and, in particular, their intentions regarding the meeting. All Member States in the region have been invited to complete a questionnaire outlining their activities and training needs.

The excellent co-ordination on TEMA activities between the IOC and the Division of Marine Sciences of Unesco has continued. Examples of this are the workshops being co-sponsored in early 1978: (1) workshop on the preparation of a syllabus for training marine technicians, Miami, Florida, U.S.A., May 22–26, 1978; (2) workshop on the preparation of a syllabus to introduce oceanography and the marine environment in secondary school curricula, Llantwit Major, South Wales, U.K., June 5–9, 1978.

TEMA Activities of IOCARIBE

The regional Secretariat of IOCARIBE is undertaking a series of TEMA activities for the countries of the region. Several missions have visited countries in Central and Northern South America to determine the availability of trained manpower, facilities, and equipment, and to identify their most pressing needs. This was done as an initial step in the development of appropriate programmes. One mission led to a feasibility study on the procedures for assessing the needs and capabilities of Caribbean countries with regard to marine science instrumentation, calibration, maintenance, and repair centres. A forthcoming mission will examine the capabilities of these countries for functioning within the framework of the ASFIS system.

Through IOCARIBE, the Commission, together with the Ocean Economics and Technology Office of the United Nations, the Division of Marine Sciences of Unesco, and the United Nations University, and with the support of UNEP, is organizing an orientation course on coastal area management, at present scheduled for early 1979, probably in Mexico.

Financial Support for Academic Studies, Etc.

Financial support was provided to scientists and technicians from Algeria, Argentina, Cameroon, Colombia, Hong Kong, Peru, and Thailand for academic studies, training, research visits, etc., in France, Germany (Federal Republic of), Mexico, and U.S.A., and to trainees from Fiji, India, Indonesia, Malaysia, Thailand, and Western Samoa, for shipboard training aboard oceanographic research vessels from Japan, United States of America, and the United Kingdom.

Marine Affairs and Management of Marine Resources
The IOC reached agreements with the University of Rhode Island and Oregon State University for awarding fellowships to students from developing countries wishing to participate in the Master of Marine Affairs and Master of Arts in Marine Affairs programmes of URI and the Marine Resources Management programme of OSU. In each case, the IOC will provide yearly stipends of US $4,000 to two fellows and the university will provide tuition fellowships. Candidates are now being selected for participation in these courses during the academic year 1978–1979 (two candidates for each university).

ANNEX I

PROGRAMME AND BUDGET OF THE BIENNIUM, 1976–1977

Breakdown of the Unesco Regular Programme funds, made available to the Commission:

		US $	
I.	Secretariat Services (Assembly, Executive Council meetings, advisory services to the Commission)	174,300	(22%)
II.	Under the Long-Term and Expanded Programme of Oceanic Exploration and Research (LEPOR):		
	(i) Ocean Science (Regional Co-operative Investigations, GIPME, IDOE projects GEBCO, GARP)	337,600	(43%)
	(ii) Ocean Services (IGOSS, Data Management, Tsunami Warning)	175,400	(22.5%)
	(iii) Training, Education, and Mutual Assistance in the marine sciences (TEMA)	87,000	(12.5%)
	TOTAL (Operational funds)	784,300	(100%)
III.	Staff costs approved under the Unesco Regular Programme	874,400	
		1,658,700	

Breakdown of the 1976–1977 total budget (in US $):

	In the form of funding	In kind; *e.g.*, for hosting meetings under the ICSPRO agreement	Salaries	Total
Unesco	784,300	. . .	874,400	1,658,700
United Nations	. . .	6,000	. . .	6,000
UNEP	465,300	. . .	. . .	465,300
FAO	. . .	8,000	106,000	114,000
WMO	. . .	. . .	43,000	43,000
IMCO	. . .	. . .	49,500	49,500
Member States *et al.* (Trust Fund)	316,000	*	500,000	816,000
TOTAL	1,565,600	14,000	1,572,900	3,152,500

NOTE—It should be noted that the above financial statement is based partly on assumptions and simplifications, and it is not an accurate representation of income or expenditure.

*No estimate attempted.

ANNEX II

[EDITORS' NOTE: Due to limitations of space, Annex II, which includes state member representatives on the Executive Council, has not been included here. The reader is referred to the original report (IOC/INF-376).]

ANNEX III

MEMBER STATES OF THE COMMISSION

At the time of the tenth session of the Assembly, 95 countries were Member States of the Commission:

+ Algeria
+ Argentina
+ Australia
  Austria
  Belgium
+ Brazil
  Bulgaria
  Cameroon
+ Canada
  Chile
  China
+ Colombia
  Congo
  Costa Rica
  Malaysia
  Malta
  Mauritania, Islamic Republic of
  Mauritius
+ Mexico
  Monaco
  Morocco
  Netherlands
  New Zealand
+ Nigeria
+ Norway
  Pakistan
  Panama
  Peru

- Cuba
- Denmark
- Dominican Republic
- Ecuador
- + Egypt, Arab Republic of
- Ethiopia
- Fiji
- Finland
- + France
- Gabon
- German Democratic Republic
- + Germany, Federal Republic of
- Ghana
- Greece
- Guatemala
- Guyana
- Haiti
- Iceland
- + India
- Indonesia
- Iran
- + Iraq
- Israel
- Italy
- Ivory Coast
- Jamaica
- + Japan
- Jordan
- + Kenya
- Korea, Republic of
- Kuwait
- Lebanon
- Libyan Arab Jamahiriya
- Madagascar
- + Philippines
- + Poland
- Portugal
- Qatar
- Romania
- + Senegal
- Sierra Leone
- Singapore
- Somalia
- South Africa, Republic of
- Spain
- Sri Lanka
- Sudan
- Surinam
- Sweden
- Switzerland
- Syrian Arab Republic
- Tanzania, United Republic of
- Thailand
- Togo
- Tonga
- Trinidad and Tobago
- + Tunisia
- Turkey
- Ukrainian SSR
- + Union of Soviet Socialist Republics
- United Arab Emirates
- + United Kingdom
- + United States of America
- Uruguay
- + Venezuela
- Viet Nam, Socialist Republic of
- Yugoslavia

+ Member States represented on the Executive Council.

ANNEX IV

LIST OF PUBLICATIONS*

1. *Intergovernmental Oceanographic Commission* (IOC)
   a) IOC Technical Series
   No. 3 Radio Communication Requirements for Oceanography (English only)
   No. 6 Perspectives in oceanography, 1968

*These publications are available in English, French, Spanish, and Russian, except where specified.

No. 7 Comprehensive outline of the scope of the long-term and expanded programme of oceanic exploration and research (English, French, and Spanish)
No. 10 Brunn memorial lectures (presented at the seventh session of the IOC, 1971)
No. 11 Brunn memorial lectures, 1973 (presented at the eighth session of the IOC Assembly)
No. 12 Oceanographic products and methods of analysis and prediction (English only)
No. 13 The International Decade of Ocean Exploration (IDOE) 1971–1980
No. 14 A Comprehensive Plan for the Global Investigation of Pollution in the Marine Environment and Baseline Study Guidelines
No. 15 Brunn memorial lectures, 1975 (presented at the ninth session of the IOC Assembly)
No. 16 Integrated Global Ocean Station System (IGOSS) General Plan and Implementation Programme 1977–1982
No. 17 Oceanographic Components of the Global Atmospheric Research Programme (GARP)
No. 18 Global Marine Pollution: An Overview
n/n Annotated Bibliography of Textbooks and Reference Materials in Marine Sciences (quadrilingual)

b) IOC Manuals and Guides
No. 1 Manual on IGOSS Data Archiving and Exchange
No. 2 International Catalogue of Ocean Data Stations (quadrilingual)
No. 2 Amend. 1 International Catalogue of Ocean Data Stations–Amendment No. 1 (quadrilingual)
No. 3 Guide to Operational Procedures for the Collection and Exchange of Oceanographic Data (BATHY and TESAC) (quadrilingual) (under revision)
No. 4 Guide to Oceanographic and Marine Meteorological Instruments and Observing Practices (English only)
No. 5 Guide to establishing a National Oceanographic Data Centre (quadrilingual)
No. 6 Wave Reporting Procedures for Tide Observers in the Tsunami Warning System (English and Spanish only)
No. 7 Guide to Operational Procedures for the IGOSS Pilot Project on Marine Pollution (Petroleum) Monitoring
No. 8 Marine Environmental Data Information Referral Catalogue (MEDI Pilot Catalogue) (English only)
No. 9 Manual on International Oceanographic Data Exchange (fourth edition)
No. 10 Operational Guide for the Marine Environmental Data Information Referral System (MEDI) (in preparation)
Catalogue on Tsunami Marigrams for Historic Tsunamis recorded at selected Tide Gauge Stations throughout the Pacific (in preparation)
Manual on Methods for Sample Collection, Preservation and Analysis (in preparation)

c) IOC Workshop Series
The Scientific Workshops of the Intergovernmental Oceanographic Commission are usually jointly sponsored with the other intergovernmental or non-governmental bodies. In each case, by mutual agreement, one of the sponsoring bodies assumes responsibility for the publication of the final report.

No.	Title	Publishing Body	Languages
1	Metallogenesis Hydrocarbons and Tectonic Patterns in Eastern Asia (Report of an IDOE Workshop); Bangkok, Thailand, September 24–29, 1973). CCOP-IOC, 1974.	Office of the Project Manager UNDP/CCOP, c/o ESCAP Sala Santitham Bangkok 2, Thailand	English
2	Ichthyoplankton, Report of the CICAR Ichthyoplankton Workshop, Mexico City, July 16–27, 1974 (Unesco Technical Paper in Marine Science, No. 20).	Division of Marine Sciences, Unesco Place de Fontenoy 75700 Paris, France	English Spanish
3	Report of the IOC/GFCM/ICSEM International Workshop on the Marine Pollution in the Mediterranean, Monte Carlo, September 9–14, 1974.	IOC, Unesco Place de Fontenoy 75700 Paris, France	English French Spanish
4	Workshop on the Phenomenon known as "El Niño," Guayaquil, Ecuador, December 4–12, 1974.	FAO Via dell Terme di Caracalla 00100 Rome, Italy	English Spanish
5	IDOE International Workshop on Marine Geology and Geophysics of the Caribbean Region and its Resources, Kingston, Jamaica, February 17–22, 1975.	IOC, Unesco Place de Fontenoy 75700 Paris, France	English Spanish
6	Report of the CCOP/SOPAC-IOC IDOE International Workshop on Geology, Mineral Resources, and Geophysics of the South Pacific, Suva, Fiji, September 1–6, 1975.	IOC, Unesco Place de Fontenoy 75700 Paris, France	English
7	Report of the Scientific Workshop to initiate planning for a co-operative investigation in the North and Central Western Indian Ocean, organized within the IDOE under the sponsorship of IOC/FAO/(OFC)/Unesco/EAC, Nairobi, Kenya, March 25–April 2, 1976.	IOC, Unesco Place de Fontenoy 75700 Paris, France	Full text (English only) Extract and Recommendations: French, Spanish, Russian

No.	Title	Publishing Body	Languages
8	Report of the IOC/FAO (IPFC)/UNEP International Workshop on Marine Pollution in East Asian Waters, Penang, April 7–13, 1976.	IOC, Unesco Place de Fontenoy 75700 Paris, France	English
9	Report of the Second International Workshop on Marine Geoscience, IOC/CMG/SCOR, Mauritius, August 9–13, 1976.	IOC, Unesco Place de Fontenoy 75700 Paris, France	English French Spanish Russian
10	Report of the Second IOC/WMO Workshop on Marine Pollution (Petroleum), Monitoring, Monaco, June 14–18, 1976.	IOC, Unesco Place de Fontenoy 75700 Paris, France	English French Spanish Russian
11	Report of the IOC/FAO/UNEP International Workshop on Marine Pollution in the Caribbean and Adjacent Regions, Port of Spain, Trinidad and Tobago, December 13–17, 1976.	IOC, Unesco Place de Fontenoy 75700 Paris, France	English Spanish
11 Suppl.	Collected contributions of invited lecturers and authors to the IOC/FAO/UNEP International Workshop on Marine Pollution in the Caribbean and Adjacent Regions, Port of Spain, Trinidad and Tobago, December 13–17, 1976.	IOC, Unesco Place de Fontenoy 75700 Paris, France	English Spanish

d) Brochures

Intergovernmental Oceanographic Commission (English, French, Spanish, Russian, Arabic, Chinese)

IDOE 1971–1980 International Decade of Ocean Exploration (English, French, Spanish, Russian)

Tsunami warning system in the Pacific (English, French, Spanish, Russian)

Guide to international marine environmental data services (English, French, Spanish, Russian)

General bathymetric chart of the oceans (English, French, Spanish, Russian)

Aquatic Sciences and Fisheries Information System (English, French, Spanish, Russian, Chinese)

Pollution in the marine environment (English only) (French and Spanish in press)

Publications of the IOC and the Division of Marine Sciences of Unesco (in preparation) (English only)

e) Ad hoc publications

International Indian Ocean Expedition (IIOE)

International Indian Ocean Expedition, Collected Reprints, Volumes VII and VIII, and Index (Volumes I to VI out of print)

International Indian Ocean Expedition Geological/Geophysical Atlas, Academy of Sciences of the USSR, 1975, Composite English/Russian

Phytoplankton Production Atlas of the International Indian Ocean Expedition, 1976, English only

Indian Ocean: Collected data on primary production, phytoplankton pigments, and some related factors (this volume includes data obtained during the International Indian Ocean Expedition co-ordinated by the IOC [1959–1965]). Compiled by B. Babenerd and J. Krey. Universitätsdruckerei Kiel, Federal Republic of Germany, 1974, English only

International Co-operative Investigations in the Tropical Atlantic (ICITA)

ICITA Atlas, Volume 1, Physical Oceanography, 1973 (quadrilingual)

ICITA Atlas, Volume 2, Chemical and Biological Oceanography, 1976 (quadrilingual)

Co-operative Study of the Kuroshio and Adjacent Regions (CSK)

CSK Atlases (7 volumes issued up to 1977), Japan Oceanographic Center, Tokyo, Japan

Co-operative Investigations of the Caribbean and Adjacent Regions (CICAR)

Progress in marine research in the Caribbean and adjacent regions CICAR-II Symposium. Abstract Volume and Annex, 1976 (composite English/Spanish)

Progress in marine research in the Caribbean and adjacent regions CICAR-II Symposium. Proceedings: Volume I—Papers on physical and chemical oceanography, marine geology and geophysics (in press); Volume II—Papers on fisheries, aquaculture, and marine biology, FAO Fisheries Report No. 200 (composite English/Spanish)

El Niño

Workshop on the Phenomenon known as "El Niño," Guayaquil, Ecuador, December 4–12, 1974, Proceedings, Spanish language papers presented, FAO Fisheries Report No. 185, 1976 (Spanish only)

Workshop on the Phenomenon known as "El Niño," Guayaquil, Ecuador, December 4–12, 1974, Proceedings, English language papers presented, Unesco (in press) (English only)

Miscellaneous

Safety Provisions of Ocean Data Acquisition Systems, Aids, and Devices (ODAS), Unesco/IMCO, published by IMCO, London, 1972

The Health of the Oceans, 1976, English (French and Spanish in press)

Marine Affairs: Register of Courses and Training Programmes (UN No. ST/ESA/54), 1976 (English only)

Oceanic Water Balance (WMO—No. 442), 1976 (English only)

2. *Unesco—Division of Marine Sciences (OCE)*

Monographs on oceanographic methodology

No. 2 Zooplankton sampling. 1968 (English only)

No. 3 A guide to the measurement of marine primary productivity under some special conditions. (SCOR WG 24), 1973 (English only)

No. 4 Zooplankton fixation and preservation. (SCOR WG 23), 1976 (English only)
No. 5 Coral reefs: research methods. (SCOR WG 35) (English only)
No. 6 Phytoplankton manual. (SCOR WG 33), 1978 (English only)

Unesco technical papers in marine science

11. An intercomparison of some current meters. (SCOR WG 21), 1969 (English only)
16. Sixth report of the joint panel on oceanographic tables and standards (SCOR WG 10), 1974 (English only)
17. An intercomparison of some current meters, II. (SCOR WG 21), 1974 (English only)
18. A review of methods used for quantitative phytoplankton studies (SCOR WG 33), 1974 (English only)
19. Marine science teaching at the university level. 1974 (English, French, Spanish, Arabic)
20. Report of the CICAR ichthyoplankton workshop. 1975 (English and Spanish) (IOC Workshop Report No. 2)
21. An intercomparison of open sea tidal pressure sensors. (SCOR WG 27), 1975
22. European subregional co-operation in oceanography. 1975 (English only)
23. An intercomparison of some current meters III. (SCOR WG 21), 1975 (English only)
24. Seventh report of the joint panel on oceanographic tables and standards. (SCOR WG 10), 1976 (English only)
25. Marine science programme for the Red Sea. 1976 (English only)
26. Marine sciences in the Gulf area. 1976 (English only; Arabic version in preparation)
27. Collected reports of the joint panel on oceanographic tables and standards, 1964–1969. Technical papers 1, 4, 8, and 14. (SCOR WG 10), 1976 (English only)
28. Eighth report of the joint panel on oceanographic tables and standards (SCOR WG 10), 1978 (English only)

Unesco reports in marine science

1. Marine ecosystem modelling in the eastern Mediterranean. 1977 (English only)
2. Marine ecosystem modelling in the Mediterranean. 1977 (English only)

IMS (International Marine Science) Newsletter

Seventeen issues had been published by the end of 1977 (English only)

Miscellaneous

Check list of the fishes of the north-eastern Atlantic and of the Mediterranean, 2 volumes, 1973 (composite English/French)

International oceanographic tables, Volume 1, second edition, revised, 1972; Volume 2, 1973 (composite English/French/Russian/Spanish)

Tsunami research symposium 1974. Published jointly by the Royal Society of New Zealand and Unesco, 1976 (English only)

Proceedings of the Symposium on warm water zooplankton. Published by the National Institute of Oceanography of India, on behalf of Unesco, 1978 (English only)

Proceedings of the second Thai national seminar on mangrove ecology. National Research Council of Thailand, 1977 (English only)

Proceedings of the Unesco seminar on benthic ecology and sedimentation of the continental shelf of the south-west Atlantic, ROSTLA (in preparation English and Spanish)

IOI: Pacem in Maribus IX[1]

Pacem in Maribus IX took place at the International Relations Institute of Cameroon (IRIC) in Yaounde, Cameroon, from January 18 to 22, 1979. IRIC was the place of the first African Symposium that formulated the concept of the Economic Zone, and Africans stressed the symbolic importance of this precedent which gave a particular relevance to Pacem in Maribus's reexamination of the new Law of the Sea and its real impact on African development.

There were 84 participants from 31 countries. Twenty-six came from Cameroon (University, IRIC, various government departments), 17 from other African countries; 41, including members of the IOI governing bodies, came from other continents.

The conference was opened by the Minister of Education of the government of Cameroon. It was closed by the Minister of Foreign Affairs.

Working sessions were organized according to regions. For each region, different uses of ocean space and resources were discussed in relation to overall development strategy, and the impact of the new Law of the Sea was examined.

The unanimous conclusion was that, whether or not a global treaty will emerge from the Third United Nations Conference on the Law of the Sea, African development will depend above all on regional cooperation and organization, on which a number of detailed recommendations were made. They are summarized in the following conclusions of the conference.

CONCLUSIONS

1. LAW OF THE SEA CONFERENCE

In the light of the relevant decisions of the U.N. General Assembly, especially the Declaration of Principles Governing the Seabed and Its Resources and the Charter of the Economic Rights and Duties of States, it was the expectation of the international community that the Third U.N. Conference on the Law of the Sea would be a universal effort to elaborate a new code of conduct and set up institutional arrangements to improve the conditions of life of the peoples on this planet by the rational use and equitable distribution of the wealth of the oceans.

Unfortunately, the prevailing trends at the Conference clearly indicate that these expectations are being increasingly frustrated. The Participants in Pacem in Maribus IX took the view that the present trend of negotiations on the so-called parallel system holds no promise of real benefits for African peoples in particular and developing countries in general.

[1]Yaounde, Cameroon, January 18–22, 1979.

With regard to the Exclusive Economic Zone, the desired objective, namely, the optimum utilization of the resources therein, for the benefits of all States, including the landlocked and geographically disadvantaged, in keeping with the spirit of the OAU Declaration on Issues Relating to the Law of the Sea, is similarly being frustrated.

2. THE AFRICAN REGION

However, the Participants are still convinced of the imperatives of a universally agreed law of the sea. They are equally aware of the need to set up effective regional machinery appropriate to African conditions in the light of internationally recognized principles. In this context, the Participants considered various issues and reached the following conclusions:

A. Africa as a Whole

a) The Development of Regional Mechanisms

i) The need to speed up the decentralization of African marine regional/subregional organizations established within the framework of the United Nations system and to relocate their secretariats in the areas concerned was recognized in order to improve their service to African States.

ii) The establishment of African regional/subregional marine science organizations within the framework of the Intergovernmental Oceanographic Commission was recommended as an appropriate mechanism to facilitate cooperation among the States concerned and to provide the scientific basis required for development, management, and protection of the marine environment and its resources. For reasons of economy and efficiency it was felt that close working relationships should be established between these organizations and those having responsibilities for development, management, and related functions, operating in the same region.

iii) Noting that in certain areas there is an undesirable duplication of efforts between marine regional/subregional organizations (e.g., in the Mediterranean) is was felt that steps should be taken by member States to improve and rationalize present arrangements.

iv) The importance of regional or subregional cooperation in the fields of surveillance and policing of marine areas within the jurisdiction of states was stressed.

b) Landlocked and Geographically Disadvantaged States

Regardless of the final outcome of the Law of the Sea Conference, regional arrangements must be made to facilitate the participation of landlocked and geographically disadvantaged States in the use of the oceans since marine resources and ocean management will play an increasingly important role in development and the exclusion of some States would widen the development gap between rich and poor States. With regard to *access,* some participants suggested in particular the establishment of free zones and harbours for landlocked States. With regard to the *living resources* it was suggested they be given equal rights with the coastal states in accordance with the OAU Declaration; in particular, it was suggested they should be given precedence, over distant-water fishing States. Landlocked States, it was suggested, would have a far better chance negotiating within the family of African countries than at the Conference on the Law of the Sea. With regard to the geographically disadvantaged States, Participants called for a better definition of the term and for a better definition of the concept of "regions" within which the rights of GDS could be implemented.

c) *Fisheries*

i) Considering the nature of marine resources, their behaviour and distribution, it is imperative that their exploration, conservation and exploitation be dealt with through regional arrangements;

ii) Bilateral or multilateral arrangements should complement regional/subregional arrangements;

iii) The regional or subregional units so established should adopt effective regulations and mechanism for their enforcement of agreed standards;

iv) Participants called for:

- the training of an adequate number of experts over the next few years;
- the protection of artisanal fisheries by reserving near-shore areas to them, to enhance both nutrition and employment;
- with regard to commercial fisheries, due regard should be given to the need of improving the nutritional conditions of African peoples;
- domestic fishing capacity should be strengthened, as dependence on foreign companies, whether through licensing or joint ventures, had not proved beneficial to African peoples.

d) *Integration of Ocean Management in Development Strategy*

Participants suggested:

i) functional regional zoning for

- recreational uses;
- transportation requiring shipping lanes;
- fishing, with provisions for spawning areas in the form of protected areas or marine parks;
- industrial or production uses;

ii) the protection of the region and its resources from pollution and abusive exploitation;

iii) the establishment of Departments of marine affairs in each country, for research, management, and regional planning;

iv) a strategy to stimulate economic, cultural, and social activities likely to increase use of the marine environment and greater integration of marine resources in the economy.

e) *Technology Transfer*

Participants attempted to define the legal and economic implications of the principle, proposed by the 1978 Arusha Symposium on UNCSTED, that technological knowledge must be considered a Common Heritage of Mankind. In this context they suggested:

i) States should endeavour to adopt laws that remove monopolies and barriers to the flow of technological knowledge;
ii) Efforts should be made to ensure that terms of transfer of technology are on equitable terms and removing burdens on recipients. Restrictions related to the production and marketing of the commodities would be removed;
iii) The foregoing terms should be packaged in regional and global treaties;
iv) Regional and global treaties are needed for the registration of various forms of technological knowledge, especially patents and copyrights, to inform parties of the stage in the development of various forms of technological knowledge and where it is obtainable ("technology banks," or "patent pools"). This would prevent companies from expropriating the ideas and devise blocking patents;
v) Programmes should be instituted to train personnel in developing countries to ensure indigenous capabilities to adopt and diffuse imported technology as well as to refine indigenous technology and adapt it to contemporary conditions;
vi) Developing countries should be alerted not to pay for patents which have expired and are no longer valid and whose life-span may have been extended by companies by spurious methods: in this connexion the work of WIPO should be taken into consideration;
vii) Collective measures should be taken against the sale of so-called "black box technology" barring recipient countries from an understanding of the plant or equipment they are acquiring. Full disclosure and full training in the techniques should be an overriding requirement by any Government inviting a foreign company;
viii) With regard to marine sciences and technology, Participants suggested the institution of *regional survey* ships, under the authority of the regional mechanisms for marine science and management. These might also become floating centres of excellence for the training of local experts.

f) *Marine Resource Research Centres*

i) Since Marine Scientific Research is the basis of any rational management of marine resources, it is imperative that countries of Africa establish and develop local capabilities for these activities;
ii) Every effort should be made at national levels to establish basic capability as nodes for a network of cooperation in marine scientific research;
iii) In view of the costly nature of facilities for marine scientific research, countries are urged to establish regional centres for research in marine affairs;
iv) For this purpose there should be one major centre to cater for each major subregion by enhancing the working capabilities of the national centres and meeting development data needs;
v) The regional research centres should be the central technical clearing-house for information on marine resource research. They should, accordingly, function as the technical advisory centres in the field;
vi) Foreign researchers interested in research should work within the framework of the programmes of the regional marine resource research centres;

vii) Foreign researchers should be under the obligation to undertake training programmes for the coastal States;

viii) Consent for foreigners to conduct research should be obtained before such research can be commenced.

g) *Data and Statistical Bureaux*

i) Every State should establish its own bureau for data and statistics on marine affairs to aid development planning;

ii) The bureaux should establish a network among themselves to ease the transfer of information with the requisite storage and retrieval mechanisms;

iii) Data on research done within a region by foreign researchers should be stored at the bureaux, as property of the bureaux's data bank.

h) *Jurisdiction over Ocean Space*

i) Uniform jurisdictional claims over ocean space would be ideal for purposes of resource management;

ii) Accordingly, African States are urged to adopt legislation for an agreed coastal jurisdiction;

iii) A limit of 200 miles for exclusive economic zones should be established at once.

i) *Training in the Multidisciplinary Skills Associated with Deep Seabed Mining*

The Conference noted with interest announced initiatives for training of personnel from developing countries to enhance their participation in the management of the resources which constitute the Common Heritage of Mankind.

i) The Chairman of the First Committee at the Third United Nations Conference on the Law of the Sea, Mr. Paul Bamela Engo of Cameroon, indicated his intention to announce at the Eighth Session of the Conference certain initiatives he has undertaken during the past two years in this field. He made it clear that this was not intended to be part of the negotiations but merely to ensure that developing countries would be in a position to participate effectively in all activities of management of seabed resources by the Authority and its organs, as soon as these commence in any part of the international area. He believed that it would be necessary for preliminary United Nations Law of the Sea Secretariat studies to be made regarding the scope of the needs in this regard.

ii) The Chairman of the Planning Council of the International Ocean Institute announced that that Institute is launching a concrete programme for the training of Third-World personnel in seabed mining technology, economics, law, administration, and contract negotiating strategy. The programme consists of a three-months interdisciplinary foundation course followed by a two-year specialized training programme. The first course is to take place starting in October, 1979. Pacem in Maribus IX called for cooperation and participated in this new and timely initiative.

B. Africa: Subregional Arrangements

a) West Africa and the Atlantic Ocean

i) Participants urged the speedy adoption of the Draft Action Plan for the Protection and Development of the Marine Environment and Coastal Areas of the West African Region and the signing and ratification of a Convention to enact the Plan.

ii) Stressing the importance of shipping for development, Participants urged action to decrease the dependence of the region upon foreign shipping services. The technical cooperation activities of IMCO, together with those of the OAU, were noted as important steps in this direction. It was urged that African States should take measures to sign and ratify the Convention on the Code of Conduct for Liner Conferences as an important measure to enhance the participation of African countries in international shipping.

b) East Africa and the Indian Ocean

i) It was stressed that the Declaration on the Indian Ocean as a Zone of Peace, adopted by the Non-Aligned Summit Conference in Lusaka in September 1970 and introduced by Ceylon and Tanzania at the 26th session of the United Nations General Assembly (1971) is of supreme importance to East African countries. Recent developments should neither discourage nor deter us from pursuing its objectives.

c) North Africa and the Mediterranean

i) In the Mediterranean area, because of its physical characteristics, historical interconnections, and the need to develop greater equality among its parts, special regional discussions are required to reduce delimitation disputes, with preference given to joint management and cooperative solutions.

ii) As pollution control is an area of clearly common interest to the coastal States, fully operational centres should be set up to enforce the standards established by the Barcelona and other Conventions and the ideals of the Blue Plan.

iii) There should be a strengthening and reorganization of existing regional institutions in the Mediterranean for fisheries and scientific research. These should be complemented by other regional institutions to manage still unregulated development of other resources (seabed, unconventional energy sources) in the collective interest of the area. The rationalization and co-ordination of the many bodies and programmes operating in the area is an urgent necessity.

iv) Because of the multiple and possibly conflicting uses of the sea and the great environmental impact of human engineering works which are becoming increasingly practicable, a forum in which such issues can be discussed with an eye to their collective management in the interest of the area as a whole, should be established.

v) The Mediterranean countries, which all (except Albania and Libya) have association agreements with the E.E.C. should attempt to negotiate them in future not on a bilateral basis but collectively, especially with regard to fisheries and transport facilities, in order to enhance inter-Mediterranean exchanges as well as North-South relations.

vi) The difficult ideal of the Mediterranean as a "lake of peace" should be fostered by cooperative measures between the littoral countries aimed at reducing dependence on the super-powers.

[EDITORS' NOTE: The International Ocean Institute and the Government of Austria will sponsor Pacem in Maribus X in Vienna, Austria, on October 27–30, 1980. The theme of the conference will be "The New Law of the Sea: Implementation and Consequences—Monitoring of Ocean Systems and Surveillance of Uses."]

SCOR Report[1]

The Thirteenth General Meeting of SCOR [Scientific Committee on Oceanic Research] was held in Edinburgh, Scotland, on the occasion of the Joint Oceanographic Assembly (JOA) in September 1976. The Twentieth Executive Meeting was held in Victoria, B.C., Canada, in May 1977 and the Twenty-first Executive Meeting in Sao Paulo, Brazil, in January 1978. The JOA is a major event in the oceanographical calendar and provides an opportunity to review progress over a wide range of marine science disciplines.[2]

The principal interdisciplinary sessions at the 1976 JOA concerned the History of the Oceans, Ocean Circulation and Marine Life, Natural Variation in the Marine Environment, Man and the Sea, and New Approaches in Oceanography. Special Symposia covered a wide range of topics and others were arranged by the International Associations which are affiliated with SCOR.

SCOR's relationships with the ICSU Unions[3] take place mainly through the International Associations—IAPSO, IABO, CMG, and IAMAP—but initial contact has recently been made with IUPAC to explore common areas of interest.

Closer contact is also being established now with SCOPE, and a new SCOR Working Group on the Carbon Cycle of the Ocean will, it is hoped, give the ocean input to the SCOPE project on the Global Carbon Cycle.

SCOR operates principally through working groups set up for specific tasks: the exchange of materials across the ocean-atmosphere interface is the topic of a new one which, as with the group on river inputs to the ocean, will tackle the subject on an interdisciplinary basis. There is also increasing attention being paid to Coastal and Estuarine Regimes, including Mangrove Ecosystems.

Activities related to air-sea interaction have, however, been dominant, and the GARP Atlantic Tropical Experiment and planning for the First GARP Global Experiment have attracted much attention. The second GARP objective is seen to be one of major interdisciplinary interest in marine science, and a special committee on Climatic Changes and the Ocean has been formed to activate this.

The Joint Physical-Biological activity on Coastal Upwelling has now been completed, and a comprehensive review of the physical oceanography of coastal upwelling will be prepared. It is likely to take longer, however, to develop all the new ideas on the biological aspects and to present them in a digested form. Equatorial upwelling continues to be studied.

Special attention is being paid to "El Niño," the catastrophic natural phenomenon which decimates the fisheries of the Peru coast and the economy of that country. Predictive techniques are being explored.

1. By R. I. Currie, Secretary, SCOR, May 1978. Report submitted to the International Council of Scientific Unions.

2. Editors' Note: See "Joint Oceanographic Assembly," *Ocean Yearbook 1,* ed. Elisabeth Mann Borgese and Norton Ginsburg (Chicago: University of Chicago Press, 1978), pp. 697–701.

3. Editors' Note: Constituent societies of the International Council of Scientific Unions.

In biology, the major interest has been concerned with the Antarctic Living Resources and, in particular, with Krill. This is a joint SCAR/SCOR activity, and following a meeting in Woods Hole in August 1976 a plan is now being proposed for a concerted effort to stimulate research on the key problems of the Southern Ocean resources.[4] Their exploitation is imminent, and yet there is much we need to know before this can be conducted in a rational manner. The programme, entitled BIOMASS (Biological Investigation of Marine Antarctic Systems and Stocks), has been published in a booklet by SCAR and SCOR.

Other new biological working groups are concerned with sampling problems and mathematical modeling and the study of mangrove ecosystems.

In the geological field interest is centering at present on marine geochronological methods, the Cenozoic history of the ocean basins, and depth indicators in marine sediments. Symposia are also being organized on the Evolution of the South Atlantic and on Oceanic Crust and Seawater Interaction. A new working group has also been formed on Sedimentation Processes at Continental Margins.

Methodological problems remain a concern of SCOR, and a number of working groups are studying these.

SCOR continues to act in an advisory capacity to the Intergovernmental Oceanographic Commission and Unesco.

4. Editors' Note: See "SCAR/SCOR Group on the Living Resources of the Southern Ocean (SCOR Working Group 54)," *Ocean Yearbook 1,* pp. 762–72.

WHO Coastal Water Quality Programme and Its Relation to Other International Efforts on Marine Pollution Control*

Richard Helmer
World Health Organization

I. BACKGROUND

During the past few years, WHO has considerably intensified its activities concerning pollution of coastal waters and the potential health hazards associated with bathing, shellfish and contamination of other marine food products. A number of projects have been started in collaboration with other international organizations as well as with national institutions in countries directly concerned with these problems.

Most of WHO's current work in this field is being pursued as part of three major international activities:

(i) the Mediterranean Action Plan
(ii) Comprehensive Action Plan for the Protection of Regional Seas
(iii) Joint Group of Experts on the Scientific Aspects of Marine Pollution (GESAMP).

These activities are carried out jointly by the United Nations Environment Programme (UNEP), the Specialized Agencies, the IAEA and other bodies of the United Nations system. UNEP has been responsible for the initiation of most of the projects on the Mediterranean and the other Regional Seas. In addition to the three Inter-Agency programmes, WHO is also developing health criteria for coastal bathing waters.

The purpose of this paper is to describe WHO's activities in the field of coastal water pollution assessment and control with particular emphasis on the progress made during the past two years. The progress in WHO activities is reported according to the three programmes mentioned above. WHO's participation in these programmes is a joint effort of the headquarters programmes (Division of Environmental Health, Legal Division and the Health Legislation Unit) and the WHO Regional Offices.

II. MEDITERRANEAN ACTION PLAN

One of the major seas being endangered by increasing pollution loads is the Mediterranean.[1] Recognizing the need for early action, the Mediterranean was chosen by UNEP as a "concentration area" for which a comprehensive programme was to be developed[2] and

*Paper presented at the WHO training course on [illegible] pollution control, held under the auspices of DANIDA, Denmark, August 1978. Updated by the author for publication in the *Ocean Yearbook* to include information through 1978. Helmer is currently Deputy Director, Regional Seas Programme Activity Centre, UNEP.

implemented. Protection of human and marine resources without hampering economic progress is the prime objective of the Mediterranean Action Plan, which consists of three major components: (1) legal (framework convention and related protocols), (2) scientific (research and monitoring), and (3) integrated planning. The Mediterranean Action Plan was developed and is being implemented through close collaboration of Member States bordering on the Mediterranean and several international organizations. To this end, a series of intergovernmental, scientific and other meetings are being held on the various aspects of the programme. The general manner in which the section plans on the regional seas are being carried out is illustrated in Annex I. The basic scheme for the implementation of the Mediterranean Action Plan is outlined in Annex II. It shows the role of the various UN bodies in this programme, and illustrates the overall approach used.

Within the Mediterranean Action Plan, WHO participates in:

(i) Preparation of a Draft Protocol for the Protection of the Mediterranean Sea against Pollution from Land-based Sources;
(ii) WHO/UNEP project on Coastal Water Quality Control in the Mediterranean (MED VII);
(iii) WHO/ECE/UNIDO/FAO/UNESCO/IAEA/UNEP Project on Pollutants from Land-Based Sources in the Mediterranean (MED X);
(iv) Integrated planning in the Mediterranean region (Blue Plan).

1. Draft Protocol for the Protection of the Mediterranean Sea against Pollution from Land-based Sources

The development of international treaties is one of the primary objectives under the Mediterranean Action Plan. The Convention for the Protection of the Mediterranean Sea against Pollution, which has already been signed by most Governments, requests Member States to "take all appropriate measures to prevent, abate and combat pollution in the Mediterranean Sea area caused by discharges from rivers, coastal establishments or outfalls, or emanating from any other land-based source within their territories." In this connection, the Conference of Plenipotentiaries[3] also adopted a resolution calling the Executive Director of UNEP "to continue the preparatory work for a Draft Protocol for the Protection of the Mediterranean Sea against Pollution from Land-based Sources."

During 1976, WHO prepared background documents concerning:

(i) Principles suggested for inclusion in the Draft Protocol for the Protection of the Mediterranean Sea against Pollution from Land-based Sources;
(ii) Technical Annexes to the Draft Protocol;
(iii) Protection of the Mediterranean Sea against Pollution from Land-based Sources: a survey of national legislation;
(iv) Compendium of Principal International Instruments relevant to the Draft Protocol.

In February 1977, UNEP, in cooperation with WHO, organized an Inter-Governmental meeting in Athens[4] to review the documentation and to discuss future plans for the preparation of the protocol. In this meeting the basic principles were revised and it was decided to convene a separate group of Government experts to deal with the technical annexes.

Based upon comments received from the countries and the UN Agencies and taking into account discussions during the WHO/UNEP Workshop on Coastal Water Pollution Control, the technical annexes to the Draft Protocol were revised and resubmitted to the Governments concerned.

Further discussions followed at a meeting of technical experts in September 1977 in Geneva[5] and a Second Inter-Governmental Consultation in Venice[6] in October 1977. Work on the Protocol continued in 1978 with the preparation of principles and guidelines on waste disposal into the marine environment.

2. WHO/UNEP Project on Coastal Water Quality Control in the Mediterranean (MED VII)

The objective of this project is to assess the degree of coastal pollution in the Mediterranean. This information will be used as a basis for the implementation of national programmes for the control of coastal pollution from land-based sources. Under the project the participating countries are to undertake sanitary and health surveillance of recreational waters, and shellfish and shellfish growing areas in selected coastal waters.

In addition, a number of epidemiological studies are to be carried out and guidelines for coastal water quality management are to be prepared. Where necessary, the participating countries will be supported by the provision of laboratory equipment and training opportunities to the national staff.

The framework for this project was prepared by a joint WHO/UNEP Expert Consultation[7] which was held in Geneva from 15 to 19 December 1975. The consultation was attended by 35 participants from 15 countries. The operational document which was prepared sets forth the administrative structure for the programme, monitoring methodology and a workplan and timetable. The WHO Regional Office for Europe is responsible for project implementation.

In 1976, the Governments were invited to designate laboratories for participation in the project. As of 30 June 1977 nine countries had responded by designating a total of 27 laboratories. Agreements are being developed with each laboratory, specifying their tasks as well as identifying their equipment and training requirements.

The Istituto Superiore di Sanita in Rome, Italy, was designated as the Regional Activity Centre. Its primary functions are to assist in the development of uniform methodology, such as data reporting procedures, and to provide on-the-job training.

In order that the monitoring activities are carried out in a uniform fashion, "Guidelines for Health-related Monitoring of Coastal Water Quality" was prepared.[8] The manual was finalized by a meeting of experts which was convened from 23 to 25 February 1977 at the Centre for Marine Research, Rudjer Boškovic Institute, in Rovinj, Yugoslavia. The English edition is available; the French edition is in preparation.

The existing criteria and standards for coastal water quality as well as procedures for carrying out epidemiological studies related to coastal water pollution were reviewed by a group of experts from 1 to 4 March 1977 in Athens, Greece.[9] Such studies are required to provide the necessary scientific evidence for standard setting. The report of the meeting, "Health Criteria and Epidemiological Studies Related to Coastal Water Pollution," is available in English and French.

The subject of waste disposal into coastal waters was addressed at a workshop on coastal water pollution control in Athens from 27 June to 1 July 1977.[10] An outline of a model code of practice was prepared by the group of experts which considers information systems, design criteria, master planning, impact assessment, and the design of treatment and discharge facilities. The final report is available in English and French.

The preparation of a Guide to Shellfish Hygiene was partly supported by the project funds.[11] It is now available in English and French.

A mid-term project review meeting was held at the Istituto Superiore di Sanita in Rome from 30 May to 1 June, 1977,[12] at which progress achieved by the participating laboratories was reviewed and plans for future action were made.

The principal investigators of the laboratories participating in the project attended in April 1978 a Seminar on Monitoring of Recreational Coastal Water Quality and Shellfish Culture Areas at the Istituto Superiore di Sanita in Rome.[13] Monitoring procedures and data handling and evaluation were discussed.

3. WHO/ECE/UNIDO/FAO/UNESCO/IAEA/UNEP Project on Pollutants from Land-based Sources in the Mediterranean (MED X)

The objectives of this project were (1) to identify the major land-based pollution sources and to estimate the total pollution load entering the Mediterranean, and (2) to assess the existing waste discharge and water pollution management practices. To this end, the cooperating organizations carried out the following tasks:

(i)	coordination with the Mediterranean Action Plan	UNEP
(ii)	assessment of municipal wastes	WHO
(iii)	assessment of industrial wastes	ECE/UNIDO
(iv)	assessment of agricultural runoff	FAO
(v)	assessment of river discharges	UNESCO
(vi)	assessment of radioactive wastes	IAEA
(vii)	project coordination	WHO

The project was started early in 1976 with the preparation of a detailed workplan and timetable. The countries bordering on the Mediterranean participate in the project by providing the required data and information.

The first major task was to develop a consistent data and information reporting format. This was accomplished by October 1976 through Inter-Agency consultations with the assistance of external experts.

Information was obtained from the national authorities concerned during the first half of 1977. Most of the countries were visited by consultants of the cooperating organizations. Detailed source inventories were prepared which were used as bases for estimating the pollution contributed by the different types of land-based sources. The final report was prepared during July and August of 1977 and submitted to a meeting of technical experts in September 1977 in Geneva.[5]

Subsequently, the report entitled Pollutants from Land-based Sources in the Mediterranean[14] was completed and provided to all Mediterranean Governments as an information paper.

4. Integrated Planning ("Blue Plan")

This component of the Mediterranean Action Plan deals with the incorporation of environmental quality considerations into economic development projects. A series of governmental consultations were held to discuss the problems involved. A special meeting was held in Split, Yugoslavia early in 1977.[15]

WHO actively participated in these deliberations and prepared a background paper on the environmental health aspects of socio-economic development in the Mediterranean Region. This paper addresses the health problems associated with water supply, waste disposal and vector control.

A variety of other sectors are now included under the Priority Action Plan such as alternative energy sources, aquaculture and freshwater resources management.

5. Future Development of the Mediterranean Action Plan

An inter-governmental review meeting of all coastal states was held in January 1978 in Monaco where the entire Action Plan was thoroughly discussed and guidance on each component provided.[16,17] Recommendations were made on the continuation of the monitoring and research part as a pilot exercise and intensive efforts to revise and finalize the Protocol on Land-based Sources were requested.[18] The next review meeting is scheduled for February 1979.

III. COMPREHENSIVE ACTION PLANS FOR THE PROTECTION OF REGIONAL SEAS

Following the pattern developed within the Mediterranean Action Plan outlined in Annexes I and II, a series of activities have been undertaken concerning other regional seas. Programmes have been initiated in the East Asian Waters, the Caribbean Sea, the Persian/Arabian Gulf and the West African Coast (Gulf of Guinea). An Action Plan is to be developed for each of these areas which will include legal, scientific and institutional components. WHO is to be a cooperating agency.

Joint programming within the UN system was achieved through an Inter-Agency Meeting on Regional Seas which was held from 16 to 18 June 1977 in Paris.[19] Common principles for the development of comprehensive action plans for marine regions were agreed upon and specific programmes were initiated.

In November 1976, an Inter-Agency Task Force Meeting was held in Nairobi, Kenya, in which the development of legal instruments for the protection of the marine environment were discussed with particular reference to the Persian/Arabian Gulf and the Gulf of Guinea.[20]

An International Workshop on Marine Pollution in the *East Asian Waters*[21] was organized by IOC, FAO and UNEP in Penang, Malaysia, from 7 to 13 April 1976. The major marine pollution problems were identified and recommendations to initiate a number of regional and subregional projects were made. Subsequently a FAO/IOC/UNEP project on the oil pollution problems in the Straits of Malacca has been started.

A similar workshop for the *Caribbean Region*[22] was organized by IOC, FAO and UNEP in Trinidad from 12 to 17 December 1976. The major marine pollution problems were discussed and priority projects were identified. These are to be initiated as part of a comprehensive Action Plan which is being developed in cooperation with the Economic Commission for Latin America (ECLA) of the UN. WHO participated in the workshop and presented a paper on health aspects of the disposal of human wastes into the marine environment.

As regards the *Persian/Arabian Gulf,* an Inter-Agency fact-finding mission visited the area to collect background information for the preparation of an Action Plan. Subsequently a technical meeting on coastal area development and protection of the marine environment was convened by UNEP and the Department of Economic and Social Affairs (ESA) of the UN in December 1976 in Kuwait.[23] WHO participated in the meeting.

The problems of the Persian/Arabian Gulf and the development of an Action Plan were further addressed by a UNEP convened Expert Meeting on the Protection of the Marine Environment which was held from 13 to 18 June 1977 in Nairobi.[24] It was decided that the evaluation of the effects of pollutants on human health and marine ecosystems be included in the proposed regional environmental assessment programme.[25] This will involve (1) a survey of land-based sources of industrial and municipal wastes and (2) studies on the impact of the wastes on human health. As regards environmental management, WHO interests will be primarily concerned with (1) strengthening of national public health services and (2) coordination of national water management policies.

WHO participated in a Meeting of Legal Experts on the Protection of the Marine Environment in Bahrain from 24 to 28 January 1977, where the preparation of a framework convention and a protocol dealing with oil pollution by other harmful substances in emergency situations was discussed.[26]

Preparations are being made for a Regional Conference of Plenipotentiaries on the Protection and Development of the Marine Environment and the Coastal Areas which was held 15–24 April 1978. This conference discussed and approved the Action Plan as well as the necessary legal instruments (convention and one protocol).

Work is being started on the development of a programme for the *West African Coast* (Gulf of Guinea). In 1976 a UNEP sponsored mission visited the area. An international workshop on marine pollution was held in May 1978 in Abidjan, Ivory Coast, in May 1978. Two studies were prepared by WHO for this purpose: one on liquid waste disposal[27] and one on health effects of coastal water pollution.[28]

IV. JOINT GROUP OF EXPERTS ON THE SCIENTIFIC ASPECTS OF MARINE POLLUTION (GESAMP)

GESAMP was established jointly by eight UN organizations to provide expert advice on marine pollution questions. It is comprised of approximately twenty scientists who meet periodically.

WHO participated in the Ninth Session of GESAMP which was held in New York from 7 to 11 March 1977. A variety of scientific problems were discussed and a programme of work for the next inter-sessional period was developed. The major points of discussion were:[29]

> (1) evaluation of hazards of harmful substances; (2) interchange of pollutants between atmosphere and the ocean; (3) sea-bed exploitation and coastal area development; (4) removal of harmful substances from wastewater; (5) biological effects of thermal discharges; and (6) monitoring of biological variables.

Within the framework of GESAMP, WHO's current responsibilities involve the organization of working groups on the harmful substances and on removal of such substances from wastewaters. Further discussions on these subjects were held at the Tenth Session of GESAMP in May/June 1978 and a report on scientific aspects of waste removal is in preparation by the Batelle Institute.

V. HEALTH CRITERIA FOR COASTAL BATHING WATERS

WHO has been involved in the development of health criteria for coastal bathing waters for a number of years. Meetings on the subject have been held in Ostend,[30] Bilthoven,[31] New York and Athens.[9] A comprehensive document has been prepared on health criteria for coastal bathing waters.[32] It is to be circulated to national focal points of the WHO Environmental Health Criteria Programme and other experts for comment. Publication is planned for early in 1979.

REFERENCES

1. Report of the IOC/GFCM/ICSEM International Workshop on Marine Pollution in the Mediterranean (Monaco, 9–14 September 1974), IOC Workshop Report No. 3, UNESCO, 1975.
2. Report of the Intergovernmental Meeting on the Protection of the Mediterranean (Barcelona, 28 January–4 February 1975); UNEP/WG.2/5, Annex; UNEP, 1975.
3. Conference of Plenipotentiaries of the Coastal States of the Mediterranean Region for the Protection of the Mediterranean Sea, Barcelona, 2–16 February 1976, Office of Public Information, United Nations, Geneva, March 1976.
4. Report of Intergovernmental Consultation concerning a Draft Protocol for the Protection of the Mediterranean Sea against Pollution from Land-based Sources, Athens, 7–11 February 1977; UNEP/IG.6/6, 25 February 1977.
5. Report on Meeting of Experts on Pollutants from Land-based Sources, Geneva, 19–24 September 1977; UNEP/IG.11/BD 31.
6. Report on Second Intergovernmental Consultation concerning a Draft Protocol for the Protection of the Mediterranean Sea against Pollution from Land-based Sources, Venice, 17–21 October 1977; UNEP/IC.11/BD 32.
7. Report of the WHO/UNEP Expert Consultation on Coastal Water Quality Control Programme in the Mediterranean (Geneva, 15–19 December 1975); EHE/76.1, WHO, 1976.
8. Guidelines for Health Related Monitoring of Coastal Water Quality; WHO, 1977.
9. Health Criteria and Epidemiology of Health Risks Related to Beach and Coastal Pollution; WHO, 1977.
10. Report on Principles and Methodology on Coastal Pollution Control Planning (WHO/UNEP Workshop on Coastal Water Quality Control, Athens, 27 June–1 July 1977); WHO, 1977.
11. P. C. Wood, Guide to Shellfish Hygiene, WHO Offset Publication No. 31; WHO, 1976.
12. Mid-term Review of the Joint WHO/UNEP Co-ordinated Pilot Project on Coastal Water Quality Control in the Mediterranean (MED VII), Rome, 30 May–1 June 1977; WHO, 1977.
13. Report of Seminar on Monitoring of Recreational Coastal Water Quality and Shellfish Culture Areas within the Framework of WHO/UNEP Pilot Project MED VII, Rome, 4–7 April 1978; WHO, 1978.
13. Pollutants from Land-based Sources in the Mediterranean; Report prepared in collaboration with ECE, UNIDO, FAO, UNESCO, WHO, IAEA; UNEP/IG.11/INF.5.
15. Report of the Intergovernmental Meeting of Mediterranean Coastal States on the Blue Plan (Split, 31 January–4 February 1977); UNEP, 1977.

16. Administrative Report on the Implementation of the Co-ordinated Mediterranean Pollution Monitoring and Research Programme (MED POL) and Related Projects of the Mediterranean Action Plan; UNEP/IG.11/INF.3.
17. Preliminary Report on the State of Pollution of the Mediterranean Sea; UNEP/IG.11/INF.4.
18. Report of the Intergovernmental Review Meeting of Mediterranean Coastal States on the Mediterranean Action Plan; UNEP/IG.11/4.
19. Report of the Interagency Meeting on Regional Seas; Paris, 16–18 June 1976; UNEP, 1976.
20. Report of the Task Force on Legal Instruments for Regional Seas, Nairobi, 1–5 November 1976; UNEP, 1976.
21. Report of the IOC/FAO(IPFC)/UNEP International Workshop on Marine Pollution in East Asian Waters, Penang, 7–13 April 1976; IOC Workshop report no. 8; UNESCO, 1976.
22. Report of the IOC/FAO/UNEP International Workshop on Marine Pollution in the Caribbean and Adjacent Regions, Trinidad, 13–17 December 1976; IOC Workshop report no. 11; UNESCO, 1977.
23. Report of the Joint UNEP/ESA Consultation on the Development and Protection of the Marine Environment and Coastal Areas, Kuwait, 6–10 December 1976; UNEP, 1977.
24. Report of the Experts Meeting on the Protection of the Marine Environment, Nairobi, 13–18 June 1977; UNEP/WG.10/8; 18 June 1977.
25. Outline of the Environmental Health Programme; Expert Meeting on the Protection of the Marine Environment; UNEP/WG.10/Background Paper 6; 17 May 1977.
26. Final Report of Bahrain Regional Meeting of Legal Experts on the Protection of the Marine Environment, Bahrain, 24–28 January 1977; UNEP/WG.9/1, 10 February 1977.
27. R. Helmer and U.W. Mörgeli; Review of the Liquid Waste Disposal Situation along the Gulf of Guinea and Adjacent Areas; WHO document CEP/77.10.
28. N. Meith-Avcin and R. Helmer; Health Effects of Coastal Water Pollution Pertaining to the Gulf of Guinea and Adjacent Areas; WHO document CEP/77.11.
29. Report of the Ninth Session of the Joint Group of Experts on the Scientific Aspects of Marine Pollution; GESAMP IX/14; 11 March 1977.
30. Health Criteria for the Quality of Recreational Waters with Special Reference to Coastal Waters and Beaches; WHO Report on a meeting held at Ostend, 13–17 March 1972.
31. Guides and Criteria for Recreational Quality of Beaches and Coastal Waters, Bilthoven, 28 October–1 November 1974; EURO 3125(1); Copenhagen, 1975.
32. Health Criteria for the Quality of Coastal Bathing Waters with Some Public Health and Associated Guidelines for Implementation; draft prepared by E. W. Mood and B. Moore, 30 March 1976.

ANNEX I
COMPREHENSIVE ACTION PLAN FOR THE PROTECTION OF REGIONAL SEAS

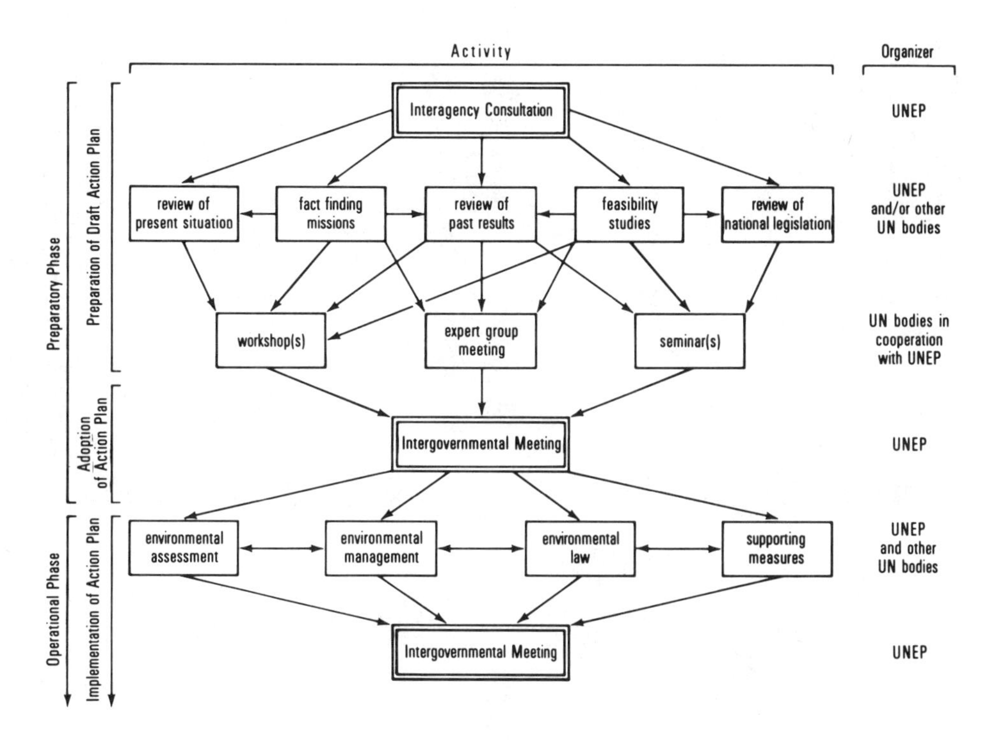

ANNEX II
MEETINGS RELEVANT TO THE MEDITERRANEAN ACTION PLAN

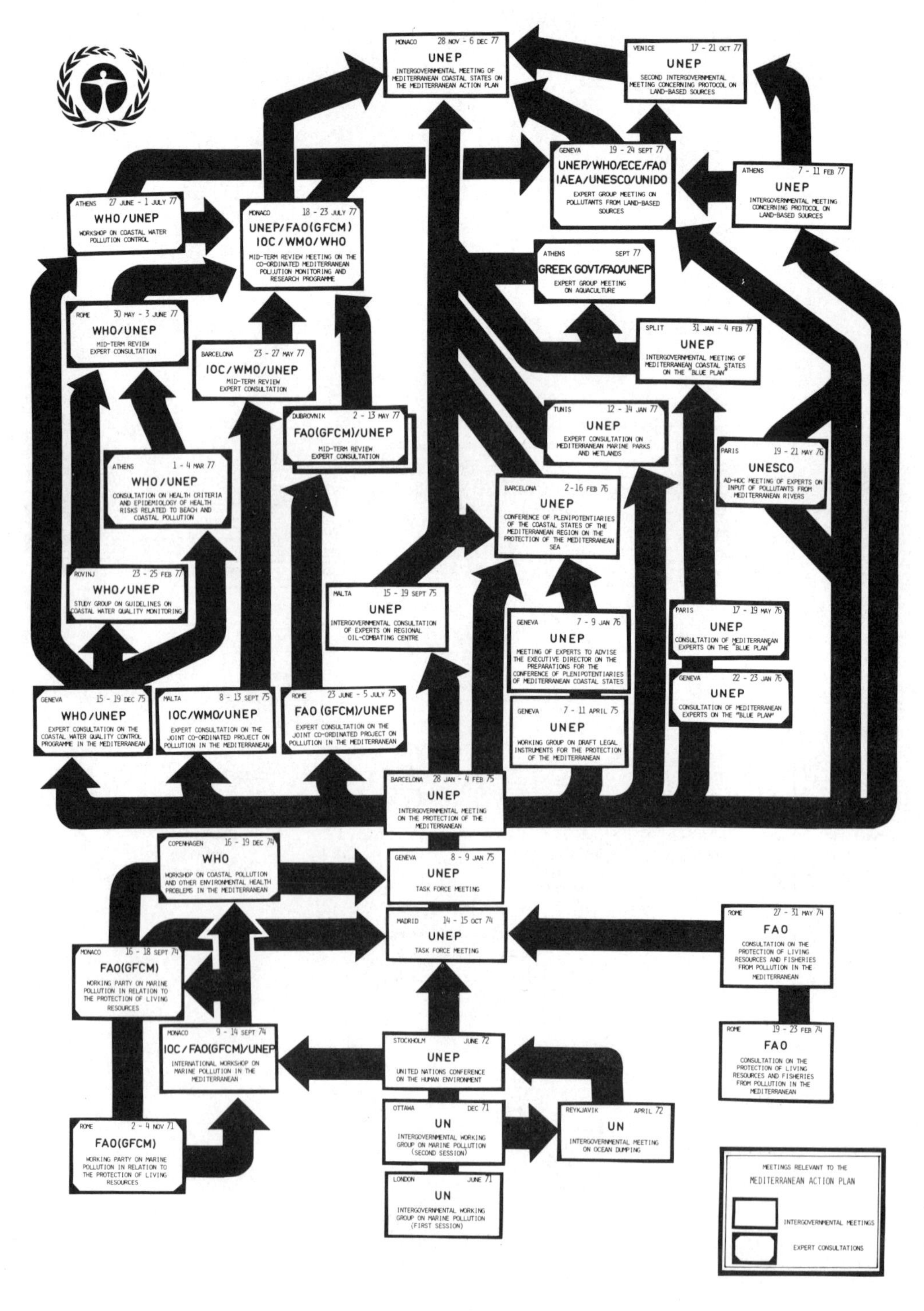

WMO: Meteorology and Ocean Affairs[1]

ACQUISITION OF MARINE ENVIRONMENTAL DATA

Mobile Ship Stations

The steady upward trend in the number of ships participating in the WMO voluntary observing ships' scheme continued. Forty-six Members participate in the scheme and contribute 4,574 selected ships, 2,071 supplementary ships and 725 auxiliary ships, making a total of 7,370 at the beginning of 1977, an increase of 114 on the corresponding total in 1976. A similar trend was reflected in the number of ships' weather reports exchanged over the GTS for the same period.

Improvement in Observational Data Coverage over the Oceans

As a first step in a specific plan to upgrade the level of real-time data availability from data-sparse ocean areas particularly in time for the First GARP Global Experiment (FGGE) and, as a basis for further action, the World Meteorological Centre (WMC) in Washington agreed to a request to prepare maps showing the distribution of ships' weather reports for the four standard times for two days in April 1977. The maps, based on the reception at the WMC Washington of ships' weather reports over the GTS, showed that there still remained many data-sparse ocean areas, especially in the tropical belt and southern hemisphere. The maps were subsequently circulated to Members, who were invited to take urgent and specific action to improve the real-time data availability from the data-sparse sea areas identified. Replies to date indicate that, within the limits of their resources, Members are endeavouring to implement the specific action proposed.

Recent studies have highlighted the fact that ship-to-shore HF [high frequency] telecommunication continues to be the principal means for the collection of meteorological and oceanographic reports from the world's oceans and Members have made increased efforts to improve the facilities and performance at certain coastal radio stations. In spite of all efforts the present state of real-time data availability remains unsatisfactory (see above) and further action had to be taken. As another step in the above-mentioned specific plan, an inquiry was addressed to ships' captains participating in the WMO voluntary observing ships' scheme to ascertain whether difficulties had been encountered in transmitting ships' weather reports and, at the same time, inviting them to provide comments or suggestions regarding possible improvements in the existing system. The enthusiastic response to this inquiry—over 1,000 replies—showed

1. Summary prepared by the Ocean Affairs Division, World Meteorological Organization, based on the WMO Annual Report for 1977 (WMO No. 502), with additional information for 1978.

that ships' officers were very willing to co-operate with WMO in seeking a more efficient ship-to-shore HF telecommunication system. A report summarizing the results of the inquiry has been prepared and circulated to Members of WMO with a request to take specific action to eliminate problem areas identified in consultation with national authorities responsible for the operation of coastal radio stations.

In order to improve the situation regarding the present availability of ships' weather reports from the southern hemisphere in time for the FGGE, the seventh session of CMM [Commission for Marine Meteorology] (Geneva, November/December 1976) adopted a recommendation which was subsequently approved by the twenty-ninth session of the Executive Committee (1977), advocating the provision of the necessary technical and financial support to developing countries for the purpose of ocean data collection and dissemination. It was felt that one way of providing this support would be to supply the services of an area marine specialist who could visit national Meteorological Services in developing countries on request and provide guidance on improvements in the collection and dissemination of ocean data and arrangements for the reception of ships' observations at coastal radio stations and their subsequent insertion into the GTS. Members were therefore asked whether such services were required in their countries or whether they would be prepared to second an expert to provide the necessary services if requested. A considerable number of Members requested such expert advice and efforts have been made to meet the requests where possible.

New Techniques and Methods for the Collection of Ships' Weather Reports

Because one of the most efficient means of sending ships' weather reports from ship to shore is the teleprinter system designed for unattended operation at coastal radio stations and on board ship, Members of WMO have been encouraged to take full advantage of developments in this system, particularly the HF radio-telex system. Replies to the inquiry directed to ships' captains mentioned above confirmed the efficiency of this mode of transmission and many captains stated they were using the system on a routine basis for the transmission of ships' weather reports.

With regard to developments in the field of ship-shore communication by means of satellites, very little can be added to the information given last year since most efforts are still at a preparatory stage.

Ocean Weather Stations

Following the withdrawal of the US North Atlantic ocean station the number of ocean stations in the global network was reduced during 1977 from seven to six. There are now four stations in the North Atlantic and two in the North Pacific, one operated by Canada and the other operated by Japan during the typhoon season. All vessels on these stations make routine surface and upper-air observations and also make and transmit BATHY and TESAC observations. In addition the majority of ocean station vessels make a wide variety of oceanographic observations.

North Atlantic Ocean Station (NAOS) System

The four ocean weather stations in the North Atlantic are all operated under the NAOS system which is supported by 16 Members of WMO as Contracting Parties and several voluntary contributing Members, and operated by five Members. Summaries concerning the operation of the network for the period July 1975—December 1976 (since the administration of the system was taken over by WMO from ICAO) and 1 January to 31 December 1977 were submitted to the WMO Secretariat by the five Operating Parties. They showed that a full programme of surface and upper-air observations had been carried out and that the observations were generally transmitted to the shore collecting centre at Bracknell.

MARINE METEOROLOGY

Marine Meteorological Services (MMS)

With the keen interest shown by nations in developing their marine activities, especially in coastal and off-shore areas, there is an increasing awareness of the need for corresponding expansion of marine meteorological services in support of these activities. To assist countries in this respect, particularly the developing countries, the seventh session of CMM formulated a number of projects. The CMM Working Group on Marine Meteorological Services, which commenced its work in May 1977, immediately took up several of these priority tasks by correspondence. These tasks included marine meteorological support to fisheries, especially in the 200-mile zone from the coastline, services to activities in and near ports which serve as important focal points in both high seas and coastal marine operations as well as to dynamically supported craft, and arrangements for the preparation of as many draft chapters as possible of the proposed Manual on Marine Meteorological Services. In particular, action was also initiated with Members concerned in arranging for the trials of the methods of graphical representation of information on radiofacsimile charts intended for marine users, commencing on 1 September 1977.

Marine Meteorological Support to Fisheries Development

WMO participated in the eleventh session of the FAO Committee on Fisheries (Rome, 22–26 April 1977). This session, recognizing the importance of marine meteorological services to safe and economic fishing operations, requested WMO to pay particular attention to the development of these services, especially in developing countries and with regard to waters off the coasts as well as in near coastal waters.

In a document presented to its twelfth session (Rome, 12–16 June 1978), the Committee was informed of the various actions taken by WMO in meeting the request of the Committee. The session further proposed that complex technical matters such as the formulation of meteorological requirements for fisheries and criteria for issuing warnings should be referred to the appropriate technical working group for detailed consideration. WMO proposes to participate at the meeting of the FAO technical group when convened.

Port Meteorological Services

The enthusiastic response by Members to an inquiry on the types of meteorological service provided for the various marine activities at ports and harbour approaches resulted in the collection of factual information in respect of about 140 ports. This information was evaluated by an informal meeting of experts of the CMM Working Group on Marine Meteorological Services, held in Geneva from 26 to 29 September 1977. Based on this evaluation, draft texts were prepared for inclusion in WMO Technical Regulations (Chapter C.1) and in the proposed Manual on Marine Meteorological Services, as well as in the Guide to Marine Meteorological Services. The draft texts of the revised Chapter C.1 of Technical Regulations and the proposed Manual, after final editing by the Secretariat, are in the process of being circulated to Members of the Commission for exchange of opinion before a formal vote is taken by correspondence for their adoption by the Commission. The modified texts in the sections of the Guide relating to port meteorological services and to the duties of port meteorological officers were, however, distributed to Members of WMO, on approval by the president of CMM, as advance information pending the issue of a regular supplement to the publication.

Meteorological Support to Maritime Search-and-Rescue Operations

WMO was represented at the fifth session of the IMCO Group of Experts on Maritime Search and Rescue (SAR), held in London from 30 May to 3 June 1977, which examined the final amendments to the draft Convention on Maritime SAR and its technical annex as well as changes to the draft Manual on Maritime SAR, for the consideration of the international conference to be held in 1979. Important amendments proposed by WMO at this session concerned the role of the NAOS vessels in SAR operations, the meteorological aspects of the operation and the terminology used in the draft Manual. In addition, the session adopted a list of meteorological parameters and phenomena considered important for the safe and efficient conduct of SAR operations.

The meteorological aspects of maritime search and rescue operations were also one of the topics considered at the twentieth session of the IMCO Sub-Committee on Safety of Navigation. WMO was represented at the session which was held in London from 5 to 9 September 1977. Since the obligations arising from the convention would be of particular importance to Meteorological Services having marine meteorological responsibilities, advance information on recent developments in this field was sent to the Services concerned, with additional explanations of possible implications of the maritime search and rescue plan. These include, for instance, the co-ordination of services in maritime search and rescue regions which usually do not coincide with the areas of responsibility under the WMO system of meteorological forecasts and warnings for the high seas. A satisfactory mechanism would be needed for obtaining the special environmental data and for developing appropriate forecasting techniques. Finally, there must be rapid and reliable communication arrangements between the meteorological forecast centre and the search and rescue co-ordination centre(s).

Co-ordination of Broadcasts of Navigational and Meteorological Warnings

Another important subject considered at the seventh session of CMM was the possible co-ordination of meteorological and navigational warning broadcasts for the benefit of the mariner. The subsequent request by WMO for an elucidation of the exact requirements for the co-ordinated system was considered by the relevant technical bodies of IMCO. According to preliminary conclusions, in view of the complex character of the well-established procedures for issuing both navigational and meteorological warnings, it was felt that the present practices should be retained. At the same time it was realized that the inclusion of certain meteorological information relating to hazardous environmental conditions in navigational warning broadcasts would contribute to safety at sea. For this purpose, it was considered sufficient if at scheduled navigational warning broadcast times the warnings were preceded by urgent meteorological messages (e.g. significant storm warnings) originated by the respective meteorological authorities and sent direct to the transmitting station, or via the area co-ordinator if so preferred. As regards coastal warnings, it was agreed that, at this stage, Members concerned should be encouraged to promote the necessary co-ordination in their country between the meteorological and navigational warnings. The Executive Committee at its thirtieth session (Geneva, May–June 1978) noted the views of IMCO on the matter and decided that suitable guidelines should be prepared by CMM for implementation of such a co-ordinated system as envisaged.

Third Session of CMM Working Group on Marine Meteorological Services

The Working Group on Marine Meteorological Services held its third session in Geneva from 10 to 14 April 1978, which was attended by 24 experts including representatives from IMCO and IOC. A major achievement of the session is the finalization of the draft Manual on Marine Meteorological Services and the draft revised Chapter C.1 of WMO Technical Regulations. Amongst other important subjects considered by the session are the identification of parameters and phenomena to be included in the services for the high seas coastal and off-shore areas, for ports and harbour areas as well as the formulation of criteria for warnings. The relevant regulations of the International Convention for the Safety of Life at Sea were reviewed and modifications to the texts were proposed in order to bring them in line with the technical terminology developed in WMO Technical Regulations. These proposals have subsequently been forwarded to IMCO for necessary action at appropriate time. The session also proposed a recommendation on the need for organizing marine meteorological services assisting in the prevention of marine pollution incidents and in the reduction of severe consequences when they occur. This proposal was accepted by the president of CMM and action has been initiated in June 1978 for adoption of the recommendation by the Commission by correspondence.

Regional Development of Marine Meteorological Activities

Marine meteorology continued to play an important role in regional development activities. A major event in this regard was the Regional Marine Meteorology Conference held in Jeddah from 24 to 30 September 1977 at the initiative of the countries concerned in the Region. WMO was represented at the conference. In a recommendation, the conference agreed in principle to establish a Regional Marine Meteorology Programme and adopted an outline of the programme. In inviting the countries concerned to participate in the development of the regional agreement for the implementation of the programme, WMO was invited to convene a meeting of experts for the preparation of the agreement in early 1978 and to convene a Plenipotentiary Conference to adopt the agreement by June 1978.

The meeting of experts took place in Tehran from 29 April to 4 May 1978 at the invitation of the Government of Iran. WMO participated in the meeting and assisted in the proceedings. The meeting considered and finalized a draft agreement, submitted by Saudi Arabia, according to which the Contracting States undertake to establish and finance jointly a Regional Marine Meteorological Programme to provide adequate services to the various marine and related activities in the region. The programme consists of five major components, namely, the observation system, the data processing and services system, the telecommunications system, marine meteorological research and training. The operational core of the programme will be a Regional Marine Meteorological Centre and National Marine Meteorological Offices to be established by each of the Contracting States.

Several countries in other regions have also expressed an interest in developing their marine meteorological services. A request has been received from Nigeria for the assignment of a marine meteorological expert to initiate and develop the marine meteorological services. The United Republic of Cameroon has also shown an active interest in developing its marine meteorological services. WMO marine meteorological experts have recently been appointed to Saudi Arabia and Malaysia to assist in the establishment and/or development of national marine meteorological services.

In the Caribbean region, a Joint Programme for Sound Environmental Management in the Wider Caribbean Areas has been initiated by the United Nations Environment Programme (UNEP) and the UN Economic Commission for Latin America (ECLA) and a project to this effect, with a duration of three years, has accordingly been put into operation. In response to a request, WMO agreed to co-operate in this joint UNEP/ECLA programme within the framework of the approved WMO projects and activities in the Region.

Guide to Marine Meteorological Services

The WMO Guide to Marine Meteorological Services was published in September 1977 (WMO-No. 471). This publication is intended for national Meteorological Services as well as for the marine users, to serve as a compact guidebook on the types of meteorological service provided to various marine activities and the international standard and recommended procedures and practices governing them.

Handbook on Wave Analysis and Forecasting

The WMO Handbook on Wave Analysis and Forecasting (WMO No. 446) was published in English in December 1976 and in French in March 1978. Arrangements have also been made for the translation of the Handbook into Russian and Spanish.

Meteorological Assistance in the Preparation of IAEA Safety Guides

WMO has been assisting the International Atomic Energy Agency in the preparation of some of the IAEA safety guides on the siting of nuclear power plants. Marine meteorological advice was given in particular to the drafting of a safety guide concerning nuclear power plants on coastal sites. This guide describes the principles of models used in the computation of probable maximum storm surges, waves, tsunami and seiches. Design studies of this kind again show the importance of a suitable network of coastal and off-shore observing stations measuring parameters such as water-level changes, waves, wind and air pressure.

Marine Climatology

A major scheme in the field of marine climatology is the preparation of marine climatological summaries for all ocean areas of the world, under a co-operative programme of WMO. In this scheme, which was initiated by the Fourth Congress of WMO, primary data are collected by various Member countries, scrutinised, quality-controlled and placed on punched cards or on magnetic tapes and sent to those Member countries which have undertaken to prepare and publish the summaries for specific areas of responsibility. The respective areas of responsibility of Members preparing the summaries are shown in the map included in Chapter 7 of WMO Guide to Marine Meteorological Services (WMO No. 471). The summaries are prepared on an annual basis starting for the year 1961 onwards. As of 1 June 1978, 46 volumes of the summaries have been published as follows: Federal Republic of Germany (1961–1969), Hong Kong (1961, 1964), India (1961, 1964, 1965), Japan (1961–1970), Netherlands (1961–1967), United Kingdom (1963–1967) and United States of America (1961–1970). Copies of the summaries can be obtained from the Members responsible for their publication, on request, at costs notified by them and the requests should be addressed direct to the Members concerned.

Fourth Session of CCM Working Group on Marine Climatology

The Working Group on Marine Climatology of the Commission for Marine Meteorology held its fourth session in the WMO Secretariat from 14 to 18 November 1977. The work of this group is of significance to the World Climate Programme since it is concerned with the international arrangements required for the establishment of basic data sets to be used in the various studies of the climate and climatic variations in oceanic areas.

The session considered a number of problems related to marine climatological summaries scheme, such as the data exchange procedures, quality control, extended IMMPC format for use on magnetic tapes, and the use of non-synchronous data in the summaries. Further, the important subject of coastal marine climatology received major attention of the session, since climatological statistics of coastal areas are increasingly being required in several regions for industrial and other applications. Accordingly the session decided on the layout of a technical paper to be prepared on the climatological requirements of coastal zones, as guidance on the subject. Based on the wide range of subjects discussed, the session also drew up a detailed work programme which included a number of projects to be carried out during the period before the eighth session of CMM in 1980.

Plan for Collection and Storage of Sea Surface Current Data

The United Kingdom agreed to assume the functions of the international surface current data centre (ISCDC), which include collection and storage of the data and supply of stored data to users on request. The detailed plan arranging for the international exchange of the data was circulated to Members requesting for their participation. There has already been a good response to the request.

Plan for Non-real Time Collection of Ships' Observations for FGGE Purposes

Several enthusiastic replies were received in response to the request for participation by Members in the plan for non-real time collection and dispatch of ships' observations for FGGE purposes. The Mobile Ship Data Centre in Hamburg, specially set up for collecting and processing this type of data, under the FGGE plan, was informed of the proposed arrangements by Members for the dispatch of data within the stipulated time limits.

Coastal Area Development

The subject of coastal area development continued to receive great attention in international circles, particularly in those regions where joint programmes are under way aimed at the integrated development of the coastal zones in the region, or protection of the environment or a better understanding of the ocean-atmosphere interaction processes. For the Manual on Coastal Area Development and Management, under preparation by the United Nations, WMO forwarded material on the climates of coastal zones, specially prepared by a WMO consultant.

Sea-Ice Services

The need for the development of uniform sea-ice symbols to be used on operational ice charts was expressed during the seventh session of CMM, when a suggestion was made that a standard operational symbology could be developed most effectively by means of operational trials conducted in association with ice reconnaissance missions. Such trials were arranged by Canada from 2 to 11 March 1977. Experts from the three Member countries, Sweden, the U.S.A. and the U.S.S.R., accompanied by a representative from the WMO Secretariat, participated in the trials, which were conducted from Gander, Newfoundland. In conjunction with operational reconnaissance flights, a number of workshop sessions were held during which a new ice symbology for international use was developed. The proposed symbology was subsequently tested in various other regions, such as the Arctic and the Baltic Sea, by members of the CMM Working Group on Sea Ice. Several details arising from the tests are now under consideration by the group.

WMO was represented at the eleventh Baltic Sea Ice Meeting which was held at Norrköping in Sweden from 3 to 6 October 1977. Co-operation between the Baltic countries with regard to sea-ice services started early this century. A common system of regional sea-ice codes and symbols has been developed during this period so as to facilitate the exchange of ice reports and charts depicting ice in the area. Recent developments in the observation of sea ice by remote-sensing techniques and the fact that in modern times ships which visit the Baltic area usually frequent other northern seas where information on ice conditions is also necessary, call for the rationalization on an international basis of regional sea-ice practices. The seventh session of CMM (Geneva, November/December 1976) took measures for the development of an international sea-ice symbology which could be used on ice charts. One of the items on the agenda of the Baltic Sea Ice Meeting was the consideration of tests made in the Baltic area of the draft international ice symbols suggested through CMM. At the same time, the meeting reviewed the present Baltic Ice Code with a view to bringing it into line with international sea-ice codes. This proved to be a difficult problem in view of the special ice features encountered in the area. A further meeting was considered desirable to solve the problem.

Workshop on Remote Sensing of Sea Ice

A Steering Committee consisting of four members of the CMM Working Group on Sea Ice was formed for the Workshop on Remote Sensing of Sea Ice which is proposed to be held in Washington, D.C., in October 1978. The Steering Committee held its first session in the WMO Secretariat from 2 to 5 November 1977, when an expert on satellites from the U.S.A. also participated. A comprehensive technical programme for the Workshop was drawn by the committee. The programme includes several laboratory sessions to acquaint participants with operational practices in the interpretation of space-borne and airborne remote-sensed data and their processing. Notification of the workshop and relevant particulars including the technical programme has been circulated to Members of WMO and others concerned, to enable their participation.

Technical Problems concerning Observing Methods

With the nomination of individual rapporteurs on various tasks enumerated by the seventh session of CMM, the chairman of the Group of Rapporteurs on Technical Problems activated the work to be carried out by further outlining the work programme of each. Proposals were made for field measurements for surface wind measurements as well as precipitation measurements at sea. These were submitted to the Advisory Working Group for CMM, which recommended that, in the case of the former, the information on available studies on the subject should be evaluated by an informal expert meeting for preparing a detailed international programme for the field tests. As regards the latter, the group advised the rapporteur to prepare a programme for the proposed field experiments.

Advisory Working Group for CMM

The fourth session of the Advisory Working Group for CMM was held in the WMO Secretariat from 8 to 11 November 1977. In addition to reviewing the work of the working groups and the rapporteurs and giving further guidance in carrying out the tasks assigned to them by CMM, the session formulated the technical basis of the main work programme for the Commission for the next financial period, 1980–1983.

A detailed draft policy statement on marine meteorological and related oceanographic activities for the eighth financial period (1980–1983) was accordingly prepared and the Executive Committee at its thirtieth session (Geneva, May–June 1978) agreed for its submission to the Eighth Congress next year for adoption. The material in the draft statement is presented in two major parts, namely, Part A—Present trends in the field of marine human activities and Part B—Major tasks during 1980–1983 and their implementation. Part A describes the recent developments in the field of shipping, fisheries, oil and gas industry, coastal area activities and marine scientific and research programmes. Part B outlines the corresponding developments needed in the coming years in the field of marine meteorological services and other related ocean activities and the programmes envisaged to be undertaken by WMO through its Commission for Marine Meteorology as well as in joint co-operation with other international organizations concerned.

OCEAN ACTIVITIES

The Integrated Global Ocean Station System (IGOSS)

The IGOSS General Plan and Implementation Programme for 1977–1982 was printed as a joint IOC/WMO publication in the four official languages and distributed to Members. In connexion with the implementation of the plan, the attention of Members was invited to the resolution of the Seventh Congress stressing the need for Members to implement the successive phases of IGOSS BATHY and TESAC data collection and exchange programme. Both the WMO Executive Committee and the IOC Executive Council urged Member States to participate actively in the implementation of IGOSS.

The BATHY/TESAC Operational Programme

A joint IOC/WMO Meeting of Governmental Experts on the Evaluation of the BATHY/TESAC Operational Programme and Preparation for the Workshop on Ocean Products and the IGOSS Data Processing and Services System (IDPSS) took place in Ottawa (22–26 August 1977). In their evaluation and appraisal of the BATHY/TESAC Operational Programme, the experts studied statistical summaries prepared by WMO which showed a considerable increase in the number of BATHY/TESAC reports exchanged over the GTS over the period 1 January 1975–30 June 1977, i.e. 16,500 in 1975, 33,500 in 1976 and 16,500 up to 30 June 1977. These statistics were based on monthly figures submitted by the following countries: Argentina, Canada, the Federal Republic of Germany, France, Japan, the United Kingdom, the U.S.A. and the Union of Soviet Socialist Republics. Semi-annual statistical summaries for the period between January

1976 and June 1977 were prepared by France and the U.S.A. Responsible National Oceanographic Data Centres were particularly encouraging in that they revealed contributions from countries in addition to those supplying monthly statistics, namely Australia, Denmark, Iceland, India, the Netherlands, Norway and Sweden.

Further Development of the IGOSS Data Processing and Services System (IDPSS)

The above-mentioned group of governmental experts prepared a preliminary plan to organize a Seminar and Workshop on Ocean Products and IDPSS in the first half of 1979, taking into account the guidelines on the implementation of the IDPSS as set out in the IGOSS General Plan and Implementation Programme 1977–1982. An important objective of the seminar and workshop will be to bring to the attention of potential user groups the availability of existing products and probable development of future products and to identify the possible scientific, economic and social benefits to be derived from their effective application to ocean activities. The workshop and seminar will be held under the joint auspices of IOC and WMO with the participation of other international organizations including FAO, IMCO, UNEP, ICES, etc. and will be planned in detail by an organizing committee.

Another Joint IOC/WMO Meeting of Governmental Experts on the IGOSS Data Processing and Services System took place in Hamburg 6–10 March 1978. This meeting concentrated on the identification of IGOSS product users, their specific requirements and the data needed for the preparation of real- and near-real-time products.

IDPSS Arrangements for the First GARP Global Experiment (FGGE)

The group considered the need for the IDPSS to meet the special requirements of the FGGE in the light of deadlines imposed by the FGGE timetable. Priorities for IGOSS products in support of the FGGE had already been identified and it was known that tropical regions and the southern oceans are areas where additional sea-surface temperature data are needed. Regional IGOSS support for projects such as the Polygon MODE Experiment and the Joint Air/Sea Interaction Experiment was providing experience in the preparation of products such as sea-surface temperature charts and this was, in effect, the groundwork for IGOSS/FGGE support. It was acknowledged that the special observing periods during the FGGE would be the times when maximum effort was needed for the collection of data and the preparation of IGOSS products. The group recommended that sea-surface temperature analysis be undertaken on a global scale and mixed-layer depth analysis be undertaken, although this may be possible only in selected areas. Canada, as a major contributor to the FGGE Southern Hemisphere Drifting Buoy Programme, agreed to consider the possibility of producing sea-surface temperature charts specifically from buoy data. Such a product would be of considerable interest when compared, as a "ground-truth" exercise, with sea-surface temperature charts produced from satellite data. It might also contribute to the requirement for additional sea-surface temperature data in the tropical regions and southern oceans. A large amount of BATHY/TESAC data will be collected by research ships participating in the oceanographic programme for the FGGE. To ensure the maximum availability of these data, the group recommended that Member States participating in the FGGE be requested to submit all their BATHY/TESAC data to IGOSS.

Institutional Arrangements for IGOSS

The Executive Committee, at its twenty-ninth session, noted that the General Plan and Implementation Programme 1977–1982 reaffirms the principle that IGOSS should be planned and operated closely with the World Weather Watch. Following this principle and in order to ensure that the implementation of IGOSS is well planned and co-ordinated, the Committee felt that there was an urgent need for a joint WMO/IOC mechanism to deal exclusively with this operational programme. The Executive Committee decided to establish a Joint IOC/WMO working Committee for IGOSS subject to a similar decision by IOC. The tenth session of the IOC Assembly in turn also decided to establish such a committee.

Information Service Bulletin on Ocean Data Buoys

In response to a Joint IOC/WMO circular letter, 21 Member States supplied information on their ocean data buoys, including information on protective measures taken or proposed in order to minimize their loss or willful disablement. The information received was compiled and circulated to Member States of IOC and WMO in August 1977 and was regarded as the first issue of a regular information service bulletin on ocean data buoys. Member States were requested to supply further information when changes occur or when new ocean data buoys are deployed so that bulletins can be updated as necessary. Members were also requested to transmit the information contained in the bulletin to national authorities concerned, not only to ensure the safety of navigation and the protection of buoys against collision, but also to inform the maritime community of the great scientific value of and the immediate benefits which can be derived from ocean data buoys.

Establishment of a Joint IOC/WMO/CPPS Working Group on the Investigation of "El Niño"

In response to requests made by the IOC Assembly at its eighth and ninth sessions, WMO has been co-operating with IOC in the implementation of a regional study of the phenomenon known as "El Niño" (ERFEN). At the tenth session of the IOC Assembly, it was proposed to intensify this co-operation through the establishment of a joint IOC/WMO/CPPS (Comisión Permanente del Pacífico Sur) Working Group on the investigation of "El Niño"; this proposal was accepted by WMO and CPPS. The main tasks of this group are:

(a) To promote activities and investigations with a view to reinforcing the co-ordination of projects related to "El Niño," emphasizing those activities undertaken in support of ERFEN;

(b) To elaborate a well-cordinated plan of action and ensure its implementation.

The Assembly also adopted a resolution inviting WMO to assist in the improvement of the real-time collection and dissemination of meteorological and oceanographic data from the "El Niño" area. The president of WMO agreed on behalf of the Executive Committee that WMO should participate in the work of this group.

Appendix B

Selected Documents and Proceedings

The documents and proceedings included in this Appendix represent a selection of documents bearing upon important ocean-related international developments. Commentaries on certain of these documents appear in the *Yearbook* text and are noted in the Index. Future editions of the *Yearbook* also will include selected international agreements, legislation, and proceedings of international conferences.

THE EDITORS

A Selection of Documents from the Eighth (Geneva) Session of UNCLOS III

UNCLOS III, EIGHTH SESSION

Volume 1 of *Ocean Yearbook* reproduced the Informal Composite Negotiating Text (ICNT)[1] as well as analysis of the major issues it raised.[2] The INCT was released after the sixth session of the conference and represented a major breakthrough in the course of these long and difficult negotiations.

The seventh session (Geneva, March 28–May 19; New York, August 23–September 15), reviewing the ICNT, agreed on a list of seven "hard-core" issues[3] crucial to the success of the conference and established seven negotiating groups (NG 1-7) to deal with them. While not reaching consensus on any of these issues, the conference, during the seventh session, initiated a great deal of useful technical work, especially with regard to the first three (questions relating to the International Seabed Authority and its system of production). With regard to production limitation, an elaborate agreement was reached which was acceptable to the largest producer country (Canada) and the largest consumer country (United States) but not to many other countries. Major difficulties with the formula surfaced during the second half of the eighth session. Detailed schemes were produced on financial arrangements between the Authority and contracting parties and on the question of technology transfer, issues a solution of which is essential if the Enterprise (the operational arm of the Authority) is to be enabled to start when commercial production begins and to compete with established industry. These questions remained intractable.

The eighth session (Geneva, March 19–April 27; New York, July 19–August 24, 1979) continued negotiations in the seven negotiating groups. A "Group of 21," representing an almost equal number of developing and developed countries, was established during the fourth week to deal comprehensively with all questions relating to the Seabed Authority and to report to the First Committee. The three main conference committees

1. UN document A/CONF.62/WP.10 (July 15, 1977).

2. Arvid Pardo, "The Evolving Law of the Sea: A Critique of the Informal Composite Negotiating Text (1977)," in *Ocean Yearbook 1,* ed. Elisabeth Mann Borgese and Norton Ginsburg (Chicago: University of Chicago Press, 1978), pp. 9–37.

3. The seven core issues are: (1) system of exploration and exploitation of the International Seabed Authority; (2) financial arrangements; (3) organs of the Authority, their composition, powers, and functions; (4) right of access of land-locked states and certain developing coastal states in a subregion or region to the living resources of the economic zone; (5) the question of the settlement of disputes relating to the exercise of the sovereign rights of coastal states in the exclusive economic zone; (6) definition of the outer limits of the continental shelf and the question of payments and contributions with respect to the exploitation of the continental shelf beyond 200 miles (question of revenue sharing); and (7) delimitation of maritime boundaries between adjacent and opposite states and settlement of disputes thereon.

continued their work. A new text (ICNT/Rev.1) was released between the two parts of the eighth session. This served as basis of discussion for the second part of the session whose purpose was to "formalize" the text, that is, to adopt it officially as a draft convention. This goal was not reached, and no new draft was agreed upon. However, a strict schedule was adopted for 1980. There are to be two 5-week sessions, in March/April and July/August, in 1980. Formalization is to be completed by the end of the first period, at which time formal amendments will be introduced. More amendments may be introduced on the first day of the resumed ninth session in July. Voting on the amendments is to be completed by the end of the second period, opening the way for the solemn signing of the convention in Caracas late in 1980 or early in 1981, thus bringing to a conclusion this unique exercise in the codification and progressive development of international law. Meanwhile, the conference has begun to look beyond the end of its mandate and toward the continuation of the development of the law of the sea and the institutions required to enact it. On the initiative of Portugal and Peru, a proposal was introduced during the eighth session to include in the "final clauses" of the convention provisions for such a continuation.

In this volume of *Ocean Yearbook* a number of documents are reproduced which convey the essence of the work of the eighth session. To begin with, a message received by the conference from UN Secretary-General Kurt Waldheim is reproduced. It stresses once more the unique importance and responsibility of the conference. This is followed by two documents. The first, drafted by a group of jurists from the developing countries and circulated by the Group of 77, argues the illegality of unilateral action with regard to mining in the international seabed area. The second is the U.S. response to this position. The prospect of such action has been hanging over the conference like a sword of Damocles. These three documents give a flavor of the political context in which the session took place.

Next comes a detailed report on the negotiations, released by the Secretariat (SEA/360, April 30, 1979), followed by the reports of the chairmen of the three main committees on which the revision of the text is based. In addition some material is included which is apt to shed new light on some of the difficulties of the hard-core issues: the questions of the outer limits of the continental shelf and of revenue sharing, and the production system of the Seabed Authority.

In connection with the first, a document is included which was released by the Intergovernmental Oceanographic Commission (IOC) on the extreme difficulties of producing a map on the basis of the criteria of the so-called Irish formula. Similar arguments were advanced during the conference by the delegation of the USSR, which produced a set of maps showing the ambiguities of this formula whose adoption, in the opinion of the Soviet experts, would give rise to many uncertainties and conflicts.

On revenue sharing, a proposal by the delegation of Nepal is included. Whereas the ICNT provides for a system of contributions or taxes on the extraction of nonliving resources from the continental margin beyond the 200-mile limit of the economic zone where national jurisdiction extends beyond that limit, the Nepalese proposal applies this system to the economic zone as well. From a legal point of view no objection can be raised against the Nepalese proposal if one accepts the provisions now in the ICNT on revenue sharing on the continental shelf beyond 200 miles, since the legal status of the continental shelf and that of the economic zone are the same. There are, nevertheless, fundamental political difficulties and problems of timing. This is a proposal that shows that the original spark of the great conference is not dead. It is a proposal whose time will come. It is remarkable that, even at this time, the proposal found as many as fifteen cosponsors!

With regard to production by the International Seabed Authority, during the last week of the spring session the delegation of the Netherlands came through with a proposal for a unitary joint venture system which conceivably might break the deadlock on the negotiations concerning the so-called parallel system. Included in this selection of documents are the text of the Netherlands's proposal and the response by the delegation of Austria stressing its basic advantages. The proposal was supported by various developing countries; and the Soviet Union declared its readiness to discuss it, provided it was adopted by the Group of 77 and was elaborated in such a way as not to be prejudicial in favor of private companies in the capitalist countries and against the principles and interests of the socialist states.

Finally, a statement by Paul Engo, chairman of the First Committee, is included. Though the title of the statement indicates a concern with technology transfer, the substance of the statement deals with the crucial and practical problem of obtaining skilled manpower for the proposed International Seabed Authority while maintaining geographical balance in staffing. This is another area in which the conference attempts to look beyond its mandate and respond to problems just over the horizon.

THE EDITORS

MESSAGE FROM THE SECRETARY-GENERAL TO THE THIRD UNITED NATIONS CONFERENCE ON THE LAW OF THE SEA[1]

Mr. Zuleta (Special Representative of the Secretary-General) read out the following message from the Secretary-General:

> As in past sessions of the Conference on the Law of the Sea I have been following with the greatest interest and attention the developments in this eighth session and the process of your negotiations.
>
> I am gratified to see that you have agreed to the same plan of work as the one adopted during your last session, namely to deal with the 'hard core' remaining issues in informal groups, and that there seems to be general agreement that this must be the final negotiating session.
>
> I wish to express my appreciation to the President of the Conference as well as the Chairmen of the Main Committees and of the Negotiating Groups for their efforts to bring this Conference to the point where a consensus on a universal treaty for a new legal order for the seas may prove feasible.
>
> The international community has given this Conference the great responsibility of adopting a legal régime to cover the traditional uses of the sea, the rational management of living and non-living resources, the preservation of the marine environment, the new scientific frontier, and the exploration and exploitation of the sea-bed beyond the limits of national jurisdiction as the common heritage of mankind. It has been agreed that the problems of ocean space are closely interrelated and need to be considered as a whole.
>
> On the outcome of this Conference depends whether all these problems can be solved under the rule of law or whether they will be left in a legal vacuum that can

1. Contained in the United Nations Third Conference on the Law of the Sea, Eighth Session, *General Committee, Provisional Summary Record of the 45th Meeting* (A/CONF.62/BUR/SR.45) (April 12, 1979), held at the Palais des Nations, Geneva, April 9, 1979, pp. 2–3.

only increase inequities and widen the gap between developing and developed nations. But there is still more involved in this Conference. You know how many vital principles and interests are at stake. Should the Conference not succeed, world public opinion will further question whether Governments have the resolve to use fully the machinery of the United Nations to achieve international understanding on global issues.

After seven sessions and a total of 54 weeks it is clear to me that the process of negotiation must be concluded. This Conference has reached a point where definitive positions have to be taken and difficult decisions have to be made. This point of view is shared by practically all the States represented in this Conference; some have even made public their position to the effect that no efforts should be spared to bring this Conference to a conclusion as soon as possible.

If we do not act now and thus lose the moment to complete the decision-making process leading to the adoption of a comprehensive law of the sea convention, we risk being overtaken by events that will make it more difficult, if not impossible, to reach agreement at a later stage on a new legal order for the oceans.

I therefore strongly appeal to all participants in this Conference to make every effort to conclude negotiations on those outstanding issues on which consensus has still to be reached. We have come to the moment of decision and we cannot afford to fail.

The Chairman requested the Special Representative of the Secretary-General to convey to the Secretary-General the Conference's gratitude for his message. He was sure that all participants in the Conference fully agreed with the sentiments which had been expressed by the Secretary-General.

LETTER DATED 24 APRIL 1979 FROM THE CHAIRMAN OF THE GROUP OF 77 ADDRESSED TO THE PRESIDENT OF THE CONFERENCE[1]

I have the honour to enclose with this letter the document concerning the question of unilateral legislation which was drawn up by the Group of Legal Experts of the Group of 77 under the chairmanship of Ambassador Roberto Herrera Caceres of Honduras. I would be grateful if you would arrange for it to be circulated as a document of the Conference to the participating States.

The Group of Legal Experts has held various meetings during the present session and will continue its work during the coming months, with the object of contributing to the definition of the legal position of the Group of 77 in defence of the common heritage of mankind.

Accept, Sir, the assurances of my highest consideration.

Mario Carias

Ambassador, Head of the Delegation of Honduras
Chairman of the Group of 77

1. United Nations Third Conference on the Law of the Sea, Eighth Session (A/CONF.62/77) (April 25, 1979), Geneva, March 19–April 27, 1979.

LETTER DATED 23 APRIL 1979 FROM THE GROUP OF LEGAL EXPERTS ON THE QUESTION OF UNILATERAL LEGISLATION ADDRESSED TO THE CHAIRMAN OF THE GROUP OF 77

In various forums, both regional and world-wide, the Group of 77 has repeatedly declared its clear legal conviction concerning the binding nature of the principles set out in resolution 2749 (XXV) of 17 December 1970, and its position regarding the unilateral initiatives and proposals of a small group of States which are seeking to explore and exploit the resources of the Area by means of so-called provisional or transitional measures.

For this reason, and in order once again to express these convictions in the most precise form possible, it was decided to establish a Group of Legal Experts composed of 12 jurists from all regions of the developing world, including members of the International Law Commission. The members of this group are: Dr. Roberto Herrera Caceres (Honduras), Ambassador to Belgium, the Netherlands, and the EEC (chairman); Professor Madjid Bencheickh (Algeria), Professor of Law; Professor Mohamed Bennouna (Morocco), Dean of the Faculty of Law, Rabat; Dr. Jorge Castañeda (Mexico), Ambassador, member of the International Law Commission; Dr. S. P. Jagota (India), Ambassador, Under-Secretary, and Legal Adviser to the Ministry of Foreign Affairs, member of the International Law Commission; Dr. Julio Cesar Lupinacci (Uruguay), Under-Secretary, Ministry of Foreign Affairs; Mr. Biram Ndiaye (Senegal), Professor of Law, University of Dakar; Dr. Frank X. Njenga (Kenya), Under-Secretary, member of the International Law Commission; Dr. Christopher Pinto (Sri Lanka), Ambassador to the Federal Republic of Germany, member of the International Law Commission; Dr. K. Rattray (Jamaica), State Counsel, Department of Public Prosecutions; Dr. S. Suchariktul (Thailand), Legal Adviser, member of the International Law Commission; and Dr. Mustafa Yasseen (United Arab Emirates), Counsellor, Permanent Mission at Geneva. The Group of Legal Experts has worked throughout the present session and will continue to work in defence of the common heritage of mankind.

The Group of Legal Experts has noted the following basic points:

1. *Development of the International Law of the Sea*

Neither the 1958 Convention on the High Seas nor general international law includes among the freedoms of the high seas the exploration and exploitation of the mineral resources of the sea-bed and the ocean floor beyond the limits of national jurisdiction.

According to Malta's proposal of 1967 envisaging a declaration and a treaty designed to reserve exclusively for peaceful purposes the sea-bed and ocean floor and the subsoil thereof beyond the limits of national jurisdiction and the use of their resources to the benefit of mankind, this area was to be declared the common heritage of mankind, the exploration and exploitation of which would be to the benefit of mankind as a whole, with special regard for the need to promote the economic development of the developing countries, on a basis of true equality.

General Assembly resolutions 2340 (XXII) of 18 December 1967, and 2467 (XXIII) of 1968 continued to define this concept, by then with the general consensus. It was decided, therefore, to establish an *Ad Hoc* Committee, and later a Committee on the Peaceful Uses of the Sea-Bed and the Ocean Floor beyond the Limits of National

Jurisdiction, one of whose special functions was to study the elaboration of the legal principles and norms which would promote international co-operation in the exploration and use of the area and ensure the exploitation of its resources for the benefit of mankind. In one of these resolutions, the Assembly expressed its conviction that the exploitation of the Area should be carried out under an international regime including appropriate machinery, considering that, pending the establishment of such a regime, States and persons, natural or juridical, were bound to refrain from any activities involving exploitation of the Area's resources.

By resolution 2574 (XXIV) of 1969, the General Assembly requested the Secretary-General of the United Nations to conduct the necessary consultations with a view to convening a conference on the law of the sea to review the regimes of the high seas, the continental shelf, and other areas, particularly in order to arrive at a clear, precise, and internationally accepted definition of the area of the sea-bed and ocean floor and the subsoil thereof lying beyond the limits of national jurisdiction, in the light of the international regime to be established for that area. The same resolution also provides that States and all persons are bound to refrain from all activities of exploitation of the resources of the area pending the establishment of an international regime, including appropriate international machinery. This provision was reiterated, *inter alia,* by resolution 52 (III) of 19 May 1972, of the United Nations Conference on Trade and Development.

General Assembly resolution 2749 (XXV) of 17 December 1970 (Declaration of Principles Governing the Sea-Bed and the Ocean Floor, and the Subsoil Thereof, Beyond the Limits of National Jurisdiction), incorporated the mandatory principles which basically regulate activities in the area.

By resolution 2750 (XXV) of 17 December 1970, and subsequent resolutions, the General Assembly decided to convene and hold a Conference on the Law of the Sea which would deal with the establishment of an equitable international regime—including international machinery—for the area and its resources, and a precise definition of that area and other interrelated issues.

Consequently, the Third United Nations Conference on the Law of the Sea was convened and is still in progress, now being in its eighth session; the broad consensus on the legal principles underlying the regime being negotiated for the exploration and exploitation of the area, as reiterated in part XI (the Area) of the Integrated Consolidated [*sic*] Negotiating Text, is once again evident.

2. *The Binding Nature of the Fundamental Principles Governing the Area*

The principles set out in resolution 2749 (XXV) (Declaration of Principles Governing the Sea-Bed and the Ocean Floor, and the Subsoil Thereof, Beyond the Limits of National Jurisdiction) are legally binding principles which were proclaimed in this Declaration and upheld by the affirmative vote of 108 States. It should be added that a number of the few States (14) which abstained on that occasion, although without formulating any objection, subsequently expressed, either explicitly or implicitly, their support for those principles, as did other States members of the international community, thus recognizing by their attitude the force of international custom as expressed in resolution 2749.

This custom has given rise to new general principles of public international law which are the basis or legal foundation of any substantive norms regulating the exploration of the area of the sea-bed and the ocean floor and the subsoil thereof and the exploitation of their resources.

3. *Normative Relationship of the Principles Applicable to the Area*

The principle that the sea-bed and ocean floor and the sub-soil thereof beyond the limits of national jurisdiction and the resources of the Area are the heritage of mankind, and the complementary principles according to which the Area is incapable of being appropriated, the need for an international regime including international machinery which would guarantee the activities carried on in the Area for the benefit of all mankind and not only for that of some States, its peaceful use and other principles contained in the said Declaration—all these form a normative unity that is indivisible and applicable to the Area. This normative unity consolidates the applicable principles laid down in the Charter of the United Nations and the Declaration on Principles of International Law concerning Friendly Relations and Co-operation among States.

4. *Legal Status of the Area*

The customary principle of the freedom of the high seas is not an absolute principle; it does not apply to the exploitation of the sea-bed and ocean floors beyond national jurisdiction, because the exploitation thereof was beyond the capacity of States at the time when that principle came into being.

But even on the assumption that this customary principle would be applicable to this exploitation, it would certainly have ceased to be applicable in consequence of the Declaration of Principles of 1970, not only because the Declaration is a resolution adopted by the General Assembly but also because that Declaration is an event reflecting a conviction incompatible with *opinio juris sive necessitatis* indispensable to the operation of the principle as an international custom in the exploitation of the sea-bed or ocean floor beyond national jurisdiction.

There is an obvious difference in legal status as regards the superjacent waters of the Area and as regards the sea-bed, sub-soil and resources of the Area.

Whereas the legal status of the superjacent waters is that of a *res communis,* the legal status of the sea-bed, sub-soil and resources thereof is that of an indivisible and inalienable common heritage of mankind to be explored and exploited for the benefit of mankind as a whole through the equitable participation of the States in the benefits to be derived therefrom, with special regard for the interests and needs of the developing countries, whether coastal or land-locked countries. The foregoing proposition is reaffirmed by the principles according to which the Area is incapable of being appropriated and that there is no such thing as acquired rights in the Area, the responsibility of States for whatever is done in the Area that is harmful to all mankind and the principle of the equitable participation of States in the economic benefits derived from the exploitation of the Area.

5. *The Legal Principles Applicable to the Area and Unilateral Acts or Limited Agreements for Its Exploration and Exploitation*

The principles of law laid down in resolution 2749 (XXV) form the basis of any international regime applicable to the Area and its resources.

All activities connected with the exploration and exploitation of the Area and other related activities will be governed by the international regime to be established by the conclusion of an international treaty that is generally acceptable and includes appropriate international machinery for implementing the principles of law referred to.

Consequently, any unilateral act or mini-treaty is unlawful in that it violates these principles, for the legal regime, whether provisional or definitive, can only be established with the consent of the international community as the sole representative of mankind and in conformity with the system determined by the international community.

The adoption of unilateral measures, draft legislation, and limited agreements would merely be an event without international legal effect and hence incapable of being invoked vis-à-vis the international community.

The great majority of States would not admit the validity of such legislation, nor could such legislation constitute valid grounds for any juridical claim to explore or exploit the Area. Furthermore, if such unilateral legislation or mini-treaty should be put into operation, the international responsibility of the States concerned would be engaged in respect of damage caused by such activities incompatible with the principles applicable in the Area.

It should be stressed that no investor would have any legal guarantee for his investments in such activities, for he would likewise be subject to individual or collective action by the other States in defence of the common heritage of mankind, and no purported diplomatic protection would carry any legal weight whatsoever.

6. *Rule of Law*

It is a function of international law to avoid the possibility that, through relationships of strength, a State might endeavor to settle by force what cannot be settled by means of the law. This can happen in cases where a claim is made subsequently to repudiate a rule that was accepted when it was formulated.

More than 119 States have reaffirmed their constant support for the respect of customary international law as the basis for the general principles of law that fundamentally apply in the Area declared as the common heritage of mankind, and their support for the principles and rules referred to above. This largely representative body of mankind should not be ignored by any one State or by a small number of States purporting to claim a *de facto* authority over all humanity.

The binding legal nature of the applicable principles and rules of international law, including those laid down in the Charter of the United Nations and those proclaimed in the Declaration of Principles of International Law concerning Friendly Relations and Co-operation among States in accordance with the Charter of the United Nations adopted by the General Assembly on 24 October 1970, emphasize the duty and the interest of maintaining international peace and security and promoting co-operation and mutual understanding among nations. Special emphasis should be given to the duty to perform fully and in good faith the obligations entered into by States by virtue of the generally recognized principles and rules of international law.

The conclusion of a mini-treaty or the adoption of unilateral legislation and any attempt to carry them into effect would likewise be inconsistent with the principles of good faith in the conduct of negotiations at international plenipotentiary conferences like the Third United Nations Conference on the Law of the Sea, which has been engaged since 1973 in efforts to work out a treaty on the exploration and exploitation of the seabed and its resources on the basis of the principles laid down in resolution 2749 (XXV) referred to earlier and in the resolutions convening the sessions of these Conferences. Accordingly, it should be the objective of all States Members that constitute humanity to ensure that the Third Conference on the Law of the Sea achieves a satisfactory result as soon as possible.

DR. H. ROBERTO HERRERA CÁCERES
Chairman

STATEMENT BY THE VICE-CHAIRMAN OF THE DELEGATION OF THE UNITED STATES OF AMERICA IN RESPONSE TO THE STATEMENT BY THE CHAIRMAN OF THE GROUP OF 77 REGARDING DEEP SEA-BED MINING LEGISLATION (A/CONF.62/89)[1]

It is regrettable that controversy has been introduced once again into the deliberations of this Conference, which can ill afford distraction from its goal of forging consensus on a comprehensive legal regime for the use and management of the oceans and their resources. In light of the full and repeated explanations of views and positions to which the Conference has already been exposed, most recently on 28 August and 15 September 1978, and 19 March 1979, I shall respond as briefly as possible to the contention that the enactment of national legislation designed to regulate the conduct of deep sea-bed mining, and exploration and exploitation activities undertaken beyond the limits of national jurisdiction, would be illegal and potentially disruptive to this Conference.

My Government rejects outright the notion that United Nations General Assembly resolutions, including United Nations General Assembly resolutions 2574D (XXIV) and 2749 (XXV) and irrespective of the majorities by which such resolutions are adopted, are legally binding on any State in the absence of an international agreement that gives effect to such resolutions and that is in force for that State. Clear statements of our position are on public record, including those made in the course of debate accompanying the passing of the United Nations General Assembly resolutions just mentioned and those made in the course of the unfortunate exchanges on this subject that have taken place during the Law of the Sea Conference and in UNCTAD.

There exists nothing in customary or conventional international law that precludes governments from acting to regulate the activities of their citizens or that forbids governments or private persons or entities access to the sea-bed beyond the limits of national jurisdiction for the purposes of exploring for and exploiting the resources there. Should the Conference succeed in producing a treaty that establishes an international regime for the regulation of such exploration and exploitation, those States for which that treaty is in force will forego the exercise of these high seas freedoms. But for States not bound by such a treaty, there are no legal impediments to these activities.

EIGHTH SESSION (FIRST PART) OF THIRD UNITED NATIONS CONFERENCE ON LAW OF SEA, GENEVA, 19 MARCH – 27 APRIL

NEGOTIATING TEXT REVISION TO BE BASIS FOR RESUMED NEW YORK SESSION[2]

Consensus on some "hard-core" issues and progress towards a compromise on sea-bed exploitation were reported at six weeks of meetings by the Third United Nations Conference on the Law of the Sea which ended Friday night, 27 April, at the Palais des Nations in Geneva.

1. Circulated at the request of the Representative of the United States of America. UN Third Conference on the Law of the Sea, Eighth Session (A/CONF.62/93) (October 1, 1979).

2. UN Department of Public Information, Press Section, United Nations Press Release SEA/360 (April 30, 1979).

Delegates from all groups of countries—developing, industrialized, and socialist—hailed what many of them described as substantial progress on the sea-bed question. But the Conference fell short of consensus on this crucial element for the law of the sea convention which is the aim of the Conference. It decided to resume the session for another six weeks, from 16 July to 24 August, at United Nations Headquarters in New York.

Highlights of the session were the completion of six years of negotiations on a legal code for the prevention of marine pollution and protection of the ocean environment, as well as an "improved prospect for consensus" on issues of particular importance to land-locked and geographically disadvantaged States, including access of the land-locked to the economic zone fisheries of nearby coastal States. The Third Committee of the Conference, which handled the pollution question, also finalized a text on the development and transfer of marine technology, but left work still to be done on rules to govern marine scientific research in the economic zones and on the continental shelves of foreign States.

Some narrowing of differences was reported on two critical questions—the breadth of the continental shelf and the delimitation of maritime boundaries where economic zones or continental shelves overlap. At the final meeting a compromise proposal on the continental shelf was presented by the Chairman of the group dealing with that question. But both these questions remain on the list of unresolved "hard-core" topics on which the Conference has been concentrating since the spring of 1978, when it last met at Geneva.

A text revision incorporating some of the results of negotiations since March 1978 is to be issued by the Conference's main officers as a basis for negotiations at the New York part of the session. The President, the Chairmen of the three main committees, and other officers met early Saturday morning, 28 April, following the end of the Conference's meetings at Geneva, and unanimously decided on what new elements to include in the text, which is a revision of the Informal Composite Negotiating Text (ICNT) prepared for the Conference in 1977.

The progress reported on sea-bed issues took the form of revised—and in some cases considerably expanded—proposals for treaty provisions on the system that is to govern future exploitation of the untapped mineral wealth on the deep ocean bottom in areas beyond the jurisdiction of any State. The focus of the negotiations was on detailed arrangments to ensure the workability of the system contemplated for this area, in which State and private enterprises would exploit the manganese nodules on the ocean bottom in parallel with an unprecedented International Sea-bed Authority that would have its own deep-sea mining organ, named the Enterprise.

Compromise proposals on this subject, based on the results of six weeks of intensive private negotiations since the Conference's eighth session opened on 19 March, sought to balance the concerns of industrialized and developing countries. Emerging from the negotiations, with varying degree of support falling short of consensus, were proposals that concentrated on ways to strengthen the two legs on which the Enterprise would have to stand—finance and access to technology. The authors of those proposals were the Chairmen of three negotiating groups set up last year to deal with sea-bed issues.

The Conference embarked on a procedural innovation at its Geneva session by setting up a Working Group of 21, divided almost equally between developed and developing countries, to seek a negotiated package on the sea-bed. Heretofore, most negotiating bodies established by the Conference have been open to all States participating. This body was unable to resolve most of the issues, but at a meeting on 26 April, which the Conference devoted to assessing its work, many speakers concluded that there had been considerable progress in sketching in the details of the future system.

On the opening day of the session, the issue of unilateral sea-bed mining legislation was raised by Honduras on behalf of the Group of 77 developing countries. It described plans for such national laws as "in violation of international law" and "a grave danger for the work and indeed the very future of the Conference." Disagreeing with this view, the United States said the legislation it contemplated was aimed at providing a legal framework so that sea-bed exploitation would not be delayed. It added that national laws would be superseded when the law of the sea convention went into force.

During the final week of the Geneva meetings, a letter on this subject from the Group of Legal Experts of the Group of 77 was circulated. It set out six points to buttress the Group's position that such unilateral legislation would be invalid and inconsistent with the principle of good faith in international negotiations.

Attending the Geneva meetings were some 1,100 delegation members from 139 States. (The participating countries are listed in Press Release SEA/355 of 30 April.)

The Conference agreed that the present structure of three main committees, seven negotiating groups and the Working Group of 21 should be maintained at the resumed session in New York, and that arrangements would be made for all outstanding questions to be discussed. The work of the first three weeks is to concentrate on sea-bed matters.

The resumed session is to be preceded by three days of preparatory meetings by the Group of 77 and by any other regional groups which wish to meet, from 16 through 18 July. The formal resumption of the session will take place on 19 July.

The aim of the Conference, fixed by the General Assembly, is to draw up a convention covering all aspects of the law of the sea. (For a brief history of the Conference, and background on the eighth session, see Press Release SEA/339 of 14 March.)

REVISION OF TEXT

The decisions on what to include in the revision of the negotiating text were taken unanimously shortly after the close of the Geneva part of the session at a meeting of the President and the Chairmen of three main committees, joined by the Chairman of the Drafting Committee and the Rapporteur-General of the Conference. These are the officials entrusted by the Conference with drawing up a revised text, on the basis of the results of negotiations.

This group decided to incorporate the following in the new revision:

—All of the compromise proposals suggested by the Chairmen of the three negotiating groups on the sea-bed;

—The compromise on access by land-locked developing countries and "States with special geographical characteristics" to the living resources of the exclusive economic zones of States in the same subregion or region, presented in May 1978 by the Chairman of Negotiating Group 4 on this subject, Satya W. Nandan (Fiji);

—The new suggested compromise by Negotiating Group 6 Chairman Andrés Aguilar (Venezuela) on the breadth of the continental shelf and on revenue sharing of shelf resources beyond 200 miles from shore;

—The compulsory conciliation formula for the settlement of disputes pertaining to the sovereign rights of States in the exclusive economic zone, presented in May 1978 by the Chairman of Negotiating Group 5 on this subject, Constantine Stavropoulos (Greece);

—Proposals informally agreed to this year by the Third Committee on protection and preservation of the marine environment and on the development and transfer of marine scientific research;

—The slightly revised article on delimitation of the territorial sea between adjacent and opposite States, presented last May by Eero J. Manner, Chairman of Negotiating Group 7 (but not the compromises suggested by Chairman Manner on delimitation of economic zones and continental shelves and on the settlement of disputes thereon);

—A proposal by Yugoslavia, made in September, on general provisions for the proposed Law of the Sea Tribunal.

The officers of the Conference met on this matter in accordance with a decision taken by the Conference last April that any revisions of the negotiating text "should emerge from the negotiations themselves and should not be introduced on the initiative of any single person . . . unless presented to plenary and found, from the widespread and substantial support prevailing in plenary, to offer a substantially improved prospect of a consensus." The Conference had decided at the same time that revision of the text should be the "collective responsibility" of the President and the Chairmen of the three main committees, "acting together as a team headed by the President."

In announcing the decision on the revision of the text (see Press Release SEA/358 of 28 April), President Amerasinghe said the new text would replace the former one as a basis for negotiations. He added that nothing in the text was binding on any delegation; it was a negotiating text, not a negotiated one.

The next revision of the negotiating text will be the fourth in a series of progressively modified negotiating documents, the first of which appeared in 1975. All have been set out in the form of treaty articles covering virtually the entire gamut of topics to be included in the future convention. The previous versions were the "single negotiating text" (1975), "revised single negotiating text" (1976), and "informal composite negotiating text" (1977).

ORGANIZATION OF WORK

The Conference continued to work through its three main committees, set up at its first session in 1973, as well as through its seven negotiating groups, established last April. (Negotiating Group 5, on settlement of economic zone fisheries disputes, did not meet, having substantially completed its task last year.) It also set up two new bodies, the Working Group of 21 on First Committee Matters and the Group of Legal Experts on settlement of sea-bed disputes. An informal plenary meeting was also held on dispute settlement. The Drafting Committee continued wording of various texts in the six official languages of the Conference.

All of the negotiating groups as well as the two new bodies met in informal, private session. All meetings of main committees were public and were held during the closing week of the session to consider reports on the negotiations. The Conference did the same in plenary session with regard to Negotiating Group 5, whose Chairman reported on the results of consultations outside the Group.

The Conference initially set the first three weeks for negotiations and then extended them through the first two days of the final week. Negotiations on the sea-bed were conducted at first in the three negotiating groups on this subject, but were then moved to the Working Group of 21, which held 14 meetings between 9 and 24 April. The membership of this group, divided almost equally between developed and developing countries, was chosen by the respective groups of States at the Conference and varied somewhat depending on the issue being discussed. Its Chairman was Paul Bamela Engo (United Republic of Cameroon), who was also Chairman of the First Committee (international sea-bed area).

The Chairmen of the main committees each reported to the Conference during the final two days of the Geneva session on the work done on matters of concern to his committee, including the work of the various Conference groups. The two other Chairmen are Andrés Aguilar (Venezuela), Second Committee (general aspects of the Law of the Sea), and Alexander Yankov (Bulgaria), Third Committee (marine environment, research, and technology).

The composition of other bodies remained as it was last year, except that Belgium replaced Ireland as a Vice-President and member of the General Committee.

SEA-BED MINING SYSTEM

Negotiations on the future system for exploiting the international sea-bed area focused on filling in the many controversial details of how the system would work in practice. The problem was described in this way by Frank X. Njenga (Kenya), Chairman of Negotiating Group 1, which dealt with the exploitation system: "This was an enormous task since we had to imagine ways and means to equalize two different kinds of entities—the powerful consortia with all their capital and credit, technology and organization, some of which are already engaged in sea-bed operations; and the Enterprise, an entity so far existing only in our imagination, a creature to be born without the necessary tools to fulfil the purpose for which it is created."

The negotiations concentrated on two main aspects of the problem: how to ensure that the International Sea-Bed Authority, and especially its Enterprise, would have enough money to compete effectively with governmental and private sea-bed miners, and how to ensure that it would be able to obtain the technology it needed.

The basics of the system had been generally agreed upon at earlier sessions of the Conference. They call for the creation of what is known as a "parallel system," in which both the Enterprise and other entities—governmental and private—would be entitled to mine the deep sea-bed, defined as the part of the ocean bottom beyond national jurisdiction. The entire system would be under the control of the Authority, an intergovernmental body which would operate it in the interests of all mankind, on the fundamental premise that the sea-bed is the common heritage of mankind and that all should share in its benefits.

Everyone who wanted to explore or exploit the deep sea-bed could do so only if the Authority granted him a contract. Mining entities which met the financial and technical criteria to be specified in the Law of the Sea Convention would be guaranteed access to the area, except when the Authority decided for economic reasons to limit the production of sea-bed minerals.

To obtain a contract, an applicant would have to prospect the area (or have it prospected for him) and then present the Authority with two mineral-bearing areas of equal commercial value. The Authority would select one, leaving the other for the successful applicant, who could mine it in accordance with a plan of operations which the Authority would be empowered to approve. The Authority's mine sites would constitute the reserved area of the sea-bed—for use either by the Enterprise or by developing countries. Other miners would operate in the non-reserved area as contractors.

This parallel system would be established for an initial period of 20 years. At the end of that time, a review conference would be called to decide whether the system needed to be changed.

These features of sea-bed exploitation being generally agreed, the Conference has been trying for the past couple of years to fill in the details in binding treaty language.

Financing the Enterprise

On the Enterprise side of the parallel system, financial arrangements have been among the main stumbling-blocks thus far. There is general agreement that the Enterprise must be provided with funds for at least its first mining project. Substantial sums would be involved, to the tune of hundreds of millions of dollars for each mine site. According to proposals before the Conference prior to this year, a large part of this would be obtained from loans floated by the Enterprise, to be guaranteed by all States parties to the convention. Other funds would come eventually from the Authority's receipts from taxation of the proceeds of miners other than the Enterprise. During the latest Geneva session a third avenue was explored—cash to be supplied directly by States parties to the convention.

According to Tommy T. B. Koh (Singapore), Chairman of Negotiating Group 2 on financial arrangements, the Group of 77 developing countries were not satisfied that proposals he had made last year provided adequately for the financing of the Enterprise's first project. Reporting to the First Committee on 25 April, he said they had criticized his proposal that the capital structure of the first project should be one-third cash and two-thirds debt. As a new organization with no assets, they felt the Enterprise should not be burdened by an undesirably large debt component. They also argued that, instead of spreading the financial burden among States according to an assessment scale similar to that used for contributions to the United Nations regular budget, the States whose nationals or firms were engaged in sea-bed mining should make an extra contribution to the Enterprise.

The response of the major industrialized countries, said Mr. Koh, was that under his 1978 proposals the Enterprise would be assured of the capital needed for its first project. They had pointed out that a ratio of one part cash or equity to two parts debt was normal in commercial practice. They had also argued that the Enterprise should not be treated as an "object of charity" but should be expected to manage its affairs efficiently and in accordance with normal commercial practice.

In the final week of the session Mr. Koh made two changes in his proposals on this matter. First, he raised the cash/debt ratio from 1:2 to 1:1, as the Group of 77 had wanted. Second, he proposed that this cash component, now amounting to half rather than a third of the total funds needed for the first mining project, should be in the form of long-term, interest-free loans, to be refunded when the Authority's Assembly so decided.

As to who should pay these cash contributions and on what basis, he said he had been unable to secure general agreement on the idea that they should be divided into two parts, one payable by all States parties and the other by States engaged in sea-bed mining and possibly also those which were major importers of the minerals found on the sea-bed. He suggested that this issue be deferred for further negotiation.

As another approach to this problem, Norway suggested during the negotiations the creation of an Establishment Fund equivalent to 20 percent of the capital required by the Enterprise. The money would come from mandatory assessments on all States parties.

Much of the work of Negotiating Group 2 was devoted to another major source of the Authority's funds—the fees and charges it would impose on sea-bed miners. Chairman Koh proposed last year a plan envisaging two main schemes of payment, either of which could be chosen by a contractor. The first would be a production charge, similar to a royalty, fixed at a certain percentage of the market value of metals processed from sea-bed nodules. This type of payment scheme is favoured by the socialist countries. The alternative choice, preferred by market-economy countries, would be a mixed system combining a production charge with payment of a share of the contractor's net proceeds.

Mr. Koh made new proposals early in the 1979 session adhering to the same scheme but revising the rates. Those proposals, he reported, were acceptable to the Soviet Union as a compromise, but not, for opposite reasons, to the other industrialized countries or to the developing ones. The Group of 77, believing they would not provide adequate revenues to the Authority, at first proposed that sea-bed contractors also be required to make a lump-sum payment of $60 million to the Authority for each mine site—half payable just after they signed the contract and the rest when they began commercial production. Later, the Group suggested the imposition of a "super-tax" on contractors who earned high profits. The industrialized countries, on the other hand, had criticized Mr. Koh's figures as too high and had complained that his fixed-rate scheme was not sensitive enough to variation in the contractor's profitability.

Mr. Koh announced revised proposals late in the session, containing what he described as two minor changes in rates which would reduce the burden on the contractors and the receipts of the Authority by a total of only $30 million.

Reviewing the situation in a report to the Conference on 26 April, First Committee Chairman Engo said the proposals before the negotiators were "designed to achieve a balance among such elements as the level of revenues accruing to the Authority and the need to devise a system of taxation that would be acceptable to countries with different economic and social systems, and also one that would respond sensitively to the contractor's profitability, thus enhancing its attractiveness to the investor." In his view, Mr. Koh's latest proposals "substantially enhance the prospects for achieving a consensus."

Consensus was announced during the session on one aspect of financing—the provisions covering the Authority's administrative operations.

Technology Transfer

Another major requirement for putting the Enterprise on its feet is access to equipment and know-how for sea-bed mining. It was generally agreed at previous sessions that sea-bed contractors would be obliged to help the Enterprise in this regard by assisting it in obtaining such technology. Exactly what this entailed was the subject of intensive negotiations at the Geneva session.

Mr. Njenga, Chairman of Negotiating Group 1, expressed the view that the most important outcome of the work of his Group had been a considerable expansion of his 1978 text on technology transfer. A number of details were added: most recently, in the last week of the session, these included a clause obliging the applicant for a mining contract to inform the Authority where the equipment he was to use was available on the open market, and a proviso that the Enterprise might require a contractor to supply a particular item of technology only if it could not buy a similar item on reasonable terms. Also added was a definition of technology.

As Chairman Engo pointed out, the fact that the Enterprise would be empowered not only to mine the sea-bed but also to process the minerals it recovered there gave rise to a discussion as to whether the obligation to transfer technology ought to cover processing technology as well. He noted that the nature and scope of such transfers had not yet been agreed. On this point, one of Chairman Njenga's latest proposals was a clause that would impose concrete obligations on States parties engaged in mining with regard to the transfer of processing technology.

Although some measure of agreement was emerging as to the contractor's obligations in respect of technology transfer, Mr. Engo reported, there was still no agreement on the transfer of technology to developing countries or on a dispute settlement and enforcement procedure in cases where a contractor and the Enterprise differed over obligations in this area.

Joint Ventures

The Netherlands advanced a suggestion during the last week of the session which was described as an additional way for the Enterprise to take part in sea-bed mining operations and secure both income and technology. It would give the Authority the option to enter into a joint mining venture with any outside contractor, including the right to participate in up to 20 percent of the capital of such an undertaking. If the Authority exercised that option, the contractor would have a similar option to join in a mining venture undertaken by the Enterprise. Some developed and developing countries welcomed the proposal as deserving of further study.

Production Policies

One article in the 1977 negotiating text contains a formula empowering the Authority to control sea-bed mining in such a way that enough minerals from that source will be produced to enable consumers to benefit, while developing countries which mine the same minerals on land would be protected from adverse effects of sea-bed production. Last year at Geneva, an expert group came up with a revised formula that would permit the application of a production ceiling linked to the growth of world nickel consumption.

This year, an informal group chaired by Satya N. Nandan (Fiji) produced a text filling in further details of this scheme. This was hailed by Canada, a major nickel producer, as a breakthrough. Chairman Engo said proposals were emerging which would make the Authority's production policies more flexible while taking the just concerns of land-based producers fully into account. The proposals which he read into the record would permit the Authority to allow sea-bed miners to exceed their planned annual production levels under certain circumstances, and to permit temporary adjustments to production levels when "a *force majeure* exists among other producers creating a significant imbalance in the world supply/demand relationship."

Mr. Nandan identified the most difficult issue as the level of the production ceiling and the number of mine sites that would be permitted to operate. He expressed confidence that the problems would be solved.

Other Features

Many other aspects of the workings of the parallel sea-bed exploitation system were dealt with in the negotiations and in Chairman Njenga's proposals. Among them are new provisions aimed at preventing monopolization of sea-bed mining by a single country, as well as new rules and criteria for the selection of applicants for mining contracts. The latter would lay down objective standards to permit the Authority to choose among applicants when a production ceiling limited the number whose requests could be approved.

Still other clauses dealt with the extent to which the Enterprise's applications to mine should be given priority over those of other entities—an approach favoured by developing countries but opposed by the industrial ones.

Mr. Njenga's assessment was that his proposals brought the negotiations "considerably closer to a final solution." He described them as "the fairest, most balanced and most feasible compromise attainable at this stage."

Organs of the Authority

No progress was reported on the long-standing issue of the voting system for decision making in the Council, the Authority's executive organ. But changes were agreed in the wording used to define two of the special interest groups of States that would be entitled to representation on the Council—those engaged in sea-bed mining, and large mineral consumers or importers.

The first category, which would be entitled to four seats on the 36-member Council, is now defined as "the eight States Parties which have the largest investments in preparation for and in the conduct of activities in the (sea-bed) Area, either directly or through their nationals, including at least one State from the Eastern (Socialist) European region." The second category, also entitled to four seats, is defined as "those States Parties which, during the last five years for which statistics are available, have either consumed more than 2 percent of total world consumption or have had net imports of more than 2 percent of total world imports of the commodities produced from the categories of minerals to be derived from the Area, and in any case one State from the Eastern (Socialist) European region."

Consensus was reported on a number of non-controversial articles dealing with other organs, including a clause empowering the Enterprise not only to mine the sea-bed but also to transport, process, and market minerals recovered from the area.

Site of the Authority

In previous years, Fiji, Jamaica, and Malta put forward bids to serve as headquarters for the Sea-bed Authority. This week, in letters to President Amerasinghe, the candidature of Jamaica was endorsed by the Latin American Group and that of Malta by the Arab Group. A request that the revised negotiating text should place the three candidates on an equal footing came from the Asian Group and the Western European and Other States Group.

Dispute Settlement

Several proposals for changes in provisions on the settlement of disputes relating to the sea-bed emerged from the newly created Group of Legal Experts on this subject, headed by Harry Wünsche (German Democratic Republic).

Chairman Wünsche reported to the First Committee that his Group had moved forward in efforts to reword the article that defines the kinds of disputes to be handled by the Sea-Bed Disputes Chamber of the proposed Law of the Sea Tribunal, and who would have access to that Chamber. Another topic discussed was the question of what limits should be imposed on the Chamber's power to question decisions by the Authority. General agreement had been reached to allow only the Assembly and the Council, among the Authority's organs to request advisory opinions of the Tribunal.

He suggested the addition of a clause prohibiting staff members of the Authority from disclosing, even after they left the staff, any industrial secret they might have picked up while working for the Authority. He noted that a number of other matters required further attention, including the procedure for selecting members of the Sea-bed Chamber.

ACCESS TO ECONOMIC ZONE FISHERIES BY LAND-LOCKED AND OTHER STATES

A "substantially improved prospect for a consensus" was reported to the Conference on issues dealing with the access of developing land-locked States and "States with special geographical characteristics" to the fisheries in the economic zones of nearby coastal States.

The 1977 negotiating text contained an article that would give land-locked countries, as well as certain developing coastal States, the right to participate equitably in the exploitation of the living resources of the exclusive economic zones of adjoining States. The wording of this provision had been endorsed by the Group of Land-Locked and Geographically Disadvantaged States, but had not been accepted by coastal States.

Last April, the Conference gave Negotiating Group 4 the task of elaborating a compromise formula on the issue. Work done at that time resulted in proposals by the Group's Chairman, Satya N. Nandan (Fiji), made in May, for major changes in the articles on utilization of the living resources of the economic zone and the rights of land-locked and other States within the zone. It was these proposals which were placed this month in the "prospect for consensus" category.

Reporting to the Second Committee on 24 April, Mr. Nandan stated that despite consultations held during the month, no substantive changes had been possible in the proposals he had made last year. It was therefore his intention to submit those proposals for inclusion in the revised negotiating text.

These proposals describe in greater detail the circumstances under which the right of access may be exercised. The general principle is that land-locked States and States with special geographical characteristics "shall have the right to participate, on an equitable basis, in the exploitation of an appropriate part of the surplus of the living resources of the exclusive economic zone of coastal States of the same subregion or region, taking into account the relevant economic and geographical circumstances." The terms and modalities of such participation would be established by the States concerned through bilateral, subregional, or regional agreements.

Another article of the text specifies that the "surplus" referred to is that part of the allowable catch in a State's economic zone which that State does not have the capacity to harvest. The land-locked States have reservations on this point. The point is still subject to negotiations as the land-locked and geographically disadvantaged States object to tying their rights to a "surplus" determined by the coastal State on its own.

The text details various elements to be considered when States work out agreement on the issue, including the need to protect the fishing communities or industries of the coastal State, the extent to which States concerned have access to other fisheries, the nutritional needs of the populations of the respective States, and the need to avoid placing a particular burden on any single coastal State.

"States with special geographical characteristics" are defined in the text as "coastal States, including States bordering enclosed or semi-enclosed seas, whose geographical situation makes them dependent upon the exploitation of the living resources of the exclusive economic zones of other States in the subregion or region, for adequate supplies of fish for the nutritional purposes of their populations or parts thereof, and coastal States which can claim no exclusive economic zone of their own."

Most delegations commenting on the Nandan proposals, including Mexico, on behalf of the Group of Coastal States, felt that they balanced to an important degree the interests of coastal States with those of land-locked States and States with special geographical characteristics. The text was thus the best possible compromise formula attainable at the present stage and a significant improvement over the existing provisions.

When the Second Committee disucssed this matter on 24 April, objections to the Nandan proposals were voiced by Ecuador, Pakistan, the Republic of Korea, and Spain among the coastal States and by Iraq, Nepal, the United Arab Emirates, and Zambia among the land-locked and geographically disadvantaged group.

One argument of the coastal State objectors was that the use of the word "right" in reference to access to fisheries by outside States contravened the jurisdiction of the coastal State in the zone. Another objection was that the term "States with special geographical characteristics" lacked precision. From the land-locked States came the objection that the coastal State would have too much latitude in determining its surplus and that inadequate attention had been given to the status of developed land-locked countries.

SETTLEMENT OF ECONOMIC ZONE DISPUTES OVER FISHERIES

A compromise formula for the settlement of disputes arising from disagreements over the exercise of sovereign rights of coastal States in their economic zones was characterized this week by Conference President Amerashinghe as satisfying the criteria for inclusion in a revised negotiating text.

The formula had been submitted to the Conference last May by the Chairman of Negotiating Group 5, Constantine Stavropoulos (Greece). It provides for compulsory conciliation in the event of such disputes. In introducing the text a year ago, Mr. Stavropoulos said that the compromise would meet the interest of two opposing groups of delegations—those who wanted all rights granted under the Law of the Sea Convention protected by effective dispute settlement provisions, and those who felt that their sovereign rights and discretions could not be effectively exercised "if they were to be harassed by an abuse of legal process and a proliferation of applications to dispute settlement prodedures." Compulsory conciliation, he added, had thus emerged as a compromise.

The compromise provides that there would be no compulsory and binding adjudication—involving judgement by a tribunal—for disputes relating to the exercise of the sovereign rights of the coastal States in the economic zone. However, a coastal State would be obliged to submit to conciliation—a more flexible procedure involving nonmandatory recommendations by a panel of conciliators—in three categories of disputes: when the living resources of the zone were endangered by inadequate conservation and management measures, when there was a refusal to determine the surplus of the living resources that may be allocated to other States beyond the capacity of the coastal State to exploit them, or when there was an arbitrary refusal to allocate the surplus to others.

In discussion in the plenary this week, Jorge Castañeda (Mexico), spokesman for the Group of Coastal States, said that while those states believed that disputes arising through the exercise of their sovereign rights in the economic zone should not be subject to any dispute settlement procedure, efforts at compromise had resulted in a solution acceptable both to the coastal States and to land-locked and geographically disadvantaged States.

The agreed text, he added, "represents the maximum consensus we can make" and it was "totally pointless" to reopen negotiations in the issue. The Group of Coastal States, he said, "would be prepared to have this important agreement included in a revised ICNT."

Also supporting inclusion of the Stavropoulos formula in a revised negotiating text were Austria (the spokesman of the Group of land-locked and geographically disadvantaged States), Ecuador, Peru, France, Switzerland, Canada, and Argentina.

Opposing this procedure was the Soviet Union, whose representative recalled that it had not accepted the report of Mr. Stavropoulos when it was submitted last year.

CONTINENTAL SHELF

No consensus resulted from negotiations on how to define the outer limits of the continental shelf and on what arrangements should be made for sharing with the international community part of the revenues derived from its exploitation beyond 200 miles. However, several new proposals were received by Negotiating Group 6 on those questions and its Chairman, Andrés Aguilar (Venezuela), reported that the positions of delegations had come closer.

On the final day of the Geneva meetings, Mr. Aguilar submitted to the plenary his compromise suggestions for a definition of the continental shelf, the exercise of coastal State rights over the shelf and revenue sharing.

In the discussion of the new Aguilar text at the final plenary meeting in Geneva, Ireland and the Soviet Union, authors of earlier proposals on this subject, characterized the new document as a step forward that could help in achieving a consensus. However, the Arab Group maintained its previous position that the shelf should be limited to 200 miles.

The new proposal is more restrictive in its definition of the continental shelf than the 1977 negotiating text, which provided for a minimum breadth of 200 miles but would permit an unlimited extension beyond that distance to the outer edge of the continental margin. In some cases this is several hundred miles from shore.

The new Aguilar proposal combines a geomorphological definition of the shelf—related to the foot of the continental slope—with a limiting clause based on distance and depth. It thus seeks to marry the positions of countries which stress the character of the shelf as the natural prolongation of a State's land territory, with the attitude of other States that feel there must be some limit to the extent of national jurisdiction of States with broad shelves.

The foot of the continental slope is the line at the place where the sea-bed begins to rise steeply from the deep ocean to the shallower off-shore waters. A State would have two options in establishing the outer edge of its continental margin: one based on the thickness of sedimentary rocks beyond the foot of the continental slope and the other by drawing a line 60 miles seaward of the foot of the slope.

Two other limits are proposed in the Aguilar formula for States whose shelves extend beyond 200 miles; 350 miles from shore, or an ocean depth of 2,500 metres plus 100 miles. Each State could choose between these two limits when defining the outer margin of its shelf.

Mr. Aguilar's proposal goes on to provide that information on the limits of the continental shelf when it extends beyond the 200-mile exclusive economic zone would be submitted by the coastal State to a Commission on the Limits of the Continental Shelf, set up on the basis of equitable geographical representation. The Commission would make recommendations to coastal States on matters related to the establishment of the outer limits of the shelf. The limits established by the coastal State, taking into account those recommendations, "shall be final and binding."

A new article proposed by Mr. Aguilar provides that the coastal State's exercise of its rights over the continental shelf "must not infringe, or result in any unjustifiable interference with navigation and other rights and freedoms of other States."

The 1977 negotiating text proposed a plan to have coastal States share with the international community, through the Sea-Bed Authority, part of the revenue which they derive from exploiting oil and other resources from their continental shelf beyond 200 miles. It provided for a sharing of revenue at a rate rising from 1 percent in the sixth year of production to a maximum of 5 percent in the tenth year and thereafter. The new formula proposes a rate that would rise from 1 percent in the sixth year to a maximum of 7 percent in the twelfth year and thereafter. In both proposals there would be a five-year grace period.

Among the new proposals received earlier in the session was one by the Soviet Union dealing with the breadth of the shelf, the emplacement of artificial islands, revenue sharing, and marine scientific research.

A proposal by Ireland, presented in 1976, seeks to define the outer limits of the shelf by reference to the structure of the sea-bed, without setting any outer limit in terms of distance from the shore. A previous Soviet proposal, presented last year, would have fixed an outer limit of 300 miles for the shelf. Both proposals, and Mr. Aguilar's new one, would allow the shelf to extend at least 200 miles from shore, regardless of the configuration of the sea bottom or the depth of water.

This year Libya supported the position set out by the Arab Group in 1978 that the continental shelf should be restricted to 200 miles.

Mr. Aguilar told the Second Committee on 24 April that other informal proposals had been submitted to the private meetings of his Negotiating Group. Two of these were amendments, by Denmark and Sri Lanka, to the Irish formula. The other two, by Sri Lanka and the Netherlands, concerned revenue sharing, and specifically the proportion of revenue derived from the continental shelf beyond 200 miles which would go to the international community.

Several delegations expressed regret that it had not been possible to set up a small group at the session this month to negotiate on the continental shelf issue and hoped that such a group would still be established. Mr. Aguilar reported that while there was unanimous agreement to setting up a small group, delegations could not agree on its composition.

DELIMITATION OF MARITIME BOUNDARIES

The problem of how to draw maritime boundaries between States situated opposite one another across a narrow body of water or adjacent to one another along the same coastline, as well as the related issue of how to settle disputes arising from boundary delimitation, continues unresolved.

In the view of Eero J. Manner (Finland), Chairman of Negotiating Group 7, which has been dealing with those questions, "none of the proposals made during the work of the Group for the modification of revision of the ICNT either secured a consensus within the Group or seemed to offer a substantially improved prospect for a consensus in the plenary."

Last year, agreement was reported on boundary delimitation as it affects the territorial seas of adjacent or opposite States. The principle, as set out in the 1977 negotiating text, is that neither State is entitled to draw the line beyond the midway position between the two coasts, except where "historic title or other special circumstances are involved."

But the problem of where to draw the line when two States' economic zones or continental shelves overlap has proved unsolvable so far, though moves towards a compromise were reported this month. Negotiating Group 7 has sought such a compromise formula to accord with the interests of those States which argue that a line should be drawn equidistant between the two countries in question, and those which insist that emplasis be placed on "equitable principles" in delimiting the boundary.

The 1977 negotiating text contains a provision stating that delimitation "shall be effected by agreement in accordance with equitable principles, employing, where appropriate, the median and equidistance line, and taking account of all the relevant circumstances."

Reporting to the Second Committee on the work of his Group, Chairman Manner expressed doubt whether, "in view of our lengthy deliberations and taking into account the controversies still prevailing, the Conference may ever be in a position to produce a provision which would offer a precise and definite answer to the question of delimitation criteria."

Assuming negotiations on the matter would continue at the next stage of the conference, Mr. Manner submitted the following text as a possible basis for compromise on delimitation criteria: "The delimitation of the exclusive economic zone (or of the continental shelf) between States with opposite or adjacent coasts shall be effected by agreement between the parties concerned, taking into account all relevant criteria and special circumstances in order to arrive at a solution in accordance with equitable principles, applying the equidistance rule or such other means as are appropriate in each specific case." Mr. Manner had reported to the conference last May that there was general agreement that the convention should provide for interim measures to be applied pending agreement or settlement in delimitation cases.

In his latest report to the Second Committee, delivered on 24 April, Mr. Manner observed that while proposals submitted to the Negotiating Group at the current session "seemed to signify a step forward in the search for a compromise, they did not gain such widespread and substantial support that would justify a revision of the ICNT." He added the observation that the most serious difficulty with the proposals concerned what some delegations regarded as "prohibitive references" to activities or measures potentially to be taken during the transitional period. A number of delegations, he said, had criticized proposals which introduced a moratorium on any economic activities in the disputed area pending a settlement.

He suggested a text under which States "shall make every effort with a view to entering into provisional arrangements" and "shall refrain from aggravating the situation or hampering in any way the reaching of a final agreement."

As to procedures for settling boundary disputes, Mr. Manner noted that, in view of opposing arguments on the nature of such procedures, the Group had been unable to resolve the issue and it remained open. There did not seem to be much prospect for finding compromise on the basis of a compulsory procedure entailing a binding decision. "Although it was abundantly clear that several delegations still remain determined to advocate compulsory and binding procedures, it would seem similarly clear that a consensus may not materialize as based on such a solution," he remarked.

Mr. Manner proposed a text calling for compulsory resort to a conciliation commission. If the parties failed to reach agreement on the basis of the commission's report, they would be obliged to use other peaceful settlement procedures of their choice as provided for under the convention.

MARINE ENVIRONMENT, RESEARCH, AND TECHNOLOGY

The Third Committee completed at this session all substantive negotiations on the parts of the convention dealing with protection and preservation of the marine environment and with the development and transfer of marine technology. But it still had proposals outstanding on the rules to govern marine scientific research by foreign vessels in the economic zone and on the continental shelves of coastal States.

In its six years of negotiations on the marine environment, the Committee had to resolve complex questions involving the potentially overlapping jursidiction of coastal States, flag States (those in which vessels are registered), and port States (those in whose harbours ships may be berthed). It did this by tentatively agreeing on a set of provisions spelling out which States are responsible for preventing and punishing polluters, particularly when pollution originates on board vessels, and what kinds of enforcement action are allowable.

Most of this work was done prior to this year. However, at the start of the 1979 session, the Conference still had to deal with several proposals, some of them aimed at enhancing the legal powers of States to protect their coasts and waters against pollution. Several measures to expand the authority of coastal States had been agreed to last year. This year Third Committee Chairman Yankov reported that, following exhaustive discussion, none of the remaining proposals could be considered as commanding widespread and substantial support so as to offer an improved prospect of consensus—the criterion established by the Conference to determine which proposals should go into its revised negotiating text.

In the Third Committee's discussion of these results on 23 April, at its only formal meeting of the session, France expressed regret that its proposal to permit the imprisonment of polluters had not been accepted. The Philippines and Somalia said they were sorry that consensus had not been reached on a proposal sponsored by their delegations and others which would have authorized States, in the interest of protecting themselves from pollution in their territorial sea, to promulgate laws and regulations concerning the design, construction, manning, or equipment of foreign ships, as long as such national rules were in conformity with generally accepted international rules. Spain said the existing text was unacceptable insofar as it left coastal States vulnerable to pollution from ships passing through straits.

The one revised article on which agreement was reached this month concerns the responsiblity and liability of States on environmental matters. It carries over from the 1977 negotiating text provisions making States "responsible for the fulfilment of their international obligations concerning the protection and preservation of the marine environment" and "liable in accordance with international law." States are to ensure that "recourse is available . . . for prompt and adequate compensation or other relief" whenever their nationals or firms damage the marine environment by pollution.

The new element in this article read as follows: "With the objective of assuring prompt and adequate compensation or other relief in respect of damage caused by pollution of the marine environment, States shall co-operate in the implementation of existing international law and the further development of international law relating to responsibility and liability for the assessment of and compensation for damage and the settlement of related disputes, as well as, where appropriate, development of criteria and procedures for payment of adequate compensation such as compulsory insurance or compensation funds."

This provision amalgamates a provision in the 1977 text with a proposal by several Arab delegations and Portugal, made last year, that called for a number of measures to ensure the payment of compensation. These included the establishment of regional and international financial and technical institutions to which victims of pollution could present claims when the culprit was unknown or when he could not provide compensation himself.

Chairman Yankov promised to consult with the Second Committee Chairman on how to deal with proposals made last year by the Soviet Union for a series of safeguards to ensure that the enforcement powers authorized in the convention were not abused. These proposals were offered as a separate part of the convention and thus do not relate exclusively to environmental matters.

As regards marine technology, the text which the Third Committee has prepared over the past several years would have States commit themselves to co-operate in promoting the transfer of such technology "on fair and reasonable terms and conditions." It mentions several methods for promoting such co-operation, including bilateral and multilateral programmes and the establishment of regional marine scientific and technological research centres, especially in developing countries.

This month the Third Committee added to these provisions an article to encourage the establishment and strengthening of national marine scientific and technological research centres. States would be obligated to promote such centres, "to stimulate and advance the conduct of marine scientific research by developing coastal States and for strengthening their national capabilities to utilize and preserve their marine resources for their economic benefit." States would also be required to "give adequate support to facilitate the establishment and strengthening of such national centres; for the provision of advance training facilities and necessary equipment, skills and know-how as well as to provide technical experts to such States which may need and request such assistance." This provision originated in a proposal made by Pakistan last year.

On marine scientific research, Chairman Yankov reported that the Third Committee had devoted to this topic a substantial part of its 10 informal meetings this month but that the discussions "could not be considered as conclusive." He added, "I consider that it is very important that we do not preclude the option for another attempt to improve the prospect for a consensus."

The Chairman noted that the Committee had considered a newly revised set of proposals by the United States on this subject, as well as other proposals by the Soviet Union and France. The revised proposals were not made public; however, the proposals which the United States made on this topic last year were aimed at facilitating the conduct of research by foreign ships in the economic zones and on the continental shelves of coastal States.

The 1977 negotiating text on this subject is based on the principles that research in these areas is permissible only with the consent of the coastal State and that such consent must be granted when the research project is conducted for peaceful purposes and fulfills other criteria laid down in the convention. Many delegations have maintained—and they repeated this stand in the Third Committee this month—that the careful balance in this text between the interests of coastal States and of researching States must not be upset by substantial changes such as some of those proposed by the United States.

REPORT OF THE CHAIRMAN OF THE FIRST COMMITTEE, MR. PAUL BAMELA ENGO (UNITED REPUBLIC OF CAMEROON), TO THE PLENARY[1]

In accordance with the recommendations included in a report of the General Committee contained in document A/CONF.62/69, adopted by the Plenary meeting at its 108th meeting on 15 September 1978, the three Negotiating Groups dealing with matters before the First Committee (Negotiating Groups 1, 2, and 3) resumed their work at the very outset of the eighth session. They concluded their work at the end of the third week as recommended.

The results of the negotiations up to that point in time are contained in the documents NG1/16/Rev. 1 and Corr. 1, NG1/17, NG2/4, NG2/5, NG2/12, and NG3/6.

In the light of the issue raised in Plenary, I, in consultation with the President, established a Group of Legal Experts to examine legal questions insofar as they related to Part XI of the Informal Composite Negotiating Text. The result of the negotiation in that Group is contained in document GLE/2.

The First Committee yesterday held its only meeting and it was formal. Our task was pursued in informal fora, created to encourage serious negotiations on the complex issues outstanding in the mandate of the First Committee. Reports were made yesterday to the Committee by all who bear the responsibilities for these subsidiary bodies created, notably the Chairmen of the Negotiating Groups and of the Group of Legal Experts, in order to record those endeavours formally. I also personally gave an account of the activities in the Working Group of 21, to bring them up to date.

As all of these were made in formal session and are being reproduced as part of my present report to Plenary, I shall refrain from embarking on a narrative concerning the committee's program of work.

A significant development in our labours was the establishment of the Working Group of 21—an idea launched by the developing countries who might ordinarily be expected to oppose that approach, given the great diversity of culture, in legal political and economic systems, in religions, and in levels of economic development. For the first time, it was possible to have a limited number of speakers, all accredited representatives of definite interests. The system of appointing Alternates allowed changes in representation at the front bench, as it were, and thus selections were made for the most interested on an issue to participate directly.

This, clearly, was an advance in the right direction and away from the unruly system of so-called open-ended meetings. Sub-Committees and Negotiating Groups of the whole (or limited but open-ended) did serve their purpose in providing opportunity for informal exchange of views, without the debauchery of publicity and records. Yet, they had only limited success owing to the tendency to address a large audience and attract a long list of speakers.

The Working Group of 21 addressed specific subjects and encouraged direct exchange between the opposing sides. It is my view that given a definite agenda early and time to prepare for each subject, the establishment of this Group may well prove to be an announcement of the much-awaited final stage of our endeavours.

1. UN Third Conference on the Law of the Sea, Eighth Session (A/CONF.62/L.36) (April 26, 1979), Geneva, March 19–April 27, 1979.

The Group of Legal Experts was another new forum for discussions. The presence of the word "legal" helped eliminate large numbers of participants, who do not have the fortune or misfortune of belonging to the legal profession. Even among those who belong to it, some were deterred by the word "expert." Thus the same phenomenon of limited membership induced thereby stimulated progress.

Mr. President, it cannot be news for anyone the fact that the First Committee's mandate contains the most complex and difficult issues at this Conference. Apart from sharing with other Committees many global realities and sentiments of national interests, much stronger than international, negotiations in the First Committee must contend with the absence of precedents. They must attempt to work out rules and regulations on the basis of assumptions, many of which may prove to be wrong and the basic approach to which even experts find little common ground. The limits of human knowledge and experience, advancements in science and technology notwithstanding, spell the scope of his wisdom. Our advance has inevitably been slow, because of those limits. What we often put down to lack of political will may consequently be a wrong diagnosis of many of our problems.

That said, I believe that we can therefore look with some satisfaction and hope on our attainments this session. For the first time in a long time, I am able to state with some degree of confidence that we have broken new grounds and I can venture to speak of consensus on some issues. Even with regard to some of the hard-core issues that have menaced us in the past have taken on new and less disagreeable dimensions. The resolution of some aspects of these have lightened the burden with regard to others.

I do not, however, wish to be interpreted as saying that all hard-core issues have been or are about to be resolved with an equal degree of success. We had time to negotiate only a few of these and what I am communicating is my impression that, if the experience of my long involvement with this effort has not led me to a misconception, the results so far demonstrate considerable progress. It is this aspect that I wish to address today.

Mr. President, perhaps the most complex and difficult task we have had to grapple with relates to the financial arrangements with regard to the Authority and the Enterprise and also the terms of contracts for exploration and exploitation. On this issue, the Conference found itself dumped in a dark tunnel when it met in Caracas. Thereafter we were to wander, almost aimlessly, in darkness. When the Revised Single Negotiating Text was worked out, we were still at a loss to truly identify what was involved. With the Informal Composite Negotiating Text, we were finally able to see some light beyond the tunnel and commenced focusing on a definite programme as a common basis for discussion.

Thanks to the indefatigable efforts of my fraternal friend, Ambassador Tommy Koh of Singapore, and the devotion of experts and nonexperts alike, we now have the clear proposals contained in documents NG2/4, NG2/5/Rev. 1 and NG2/12/Rev. 1. We are finding our way out of the tunnel. Ambassador Koh's report to the First Committee is detailed, and I need not insult him by unnecessary duplication.

With regard to NG2/4, which deals with the financial arrangements of the Authority, it is generally felt that a consensus has been reached.

The financial arrangments to be entered into between the Authority and contractors undertaking sea-bed mining activities, were discussed in great detail on the basis of NG2/12. Proposals before the negotiators are designed to achieve a balance among such elements as the level of revenues accruing to the authority and the need to devise a system of taxation that would be acceptable to countries with different social and economic systems, and also one that would respond sensitively to the contractor's profitability, thus enhancing its attractiveness to the investor.

Considerable progress was made in clarifying elements in the proposals. It is probably only on the question of the level of revenues accruing to the Authority that agreement has yet to be achieved. Document NG2/12/Rev. 1 contains the latest situation which reveals tremendous progress.

However, two paragraphs will probably require further negotiations: (i) Under paragraph 7 (*sexies*) there appear to be three interrelated questions which must be taken together. These are the two production charge rates, the ANP figure and the two tax rates. (ii) Under paragraph 7 (*tertius*) the question is whether or not it is possible and desirable to spell out the financial terms of contracts for contractors which will mine the nodules but will not undertake a fully integrated project, e.g., that will engage only in nodule recovery and perhaps transportation.

In spite of this observation, it is widely accepted that the suggestions in NG2/12/Rev. 1, compared to the ICNT, substantially enhance the prospects for achieving a consensus.

During the negotiations, one particular element evoked wide discussion: the need for the level of the Authority's revenues to be such as "to enable the Enterprise to engage in sea-bed mining effectively at the same time as the entities referred to in Article 151 (2) (ii)." [See paragraph 7 (e) of Annex II.]

While it was recognized that the Authority's revenues would not be the sole, or even the principal source of initial financing for the Enterprise, the developing countries sought to ensure that the levels of revenues from mining contracts would make a substantial contribution to that end. Resulting from this discussion, proposals were made regarding the separate but related question of *how*, in fact, the enterprise and its initial mining and other related activities might be financed, pursuant to earlier assurances given by responsible spokesmen of the industrialized countries.

Negotiation of the level of the Authority's revenues, including aspects relating to national taxation, still presents difficulties, while serious consideration must continue to be given to the financial structure and the network of commitments by States Parties for ensuring the timely execution of the Enterprises first mining operations.

On the whole, with regard to NG2/5/Rev. 1, the one real outstanding issue still to be negotiated, I must emphasize, is how the cash capital of the first project of the enterprise should be raised. Ambassador Koh gave valuable leadership in his statement yesterday. Two basic ideas exist: (i) that the cash capital should be raised from all States Parties in accordance with the schedule referred to in Article 158 (2) (vi); (ii) that the cash capital should be divided into two parts: the first part should be raised from all States in accordance with the said schedule and the second from one of three ways: (*a*) States Parties referred to in Article 159 (1) (a); (*b*) States Parties referred to in Article 159 (1) (a) and (b); and (*c*) States Parties engaged in activities of exploration and exploitation in the Area and by the States Parties sponsoring applicants for contracts.

It is perhaps appropriate to say that the industrialized countries bear basic responsibility for ensuring the capabilities of the Enterprise over the preliminary phase of its activities. I say this because, as one representative of the industrialized countries admitted during the negotiations, they have a fundamental interest in seeing this realized because if the Enterprise does not work, especially at the beginning, no activities would, in fact, commence. Secondly, some sacrifice in financial terms on the part of industrialized States would ease financial burdens on contractors and the Enterprise, as well as reassure the developing countries that the Enterprise will not turn out to be a mere paper institution.

Here again, I am of the opinion that the ideas reflected in document NG2/5/Rev. 1, compared with the ICNT, also substantially enhance the propects for achieving a consensus.

SYSTEM OF EXPLORATION AND EXPLOITATION

Mr. President, I shall now turn to another area in our mandate which is crucial: the system of exploration and exploitation.

Considerable progress was made at this session on the elaboration of the "parallel system" of exploration and exploitation of sea-bed mineral resources—a system which is viewed as the tentative compromise arrangement for initiating mining of the deep sea-bed. The implications of this approach transformed the subject into one of central importance for many delegations.

There was, therefore, concern to ensure that at the end of the interim period, specified as 20 years, there would be a review of the working of the system. This would be done by reference to stated criteria acceptable to all, in accordance with Article 153, and a fair opportunity for alteration of the system if it proves unsatisfactory or if a better system can be identified by a future generation more knowledgeable than ours in the light of known data and ongoing truths.

The terms of Article 153 relating to the procedures to be followed, as well as the scope and implementation of the decisions to be taken, were discussed. However, agreement still eludes us on this rather difficult issue. I am unable to recommend at this stage any revision of the ICNT relating to Article 153(6), although I feel strongly that the relevant suggestion contained in document NG1/16/Rev. 1 and Corr. 1 must be regarded as a helpful alternative. The decision here is for the Plenary. The application of a moratorium if the Review Conference fails to reach agreement before the fifth year of the Conference is still an issue under serious discussion.

Priority for the Enterprise

The Working Group of 21 discussed at some length a possible device for ensuring the continuance, on a viable and effective basis, of the activities of the Enterprise, viz., by according a certain priority to activities of the Enterprise. In this way, it was felt, undue concentration on contractual activites and neglect of sites reserved for the Enterprise would be avoided, and a balance maintained in the working of the "parallel system." Central to this discussion were the provisions of sub-paragraph 5 *bis* (d) which, in reference to the limit on production imposed by article 150 *bis*, would accord to the Enterprise's activites on sites reserved to it, a priority over any applicant for a contract, provided only that the Enterprise were ready to engage in mining and its application to carry out activities met the Authority's requirements in all other respects. While it seemed generally agreed that activities in sites reserved pursuant to paragraph 5 *tertius* should be accorded due weight in any selection process, there are still substantial divergences of view regarding the nature and scope of any priority to be accorded to the Enterprise. This remains, therefore, one of the most critical of the outstanding issues.

I am of the opinion that it would be undesirable for me to submit any ideas for review of the ICNT on this subject at this time. The suggestions contained in document NG1/16/Rev. 1 must be studied with a view to some serious negotiations in the Working Group of 21 at the next session. The Plenary has responsbility for decision on the procedure.

Transfer of Technology

This subject is well known and the issues involved are now fairly clear. The Chairman of the Negotiating Group 1, in his statement yesterday, ventured a definition of what is meant by technology for the specific purposes of sea-bed mining. The general public tends to conclude that our efforts here on the subject are an extension of the North-South dialogue or the UNCTAD endeavours. The definition advanced should provide a helpful basis for greater understanding.

It was clear from the negotiations in the Working Group of 21 that this issue is very closely connected with ensuring the attainment and continuance, on a viable and effective basis, of the activities of the Enterprise.

Here there was substantial agreement on the need, not merely to finance the Enterprise, but also to ensure its access to the technology it would require over the whole range of its activities.

As I pointed out yesterday, in reporting on the work of Negotiating Group 3, there is now happy agreement that the activities of the Enterprise would comprehend such stages as transportation and processing of sea-bed minerals. This gave rise to a discussion as to whether or not paragraph 4 (*bis*) on the transfer of technology ought to include specific provisions on the transfer of technologies in those and related fields—i.e., not merely recovery technology or technology connected with sea-bed mining operations, on which there now appears to exist a consensus.

Although some measure of agreement is emerging as to the contractor's obligations in respect of technology transfer, there appears to be, as yet, no agreement on important aspects, such as the nature and scope of the technology to be transferred, transfer of technology to developing country applicants for reserved sites, and the dispute settlement and enforcement procedure.

The Working Group of 21 touched upon the settlement of disputes in relation to transfer of technology, notably the question whether binding commercial arbitration is a sufficient device, where the negotiations on the terms and conditions of transfer of technology fail to reach agreement. The question was whether issues not relating strictly to price could not be referred to the Sea-Bed Tribunal. It would appear that some tentative consensus has emerged that there should be a division of jurisdiction. Extensive and inconclusive discussions took place on the issue of a possibility of the transfer of technology provisions where such technology was owned by a third party.

Production Policies

Much important work has been done on the production policies set forth in Article 150 (b). An informal group consisting of interested delegations, under the Chairmanship of Ambassador Nandan, studied the issue, with a view to attaining greater agreement on an improvement on the provision of the ICNT. He gave a detailed report yesterday, as a supplement to the report of Chairman Frank Njenga of Negotiating Group 1.

It would appear that very interesting proposals are emerging which would make the Authority's production policies more flexible in regard to the award of contracts and plans of work, while at the same time taking fully into account the just concerns of the land-based producers.

Indications reaching me this morning reveal that the interested delegations working with Ambassador Nandan have assembled textual ideas as an improved basis for further discussion. I feel duty-bound to report their content here because of the belief that the Plenary ought to find out if such a text could, in fact, improve the prospects of consensus on this issue.

> 1. The nickel production ceiling referred to in Article 150 (*bis*) (2) establishes the level of production of other metals that could be extracted from that tonnage of nodules which would allow for the production of the quantity of nickel specified by the ceiling, irrespective of whether the production of nickel takes place.
> 2. The Review Conference shall begin 15 years after the date of first commercial production.
> 3. The Authority shall reserve the level of planned production referred to in Article 150 (*bis*) (2) only when an operator informs the Authority that a mineral deposit exists, that adequate financial resources are available for its development, and that it is technologically feasible to achieve commercial production from that deposit by a specified date, not later than five years following the approval of that plan of work covering exploitation. In the event that commercial production from that deposit is not achieved by the date specified, the Authority shall review the situation and if no sufficient cause can be shown for the delay, the reservation of that level of planned production shall be cancelled.
> 4. Any year during the Interim Period the Authority may allow any contractor to go over his planned level of annual production as approved under a plan of work covering exploitation, if in so doing the total production ceiling for that year as calculated in Article 150 (*bis*) (2) is not exceeded.
> Alternatively,
> An operator may in some years produce up to ________ more than that level of annual production of minerals from nodules as specified in his approved plan of work covering exploitation. Any increase over ________ and up to ________ or a continuous increase for more than two years of any percent exceeding that specified in his plan of work shall be negotiated with the Authority which will be guided by the principle of not exceeding the total production allowed under calculations in Article 150 (*bis*) (2).
> 5. The Authority may allow a temporary increase or decrease in the level of production of existing sea-bed operators provided that, and only so long as a cause of *force majeure* exists among other producers creating a significant imbalance in the world supply/demand relationship.

Regarding paragraph 3, it is understood that only actual production counts towards the filling of the ceiling, irrespective of whether it comes from the operation under the contract or under the plan of work of the Enterprise.

Nationality and Sponsorship

Some delegations have pointed out that the legal issues of nationality and sponsorhip have yet to be discussed. These are complex matters that are of practical importance and would be of critical significance in the operation of "anti-dominance" (anti-monopoly) provisions such as paragraphs 5 (c) and (d), and 5 *bis* (c), as well as any corresponding provisions which may be considered for application to activities in the "reserved" sites. The issues arising in this connection will have to be discussed and resolved before appropriate provision is made in the text.

Reserved Sites

Reference was made in the course of the negotiations to the lack of clear provisions relating to activities in reserved sites. The award of contracts by the Authority for activities on reserved sites, the basic contractual conditions (including transfer of technology and financial arrangements) to be incorporated, and the extent of, and conditions for, non-developing country participation in such activities as foreseen in paragraph 5 *tertius* (c) would all need to be covered in an appropriate manner.

Joint Ventures

Finally, I would like to mention a proposal for introduction of more specific provisions on joint ventures, and more particularly joint ventures in which the Enterprise would participate. Many countries, both developing and developed, found this proposal interesting. At the very least, the proposal should have the effect of encouraging the negotiators to pay greater attention to the mechnaism of the joint ventures which is referred to frequently in Annex II, to ensure clarity and dispel doubts or apprehensions, if any, regarding such arrangements.

Conclusions on the System

It is difficult to make definite recommendations on the prospects of all suggestions made in document NG1/16/Rev. 1. It may rightly be said that it shows progress, even though this cannot be identified in as concrete a form as those on financial questions. The Working Group of 21 did not have enough time to negotiate all the issues in the light of the distinguished Chairman's suggestions submitted following consultations and open debate in Negotiating Group 1.

Three matters were discussed in the Working Group of 21 relating to NG1/16 and NG1/16/Rev. 1 was submitted by the Chairman of Negotiating Group 1 following them.

They were: (i) transfer of technology; (ii) review conference; (iii) priority for the Enterprise.

The first of these was considered thoroughly, and it is my view that although it is not safe to declare that consensus exists on the content of Rev. 1, I am persuaded that it offers an improved basis for further negotiations and may well be considered in the process of revision.

The other two, as I have explained, were still in the elementary stages of negotiations in the Working Group of 21. The views of the author of the suggestions are contained in his report presented to the first Committee yesterday.

However, the value of that document must not be lost. It was produced by the Chairman of Negotiating Group 1 after strenuous efforts. The suggestions may not all attract agreement on all sides, but it must be remembered that they are currently the ones which lie close to those of the ICNT and that the latter is being examined in the light of the former. It would be undesirable to exclude obvious consensus and improvements in a revision on the ground that the Working Group of 21 had not fully discussed them or at all.

ORGANS OF THE AUTHORITY, ETC.

Mr. President, I gave a full review of the situation with regard to subjects within the mandate of Negotiating Group 3. There are consensus suggestions, in document NG3/6, I would recommend should be incorporated in any revision. No one has challenged my declarations on them. Further consultations reveal that a further change to NG3/6 regarding Article 160 (2) (xxi) and (xxii) offer good propsect for consensus. The text of Article 160 (2) (xxi) should be modified to read as follows: "Issue emergency orders, which may include order for the suspension *or adjustment* of operations, to prevent serious harm to the marine environment arising out of any activity in the Area." In Article 160 (2) (xxii) revise the phrase "irreparable harm to a unique environment" to read "serious harm to the marine environment."

However, Article 159 will have to take its turn in the list of outstanding issues in spite of my stated opinion that the issues involved are, indeed, ripe for resolution.

LEGAL ISSUES (GLE)

The deliberations of the Group of Experts were fully reported by Dr. Wünsche yesterday, and I do not have much to add because of its clarity on the issues. However, he has informed me of a desirable amendment to his suggestions under Article 188, relating to the submission of disputes to *ad hoc* Chambers of the Sea-Bed Disputes Chamber and to binding arbitration. I would invite the Plenary to amend paragraph 1 by deletion in the last line of the words "any party" and substitute therefor "the parties."

With regard to the suggestions from the Group of Legal Experts, I am satisfied from consultations that there is widespread feeling that they offer excellent prospects for consensus or at least a much better basis for further negotiations than does the ICNT.

Mr. President, I wish to draw attention to a document which as been issued as WG21/1. I should like to point out that as such it in no way attempts or relates to a revision of the ICNT. As I explained in the Working Group of 21 and indicated in the First Committee, the idea is merely to give delegations a picture of how the suggestions from the Chairmen of the various negotiating fora dealing with First Committee matters would, if adopted, fit into the scheme of things proposed by the ICNT. It assembles all the suggestions in a common document and gets rid of multifarious numbering systems. It is also my view that it will aid the negotiating efforts in the First Committee and will ease the revision of Part XI in areas where a revision is considered by the Plenary to be necessary.

Mr. President,

I sincerely hope that in spite of the length of this report I have managed to give you a full account of the deliberations in the First Committee during this session and in a way that enables the Plenary to take the vital decisions incumbent on it. My opinions cannot attract more weight than the consensus of all the distinguished representatives of sovereign States assembled here. What must dominate our thinking is the will to resolve problems and to attain a viable universal treaty in which all of mankind will gain.

I should like, in closing, to express my profound gratitude to the delegations who have worked so hard to achieve progress in the First Committee. As I said at the beginning, I have no doubt that they have demonstrated greater political will at this session than at any other.

I should also like, once again, to express the tremendous satisfaction I have had in observing the continuing co-operation of my fraternal friends, Frank Njenga, Tommy Koh, and Harry Wünsche. The reports they have submitted testify to their dedication and capacity for hard work. It is not as a matter of mere formality that I express my profound gratitude to the distinguished Representative of the Secretary-General and his able staff who loyally serve us and whose presence in our official life makes the burden of office far less difficult than it otherwise could have been.

I should also like to thank all the others, the interpreters, the secretaries, the précis writers, etc., who have, as always, made a very valuable contribution to our success.

Thank you.

REPORT TO THE PLENARY BY AMBASSADOR ANDRES AGUILAR (VENEZUELA), CHAIRMAN OF THE SECOND COMMITTEE[1]

1. Three of the Negotiating Groups established in accordance with the organization of work adopted by the Plenary (A/CONF.62/62) deal with issues falling wholly or partly within the mandate of the Second Committee. They are Negotiating Groups 4, 6, and 7.
2. Negotiating Group 4, under the chairmanship of Ambassador Satya Nandan (Fiji), deals with the right of access of land-locked States and certain developing coastal States in a subregion or region—or geographically disadvantaged States—to the living resources of the exclusive economic zone.
3. Negotiating Group 7, under the chairmanship of Judge E. J. Manner (Finland), deals with the definition of maritime boundaries between adjacent and opposite States—a Second Committee issue—and settlement of disputes thereon which is dealt with by the Plenary Conference.
4. Negotiating Group 6, under my chairmanship, deals with the definition of the outer limit of the continental shelf and the question of payments and contributions with respect to the exploitation of the continental shelf beyond 200 miles or the question of revenue sharing.
5. At its 57th and 58th meetings, on 27 April 1979, the Second Committee received the reports relating to the eighth session at Geneva from the Chairmen of the three above-mentioned Negotiating Groups.

1. UN Third Conference on the Law of the Sea, Eighth Session (A/CONF.62/L.38) (April 27, 1979), Geneva, March 19–April 27, 1979.

6. In the case of Negotiating Group 4, the debate focused on the question whether the Chairman's proposals (document NG4/9/Rev.2) concerning article 62, paragraph 2, article 69 and article 70 met the requirement laid down in document A/CONF.62/62, paragraph 10, by offering a substantially improved prospect of a consensus. The debate showed that the general feeling in the Committee was that they did, and I would accordingly recommend their inclusion in any revision or modification of the Informal Composite Negotiating Text. The comments on and objections voiced with regard to the compromise proposals of the Chairman of Negotiating Group 4 are, of course, reported in the summary records of the meeting.

7. The report of the Chairman of Negotiating Group 7 on the work of that Group is contained in document NG7/39. As stated there, except for two drafting amendments to article 15, none of the proposals for revision of the Informal Composite Negotiating Text offered a substantially improved prospect of a consensus.

8. I wish to place on record my gratitude to Ambassador Nandan and Judge Manner for their dedication and their contribution to the Committee's work.

9. With regard to Negotiating Group 6, I shall not repeat here my report to the Second Committee, which appears in document A/CONF.62/C.2/L.100. As I said in the closing paragraph of that report, I held intensive consultations over the last few days with a number of delegations which had been most active in the Group's discussions.

The results of these efforts to achieve what may be a decisive break-through resolving Second Committee issues, and perhaps issues of the Conference itself, are set out in document A/CONF.62/L.37.

I know that there are delegations which will have reservations or objections to my suggestions, but I hope they understand that the negotiations are open-ended and that they will have an opportunity to put forward their own views in our future negotiations.

The suggestions I am making for inclusion in any revision of the Composite Negotiating Text are self-explanatory. I am convinced that they will improve the prospect of a consensus.

I shall confine myself to two comments on my proposal:

a) The question of establishing the starting-point from which the distance specified in paragraph 3 *bis* applies, that is to say, the baseline from which the breadth of the territorial sea or the outer limit of the exclusive economic zone is measured, is still a controversial issue which may require further negotiations.

b) It is clear to me from the negotiations and consultations on the subject of the continental shelf that the question of scientific research is an important element in any over-all compromise on the issue of jurisdiction of the coastal State vis-à-vis other States. Judging from the consultations I have held, I believe that final acceptance of the formulae I am now presenting will depend on a settlement of this question. I therefore hope that the Third Committee will find a suitable formula concerning scientific research on the continental shelf so that a solution satisfactory to all delegations concerned may be achieved.

10. I reiterate the recommendation I made in paragraph 13 of my report to the Plenary at the conclusion of the first part of the seventh session, which appears on page 85 of the English text of volume 10 of the Official Records of the Conference, concerning suggestions which had widespread support in the Second Committee. To these may be added the Belgian delegation's suggestion that the words "or for the safety of ships" should be added at the end of the first sentence of paragraph 3 of article 25.

11. With regard to another issue, I have continued to hold consultations on the provisions for improved protection of marine mammals.

12. In conclusion, I wish to reiterate my thanks to all delegations which have participated in the work of the Second Committee and to the efficient Secretariat staff. I am gratified at the will displayed by participating States in successfully completing consideration of the long list of very complex and important items and issues assigned to the Second Committee.

REPORT BY THE CHAIRMAN OF THE THIRD COMMITTEE, AMBASSADOR A. YANKOV (BULGARIA)[1]

RESULTS OF NEGOTIATIONS ON PART XIII AND XIV ON THE ICNT DURING THE EIGHTH SESSION

1. I have the honour to submit for your consideration this report on the work of the Third Committee during this session. The report was considered at the 40th meeting of the Committee held on 23 April. But having in mind the stage of the Conference, this report indeed reflects the results which have been achieved until now. We have opted since Caracas, to negotiating fairly in open-ended meetings with the flexible use of all available means of negotiations, but always on the condition that the results should be brought to the attention of the Committee as a whole. The negotiations and discussions which took place during this Session were concentrated on the main pending issues in all parts within the mandate of the Third Committee namely, Part XII—the Protection and Preservation of the Marine Environment; Part XIII—Marine Scientific Research; and Part XIV—Development and Transfer of Technology.

2. During the first meeting of the Third Committee at this session of the Conference, held on 2 April, I reviewed the outstanding issues, enumerating all the pending informal proposals. Although most of them have been the subject of extensive consideration at the previous sessions, we agreed to provide the sponsors with an additional opportunity to present them to the Committee and hear the reactions of the interested delegations. This was to allow the Committee to assess the chances of acceptability and enable the sponsors to consider how to pursue matters of special interest to them in the future. We agreed also to provide adequate opportunity to discuss those informal proposals which, owing to lack of time during previous sessions, were not thoroughly examined so that they could be taken up again and negotiated at this session.

3. In the course of the present session the Third Committee held 10 meetings in which we heard over 220 interventions. Since during the previous session priority was accorded to discussions and negotiations on matters pertaining to the protection and preservation of the marine environment, this time we made an effort to also give some priority to the pending issues within Part XIII—Marine Scientific Research.

4. A number of meetings were scheduled to discuss marine scientific research and to provide to the sponsors an opportunity to present their informal proposals and also to give a chance to the members of the Committee to comment on those proposals. It was my belief that a further consideration of them would help us to ascertain the possibilities of broadening the area of agreement.

1. UN Third Conference on the Law of the Sea, Eighth Session (A/CONF.62/L.34) (April 26, 1979), Geneva, March 19–April 27, 1979.

5. Sensing the feeling in the Committee during this session of wishing to conclude the discussion of Part XII, it was agreed that I would chair meetings on some of the pending amendments on the protection and preservation of the marine environment which were considered during the previous informal negotiations under the chairmanship of Señor José Luis Vallarta as well as all the meetings on marine scientific research.

RESULTS OF NEGOTIATIONS ON PART XII—PROTECTION AND PRESERVATION OF THE MARINE ENVIRONMENT

6. As already stated, some of the meetings on this part were chaired by me and the others, as agreed during our meeting on the organization of work, by Señor José Luis Vallarta of Mexico. The basic aim of these negotiations and the procedure followed was to broaden the area of compromise and to try to retain and improve those texts and amendments which after prolonged and exhaustive negotiations have proved to command a substantially improved prospect of consensus thus alleviating the need to repeatedly come back to the same proposals.
7. I wish to point out further that the present report follows the general lines of the same pattern of reporting as the previous reports submitted to you during the past sessions. However, in view of paragraph 10 of Document A/CONF.62/62, paragraph 7 of Document A/CONF.62/69 and recommendations 6 and 7 of Document A/CONF.62/BUR.11/Rev.1, taking into account the requirements contained therein at this stage, we have to try in the document reflecting the results of the session to incorporate those provisions which have emerged from intensive negotiations and which offer substantially improved prospects of consensus compared to the ICNT.
8. Under my chairmanship, the Committee discussed the Brazilian proposal on Article 209, paragraphs 1 and 5; the informal proposals of Canada, Bahamas, Barbados, Iceland, Kenya, New Zealand, Philippines, Portugal, Somalia, Spain, and Trinidad and Tobago on Article 212 paragraph 3; the informal proposal of Spain on Article 234; the informal proposals submitted by the United Republic of Tanzania on Article 212, paragraph 5, Article 229; and a general proposal for the substitution of the expression "competent international organization" by the expression "competent international organizations" wherever it appears in the text. The Committee also addressed itself to the French proposal as contained in informal document MP/29 related to paragraph 1 of Article 231.
9. These negotiations, in my personal view, were exhaustive, and under the existing guidelines as contained in document A/CONF.62/62 those informal proposals could not be considered as commanding a widespread and substantial support to offer an improved prospect of consensus.
10. Under the chairmanship of Señor José Luis Vallarta, the Committee held four informal meetings trying to amalgamate the provisions contained in Article 236 of the ICNT with those contained in document MP/18/Rev.1 as proposed by several of the Arab delegations and Portugal. I am pleased to inform you that negotiations were successful and due to the able and flexible chairmanship of Señor José Luis Vallarta, and the sense of co-operation and moderation demonstrated by the co-sponsors of the amendment to Article 236 a revision has been successfully agreed upon. The new text of Article 236 thus will read as follows:

Article 236. Responsibility and Liability

1. States are responsible for the fulfilment of their international obligations concerning the protection and preservation of the marine environment. They shall be liable in accordance with international law.
2. States shall ensure that recourse is available in accordance with their legal systems for prompt and adequate compensation or other relief in respect of damage caused by pollution of the marine environment by persons, natural or juridical, under their jurisdiction.
3. With the objective of assuring prompt and adequate compensation in respect of all damage caused by pollution of the marine environment, States shall co-operate in the implementation of existing international law and the further development of international law relating to responsibility and liability for the assessment of and compensation for damage and the settlement of related disputes, as well as, where appropriate, development of criteria and procedures for payment of adequate compensation such as compulsory insurance or compensation funds."

11. The only remaining proposal on this part is the proposal submitted by the Soviet Union, for a new "Part XIV *bis*—General Safeguards," as appears on page 186 of Volume 10 of the Official Records of the Conference. From my personal contacts with various interested delegations and from the discussions on the proposals held during the last session, I got the feeling that the Committee would prefer not to discuss this matter in this Committee because of the close link existing between the Soviet proposal and matters pertaining to the Second Committee. In these circumstances, the Committee entrusted me to discuss the matter with the Chairman of the Second Committee and jointly to agree as to the best procedure to be suggested for dealing with the Soviet proposal.
12. In view of the progress of the negotiations made during this Session, and the very important positive results that were achieved, I would venture to state that the substantive negotiations on Part XII—Protection and Preservation of the Marine Environment, could be considered completed.

In this connexion, I wish to reiterate the assessment contained in my report on 13 September 1978, that "with respect to matters relating to the protection and preservation of the marine environment, we have reached a stage where the ICNT thus constitutes a good basis for a consensus. This does not mean that there is no room for further negotiations aiming at improving the texts. But at the same time, we should take into account the fact that we have reached a balance which should not be disturbed."

RESULTS OF NEGOTIATIONS ON PART XIII—MARINE SCIENTIFIC RESEARCH

13. A substantial time of our negotiations was devoted to marine scientific research at this session.

As all of us are well aware, some differences of opinion as to the regime of marine scientific research still persist. The Committee addressed itself to the revised version of the proposals presented by the United States delegation as contained in document MSR/2/Rev.1. During the discussion two new proposals were tabled: by the USSR on

Article 256 as contained in document MSR/3, and by France on Article 248 as contained in document MSR/4.

14. The discussions were exhaustive, although in the view of several delegations they could not be considered as conclusive. I felt that there was substantial support for the ICNT, and for the maintenance of the delicate balance achieved so far in the overall package with regard to Part XIII. However, it is well known that several delegations maintained that they should have the opportunity to continue the negotiations on this vitally important issue, considering that all efforts to reach a compromise on some of the outstanding questions in this part have not been exhausted.

15. More than 50 interventions were made on the United States proposals—some of them opposing any change in the ICNT; others advocating the need for certain drafting, stylistic, or substantive modifications which would improve the text. It is my submission and without prejudice to the interpretation given by the sponsors, some of the United States proposals, especially those referring to the conduct of marine scientific research on the continental shelf, were of a substantive nature while others entailed drafting modifications, further clarification of existing provisions, or their interpretation. Therefore, it is my personal view that at the later stage of our negotiations and in the light of negotiations in the other Committees, we might at an appropriate time try to broaden the basis for agreement on those other pending issues. I would then venture to conclude that, since we have not attained all the required elements to enable us to proceed to a revision of this part of the ICNT, I consider that it is very important that we do not preclude the option for another attempt to improve the prospect for a consensus.

16. We had a proposal on this part referring to Article 264 submitted by a number of Arab States and Portugal. In the light of the results reached on Article 236, no modification of Article 264 is needed since paragraph 3 of this Article contains an explicit reference to Article 236. At the meeting of the Third Committee, the sponsors of the proposal on Article 264 agreed with my assumption and withdrew their proposals.

RESULTS OF NEGOTIATIONS ON PART XIV—DEVELOPMENT AND TRANSFER OF MARINE TECHNOLOGY

17. As you will recall during a previous session, Pakistan submitted an informal proposal for the inclusion in the ICNT of a new Article 275 *bis*. We had an exhaustive discussion on this part during this session and my impression was that Pakistan's proposal was overwhelmingly supported. There were some suggestions for changes which were favourably considered by the Committee on the basis of those proposals and the comments which were made. I would suggest therefore to include in Part XIV of the following article as amended:

Section 3. National and Regional Marine Scientific and Technological Centres

Article 275 bis. Establishment of National Centres

1. States, through competent international organizations, and the Authority shall, individually or jointly, promote the establishment, especially in developing coastal

States, of national marine scientific and technological research centres and strengthening of the existing national centres, in order to stimulate and advance the conduct of marine scientific research by developing coastal States and for strengthening their national capabilities to utilize and preserve their marine resources for their economic benefit.

2. States, through competent international organizations, and the Authority shall give adequate support to facilitate the establishment and strengthening of such national centres: for the provision of advance training facilities and necessary equipment, skills and know-how as well as provide technical experts to such States which may need and request such assistance.

18. During the Seventh Session, the delegation of the United States of America submitted a set of informal suggestions which contained revisions to Articles 274 and 276. When submitting the revised version of their amendments, contained in informal document MSR 2/Rev.1, those articles did not appear and I got the impression that the United States delegation would not press on maintaining those proposals. In this case the negotiations on Part XIV could also be considered as completed.

19. In conclusion, I would like to reiterate my understanding that with regard to the provisions of the ICNT within the terms of reference of the Third Committee, further progress has been made to broaden the areas of agreement and that the basis for a reasonable compromise offering us a substantially improved prospect of consensus has been set.

20. I should like to add for the record that this report was considered at the fortieth meeting of the Third Committee on 23 April 1979. I am pleased to inform you that the report as well as my conclusions were received with general approval by the Committee. I will go even further and say that the support expressed by the members of the Committee was so significant and clear that I would venture to consider this report not only as a mere information on our work, but as an important summing up of our deliberations which have taken place until now and also as a collective assessment of the results of the negotiations which have been achieved so far. Although I would refrain from saying that we had completed our mandate since there are some pending proposals on Part XIII, nevertheless it should be assumed that the considerations of Parts XII and XIV have been concluded. Accordingly I would suggest that all provisions on which consensus was reached or emerged from intensive negotiations during the Seventh and the present sessions and which offer a substantially improved prospect of consensus, be incorporated in a revised edition of the ICNT as agreed by the Conference.

21. Finally, I should like once again to express my most sincere thanks and appreciation to all members of the Committee for their co-operation and sense of goodwill which enabled us to arrive at a successful conclusion of our work at this session. I should also like to pay special tribute to the Secretariat for their dedication, competence, and most valuable assistance in fulfilling the mandate of this Committee.

A STUDY OF THE IMPLICATIONS OF PREPARING LARGE-SCALE MAPS FOR THE THIRD UNITED NATIONS CONFERENCE ON THE LAW OF THE SEA (UNCLOS)[1]

1. Background

1.1 At its 106th plenary meeting, held on 19 May 1978, the Third United Nations Conference on the Law of the Sea decided that the Intergovernmental Oceanographic Commission (IOC) and other competent international bodies should be asked to study the financial and other implications, including in particular the time required, for preparing large-scale maps (1:10 million) of the Atlantic, Indian, Pacific, and Arctic Oceans, showing the effects of different formulae for defining the outer limits of the continental shelf (ref. doc. A/CONF.62/SR.106).

1.2 Following a request from the President of the Conference, the Special Representative of the Secretary-General for UNCLOS invited the IOC and other competent bodies to assist in this task.

1.3 The IOC Executive Council, at its tenth session (June 1978), accepted this task and instructed the Secretary to arrange for this task to be undertaken with the least possible delay, working in close co-operation with the International Hydrographic Organization (IHO).

2. The Task

2.1 A study of the implications of preparing suitable maps on 1:10 million scale of the Atlantic, Indian, Pacific, and Arctic Oceans showing the effects of three different formulae for defining the outer limits of the legal Continental Shelf, namely: the 200-mile limit (proposed in particular by the Arab Group)—ref. doc. NG6/2 (Annex I); the two parts of the Irish formula—ref. doc. NG6/1 (Annex II); the proposal submitted by the USSR—ref. doc. C.2/Informal Meeting/14 (Annex III).

3. Commentary on the "Preliminary Study Illustrating Various Formulae for the Definition of the Continental Shelf" with Associated Map at 1:30,000,000 (A/CONF.62/C.2/L.98-Add.1)

3.1 This study was requested by the Third United Nations Conference on the Law of the Sea in June 1977 and contracted by the United Nations Secretariat in January 1978 to provide an assessment of the impact of various formulae proposed to define the edge of the Continental Shelf.

3.2 The scale of 1:30,000,000 was chosen for this map in order to provide a visual impression on a single sheet of paper of the margins of the world interpreted according to the different formulations and to demonstrate the areas that might accrue to coastal

1. United Nations Third Conference on the Law of the Sea, Eighth Session (A/CONF.62/C.2/L.99) (April 9, 1979). Where reference has been made to a limit or line at a set distance from the coastline (or baseline), the word "mile" has been used hyphenated to the distance in figures (e.g., 300-mile line). In all cases these are nautical miles.

States or to the international sea-bed authority. In the time available for the preparation of this map only immediately available data sources could be used and many generalizations had to be made. Definitive interpretations of the formulae were made without detailed supporting arguments or justifications and precision was sacrificed for speed. The result, which contained a number of errors, omissions, and wrong evaluations, many of which have subsequently been notified to the United Nations Secretariat and to the Conference (doc. A/CONF.62/C.2/L.98/Add. 3), was however intended to be illustrative only.

3.3 In particular the line depicting the outer edge of the continental margin was extremely uncertain in places because of the lack of an accepted definition of this term which could be universally applied.

3.4 The map did not include a presentation of the Informal Suggestion by the USSR (doc. C2/Informal Meeting/14 of 27 April 1978) since it had not at that time been presented. The IOC was requested to arrange for a line illustrating the USSR proposal to be added to the Preliminary Study. However after several versions were drafted, it was found not to be possible (cf. Statement by the Representative of the Intergovernmental Oceanographic Commission to Negotiating Group 6, 29 August 1978).

4. Study of the Proposal to Prepare Larger-Scale Maps—General Comments

4.1 The study avoids commenting on the relative value of the different proposals, recognizing that this may depend on issues outside the present task. However, in order to assess the technical implications of preparing larger-scale maps showing the different formulae, it is necessary to analyse the formulae themselves and point out technical difficulties which affect the production of these maps.

4.2 There is clearly a danger that if a series of 1:10,000,000 maps is produced under the auspices of an international authority, it could be considered as a definitive document and be used in support of national claims. It is therefore essential to maintain a clear distinction between small-scale maps such as the 1:30,000,000 scale Preliminary Study prepared for illustrative purposes only, based on current knowledge of the morphology and geology of the continental margins, and large-scale definitive charts which could vary in scale from 1:10,000,000 to 1:50,000. These large-scale maps will be necessary for ultimate determination of limits by coastal States and for possible negotiations, and may include data obtained especially for the purpose.

4.3 The analysis that follows considers the suitability of map scales in relation to the quantity and precision of the information to be displayed. At 1:30,000,000 the quantity of information displayed does not lead to overcrowding. The precision with which the position of the lines requested can be drawn at any scale depends on: (*a*) the practical interpretation of the formulae; (*b*) the quality and quantity of the geographical, morphological, and geological data. These factors will be discussed in relation to each of the formulae.

5. The 200-Nautical Mile Limit

5.1 Document NG6/2 of 11 May 1978 (Annex I) describes the proposal to use a line drawn at 200 nautical miles offshore from the baselines from which the breadth of the territorial sea is measured.

5.2 In principle a limit drawn 200 miles from the coastline can readily be depicted with an accuracy appropriate to any chosen scale. However, it would be necessary to carry out a careful and detailed review of the extent to which features affecting the baselines under the provisions of the Informal Composite Negotiating Text (ICNT) would affect the drawing of such lines. No estimate has been made of the cost, time, or other requirements of drawing the 200-mile line from baselines as provided for in the ICNT, since doing so would presuppose knowledge of the manner in which coastal states would apply the relevant provisions.
5.3 The data required are the world coastline, already available in suitable graphic form on Mercator's projection on a scale of 1:10,000,000 at the equator in the Carte général du monde (Institut géographique national, France) which is being used on sheets of the 5th dition of the General Bathymetric Chart of the Oceans (GEBCO) (see Annex IV). The coastline also exists in digitized format. Some information on baselines is already available from national declarations made by coastal states.
5.4 There are several areas of uncertainty. The above coastline has been compiled for specific purposes. It is not known what off-shore features, islands, etc., have been omitted because of scale considerations. To check for missing information would be a considerable task (see 5.2 above). Whereas the use of baselines already declared presents no cartographic problem, it would be inappropriate to make assumptions about undeclared baselines. Even the declared baselines may not all be acceptable internationally, and there is some doubt as to the propriety of using these on an international document.
5.5 In view of these uncertainties, the only immediately practical procedure is to draw the 200-mile limit based on the coastline. Two methods of preparation are possible—by computer or by hand.
5.6 The following are estimates of the time needed for preparation and of production costs: (*a*) The maps prepared by computer using the coastline as a base would take 12 to 14 months to produce, taking into account programme and tape preparation, printout, scribing, and printing time, including the initial time to arrange contracts. Conversion to equal-area or other projections would take two months more. The cost would total approximately $50,000, including the price of the original tape, plotting time on the computer, preparation and production, printing, management, and overheads. (*b*) To prepare the series by hand would take a minimum time of 6–8 months and would not cost appreciably less than (*a*). Both these figures depend on the availability of a large number of competent staff.

6. The Irish Formula

6.1 The Irish formula (doc. NG6/1 of 1 May 1978 [Annex II]) contains two options for establishing the outer edge of the continental margin where it extends beyond 200 nautical miles from the baselines, both options requiring the establishment of the position of the "foot of the continental slope."
6.2 Determination of the "foot of the continental slope" presents some interpretational difficulty. The formula states that "in the absence of evidence to the contrary, the foot of the continental slope shall be determined as the maximum change of gradient at its base." This allows considerable flexibility in the positioning of the line based on the nature of the "evidence." This "evidence" could be morphological or geological. "In the absence of evidence to the contrary," the definition is purely morphological and is based

on geometrical concepts associated with a margin profile, and this is the only aspect that has been considered in this study. The "maximum change of gradient at its base" suggests that if there are several large changes of gradient, only that at the base (a word which is synonymous with foot) should be chosen to determine the foot of the continental slope.

6.3 The drafting of foot of slope lines beyond 200 nautical miles on a world map at any scale thus raises a variety of difficulties. There are basically three types of slope: (*a*) Slopes connecting a shallow shelf and normal deep ocean floor. These may cross terraces, ridges, canyons, etc., which give rise to rapid changes of gradient. (*b*) Slopes connecting a shallow shelf and extra-deep ocean floor in oceanic trenches. The region between the shelf edge and the axis of the trench may contain ridges parallel to the trench axis giving rise to several reversals of gradient in profiles. (*c*) Slopes from shallow shelves to normal deep ocean floor, interrupted by areas of intermediate or shallow depth. The Preliminary Study indicates that examples of these types, especially of (*c*), occur outside the 200-mile limit.

6.4 In order to enhance significantly the illustrative value and the accuracy of the foot of the slope line in a new study at a larger scale, it will be necessary to examine several tens of thousands of individual echo-sounding profiles. In the Preliminary Study, the foot of the slope line was obtained for convenience directly from the World Ocean Floor Physiographic Diagram (which was provided to all delegations at the seventh session of UNCLOS-III in May 1978). This Diagram, however, used only a part of the total slope data now available and furthermore has inherent positional inaccuracies. Nevertheless it took ten years to complete at a total cost of over a million dollars.

6.5 In addition to profiles, it is necessary to establish, from other depth data and geological knowledge, the topography between the profiles, and this can best be done using detailed contouring. Such contours are currently being prepared in the international project, the General Bathymetric Chart of the Oceans (GEBCO), the 5th edition of which, providing world cover, is due to be completed in 1982 (see Annex IV).

6.6 The over-all accuracy and usefulness of a presentation of the line depicting the foot of the continental slope will depend on the quantity of data, its distribution in space and its availability.

6.7 It is estimated that at least three years will be necessary to complete a delineation of the foot of the slope line to the accuracy appropriate to a scale of 1:10,000,000 and that the cost could exceed a million dollars.

7. Irish Formula 3a

7.1 This option requires the measurement of sediment thickness seaward of the foot of the slope line to be obtained initially from geophysical data and later tested by drilling.

7.2 The Preliminary Study used a composite of published maps of sediment thickness, contoured initially in units of the reflection time for seismic waves, and then recontoured in units of thickness using some measured and some assumed seismic velocities. Incompleteness of data coverage and quality necessitated considerable interpolation and extrapolation.

7.3 The Preliminary Study also showed that the Irish 3a line is seldom further than 120 nm from the base of the slope (equivalent to 1.2 nm sediment thickness) and often is less than 60 nm (equivalent to 0.6 nm sediment thickness). In such thicknesses the base of the sediments is not difficult to identify and the accuracy of the measurement of the sediment thickness is such that the ambiguity of the 1 percent line is no greater than that of the foot of the slope line. Very few drilled holes are at present available to test the thickness profiles.

7.4 However, although a considerable body of data exists, or is presumed to exist, that will permit identification of the base of the sediments in this manner, a considerable amount of processing and interpretation will be needed before they can be compiled in map form. Even so, there may be insufficient profiles in some areas to satisfy the need for data points at 60-mile intervals required by the formula.

7.5 Even for the purpose of preparing 1/10 million scale maps, the task of locating, gaining access to, sorting and co-ordinating the many thousands of individual seismic reflection profiles that exist in the public domain will be enormous. Reflection profiles are not archived in data centres, and most profiles reside in private or national libraries not accessible to the public. The research task can thus not be accomplished by a single agency so that a multinational and multi-institutional study would be required. This would involve dozens of experts who may already be committed to other ongoing research projects. The task could take many years and could cost over one million dollars.

8. Irish Formula 3b

8.1 Since Irish formula 3b incorporates the same foot of the continental slope line as formula 3a, it is subject to similar difficulties of interpretation (see pars. 6.2 and 6.3 above) that will not be resolved to any significant extent by increasing the map scale.

8.2 Once the foot of the continental slope is established on a map series at 1:10,000,000, the addition of a line 60 nm from the foot of the slope would be a relatively small additional task. Its accuracy would be dependent on the accuracy of the foot of the slope line.

9. USSR Proposal

9.1 As with the Irish formulae, the USSR proposal (ref. doc. C.2/Informal Meeting/14 dated 27 April 1978 [Annex III]) is concerned with wide margins exceeding 200 nm. The proposal suggests that the words "but not further than 100 nautical miles from the outer limit of the 200-mile economic zone" should be inserted in the existing text of article 76 of the ICNT. There would then be no circumstances where the outer limit of the Continental Shelf of a coastal state would exceed 300 nm from baselines for the establishment of the territorial sea. The addition of the 300-mile line to charts at a scale of 1:10,000,000 already showing the 200-mile line discussed in section 5 above, would require relatively small additional effort and cost.

9.2 Difficulty with implemetation of the USSR proposal lies with the fact that its application in specific geographic areas requires delimitation of "the outer edge of the continental margin." No practical guidelines are provided in the text of the USSR proposal to define "the outer edge of the continental margin." Since considerable flexibility is given in the use of "scientifically sound geological and geomorphological data" for this purpose, determination of this limit, as with the Irish formulae (see pars.

6.2, 6.3, and 8.1 above), would not be significantly improved by an increase of the map scale.

10. Conclusions

10.1 With the information available at the present time, only the 200-mile limit drawn from the coastline (as distinct from baselines) can be more accurately displayed on a 1:10,000,000 scale than on the 1: 30,000,000 scale of the Preliminary Study (see par. 5.5 above).
10.2 The preparation of a map series at 1:10,000,000 showing with appropriate precision the 200-mile line based on a published coastline would take between 6 and 16 months and would cost approximately $50,000.
10.3 Difficulties of interpretation of the "foot of the continental slope" in the Irish formulae and "the outer edge of the continental margin" in the USSR proposal prevent any greater precision and clarity being achieved at present by producing a map on a scale of 1:10,000,000.
10.4 If these difficulties were to be removed, the preparation of a map series at 1:10,000,000 showing with appropriate precision (limited by variations in the availability of data) the 200-mile line (if drawn from baselines), the two parts of the Irish formula, and the 300-mile line from the USSR proposal would take at least three years and cost about two million dollars.
10.5 The work could only start after a decision has been made, funding assured, and contractors found. A crucial limitation on the time required may be the lack of sufficient specialized personnel available to carry out this task. It could therefore not be completed within the period of 6 to 10 months (from May 1978) specified by the Conference.

INFORMAL SUGGESTION BY THE ARAB GROUP[2]

Article 76. Definition of the Continental Shelf

The continental shelf of a coastal State comprises the sea-bed and subsoil of the submarine areas that extend beyond its territorial sea throughout the natural prolongation of its land territory to a distance of 200 nautical miles from the baselines from which the breadth of the territorial sea is measured.

INFORMAL SUGGESTION BY IRELAND[3]

Article 76. Definition of the Continental Shelf

1. Same as ICNT, *viz.*, The continental shelf of a coastal State comprises the sea-bed and subsoil of the submarine areas that extend beyond its territorial sea throughout the natural prolongation of its land territory to the outer edge of the continental margin, or

2. A/CONF.62/C.2/L.99, Annex I (NG6/2) (May 1, 1978).
3. A/CONF.62/C.2/L.99, Annex II (NG6/1) (May 1, 1978).

to a distance of 200 nautical miles from the baselines from which the breadth of the territorial sea is measured where the outer edge of the continental margin does not extend up to that distance.

2. The continental margin comprises the submerged prolongation of the land mass of the coastal State and consists of the sea-bed and subsoil of the shelf, the slope, and the rise. It does not include the deep ocean floor nor the subsoil thereof.

3. For the purpose of this Convention, the coastal State shall establish the outer edge of the continental margin wherever the margin extends beyond 200 nautical miles from the baselines from which the breadth of the territorial sea is measured, by either: (*a*) a line delineated in accordance with paragraph 4 by reference to the outermost fixed points at each of which the thickness of sedimentary rocks is at least 1 percent of the shortest distance from such point to the foot of the continental slope; or (*b*) a line delineated in accordance with paragraph 4 by reference to fixed points not more than 60 nautical miles from the foot of the continental slope. In the absence of evidence to the contrary, the foot of the continental slope shall be determined as the point of maximum change in the gradient at its base.

4. The coastal State shall delineate the seaward boundary of its Continental Shelf where that Shelf extends beyond 200 nautical miles from the baselines from which the breadth of the territorial sea is measured by straight lines not exceeding 60 nautical miles in length, connecting fixed points, such points to be defined by co-ordinates of latitude and longitude.

5. Every delineation pursuant to this Article shall be submitted to the Continental Shelf Boundary Commission for certification in accordance with Annex____. Acceptance by the Commission of a delineation so submitted in accordance with Annex______and the seaward boundary so fixed, shall be final and binding.

6. The coastal State shall deposit with the Secretary-General of the United Nations charts and relevant information, including geodetic data, permanently describing the outer limit of its Continental Shelf. The Secretary-General shall give due publicity thereto.

7. The provisions of this Article are without prejudice to the question of delimitation of the Continental Shelf between opposite or adjacent States.

INFORMAL SUGGESTION BY THE USSR[4]

Part VI

Article 76

The continental shelf of a coastal State comprises the sea-bed and subsoil of the submarine areas that extend beyond its territorial sea throughout the natural prolongation of its land territory to the outer edge of the continental margin, *but not further than 100 nautical miles from the outer limit of the 200-mile economic zone,* or to a distance of 200 nautical miles from the baselines from which the breadth of the territorial sea is measured where the outer edge of the continental margin does not extend beyond the outer limit of the 200-mile zone.

4. A/CONF.62/C.2/L.99, Annex III (C.2/Informal Meeting/14) (April 27, 1978).

The Soviet delegation deems it necessary to propose that the outer edge of the continental shelf should be defined with reference to a precise distance criterion, by fixing a specific maximum distance of up to 100 miles beyond the limit of the 200-mile economic zone. This would make it possible to determine exactly where the continental shelf of a particular State ends and where the international area, i.e., the area proclaimed to be the common heritage of mankind, begins.

For this reason it is suggested that the words *"but not further than 100 nautical miles from the outer limit of the 200-mile economic zone"* should be inserted in the existing text of article 76 after the words "to the outer edge of the continental margin."

Within the indicated 100-mile strip beyond the limit of the economic zone, any scientifically sound geological and geomorphological data could be used to determine the precise limits of the continental shelf of a particular State, and in cases where such data are not available, paragraph 3(b) of the Irish amendment submitted at the fourth session of the Conference could be applied.

Thus, according to the proposed formulation the outer edge of the continental shelf would be determined in the following manner:

1. Where the continental margin does not extend beyond the confines of the 200-mile economic zone, the edge of the continental shelf will lie along the outer limit of the economic zone.

2. In cases where the edge of the continental margin extends less than 100 miles beyond the outer limit of the 200-mile economic zone, the continental shelf of the coastal State will be determined on the basis of scientifically sound geological and geomorphological data. If such data are not available, the outer edge of the continental shelf will be determined in accordance with paragraph 3(b) of the Irish amendment ("not more than 60 nautical miles from the foot of the continental slope"), on the understanding, however, that the edge of the continental shelf shall not under any circumstances be fixed at more than 100 miles beyond the outer limit of the 200-mile economic zone.

3. Where the continental margin extends beyond the 100-mile strip adjacent to the 200-mile economic zone, the edge of the continental shelf will be fixed at a distance of 100 miles from the outer limit of the economic zone.

Consequently, according to the suggested formula the 100-mile extension of the continental shelf beyond the outer limit of the 200-mile economic zone represents a maximum limit beyond which no State may exercise its sovereign rights over the continental shelf.

THE GENERAL BATHYMETRIC CHART OF THE OCEANS (GEBCO)[5]

GEBCO is the only internationally produced series of bathymetric maps which covers the entire world. The series is produced under the auspices of the IOC and the IHO and represents a successful blending of marine geoscientists and hydrographers working in concert. The series covers the world in 18 sheets, 16 of which are at a scale of 1:10,000,000 at the equator, Mercator projection, while the two polar sheets are at 1:6,000,000 on Polar Stereographic projection. Contour interval is 500 metres in the worst case, being reduced to 100 metres in many areas. Data control is shown on the body of maps so that users can perceive the framework within which the contours were interpreted.

5. A/CONF.62/C.2/L.99, Annex IV.

The participants in this programme do so on a voluntary basis and consequently the costs to IOC have been minimal. To date, four sheets have been printed, with another three scheduled for press in early 1979. Three other sheets are in drafting stage and it is projected that the remaining sheets will be printed by May 1982.

AMENDMENTS TO ARTICLE 173 OF THE ICNT

PROPOSAL BY THE DELEGATION OF NEPAL

A. The name of "Article 173. Special Fund" shall be changed to "Article 173. Common Heritage Fund."

B. Paragraph 1 of Article 173 shall be amended to read as follows:

1. *There shall be established a Common Heritage Fund to which shall be credited:* any excess of revenues of the Authority over its expenses and costs to an extent determined by the Council; all payments received pursuant to Article 170; *and contributions which shall be made by States Members of the Authority, taking into account the net proceeds accruing to them from the exploitation of non-living resources beyond the territorial sea as defined in Article 3 of this Convention. The Council shall make recommendations from time to time to the Assembly in regard to the criteria for determining such contributions.*

C. Paragraph 2 of Article 173 shall be amended to read as follows:

2. *Disbursements from the Common Heritage Fund* shall be apportioned and made available equitably to *developing States Members* in such manner and in such currencies, and otherwise in accordance with criteria, rules, regulations, and procedures adopted by the Assembly pursuant to subparagraph (xii) of paragraph 2 of Article 158. *In establishing such criteria the Assembly should ensure that the Fund's disbursements are used to promote human welfare and world peace, to protect the marine environment, to foster the transfer of marine technology and to assist United Nations activities in these fields.*

STATEMENT BY PROFESSOR W. RIPHAGEN IN THE WORKING GROUP OF 21 ON FIRST COMMITTEE MATTERS ON APRIL 20, 1979[1]

Mr. Chairman,

At the outset I must make it clear that I speak now for the Netherlands only. I should also make it clear that what I am going to say is not a proposal in any formal sense of the word, but only a suggestion, thrown out with the purpose of possibly helping our Group to reach an overall consensus. For that same purpose this suggestion is, so to speak, automatically withdrawn if it should appear not to be helpful.

Mr. Chairman, our discussions up till now seem to center around the position of the Enterprise and its actual possibilities to operate speedily and effectively. We have discussed at length the financial and technological means to be put at its disposal.

1. Delegation of the Kingdom of the Netherlands to the Third United Nations Conference on the Law of the Sea, Eighth Session, Geneva.

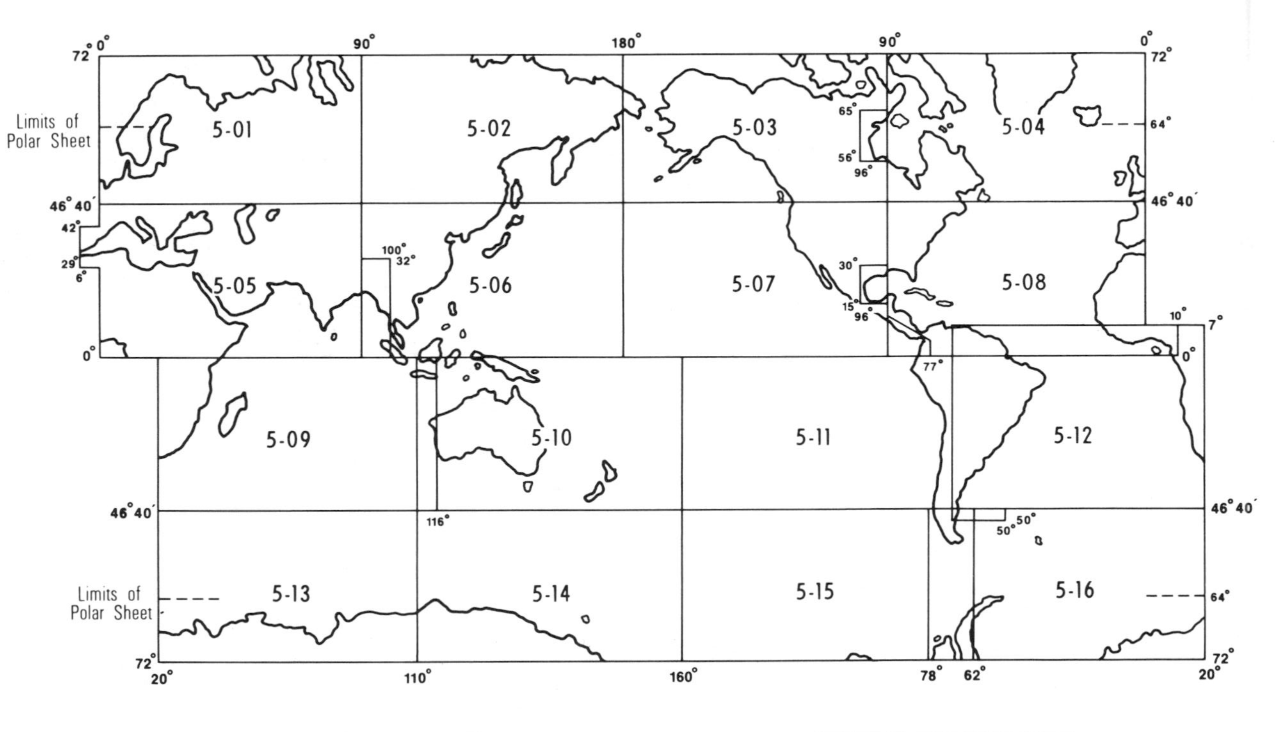

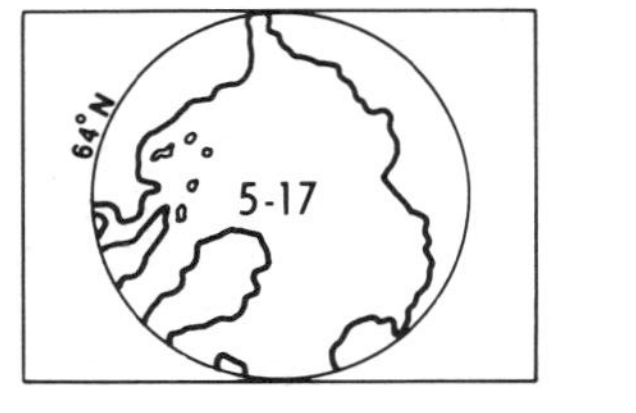

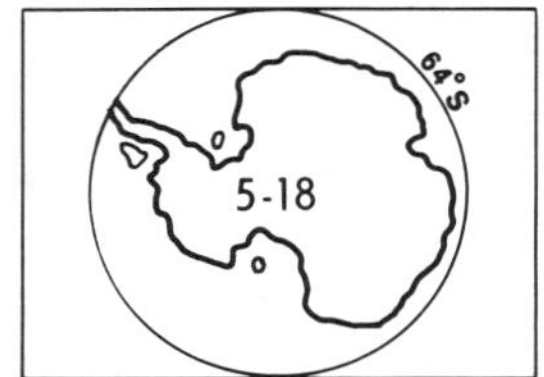

Assembly diagram for GEBCO sheets, 5th ed. (A/CONF.62/C.2/L.99, Annex IV)

It is within this context that my Delegation feels that perhaps some thought might be given to the *way* in which the Enterprise might start to operate.

Basing ourselves on the assumption that we have found a solution for the financing of the Enterprise—and my Delegation at least is hopeful that such a solution will be found—the next question is obviously what the Enterprise is going to do with the money put at its disposal.

Now this is clearly a matter for the Enterprise itself to decide upon. It could start on its own *or* it might wish first to participate in the operations of private contractors or state enterprises willing to engage in seabed mining.

Now the second branch of this alternative would seem to my Delegation the best choice, but again this is a matter for the Enterprise itself to decide, However, *if* the Enterprise should decide that the best way to start its work would be to participate in the operations of private companies or state enterprises, we should, in the opinion of my Delegation build into the Law of the Sea Convention some guarantee that the Enterprise can in actual fact realize that wish.

With this purpose in view, my Delegation ventures to suggest that possibly the following system might be introduced in the Convention.

At the time of the granting by the Authority of a contract with respect to activities for exploration and exploitation, the Enterprise should be offered the option to enter into a joint-venture arrangement with the applicant. To avoid undue uncertainties for the applicant the Enterprise should exercise its optional right within a limited time period.

If the Enterprise decides to exercise its optional participation right and enters into such arrangement with the contractor, the latter should subsequently have a similar and equivalent optional right for entering into a joint-venture arrangement with the Enterprise in the exploration and exploitation of the corresponding reserved area.

The participation by either party in such joint-venture arrangement should not exceed 20%. This leaves open the possibility for participation with a lower percentage, which, then, should be applied in both areas. If the Enterprise would, for instance, choose to participate to only 10% in a particular joint-venture arrangement, it would follow that the contractor would have an optional participation right for a joint-venture arrangement in the corresponding reserved area to the same extent, i.e., 10%.

A share of 20% in a seabed mining project makes the Enterprise a significant partner in the joint-venture arrangement. The Enterprise may, for example, participate in each of the existing 5 consortia at a percentage of 20. This, of course, is only theory—for it does not go without saying that the Enterprise will spend all its money at once. By participating in one or two consortia only, the Enterprise will be able to gain access to technology and the skill to operate it. After it has gained enough experience, the Enterprise can make a choice between the different technologies for its own operation.

The draft convention should provide for the optional right as such, while the modalities of any such joint arrangements could be negotiated—in the operating agreement—by the parties (i.e., Enterprise and contractor) at the appropriate time. It should, however, be mentioned that the contractual arrangements either way would be on normal commercial terms, giving both parties the same rights and obligations in the same manner as would apply in joint-venture arrangements entered into between commercial undertakings.

The exercise of such optional participation rights, to the extent of this participation, results in a unitary system of exploration and exploitation. It would, however, not eliminate the parallel system inasmuch as it may be expected that the Enterprise would not wish to exercise *all* options offered to it. When the Enterprise opts for participation, there is a unitary system coming into existence; when the Enterprise decides *not* to opt for that, the parallel system shall be applied to that extent.

The basic structure of any system of exploration and exploitation should be conducive to the direct participation of the Enterprise in seabed mining operations from the outset, to acquire adequate insight in the technical, financial, and managerial means that are essential in a mining project.

The specialized technology that is going to be used in future operations is nowadays concentrated in a few seabed mining consortia. For the Enterprise to get access to the relevant technology there are in essence two possibilities: firstly, the Enterprise could participate, as a partner, in an already established consortium. This means that the share of each party in the consortium will be diminished in relation to the share of the Enterprise. Secondly, the Enterprise could create a joint arrangement with a consortium and establish thereby a new consortium, in which the Enterprise has a share and the existing consortium as a whole has a share.

Taking into account the unanimously accepted objective that the Enterprise should be enabled to explore and exploit the Area for the benefit of all, the system in the draft Convention should be supplemented so as to ensure this objective.

It is in order to guarantee to the Enterprise the effective realisation of these possibilities of participation that we make the suggestions as outlined here, thereby trying to offer a modest contribution to the success of our work in this Group.

New Article 151 para. 3 *(bis)*[1]

The Enterprise has the option to enter into a joint venture arrangement with the contractor to a maximum of 20% participation under terms and conditions to be agreed upon between the Enterprise and the contractor.

If the Enterprise decides to exercise this optional right and enters into such an arrangement with the contractor, the latter has an equal optional right for entering into a joint-venture arrangement with the Enterprise in the exploration and exploitation of the corresponding reserved area.

The contractual arrangements either way shall be on commercial terms and conditions as customarily applied to joint ventures freely entered into between independent parties.

STATEMENT BY THE DELEGATION OF AUSTRIA[2]

Mr. Chairman:

The Austrian Delegation has listened with keen interest to the distinguished Delegate of the Netherlands, on April 20 in the Working Group of 21 and now again, on his timely, conceptually bold and flexible proposal, and I should like to congratulate him on his initiative. At this moment, when the Conference is faced with serious difficulties in finding a *modus operandi* acceptable to both the industrialized and the developing

1. Details, including time limit for exercising of option and conciliation and arbitration in accordance with NG.1/16, to be inserted in Annex II.

2. Presented to the United Nations Third Conference on the Law of the Sea, Eighth Session, Geneva.

countries and, at the same time, ensuring that the Enterprise can function on an equal footing with the Contractors—the Netherlands' proposal may indeed open the way towards a solution, a way out of our deadlock.

As was pointed out by various delegations the idea of a unitary joint-venture system is not new to this Conference. The Delegations of Nigeria, Sri Lanka and others have introduced it into our debates on various occasions. I may remind you that a proposal in this direction was also make by my own Delegation, albeit during an informal working session in the Spring of 1977. The text of this proposal is available in the Report of Informal Consultations in Geneva Annex 6, which was distributed by the Secretariat.

I should like to summarize here very succinctly the advantages of a unitary joint venture system, with particular reference to the version proposed by the Netherlands.

It should be stressed at the outset that the Netherlands proposal introduces a unitary system *only to the extent* that the Enterprise exercises its option for joint venture with the Contractor in the non-reserved area and the Contractor exercises his option to enter into a joint venture with the Enterprise in the reserved area. To the extent that these options are *not* exercised, the parallel system is retained. This means that the changes required in the ICNT are relatively minimal. They could, conceivably, be contained in an additional single Article 151 *bis* and some changes in Annexes II and III. If the Conference could agree on financial terms, terms and conditions of technology transfer, etc., all these paragraphs and articles could be included in the Text and remain the basis for the parallel system. Should, however, the Conference fail to agree on these detailed provisions, we need not despair: for we could then assume that Enterprise and Contractors would fall back on exercising the option for joint ventures. The availability of the option cools the burning importance of the provisions for financial arrangements and technology transfer.

It should also be stressed at the outset that the Netherlands' proposal does not detract one iota from the rights and aspirations of the Enterprise as conceived by the Developing countries. It merely *adds* to these rights. The Enterprise retains its full rights to operate by itself, but *in addition* it has the right to *share in all seabed production operations.* Theoretically, it has this option also under the ICNT; practically, however, it was not assured that there would be State or private partners for the Enterprise for joint ventures. The Netherlands' proposal assures that the option can be exercised.

Let us assume now that the options are exercised. What are the advantages of the system? I should like to develop the following 9 points:

1. *The system ensures that the Enterprise can initiate its operations at the same time as the private sector.* It is the only system that gives this assurance.

2. The problem of the *financial terms of contracts* becomes far simpler. They can be solved in accordance with standard commercial practice: The share of the produce, the share of the profits, and the share of decision-making power are proportionate to the Enterprise's investment share, which, according to the Netherlands' proposal, would be up to 20 percent in the non-reserved, and at least 80 percent in the reserved areas, that is, an average of 50 percent, if all options were exercised.

3. The system thus *maximizes financial benefits* for the Enterprise and the Authority (in the optimal case, 50 percent of the total seabed production); at the same time, it is financially advantageous to States and companies since it *reduces their investment* up to 50 percent average while providing for the kind of flexible profit and risk sharing system which the industrialized countries have been advocating during our discussions here.

4. The system *solves the problems of technology transfer* which is automatically assured in a joint venture.

5. Joint ventures may cover one or more or all stages of an integrated operation, from Research and Development through Prospecting, Exploration, Exploitation, Processing, and Marketing. *The untractable problem of calculating the ANP (attributable net proceeds) is thus avoided under a joint venture system.*

6. The *banking system,* which caused a great many difficulties, some of which have not even been fully discussed at this Conference, is greatly simplified under the Netherlands' proposal. Under the ICNT it was indeed difficult to decide at what point the two mine sites under consideration could be deemed to be of equal commercial value, and what this value was to be. The question of who was to be responsible for the costs of exploration up to the point of this decision had indeed not been solved satisfactorily to all parties. Under the Netherlands' proposal this difficulty is avoided. In practice, the banking system, under the Netherlands' proposal, would work as follows: For each Contract A, in which the Enterprise has the option for a 20 percent participation, there is a Contract B, in which States and companies have an option for an equal participation up to 20 percent. Exploration, in each case, is to be carried out by the joint venture.

7. The problem of *discrimination between the Enterprise and States and companies with regard to taxation* is avoided. All joint ventures will be treated in the same way, without discrimination.

8. The most important and basic advantage of the system is that the established industry is *built into it on the basis of cooperation rather than competition.* The Netherlands' proposal introduces this principle in a most flexible way, without shaking the basis of the parallel system. It opens options. Commercial practice and experience themselves will decide to what extent these options will be exercised.

9. The problems of the *Review Conference,* therefore, lose much of their pungency and become far more tractable. For if the system of exploration and exploitation is built in such a way that the most efficient form of cooperation is allowed to emerge during the first 20 years, the task of the Review Conference will be greatly facilitated. Rather than a consolation prize for those who did not really want to accept the parallel system in the present Convention, the Review Conference will be a normal occurrence, faced with the normal tasks of consolidating the system and making minor improvements, not of basically changing the system, under the two-edged sword of Damocles of a moratorium.

The Conference owes a debt of gratitude to the Delegation of the Netherlands for this proposal and we are looking forward to deliberations on it in the resumed session.

TRANSFER OF TECHNOLOGY: EMERGENCY TRAINING OF PERSONNEL FOR THE SEA-BED AUTHORITY[1]

STATEMENT IN PLENARY BY PAUL BAMELA ENGO (CAMEROON), CHAIRMAN OF THE FIRST COMMITTEE

I take the floor to raise an important question relating to the hard core issue of the transfer of technology. One of the main conditions demanded by the developing countries for accepting the so-called dual or parallel system proposed by the

1. Statement in Plenary on March 19, 1979, UN Third Conference on the Law of the Sea, Eighth Session, Geneva, March 19–April 27, 1979.

industrialised countries, is that the latter should accept, in the words of a recent OAU Declaration, "provisions for adequate financing, transfer of technology and training of personnel" as means of ensuring that the "Enterprise shall be an effective organ capable of undertaking activities in the Area at the same time as other entities." The idea as I see it is to make both sides of the system equally viable.

Technology for the purposes of our negotiations would appear to consist of two main areas: (1) the actual equipment or machinery developed or designed for the activities; and (2) skills or "know-how" in the technical fields involved in sea-bed mining operations: from negotiating a contract to processing and perhaps beyond.

The success or failure in reaching agreement on the issue of the transfer of these for the interim period of 20 years may well dictate the fate of the system we now contemplate.

The acquisition of technology in terms of machines or equipment is comparatively easier to negotiate, for it is now common ground that the Enterprise must be seen to be effectively operative—a point which breathes limited confidence into our labours. The Enterprise must possess technology. The issue thus becomes "how?" The imperative of ensuring a financial capability to acquire machines is in principle agreed, even if the scope and source of this must still be worked out. For the source, an applicant could buy double his needs and arrange for the Authority to have half of this at the same price he paid for them. A danger may exist in his buying too many over-sophisticated models to meet a high standard, while something just as effective may be obtained for less cost. The product may, in fact, prove to be poor and untested. An optional clause (discretion on the part of the Authority to receive or not to receive) becomes imperative. In the unlikely event that technology is developed by a State, states enterprise or private entity for purposes not of sale but of its own use, some difficulty may arise as to who pays for cost of research and development. All these are concrete situations which can be negotiated.

I do not believe that this type of technology presents a real problem either, with regard to the fundamental question: "to whom must it be transferred?" In effect it means: "who must be entitled to obtain a licence under the Convention." The Enterprise appears to be the undisputed beneficiary. Some begrudge provisions for a direct transfer of this technology to the developing countries. I believe that we ought all to agree that with regard to the Reserved Area, any entity given access, including States, *must,* if it is to be productive, have technology (in terms of machines, etc.) on the same conditions as the Enterprise itself. Otherwise, there would be a risk of mismanagement of that part of the common heritage. No one could in all seriousness contemplate that.

All of these are still comparatively easier to negotiate, given the proper political will.

A far more complex question is posed by the second categorisation of mining technology, i.e., in terms of skills and what is popularly known as "know-how." How does an applicant fulfill the obligations of transfer here? Skilled personnel in this specialized field is not easy to come by and, under the existing global high demand, expensive.

Yet it is a crucial element in the business of technology transfer. Of what use would sophisticated machinery be to the Enterprise and others who carry out activities in the Reserved Area during the 20-year interim period, if the necessary know-how for their operations is either deficient or non-existent? I do not see that this important aspect could divide the Conference. We must together examine ways and means of increasing the availability of this category of technology because we are all committed: (*a*) to a viable Enterprise coming into being as soon as activities commence in the area, as well as to its ability to exploit efficiently thus generating maximum benefits for the Authority; (*b*) to the Authority fulfilling the important function of ensuring effective participation in activities by all of mankind, no matter what their geographical location or standard of economic development; and (*c*) to ensuring that the contract Area is efficiently exploited to generate maximum profit to the Authority as well as to the Contractor, in terms of financial and other benefits to be derived from the Area.

I have been greatly troubled over the past few years about the thought of how the international community can find a solution to this problem in times of grave shortages and uneven distribution of the benefits of sea-bed technology, which many now regard as the common heritage of mankind. I have dreaded the thought of our labours, in producing a historic Convention, being frustrated by the tragedy of lack of means to implement all its aspects, especially having regard to the nature of the cruel world of subjectivity we live in, and the character of our young, fragile international community still plagued by conditions of war and ultra-nationalism.

The argument of some industrialized countries and the international press, that the Authority and the Enterprise will belong to the developing countries, does not reduce one's concern. The Authority is being created for mankind as a whole; we must not treat it patronisingly or in terms of dealing with a representative of the developing world. Yet, we must not create, even indirectly, conditions of monopoly for those in whose hands lie the technological and financial capabilities to carry out activities in the Area in contemporary times. I recently came to the conclusion that it has become timely to have this matter discussed openly.

It is heartening that sympathetic reactions have come from some of the industrialized countries with whom I have consulted informally for combating a serious problem *now.* The OAU Council of Ministers, meeting earlier this month in Nairobi in Special Session on the issues of Law of the Sea, called for an immediate programme for training of personnel, while expressing concern for probable monopoly by nationals of industrialized countries in the technical fields of mining and processing.

A similar call had been made in the Conclusions of an important convocation held in Yaounde recently, which was sponsored by the Government of the United Republic of Cameroon and organized as the annual conference *Pacem in Maribus* by the International Ocean Institute of Malta. That Institute announced concrete plans in the field. The 80 experts assembled adopted the theme "Africa and the Law of the Sea," examining, *inter alia,* what that continent would gain or lose under the Convention.

Even though progress in reaching agreement on a treaty governing exploration and production of minerals from the deep ocean floor has seemed slow to all of us, our negotiations have nevertheless reached a stage where we can be reasonably confident that they will be completed successfully and that within a period of four or five years or so sea-bed mining will be taking place under an international regime. We can also foresee much of the framework under which it will take place, and prominent in this is the key role that will be played by the International Sea-Bed Authority. It will be the resource management organization, undertaking the diverse and complex range of activities entailed in that broad responsibility. For example, it will carry out scientific research. It will review and act on applications for exploration and mining and examine proposed work plans. Under the proposed banking system it will select the mining sites to be reserved for the Authority. It will develop regulations governing the conduct of activities in the area and see to it that operations meet the requirements so established. It will be deeply involved in the process of technology transfer to the international enterprise. And, of course, it will also collect revenues.

The Authority must be ready to do its job when sea-bed mining under the Law of the Sea treaty begins. We should plan therefore on making the Authority operational as soon as possible after the treaty is ratified. Perhaps it is too early to begin organizational planning now, but it is *not* too early to start thinking about how to staff the Authority with the full range of experts that its responsibilities will require and where those experts will come from.

As an international organization, the staff of the Authority should, of course, be geographically balanced in national origins. Achieving this equitable distribution in the early years of the Authority poses a problem that deserves our attention *now*. Many of the developing countries do not now have a surplus of people with the education and training that will be required for their participation in the work of the Authority. Some have none at all. This situation threatens effective participation by most young countries in activities in the Area. This does not apply to all positions that will need to be filled. Nearly all developing countries, for example, have sufficient numbers of lawyers and people trained in non-scientific and non-technical fields. But I am sure none of us would want to see an Authority in which the lawyers and typists are all from developing countries and the geologists, engineers, and oceanographers are all from the industrialized countries. Unless we take the steps *now* to educate and train people from the developing countries in the relevant scientific and technical fields, we will be a long time in achieving across-the-board geographic balance in the staff of the Authority.

Fortunately, there are two circumstances that should make it possible to provide a balanced staff for the authority early in its operation. One is that we have time, if we take action *now,* to pursue an orderly process of education and training of citizens of developing countries, for it will be several years before sea-bed mining begins. The other is that there are already in existence technical training programmes designed to assist students from the developing countries. While they have not before now focused on fields directly related to deep ocean mining, it would seem likely that some of them could accommodate a shift in this direction. If we define the grave needs of this generation, they may be expected to react favourably. As it happens, such technical assistance programmes exist in one form or another in all the countries that have the greatest current involvement in preparations for deep ocean mining. Most of the expertise in sea-bed mining *per se* doubtless exists in the mining consortia, but oceanography, marine geology, ocean engineering and other relevant and necessary disciplines are a part of the ongoing curricula of several universities in the industrialized countries.

Support could come not only from the industrialized countries for assistance in training students in sea-bed resource management, but also from the more fortunate developing countries which also themselves are already carrying out such activities in conjunction with the production of their own resources. Some of them may be able to provide on-the-job training and some, such as OPEC members, might also be able to give some financial support as well. Considering what we already have to work from, it would not seem difficult to prepare students from the developing countries for full participation in the work of the Authority, if we set about the task *now.*

Several steps would seem appropriate to get this programme under way. One would be to request the Secretary-General of the UN to make an up-to-date analysis of the probable composition of the Authority's staff, and that of its Secretariat in particular, in the light of our negotiations thus far, to give us an idea of the number of people in various disciplines that may be needed during, say, the first five years of its existence. We might also ask the Secretary-General to compile a list of the institutions which might appropriately give financial assistance for education and training in this field. The solid support of developed countries, especially the industrialized ones among them, is indispensable. This support must be active if the programme is to succeed. Finally, the UN could serve as a clearing house for requests from the developing countries for assistance in this area, helping both to find support for trainees as well as to place them in the appropriate programme. Thought must also be given to establishing some of these in the developing world where possible.

As another step, it would be well for the developing countries to take stock of their manpower pools and to take the initiative in seeking education or training for their nationals in this area. Some developing countries already have trained personnel in this or closely related fields, and will not need such assistance. Many others will need it, however, but they will need to take the initiative in seeking the help they need.

For the longer term, it will be desirable for many of the developing countries to increase the capabilities of their own educational institutions, and that also will require some initial and substantial outside help. Here again, however, there is already some effort ongoing and some prospect of directing new efforts toward the sea-bed problem from existing programmes. For example, I am informed that the United States' sea-grant college programme is beginning support for the development of a "sister universities" programme in which US institutions will help counterpart educational institutions in developing countries to begin or to strengthen educational facilities in marine sciences. Such programmes are not needed just to provide staff for the Authority, but for the even more important long-term objective of helping prepare developing countries to participate in the full range of activities involved in the the use of marine resources.

In summary, I can foresee a problem in preparing personnel from developing countries to participate in the work of the Authority and in other marine activities. But I can also see reasonable and feasible ways of surmounting the problem if we take action *now*.

I would consequently request the distinguished Representative of the United Nations Secretary-General to carry out the studies of which I have spoken. I have already sent him a letter indicating the full scope of such studies. If an immediate programme is to succeed, we must know the type of manpower that is needed to make the Authority fully effective in the technical field. Planning depends on it and so does success. I am sure we can, as always, count on the characteristic dedication and expertise of the Secretariat in this matter.

Kuwait Regional Conference of Plenipotentiaries on the Protection and Development of the Marine Environment and the Coastal Areas[1]

FINAL ACT
OF THE KUWAIT REGIONAL CONFERENCE
OF PLENIPOTENTIARIES
ON THE PROTECTION AND DEVELOPMENT OF THE MARINE
ENVIRONMENT AND THE COASTAL AREAS

1. On the initiative of the States of the Region, the Kuwait Regional Conference of Plenipotentiaries on the Protection and Development of the Marine Environment and the Coastal Areas was convened by the Executive Director of the United Nations Environment Programme in pursuance of decision 58 (IV) of the Governing Council of UNEP.
2. The Conference met at the Messilah Beach Hotel, Kuwait, at the kind invitation of the Government of Kuwait, from 15 to 23 April 1978.
3. The following States of the Region were invited to participate in the Conference: Bahrain, Iran, Iraq, Kuwait, Oman, Qatar, Saudi Arabia and the United Arab Emirates. All of these States accepted the invitation and participated in the Conference.
4. The following secretariat, bodies, agencies and organizations were represented by observers at the Conference:

United Nations Secretariat
Department of Economic and Social Affairs
Economic Commission for Western Asia
United Nations Industrial Development Organization
United Nations bodies
United Nations Development Programme
Specialized agencies
Food and Agriculture Organization of the United Nations
United Nations Educational, Scientific and Cultural Organization
World Health Organization
Inter-Governmental Maritime Consultative Organization
Intergovernmental organizations
Organization of Arab Petroleum Exporting Countries

5. At the inaugural ceremony, Dr. Abdul Rahman Al-Awadi, Minister of Health, President of the High Committee for the Protection of the Environment in Kuwait, and Representative of His Highness, the Amir of Kuwait, made a welcoming address to the

1. Kuwait, April 15–23, 1978.

participants on behalf of the host Government. He informed the Conference that His Highness attributes special importance to the protection and preservation of the environment. His Highness had demonstrated his concern for the environment by taking the Conference under his personal auspices.

6. Dr. Al-Awadi emphasized the necessity of enhancing the quality of the environment to safeguard the social and economic well-being of the population of the Region. He pointed out that the Region is witnessing rapid development which might entail adverse effects on the quality of the environment. He urged that initiatives be taken by the States of the Region to eliminate every threat to the environment. He noted that efforts to protect the environment and preserve the natural resources of the Region must be based on regional and international co-operation and co-ordination. He warned that in the absence of such co-operation and co-ordination national efforts might go to waste.

7. Dr. Al-Awadi expressed his hope that the discussions at the Conference would lead to a comprehensive agreement on the positive steps which must be taken to preserve, protect, and enhance the quality of the environment of the Region. In concluding his welcoming address, Dr. Al-Awadi said that Kuwait would be honored to host the proposed Regional Organization for the protection of the Marine Environment and to provide it with all the necessary facilities.

8. Mr. Mostafa Kamal Tolba, Executive Director of UNEP, made a statement expressing his gratitude to the Government of Kuwait for its hospitality in hosting the Conference. He then expressed his special appreciation to His Highness, the Amir of Kuwait, for placing the Conference under his own auspices.

9. The Executive Director congratulated the States of the Region for the work they had done in preparing for the Conference. He applauded their efforts as a clear indication of their commitment to preserve the environment of the Region. In addition, he expressed his appreciation for the contribution that had been made by other bodies in the United Nations system which had co-operated in the preparation of the substantive programme to be considered by the Conference.

10. The Executive Director called attention to the important task that lay before the Conference in adopting a Final Act which would include an Action Plan for the Protection and Development of the Marine Environment and the Coastal Areas, the Kuwait Regional Convention for Co-operation on the Protection of the Marine Environment from Pollution, and a Protocol concerning Co-operation in Combating Pollution by Oil and other Harmful Substances in Cases of Emergency.

11. In concluding his address, the Executive Director pledged the full support of UNEP in servicing the Conference as well as in implementing the regional Action Plan.

12. In his address to the Conference, Mr. C. P. Srivastava, Secretary-General of the Inter-Governmental Maritime Consultative Organization, stated that IMCO attached great importance to the objectives of the Conference and pledged full support to the Conference and its follow-up action in the fields of IMCO's competence, particularly in the implementation of the Protocol concerning Co-operation in Combating Pollution and in the establishment of a Marine Emergency Mutual Aid Centre. Mr. Srivastava assured the Conference of IMCO's willingness to co-operate with UNEP in the implementation of the comprehensive Action Plan for the Region.

13. After the inaugural ceremony, the Executive Director opened the Conference in his capacity as Secretary-General of the Conference. By unanimous vote, the Conference then elected Dr. Abdul Rahman Al-Awadi, head of the Kuwait delegation, President of the Conference.

14. The Conference then adopted its agenda as follows:

1. Opening of the Conference
2. Election of the President
3. Organizational Matters:
   (a) adoption of the agenda
   (b) adoption of the Rules of Procedure
   (c) election of two Vice-Presidents and a Rapporteur-General of the Conference, and a Chairman and a Rapporteur for each of the two Main Committees
   (d) appointment of the Credentials Committee
   (e) appointment of the Drafting Committee
   (f) organization of the work of the Conference
4. Examination of the Draft Action Plan comprising chapters on:
   (a) environmental assessment
   (b) environmental management
   (c) legal aspects
   (d) institutional and financial arrangements
5. Considerations of the report of the Credentials Committee
6. Adoption of the Final Act of the Conference
7. Signature of the Final Act
8. Closing of the Conference

15. The Conference adopted its Rules of Procedure and then proceeded to unanimously elect Mr. Farrokh Parsi, deputy head of the delegation of Iran, and Mr. Abdular Al-Gain, head of the delegation of Saudi Arabia, Vice-Presidents of the Conference. Mr. Amer Araim, delegate from Iraq, was unanimously elected Rapporteur-General of the Conference.

16. In conformity with the Rules of Procedure the Conference established the following committees:

General Committee
Chairman: President of the Conference
Members: Vice-Presidents and Rapporteur-General of the Conference, and the Chairmen and Rapporteurs of the two Main Committees

Credentials Committee
Chairman: Vice-President of the Conference (Iran)
Members: Rapporteur-General of the Conference, Chairmen of Committees I and II, and Rapporteurs of Committees I and II

Main Committee I
Chairman: Mr. Rifat M. Ali (United Arab Emirates)
Rapporteur: Mr. Babakar Al-Noor (Oman)

Main Committee II
Chairman: Mr. Salah Al-Madani (Bahrain)
Rapporteur: Mr. Saad Allam (Qatar)

Drafting Committee

Chairman:	Vice-President of the Conference (Saudi Arabia)
Members:	Rapporteur-General of the Conference and the Rapporteurs of Committees I and II

Joint Committee

Chairman:	Vice-Presidents of the Conference
Rapporteur:	Rapporteur-General of the Conference

17. The Conference referred agenda items 4 (a), (b) and (d) to Main Committee I and item 4 (c) to Main Committee II and requested that the Committees report the results of their deliberations to the Plenary of the Conference.
18. The General Committee decided to establish a Joint Committee, referred the Protocol concerning Co-operation in Combating Pollution by Oil and other Harmful Substances in Cases of Emergency under agenda item 4 (c) to the Committee, and requested that the Committee report the result of its deliberations to the Plenary of the Conference.
19. The primary documents which served as the basis for the deliberations of the Conference were:

—Draft Action Plan for the Protection and Development of the Marine Environment and the Coastal Areas;
—Draft Kuwait Regional Convention for Co-operation on the Protection of the Marine Environment from Pollution;
—Draft Protocol concerning Co-operation in Combating Pollution by Oil and other Harmful Substances in Cases of Emergency;
—Draft Resolutions on the Marine Emergency Mutual Aid Centre and on Interim Arrangements.

20. The primary as well as the secondary documents which were considered by the Conference are identified by title and number in the Appendix.
21. The Conference approved that the credentials of the representatives of the participating States be recognized as in order for signing the Final Act.
22. On the basis of the deliberations of the two Main Committees and the Joint Committee as embodied in their reports,[2] the Conference, on 23 April 1978, adopted the Action Plan for the Protection and Development of the Marine Environment and the Coastal Areas, the Kuwait Regional Convention for Co-operation on the Protection of the Marine Environment from Pollution together with the Protocol concerning Co-operation in Combating Pollution by Oil and other Harmful Substances in Cases of Emergency. The Convention and the Protocol which are annexed to this Final Act will be opened by the Depositary, the Government of Kuwait, for signature by the States of the Region on 24 April 1978 in Kuwait. During the Plenary's consideration of the report of Committee I concerning institutional and financial arrangements, it was emphasized by

2. UNEP/IG.8/L.1, UNEP/IG.8/L.2, UNEP/IG.8/L.3.

the Conference that it was not the intention to limit recruitment for the interim secretariat solely to nationals from the States of the Region.
23. The Conference also adopted the following resolutions which are annexed to this Final Act:

1. Resolution on the Interim Secretariat
2. Resolution on Financial Arrangements
3. Resolution on Steps to be taken for the Establishment of the Marine Emergency Mutual Aid Centre
4. Resolution on Co-ordination between the Regional Marine Meteorological and Environmental Programmes
5. Tribute to the Government of Kuwait

IN WITNESS WHEREOF the undersigned Plenipotentiaries, being duly authorized by their respective Governments, have signed this Final Act:
DONE AT KUWAIT this twenty-third day of April, in the year one thousand nine hundred and seventy-eight in the Arabic, English and Persian languages.

APPENDIX

LIST OF DOCUMENTS BEFORE THE CONFERENCE

UNEP/IG.8/1	Provisional agenda
UNEP/IG.8/2 and Rev.1	Provisional annotated agenda
UNEP/IG.8/3 and Cor.1/Rev.1	Draft rules of procedure
UNEP/IG.8/4	Report of the Experts Meeting on the Protection of the Marine Environment, Nairobi, 13–18 June 1977 with the following Annexes: Annex I: Draft Action Plan Annex II: Draft Kuwait Regional Convention for Cooperation on the Protection of the Marine Environment Annex III: Draft Protocol on the Cooperation in Cases of Pollution Emergencies.
UNEP/IG.8/5	Not issued
UNEP/IG.8/6	Outline of Environmental Assessment Programmes included in the Action Plan
UNEP/IG.8/7	Outline of Environmental Management Programmes included in the Action Plan
UNEP/IG.8/8	Institutional and Financial Implications for the Action Plan for the two-year period 1978–1979
UNEP/IG.8/Inf.1 and Rev.1	Information for Participants
UNEP/IG.8/Inf.2	List of documents
UNEP/IG.8/Inf.3	List of participants

ACTION PLAN FOR THE PROTECTION AND DEVELOPMENT OF THE MARINE ENVIRONMENT AND THE COASTAL AREAS OF BAHRAIN, IRAN, IRAQ, KUWAIT, OMAN, QATAR, SAUDI ARABIA AND THE UNITED ARAB EMIRATES

CONTENTS

INTRODUCTION

1. The Region has been recognized by the Governments concerned and by the Governing Council of the United Nations Environment Programme (UNEP) as a "concentration area" in which UNEP, in close collaboration with the relevant components of the United Nations system, will attempt to fulfill its catalytic role in assisting States of the Region to develop and implement, in a consistent manner, an Action Plan commonly agreed upon.
2. The protection and development of the marine environment and the coastal areas of the Region for the benefit of present and future generations will be the central objective of the Action Plan. This Action Plan sets forth a framework for an environmentally sound and comprehensive approach to coastal area development, particularly appropriate for this rapidly developing Region.
3. Recognizing the complexity of the problem and the numerous ongoing activities, the Action Plan has been based upon:

3.1 findings of an interagency mission[1] organized by UNEP in co-operation with UN/ESA which visited Bahrain, Iran, Iraq, Kuwait, Oman, Qatar, Saudi Arabia and the United Arab Emirates from 15 March to 25 May 1976;

3.2 Consultative Meeting on Marine Sciences in the Region convened by UNESCO in Paris, 11–14 November 1975;

3.3 recommendations for a marine science project endorsed by the Conference of Ministers of Arab States Responsible for the Application of Science and Technology for Development, CASTARAB, convened by UNESCO in Rabat, 16–25 August 1976;

3.4 Meeting of a Group of Experts on Coastal Area Development convened by UN/ESA in New York, November 1974;

1. UNEP, United Nations Department of Economic and Social Affairs (UN/ESA), Inter-Governmental Maritime Consultative Organization (IMCO), Food and Agriculture Organization of the United Nations (FAO) and United Nations Educational, Scientific and Cultural Organization (UNESCO).

3.5 recommendations of the Kuwait Technical Meeting on Coastal Area Development and Protection of the Marine Environment co-sponsored by UNEP and UN/ESA in Kuwait, 6–9 December 1976;

3.6 a feasibility study for a co-ordinated applied marine science and basic marine science programme conducted by UNEP and UNESCO in co-operation with the Intergovernmental Oceanographic Commission (IOC) and FAO;

3.7 Regional Meeting of Legal Experts on the Protection of the Marine Environment held by UNEP in Bahrain, 24–28 January 1977;

3.8 Experts Meeting on the Protection of the Marine Environment, Nairobi, 13–18 June 1977;

3.9 additional suggestions and proposals received from the United Nations system.

4. The Action Plan aims to achieve the following:

4.1 assessment of the state of the environment including socio-economic development activities related to environmental quality and of the needs of the Region in order to assist Governments to cope properly with environmental problems, particularly those concerning the marine environment;

4.2 development of guidelines for the management of those activities which have an impact on environmental quality or on the protection and use of renewable marine resources on a sustainable basis;

4.3 development of legal instruments providing the legal basis for co-operative efforts to protect and develop the Region on a sustainable basis;

4.4 supporting measures including national and regional institutional mechanisms and structure needed for the successful implementation of the Action Plan.

5. For this document, it is assumed that the Region includes the marine area bounded in the south by the following rhumb lines:

From Ras Dharbat Ali
Lat. 16°39′ N — Long. 53°3′30″ E; then
to a position in:
Lat. 16°00′ N — Long. 53°25′ E; then
to a position in:
Lat. 17°00′ N — Long. 56°30′ E; then
to a position in:
Lat. 20°30′ N — Long. 60°00′ E; then
to Ras Al-Fasteh in:
Lat. 25°04′ N — Long 61°25′ E

The coastal area to be considered as part of the Region will be identified by the relevant Governments of the Region on an ad hoc basis depending on the type of activities to be carried out within the framework of the Action Plan. Nevertheless, coastal areas not included in the Region as defined above, should not be a source of marine pollution.

6. All components of the Action Plan are interdependent and provide a framework for comprehensive action to contribute to both the protection and the continued development of the ecoregion. No component will be an end in itself. Each activity is intended to assist the Governments of the Region to improve the quality of the information on which environmental management policies are based.

7. The protection of the marine environment is considered as the first priority of the Action Plan, and it is intended that measures for marine and coastal environmental protection and development should lead to the promotion of the human health and well-being as the ultimate goal of the Action Plan.

8. The Action Plan is intended to meet the environmental needs and enhance the environmental capabilities of the Region and is aimed primarily toward implementation by way of co-ordinated national and regional activities. To achieve this goal, an intensive training programme should be formulated in the early phases of the implementation of the Action Plan.
9. A general description of the various components of the Action Plan is given in the following paragraphs.

I. ENVIRONMENTAL ASSESSMENT

10. Environmental assessment is one of the basic activities which will underlie and facilitate the implementation of the other components of the Action Plan.
11. The identification of the present quality of the marine environment and the factors currently influencing its quality and having an impact on human health will be given priority together with an assessment of future trends.
12. Due to the lack or inadequacy of available basic data on the marine environment, a co-ordinated basic and applied regional marine science programme and marine meteorological program will be formulated as a basis for the protection of the marine environment of the Region. In formulating the operational details of these programmes, planned and ongoing national and regional programmes will be taken into account.
13. The following programmes are recognized as components of the co-ordinated regional environmental assessment programme:

13.1 survey of national capabilities of the Region in the field of marine sciences including marine meteorology covering:

(a) scientific and administrative institutions;
(b) information centres and data sources;
(c) research facilities and equipment;
(d) manpower;
(e) existing environmental laws and regulations;
(f) ongoing and planned activities;
(g) publications.

13.2 assessment of the origin and magnitude of oil pollution in the Region comprising:

(a) baseline studies on the sources, transport and distribution of oil and petroleum hydrocarbon pollution in the Region;
(b) physical, chemical and biological oceanography of the Region relevant to the transport, distribution and fate of oil as a pollutant;
(c) marine meteorology relevant to the transport and distribution of oil as a pollutant.

13.3 assessment of the magnitude of pollutants affecting human health and marine ecosystems of the Region consisting of:

(a) survey of land-based sources of industrial and municipal wastes discharged directly or indirectly into the sea or reaching it through the atmosphere;
(b) studies on the impact of industrial and municipal waste, including microbiological agents, on human health;

(c) research on effects of pollutants and other human activities, such as dredging and land reclamation on important marine species, communities and ecosystems;

(d) baseline studies and monitoring of the levels of selected pollutants, in particular heavy metals, in marine organisms.

13.4 assessment of factors relevant to the ecology of the Region and to the exploitation of its living resources including:

(a) biology of commercially important species of crustaceans, molluscs and fish in the Region, including their stock assessment;

(b) plankton productivity and distribution in the Region;

(c) ecological studies of important natural habitats in the intertidal and subtidal zones, including creeks (khores) in the Region.

13.5 assessment of geological processes such as sedimentation contributing to, or modifying, the fate of pollutants in the Region, and their impact on human health, marine ecosystems and human activities, as well as effects of coastal engineering and mining.

14. The programmes listed in paragraph 13 are interdisciplinary and interrelated in nature. Therefore, while preparing the operational details of each programme, due attention should be paid to their close co-ordination in order to avoid duplication.

15. The priorities to be assigned to the activities listed in paragraph 13 will be determined by the Governments of the Region taking into account the present level of development in the Region and the pressing need to provide reliable and comparable data on which sound management decisions can rest.

16. The agreed programme will be executed primarily through existing national institutions within the framework of regional co-operation keeping in mind that for some projects a training programme should be formulated and that the assistance of experts from outside the Region might be required in the initial phase of some projects.

17. Operational details of each programme will be developed primarily by experts nominated by the Governments of the Region.

II. ENVIRONMENTAL MANAGEMENT

18. The countries of the Region have experienced unprecedented rates of growth during recent years, particularly in areas such as urbanization, industrialization, agriculture, transport, trade, and exploration and exploitation of the Region's resources. Continuous socio-economic development can be achieved on a sustainable basis if environmental considerations are taken into account.

19. To achieve the objectives of the development and environmental management component of the Action Plan the following preparatory activities should be undertaken:

19.1 preparation and up-dating of a directory of Government-designated institutions available in the Region and active in fields related to the environmental management components of the Action Plan;

19.2 assessment of present and future development activities and their major environmental impact in order to evaluate the degree of their influence on the environment and to find appropriate measures to either eliminate or reduce any damaging effects which they may have;

19.3 identification of the most relevant ongoing national, regional or internationally supported development projects which have beneficial environmental effects such as the various fisheries projects of FAO, the environmental sanitation activities of the World Health Organization, and the assistance in industrial waste treatment provided through the United Nations Industrial Development Organization. The most significant of these projects should be strengthened and expanded to serve as demonstrations and training sites on a regional basis.

20. Furthermore, in view of the priorities and needs of the Region, the following co-operative programmes relevant to the management of regional environmental problems stemming from national development activities will be undertaken:

20.1 formulation of regional contingency plans for accidents involving oil exploration, exploitation and transport, and strengthening the meteorological services contributing to the development of contingency plans and to their execution in co-ordination with existing or future marine regional meteorological programmes;

20.2 assistance in development of national capabilities in engineering knowledge needed for regional environmental protection;

20.3 strengthening the national public health services and their co-ordination whenever transboundary interests require it;

20.4 rational exploitation and management of marine living resources, including aquaculture, on a sustainable basis, and the establishment of protected aquatic and terrestrial areas, such as marine parks, wetlands and others;

20.5 co-ordination of marine and land transport activities and the creation of a regional transport co-ordinated programme with special emphasis on port-generated pollution;

20.6 development of principles and guidelines for coastal area development and management through workshops;

20.7 co-ordination of national water management policies including community water supply and water quality control, whenever they may have impact on the marine environment of the Region;

20.8 upkeep of records of oil pollution incidents in the Region with relevant information on the impact of such pollution on the marine environment.

21. As part of the activities and regional co-operative programmes mentioned in paragraphs 19 and 20 a vast training programme should be developed for personnel from the Region. Such a programme may be executed through training at existing national, regional or international institutions ready to offer their facilities.

22. Marine and coastal area environmental protection and enhancement cannot be achieved without the full support and co-operation of all those concerned. Therefore, adequate resources should be devoted to systematic and regular campaigns for public awareness of environmental issues in the Region.

III. LEGAL COMPONENT

23. Regional legal agreements provide a fundamental basis for regional co-operation to protect the marine environment in the Region. Recognizing the importance of sound environmental development of the Region, the Governments agree to the need for early ratification of the Kuwait Regional Convention for Co-operation on the Protection of the Marine Environment from Pollution, and the Protocol concerning Regional Co-operation in Combating Pollution by Oil and other Harmful Substances in Cases of Emergency, which are adopted by the Kuwait Regional Conference of Plenipotentiaries on the Protection and Development of the Marine Environment and the Coastal Areas.

24. It is recommended that UNEP should, in co-operation with the Governments and United Nations bodies concerned, convene intergovernmental groups to prepare additional protocols which will include:

24.1 scientific and technical co-operation;

24.2 pollution resulting from exploration and exploitation of the continental shelf and the sea bed and its subsoil;

24.3 development, conservation, protection and harmonious utilization of the marine living resources of the Region;

24.4 liability and compensation for damage resulting from pollution of the marine environment;

24.5 pollution from land-based sources.

25. Aware of the need to give special protection to the Region against pollution from ships through normal operations or dumping activities, an appeal is made to Governments of the Region to strengthen the measures for the protection of the Region through ratification and implementation of the relevant international conventions, particularly:

25.1 1954 International Convention for the Prevention of Pollution of the Sea by Oil, and its amendments;

25.2 1972 Convention on Prevention of Marine Pollution by Dumping of Wastes and Other Matter;

25.3 1973 International Convention for the Prevention of Pollution from Ships as modified by Protocol of 1978.

IV. INSTITUTIONAL AND FINANCIAL ARRANGEMENTS

26. In establishing institutional arrangements for carrying out the Action Plan, a mechanism should be established which uses, to the greatest possible extent, the national capabilities available in the Region and the capabilities of existing international organizations and co-ordinating bodies and which would deal with national institutions through the appropriate national authorities of the States concerned. Where necessary, national institutions should be strengthened so that they may participate actively and efficiently in the various programmes.

27. Subject to the approval of the Governments of the Region and in close co-operation with the international bodies concerned, UNEP should make such interim arrangements as may be required for the achievement of the objectives of the Action Plan, including the establishment of an interim secretariat, until the permanent Regional Organization for the Protection of the Marine Environment is established. In order to fulfill this task the interim secretariat should have adequate professional and supporting staff recruited mainly from the signatory States in consultation, as far as possible, with the Governments of the Region. The interim secretariat shall be responsible for the overall co-ordination of the Action Plan and of matters arising out of the Convention and any protocol thereto. The interim secretariat should convene annual meetings of the States of the Region and, as necessary, working groups of regional experts to review progress achieved pursuant to recommendations set forth in the Action Plan and to advise the Executive Director of UNEP on the development of additional activities.

28. In addition to the functions assigned to it by the States of the Region, the interim secretariat or the secretariat should establish and maintain liaison with competent bodies responsible for similar activities in the Region and in other regions of the world so that each region may benefit from the experience of others and data generated in all regions may be compatible and may contribute to an overall view of the marine environment.

29. The Governments of the Region agree to the necessity of establishing a Marine Emergency Mutual Aid Centre. The Centre should have primarily a co-ordinating role in exchange of information, training programmes and monitoring. The possibility of the Centre initiating operations to combat pollution by oil and other harmful substances may be considered at a later stage in accordance with Article III of the Protocol

concerning Regional Co-operation in Combating Pollution by Oil and other Harmful Substances in Cases of Emergency.
30. Responsibilities should be transferred from the interim secretariat to the Regional Organization for the Protection of the Marine Environment as soon as this Organization is established.
31. It is proposed that the programme be financed by proportional contributions by the Governments to be assessed on the basis of a mutually agreed scale of contributions and supplemented especially in the initial stages by assistance that could be available from international bodies. The ultimate aim should be to make the programme self-supporting within the regional context, not only by developing institutional capabilities to perform the required tasks, but also by supporting training, provision of equipment and other forms of assistance from within the Region.

KUWAIT REGIONAL CONVENTION FOR CO-OPERATION ON THE PROTECTION OF THE MARINE ENVIRONMENT FROM POLLUTION

The Government of the STATE OF BAHRAIN, The Imperial Government of IRAN, The Government of the REPUBLIC OF IRAQ, The Government of the STATE OF KUWAIT, The Government of the SULTANATE OF OMAN, The Government of the STATE OF QATAR, The Government of the KINGDOM OF SAUDI ARABIA, The Government of the UNITED ARAB EMIRATES,

Realizing that pollution of the marine environment in the Region shared by Bahrain, Iran, Iraq, Kuwait, Oman, Qatar, Saudi Arabia and the United Arab Emirates, by oil and other harmful or noxious materials arising from human activities on land or at sea, especially through indiscriminate and uncontrolled discharge of these substances, presents a growing threat to marine life, fisheries, human health, recreational uses of beaches and other amenities,

Mindful of the special hydrographic and ecological characteristics of the marine environment of the Region and its particular vulnerability to pollution,

Conscious of the need to ensure that the processes of urban and rural development and resultant land use should be carried out in such a manner as to preserve, as far as possible, marine resources and coastal amenities, and that such development should not lead to deterioration of the marine environment,

Convinced of the need to ensure that the processes of industrial development should not, in any way, cause damage to the marine environment of the Region, jeopardize its living resources or create hazards to human health,

Recognizing the need to develop an integrated management approach to the use of the marine environment and the coastal areas which will allow the achievement of environmental and development goals in a harmonious manner,

Recognizing also the need for a carefully planned research, monitoring and assessment programme in view of the scarcity of scientific information on marine pollution in the Region,

Considering that the States sharing the Region have a special responsibility to protect its marine environment,

Aware of the importance of co-operation and co-ordination of action on a regional basis with the aim of protecting the marine environment of the Region for the benefit of all concerned, including future generations,

Bearing in mind the existing international conventions relevant to the present Convention,

HAVE AGREED AS FOLLOWS:

ARTICLE I
DEFINITIONS

For the purpose of the present Convention:

(a) marine pollution means the introduction by man, directly or indirectly, of substances or energy into the marine environment resulting or likely to result in such deleterious effects as harm to living resources, hazards to human health, hindrance to marine activities including fishing, impairment of quality for use of sea and reduction of amenities;

(b) "National Authority" means the authority designated by each Contracting States as responsible for the co-ordination of national efforts for implementing the Convention and its protocols;

(c) "Organization" means the organization established by the Contracting States in accordance with Article XVI;

(d) "secretariat" means the organ of the Organization established in accordance with Article XVI;

(e) "Action Plan" means the Action Plan for the Development and Protection of the Marine Environment and the Coastal Areas of Bahrain, Iran, Iraq, Kuwait, Oman, Qatar, Saudi Arabia and the United Arab Emirates adopted at the Kuwait Regional Conference of Plenipotentiaries on the Protection and Development of the Marine Environment and the Coastal Areas, convened from 15 to 23 April 1978.

ARTICLE II
GEOGRAPHICAL COVERAGE

(a) The present Convention shall apply to the sea area in the Region bounded in the south by the following rhumb lines: from Ras Dharbat Ali in (16°39′ N, 53°3′30″ E) then to a position in (16°00′ N, 53°25′ E) then to a position in (17°00′ N, 56°30′ E) then to a position in (20°30′ N, 60°00′ E) then to Ras Al-Fasteh in (25°04′ N, 61°25′ E). (Hereinafter referred to as the "Sea Area");

(b) The Sea Area shall not include internal waters of the Contracting States unless it is otherwise stated in the present Convention or in any of its protocols.

ARTICLE III
GENERAL OBLIGATIONS

(a) The Contracting States shall, individually and/or jointly, take all appropriate measures in accordance with the present Convention and those protocols in force to which they are party to prevent, abate and combat pollution of the marine environment in the Sea Area;

(b) In addition to the Protocol concerning Regional Co-operation in Combating Pollution by Oil and other Harmful Substances in Cases of Emergency opened for signature at the same time as the present Convention, the Contracting States shall co-operate in the formulation and adoption of other protocols prescribing agreed measures, procedures and standards for the implementation of the Convention;

(c) The Contracting States shall establish national standards, laws and regulations as required for the effective discharge of the obligation prescribed in paragraph (a) of this article, and shall endeavour to harmonise their national policies in this regard and for this purpose appoint the National Authority;

(d) The Contracting States shall co-operate with the competent international, regional and sub-regional organizations to establish and adopt regional standards, recommend practices and procedures to prevent, abate and combat pollution from all sources in conformity with the objectives of the present Convention, and to assist each other in fulfilling their obligations under the present Convention;

(e) The Contracting States shall use their best endeavour to ensure that the implementation of the present Convention shall not cause transformation of one type of pollution to another which could be more detrimental to the environment.

ARTICLE IV
POLLUTION FROM SHIPS

The Contracting States shall take all appropriate measures in conformity with the present Convention and the applicable rules of international law to prevent, abate and combat pollution in the Sea Area caused by intentional or accidental discharges from ships, and shall ensure effective compliance in the Sea Area with applicable international rules relating to the control of this type of pollution, including load-on-top, segregated ballast and crude oil washing procedures for tankers.

ARTICLE V
POLLUTION CAUSED BY DUMPING FROM SHIPS AND AIRCRAFT

The Contracting States shall take all appropriate measures to prevent, abate and combat pollution in the Sea Area caused by dumping of wastes and other matter from ships and aircraft, and shall ensure effective compliance in the Sea Area with applicable international rules relating to the control of this type of pollution as provided for in relevant international conventions.

ARTICLE VI
POLLUTION FROM LAND-BASED SOURCES

The Contracting States shall take all appropriate measures to prevent, abate and combat pollution caused by discharges from land reaching the Sea Area whether water-borne, air-borne, or directly from the coast including outfalls and pipelines.

ARTICLE VII
POLLUTION RESULTING FROM EXPLORATION AND EXPLOITATION OF THE BED OF THE TERRITORIAL SEA AND ITS SUB-SOIL AND THE CONTINENTAL SHELF

The Contracting States shall take all appropriate measures to prevent, abate and combat pollution in the Sea Area resulting from exploration and exploitation of the bed of the territorial sea and its sub-soil and the continental shelf, including the prevention of accidents and the combating of pollution emergencies resulting in damage to the marine environment.

ARTICLE VIII
POLLUTION FROM OTHER HUMAN ACTIVITIES

The Contracting States shall take all appropriate measures to prevent, abate and combat pollution of the Sea Area resulting from land reclamation and associated suction dredging and coastal dredging.

ARTICLE IX
CO-OPERATION IN DEALING WITH POLLUTION EMERGENCIES

(a) The Contracting States shall, individually and/or jointly, take all necessary measures, including those to ensure that adequate equipment and qualified personnel are readily available, to deal with pollution emergencies in the Sea Area, whatever the cause of such emergencies, and to reduce or eliminate damage resulting therefrom;

(b) Any Contracting State which becomes aware of any pollution emergency in the Sea Area shall, without delay, notify the Organization referred to under Article XVI and, through the secretariat, any Contracting State likely to be affected by such emergency.

ARTICLE X
SCIENTIFIC AND TECHNOLOGICAL CO-OPERATION

(a) The Contracting States shall co-operate directly, or, where appropriate, through competent international and regional organizations, in the field of scientific research, monitoring and assessment concerning pollution in the Sea Area, and shall exchange data as well as other scientific information for the purpose of the present Convention and any of its protocols;

(b) The Contracting States shall co-operate further to develop and co-ordinate national research and monitoring programmes relating to all types of pollution in the Sea Area and to establish in co-operation with competent regional or international organizations, a regional network of such programmes to ensure compatible results. For this purpose, each Contracting State shall designate the National Authority responsible for pollution research and monitoring within the areas under its national jurisdiction. The Contracting States shall participate in international arrangements for pollution research and monitoring in areas beyond their national jurisdiction.

ARTICLE XI
ENVIRONMENTAL ASSESSMENT

(a) Each Contracting State shall endeavour to include an assessment of the potential environmental effects in any planning activity entailing projects within its territory, particularly in the coastal areas, which may cause significant risks of pollution in the Sea Area;

(b) The Contracting States may, in consultation with the secretariat, develop procedures for dissemination of information of the assessment of the activities referred to in paragraph (a) above;

(c) The Contracting States undertake to develop, individually or jointly, technical and other guidelines in accordance with standard scientific practice to assist the planning of their development projects in such a way as to minimize their harmful impact on the marine environment. In this regard international standards may be used where appropriate.

ARTICLE XII
TECHNICAL AND OTHER ASSISTANCE

The Contracting States shall co-operate directly or through competent regional or international organizations in the development of programmes of technical and other assistance in fields relating to marine pollution in co-ordination with the Organization referred to in Article XVI.

ARTICLE XIII
LIABILITY AND COMPENSATION

The Contracting States undertake to co-operate in the formulation and adoption of appropriate rules and procedures for the determination of:

(a) civil liability and compensation for damage resulting from pollution of the marine environment, bearing in mind applicable international rules and procedures relating to those matters; and

(b) liability and compensation for damage resulting from violation of obligations under the present Convention and its protocols.

ARTICLE XIV
SOVEREIGN IMMUNITY

Warships or other ships owned or operated by a State, and used only on Government non-commercial service, shall be exempted from the application of the provisions of the present convention. Each Contracting State shall, as far as possible, ensure that its warships or other ships owned or operated by that State, and used only on Government non-commercial service, shall comply with the present Convention in the prevention of pollution to the marine environment.

ARTICLE XV
DISCLAIMER

Nothing in the present Convention shall prejudice or affect the rights or claims of any Contracting State in regard to the nature or extent of its maritime jurisdiction which may be established in conformity with international law.

ARTICLE XVI
REGIONAL ORGANIZATION FOR THE PROTECTION OF THE MARINE ENVIRONMENT

(a) The Contracting States hereby establish a Regional Organization for the Protection of the Marine Environment, the permanent headquarters of which shall be located in Kuwait.

(b) The Organization shall consist of the following organs:

(i) a Council which shall be comprised of the Contracting States and shall perform the functions set forth in paragraph (d) of Article XVII;

(ii) a secretariat which shall perform the functions set forth in paragraph (a) of Article XVIII; and

(iii) a Judicial Commission for the Settlement of Disputes whose composition, terms of reference and rules of procedure shall be established at the first meeting of the Council.

ARTICLE XVII
COUNCIL

(a) The meetings of the Council shall be convened in accordance with paragraph (a) of Article XVIII and paragraph (b) of Article XXX. The Council shall hold ordinary meetings once a year. Extraordinary meetings of the Council shall be held upon the request of at least one Contracting State endorsed by at least one other Contracting State, or upon the request of the Executive Secretary endorsed by at least two Contracting States. Meetings of the Council shall be convened at the headquarters of the Organization or at any other place agreed upon by consultation amongst the Contracting States. Three-fourths of the Contracting States shall constitute a quorum.

(b) The Chairmanship of the Council shall be given to each Contracting State in turn in alphabetical order of the names of the States in the English language. The Chairman shall serve for a period of one year and cannot during the period of chairmanship serve as a representative of his State. Should the chairmanship fall vacant, the Contracting State chairing the Council shall designate a successor to remain in office until the term of chairmanship of that Contracting State expires.

(c) The voting procedure in the Council shall be as follows:

(i) each Contracting State shall have one vote;

(ii) decisions on substantive matters shall be taken by a unaminous vote of the Contracting States present and voting;

(iii) decisions on procedural matters shall be taken by three-fourths majority vote of the Contracting States present and voting.

(d) The functions of the Council shall be:

(i) to keep under review the implementation of the Convention and its protocols, and the Action Plan referred to in paragraph (e) of Article I;

(ii) to review and evaluate the state of marine pollution and its effects on the Sea Area on the basis of reports provided by the Contracting States and the competent international or regional organizations;

(iii) to adopt, review and amend as required in accordance with procedures established in Article XXI, the annexes to the Convention and to its protocols;

(iv) to receive and to consider reports submitted to by Contracting States under Articles IX and XXIII;

(v) to consider reports prepared by the secretariat on questions relating to the Convention and to matters relevant to the administration of the Organization;

(vi) to make recommendations regarding the adoption of any additional protocols or any amendments to the Convention or to its protocols in accordance with Articles XIX and XX;

(vii) to establish subsidiary bodies and ad hoc working groups as required to consider any matters related to the Convention and its protocols and annexes to the Convention and its protocols;

(viii) to appoint an Executive Secretary and to make provision for the appointment by the Executive Secretary of such other personnel as may be necessary;
(ix) to review periodically the functions of the secretariat;
(x) to consider and to undertake any additional action that may be required for the achievement of the purposes of the Convention and its protocols.

ARTICLE XVIII
SECRETARIAT

(a) The secretariat shall be comprised of an Executive Secretary and the personnel necessary to perform the following functions:
(i) to convene and to prepare the meetings of the Council and its subsidiary bodies and ad hoc working groups as referred to in Article XVII, and conferences as referred to in Articles XIX and XX;
(ii) to transmit to the Contracting States notifications, reports and other information received in accordance with Articles IX and XXIII;
(iii) to consider enquiries by, and information from, the Contracting States and to consult with them on questions relating to the Convention and its protocols and annexes thereto;
(iv) to prepare reports on matters relating to the Convention and to the administration of the Organization;
(v) to establish, maintain and disseminate an up-to-date collection of national laws of all States concerned relevant to the protection of the marine environment;
(vi) to arrange, upon request, for the provision of technical assistance and advice for the drafting of appropriate national legislation for the effective implementation of the Convention and its protocols;
(vii) to arrange for training programmes in areas related to the implementation of the convention and its protocols;
(viii) to carry out its assignments under the protocols to the Convention;
(ix) to perform such other functions as may be assigned to it by the Council for the implementation of the Convention and its protocols.

(b) The Executive Secretary shall be the chief administrative official of the Organization and shall perform the functions that are necessary for the administration of the present Convention, the work of the secretariat and other tasks entrusted to the Executive Secretary by the Council and as provided for in its rules of procedure and financial rules.

ARTICLE XIX
ADOPTION OF ADDITIONAL PROTOCOLS

Any Contracting State may propose additional protocols to the present Convention pursuant to paragraph (b) of Article III at a diplomatic conference of the Contracting States to be convened by the secretariat at the request of at least three Contracting States. Additional protocols shall be adopted by a unanimous vote of the Contracting States present and voting.

ARTICLE XX
AMENDMENTS TO THE CONVENTION AND ITS PROTOCOLS

(a) Any Contracting State to the present Convention or to any of its protocols may propose amendments to the Convention or to the protocol concerned at a diplomatic conference to be convened by the secretariat at the request of at least three Contracting States. Amendments to the convention and its protocols shall be adopted by a unanimous vote of the Contracting States present and voting.

(b) Amendments to the Convention or any protocol adopted by a diplomatic conference shall be submitted by the Depositary for acceptance by all Contracting States. Acceptance of amendments to the Convention or to any protocol shall be notified to the Depositary in writing. Amendments adopted in accordance with this article shall enter into force for all Contracting States, except those which have notified the Depositary of a different intention, on the thirtieth day following the receipt by the Depositary of notification of their acceptance by at least three-fourths of the Contracting States to the Convention or any protocol concerned as the case may be.

(c) After the entry into force of an amendment to the Convention or to a protocol, any new Contracting State to the Convention or such protocol shall become a Contracting State to the instrument as amended.

ARTICLE XXI

ANNEXES AND AMENDMENTS TO ANNEXES

(a) Annexes to the Convention or to any protocol shall form an integral part of the Convention or such protocol.

(b) Except as may be otherwise provided in any protocol, the following procedure shall apply to the adoption and entry into force of any amendments to annexes to the Convention or to any protocol:

(i) any Contracting State to the Convention or to a protocol may propose amendments to the annexes to the instrument in question at the meetings of the Council referred to in Article XVII;
(ii) such amendments shall be adopted at such meetings by a unanimous vote;
(iii) the Depositary referred to in Article XXX shall communicate amendments so adopted to all Contracting States without delay;
(iv) any Contracting State which has a different intention with respect to an amendment to the annexes to the convention or to any protocol shall notify the Depositary in writing within a period determined by the Contracting States concerned when adopting the amendment;
(v) the Depositary shall notify all Contracting States without delay of any notification received pursuant to the preceding sub-paragraph;
(vi) on the expiry of the period referred to in sub-paragraph (iv) above, the amendment to the annex shall become effective for all Contracting States to the Convention or to the protocol concerned which have not submitted a notification in accordance with the provisions of that sub-paragraph.

(c) The adoption and entry into force of a new annex to the Convention or to any protocol shall be subject to the same procedure as for the adoption and entry into force of an amendment to an annex in accordance with the provisions of this article, provided that, if any amendment to the Convention or the protocol concerned is involved, the new annex shall not enter into force until such time as the amendment to the Convention or the protocol concerned enters into force.

ARTICLE XXII
RULES OF PROCEDURE AND FINANCIAL RULES

(a) The Council shall, at its first meeting, adopt its own rules.
(b) The Council shall adopt financial rules to determine, in particular, the financial participation of the Contracting States.

ARTICLE XXIII
REPORTS

Each Contracting State shall submit to the secretariat reports on measures adopted in implementation of the provisions of the Convention and its protocols in such form and at such intervals as may be determined by the Council.

ARTICLE XXIV
COMPLIANCE CONTROL

The Contracting States shall co-operate in the development of procedures for the effective application of the Convention and its protocols, including detection of violations, using all appropriate and practicable measures of detection and environmental monitoring, including adequate procedures for reporting and accumulation of evidence.

ARTICLE XXV
SETTLEMENT OF DISPUTES

(a) In case of a dispute as to the interpretation or application of this Convention or its protocols, the Contracting States concerned shall seek a settlement of the dispute through negotiation or any other peaceful means of their own choice.
(b) If the Contracting States concerned cannot settle the dispute through the means mentioned in paragraph (a) of this article, the dispute shall be submitted to the Judicial Commission for the Settlement of Disputes referred to in paragraph (b) (iii) of Article XVI.

ARTICLE XXVI
SIGNATURE

The present Convention together with the Protocol concerning Regional Co-operation in Combating Pollution by Oil and other Harmful Substances in Cases of Emergency shall be open for signature in Kuwait from 24 April to 23 July 1978 by any State invited as a participant in the Kuwait Regional Conference of Plenipotentiaries on the Protection and Development of the Marine Environment and the Coastal Areas, convened from 15 to 23 April 1978 for the purpose of adopting the Convention and the Protocol.

ARTICLE XXVII
RATIFICATION, ACCEPTANCE, APPROVAL OR ACCESSION

(a) The present Convention together with the Protocol concerning Regional Co-operation in Combating Pollution by Oil and other Harmful Substances in Cases of Emergency and any other protocol thereto shall be subject to ratification, acceptance, or approval by the States referred to in Article XXVI.

(b) As from 24 July 1978, this Convention together with the Protocol concerning Regional Co-operation in Combating Pollution by Oil and other Harmful Substances in Cases of Emergency shall be open for accession by the States referred to in Article XXVI.

(c) Any State which has ratified, accepted, approved or acceded to the present Convention shall be considered as having ratified, accepted, approved or acceded to the Protocol concerning Regional Co-operation in Combating Pollution by Oil and other Harmful Substances in Cases of Emergency;

(d) Instruments of ratification, acceptance, approval or accession shall be deposited with the Government of Kuwait which will assume the functions of Depositary.

ARTICLE XXVIII
ENTRY INTO FORCE

(a) The present Convention together with the Protocol concerning Regional Co-operation in Combating Pollution by Oil and other Harmful Substances in Cases of Emergency shall enter into force on the ninetieth day following the date of deposit of at least five instruments of ratification, acceptance or approval of, or accession to, the Convention;

(b) Any other protocol to this Convention, except as otherwise provided in such protocol, shall enter into force on the ninetieth day following the date of deposit of at least five instruments of ratification, acceptance or approval of, or accession to, such protocol;

(c) After the date of deposit of five instruments of ratification, acceptance or approval of, or accession to, this Convention or any other protocol, this Convention or any such protocol shall enter into force with respect to any State on the ninetieth day following the date of deposit by that State of the instrument of ratification, acceptance, approval or accession.

ARTICLE XXIX
WITHDRAWAL

(a) At any time after five years from the date of entry into force of this Convention, any Contracting State may withdraw from this Convention by giving written notification of withdrawal to the Depositary;

(b) Except as may be otherwise provided in any other protocol to the Convention, any Contracting State may, at any time after five years from the date of entry into force of such protocol, withdraw from such protocol by giving written notification of withdrawal to the Depositary;

(c) Withdrawal shall take effect ninety days after the date on which notification of withdrawal is received by the Depositary;

(d) Any Contracting State which withdraws from the Convention shall be considered as also having withdrawn from any protocol to which it was a party;

(e) Any Contracting State which withdraws from the Protocol concerning Regional Co-operation in Combating Pollution by Oil and other Harmful Substances in Cases of Pollution Emergency shall be considered as also having withdrawn from the Convention.

ARTICLE XXX
RESPONSIBILITIES OF THE DEPOSITARY

(a) The Depositary shall inform the Contracting States and the secretariat of the following:

(i) signature of this Convention and of any protocol thereto, and of the deposit of the instruments of ratification, acceptance, approval or accession in accordance with Article XXVII;

(ii) date on which Convention and any protocol will enter into force in accordance with the provision of Article XXVIII;

(iii) notification of a different intention made in accordance with Articles XX and XXI;

(iv) notification of withdrawal made in accordance with Article XXIX;

(v) amendments adopted with respect to the Convention and to any protocol, their acceptance by the Contracting State and the date of entry into force of those amendments in accordance with the provisions of Article XX;

(vi) adoption of new annexes and of the amendment of any annex in accordance with Article XXI;

(b) The Depositary shall call the first meeting of the Council within six months of the date on which the Convention enters into force.

The original of this Convention, of any protocol thereto, of any annex to the Convention or to a protocol, or of any amendment to the Convention, to a protocol or to an annex of the Convention or of a protocol shall be deposited with the Depositary, the Government of Kuwait who shall send copies thereof to all States concerned and shall register all such instruments and all subsequent actions in respect of them with the Secretariat of the United Nations in accordance with article 102 of the Charter of the United Nations.

IN WITNESS WHEREOF the undersigned Plenipotentiaries, being duly authorized by their respective Governments, have signed the present Convention.

DONE AT KUWAIT this twenty-fourth day of April, in the year one thousand nine hundred and seventy-eight in the Arabic, English and Persian languages, the three texts being equally authentic. In case of a dispute as to the interpretation or application of the Convention or its protocols, the English text shall be dispositively authoritative.

PROTOCOL CONCERNING REGIONAL CO-OPERATION IN COMBATING POLLUTION BY OIL AND OTHER HARMFUL SUBSTANCES IN CASES OF EMERGENCY

THE CONTRACTING STATES

Being Parties to the Kuwait Regional Convention for Co-operation on the Protection of the Marine Environment from Pollution (hereinafter referred to as "the Convention");

Conscious of the particular urgency to realize the ever present potentiality of emergencies which may result in substantial pollution by oil and other harmful substances and to provide co-operative and effective measures to deal with them;

Being aware that existing measures for responding to pollution emergencies need to be enhanced on a national and regional basis to deal with this problem in a comprehensive manner for the benefit of the Region;

HAVE AGREED AS FOLLOWS:

ARTICLE I

For the purposes of this Protocol:
(1) "Appropriate Authority" means either the National Authority defined in Article I of the Convention, or the authority or authorities within the Government of a Contracting State, designated by the National Authority and responsible for:
(a) combating and otherwise operationally responding to marine emergencies;
(b) receiving and co-ordinating information of particular marine emergencies;
(c) co-ordinating available national capabilities, for dealing with marine emergencies in general within its own Government and with other Contracting States.
(2) "Marine Emergency" means any casualty, incident, occurrence or situation, however caused, resulting in substantial pollution or imminent threat of substantial pollution to the marine environment by oil or other harmful substances and includes, *inter alia,* collisions, strandings and other incidents involving ships, including tankers, blow-outs arising from petroleum drilling and production activities, and the presence of oil or other harmful substances arising from the failure of industrial installations;
(3) "Marine Emergency Contingency Plan" means a plan or plans, prepared on a national, bilateral or multilateral basis, designed to co-ordinate the deployment, allocation and use of personnel, material and equipment for the purpose of responding to marine emergencies;
(4) "Marine Emergency Response" means any activity intended to prevent, mitigate or eliminate pollution by oil or other harmful substances or threat of such pollution resulting from marine emergencies;
(5) "Related Interests" means the interests of a Contracting State directly or indirectly affected or threatened by a marine emergency, such as:
(a) Maritime, coastal, port or estuary activities, including fisheries activities, constituting an essential means of livelihood of the persons concerned;
(b) historic and tourist attractions of the area concerned;
(c) the health of the coastal population and the wellbeing of the area concerned, including conservation of living marine resources and of wildlife;
(d) industrial activities which rely upon intake of water, including distillation plants, and industrial plants using circulating water;
(6) "Convention" means the Kuwait Regional Convention for Co-operation on the Protection of the Marine Environment from Pollution;
(7) "Sea Area" means the area specified in paragraph (a) of Article II of the Convention;
(8) "Council" means the organ of the Regional Organization for the Protection of the Marine Environment established under Article XVI of the Convention;
(9) "Centre" means the Marine Emergency Mutual Aid Centre established under Article III, paragraph 1 of the present Protocol.

ARTICLE II

1. The Contracting States shall co-operate in taking the necessary and effective measures to protect the coastline and related interests of one or more of the States from the threat and effects of pollution due to the presence of oil or other harmful substances in the marine environment resulting from marine emergencies.

2. The Contracting States shall endeavour to maintain and promote, either individually or through bilateral or multilateral co-operation, their contingency plans and means for combating pollution in the Sea Area by oil and other harmful substances. These means shall include, in particular, available equipment, ships, aircraft and manpower prepared for operations in cases of emergency.

ARTICLE III

1. The Contracting States hereby establish the Marine Emergency Mutual Aid Centre.
2. The objectives of the Centre shall be:

(a) to strengthen the capacities of the Contracting States and to facilitate co-operation among them in order to combat pollution by oil and other harmful substances in cases of marine emergencies;

(b) to assist Contracting States, which so request, in the development of their own national capabilities to combat pollution by oil and other harmful substances and to co-ordinate and facilitate information exchange, technological co-operation and training.

(c) a later objective, namely the possibility of initiating operations to combat pollution by oil and other harmful substances at the regional level, may be considered. This possibility should be submitted for approval by the Council after evaluating the results achieved in the fulfilment of the previous objectives and in the light of financial resources which could be made available for this purpose.

3. The functions of the Centre shall be:

(a) to collect and disseminate to the Contracting States information concerning matters covered by this Protocol, including:

(i) laws, regulations and information concerning appropriate authorities of the Contracting States and marine emergency contingency plans referred to in Article V of this Protocol;
(ii) information concerning methods, techniques and research relating to marine emergency response referred to in Article VI of this Protocol; and
(iii) list of experts, equipment and materials available for marine emergency responses by the Contracting States;

(b) to assist the Contracting States, as requested:

(i) in the preparation of laws and regulations concerning matters covered by this Protocol and in the establishment of appropriate authorities;
(ii) in the preparation of marine emergency contingency plans;
(iii) in the establishment of procedures under which personnel, equipment and materials involved in marine emergency responses may be expeditiously transported into, out of, and through their respective countries;
(iv) in the transmission of reports concerning marine emergencies; and
(v) in promoting and developing training programmes for combating pollution.

(c) to co-ordinate training programmes for combating pollution and prepare comprehensive anti-pollution manuals;

(d) to develop and maintain a communication/information system appropriate to the needs of the Contracting States and the Centre for the prompt exchange of information concerning marine emergencies required by this Protocol;

(e) to prepare inventories of the available personnel, material, vessels, aircraft, and other specialized equipment for marine emergency responses;

(f) to establish and maintain liaison with competent regional and international organizations, particularly the Inter-Governmental Maritime Consultative Organization, for the purposes of obtaining and exchanging scientific and technological information and data, particularly in regard of any new innovation which may assist the Centre in the performance of its functions;

(g) to prepare periodic reports on marine emergencies for submission to the Council; and

(h) to perform any other functions assigned to it either by this Protocol or by the Council.

4. The Centre may fulfill additional functions necessary for initiating operations to combat pollution by oil and other harmful substances on a regional level, when authorized by the Council, in accordance with paragraph 2 (c) above.

ARTICLE IV

1. The present Protocol shall apply to the Sea Area specified in paragraph (a) of Article II of the Convention.

2. For the purposes of dealing with a marine emergency, ports, harbours, estuaries, bays and lagoons may be treated as part of the Sea Area if the concerned Contracting State so decides.

ARTICLE V

Each Contracting State shall provide the Centre and the other Contracting States with information concerning:

(a) its appropriate authority;

(b) its laws, regulations, and other legal instruments relating generally to matters addressed in this Protocol, including those concerning the structure and operation of the authority referred to in Paragraph (a) above;

(c) its national marine emergency contingency plans.

ARTICLE VI

Each Contracting State shall provide to other Contracting States and the Centre information concerning:

(a) existing and new methods, techniques, materials, and procedures relating to marine emergency response;

(b) existing and planned research and developments in the areas referred to in Paragraph (a) above; and

(c) results of research and developments referred to in Paragraph (b) above.

ARTICLE VII

1. Each Contracting State shall direct its appropriate officials to require masters of ships, pilots of aircraft and persons in charge of offshore platforms and other similar structures operating in the marine environment and under its jurisdiction to report the existence of any marine emergency in the Sea Area to the appropriate national authority and to the Centre.

2. Any Contracting State receiving a report pursuant to paragraph 1 above shall promptly inform the following of the marine emergency:

(a) the Centre;

(b) all other Contracting States;

(c) the flag State of any foreign ship involved in the marine emergency concerned.

3. The content of the reports, including supplementary reports where appropriate, referred to in paragraph 1 above should conform to Appendix A of this Protocol.

4. Any Contracting State which submits a report pursuant to paragraphs 2 (a) and (b) above, shall be exempted from the obligations specified in paragraph (b) of Article IX of the Convention.

ARTICLE VIII

The Centre shall promptly transmit information and reports which it receives from a Contracting State pursuant to Article V, VI and paragraph 2 of Article VII of this Protocol to all other Contracting States.

ARTICLE IX

Any Contracting State which transmits information pursuant to this Protocol may specifically restrict its dessemination. In such a case, any Contracting State or the Centre to whom this information has been transmitted shall not divulge it to any other person, government, or to any public or private organization without the specific authorization of the former Contracting State.

ARTICLE X

Any Contracting State faced with a marine emergency situation as defined in Paragraph 2 of Article I of this Protocol shall:

(a) take every appropriate measure to combat pollution and/or to rectify the situation;

(b) immediately inform all other Contracting States, either directly or through the Centre, of any action which it has taken or intends to take to combat the pollution. The Centre shall promptly transmit any such information to all other Contracting States;

(c) make assessment of the nature and extent of the marine emergency, either directly or with the assistance of the Centre;

(d) determine the necessary and appropriate action to be taken with respect to the marine emergency, in consultation, where appropriate, with other Contracting States, affected States and the Centre.

ARTICLE XI

1. Any Contracting State requiring assistance in a marine emergency response may call for assistance directly from any other Contracting State or through the Centre. Where the services of the Centre are utilized, the Centre shall promptly transmit requests received to all other Contracting States. The Contracting States to whom a request is made pursuant to this paragraph shall use their best endeavours within their capabilities to render the assistance requested.
2. The assistance referred to in paragraph 1 above may include:

(a) personnel, material, and equipment, including facilities or methods for the disposal of recovered pollutant;

(b) surveillance and monitoring capacity;

(c) facilitation of the transfer of personnel, material, and equipment into, out of, and through the territories of the Contracting States.

3. The services of the Centre may be utilized by the Contracting States to co-ordinate any marine emergency response in which assistance is called for pursuant to paragraph 1 above.

4. Any Contracting State calling for assistance pursuant to paragraph 1 above shall report the activities undertaken with this assistance and its results to the Centre. The Centre shall promptly transmit any such report to all other Contracting States.
5. In cases of special emergencies, the Centre may call for the mobilization of resources made available by the Contracting States to combat pollution by oil and other harmful substances.

ARTICLE XII

1. Having due regard to the functions assigned to the Centre under this Protocol, each Contracting State shall establish and maintain an appropriate authority to carry out fully its obligations under this Protocol. With the assistance of the Centre, where appropriate, the appropriate authority of each Contracting State shall co-operate and co-ordinate its activities with counterparts in the other Contracting States.
2. Among other matters with respect to which co-operation and co-ordination efforts shall be directed under paragraph 1 above are the following:

(a) distribution and allocation of stocks of material and equipment;

(b) training of personnel for marine emergency response;

(c) marine pollution surveillance and monitoring activities;

(d) methods of communication in respect of marine emergencies;

(e) facilitation of the transfer of personnel equipment and materials involved in marine emergency responses into, out of, and through the territories of the Contracting States;

(f) other matters to which this Protocol applies.

ARTICLE XIII

The Council shall:

(a) review periodically the activities of the Centre performed under this Protocol;

(b) decide on the degree to which, and stages by which, the functions of the Centre set out in Article III will be implemented; and

(c) determine the financial, administrative and other support to be provided by the Contracting States to the Centre for the performance of its functions.

IN WITNESS WHEREOF the undersigned Plenipotentiaries, being duly authorized by their respective Governments, have signed this Protocol:

DONE AT KUWAIT this twenty-fourth day of April, in the year one thousand nine hundred and seventy-eight in the Arabic, English, and Persian languages, the three texts being equally authentic. In case of a dispute as to the interpretation or application of this Protocol, the English text shall be dispositively authoritative.

APPENDIX A
GUIDELINES FOR THE REPORT TO BE MADE PURSUANT TO ARTICLE VII OF THE PROTOCOL

1. Each report shall, as far as possible, contain, in general:

(a) the identification of the source of pollution (e.g. identity of the ship), where appropriate;

(b) the geographic position, time and date of the occurrence of the incident or of the observation;

(c) the marine meteorological conditions prevailing in the area;

(d) where the pollution originates from a ship, relevant details respecting the conditions of the ship.

2. Each report shall contain, whenever possible, in particular:

(a) a clear indication or description of the harmful substances involved, including the correct technical names of such substances (trade names should not be used in place of the correct technical names);

(b) a statement or estimate of the quantities, concentrations and likely conditions of harmful substances discharged or likely to be discharged into the sea;

(c) where relevant, a description of the packaging and identifying marks; and

(d) the name of the consignor, consignee or producer.

3. Each report shall clearly indicate, whenever possible, whether the harmful substance discharged or likely to be discharged is oil or a noxious liquid, solid or gaseous substance, and whether such substance was or is carried in bulk or contained packaged form, freight containers, portable tanks, or submarine pipelines.

4. Each report shall be supplemented, as necessary, by any relevant information requested by a recipient of the report or deemed appropriate by the person sending the report.

5. Any of the persons referred to in Article VII, paragraph 1 of this Protocol shall:

(a) supplement as far as possible the initial report, as necessary, with information concerning further developments; and

(b) comply as fully as possible with requests from affected States for additional information.

INTERIM SECRETARIAT

THE CONFERENCE

Having adopted the Action Plan for the Protection and Development of the Marine Environment and the Coastal Areas of Bahrain, Iran, Iraq, Kuwait, Oman, Qatar, Saudi Arabia and the United Arab Emirates;

Having adopted the Kuwait Regional Convention for Co-operation on the Protection of the Marine Environment and the Protocol to that Convention concerning Regional Co-operation in Combating Pollution by Oil and other Harmful Substances in Cases of Emergency;

Noting the statement of the Executive Director of the United Nations Environment Programme to the effect that UNEP is willing to accept responsibility in the interim period before the entry into force of the Convention for the overall co-ordination of the development of activities under the Action Plan;

Recognizing the importance of co-operation with other bodies in the Region, the United Nations system and other international organizations and expert bodies in the field of marine pollution;

Calls upon the Executive Director of UNEP, in consultation with the Governments of the Region and in close co-operation with relevant United Nations bodies, to make such interim arrangements as may be required until the establishment of the Regional Organization for the Protection of the Marine Environment in order to achieve the objectives of the Action Plan and to convene annual meetings of the States of the Region and, as necessary, working groups of regional experts to review progress achieved pursuant to recommendations set forth in the Action Plan and to advise on the development of additional activities.

FINANCIAL ARRANGEMENTS

THE CONFERENCE

Having adopted the Action Plan for the Protection and Development of the Marine Environment and the Coastal Areas of Bahrain, Iran, Iraq, Kuwait, Oman, Qatar, Saudi Arabia and the United Arab Emirates;

Having adopted the Kuwait Regional Convention for Co-operation on the Protection of the Marine Environment from Pollution and the Protocol concerning Co-operation in Combating Pollution by Oil and other Harmful Substances in Cases of Emergency as part of the legal component of the Action Plan;

Welcoming the willingness of the Executive Director of the United Nations Environment Programme to assume responsibility for such interim arrangements as may be required for the achievement of the objectives of the Action Plan prior to the establishment of the Regional Organization for the Protection of the Marine Environment;

Further welcoming the offer of the Executive Director of UNEP to contribute toward the costs of the interim secretariat up to a maximum of U.S. $500,000 for the initial two and one-half years;

Having regard to the cost estimates for the implementation of the Action Plan totalling U.S. $6.3 million for the initial two and one-half year operating period during which the projects stipulated in the Action Plan will be implemented;

Agrees to establish a Regional Trust Fund to cover the costs of implementing the Action Plan for the Protection and Development of the Marine Environment and the Coastal Areas;

Decides that the Regional Trust Fund be financed for the initial two and one-half year period by proportional contributions from the Governments to be assessed as shown in Table 1;

TABLE 1.—EXPECTED CONTRIBUTIONS

Area	%	U.S.$
Bahrain	2.00	116,400
Iran	28.04	1,631,928
Iraq	12.66	736,812
Kuwait	15.46	899,772
Oman	2.00	116,400
Qatar	8.93	519,726
Saudi Arabia	19.18	1,116,276
United Arab Emirates	11.73	682,686
Subtotal	...	5,820,000
UNEP	...	500,000
Total	...	6,320,000

Requests that the Executive Director of UNEP assume responsibility for administering the Regional Trust Fund in the interim period prior to the establishment of the Regional Organization for the Protection of the Marine Environment.

STEPS TO BE TAKEN FOR THE ESTABLISHMENT OF THE MARINE EMERGENCY MUTUAL AID CENTRE

THE CONFERENCE

Having adopted the Kuwait Regional Convention for Co-operation on the Protection of the Marine Environment from Pollution and the Protocol concerning Regional Cooperation in Combating Pollution by Oil and other Harmful Substances in Cases of Emergency;

Noting that Article III of the Protocol provides for the establishment of the Marine Emergency Mutual Aid Centre;

Confirming the desirability of taking the necessary steps as soon as possible to bring the Centre into operation upon entry into force of the Protocol;

Takes note of the kind invitation of the State of Bahrain to act as host to the Centre;

Requests that the Executive Director of the United Nations Environment Programme, in co-operation with the Secretary-General of the Inter-Governmental Maritime Consultative Organization, convene a meeting of governmental experts of the States signatory to the Protocol to consider the steps to be taken for the establishment of the Marine Emergency Mutual Aid Centre following the entry into force of the Protocol, and prepare the necessary documentation for consideration by the meeting with a view to making recommendations on the following to be submitted for consideration by the Governments and, subsequently, by the Council of the Regional Organization for the Protection of the Marine Environment at its first meeting:

(a) the facilities to be offered by the host Government to the Centre;
(b) the staff necessary for the Centre to fulfill its functions;
(c) the character and legal status to be accorded to the Centre and its staff;
(d) the facilities and services which could be made available to the Centre;
(e) the financial requirements for the establishment and operation of the Centre; and
(f) the financial and other support to be rendered in order to meet these requirements.

CO-ORDINATION BETWEEN THE REGIONAL MARINE METEOROLOGICAL AND ENVIRONMENTAL PROGRAMMES

THE CONFERENCE

Aware of the current plans for developing a marine meteorological programme by the States of the Region in co-operation with the World Meteorological Organization;

Recognizing the vital role that the regional marine meteorological programme will have in the future environmental assessment and management programmes in the Region;

Emphasizing that an Action Plan for the Protection and Development of the Marine Environment and the Coastal Areas was adopted by the Kuwait Regional Conference of Plenipotentiaries on the Protection and Development of the Marine Environment and the Coastal Areas held from 15 to 23 April 1978;

Welcoming the fact that the first Regional Marine Meteorological Conference held in Jeddah from 24 to 30 September 1977 identified marine meteorological support services for the monitoring of pollution in the marine environment and environmental protection as among the main functions of the regional marine meteorological programme;

Requests that both the Regional Meeting of Marine Meteorological Experts to be held in Tehran from 29 April to 4 May 1978 and the Conference of Plenipotentiaries on the Regional Marine Meteorological Programme to be held in Jeddah in 1978 take into consideration the requirements of the Action Plan and maintain continuous consultation with the interim secretariat of the Regional Organization for the Protection of the Marine Environment in order to ensure maximum efficiency and benefit for the two regional programmes by joint planning and sharing of resources;

Requests also that the Executive Director of UNEP and the Secretary-General of WMO keep in view the same considerations and promote maximum co-ordination between the two programmes.

TRIBUTE TO THE GOVERNMENT OF KUWAIT

THE CONFERENCE

Having met in Kuwait from 15 to 23 April 1978 at the gracious invitation of the Government of Kuwait and under the auspices of His Highness, the Amir of Kuwait;

Convinced that the efforts made by the Government of Kuwait in providing facilities, premises and other resources contributed significantly to the efficient conduct of its proceedings;

Deeply appreciative of the courtesy and hospitality extended by the Government of Kuwait to the members of the delegations, observers and the secretariat attending the Conference;

Expresses its sincere gratitude to His Highness, the Amir of Kuwait, and through him, to the Government and people of Kuwait, for the cordial welcome which they accorded to the Conference and to those associated with its work and for their contribution to the success of the Conference.

Recommendations for the Future Development of the Mediterranean Action Plan[1]

I. GENERAL RECOMMENDATIONS

1. As an expression of their full support for the protection and harmonious development of the Mediterranean Basin and the activities launched as part of the agreed Action Plan, the Governments of the Mediterranean States and the EEC should ratify, with the shortest possible delay, the Convention for the Protection of the Mediterranean Sea against Pollution, the Protocol for the Prevention of Pollution of the Mediterranean Sea by Dumping from Ships and Aircraft and the Protocol concerning Co-operation in Combating Pollution of the Mediterranean Sea by Oil and Other Harmful Substances in Cases of Emergency, and should continue the negotiations on the Protocol for the Protection of the Mediterranean Sea against Pollution from Land-Based Sources, leading to its final adoption and early signature, ratification and implementation.

2. The pilot phase of the various activities undertaken as part of the assessment of the sources, amounts, pathways, levels and effects of pollutants should be continued. Using the experience and results obtained during the pilot phase, as well as the established network of collaborating national institutions, a long-term monitoring programme should be prepared in consultation with Governmental experts and adopted by Governments and the EEC. By analysing the trends in levels and effects of pollutants in the Mediterranean region, this programme should serve as the basis on which to take environmentally-sound management devisions essential for the future socio-economic development of the region; these trends constitute the most objective indicator of the effectiveness of the measures taken by Governments under the Convention and protocols.

3. Recognizing the importance of environmental management for sustained socio-economic development, the Governments of the Mediterranean Region and the EEC should, through their institutions, play an active role in all the activities relevant to the integrated planning and management of natural resources. In particular, they should give firm, substantive and financial support to and participate in the implementation of the Blue Plan and the Priority Actions Programme as developed in the framework of the Mediterranean Action Plan.

4. Institutions having a regional role should be strengthened and used more efficiently for the benefit of the Mediterranean States. In particular, Governments are invited to provide to the extent possible, support and co-operation to the Regional Oil

1. Adopted by the Intergovernmental Review Meeting of Mediterranean Coastal States on the Mediterranean Action Plan, Monaco, January 9–14, 1978. United Nations Environment Programme, *Report of the Intergovernmental Review Meeting of Mediterranean Coastal States on the Mediterranean Action Plan* (UNEP/IG.11/4), annex IV. For related documents see *Ocean Yearbook 1,* ed. Elisabeth Mann Borgese and Norton Ginsburg (Chicago: University of Chicago Press, 1978), pp. 702–33.

Combating Centre in Malta. The feasibility of establishing subregional oil combating centres may be reviewed at a later stage after more experience has been gained through the operation of the Malta Centre.

5. As in the past, the activities agreed upon as part of the Action Plan should be executed by national institutions of the Governments which have been involved in formulating the Action Plan. UNEP, in close collaboration with the relevant parts of the United Nations system and under the guidance of Governments and the EEC, should act as the overall co-ordinator of these activities.

II. SPECIFIC RECOMMENDATIONS

Environmental Assessment

6. The pilot project phase of the environmental assessment component of the Mediterranean Action Plan should be extended until a reasonable amount of data is collected with a view to transforming it, as soon as feasible, into a permanent monitoring system.

7. The various projects of the environment assessment component of the Mediterranean Action Plan should be more strongly integrated and efficiently co-ordinated to make possible a comprehensive contribution to the other components of the Mediterranean Action Plan, and thus provide the indispensable scientific basis for management activities and for the legislative initiatives the Contracting Parties to the Barcelona Convention may wish to take.

8. Hazard profiles should be reassembled and updated for substances identified by the Convention and the annexes to the protocols.

9. Based on research centres and institutions nominated by the Mediterranean Governments and the EEC as participants in the various pilot projects, the network of institutions needed for systematic and comparable Mediterranean-wide data-reporting on the levels and effects of pollutants should be completed by Governments.

10. With a view to facilitating the implementation of Article 10 of the Convention, and taking into account the experience gained and the results obtained up to the present, UNEP should prepare, in consultation with the Governments of the Mediterranean States and the EEC and in collaboration with the relevant specialized parts of the United Nations system, a draft outline of a medium-term monitoring programme which, after the approval of Governments and the EEC, would replace the present pilot programme. To this end the Mediterranean Governments and the EEC should provide UNEP with information concerning their monitoring programmes and the measures they have already taken to analyse the sources, amounts, levels, trends, pathways and effects of pollutants in the Mediterranean.

11. Taking into account existing national provisions and international arrangements and agreements, proposals for criteria applicable to the quality of recreational waters and seafood should be collected and eventually developed.

12. Principles and guidelines should be prepared by UNEP allowing the Mediterranean countries to select, establish and manage specially protected Mediterranean areas.

13. A model code of practice for the disposal of liquid wastes into the Mediterranean should be developed, covering initially criteria and guidelines essential for the implementation of the protocol on pollutants from land-based sources.

14. A report on the state of pollution of the Mediterranean Sea should be prepared using reliable and comparable data, primarily those supplied by the competent national institutions and by studies which may be carried out during the pilot projects of the environmental assessment of the Mediterranean Action Plan. The final form of this report should be prepared and released in consultation with the Mediterranean Governments and the EEC.

15. UNEP should aid Governments that request assistance in defining the nature and extent of help they might need and to favourably respond to it in order to participate in the pollution monitoring and research projects through their national institutions.

16. Governments, with the assistance of UNEP if necessary, should further strengthen those national research centres that do not have either sufficiently trained personnel or the equipment for their effective participation in the programme. UNEP, if necessary, should assist Governments in the installation of new research centres.

17. Additional research centres should be designated by Governments and the EEC to participate in the pollution monitoring and research projects in geographic zones at present inadequately covered.

18. Under UNEP's overall co-ordination, and with the assistance of the relevant organizations of the United Nations system, the collaboration between research centres should be reinforced and, in view of the complementary nature of the data generated by the various pilot projects, further efforts should be made to make an interdisciplinary assessment of the origin, amounts, levels, pathways and effects of pollutants of the Mediterranean.

19. The methods used by participants in the various pilot projects assessing the levels and effects of pollutants are already well harmonized and, whenever necessary, unified. Nevertheless, as the results of the monitoring and research activities may have legislative implications for the Contracting Parties of the Barcelona Convention, UNEP, as the Secretariat of the Convention, should assist in elaborating reference methods for Mediterranean marine pollution studies and submit them for approval to the Governments and the EEC.

20. Subject to further evaluation and to the approval of the Mediterranean coastal States, a joint oceanographic cruise (MED CRUISE) could be considered by UNEP, in collaboration with the relevant organizations of the United Nations system and the national institutions of the region, to increase the number and quality of data on the open waters of the Mediterranean.

21. The input of riverborne and airborne pollutants into the Mediterranean may turn out to belong to the major groups of unknown parameters needed to assess the state of pollution in the Mediterranean Basin and UNEP should organize their assessment.

22. The facilities of the Geneva-based United Nations International Computing Centre (ICC) should be selected and used on a trial basis as the central data repository and processing facility satisfying the requirements of the entire Mediterranean Action Plan. Data reported to this facility, directly or through the organizations co-operating in the implementation of the various activities, should be considered as unclassified, unless stated otherwise. Data should be collected, handled and disseminated according to existing, standard practices, making full use of the existing mechanisms for data exchange.

23. The build-up of modelling capabilities of the Mediterranean scientists, particularly those in developing countries, should be promoted by UNEP. Initial targets for

modelling may include biogeochemical cycles of heavy metals, oil, chlorinated hydrocarbons, and ecosystems; they should be integrated with hydrodynamic models because they constitute the common basis for such models.

24. Without prejudice to the development of water pollution standards, technical principles and methodological guidelines for the scientific assessment of possible waste-absorptive capacity of the marine environment should be developed.

Integrated Planning (Environmental Management)

25. The Meeting took note of the progress made in the implementation of the Blue Plan and recommended that, in order to allow the Executive Director to proceed with the implementation of the first phase of the Blue Plan, as agreed at the 1977 Split consultation, the Governments which have not done so should, as soon as possible, and not later than the end of March 1978:

(i) designate national focal points for the Blue Plan;
(ii) nominate national participating institutions and experts for the Blue Plan activities;
(iii) indicate the surveys and prospective studies of the Blue Plan in which their institutions and experts would be ready to play an active role;
(iv) forward to UNEP Fund their financial contributions to the agreed Blue Plan budget.

26. The Meeting, having learned of the activities directly related to PAP scheduled for 1978, recommends that Governments, through their institutions and experts, take an active part in such activities. This applies, in particular, to the following:

(i) a seminar on fresh water resources management in the Mediterranean region, to be held in France in April 1978;
(ii) the expert consultation on aquaculture development in the Mediterranean region, being convened and hosted by the Greek Government in Athens from 13 to 18 March 1978 and sponsored by UNEP and GFCM of FAO;
(iii) training and information exchange in urban, environmental pollution control, including tourist resorts, Athens;
(iv) the Government of Italy/UNEP International Training Programme in Environmental Management, Urbino, Italy, 3 April - 27 July 1978;
(v) seminar on the geographic and socio-economic framework of the Blue Plan, scheduled to take place in Yugoslavia next spring.

27. The Meeting took note of the significance of protected areas from the socio-economic, scientific and conservation points of view and recommended that Governments should support the protection and rational management of existing marine parks, wetlands and other protected areas. They should also promote the creation of new protected areas in the region. In particular, Governments should:

(i) support the efforts of the Executive Director to create an Association of Protected Mediterranean Areas and advise him on the designation of one member of the Association to act as the co-ordinator of the Association's activities;
(ii) request the Executive Director to convene periodic meetings of representatives of Mediterranean protected areas to compare and develop their experiences and problems;

(iii) expand the research projects on ecological problems of protected areas and relate them to UNEP MED POL activities;

(iv) ask the Executive Director to convene an intergovernmental meeting to consider and adopt guidelines and technical principles for the establishment and management of Mediterranean protected areas. The Meeting should also consider the development of a protocol on the protection and management of Mediterranean Protected Areas (see paragraph 34 below);

(v) contribute to the preparation and periodic updating of a Directory of Mediterranean protected areas.

28. In relation to sub-paragraph (i) above, the Meeting welcomed the offer of Tunisia to serve as the co-ordinator of the Association of Protected Mediterranean Areas.

29. The Meeting noted the appropriateness of the various projects relating to evaluation of the sources, amounts, levels and effects of pollutants (see UNEP/IG.11/3/Annex I), in particular MED VII and MED X, for the management of environmental problems, which had been clearly brought out. It therefore invited Governments to take such measures as to ensure that the results of these projects could help them to develop the administrative, economic and other measures needed for environmental management.

Environmental Legislation

30. As an expression of their full support for the protection and harmonious development of the Mediterranean Basin and the activities launched as part of the agreed Action Plan, the Governments of the Mediterranean States and the EEC are urged to ratify, with the shortest possible delay, the Convention for the Protection of the Mediterranean Sea against Pollution, the Protocol for the Prevention of Pollution of the Mediterranean Sea by Dumping from Ships and Aircraft, and the Protocol concerning Co-operation in Combating Pollution of the Mediterranean Sea by Oil and Other Harmful Substances in Cases of Emergency.

31. UNEP, as the Organization responsible for the Secretariat functions under Article 13 of the Convention, should convene the first meeting of the Contracting Parties to the Convention and protocols within one year of the entry into force of the Convention. By that time it is hoped that the number of Contracting Parties will include a large majority of the Mediterranean coastal States.

32. In preparation for the first meeting of the Contracting Parties, UNEP should prepare, in consultation with the Governments of the region, the EEC and relevant international organizations, a draft of the rules of procedure and financial rules to be presented for consideration to the Contracting Parties as provided for in Article 18 of the Convention.

33. Recognizing that pollution from man's activities on land represents the most significant source of pollution in the Mediterranean Basin, the Governments of the Mediterranean States and the EEC should continue their consultations on the Protocol for the Protection of the Mediterranean Sea against Pollution from Land-Based Sources, leading to the adoption of the Protocol at a diplomatic conference. UNEP should assist the States in this task by providing as complete technical data on land-based pollutants as possible.

34. Recognizing the activities already under way within the Action Plan on specially protected areas, UNEP should, in co-operation with FAO, UNESCO and IUCN, prepare background material on existing legislation and regional legal alternatives for the protection of such marine and coastal areas. UNEP should convene a meeting of Government experts to review this material and to advise on the feasibility of developing a protocol on specially protected marine and coastal areas.

35. Taking note of the work already under way within the UNEP Working Group on Environmental Law regarding corrective and preventive measures for pollution damage arising from offshore mining and drilling carried out in the areas within national jurisdiction and of the forthcoming IJO meeting of experts on Legal Aspects of Pollution Resulting from Exploration and Exploitation of the Continental Shelf, the Seabed and its Subsoil in the Mediterranean, UNEP is requested to report to the first meeting of the Contracting Parties on the progress achieved in those fora so that a decision may be taken as to the feasibility of developing a protocol in this respect.

36. The Mediterranean States, taking note of the forthcoming Conference of Plenipotentiaries on Tanker Safety and Pollution Prevention to be convened in February 1978, should become Parties to the 1973 International Convention on the Prevention of Pollution from Ships, and at the appropriate time, should study the advisability of using their concerted efforts, within the framework of IMCO, to have the Mediterranean designated as a special area for the purposes of Annex II of that Convention.

37. Within the perspective of the application of Article 12 of the Barcelona Convention and in order to implement Resolution 4 adopted by the Barcelona Conference of Plenipotentiaries, the Executive Director should be prepared to propose to the first meeting of the Contracting Parties that a study be made of:

(a) appropriate procedures for the determination of liability and compensation for damage resulting from pollution of the marine environment deriving from violations of the provisions of the above-mentioned Convention and applicable protocols;

(b) an Interstate Guarantee Fund for the Mediterranean Sea Area.

This study should be entrusted to a committee of Government experts.[2]

38. The Mediterranean coastal States should provide, to the extent possible, support and co-operation to the Regional Oil Combating Centre so that it may effectively fulfil the objectives assigned to it. Each State should develop its national contingency plans and capabilities for dealing with oil pollution emergencies. Sectoral and subregional contingency plans for neighbouring countries should be promoted through bilateral or multilateral agreements for the above plans, technical arrangements should be agreed and assistance could be provided. When experience has been gained through the operation of the regional centre, the feasibility of establishing subregional oil combating centres may be considered.

2. One delegation recalled its government's reservations in regard to Resolution 4.

Institutional and Financial

39. The delegations convened at Monaco took note of the policy directives of the Governing Council of UNEP under which the Executive Director is carrying out the Mediterranean Action Plan, especially decision 47 (paragraph 9) and 50 (paragraph 7) adopted at the fourth session of the Council in 1976.[3]

40. Considering the exemplary nature of the Mediterranean Action Plan, the Executive Director is requested to carry out this Plan as a pilot project for other seas of the world. This long-term pilot project should be conducted under the direction of UNEP, with the assistance of the specialized international organizations concerned with the development of the Mediterranean Action Plan.

41. As in the past, activities agreed upon as part of the Action Plan should be carried out with the assistance of national institutions designated by their Governments. In this task the institutions shall be assisted by UNEP and relevant specialized United Nations organizations. UNEP, in close collaboration with the relevant parts of the United Nations system and under the guidance of Governments, will continue the role it has thus far assumed as the Secretariat of the Action Plan and of the Convention, which is an integral part of the Action Plan. Consequently, upon entry into force of the Convention, the Executive Director will make arrangements to carry out the secretariat responsibility on a continuing basis.

42. For reasons of administrative and operational efficiency, and taking into account the use of the Mediterranean programme as a model for UNEP's work in the global regional seas programme, the Executive Director will maintain the staff responsible for all main components of the Mediterranean Action Plan in a single secretariat at Geneva on an interim basis. Since the Governments convened in Monaco were not able to take a decision on the future location of the final headquarters of this co-ordination centre of the Mediterranean Action Plan, the Governments of Greece, Lebanon, Monaco and Spain repeated their offers to host this co-ordination centre on their territories; it being considered, *inter alia,* that the centre may most appropriately be situated in one of the countries of the Mediterranean Basin. Any other Governments wishing to make proposals to host the centre were invited to submit their offers to UNEP.

43. With regard to the assessment component, the Executive Director will continue, with the assistance of the United Nations system, to strengthen, during the whole pilot phase, Regional Activity Centres of the research and monitoring programme and other national scientific institutions duly nominated by their Governments.

3. Decision 47 (IV), paragraph 9, "*Considers* that the successful achievements of the United Nations Environment Programme in the field of protection of the environment in the Mediterranean region afford a concrete example of both the integrated approach and the proper co-ordinating role that should be the major concern of the Programme in its activities, and requests the Executive Director to ensure that the catalytic function, co-ordination and integration, as opposed to involvement in longer-term activities of a primarily executive character, always constitute the main contribution of the Programme in its endeavours to ensure the protection and improvement of the environment"; and Decision 50 (IV), paragraph 7, "*Notes* the Executive Director's account of how the concepts of environmental assessment and environmental management, as well as supporting activities, have been applied in the Mediterranean, and requests the Executive Director further to develop work in the Mediterranean in accordance with this framework, while taking steps towards the progressive transfer of executive responsibility to the Governments of the region."

44. The Governments of France and Yugoslavia have put at the disposal of UNEP the necessary facilities for units intended to assist in the co-ordination under the integrated planning component of the Action Plan, of the Blue Plan and PAP respectively. The Government of Spain undertook similar actions with regard to a legal unit. With a view to ensuring a balanced distribution of institutions between the countries of the region, it was agreed to establish a unit in one of the countries of the southern Mediterranean within the framework of the implementation of the Action Plan. The Government of Tunisia offered to host this unit. In view of the arrangements already made for the organization and the financing of activities under the Blue Plan, the meeting considered that UNEP should attempt to mobilize additional resources in order to strengthen and accelerate the activities under the PAP, including organizational steps to launch and co-ordinate the specific activities which are the subject of earlier recommendations. For the same reasons mentioned above, one delegation requested that the establishment of a unit in the eastern Mediterranean should be considered and offered to host such a unit.

45. Having been informed of decision 98 (V)[4] adopted at the last session of the Governing Council concerning the total commitment authority for UNEP, and taking into account the exemplary nature of the Mediterranean Action Plan, which is a pilot action plan, the delegations present in Monaco request the Executive Director to continue the effort undertaken for a substantial period.

46. The Governments convened at Monaco requested the Executive Director to prepare a report on the budget provided for the Mediterranean Action Plan. They endorsed the principle of a separate trust fund to ensure the harmonious development and effective co-ordination of jointly agreed activities. This fund could be financed as follows:

> 50 per cent to be covered by Governments of the region and the EEC. Contributions from Governments will be determined by the United Nations assessment scale and for the EEC by agreement between it and UNEP.
>
> 50 per cent by UNEP and the international organizations concerned.

47. The Meeting welcomed the intention of the Executive Director to convene in 1978 a meeting of Government-nominated representatives, to examine the Executive Director's report on the budget for the 1979/1980 biennium.

4. The total commitment authority for the Environment Fund stands as follows: 1978, $31.6 million; 1979, $30 million. Of this amount the allocation for oceans has been fixed at: 1978, $4.0 million, or 13 percent of the total; 1979, $3.19 million, or 11 percent of the total.

Final Declaration of the Review Conference of the Parties to the Treaty on the Prohibition of the Emplacement of Nuclear Weapons and Other Weapons of Mass Destruction on the Seabed and the Ocean Floor and in the Subsoil Thereof[1]

PREAMBLE

The States Parties to the Treaty on the Prohibition of the Emplacement of Nuclear Weapons and Other Weapons of Mass Destruction on the Seabed and the Ocean Floor and in the Subsoil Thereof which met in Geneva in June 1977 in accordance with the provisions of Article 7 to review the operation of the Treaty with a view to assuring that the purposes of the preamble and the provisions of the Treaty are being realized:

Recognizing the continuing importance of the Treaty and its objectives,

Affirming their belief that universal adherence to the Treaty would enhance international peace and security,

Recognizing that an arms race in nuclear weapons or any other types of weapons of mass destruction on the seabed would present a grave threat to international security,

Recognizing also the importance of continuing negotiations concerning further measures in the field of disarmament for the prevention of an arms race on the seabed, the ocean floor and the subsoil thereof,

Considering that the continuation of the trend towards a relaxation of tension in international relations provides a favourable climate in which more significant progress can be made towards the cessation of the arms race,

Reaffirming their conviction that the Treaty constitutes a step towards the exclusion of the seabed, the ocean floor and the subsoil thereof from the arms race,

Emphasizing the common interest of mankind in the progress of the exploration and use of the seabed and the ocean floor for peaceful purposes,

Recognizing that the natural resources of the seabed and ocean floor beyond the limits of national jurisdiction, will have an increasing role in assuring the economic progress of States, particularly of developing countries, and recalling in this connexion General Assembly resolution 2749 (XXV),

Appealing to States to refrain from any action which might lead to the extension of the arms race to the seabed and ocean floor, and might impede the exploration and exploitation by States of the natural resources of the seabed and ocean floor for their economic development,

1. United Nations document SBT/CONF/25, pt. 2. The text of the treaty will be found in *Ocean Yearbook I*, ed. Elisabeth Mann Borgese and Norton Ginsburg (Chicago: University of Chicago Press, 1978), pp. 402–5.

Affirming that no measures which may be decided upon in the context of international negotiations on the Law of the Sea will affect the rights and obligations assumed by the States Parties under this Treaty,

Declare as follows:

PURPOSES

The States Parties to the Treaty reaffirm their strong common interest in avoiding an arms race on the seabed in nuclear weapons or any other types of weapons of mass destruction. They reaffirm their strong support for the Treaty, their continued dedication to its principles and objectives and their commitment to implement effectively its provisions.

Article 1

The review undertaken by the Conference confirms that the obligations assumed under Article 1 of the Treaty have been faithfully observed by the States Parties. The Conference is convinced that the continued observance of this Article remains essential to the objective which all States Parties share of avoiding an arms race in nuclear weapons or any other type of weapons of mass destruction on the seabed.

Article 2

The Conference reaffirms its support for the provisions of Article 2 which define the zone covered by the Treaty.

Article 3

The Conference notes with satisfaction that no State Party has found it necessary to invoke the provisions of Article 3, paragraphs 2, 3, 4 and 5 dealing with international complaints and verifications procedures. The Conference considers that the provisions for consultation and co-operation contained in paragraphs 2, 3 and 5 include the right of interested States Parties to agree to resort to various international consultative procedures, such as *ad hoc* consultative groups of experts and other procedures.

The Conference reaffirms in the framework of Article 3 and Article 4 that nothing in the verification provisions of this Treaty should be interpreted as affecting or limiting, and notes with satisfaction that nothing in these provisions has been identified as affecting or limiting, the rights of States Parties recognized under international law and consistent with their obligations under the Treaty, including the freedom of the high seas and the rights of coastal States.

The Conference reaffirms that States Parties should exercise their rights under Article 3 with due regard for the sovereign rights of coastal States as recognized under international law.

Article 4

The Conference notes the importance of Article 4 which provides that nothing in this Treaty shall be interpreted as supporting or prejudicing the position of any State Party with respect to existing international conventions, including the 1958 Convention on the Territorial Sea and Contiguous Zone, or with respect to rights or claims which such State Party may assert, or with respect to recognition or non-recognition of rights or claims asserted by any other State, related to waters off its coast, including, *inter alia,* territorial seas and contiguous zones, or to the seabed and the ocean floor, including continental shelves. The Conference also noted that obligations assumed by States Parties to the Treaty arising from other international instruments continue to apply. The Conference agrees that the zone covered by the Treaty reflects the right balance between the need to prevent an arms race in nuclear weapons and any other types of weapons of mass destruction on the seabed and the right of States to control verification activities close to their own coasts.

Article 5

The Conference affirms the commitment undertaken in Article 5 to continue negotiations in good faith concerning further measures in the field of disarmament for the prevention of an arms race on the seabed, the ocean floor and the subsoil thereof. To this end, the Conference requests that the Conference of the Committee on Disarmament in consultation with the States Parties to the Treaty, taking into account the proposals made during this Conference and any relevant technological developments, proceed promptly with consideration of further measures in the field of disarmament for the prevention of an arms race on the seabed, the ocean floor and the subsoil thereof.

Article 6

The Conference notes that over the five years of the operation of the Treaty no State Party proposed any amendments to this Treaty according to the procedure laid down in this Article.

Article 7

The Conference notes with satisfaction the spirit of co-operation in which the Review Conference was held.

The Conference takes note of the fact that no information has been presented to it indicating that major technological developments have taken place since 1972 which affect the operation of the Treaty. The Conference, nevertheless, recognizes the need to keep such developments under continuing review and invites the Conference of the Committee on Disarmament, in consultation with the States Parties to the Treaty, to consider establishing an *ad hoc* expert group under its auspices for this purpose. Such a group might facilitate the implementation of the purposes stated in the section dealing with Article 5. It might also contribute to the orderly preparation of the next Review Conference.

In order further to facilitate the dissemination of information relevant to the Treaty to States for their assessment, the Conference invites the Secretary-General of the United Nations to collect such information from officially available sources and publish it in the United Nations Yearbook on Disarmament.

The Conference, recognizing the importance of the review mechanism provided in Article 7, decides that a further review conference should be held in Geneva in 1982 unless a majority of the States Parties indicate to the Depositaries that they wish it to be postponed. In any case a further review conference shall be convened not later than 1984. The next conference shall determine in accordance with the views of a majority of those States Parties attending whether and when an additional review conference shall be convened.

Article 8

The Conference notes with satisfaction that no State Party has exercised its rights to withdraw from the Treaty under Article 8.

Article 9

The Conference reaffirms its conviction that nothing in the Treaty affects the obligations assumed by States Parties to the Treaty under international instruments establishing zones free from nuclear weapons.

Article 10

The Conference stresses that the five years that have elapsed since the date of entry of the Treaty into force have demonstrated its effectiveness. At the same time the Conference notes with concern that the Treaty has not yet achieved universal acceptance. Therefore the Conference calls upon the States that have not yet become Parties, particularly those possessing nuclear weapons or any other types of weapons of mass destruction, to do so at the earliest possible date. Such adherence would be a significant contribution to international confidence.

Treaty between Australia and the Independent State of Papua New Guinea concerning Sovereignty and Maritime Boundaries in the Area between the Two Countries, including the Area Known as Torres Strait, and Related Matters[1]

Australia and Papua New Guinea, desiring to set down their agreed position as to their respective sovereignty over certain islands, to establish maritime boundaries, and to provide for certain other related matters in the area between the two countries including the area known as Torres Strait; recognising the importance of protecting the traditional way of life and livelihood of Australians who are Torres Strait Islanders and of Papua New Guineans who live in the coastal area of Papua New Guinea in and adjacent to the Torres Strait; recognising also the importance of protecting the marine environment and ensuring freedom of navigation and overflight for each other's vessels and aircraft in the Torres Strait area; desiring also to cooperate with one another in that area in the conservation, management, and sharing of fisheries resources and in regulating the exploration and exploitation of seabed mineral resources; as good neighbours and in a spirit of cooperation, friendship, and goodwill; have agreed as follows:

PART 1: DEFINITIONS

Article 1: Definitions

1. In this treaty,

a) "adjacent coastal area" means, in relation to Australia, the coastal area of the Australian mainland and the Australian islands near the Protected Zone; and, in relation to Papua New Guinea, the coastal area of the Papua New Guinea mainland and the Papua New Guinea islands near the Protected Zone;

b) "fisheries jurisdiction" means sovereign rights for the purpose of exploring and exploiting, conserving, and managing fisheries resources other than sedentary species;

c) "fisheries resources" means all living natural resources of the sea and seabed, including all swimming and sedentary species;

d) "free movement" means movement by the traditional inhabitants for or in the course of traditional activities;

e) "indigenous fauna and flora" includes migratory fauna;

Editors' Note: Due to limitations of space, the seven annexes to this treaty, which list the boundary points, have been omitted. Instead, the reader is referred to the map found in Mr. Prescott's article (p. 339).

1. Signed at Sydney, December 18, 1978. Text published by the Australian Department of Foreign Affairs, Canberra, 1978.

f) "mile" means an international nautical mile, being 1,852 metres in length;

g) "Protected Zone" means the zone established under article 10;

h) "Protected Zone commercial fisheries" means the fisheries resources of present or potential commercial significance within the Protected Zone and, where a stock of such resources belongs substantially to the Protected Zone but extends into an area outside but near it, the part of that stock found in that area within such limits as are agreed from time to time by the responsible authorities of the parties;

i) "seabed jurisdiction" means sovereign rights over the continental shelf in accordance with international law, and includes jurisdiction over low-tide elevations and the right to exercise such jurisdiction in respect of those elevations in accordance with international law;

j) "sedentary species" means living organisms which, at the harvestable stage, either are immobile on or under the seabed or are unable to move except in constant physical contact with the seabed or the subsoil;

k) "traditional activities" means activities performed by the traditional inhabitants in accordance with local tradition, and includes, when so performed, (i) activities on land, including gardening, collection of food, and hunting; (ii) activities on water, including traditional fishing; (iii) religious and secular ceremonies or gatherings for social purposes, for example, marriage celebrations, and settlement of disputes; and (iv) barter and market trade. In the application of this definition, except in relation to activities of a commercial nature, "traditional" shall be interpreted liberally and in the light of prevailing custom;

l) "traditional fishing" means the taking by traditional inhabitants, for their own or their dependants' consumption or for use in the course of other traditional activities, of the living natural resources of the sea, seabed, estuaries, and coastal tidal areas, including dugong and turtle;

m) "traditional inhabitants" means, in relation to Australia, persons who (i) are Torres Strait Islanders who live in the Protected Zone or the adjacent coastal area of Australia, (ii) are citizens of Australia, and (iii) maintain traditional customary associations with areas or features in or in the vicinity of the Protected Zone in relation to their subsistence or livelihood or social, cultural, or religious activities; and, in relation to Papua New Guinea, persons who (i) live in the Protected Zone or the adjacent coastal area of Papua New Guinea, (ii) are citizens of Papua New Guinea, and (iii) maintain traditional customary associations with areas or features in or in the vicinity of the Protected Zone in relation to their subsistence or livelihood or social, cultural, or religious activities.

2. Where for the purposes of this treaty it is necessary to determine the position on the surface of the Earth of a point, line, or area, that position shall be determined by reference to the Australian Geodetic Datum, that is to say, by reference to a spheroid having its centre at the centre of the Earth and a major (equatorial) radius of 6,378,160 metres and a flattening of $\frac{100}{29,825}$ and by reference to the position of the Johnston Geodetic Station in the Northern Territory of Australia. That station shall be taken to be situated at latitude 25°56′54.5515″ south and at longitude 133°12′30.0771″ east and to have a ground level of 571.2 metres above the spheroid referred to above.

3. In this treaty, the expression "in and in the vicinity of the Protected Zone" describes an area the outer limits of which might vary according to the context in which the expression is used.

PART 2: SOVEREIGNTY AND JURISDICTION

Article 2: Sovereignty over Islands

1. Papua New Guinea recognises the sovereignty of Australia over *(a)* the islands known as Anchor Cay, Aubusi Island, Black Rocks, Boigu Island, Bramble Cay, Dauan Island, Deliverance Island, East Cay, Kaumag Island, Kerr Islet, Moimi Island, Pearce Cay, Saibai Island, Turnagain Island, and Turu Cay; and *(b)* all islands that lie between the mainlands of the two countries and south of the line referred to in paragraph 1 of article 4 of this treaty.

2. No island over which Australia has sovereignty, other than those specified in subparagraph 1*(a)* of this article, lies north of the line referred to in paragraph 1 of article 4 of this treaty.

3. Australia recognises the sovereignty of Papua New Guinea over *(a)* the islands known as Kawa Island, Mata Kawa Island, and Kussa Island; and *(b)* all the other islands that lie between the mainlands of the two countries and north of the line referred to in paragraph 1 of article 4 of this treaty, other than the islands specified in subparagraph 1*(a)* of this article.

4. In this treaty, sovereignty over an island shall include sovereignty over *(a)* its territorial sea; *(b)* the airspace above the island and its territorial sea; *(c)* the seabed beneath its territorial sea and the subsoil thereof; and *(d)* any island, rock, or low-tide elevation that may lie within its territorial sea.

Article 3: Territorial Seas

1. The territorial sea boundaries between the islands of Aubusi, Boigu and Moimi, and Papua New Guinea and the islands of Dauan, Kaumag and Saibai, and Papua New Guinea shall be the lines described in Annex 1 to this treaty, which are shown on the map annexed to this treaty as Annex 2, together with such other portion of the outer limit of the territorial sea of Saibai described in Annex 3 to this treaty that may abut the territorial sea of Papua New Guinea.

2. The territorial seas of the islands specified in subparagraph 1*(a)* of article 2 of this treaty shall not extend beyond three miles from the baselines from which the breadth of the territorial sea around each island is measured. Those territorial seas shall not be enlarged or reduced, even if there were to be any change in the configuration of a coastline or a different result from any further survey.

3. The provisions of paragraph 2 of this article shall not apply to that part of the territorial sea of Pearce Cay which lies south of the line referred to in paragraph 1 of article 4 of this treaty.

4. The outer limits of the territorial seas of the islands specified in subparagraph 1*(a)* or article 2 of this treaty, except in respect of that part of the territorial sea of Pearce Cay which lies south of the line referred to in paragraph 1 of article 4 of this treaty, shall be as described in Annex 3 to this treaty. The limits so described are shown on the maps annexed to this treaty as Annexes 2 and 4.

5. Australia shall not extend its territorial sea northwards across the line referred to in paragraph 1 of article 4 of this treaty.

6. Papua New Guinea shall not *(a)* extend its territorial sea off its southern coastline between the meridians of longitude 142°03′30″ east and of longitude 142°51′00″ east, beyond three miles from the baselines from which the breadth of the territorial sea is measured; *(b)* extend its territorial sea or archipelagic waters into the area bounded by that portion of the line referred to in paragraph 2 of article 4 of this treaty running from the point of latitude 9°45′24″ south, longitude 142°03′30″ east to the point of latitude 9°40′30″ south, longitude 142°51′00″ east, and that portion of the line referred to in paragraph 1 of article 4 of this treaty which runs between those two points; *(c)* establish an archipelagic baseline running in or through the area referred to in subparagraph *(b)* of this paragraph; or *(d)* extend its territorial sea southwards across the line referred to in paragraph 1 or article 4 of this treaty.

Article 4: Maritime Jurisdiction

1. Subject to the provisions of article 2 of this treaty, the boundary between the area of seabed and subsoil that is adjacent to and appertains to Australia and the area of seabed and subsoil that is adjacent to and appertains to Papua New Guinea, and over which Australia and Papua New Guinea, respectively, shall have seabed jurisdiction, shall be the line described in Annex 5 to this treaty. The line so described is shown on the map annexed to this treaty as Annex 6 and, in part, on the map annexed to this treaty as Annex 7.

2. Subject to the provisions of article 2 of this treaty, the boundary between the area of sea that is adjacent to and appertains to Australia and the area of sea that is adjacent to and appertains to Papua New Guinea, and in which Australia and Papua New Guinea, respectively, shall have fisheries jurisdiction, shall be the line described in Annex 8 to this treaty. The line so described is shown on the map annexed to this treaty as Annex 6 and, in part, on the maps annexed to this treaty as Annexes 2 and 7.

3. In relations to the area bounded by the portion of the line referred to in paragraph 2 of this article running from the point of latitude 9°45′24″ south, longitude 142°03′30″ east to the point of latitude 9°40′30″ south, longitude 142°51′00″ east and that portion of the line referred to in paragraph 1 of this article which runs between those two points, exclusive of the territorial seas of the islands of Aubusi, Boigu, Dauan, Kaumag, Moimi, Saibai, and Turnagain, *(a)* neither party shall exercise residual jurisdiction without the concurrence of the other party; and *(b)* the parties shall consult with a view to reaching agreement on the most effective method of application of measures involving the exercise of residual jurisdiction.

4. In paragraph 3 of this article, "residual jurisdiction" means *(a)* jurisdiction over the area other than seabed jurisdiction or fisheries jurisdiction, including jurisdiction other than seabed jurisdiction or fisheries jurisdiction insofar as it relates to, *inter alia*, (i) the preservation of the marine environment; (ii) marine scientific research; and (iii) the production of energy from the water, currents, and winds; and *(b)* seabed and fisheries jurisdiction to the extent that the exercise of such jurisdiction is not directly related to the exploration or exploitation of resources or to the prohibition of, or refusal to authorise, activities subject to that jurisdiction.

PART 3: SOVEREIGNTY AND JURISDICTION—RELATED MATTERS

Article 5: Existing Petroleum Permit

1. Where prior to September 16, 1975, Australia has granted an exploration permit for petroleum under Australian law in respect of a part of the seabed over which it ceases by virtue of this treaty to exercise sovereign rights, and a permittee retains rights in respect

of that permit immediately prior to the entry into force of this treaty, Papua New Guinea, upon application by that permittee, shall offer to that permittee a petroleum-prospecting licence or licences under Papua New Guinea law in respect of the same part of the seabed on terms that are not less favourable than those provided under Papua New Guinea law to any other holder of a seabed petroleum-prospecting licence.

2. An application for a licence under paragraph 1 of this article shall be made *(a)* in respect of a part of the seabed lying outside the Protected Zone, within six months after the date of entry into force of this treaty; *(b)* in respect of a part of the seabed lying within the Protected Zone, during the period referred to in article 15 and any extension of that period to which the parties may agree.

Article 6: Exploitation of Certain Seabed Deposits

If any single accumulation of liquid hydrocarbons or natural gas, or if any other mineral deposit beneath the seabed, extends across any line defining the limits of seabed jurisdiction of the parties, and if the part of such accumulation or deposit that is situated on one side of such a line is recoverable in fluid form wholly or in part from the other side, the parties shall consult with a view to reaching agreement on the manner in which the accumulation or deposit may be most effectively exploited and on the equitable sharing of the benefits from such exploitation.

Article 7: Freedoms of Navigation and Overflight

1. On and over the waters of the Protected Zone that lie *(a)* north of the line referred to in paragraph 1 of article 4 of this treaty and seaward of the low water lines of the land territory of either party, and *(b)* south of that line and beyond the outer limits of the territorial sea, each party shall accord to the vessels and aircraft of the other party, subject to paragraphs 2 and 3 of this article, the freedoms of navigation and overflight associated with the operation of vessels and aircraft on or over the high seas.

2. Each party shall take all necessary measures to ensure that, in the exercise of the freedoms of navigation and overflight accorded to its vessels and aircraft under paragraph 1 of this article, *(a)* those vessels observe generally accepted international regulations, procedures, and practices for safety at sea and for the prevention, reduction, and control of pollution from ships; *(b)* those civil aircraft observe the Rules of the Air established by the International Civil Aviation Organization as they apply to civil aircraft, and state aircraft normally comply with such of those rules as relate to safety and at all times operate with due regard for the safety of navigation; *(c)* those vessels and aircraft north of the line referred to in paragraph 1 of article 4 of this treaty do not engage in the embarking or disembarking of any commodity, currency, or person contrary to the customs, fiscal, immigration, or sanitary laws and regulations of the other party, provided that the relevant laws and regulations of that party do not have the practical effect of denying, hampering, or impairing the freedoms of navigation and overflight accorded under paragraph 1 of this article; and *(d)* those vessels and aircraft, north of the line referred to in paragraph 1 of article 4 of this treaty, do not act in a manner prejudicial to the peace, good order, or security of the other party.

3. Vessels of a party engaged in the exploration or exploitation of resources in an area of jurisdiction of the other party shall remain subject to the laws and regulations of the other party made in the exercise of its resources jurisdiction consistently with this treaty and with international law, including the provisions of those laws and regulations concerning the boarding, inspection, and apprehension of vessels.

4. In those areas of the Protected Zone north of the line referred to in paragraph 1 of article 4 of this treaty to which paragraph 1 of this article does not apply, civil aircraft of a party engaged in scheduled or nonscheduled air services shall have the right of overflight and the right to make stops for nontraffic purposes without the need to obtain prior permission from the other party, subject to compliance with any applicable laws or regulations made for the safety of air navigation.

5. In areas of the Protected Zone to which paragraph 1 of this article does not apply, the vessels of a party shall enjoy the right of innocent passage. There shall be no suspension of that right, and neither party shall adopt laws or regulations applying to those areas that might impede or hamper the normal passage of vessels between two points both of which are in the territory of one party.

6. In cases where the provisions of neither paragraph 1 nor paragraph 5 of this article apply, a regime of passage over routes used for international navigation in the area between the two countries, including the area known as Torres Strait, shall apply in respect of vessels that is no more restrictive of passage than the regime of transit passage through straits used for international navigation described in articles 34–44 inclusive of document A/Conf. 62/WP.10 of the Third United Nations Conference on the Law of the Sea, provided that, before a party adopts a law or regulation that might impede or hamper the passage over these routes of vessels proceeding to or from the territory of the other party, it shall consult with the other party. If the provisions of those articles are revised, are not included in any Law of the Sea Convention, or fail to become generally accepted principles of international law, the parties shall consult with a view to agreeing upon another regime of passage that is in accordance with international practice to replace the regime of passage applying under this paragraph.

7. The rights of navigation and overflight provided for in this article are in addition to, and not in derogation of, rights of navigation and overflight in the area concerned under other treaties or general principles of international law.

Article 8: Navigational Aids

With a view to maintaining and improving the safety of navigation through the waters in the area between the two countries, the parties shall cooperate and, with due regard to the technical and other means available to each of them, shall, where appropriate and as may be agreed between them, provide mutual assistance in the provision and maintenance of navigational aids and in the preparation of charts and maps.

Article 9: Wrecks

1. Wrecks of vessels and aircraft which lie on, in, or under the seabed in an area of seabed jurisdiction of a party shall be subject to the jurisdiction of that party.

2. If a wreck of historical or special significance to a party is located or found in an area between the two countries under the jurisdiction of the other party, the parties shall consult with a view to reaching agreement on the action, if any, to be taken with respect to that wreck.

3. The provisions of this article shall be without prejudice to the competence of the courts of a party, for the purposes of the laws of that party, in relation to maritime causes of action in respect of wrecks coming within the provisions of this article.

4. This article shall not apply to any military vessel or aircraft of either party wrecked after the date of entry into force of this treaty.

PART 4: THE PROTECTED ZONE

Article 10: Establishment and Purposes of the Protected Zone

1. A Protected Zone in the Torres Strait is hereby established comprising all the land, sea, airspace, seabed, and subsoil within the area bounded by the line described in Annex 9 to this Treaty. The line so described is shown on the maps annexed to this treaty as Annexes 6 and 7 and, in part, on the map annexed to this treaty as Annex 2.

2. The parties shall adopt and apply measures in relation to the Protected Zone in accordance with the provisions of this treaty.

3. The principal purpose of the parties in establishing the Protected Zone, and in determining its northern, southern, eastern, and western boundaries, is to acknowledge and protect the traditional way of life and livelihood of the traditional inhabitants, including their traditional fishing and free movement.

4. A further purpose of the parties in establishing the Protected Zone is to protect and preserve the marine environment and indigenous fauna and flora in and in the vicinity of the Protected Zone.

Article 11: Free Movement and Traditional Activities including Traditional Fishing

1. Subject to the other provisions of this treaty, each party shall continue to permit free movement and the performance of lawful traditional activities in and in the vicinity of the Protected Zone by the traditional inhabitants of the other party.

2. Paragraph 1 of this article shall not be interpreted as sanctioning the expansion of traditional fishing by the traditional inhabitants of one party into areas outside the Protected Zone under the jurisdiction of the other party not traditionally fished by them prior to the date of entry into force of this treaty.

3. The provisions of this article and the other provisions of this treaty concerning traditional fishing are subject to article 14 and paragraph 2 of article 20 of this treaty.

Article 12: Traditional Customary Rights

Where the traditional inhabitants of one party enjoy traditional customary rights of access to and usage of areas of land, seabed, seas, estuaries, and coastal tidal areas that are in or in the vicinity of the Protected Zone and that are under the jurisdiction of the other party, and those rights are acknowledged by the traditional inhabitants living in or in proximity to those areas to be in accordance with local tradition, the other party shall permit the continued exercise of those rights on conditions not less favourable than those applying to like rights of its own traditional inhabitants.

Article 13: Protection of the Marine Environment

1. Each party shall take legislative and other measures necessary to protect and preserve the marine environment in and in the vicinity of the Protected Zone. In formulating those measures each party shall take into account internationally agreed rules, standards, and recommended practices which have been adopted by diplomatic conferences or by relevant international organisations.

2. The measures that each party shall take in accordance with paragraph 1 of this article shall include measures for the prevention and control of pollution or other damage to the marine environment from all sources and activities under its jurisdiction

or control and shall include, in particular, measures to minimise to the fullest practicable extent *(a)* the release of toxic, harmful, or noxious substances from land-based sources, from rivers, from or through the atmosphere, or by dumping at sea; *(b)* pollution or other damage from vessels; and *(c)* pollution or other damage from installations and devices used in the exploration and exploitation of the natural resources of the seabed and subsoil thereof.

3. The measures taken by each party in accordance with paragraph 1 of this article shall be consistent with its obligations under international law, including obligations not to prejudice the rights of foreign ships and aircraft, and shall be subject to the provisions of article 7 of this treaty.

4. The parties shall consult, at the request of either, for the purpose of *(a)* harmonising their policies with respect to the measures that each shall take pursuant to this article, and *(b)* ensuring the effective and coordinated implementation of those measures.

5. If either party has reasonable grounds for believing that any planned activity under its jurisdiction or control may cause pollution or other damage to the marine environment in or in the vicinity of the Protected Zone, that party shall, after due investigation, communicate to the other party its assessment of the potential impact of that activity on the marine environment.

6. If either party has reasonable grounds for believing that any existing or planned activity under the jurisdiction or control of the other party is causing or may cause pollution or other damage to the marine environment in or in the vicinity of the Protected Zone, it may request consultations with the other party, and the parties shall then consult as soon as possible with a view to adopting measures to prevent or control any pollution or other damage to that environment from that activity.

Article 14: Protection of Fauna and Flora

1. Each Party shall, in and in the vicinity of the Protected Zone, use its best endeavours to *(a)* identify and protect species of indigenous fauna and flora that are or may become threatened with extinction; *(b)* prevent the introduction of species of fauna and flora that may be harmful to indigenous fauna and flora; and *(c)* control noxious species of fauna and flora.

2. Notwithstanding any other provision of this treaty except paragraph 4 of this article, a party may implement within its area of jurisdiction measures to protect species of indigenous fauna and flora which are or may become threatened with extinction or which either party has an obligation to protect under international law.

3. The parties shall as appropriate and necessary exchange information concerning species of indigenous fauna and flora that are or may become threatened with extinction and shall consult, at the request of either of them, for the purpose of *(a)* harmonising their policies with respect to the measures that each may take to give effect to paragraphs 1 and 2 of this article; and *(b)* ensuring the effective and coordinated implementation of those measures.

4. In giving effect to the provisions of this article, each party shall use its best endeavours to minimise any restrictive effects on the traditional activities of the traditional inhabitants.

Article 15: Prohibition of Mining and Drilling of the Seabed

Neither party shall undertake or permit within the Protected Zone mining or drilling of the seabed or the subsoil thereof for the purpose of exploration for or exploitation of liquid hydrocarbons, natural gas, or other mineral resources during a period of ten years from the date of entry into force of this treaty. The parties may agree to extend that period.

Article 16: Immigration, Customs, Quarantine, and Health

1. Except as otherwise provided in this treaty, each party shall apply immigration, customs, quarantine, and health procedures in such a way as not to prevent or hinder free movement or the performance of traditional activities in and in the vicinity of the Protected Zone by the traditional inhabitants of the other party.

2. Each party, in administering its laws and policies relating to the entry and departure of persons and the importation and exportation of goods into and from areas under its jurisdiction in and in the vicinity of the Protected Zone, shall act in a spirit of mutual friendship and good neighbourliness, bearing in mind relevant principles of international law and established international practices and the importance of discouraging the occurrence, under the guise of free movement or performance of traditional activities, of illegal entry, evasion of justice, and practices prejudicial to effective immigration, customs, health, and quarantine protection and control.

3. Notwithstanding the provisions of paragraph 1 of this article, *(a)* traditional inhabitants of one party who wish to enter the other country, except for temporary stay for the performance of traditional activities, shall be subject to the same immigration, customs, health, and quarantine requirements and procedures as citizens of that party who are not traditional inhabitants; *(b)* each party reserves its right to limit free movement to the extent necessary to control abuses involving illegal entry or evasion of justice; and *(c)* each party reserves its right to apply such immigration, customs, health, and quarantine measures, temporary or otherwide, as it considers necessary to meet problems which may arise. In particular each party may apply measures to limit or prevent free movement, or the carriage of goods, plants, or animals in the course thereof, in the case of an outbreak or spread of an epidemic, epizootic, or epiphytotic in or in the vicinity of the Protected Zone.

Article 17: Implementation and Coordination

In order to facilitate the implementation of the provisions of this treaty relating to the Protected Zone, the authorities of each party shall, at the request of the authorities of the other party, as may be appropriate and necessary, *(a)* make available to the authorities of the other party information on the relevant provisions of its laws, regulations, and procedures relating to immigration, citizenship, customs, health, quarantine, fisheries, the protection of the environment, and other matters; and *(b)* consult with the authorities of the other party with a view to making appropriate administrative or other arrangements to resolve any problems arising in the implementation of those provisions.

Article 18: Liaison Arrangements

1. Each party shall designate a representative who shall facilitate the implementation at the local level of the provisions of this treaty.

2. The two designated representatives shall *(a)* exchange information on relevant developments in and in the vicinity of the Protected Zone; *(b)* consult together and take such action as is appropriate to their respective functions to facilitate the practical operation at the local level of the provisions of this treaty and to resolve any problems arising therefrom; *(c)* keep under review free movement by the traditional inhabitants

of one party into areas under the jurisdiction of the other party and the local arrangements applying in respect of such free movement; and *(d)* draw to the attention of their governments, and make recommendations as appropriate on, any matters affecting the implementation of the provisions of this treaty or arising therefrom which are not capable of resolution at the local level or which may otherwise require consideration by both parties.

3. In the exercise of his functions, each representative shall *(a)* consult closely with representatives of the traditional inhabitants of his country, particularly in relation to any problems which may arise in respect of free movement, traditional activities, and the exercise of traditional customary rights as provided for in this treaty, and convey their views to his government; and *(b)* maintain close liaison with national, state, provincial, and local authorities of his country on all matters falling within their respective responsibilities.

4. Unless a different location is required by the circumstances, the representative of Australia shall be based at Thursday Island and the representative of Papua New Guinea shall be based at Daru.

Article 19: Torres Strait Joint Advisory Council

1. The parties shall jointly establish and maintain an advisory and consultative body which shall be known as the Torres Strait Joint Advisory Council (called in this article "the Advisory Council").

2. The functions of the Advisory Council shall be *(a)* to seek solutions to problems arising at the local level and not resolved pursuant to article 18 of this treaty; *(b)* to consider and to make recommendations to the parties on any developments or proposals which might affect the protection of the traditional way of life and livelihood of the traditional inhabitants, their free movement, performance of traditional activities, and exercise of traditional customary rights as provided for in this treaty; and *(c)* to review from time to time as necessary, and to report and to make recommendations to the parties on, any matters relevant to the effective implementation of this treaty, including the provisions relating to the protection and preservation of the marine environment, and fauna and flora, in and in the vicinity of the Protected Zone.

3. The Advisory Council shall not have or assume responsibilities for management or administration. These responsibilities shall, within the respective areas of jurisdiction of each party, continue to lie with the relevant national, state, provincial, and local authorities.

4. In the exercise of its functions, the Advisory Council shall ensure that the traditional inhabitants are consulted, that they are given full and timely opportunity to comment on matters of concern to them, and that their views are conveyed to the parties in any reports and recommendations made by the Advisory Council to the parties.

5. The Advisory Council shall transmit its reports and recommendations to the foreign ministers of the parties. After consideration by appropriate authorities of the parties, consultations may be arranged with a view to the resolution of matters to which the Advisory Council has invited attention.

6. Unless otherwise agreed by the parties, the Advisory Council shall consist of eighteen members, that is, nine members from each party who shall include *(a)* at least two national representatives; *(b)* at least one member representing the government of Queensland in the case of Australia and one representing the Fly River provincial government in the case of Papua New Guinea; and *(c)* at least three members representing the traditional inhabitants, with each party being free to decide from time to time from which of the aforementioned categories any other of its members will be drawn.

7. The Advisory Council shall meet when necessary at the request of either party.

Consecutive meetings of the Advisory Council shall be chaired alternately by a representative of Australia and a representative of Papua New Guinea. Meetings shall be held alternately in Australia and Papua New Guinea or as may from time to time be otherwise arranged.

PART 5: PROTECTED ZONE COMMERCIAL FISHERIES

Article 20: Priority of Traditional Fishing and Application of Measures to Traditional Fishing

1. The provisions of this part shall be administered so as not to prejudice the achievement of purposes of part 4 of this treaty in regard to traditional fishing.

2. A party may adopt a conservation measure consistent with the provisions of this part which, if necessary for the conservation of a species, may be applied to traditional fishing, provided that the party shall use its best endeavours to minimise any restrictive effects of that measure on traditional fishing.

Article 21: Conservation, Management, and Optimum Utilisation

The parties shall cooperate in the conservation, management, and optimum utilisation of Protected Zone commercial fisheries. To this end, the parties shall consult at the request of either and shall enter into arrangements for the effective implementation of the provisions of this part.

Article 22: Conservation and Management of Individual Fisheries

1. The parties shall, where appropriate, negotiate subsidiary conservation and management arrangements in respect of any individual Protected Zone commercial fishery.

2. If either party notifies the other in writing that it regards one of the Protected Zone commercial fisheries as one to which common conservation and management arrangements should apply, the parties shall within ninety days from the date of the notification enter into consultations with a view to concluding arrangements specifying the measures to be applied by them with respect to that fishery.

3. The parties shall, where appropriate, also negotiate supplementary conservation and management arrangements in respect of resources directly related to a fishery referred to in paragraph 1 of this article, including resources invoking stocks occurring in the Protected Zone where such stocks are not otherwise subject to the provisions of this treaty.

Article 23: Sharing of the Catch of the Protected Zone Commercial Fisheries

1. The parties shall share the allowable catch of the Protected Zone commercial fisheries in accordance with the provisions of this article and of articles 24 and 25 of this treaty.

2. The allowable catch, that is to say the optimum sustainable yield, of a Protected Zone commercial fishery shall be determined jointly by the parties as part of the subsidiary conservation and management arrangements referred to in paragraph 1 of article 22 of this treaty.

3. If either party has reasonable grounds for believing that the commercial exploi-

tation of a species of Protected Zone commercial fisheries would, or has the potential to, cause serious damage to the marine environment or might endanger another species, that party may request consultations with the other party and the parties shall then consult as soon as possible with a view to reaching agreement on whether such commercial exploitation could be undertaken in a manner which would not result in such damage or endanger another species.

4. In respect of any relevant period where the full allowable catch of a particular Protected Zone commercial fishery might be taken, each party shall be entitled to a share of the allowable catch apportioned, subject to paragraphs 5, 6, and 8 of this article and to articles 24 and 25 of this treaty, as follows: *(a)* in areas under Australian jurisdiction, except as provided in *(b)* below, Australia 75%, Papua New Guinea 25%; *(b)* within the territorial seas of Anchor Cay, Black Rocks, Bramble Cay, Deliverance Island, East Cay, Kerr Islet, Pearce Cay, and Turu Cay, Australia 50%, Papua New Guinea 50%; *(c)* in areas under Papua New Guinea jurisdiction, Australia 25%, Papua New Guinea 75%.

5. Papua New Guinea shall have the sole entitlement to the allowable catch of the commercial barramundi fishery near the Papua New Guinea coast, except within the territorial seas of the islands of Aubusi, Boigu, Dauan, Kaumag, Moimi, and Saibai where, in respect of that fishery, the provisions of paragraph 4 *(a)* of this article shall not apply.

6. In apportioning the allowable catch in relation to an individual fishery, the parties shall normally consider the allowable catch expressed in terms of weight or volume. In calculating the apportionment of the total allowable catch of the Protected Zone commercial fisheries, the parties shall have regard to the relative value of individual fisheries and shall, for this purpose, agree on a common value for production from each individual fishery for the period in question, such value being based on the value of the raw product at the processing facility or such other point as may be agreed, but prior to any enhancement of value through processing, including processing at a pearl culture farm, or further transportation or marketing.

7. The parties may agree to vary the apportionment of the allowable catch determined for individual fisheries as part of the subsidiary conservation and management arrangements referred to in paragraph 1 of article 22 of this treaty but so as to maintain in respect of the total allowable catch of the Protected Zone commercial fisheries the apportionment specified in paragraph 4 of this article for each party.

8. In calculating the total allowable catch of the Protected Zone commercial fisheries, the allowable catch of the commercial barramundi fishery referred to in paragraph 5 of this article shall be disregarded.

Article 24: Transitional Entitlement

1. As part of the subsidiary conservation and management arrangements referred to in paragraph 1 of article 22 of this treaty, the level of the catch of each Protected Zone commercial fishery to which each party is entitled, provided it remains within the allowable catch, *(a)* shall not, during the period of five years immediately after the entry into force of this treaty, be reduced below the level of catch of that party before the entry into force of this treaty; but *(b)* may, during the second period of five years after the entry into force of this treaty, be adjusted progressively so that at the end of that second five-year period it reaches the level of catch apportioned in each case in article 23 of this treaty.

2. The entitlement of a party under this article shall, where the limitation of the allowable catch makes it necessary, take priority over the entitlement of the other party under article 23 of this treaty, but shall be taken into account in calculating the entitlement of the first party.

Article 25: Preferential Entitlement

If, in any relevant period, a party does not itself propose to take all the allowable catch of a Protected Zone commercial fishery to which it is entitled, either in its own area of jurisdiction or that of the other party, the other party shall have a preferential entitlement to any of the allowable catch of that fishery not taken by the first party.

Article 26: Licensing Arrangements

1. In the negotiation and implementation of the conservation and management arrangements referred to in paragraph 1 of article 22 of this treaty, *(a)* the parties shall consult and cooperate in the issue and endorsement of licences to permit commercial fishing in Protected Zone commercial fisheries; *(b)* the responsible authorities of the parties may issue licences to fish in any Protected Zone commercial fishery; and *(c)* persons or vessels which are licensed by the responsible authorities of one party to fish in any relevant period in a Protected Zone commercial fishery shall, if nominated by the responsible authorities of that party, be authorised by the responsible authorities of the other party, wherever necessary, by the endorsement of licences or otherwise, to fish in those areas under the jurisdiction of the other party in which the fishery concerned is located.

2. The persons or vessels licensed by one party which have been authorised, or are to be authorised, under the provisions of paragraph 1 of this article to fish in waters under the jurisdiction of the other party shall comply with the relevant fisheries laws and regulations of the other party except that they shall be exempt from licensing fees, levies, and other charges imposed by the other party in respect of such fishing activities.

3. In issuing licences in accordance with paragraph 1 of this article, the responsible authorities of both parties shall have regard to the desirability of promoting economic development in the Torres Strait area and employment opportunities for the traditional inhabitants.

4. The responsible authorities of both parties shall ensure that the traditional inhabitants are consulted from time to time on the licensing arrangements in respect of Protected Zone commercial fisheries.

Article 27: Third-State Fishing in Protected Zone Commercial Fisheries

1. The responsible authorities of the parties shall inform one another and shall consult, at the request of either of them, concerning the proposed exploitation of the Protected Zone commercial fisheries *(a)* by a joint venture in which there is third-state equity participation; or *(b)* by a vessel of third-state registration or with a crew substantially of the nationality of a third state.

2. Vessels the operations of which are under the control of nationals of a third state shall not be licensed to exploit the Protected Zone commercial fisheries without the concurrence of the responsible authorities of both parties in a particular case or class of cases.

Article 28: Inspection and Enforcement

1. The parties shall cooperate, including by exchange of personnel, in inspection and enforcement to prevent violations of the Protected Zone commercial fisheries arrangements and in taking appropriate enforcement measures in the event of such violations.

2. The parties shall consult from time to time, as necessary, so as to ensure that legislation and regulations adopted by each party pursuant to paragraph 1 of this article are, as far as practicable, consistent with the legislation and regulations of the other party.

3. Each party shall make it an offence under its fisheries laws or regulations for a person to use a vessel of its nationality to fish in Protected Zone commercial fisheries for species of fisheries resources in areas over which the other party has jurisdiction in respect of those species *(a)* without being duly licensed or authorised by that other party; or *(b)* in the case of a licensed or authorised vessel, in breach of the fisheries laws or regulations of the other party applying within those areas.

4. Each party will, in relation to species of fisheries resources in areas where it has jurisdiction in respect of those species, *(a)* investigate suspected offences against its fisheries laws and regulations; and *(b)* except as provided in or under this article, take corrective action when necessary against offenders against those laws or regulations.

5. In this article, "corrective action" means the action normally taken in respect of a suspected offence, after due investigation, and includes, where appropriate, the apprehension of a suspected offender, the prosecution of an alleged offender, or the execution of a penalty imposed by a court or the cancellation or suspension of the licence of an offender.

6. In accordance with the provisions of this article, and in other appropriate cases as may be agreed between the parties, corrective action in respect of offences or suspected offences against the fisheries laws or regulations of the parties shall be taken by the authorities of the party whose nationality is borne by the vessel or person concerned (called in this article "the first party") and not by the party in whose area of jurisdiction the offence or suspected offence occurs (called in this article "the second party").

7. The parties acknowledge that the principle stated in paragraph 6 of this article should not be applied so as to frustrate the enforcement of fisheries laws or regulations or to enable offenders against those laws or regulations to go unpunished.

8. Where, in the case of a suspected offence alleged to have been committed in or in the vicinity of the Protected Zone, it appears that the offence was, or might reasonably be considered to have been, committed in the course of traditional fishing, corrective action, or other measures shall be taken by the authorities of the first party and not by the authorities of the second party and, if being detained by the authorities of the second party, the alleged offenders and their vessel shall be either released or handed over to the authorities of the first party, in accordance with arrangements that will avoid undue expense or inconvenience to the authorities of the second party.

9. Where paragraph 8 of this article applies, the authorities of the second party may require assurance in a particular case that corrective action or other measures will be taken by the authorities of the first party that will adequately ensure that the activity complained of will not be repeated.

10. Where the provisions of paragraph 8 of this article do not apply, and the person or vessel alleged to have been involved or used in the commission of a suspected offence in the Protected Zone is licensed to fish in the Protected Zone by the authorities of the first party, corrective action shall be taken by the authorities of the first party and not by the authorities of the second party and, if being detained by the authorities of the second party, the alleged offenders and their vessel shall be either released or handed over to the authorities of the first party, in accordance with arrangements that will avoid undue expense or inconvenience to the authorities of the second party, and the provisions of paragraphs 13 and 14 of this article shall apply.

11. The provisions of paragraph 10 of this article shall also apply in respect of a suspected offence by a person or vessel of the first party in an area of jurisdiction of the second party outside the Protected Zone where *(a)* that person or vessel was authorised by the authorities of the second party to fish in the area where the suspected offence was

committed under the arrangements referred to in paragraph 1 of article 22 of this treaty; and *(b)* the suspected offence was committed in relation to the fishery the subject of that authorisation and did not involve the taking of other species or potential injury to another fishery.

12. Persons or vessels of the first party detained by the authorities of the second party in the circumstances described in paragraphs 8 and 10 of this article may be detained for as long as necessary to enable those authorities to conduct an expeditious investigation into the offence and to obtain evidence. Thereafter, they shall not be detained other than for the purpose of the handing over of the persons or vessels in accordance with the provisions of those paragraphs unless they are lawfully detained on some other ground.

13. If an alleged offender referred to in paragraph 10 of this article is, in respect of conduct in waters under the jursidiction of the second party, *(a)* convicted of an offence against the fisheries laws or regulations of the first party; or *(b)* found by the authorities of the first party, on the basis of sufficient available evidence, to have contravened or failed to comply with a condition of his licence or authorisation or that of his vessel, the authorities of the first party shall, where appropriate and having regard to paragraph 7 of this article, cancel or suspend the licence or authorisation of the person or his vessel so far as it relates to the Protected Zone commercial fisheries.

14. Where a person or vessel involved or used in the commission of the alleged offence referred to in paragraph 10 of this article is also currently licensed or authorised to fish in the area of the Protected Zone by the second party, the authorities of the second party may, after receiving a report and representations, if any, from the authorities of the first party, cancel or suspend that licence or authorisation in accordance with its laws for such period as is warranted by the circumstances of the case.

15. Each party shall provide the other party with any evidence obtained during investigations carried out in accordance with this article into a suspected offence involving a person or vessel of the other party. Each party shall take appropriate measures to facilitate the admission of such evidence in proceedings taken in respect of the suspected offence.

16. In this article references to persons and vessels of, or of the nationality of, a party include references to persons or vessels licensed by that party under subparagraph 1*(b)* of article 26 of this treaty, and the crews of vessels so licensed, except where such persons or vessels have a prior current licence from the other party under that subparagraph.

PART 6: FINAL ARTICLES

Article 29: Settlement of Disputes

Any dispute between the parties arising out of the interpretation or implementation of this treaty shall be settled by consultation or negotiation.

Article 30: Consultations

The parties shall consult, at the request of either, on any matters relating to this treaty.

Article 31: Annexes

The Annexes to this treaty shall have force and effect as integral parts of this treaty.

Article 32: Ratification

This Treaty shall be subject to ratification and shall enter into force on the exchange of the instruments of ratification.

In witness whereof the undersigned being duly authorised have signed the present treaty and have affixed thereto their seals.

Done in duplicate at Sydney on this eighteenth day of December, one thousand nine hundred and seventy-eight.

FOR AUSTRALIA

Signed MALCOLM FRASER
Prime Minister

Signed ANDREW PEACOCK
Minister for
Foreign Affairs

FOR PAPUA NEW GUINEA

Signed MICHAEL SOMARE
Prime Minister

Signed N. EBIA OLEWALE
Deputy Prime Minister
and Minister for
Foreign Affairs and
Trade

South Pacific Forum Fisheries Agency Convention[1]

The governments comprising the South Pacific Forum
Noting the Declaration on Law of the Sea and a Regional Fisheries Agency adopted at the 8th South Pacific Forum held in Port Moresby in August 1977;
Recognising their common interest in the conservation and optimum utilisation of the living marine resources of the South Pacific region and in particular of the highly migratory species;
Desiring to promote regional co-operation and co-ordination in respect of fisheries policies;
Bearing in mind recent developments in the law of the sea;
Concerned to secure the maximum benefits from the living marine resources of the region for the peoples and for the region as a whole and in particular the developing countries; and
Desiring to facilitate the collection, analysis, evaluation, and dissemination of relevant statistical scientific and economic information about the living marine resources of the region, and in particular the highly migratory species;
have agreed as follows:

Article 1
Agency

1. There is hereby established a South Pacific Forum Fisheries Agency.
2. The Agency shall consist of a Forum Fisheries Committee and a Secretariat.
3. The seat of the Agency shall be at Honiara, Solomon Islands.

Article 2
Membership

Membership of the Agency shall be open to:
a) members of the South Pacific Forum;
b) other states or territories in the region on the recommendation of the Committee and with the approval of the Forum.

Article 3
Recognition of coastal states' rights

1. The Parties to this Convention recognise that the coastal state has sovereign rights, for the purpose of exploring and exploiting, conserving and managing the living marine resources, including highly migratory species, within its exclusive economic zone or fishing zone which may extend 200 nautical miles from the baseline from which the breadth of its territorial sea is measured.
2. Without prejudice to Paragraph (1) of this Article the Parties recognise that effective co-operation for the conservation and optimum utilisation of the highly migratory

1. Adopted by the South Pacific Forum at Honiara, Solomon Islands, on July 10, 1979.

species of the region will require the establishment of additional international machinery to provide for co-operation between all coastal states in the region and all states involved in the harvesting of such resources.

Article 4
Committee

1. The Committee shall hold a regular session at least once every year. A special session shall be held at any time at the request of at least four Parties. The Committee shall endeavor to take decisions by consensus.
2. Where consensus is not possible each Party shall have one vote and decisions shall be taken by a two-thirds majority of the Parties present and voting.
3. The Committee shall adopt such rules of procedure and other internal administrative regulations as it considers necessary.
4. The Committee may establish such sub-committees, including technical and budget sub-committees as it may consider necessary.
5. The South Pacific Bureau for Economic Co-operation (SPEC) may participate in the work of the Committee. States, territories, and other international organisations may participate as observers in accordance with such criteria as the Committee may determine.

Article 5
Functions of the Committee

1. The functions of the Committee shall be as follows:
   a) to provide detailed policy and administrative guidance and direction to the Agency;
   b) to provide a forum for Parties to consult together on matters of common concern in the field of fisheries;
   c) to carry out such other functions as may be necessary to give effect to this Convention.
2. In particular the Committee shall promote intraregional coordination and co-operation in the following fields:
   a) harmonisation of policies with respect to fisheries management;
   b) co-operation in respect of relations with distant-water fishing countries;
   c) co-operation in surveillance and enforcement;
   d) co-operation in respect of onshore fish processing;
   e) co-operation in marketing;
   f) co-operation in respect of access to the 200-mile zones of other Parties.

Article 6
Director, staff, and budget

1. The Committee shall appoint a Director of the Agency on such conditions as it may determine.
2. The Committee may appoint a Deputy Director of the Agency on such conditions as it may determine.
3. The Director may appoint other staff in accordance with such rules and conditions as the Committee may determine.
4. The Director shall submit to the Committee for approval:
   a) an annual report on the activities of the Agency for the preceding year;
   b) a draft work programme and budget for the succeeding year.

5. The approved report, budget, and work programme shall be submitted to the Forum.
6. The budget shall be financed by contributions according to the shares set out in the Annex to this Convention. The Annex shall be subject to review from time to time by the Committee.
7. The Committee shall adopt financial regulations for the administration of the finances of the Agency. Such regulations may authorise the Agency to accept contributions from private or public sources.
8. All questions concerning the budget of the Agency, including contributions to the budget, shall be determined by the Committee.
9. In advance of the Committee's approval of the budget, the Agency shall be entitled to incur expenditure up to a limit not exceeding two-thirds of the preceeding year's approved budgetary expenditure.

Article 7

Functions of the Agency

Subject to direction by the Committee the Agency shall:

a) collect, analyse, evaluate, and disseminate to Parties relevant statistical and biological information with respect to the living marine resources of the region and in particular the highly migratory species;

b) collect and disseminate to Parties relevant information concerning management procedures, legislation and agreements adopted by other countries both within and beyond the region;

c) collect and disseminate to Parties relevant information on prices, shipping, processing and marketing of fish and fish products;

d) provide, on request, to any Party technical advice and information, assistance in the development of fisheries policies and negotiations, and assistance in the issue of licences, the collection of fees or in matters pertaining to surveillance and enforcement;

e) seek to establish working arrangements with relevant regional and international organisations, particularly the South Pacific Commission; and

f) such other functions the Committee may decide.

Article 8

Legal status, privileges, and immunities

1. The Agency shall have legal personality and in particular the capacity to contract, to acquire and dispose of movable and immovable property, and to sue and be sued.
2. The Agency shall be immune from suit and other legal process and its property shall be inviolable.
3. Subject to approval by the Committee the Agency shall promptly conclude an agreement with the Government of the Solomon Islands providing for such privileges and immunities as may be necessary for the proper discharge of the functions of the Agency.

Article 9

Information

The Parties shall provide the Agency with available and appropriate information including:

a) catch and effort statistics in respect of fishing operations in waters under their jurisdiction or conducted by vessels under their jurisdiction;

b) relevant laws, regulations, and international agreements;
c) relevant biological and statistical data; and
d) action with respect to decisions taken by the Committee.

Article 10

Signature, accession, entry into force

1. This Convention shall be open for signature by members of the South Pacific Forum.
2. This Convention is not subject to ratification and shall enter into force 30 days following the eighth signature. Thereafter it shall enter into force for any signing or acceding state thirty days after signature or the receipt by the depositary of an instrument of accession.
3. This Convention shall be deposited with the Government of the Solomon Islands (herein referred to as the depositary) who shall be responsible for its registration with the United Nations.
4. States or territories admitted to membership of the Agency in accordance with article 2(b) shall deposit an instrument of accession with the depositary.
5. Reservations to this Convention shall not be permitted.

Article 11

Withdrawal and amendment

1. Any Party may withdraw from this Convention by giving written notice to the depositary. Withdrawal shall take effect one year after receipt of such notice.
2. Any Party may propose amendments to the Convention for consideration by the Committee. The text of any amendment shall be adopted by a unanimous decision. The Committee may determine the procedures for entry into force of amendments to this Convention.

Annex

The following are the shares to be contributed by Parties to the Convention towards the budget of the Agency in accordance with article 6(6):

Australia	1/3
Cook Islands	1/30
Fiji	1/30
Gilbert Islands	1/30
Nauru	1/30
New Zealand	1/3
Niue	1/30
Papua New Guinea	1/30
Solomon Islands	1/30
Tonga	1/30
Tuvalu	1/30
Western Samoa	1/30

Appendix C

List of Acronyms and Abbreviations

This List is based upon acronyms and abbreviations found in *Ocean Yearbooks 1* and *2*. When possible, the entries have been checked against the references cited below.[1] In addition to the entry and its expansion, qualifications in parentheses and translations in brackets have been occasionally inserted to aid the reader. The sponsoring agency, parent organization, or participants follow the expansion.

THE EDITORS

1. For additional information see, among others, *The Europa Year Book 1976* (London: Europa, 1976); Food and Agriculture Organization, *Initials and Acronyms of Bodies, Activities and Projects Concerned with Fisheries and Aquatic Sciences* by Gianna Landi, FAO Fisheries Circular no. 110, Revision 2 (FIRS/C100 [Rev. 2]) (Rome: FAO, 1975); Klaus Schubert, *Internationales Abkürzungslexikon* (Munich: Wilhelm Fink Verlag, 1978); Stockholm International Peace Research Institute, *Armaments and Disarmament in the Nuclear Age* (Atlantic Highlands, N.J.: Humanities Press, 1976) and other selected SIPRI publications; and U.S. Department of Commerce, National Oceanic and Atmospheric Administration, *Annotated Acronyms and Abbreviations of Marine Science Related International Organizations,* 2d ed. (Washington, D.C.: Government Printing Office, 1976).

AAM air-to-air missile
ACMRR Advisory Committee on Marine Resources Research, FAO
ACOMR Advisory Committee on Oceanic Meteorological Research (defunct), WMO
ACV air-cushion vehicle
AEM Applications Explorer Mission (satellite), US
AEW airborne early warning (system)
AFAR Azores fixed acoustic range (sonar), NATO
AIDJEX Arctic Ice Dynamics Joint Experiment, AINA, ICSI
ALECSO Arab League Educational, Cultural, and Scientific Organization
AS antisubmarine
ASCOPE ASEAN Council on Petroleum
ASEAN Association of Southeast Asian Nations
ASEP ASEAN Sub-regional Environment Programme
ASFIS Aquatic Sciences and Fisheries Information System, FAO/IOC
ASM air-to-surface missile
ASW antisubmarine warfare
ATS Applications Technology Satellites, US
AWACS airborne warning and control system, US

BIOMASS Biological Investigations of Marine Antarctic Systems and Stocks, SCAR/SCOR
BM ballistic missile
BOD biological oxygen demand

CARPAS Comisión Asesora Regional de Pesca para el Atlántico Sudoccidental [Regional Fisheries Advisory Commission for the Southwest Atlantic], FAO
CATIE Tropical Agricultural Research and Training Centre, IICA
CCD Conference of the Committee on Disarmament (formerly ENDC)
CCOP Committee for Co-ordination of Joint Prospecting for Mineral Resources in Asian Off-Shore Areas, ESCAP
CDCC Caribbean Development and Co-operation Committee, ECLA
CECAF Fishery Committee for the Eastern Central Atlantic, FAO
CEP Caribbean Environment Project, UNEP
CEP circular error probable
CGIAR Consultative Group on International Agricultural Research, IBRD/FAO/UNDP
CICAR Co-operative Investigations of the Caribbean and Adjacent Regions, IOC
CIFA Committee for Inland Fisheries of Africa, FAO
CIM Cooperative Investigations of the Mediterranean, IOC/ICSEM/FAO/CFCM
CINCWIO Cooperative Investigation in the North and Central Western Indian Ocean, IOC
CINECA Cooperative Investigation of the Northern Part of the Eastern Central Atlantic, ICES/IOC/FAO/CECAF
CLIMAP Climate Long Range Investigation Mapping and Prediction, IDOE
CMG Commission for Marine Geology, IUGS
CMM Commission for Marine Meteorology, WMO
CNEXO Centre National pour l'Exploitation des Océans, France
COFI Committee on Fisheries, FAO
CPPS Comisión Permanente del Pacífico Sur (see PCSP)
CSIRO Commonwealth Scientific and Industrial Research Organisation, Australia
CSK Cooperative Study of the Kuroshio and Adjacent Regions, IOC
CUEA Coastal Upwelling Ecosystem Analysis, US/Canada

DANIDA Danish International Development Agency

DARPA Defense Advanced Research Projects Agency, US
DDT dichloro-diphenyl-trichloroethane
DEVIS Development Science Information System, ILO
DHI Deutsches Hydrographisches Institut [German Hydrographic Institute], F.R. of Germany
DIESA Department of International Economic and Social Affairs, UN
DIFAR directional low-frequency analyzer and ranging (ASW)
DIMUS digital multibeam steering (ASW)
DMSP Defense Meterological Satellite Program, US
DNA digital multibeam, narrow-band processing, accelerated active search rate sonar
DSDP Deep Sea Drilling Program (see IPOD)
DSIR Department of Scientific and Industrial Research, New Zealand

EASTROPAC International Cooperative Effort toward Understanding of the Eastern Tropical Pacific Ocean, EPOC/IATTC
ECA Economic Commission for Africa, UN
ECE Economic Commission for Europe, UN
ECG Ecosystem Conservation Group, Unesco/UNEP/FAO/IUCN
ECLA Economic Commission for Latin America, UN
ECM electronic countermeasures
ECOR Engineering Committee on Oceanic Resources, UIEO
ECOSOC Economic and Social Council, UN
EEC European Economic Community
EEZ exclusive economic zone
EIFAC European Inland Fisheries Advisory Commission, FAO
ENDC Eighteen Nation Disarmament Committee (see CCD)
ERFEN Estudio Regional del Fenómeno El Niño [Regional Study of the Phenomenon known as El Nino]
ESCAP Economic and Social Commission for Asia and the Pacific, UN
EW electronic warfare

FAMOUS French-American Mid-Ocean Undersea Study, France/US
FAO Food and Agriculture Organization of the United Nations
FCS fire control system (weapons)
FGGE First GARP Global Experiment, WMO/ICSU
FIBEX First International BIOMASS Experiment
FLIP floating instrument platform
FLIR forward-looking infrared (sensor)
FMIS Fisheries Management Information System, FAO
FPB fast patrol boat
FPC fish protein concentrate

GARP Global Atmospheric Research Programme, WMO/ICSU
GATE GARP Atlantic Tropical Experiment, WMO/ICSU
GEBCO General Bathymetric Chart of the Oceans, IHO/IOC
GEMS Global Environmental Monitoring System, UNEP
GEOS Geodynamic Experiment Ocean Satellite, US
GEOSECS Geochemical Ocean Section Study, IOC/IUGS
GERS Group d'Etudes et de Recherches Sous-marines, France
GESAMP Joint Group of Experts on the Scientific Aspects of Marine Pollution, IMCO/FAO/Unesco/WMO/WHO/IAEA/UN
GFCM General Fisheries Council for the Mediterranean, FAO
GIPME Global Investigation of Pollution in the Marine Environment, SCOR/ACMRR/ACOMR/ECOR/ICES/GESAMP
GOES Geostationary Operational Environmental Satellite, US
GTS Global Telecommunications System, WMO/WWW

IABO International Association of Biological Oceanography, IUBS
IAEA International Atomic Energy Agency, UN
IAHS International Association of Hydrological Sciences, IUGG
IALA International Association of Lighthouse Authorities
IAMAP International Association of Meteorology and Atmospheric Physics, IUGG
IAMS International Association of Microbiological Societies, IUBS
IAPSO International Association for the Physical Sciences of the Ocean, IUGG
ICAO International Civil Aviation Organization, UN
ICBM intercontinental ballistic missile
ICES International Council for the Exploration of the Sea
ICG International Coordination Group, IOC
ICG/ITSU International Coordination Group for the Tsunami Warning System in the Pacific, IOC
ICITA International Cooperative Investigation of the Tropical Atlantic, IOC
ICLARM International Center for Living Aquatic Resources Management
ICNAF International Commission for the Northwest Atlantic Fisheries (now NAFO)
ICNT Informal Composite Negotiating Text (UNCLOS)
ICRP International Commission on Radiological Protection
ICSEAF International Commission for the Southeast Atlantic Fisheries, FAO
ICSEM International Commission for the Scientific Exploration of the Mediterranean
ICSPRO Inter-Secretariat Committee on Scientific Programmes relating to Oceanography, UN/Unesco/WMO/FAO/IMCO/IOC
ICSU International Council of Scientific Unions
IDOE International Decade of Ocean Exploration, IOC/UN
IDPSS IGOSS Data Processing and Services System
IGOSS Integrated Global Ocean Station System, IOC/WMO
IGY International Geophysical Year
IHB International Hydrographic Bureau (see IHO)
IHD International Hydrological Decade, Unesco
IHO International Hydrographic Organization
IIASA International Institute for Applied Systems Analysis
IIOE International Indian Ocean Expedition, SCOR/IOC
IISS International Institute for Strategic Studies
IJO International Juridical Organization for Developing Countries
ILMR International Laboratory of Marine Radioactivity, IAEA
ILU Institute of London Underwriters
IMCO Inter-Governmental Maritime Consultative Organization, UN
INMARSAT International Maritime Satellite System, IMCO
INPFC International North Pacific Fisheries Commission, Canada/Japan/US
INRES Information Referral System, UNDP
IOC Intergovernmental Oceanographic Commission, Unesco
IOCARIBE IOC Association for the Caribbean and Adjacent Regions
IODE International Oceanographic Data Exchange, IOC
IOI International Ocean Institute
IPFC Indo-Pacific Fisheries Council, FAO
IPOD International Program of Ocean Drilling (formerly DSDP), France/F.R. of Germany/Japan/UK/US/USSR
IRB Instituto de Resseguros do Brasil
IRIC International Relations Institute of Cameroon
IRPTC International Register of Potentially Toxic Chemicals, UNEP

IRS International Referral System, UNEP
ISA International Seabed Authority (UNCLOS)
ISCDC international surface current data centre
ISOS International Southern Ocean Studies, IDOE
ITCWRM International Training Centre for Water Resources Management
ITIC International Tsunami Information Center, IOC
ITU International Telecommunication Union, UN
IUBS International Union of Biological Sciences, ICSU
IUCN International Union for the Conservation of Nature and Natural Resources
IUGG International Union of Geodesy and Geophysics, ICSU
IUGS International Union of Geological Sciences, ICSU
IUMI International Union of Marine Insurance
IUPAC International Union of Pure and Applied Chemistry, ICSU
IWC International Whaling Commission
IWP Indicative World Plan for Agriculture (see PSWAD)

JOA Joint Oceanographic Assembly (1976), SCOR/IAPSO/IABO/CMG/ACMRR/ACOMR/ECOR/The Royal Society of London
JOIDES Joint Oceanographic Institutions for Deep Earth Sampling

LAMPS Light Airborne Multipurpose System (ASW), US
LEPOR Long-term and Expanded Programme of Oceanic Exploration and Research, UN/IOC
LL/GDS landlocked and geographically disadvantaged states
LNG liquid natural gas

MACTECH Marine and Coastal Technology Programme, UN
MACTIS Marine and Coastal Technology Information Service, UN
MAD magnetic anomaly detector (ASW)
MAPMOPP Marine Pollution (Petroleum) Monitoring Pilot Project, IGOSS
MARISAT maritime satellite system
MARV maneuverable reentry vehicle
MEDEAS Centre d'Activités Environnement-Développement en Méditerranée
MEDI Marine Environment Data Information Referral System, UN
MED POL Coordinated Mediterranean Pollution Monitoring and Research Programme
MESA Marine Ecosystems Analysis Program, US
MIRV multiple independently targetable reentry vehicle
MMS marine meteorological services
MR maritime reconnaissance
MRV multiple reentry vehicle
MSS moored sonobuoy system (ASW)
MSY maximum sustainable yield

NACOA National Advisory Committee on Oceans and Atmosphere, US
NAFO Northwest Atlantic Fishery Organization (formerly ICNAF)
NAIS National Aquaculture Information System, US Sea Grant Program
NAOS North Atlantic Ocean Stations, WMO
NAS National Academy of Sciences, US
NASA National Aeronautics and Space Administration, US
NASCO National Academy of Science's Committee on Oceanography (replaced by OSB), US
NATO North Atlantic Treaty Organization
NERC Natural Environment Research Council, UK
NOAA National Oceanic and Atmospheric Administration, US

NODC national oceanographic data centre, IOC
NORPAC Cooperative Survey of the Northern Pacific, Canada/Japan/US
NORPAX North Pacific Experiment, US
NSF National Science Foundation, US
NTDS naval tactical data system

OAU Organization of African Unity
OECD Organization for Economic Cooperation and Development
OETO Ocean Economics and Technology Office, UN
ONR Office of Naval Research, US
OPC Ocean Policy Committee, NAS/NRC
ORICS optical ranging identification and communication system (ASW)
ORSTOM Office de la Recherche Scientifique et Technique d'Outre-mer [Office of Overseas Scientific and Technical Research]
OSB Ocean Sciences Board, NAS/NRC
OSIS Ocean Surveillance Information System, US Navy
OTEC ocean thermal energy conversion
OTH over the horizon (radar)
OTH-B over the horizon backscatter (radar)

PAP Priority Actions Programme (Mediterranean)
PC patrol craft (coastal)
PCB polychlorinated biphenyl
PCSP Permanent Commission of the Conference on the Use and Conservation of the Marine Resources of the South Pacific, Chile/Ecuador/Peru
PEPAS Centre for the Promotion of Environmental Planning and Applied Science, WHO
PGM precision guided munitions
PIM Pacem in Maribus, IOI
POOL Ad hoc Group of Experts on Pollution of the Ocean Originating on Land, IOC
PSWAD Perspective Study of World Agricultural Development (formerly IWP), FAO
PTBT Partial Test Ban Treaty
PVC polyvinyl chloride

R&D research and development (weapons)
RDSS rapidly deployed surveillance system (sonar)
RIOS Working Group on River Inputs to Ocean Systems, SCOR/ACMRR/ECOR/IAHS/Unesco
RNODC responsible national oceanographic data centre, IOC
ROMBI results of marine biological investigations (inventory form), IOC
ROMS Remote Ocean Measurement System (satellite), US
ROSCOP report of observations of samples collected by oceanographic programmes (inventory form), IOC
RSNT Revised Single Negotiating Text (UNCLOS)
RTG radionuclide thermoelectric generator (satellites)
RV reentry vehicle

SAD submarine anomaly detector (ASW)
SAM surface-to-air missile
SAR search and rescue
SAS suspended array system (ASW)
SCAR Scientific Committee on Antarctic Research, ICSU
SCOPE Scientific Committee on Problems of the Environment, ICSU
SCOR Scientific Committee on Oceanic Research, ICSU
SCSP South China Sea Fisheries Development and Coordinating Programme, FAO
SEAFDEC Southeast Asian Fisheries Development Centre

SEOS Synchronous Earth Observatory Satellite, US
SEPS Swedish Environmental Protection Service
SIPRI Stockholm International Peace Research Institute
SLBM submarine-launched ballistic missile
SMS stationary meteorological satellite
SNAP systems for nuclear auxiliary power (satellites)
SOC Southern Ocean Coordination Group, IOC
SOSUS sonar surveillance system (ASW)
SPC South Pacific Commission
SPEC South Pacific Bureau for Economic Cooperation, SPF
SPF South Pacific Forum
SS submarine (nonnuclear propulsion), US Navy
SSB fleet ballistic missile submarine, US Navy
SSBN fleet ballistic missile submarine (nuclear powered), US Navy
SSG guided-missile submarine, US Navy
SSM surface-to-surface missile
SSN nuclear-powered attack submarine, US Navy
SURTASS surveillance towed array surveillance system (sonar)

TAC total allowable catch
TASS towed array surveillance systems (sonar)
TCDC technical cooperation among developing countries, UN
TEMA Training, Education and Mutual Assistance, IOC
TESAC temperature, salinity, and currents (Ocean SYNDARC format), IOC
TTP Working Group on Tropospheric Transport of Pollutants, SCOR/ACOMR/IAMAP

UN United Nations
UNCLOS United Nations Conference on the Law of the Sea
UNCSTD United Nations Conference on Science and Technology for Development
UNCTAD United Nations Conference on Trade and Development
UNDP United Nations Development Programme
UNEP United Nations Environment Programme
Unesco United Nations Educational, Scientific and Cultural Organization
UNIDO United Nations Industrial Development Organization
UNOLS University National Oceanographic Laboratory System, US

VAP Voluntary Assistance Programme, IOC/WMO
VDS variable depth sonar
VLCC very large crude carrier
VSS VSTOL support ship
VSTOL vertical or short takeoff and landing (vehicle)

WARC World Administrative Radio Conference
WECAFC Western Central Atlantic Fishery Commission, FAO
WHO World Health Organization
WIPO World Intellectual Property Organization
WMC World Meteorological Center, WWW
WMO World Meteorological Organization, UN
WTO Warsaw Treaty Organization
WTO World Tourist Organization
WWF World Wildlife Fund
WWW World Weather Watch, WMO

Appendix D

Living Resources

The eight tables included in this Appendix follow the basic typology of the tables found in *Ocean Yearbook 1,* with some modification as to coverage and the addition of data on fishing fleets and values of catch. Due to limitations of space, serial coverage is limited to the last 5 years for which data are available plus the convenient benchmark year of 1970. A rate of change for at least the last year covered has been appended to all the serial tables. Data from years previously reported have been updated wherever possible. The reader is referred to "Marine Fisheries" by S. J. Holt and C. Vanderbilt (pp. 9–56) for additional tables—especially for human nutritional data.

THE EDITORS

TABLE 1D.--WORLD NOMINAL MARINE CATCH, BY CONTINENT* (1,000 Metric Tons)

	1970	1973	1974	1975	1976	1977	Change 1976-77 (%)
Africa	3,131	3,426.5	3,389.3	3,012.0	2,820.6	2,728.4	-3.3
America, N.	4,750	4,688.7	4,737.4	4,772.8	5,339.8	5,558.2	4.1
America, S.	14,629	4,309.2	6,540.2	5,669.7	7,118.4	5,558.1	-21.9
Asia	19,453	22,902.0	24,318.5	24,347.0	25,044.4	25,639.3	2.4
Europe	11,815	12,425.0	12,490.0	12,347.6	13,245.3	13,294.5	.4
Oceania	194	274.3	290.0	242.1	283.1	305.3	7.8
USSR	6,399	7,769.2	8,462.7	8,991.6	9,363.4	8,581.3	-8.4
World total†	61,432	57,405.7	61,030.0	60,330.6	64,107.5	62,743.4	-2.1

SOURCE.--FAO, Yearbook of Fishery Statistics.

*Nominal marine catch is the total nominal catch minus the nominal inland catch. Continental classification follows FAO usage. Data for 1974-76 have been updated from those found in the last volume.

†Exceeds the sum of the figures by continent due to the inclusion of catches not elsewhere included (see tables A-1B and A-1C in source).

TABLE 2D.--WORLD NOMINAL MARINE CATCH, BY MAJOR FISHING AREA

	Million Metric Tons						Change (%)		
	1970	1973	1974	1975	1976	1977*	1974-75	1975-76	1976-77*
Atlantic, N.W.	4.23	4.46	4.02	3.77	3.46	3.52	-6.22	-8.	1.73
Atlantic, N.E.	10.70	11.30	11.82	12.14	13.33	13.40	2.71	9.80	.53
Atlantic, N.	14.93	15.76	15.84	15.91	16.79	16.92	.44	5.53	.77
Excluding capelin	13.45	13.71	13.94	13.66	13.42	N.A.	.98	-1.76	N.A.
Atlantic, W.C.	1.42	1.40	1.50	1.59	1.55	1.43	6.00	-2.52	-7.74
Atlantic, E.C.	2.77	3.30	3.48	3.49	3.55	3.91	.29	1.72	10.14
Mediterranean and Black	1.15	1.16	1.37	1.28	1.30	1.32	-6.57	1.56	1.54
Atlantic, C.	5.34	5.86	6.35	6.36	6.40	6.66	.16	.63	4.06
Atlantic, S.W.	1.10	.95	1.00	1.00	.93	1.15	0	-7.00	23.66
Atlantic, S.E.	2.52	3.17	2.86	2.58	2.77	2.61	-9.79	7.36	-5.78
Atlantic, S.	3.62	4.12	3.86	3.58	3.70	3.76	-7.25	3.35	1.62
Atlantic	23.89	25.74	26.05	25.85	26.89	27.34	-.8	4.02	1.67
Excluding capelin	22.39	23.69	23.85	23.60	23.90	N.A.	-1.0	1.3	N.A.
Indian, W.	1.72	1.96	2.22	2.11	2.12	2.19	-5.0	.5	3.3
Indian, E.	.81	.88	1.04	1.10	1.23	1.26	5.8	11.8	2.4
Indian†	2.53	2.84	3.26	3.21	3.35	3.45	-1.5	4.4	3.0
Pacific, N.W.	13.01	16.53	16.68	17.00	18.12	18.14	1.9	6.6	.1
Pacific, N.E.	2.65	1.90	2.32	2.24	2.41	2.37	-3.4	7.6	-1.7
Pacific, N.	15.66	18.43	18.99	19.24	20.53	20.51	1.3	6.7	-.1
Excluding Alaska pollack	12.56	13.81	14.08	14.22	14.57	N.A.	1.0	2.5	N.A.
Pacific, W.C.	4.22	5.06	5.16	5.14	5.40	5.75	-.4	5.1	6.5
Pacific, E.C.	.91	1.25	1.10	1.32	1.74	1.99	20.0	31.8	14.4
Pacific, C.	5.12	6.31	6.25	6.47	6.14	7.74	3.5	-5.1	26.1

TABLE 2D. (continued)

	Million Metric Tons						Change (%)		
	1970	1973	1974	1975	1976	1977*	1974-75	1975-76	1976-77*
Pacific, S.W.	.17	.32	.36	.31	.36	.39	-13.9	16.1	8.3
Pacific, S.E.	13.76	3.07	5.33	4.41	5.79	3.85	-17.3	31.3	-33.5
Excluding anchoveta	.66	1.37	1.36	1.09	1.37	3.04	-19.9	25.7	121.9
Pacific, S.	13.93	3.39	5.70	4.72	6.15	4.24	-17.2	30.3	-31.1
Pacific	34.71	28.13	30.94	30.42	32.82	32.49	-1.7	7.9	-1.0
Excluding anchoveta	21.61	26.43	26.97	27.10	28.28	31.68	.5	4.4	12.0
World total‡	70.35	66.81	70.30	69.89	74.65	73.99	-.6	6.8	-.9
Northern regions	30.59	34.19	34.83	35.15	36.43	37.43	.9	3.6	2.7
Central regions	13.00	15.01	15.86	16.03	16.58	17.85	1.1	3.4	7.7
Southern regions	17.54	7.51	9.56	8.30	10.11	8.00	-13.2	21.8	-20.9

SOURCE.--FAO.

NOTE.--N.A. = not available.

*Provisional figures, as of October 10, 1978, and including FAO "estimates" for USSR and Japan.

†Temperate and tropical.

‡Exceeds the sum of the figures due to the inclusion of catches not elsewhere included.

TABLE 3D.--WORLD NOMINAL FISH CATCH, DISPOSITION* (Million Metric Tons)

	1970	1973	1974	1975	1976	1977	Change 1976-77 (%)
Human consumption	44.6 (100.0)	49.5 (100.0)	50.2 (100.0)	50.1 (100.0)	52.0 (100.0)	52.9 (100.0)	1.7
Marketing fresh	18.6 (41.7)	20.4 (41.2)	20.9 (41.7)	20.4 (40.7)	20.9 (40.2)	22.4 (42.4)	7.2 (5.5)
Freezing	9.8 (22.0)	11.8 (23.8)	11.7 (23.3)	11.9 (23.8)	13.0 (25.0)	12.6 (23.7)	-3.1 (-5.2)
Curing	8.1 (18.1)	8.0 (16.2)	8.2 (16.2)	8.3 (16.5)	8.4 (16.2)	8.2 (15.5)	-2.4 (-4.3)
Canning	8.1 (18.2)	9.3 (18.8)	9.4 (18.8)	9.5 (19.0)	9.7 (18.6)	9.7 (18.4)	0 (-1.1)
Other purposes	26.0 (100.0)	18.2 (100.0)	21.1 (100.0)	20.9 (100.0)	22.8 (100.0)	20.6 (100.0)	-9.6
Reduction	25.0 (96.2)	17.2 (94.5)	20.1 (95.3)	19.9 (95.2)	21.8 (95.6)	19.6 (95.1)	-10.1 (-.5)
Miscellaneous	1.0	1.0	1.0	1.0	1.0	1.0	0
World total	70.6	67.7	71.3	71.0	74.7	73.5	-1.6
World marine total	61.4	57.4	61.0	60.3	64.1	62.7	-2.1

SOURCE.--FAO, Yearbook of Fishery Statistics.

NOTE.--Percentages of subtotals shown in parentheses.

*The figures for disposition are based on "live weight" and include freshwater catches. The disposition figures for only the marine catches are not currently available. The data for total marine catch are included for comparison.

TABLE 4D.--WORLD NOMINAL MARINE CATCH, BY COUNTRY* (1,000 Metric Tons)

	1970	1973	1974	1975	1976	1977	Change 1976-77 (%)
Anglo-America:							
Canada	1,345.2	1,112.0	989.6	978.4	1,092.6	1,235.6	13.1
Greenland	39.8	44.5	50.9	47.5	44.7	59.7	33.6
U.S.A.	2,810.9	2,773.9	2,844.3	2,826.8	3,099.3	3,029.7	-2.2
Other	7.7	6.2	5.2	11.5	14.3	14.8	3.5
Total	4,203.6	3,936.6	3,890.0	3,864.2	4,250.9	4,339.8	2.1
Latin America:							
Argentina	209.4	294.5	286.2	214.2	271.9	382.4	40.6
Brazil	423.8	619.4	572.2	598.7	564.4	646.5	14.5
Chile	1,209.3	691.0	1,157.1	929.5	1,406.5	1,285.3	-8.6
Colombia	21.3	32.2	25.2	24.5	23.7	23.7†	N.A.
Costa Rica	7.0	10.7	13.4	14.0	12.8	13.0	1.6
Cuba	105.3	148.9	162.8	141.6	192.3	183.3	-4.7
Ecuador	91.4	153.9	174.4	263.4	315.0	475.5	51.0
Guyana	17.4	19.0	23.6	20.1	19.1	21.8	14.1
Jamaica	8.5	9.6	10.1	10.1	10.1	10.1	0
Mexico	379.3	464.4	427.9	481.5	554.3	651.1	17.5
Nicaragua	8.5	11.3	13.6	14.9	13.4	18.3	36.6
Panama	52.2	108.6	88.4	111.3	172.1	228.0	32.5
Peru	12,532.9	2,323.1	4,139.3	3,440.7	4,337.8	2,523.5	-41.8
Puerto Rico	46.0	80.0	76.0	80.7	80.5	58.4	-27.5
Uruguay	13.2	17.5	15.7	26.1	33.6	48.1	43.2
Venezuela	122.6	153.8	140.9	145.8	139.1	144.2	3.7
Other†	55.1	62.4	60.9	61.3	60.9	63.2	3.8
Total	15,303.2	5,200.3	7,387.7	6,578.4	8,207.5	6,776.4	-17.4
Western Europe:							
Belgium	53.0	52.7	46.4	49.0	44.4	45.4	2.3
Denmark	1,217.1	1,450.5	1,822.1	1,750.6	1,896.5	1,792.0	-5.5
Faeroe Is.	207.8	246.3	246.4	285.6	342.0	310.3	-9.3
Finland	64.0	82.2	87.7	87.1	93.7	94.1	.4
France	782.5	813.9	807.5	805.8	806.0	760.3	-5.7
Germany, F.R.	597.9	463.2	510.7	426.7	439.4	417.1	-5.1
Greece	91.5†	88.0†	85.6	86.5	97.2	97.2†	N.A.

Iceland	733.3	901.3	944.4	994.3	985.7	1,374.0	39.4
Ireland	78.9	90.5	89.5	85.1	96.3	95.5	-.8
Italy	379.3	384.0	408.6	386.7	399.0	404.0	1.3
Netherlands	298.8	340.8	322.5	346.1	281.8	309.7	9.9
Norway	2,985.7	2,987.4	2,644.9	2,550.4	3,435.3	3,562.2	3.7
Portugal	464.5	478.2	429.5	375.4	346.0	310.2	-10.3
Spain	1,527.0	1,564.1	1,496.0	1,501.1	1,457.9	1,436.5	-1.5
Sweden	284.2	218.4	203.2	204.9	202.9	182.2	-10.2
U.K.	1,113.8	1,151.2	1,103.3	995.6	1,052.5	1,003.6	-4.6
Other	1.2	1.6	1.5	1.5	1.6	1.5	-6.3
Total	10,880.5	11,314.3	11,249.8	10,932.4	11,978.2	12,195.8	1.8
Socialist Eastern Europe:							
Bulgaria	88.0	94.9	111.1	150.3	159.2	129.1	-18.9
German D.R.	308.2	351.5	349.8	361.4	266.1	196.1	-26.3
Poland	451.3	557.4	657.1	777.4	726.3	638.5	-12.1
Romania	24.8	60.9	87.5	89.9	76.9	95.8	24.6
USSR	6,399.7	7,769.1	8,462.7	8,991.6	9,363.4	8,581.3	-8.4
Yugoslavia	26.7	31.0	30.7	32.3	34.9	35.2	.9
Other†	4.0	4.0	4.0	4.0	4.0	4.0	N.A.
Total	7,302.7	8,868.8	9,702.9	10,406.9	10,630.8	9,680.0	-8.9
Near East:							
Algeria	25.7	31.2	35.7	37.7	35.1	43.5	23.9
Egypt	27.2	27.8	27.5	25.9	30.5	29.6	-3.0
Iran†	18.0	16.9	16.9	16.9	16.9	16.9	N.A.
Morocco	250.0	399.5	287.9	228.6	286.2	260.3	-9.0
Oman	180.0†	180.0†	180.0†	198.8	198.0	198.0†	N.A.
Saudi Arabia	21.7	26.4	23.6	23.0	23.3	18.4	-21.0
Tunisia	24.0	31.7	41.9	31.7	34.9	38.4	10.0
Turkey	171.1	152.9†	243.5†	184.8†	138.2†	138.2†	N.A.
United Arab Emirates	40.0†	43.0	68.0	68.0	64.4	64.4	0
Yemen Arab R.	7.6†	10.0†	12.4	14.6	16.5	17.5	6.1
Yemen, Dem.	120.0	133.5	145.5	142.8	152.6	161.7	6.0
Other†	35.8	42.7	44.2	43.3	48.2	47.6	-1.2
Total	921.1	1,095.1	1,127.1	1,016.1	1,044.8	1,034.5	-1.0

TABLE 4D. (continued)

	1970	1973	1974	1975	1976	1977	Change 1976-77 (%)
Sub-Saharan Africa:							
Angola	368.2	472.0	393.3	153.6	74.5	113.4	52.2
Cameroon	20.8	21.6†	21.6†	21.6†	21.6†	21.6†	N.A.
Congo	7.6†	15.2	15.7	15.1	17.9	15.4	-14.0
Ethiopia	16.3	25.8†	25.8†	25.8†	25.8†	25.8†	N.A.
Ghana	141.5	182.4	182.2	212.6	195.8	340.7	74.0
Ivory Coast	66.5†	59.5	69.3	62.5	72.9	79.3	8.8
Liberia†	10.7	12.5	12.6	12.6	12.6	12.5	-.8
Madagascar	13.1	21.2	25.4	14.5	13.4	14.5	8.2
Mauritania	50.2	29.4	21.2†	21.2†	21.2†	21.2†	N.A.
Mozambique	7.6	13.3	15.7	12.5	14.9	14.0	-6.0
Namibia†	711.2	709.7	840.4	760.8	574.4	404.1	-29.6
Nigeria	217.0	155.8	158.5	160.1	165.7	169.8	2.5
Senegal	169.2	303.8	347.0	352.9	350.9	282.3	-19.5
Sierra Leone	29.6	65.7	66.7	67.5	68.2	78.9	15.7
Somalia	30.0†	30.0†	32.6	32.6†	32.6†	32.6†	N.A.
South Africa	696.1	710.4	649.7	637.3	638.4	602.8	-5.6
Tanzania	18.6	23.0	28.3	31.2	48.5	50.0	3.1
Togo	6.4	7.9	8.2	11.4	9.5	7.6	-20.0
Zaire	14.7	12.2	13.4	13.4	7.9	2.4	-69.6
Other	52.7	58.8	60.2	59.8	58.4	58.6	.3
Total	2,648.0	2,930.2	2,987.8	2,679.0	2,425.1	2,347.5	-3.2
South Asia:							
Bangladesh	90.0†	88.0	89.0	89.0	90.0	95.0	5.6
India	1,085.6	1,210.4	1,472.0	1,478.0	1,525.0	1,610.0	5.6
Maldives	34.5	33.7	37.5	27.9	32.3	26.7	-17.3
Pakistan	149.3	209.1	163.3	167.8	177.2	215.4	21.6
Sri Lanka	89.8	93.7	103.0	111.4	123.3	125.7	1.9
Total	1,449.2	1,634.9	1,864.8	1,874.1	1,947.8	2,072.8	6.4
East Asia:							
China†	2,102.0	2,312.0	2,312.0	2,312.0	2,312.0	2,312.0	N.A.
Hong Kong	133.3	124.8	135.9	146.9	152.7	154.1	.9

Japan	9,198.7	10,569.1	10,625.2	10,324.9	10,461.5	10,525.5	.6
Korea, D.P.R.†	1,000.0	1,300.0	1,400.0	1,500.0	1,600.0	1,600.0	N.A.
Korea, Rep.	842.1	1,682.2	2,022.3	2,124.9	2,390.3	2,393.1	.1
Macau	9.6	10.1†	10.1†	10.1†	10.1†	10.1†	N.A.
Taiwan‡	540.5	651.0	583.4	652.4	675.1	N.A.	N.A.
Total	13,826.2	16,649.2	17,088.9	17,071.2	17,601.7	N.A.	N.A.
Southeast Asia:							
Burma	311.4	338.1	307.6	355.1	367.2	379.8	3.4
Cambodia	20.2	10.8	10.8†	10.8†	10.8†	10.8†	N.A.
Indonesia	807.2	886.4	948.6	996.9	1,081.6	1,144.0	5.8
Malaysia	338.5	440.4	522.2	471.5	514.5	616.0	19.7
Philippines	941.7	1,204.3	1,268.4	1,336.8	1,240.1	1,348.4	8.7
Singapore	17.3	17.9	18.6	16.9	15.8	14.4	-8.9
Thailand	1,343.4	1,540.2	1,355.5	1,392.3	1,512.7	1,628.1	7.6
Vietnam†	668.3	837.2	837.2	837.2	837.2	837.2	N.A.
Other	1.5	1.5	1.5	1.5	1.6	2.1	31.3
Total	4,449.5	5,276.8	5,270.4	5,419.0	5,581.5	5,980.8	7.2
Australasia:							
Australia	100.7	128.5	136.3	107.2	110.0	126.3	14.8
New Zealand	58.3	64.7	67.8	62.4	74.9	109.5	46.2
Papua New Guinea	17.7	47.3	51.5	36.2	52.4	27.9	-46.8
Solomon Islands	1.0†	7.3†	11.6	8.7	18.6	15.8	-15.1
Other†	18.6	25.6	23.0	27.6	27.2	25.8	-5.1
Total	196.3	273.4	290.2	242.1	283.1	305.3	7.8
Other NEI§	243.0	225.6	171.3	247.4	157.3	1,015.5	
World total	61,432.3	57,405.7	61,030.0	60,330.6	64,107.5	62,743.4	-2.1

SOURCE.--FAO, Yearbook of Fishery Statistics, 1977, unless otherwise indicated.

NOTE.--N.A. = not available.

*Nominal marine catch = nominal catch in marine areas. Countries which reported marine catches of less than 10,000 metric tons in 1975 are included under "Other" for all years.

†Based on FAO estimates.

‡Estimates based on Republic of China, Executive Yuan, Economic Planning Council, Taiwan Statistical Data Book, 1977, p. 70.

§NEI = not elsewhere included. These data differ from values found in the source due to the exclusion of Taiwan's catch from the sum, the inclusion of the catches for the French Southern and Antarctic Territories in the sum, and small discrepancies due to rounding.

TABLE 5D.--TRADE OF FISHERY COMMODITIES, BY MAJOR IMPORTING AND EXPORTING COUNTRIES* (1,000 Metric Tons)

	1970	1973	1974	1975	1976	1977	Change 1976-77 (%)
Imports:							
U.S.A.	1,053.1	1,109.2	1,042.9	943.3	1,106.5	1,027.9	-7.1
Germany, F.R.	971.4	766.1	825.9	848.3	798.7	790.9	-1.0
Japan	352.6	592.2	587.5	642.8	722.3	941.7	30.4
U.K.	731.2	653.5	548.3	621.4	675.8	606.1	-10.3
France	353.7	368.5	359.6	397.5	397.3	450.7	13.4
Italy	319.3	290.7	272.5	317.9	354.1	321.7	-9.1
Netherlands	340.6	257.7	246.0	284.2	309.7	287.5	-7.2
Spain	173.2	183.0	237.3	176.1	137.1	128.6	-6.2
Poland	144.4	159.5	214.1	217.2	168.6	159.1†	-5.6
Denmark	176.4	188.4	209.9	175.5	175.9	195.0	10.9
Belgium	217.8	172.7	160.5	162.9	163.9	164.2	.2
Sweden	190.9	184.7	152.1	161.8	162.0	158.4	-2.2
Singapore	111.8	132.0	123.5	137.9	131.8	146.1	10.8
Yugoslavia	143.2	96.7	120.3	79.7	90.9	109.9	20.9
Switzerland	101.6	106.6	110.8	119.7	119.7	123.0	2.8
Czechoslovakia	166.4	106.7	106.7†	113.6	133.9	95.6†	-28.6
Portugal	84.0	92.3	98.8	112.7	111.8	102.1	-8.7
Norway	61.9	47.3	95.1	34.5	38.7	39.9	3.1
Ivory Coast	13.7	51.5	90.3	96.5	111.2	66.7†	-40.0
Ghana	46.3	111.0	86.8	76.6	76.6†	76.6†	N.A.
Hong Kong	61.1	81.6	78.9	83.4	86.6	84.4	-2.5
Malaysia, W.	71.8	66.6	74.4	93.2	115.9	124.0†	7.0
Finland	60.9	74.5	73.7	64.4	81.8	64.4	-21.3
Australia	70.1	61.8	70.8	74.1	60.8	65.6	7.9
Subtotal	6,017.4	5,954.8	5,986.7	6,035.2	6,331.6	6,330.1	0
Other	1,443.8	1,142.3	1,336.4	1,651.7	1,653.6	1,556.0	-5.9
World total	7,461.2	7,097.1	7,323.1	7,686.9	7,985.2	7,886.1	-1.2

Exports:							
Peru	2,122.8	391.9	733.7	951.1	649.1	489.6	-24.6
Japan	534.3	674.2	706.2	593.4	642.4	582.9	-9.3
Norway	663.9	770.4	577.6	707.1	864.8	899.2	4.0
Denmark	414.9	506.8	572.9	611.2	671.1	611.3	-8.9
USSR	316.4	301.6	411.8	551.0	526.9	459.1	-12.9
Iceland	304.9	323.1	300.2	365.9	325.2	396.0	21.8
Canada	371.8	354.9	299.5	301.6	349.6	442.0	26.4
South Africa	218.3	211.6	230.1†	258.1†	218.9†	107.0†	-51.1
U.S.A.	141.6	253.2	221.8	196.3	220.0	226.0	2.7
Netherlands	228.4	238.4	217.1	235.0	227.5	232.9	2.4
Spain	177.6	188.2	195.7	178.4	232.1	190.1	-18.1
U.K.	133.4	187.1	186.7	156.3	165.6	185.5	12.0
Germany, F.R.	128.3	152.4	183.9	153.3	189.0	199.4	5.5
Korea, Rep.	64.3	178.2	146.7	395.7	284.8	513.9	80.4
Chile	125.8	29.6	127.0	112.3	231.0	231.0†	N.A.
Angola	98.0	162.5	111.2†	78.7†	35.4†	23.2†	-34.5
Faeroe Is.	97.3	104.4	109.4	119.3	118.6	87.5	-26.2
Morocco	81.4	135.8	102.3	88.3	98.4	68.1	-30.8
France	56.8	119.5	100.6	88.1	126.7	123.8	-2.3
Italy	31.0	71.9	97.5	87.1	78.4	64.0	-18.4
Malaysia, W.	102.9	115.1	96.0	81.2	107.3	106.9†	-.4
Thailand	44.1	105.2	87.6	96.4	131.6	180.3	37.0
Poland	59.2	80.7	87.2	87.5	96.9	96.9†	N.A.
Sweden	164.3	108.3	84.9	82.2	92.2	84.9	-7.9
Subtotal	6,681.7	5,765.0	5,987.6	6,575.5	6,683.5	6,601.5	-1.2
Other	751.1	1,126.8	1,098.7	1,037.4	1,209.0	1,361.8	12.6
World total	7,432.8	6,891.8	7,086.3	7,612.9	7,892.5	7,963.3	.9

SOURCE.--FAO, Yearbook of Fishery Statistics, 1977.

NOTE.--N.A. = not available.

*The term "Fishery Commodities" follows FAO usage and includes the seven principal fishery commodity groups. Countries are ranked by 1974 figures. Countries which reported exports or imports of less than 70,000 metric tons in 1974 are included under "Other."

†FAO estimate.

TABLE 6D.--FISHING FLEETS, BY COUNTRY, 1979

	Trawlers and Fishing Vessels*			Factory Ships and Carriers	
	grt	No.	Over 500 grt† (%)	grt	No.
Albania	300	2	0		
Algeria	2,511	22	0		
Angola	1,280	4	0		
Antigua	263	1	0		
Argentina	67,453	133	66.9	2,668	1
Australia	23,176	125	2.2	549	1
Bahamas	4,073	14	66.0		
Bahrain	532	5	0		
Bangladesh	742	4	0	263	2
Barbados	3,648	30	0		
Belgium	12,909	83	4.2		
Benin	998	8	0		
Bermuda	3,268	7	52.6	36,471	2
Brazil	11,975	61	7.2		
Bulgaria	77,950	34	99.3	32,176	6
Burma	3,260	17	0		
Cameroon	5,328	24	39.9		
Canada	144,454	491	44.6	102	1
Cape Verde	1,083	3	0		
Cayman Is.	1,222	7	0		
Chile	9,929	51	13.0		
China	5,690	16	29.4	4,523	12
Colombia	999	7	0		
Congo	6,242	13	76.4		
Costa Rica	10,275	12	82.5		
Cuba	166,718	203	90.7		
Cyprus	4,338	9	73.7		
Denmark	71,840	360	13.6		
Djibouti	156	1	0		
Ecuador	9,528	29	29.6		
Egypt	8,728	6	97.4		
El Salvador	330	3	0		
Ethiopia	327	3	0		
Faeroe Is.	44,999	141	74.8		
Fiji	420	3	0		
Finland	2,291	10	25.4		
France	166,025	519	47.2	4,523	2
Gabon	874	6	0		
German D.R.	118,096	162	88.4	54,848	9
Germany, F.R.	127,884	146	89.7		
Ghana	42,757	54	87.1	5,602	3
Greece	40,404	89	68.0	499	1
Guatemala	250	2	0		
Guinea	1,450	3	74.6		
Guinea-Bissau	151	1	0		
Guyana	3,603	34	0		
Honduras	3,303	17	40.1	2,001	1
Hong Kong	1,768	8	28.3		
Iceland	90,797	309	24.3		
India	5,323	32	0		
Indonesia	25,715	127	15.2	450	3
Iran	4,316	21	28.3		
Iraq	9,362	12	77.1	10,413	2
Ireland	7,636	37	39.7		

TABLE 6D. (continued)

	Trawlers and Fishing Vessels*			Factory Ships and Carriers	
	grt	No.	Over 500 grt† (%)	grt	No.
Israel	2,010	2	100.0		
Italy	85,562	239	59.5		
Ivory Coast	12,432	39	54.9	499	1
Jamaica	648	3	0		
Japan	892,811	2,692	26.8	188,296	104
Jordan	200	1	0		
Kenya	475	2	0		
Korea, D.R.	2,867	2	90.7	36,190	6
Korea, R.	272,820	718	38.4	54,292	13
Kuwait	12,826	92	0	788	1
Lebanon	560	4	0		
Liberia	1,012	9	0		
Libya	5,177	27	0		
Madagascar	2,983	21	0		
Malaysia	440	3	0		
Maldives	1,827	4	54.6		
Mauritania	375	2	0		
Mauritius	2,422	8	25.3		
Mexico	41,729	182	42.4		
Morocco	14,083	54	10.1		
Mozambique	7,690	45	0		
Netherlands	89,182	371	13.6		
New Zealand	6,692	28	15.8		
Nicaragua	4,803	15	67.9		
Nigeria	6,968	35	34.1		
Norway	241,852	719	34.6	1,252	5
Oman	236	1	0		
Pakistan	398	2	0		
Panama	144,030	361	35.0	21,237	10
Papua New Guinea	2,347	16	0		
Peru	130,357	611	5.0		
Philippines	31,880	124	12.7	1,616	5
Poland	279,665	335	93.0	75,340	9
Portugal	129,527	172	80.8		
Qatar	346	3	0		
Romania	111,344	37	100.0	58,512	6
St. Lucia	105	1	0		
St. Vincent	492	1	0		
Samoa, W.	213	1	0		
Saudi Arabia	364	2	0		
Senegal	18,372	68	34.0		
Sierra Leone	119	1	0		
Singapore	1,056	4	0		
South Africa	68,479	167	54.1	303	1
Spain	565,572	1,794	33.1	3,731	3
Sri Lanka	3,178	11	21.6		
Surinam	1,260	7	0		
Sweden	13,780	73	12.7		
Taiwan	72,444	248	13.5		
Tanzania	614	2	0		
Thailand	906	3	0		
Togo	134	1	0		
Tonga	289	2	0		
Trinidad & Tobago	2,145	18	0		

TABLE 6D. (continued)

	Trawlers and Fishing Vessels*			Factory Ships and Carriers	
	grt	No.	Over 500 grt† (%)	grt	No.
Tunisia	454	3	0		
Turkey	751	3	0	663	4
Turks & Caicos Is.	124	1	0		
USSR	3,580,395	3,884	88.7	2,765,042	576
United Arab Emirates	429	2	0		
U.K.	170,100	507	47.2		
U.S.A.	459,025	2,182	33.2	5,180	8
Uruguay	7,437	20	58.6		
Venezuela	9,847	51	21.4		
Vietnam	2,796	20	0		
Yemen, D.	3,546	13	42.0		
Yugoslavia	2,518	5	83.2	113	1
Zaire	4,793	14	46.5		
World total	8,891,271	19,609	62.2	3,378,687	799

SOURCE.--Lloyd's Register of Shipping Statistical Tables, 1979.

*Data exclude vessels of less than 100 grt. The smaller vessels used in artisanal fishing are therefore excluded.

†Percent of grt.

TABLE 7D.--VALUES OF 1973 FISH CATCH, BY COUNTRY

Country*	Catch (Million Metric Tons) All	Catch (Million Metric Tons) Marine	Value†/ Metric Ton (US$)	Total Value ($Million)	Value/ Capita (US$)	Catch as % GNP
1. Japan	10.75	10.57	315	3,371	31	1.4
2. USSR	8.61	7.76	350	3,017	12	.9
3. Norway	2.99	2.99	119.5	356	90	2.7
4. U.S.A.	2.72	2.64	339.9	907	4	.1
5. Peru	2.33	2.32	37.4	86	6	1.2
6. China	6.88	2.31	450	3,408	4	2.6
7. Korea, Rep.	1.68	1.68	235	389	12	4.5
8. Spain	1.58	1.57	425	667	19	1.8
9. Thailand	1.68	1.54	211.0	357	9	5.0
10. Denmark	1.46	1.45	163.1	239	48	1.4
11. India	1.96	1.21	226.4	443	1	.8
12. Philippines	1.30	1.20	411.1	513	13	6.6
13. U.K.	1.12	1.12	335.4	384	7	.3
14. Canada	1.16	1.11	260.6	300	14	.3
15. Iceland	.90	.90	140	127	598	20.5
16. Indonesia	1.27	.89	411	534	4	6.2
17. Vietnam	1.01	.83	350	355	9	7.6
18. France	.81	.81	634.4	506	10	.3
19. Korea, Dem. Rep.	.80	.80	235	188	13	4.0
20. S. Africa	.71	.71	57.3	76	3	.4
21. Chile	.69	.69	13.8	9	1	.1
22. Brazil	.70	.62	370	218	2	.5
23. Poland	.58	.56	350	203	6	.5
24. Angola	.47	.47	23.0	11	2	.5

TABLE 7D. (continued)

Country*	Catch (Million Metric Tons) All	Catch (Million Metric Tons) Marine	Value†/ Metric Ton (US$)	Total Value ($Million)	Value/ Capita (US$)	Catch as % GNP
25. Mexico	.48	.46	300.7	145	3	.4
26. Germany, F.R.	.48	.46	380.2	181	3	.1
27. Portugal	.48	.44	301.8	137	16	2.1
28. Malaysia	.44	.44	617.6	275	24	6.2
29. Morocco	.40	.40	90	36	2	.9
30. Italy	.40	.38	791.9	309	6	.3
31. Germany, D.R.	.36	.35	350	128	8	.3
32. Netherlands	.34	.34	444.7	153	11	.4
33. Burma	.46	.33	200.9	93	3	4.6
34. Senegal	.32	.31	281.6	91	22	10.5
35. Argentina	.30	.29	228.2	69	3	.3
36. Faeroe Is.	.25	.25	140	35	821	37.2
37. Sweden	.23	.22	238.8	54	7	.2
38. Pakistan	.24	.21	320.7	69	1	1.3
39. Ghana	.22	.18	291.7	57	6	2.8
40. Oman	.18	.18	250	25	35	7.7
41. Nigeria	.47	.16	200	133	2	1.9
42. Turkey	.17	.15	480	80	2	.6
43. Venezuela	.16	.15	241.9	39	4	.3
44. Cuba	.15	.15	500	70	8	1.5
45. Ecuador	.15	.15	190	20	3	1.2
46. Yemen, Dem.	.13	.13	250	33	22	15.6
47. Australia	.13	.13	979.8	121	9	.3
48. Hong Kong	.13	.13	677.1	78	19	1.9
49. Bangladesh	.82	.10	630	156	2	3.6
50. Sri Lanka	.10	.09	328.2	33	3	1.5
51. Bulgaria	.10	.09	310	32	4	.5
All	62.25	53.42	310.3	19,315	6	.7
World	66.8	56.7	310	21,000	5	.7

SOURCE.--S. J. Holt and C. Vanderbilt, in this volume, p. 16.

*Countries whose marine catches reached 100,000 metric tons in 1973 or soon thereafter.

†Landed values (first sale) per metric ton caught were obtained by dividing the total tonnage by the reported value where both are given by the FAO (20 countries). These data refer to marine plus inland catches. Where data were incomplete in 1973, values per ton have been extrapolated from previous years or estimated by comparing countries with similar economic structure, with similar fishing industries, and located in the same region, following a study by Alan Marriot.

TABLE 8D.--SPECIES GROUPS IN MARINE CATCHES

FAO Taxonomic Classification		Million Metric Tons				% of Total		% Increase	
		1973	1974	1975	1976	1975	1976	1974-75	1975-76
23, 24, 25	Diadromous fishes*	.75	.66	.74	.70	1.25	1.11	12.12	-5.41
31	Flounders, halibuts, soles, etc.	1.25	1.18	1.14	1.12	1.92	1.78	-3.39	-1.75
32	Cods, hakes, haddocks, etc.	11.97	12.70	11.88	12.12	20.05	19.27	-6.46	2.02
	Alaska pollack	4.62	4.91	5.02	5.07	8.47	8.06	2.24	1.00
	Other	7.35	7.79	6.86	7.05	11.58	11.21	-11.94	2.77
33	Redfishes, basses, congers, etc.	4.32	4.87	5.07	4.95	8.56	7.87	4.11	-2.37
34	Jacks, mullets, sauries, etc.	5.74	5.45	5.93	7.38	10.01	11.73	8.81	24.45
	Capelin	2.05	1.90	2.25	3.36	3.80	5.34	18.42	49.33
	Other	3.68	3.55	3.68	4.02	6.21	6.39	3.66	9.24
35	Herrings, sardines, anchovies, etc.	11.31	13.89	13.62	15.09	22.99	23.99	-1.94	10.79
	Anchoveta	1.70	3.97	3.32	4.28	5.60	6.80	-16.37	28.92
	Other	9.61	9.92	10.30	10.81	17.39	17.18	3.83	4.95
36	Tunas, bonitos, billfishes, etc.	2.00	2.12	1.98	2.21	3.34	3.51	-6.60	11.62
37	Mackerels, snoeks, cuttlass-fishes, etc.	3.42	3.61	3.59	3.34	6.06	5.31	-.55	-6.96
38	Sharks, rays, chimaeras, etc.	.59	.54	.57	.53	.96	.84	5.56	-7.02
	Total Fishes†	49.93	53.30	52.44	55.79	88.52	88.68	-1.61	-7.02
42, 43, 44	Large crustaceans‡	.54	.58	.56	.60	.95	.95	-3.45	7.14
45, 46	Shrimps, prawns, krill, etc.	1.25	1.31	1.28	1.32	2.16	2.10	-2.29	3.13
	Total Crustaceans†	1.86	1.95	1.91	2.00	3.22	3.18	-2.05	4.71

52	Gastropod molluscs§	.06	.05	.07	.06	.12	.10	40.00	-14.29
53-56	Bivalve molluscs	2.12	2.07	2.30	2.43	3.88	3.86	11.11	5.65
57	Squids, cuttlefishes, octopuses, etc.	1.05	1.06	1.16	1.18	1.96	1.88	9.43	1.72
	Total Molluscs†	3.41	3.38	3.72	3.86	6.28	6.14	10.06	3.76
	Other invertebrate animals‖	.14	.07	.06	.07	.10	.11	-28.57	16.67
	Total "Shellfish"#	5.41	5.40	5.69	5.93	9.6	9.0	5.4	4.2
	Algae**	1.12	1.30	1.11	1.19	1.87	1.89	-14.62	7.21
	Total of fishes, shellfish, and algae	56.46	60.00	59.24	62.91	100.00	100.00	-1.27	6.20

SOURCE.--S. J. Holt and C. Vanderbilt, in this volume, pp. 17-18.

NOTE.--Marine mammals (whales, dolphins and porpoises, seals and sea-lions) are excluded.

*Includes sea catches of salmons, trouts, smelts (except capelin), shads, milkfishes, and other diadromous species; that is, species which migrate between the seas and fresh or brackish waters. The figures in this row are less reliable than those in other rows because of uncertainties as to the distribution of catches between sea and inland waters.

†Includes miscellaneous catches not assigned to above categories.

‡Crabs, sea-spiders, lobsters, spiny rock lobsters, squat lobsters, _Nephrops_, etc.

§Abalones, winkles, conches, etc.

‖Mainly sea-urchins, sea cucumbers, and other echinoderms; also sea-squirts and sponges, etc.

#Crustaceans, molluscs, and other invertebrates.

**Brown, red and green seaweeds.

Appendix E

Tables, Nonliving Resources

The four tables included in this Appendix follow the basic typology of the tables found in *Ocean Yearbook 1* with some modification. Coverage focuses on the last five years for which data are available. A rate of change for the last year covered has been appended where applicable. Earlier data have been updated wherever possible.

THE EDITORS

TABLE 1E.--WORLD PRODUCTION OF CRUDE OIL, TOTAL AND OFFSHORE
(Barrels per Day, in Thousands)

Year	World Production	Offshore Production	Offshore as % of World
1972	49,698.00	8,858.77	17.8
1973	55,212.70	10,067.28	18.2
1974	56,722.00	9,268.62	16.3
1975	53,850.00	8,278.36	15.4
1976	57,210.00	9,431.91	16.5
1977	56,567.00	11,436.75	20.2
1978	60,337.00	11,480.75	19.0
1979	62,768.00	12,646.93	20.2

SOURCE.--*Offshore* (June 20, 1977, 1978, 1979, and 1980).

NOTE.--6.998 barrels of crude petroleum approximately equal 1.0 metric ton (ASTM-IP Petroleum Measurement Tables).

TABLE 2E.--OFFSHORE CRUDE OIL PRODUCTION, BY REGION AND COUNTRY (Barrels per Day, in Thousands)

Area and Country	1970	1974	1975	1976	1977	1978	1977-78 (% Change)
Total	7,532.0	9,268.6	8,278.4	9,431.9	11,436.8	11,480.8	.4
Anglo-America:							
U.S.A.	1,577	1,427.5	909.6	1,064.0	1,237.8	1,123.5	-9.2
Latin-America:							
Brazil	8.0	20.4	19.0	35.4	28.8	38.9	35.1
Mexico	35.0	11.9	45.0	45.4	48.4	40.2	-16.9
Peru		34.4	28.9	31.8	28.6*	28.9	1.1
Trinidad & Tobago	76	135.1	174.0	180.1	189.3	175.6	-7.3
Venezuela	2,460*	2,071.2	1,737.1	1,677.2*	1,249.8*	1,083.5	-13.3
Subtotal	2,579	2,273.0	2,004.0	1,969.9	1,544.9	1,367.1	-11.5
Western Europe:							
Denmark		1.9	3.3	8.0	10.2	8.5	-16.7
Italy	12	10.2	10.4	10.1	12.0*	4.3	-64.1
Norway		35.6	189.6	242.6	279.7	356.5	27.4
Spain		34.1	32.9	33.3	23.0	20.0	-13.0
U.K.			83.0	446.0	760.0†	1,070.0	40.8
Subtotal	12	81.8	319.2	740.0	1,084.9	1,459.3	34.5
Socialist Eastern Europe:							
USSR	258	231.0	228.0	220.0†	205.0†	200.0*	-2.4
Near East:							
Abu Dhabi	269	513.0	462.7	560.0	627.0	590.7	-5.8
Divided Zone		333.9†	315.1	247.1	165.6	240.0*	44.9
Dubai	70	132.9	249.3	308.3	317.0*	362.0	13.2
Egypt	257	147.2	165.0	231.1	399.0	396.0	-.8
Iran	322	455.2	481.2	426.5	507.4*	654.5*	29.0
Qatar	172				244.3	260.0	6.4
Saudi Arabia	1,251	2,024.6	1,385.8	1,694.8*	2,621.4*	2,621.4*	N.A.
Sharjah			38.4	37.0	32.6*	25.4	-21.9
Tunisia	0	0	43.0	37.0	45.5*	45.2	-.6
Subtotal	2,341	3,606.8	3,140.5	3,541.8	4,959.8	5,195.2	4.7
Sub-Saharan Africa:							
Cabinda	96	140.4	143.2	33.6*	130.5*	94.6	-27.5
Congo		46.0	37.3	38.0	25.0*	52.4*	109.8
Gabon	29	59.6	179.9	165.9	183.2	138.0	-24.7
Ghana						5.0	N.A.

TABLE 2E. (continued)

Area and Country	1970	1974	1975	1976	1977	1978	1977-78 (% Change)
Nigeria	275	648.9	431.3	525.0	536.4*	384.4	-28.3
Zaire				20.0	23.4*	19.0	-18.9
Subtotal	400	894.9	791.7	782.5	898.5	693.4	-22.8
South Asia:							
India				15.0	80.0*	61.7	-22.8
East Asia:							
China						2.0	N.A.
Japan	3	1.9	0.9	3.0	3.1	2.5	-17.7
Subtotal	3	1.9	0.9	3.0	3.1	4.5	45.2
Southeast Asia:							
Brunei	146‡	287.1‡	141.2	170.2	223.0	191.0	-14.3
Malaysia			84.5	151.4	178.1*	225.0	26.3
Indonesia		247.4	246.4	426.0	591.0	545.2	-7.7
Subtotal	146	534.5	472.1	747.6	992.1	961.2	-3.1
Australasia:							
Australia	216	217.3	412.5	348.0	430.6	414.8	-3.7

SOURCE.--Offshore (June 20, 1977; June 20, 1979).

*Estimate.

†Crude and condensate.

‡Brunei/Malaysia combined production.

TABLE 3E.--WORLD PRODUCTION OF NATURAL GAS, TOTAL AND OFFSHORE
(Cubic Feet, in Billions)

Year	World Production	Offshore Production	Offshore as % of World
1972	43,463.4	6,824.4*	15.7
1973	56,992.3	7,697.0	13.5
1974	47,253.3	8,088.7	17.1
1975	47,029.9	9,532.1	20.3
1976	50,407.5	10,847.0†	21.5
1977	53,883.7	6,663.3†	12.4
1978	53,859.5	9,509.0†	17.8
1979	57,194.6	9,369.0†	16.4

SOURCES.--Basic Petroleum Data Book for 1972-75, Oil and Gas Journal, and Offshore for 1976-80.

*Reflects Free World offshore production since production figures for socialist countries are not available.

†Based on extrapolation from average daily rate.

TABLE 4E.--OFFSHORE NATURAL GAS PRODUCTION, BY REGION AND COUNTRY
(Cubic Feet per Day, in Millions)

Area and Country	1970	1974	1975	1976	1977	1978	1977-78 (% Change)
Total	10,315.05	17,106.44	20,966.96	29,717.82	18,255.60	26,051.98	42.71
Anglo-America:							
U.S.A.	8,591.78	11,588.36	11,664.27	11,864.46	9,804.60	13,983.84	42.62
Latin America:							
Brazil		21.50	25.00	26.00*	25.00*	65.92	163.68
Mexico						107.90	N.A.
Peru	64.66	69.85	77.00				N.A.
Trinidad & Tobago	10.96	18.00	123.00	311.20	344.30	428.50	24.45
Subtotal	75.62	109.35	225.00	337.20	369.30	602.32	63.10
Western Europe:							
Ireland						1.79	N.A.
Italy						38.00	N.A.
Netherlands			186.00	298.40	530.00	530.00*	0
Norway			16.50	560.00†	269.00	1,375.00	411.15
Spain						850.07	N.A.
U.K.	1,086.30	3,600.00	3,600.00	13,912.10‡	3,880.00	3,925.92	1.18
Subtotal	1,086.30	3,600.00	3,802.50	14,770.50	4,679.00	6,720.78	43.64
Socialist Eastern Europe:							
USSR		725.00	774.00	897.00	996.00	1,064.00	6.82

Near East:							
Abu Dhabi				510.00	615.00	629.00	2.27
Egypt						72.00	N.A.
Saudi Arabia	494.80	711.40	3,825.39	200.00	300.00*	300.00*	0
Subtotal	494.80	711.40	3,825.39	710.00	915.00	1,001.00	9.40
Sub-Saharan Africa:							
Gabon						.60	N.A.
Nigeria			314.00	500.00*	218.80§	220.00*	.55
Subtotal			314.00	500.00	218.80	220.60	.82
East Asia:							
Japan						60.00	N.A.
South Asia:							
India						3.44	N.A.
Southeast Asia:							
Brunei						988.00	N.A.
Indonesia		155.05	158.80	345.66	621.90	561.00	-9.79
Subtotal		155.05	158.80	345.66	621.90	1,549.00	149.08
Australasia:							
Australia	60.55	217.28	203.00	293.00	651.00	847.00	30.10

SOURCE.--*Offshore*.

*Estimated.

†Estimate based on oil/gas ratios.

‡Decrease caused by difference in reporting procedure.

§Nigerian Gulf Oil Company only.

Appendix F

Tables, Transportation and Communication

The first four of the six tables in this Appendix are continuations, with some modifications, of tables on shipping included in *Ocean Yearbook 1.* Due to limitations of space, these tables focus on the last 5 years for which data are available. In addition to information on shipping, the Appendix also includes two tables describing the submarine cable network listed in ITU publications. Future editions of the *Yearbook* will include data on satellite communication and underwater pipelines.

THE EDITORS

TABLE 1F.--WORLD SHIPPING TONNAGE, BY TYPE OF VESSEL (Million grt as of July 1)

	1970	1975	1976	1977	1978	1979	Change 1978-79 (%)
Oil tankers	86.1	150.1	168.2	174.1	175.0	174.2	-.5
Liquefied gas carriers*	1.4	3.0	3.4	4.4	5.5	6.7	21.8
Chemical carriers	.5	1.0	1.3	1.8	1.9	2.1	10.5
Miscellaneous tankers	N.A.	.1	N.A.	.2	.2	.2	0
Bulk/oil carriers†	8.3	23.7	25.0	26.1	26.4	26.5	.4
Ore and bulk carriers	38.3	61.8	66.7	74.8	80.2	81.8	2.0
General cargo‡	72.4	70.4	73.6	77.1	78.0	81.7	4.7
Miscellaneous cargo ships	N.A.	.4	N.A.	N.A.	N.A.	N.A.	
Container ships (fully cellular)	1.9	6.2	6.7	7.5	8.7	10.0	14.9
Barge-carrying vessels	N.A.	.8	N.A.	N.A.	N.A.	N.A.	
Vehicle carriers	N.A.	.4	N.A.	.6	1.2	1.6	33.3
Fishing factories, carriers, and trawlers	7.8	11.3	N.A.	12.1	12.4	12.4	0
Passenger liners	3.0	2.8	N.A.	N.A.	N.A.	N.A.	
Ferries and other passenger vessels	N.A.	4.6	N.A.	7.1	6.9	7.2	4.3
All other vessels§	7.8	5.3	N.A.	8.0	9.6	8.5	-11.5

SOURCE.--Lloyd's Register of Shipping Statistical Tables.

NOTE.--grt = gross registered tons; N.A. = not available.

*I.e., ships capable of transporting liquid natural gas or liquid petroleum gas or other similar hydrocarbon and chemical products which are all carried at pressures greater than atmosphere or at subambient temperature or a combination of both.

†Including ore/oil carriers.

‡Including passenger/cargo.

§Including livestock carriers, supply ships and tenders, tugs, cable ships, dredgers, icebreakers, research ships, and others.

TABLE 2F.--ESTIMATED AVERAGE SIZE OF SELECTED TYPES OF VESSELS: EXISTING WORLD FLEETS, MIDYEAR (grt)

	1970	1975	1976	1977	1978	1979	Change 1978-79 (%)
Oil tankers (100 grt and above)	14,110	21,363	23,955	25,191	25,434	25,067	-1.4
Ore/bulk carriers* (6,000 grt and above)	18,450	23,052	23,331	23,399	23,381	23,356	-.1
Container ships (100 grt and above)	11,420	14,859	15,091	14,878	16,335	16,828	3.0
Liquefied gas carriers (grt)	4,690	7,123	7,799	8,947	15,711	17,799	13.3

SOURCE.--Lloyd's Register of Shipping Statistical Tables.
NOTE.--grt = gross registered tons.
*Including bulk/oil carriers.

TABLE 3F.--WORLD MERCHANT FLEETS, BY REGION AND COUNTRY (grt as of July 1)

	1975	1976	1977	1978	1979	Change 1978-79 (%)
Anglo-America:						
Canada	988,726	2,638,692	2,822,948	2,954,499	3,015,752	2.1
U.S.A.	10,931,002	14,908,445	15,299,681	16,187,636	17,542,220	8.4
Subtotal	11,919,728	17,545,137	18,122,629	19,142,135	20,557,972	7.4
Latin America:						
Anguilla	N.A.	N.A.	N.A.	399	399	0
Argentina	1,447,165	1,469,754	1,677,169	2,000,879	2,343,671	17.1
Bahamas	189,890	147,817	106,317	84,269	120,581	43.1
Barbados	3,897	3,897	4,448	4,448	5,107	14.8
Belize	620	620	620	620	620	0
Bolivia	N.A.	N.A.	N.A.	N.A.	15,130	N.A.
Brazil	2,691,408	3,096,293	3,329,951	3,701,731	4,007,498	8.3
Cayman Is.	49,320	78,251	123,787	169,100	229,973	36.0
Chile	386,322	409,756	405,971	466,319	536,616	15.1
Colombia	208,407	211,691	247,240	271,953	291,702	7.3
Costa Rica	5,102	6,257	6,811	10,462	19,270	84.2
Cuba	476,279	603,750	667,518	779,187	852,604	9.4
Dominican Rep.	9,920	8,469	569	18,313	25,140	37.3
Ecuador	142,356	180,623	197,244	201,244	234,240	16.4
El Salvador	1,957	2,128	1,987	1,987	2,317	16.6
Falkland Is.	7,931	6,937	6,937	7,937	7,937	0
Grenada	226	226	226	226	226	0
Guatemala	9,584	8,197	11,854	11,645	9,294	-20.2
Guyana	16,828	19,105	16,274	16,733	17,243	3.0
Haiti	N.A.	N.A.	N.A.	394	394	0
Honduras	67,923	71,042	104,903	130,831	193,256	47.7
Jamaica	6,740	6,892	7,075	10,430	12,927	23.9
Mexico	574,857	593,875	673,964	727,201	914,898	25.8
Monserrat	949	1,130	1,248	1,248	1,248	0
Nicaragua	32,700	26,415	34,588	34,588	13,241	-61.7
Panama	13,667,123	15,631,180	19,458,419	20,748,679	22,323,931	7.6
Paraguay	21,930	21,930	21,930	21,930	23,019	5.0
Peru	518,316	525,137	555,419	574,718	646,380	12.5
St. Kitts-Nevis	405	405	256	256	256	0

St. Lucia	904	904	928	1,243	1,243	0
St. Vincent	5,507	5,663	8,428	11,523	12,718	10.4
Surinam	N.A.	4,890	7,277	8,847	11,171	26.3
Trinidad	13,864	13,603	17,192	15,890	17,165	8.0
Turks & Caicos Is.	1,572	2,405	2,405	2,408	2,408	0
Uruguay	130,998	151,255	192,792	174,357	198,169	13.7
Venezuela	515,661	543,446	639,396	823,543	882,098	7.1
Virgin Is. (U.K.)	2,420	2,409	4,057	4,158	4,726	13.7
Subtotal	21,210,181	23,856,352	28,543,100	31,039,696	33,978,816	9.5*
Western Europe:						
Austria	75,396	82,982	53,284	46,148	81,437	76.5
Belgium	1,358,425	1,499,431	1,595,489	1,684,692	1,788,538	6.2
Denmark	4,478,112	5,143,022	5,331,165	5,530,408	5,524,416	-.1
Faeroe Is.	N.A.	N.A.	N.A.	60,939	63,293	3.9
Finland	2,001,618	2,115,322	2,262,095	2,358,623	2,508,764	6.4
France	10,745,999	11,278,016	11,613,859	12,197,354	11,945,837	-2.1
Germany, F.R.	8,516,567	9,264,671	9,592,314	9,736,667	8,562,780	-12.1
Gibraltar	28,850	21,526	10,549	832	2,291	175.4
Greece	22,527,156	25,034,585	29,517,059	33,956,093	37,352,597	10.0
Iceland	154,381	162,268	166,702	175,097	180,442	3.1
Ireland	210,389	201,965	211,872	212,143	200,714	-5.4
Italy	10,136,989	11,077,549	11,111,182	11,491,873	11,694,872	1.8
Malta	45,950	39,140	100,420	101,541	116,299	14.5
Monaco	14,588	3,998		3,268	31,422	861.5
Netherlands	5,679,413	5,919,892	5,290,360	5,180,392	5,403,350	4.3
Norway	26,153,682	27,943,834	27,801,471	26,128,428	22,349,337	-14.5
Portugal	1,209,701	1,173,710	1,281,439	1,239,963	1,205,478	-2.8
Spain	5,433,354	6,027,763	7,186,081	8,056,080	8,313,658	3.2
Sweden	7,486,196	7,971,246	7,429,394	6,508,255	4,636,662	-28.8
Switzerland	193,657	212,526	252,746	230,762	265,336	15.0
U.K.	33,157,422	32,923,308	31,646,351	30,896,606	27,951,342	-9.5
Subtotal	139,607,845	148,096,754	152,453,832	155,796,164	150,178,865	-3.6
Socialist Eastern Europe:						
Albania	57,368	57,368	55,870	55,870	56,127	.5
Bulgaria	937,458	933,361	964,156	1,082,477	1,150,299	6.3
Czechoslovakia	116,148	148,689	148,689	150,770	154,819	2.7
German D.R.	1,389,000	1,437,054	1,486,838	1,539,994	1,552,148	.8
Hungary	47,943	54,926	63,016	77,738	77,738	0
Poland	2,817,129	3,263,206	3,447,517	3,490,587	3,580,294	2.6

TABLE 3F. (continued)

	1975	1976	1977	1978	1979	Change 1978-79 (%)
Romania	777,309	994,184	1,218,171	1,428,041	1,797,108	25.8
USSR	19,235,973	20,667,892	21,438,291	22,261,927	22,900,201	2.9
Yugoslavia	1,873,482	1,943,750	2,284,526	2,365,630	2,407,221	1.8
Subtotal	27,251,810	29,500,430	31,107,074	32,453,034	33,675,955	3.8
Sub-Saharan Africa:						
Angola	N.A.	N.A.	N.A.	21,820	64,312	194.7
Benin	656	656	912	1,074	4,446	314.0
Cameroon	3,199	19,045	78,180	83,777	38,580	-53.9
Cape Verde	N.A.	N.A.	N.A.	5,516	7,510	36.1
Comoro	N.A.	N.A.	N.A.	765	467	-39.0
Congo	1,846	2,453	4,172	6,942	6,784	-2.3
Djibouti	N.A.	N.A.	N.A.	1,971	3,291	67.0
Equatorial Guinea	N.A.	N.A.	N.A.	3,070	6,412	108.9
Ethiopia	24,953	24,953	23,989	23,490	23,999	2.2
Gabon	106,738	98,285	98,645	77,520	77,095	-.5
Gambia	1,337	1,337	1,608	4,224	3,907	-7.5
Ghana	180,351	183,089	182,696	186,079	196,976	5.9
Guinea	15,054	15,280	12,597	15,041	16,412	9.1
Guinea-Bissau	N.A.	N.A.	219	370	560	51.4
Ivory Coast	119,215	114,191	115,717	156,749	180,639	15.2
Kenya	17,331	15,469	15,192	15,224	26,174	71.9
Liberia	65,820,414	73,477,326	79,982,968	80,191,329	81,528,175	1.7
Madagascar	44,273	49,738	39,850	40,303	55,508	37.7
Mali	N.A.	N.A.	N.A.	N.A.	200	N.A.
Mauritania	1,681	1,113	1,113	489	375	-23.3
Mauritius	33,105	35,146	37,288	40,732	40,390	-.8
Mozambique	149	13,825	27,618	36,169	36,704	1.5
Nigeria	142,050	181,565	335,540	324,024	382,879	18.2
St. Helena	N.A.	N.A.	N.A.	N.A.	3,150	N.A.
Senegal	23,261	26,621	28,044	29,404	33,752	14.8
Seychelles	1,901	1,901	59,140	53,646	3,426	-93.6
Sierra Leone	17,209	17,209	7,298	4,689	2,256	-51.9
Somalia	1,813,313	1,792,900	158,166	72,961	54,895	-24.8
South Africa	565,575	477,011	476,324	660,735	741,469	12.2

Sudan	45,578	45,578	43,375	43,375	43,375	0
Tanzania	33,449	34,934	35,613	36,968	57,728	56.2
Togo	N.A.	N.A.	N.A.	15,498	15,501	0
Uganda	5,510	5,510	5,510	5,510	5,510	0
Zaire	85,232	107,278	109,785	109,785	91,784	-16.4
Zambia	5,513	5,513	5,513	5,513	5,513	0
Subtotal	69,108,793	76,747,926	81,887,072	82,274,762	83,760,154	1.8†
Near East:						
Algeria	246,432	463,094	1,055,962	1,152,086	1,258,081	9.2
Bahrain	3,670	25,096	6,409	7,161	8,795	22.8
Cyprus	3,221,070	3,114,263	2,787,908	2,599,529	2,355,543	-9.4
Egypt	301,383	376,066	407,818	456,291	541,721	18.7
Iran	479,718	683,329	1,002,061	1,194,675	1,207,372	1.1
Iraq	310,594	748,774	1,135,245	1,305,907	1,328,256	1.7
Israel	451,323	481,594	404,651	420,933	435,394	3.4
Jordan	200	200	696	2,295	696	-69.7
Kuwait	990,857	1,106,816	1,831,194	2,240,030	2,428,200	8.4
Lebanon	167,490	213,572	227,009	277,846	260,125	-6.4
Libya	241,725	458,805	673,969	885,362	885,247	0
Morocco	79,863	136,596	270,295	341,410	364,364	6.7
Oman	3,159	3,374	6,137	5,630	6,954	23.5
Qatar	1,389	75,747	84,710	87,767	90,586	3.2
Saudi Arabia	180,246	588,745	1,018,713	1,246,112	1,442,952	15.8
Syria	7,531	10,192	20,679	26,518	31,829	20.0
Tunisia	40,827	62,941	100,128	112,303	127,968	13.9
Turkey	994,668	1,079,347	1,288,282	1,358,779	1,421,715	4.6
United Arab Emirates	50,638	143,109	152,100	156,479	156,120	-.2
Yemen, A.R.	1,260	1,260	1,436	1,436	1,956	36.2
Yemen, Dem.	5,860	6,654	6,390	10,061	10,775	7.1
Subtotal	7,779,893	9,779,574	12,481,792	13,888,610	14,364,649	3.4
South Asia:						
Bangladesh	133,016	146,818	244,314	284,496	298,524	4.9
India	3,869,187	5,093,984	5,482,176	5,759,224	5,854,285	1.7
Maldives	95,154	121,462	110,681	96,218	91,786	-4.6
Pakistan	479,358	483,433	475,600	442,401	442,694	.1
Sri Lanka	80,862	91,031	92,581	92,528	92,941	.4
Subtotal	4,657,577	5,936,728	6,405,352	6,674,867	6,780,230	1.6

TABLE 3F. (continued)

	1975	1976	1977	1978	1979	Change 1978-79 (%)
East Asia:						
China, P.R.	2,828,290	3,588,726	4,245,446	5,168,898	6,336,747	22.6
Hong Kong	418,512	423,218	609,679	874,850	1,469,623	68.0
Japan	39,739,598	41,663,188	40,035,853	39,182,079	39,992,925	2.1
Korea, D.P.R.	81,782	89,482	89,482	90,078	162,261	80.1
Korea, Rep.	1,623,532	1,796,106	2,494,724	2,975,389	3,952,946	32.9
Taiwan	1,450,000	1,483,981	1,558,713	1,619,595	2,011,311	24.2
Subtotal	46,141,714	49,044,701	49,033,897	49,910,889	53,925,813	8.0
Southeast Asia:						
Brunei	283	899	899	899	899	0
Burma	54,548	68,867	67,502	70,848	64,400	-9.1
Cambodia	1,208	1,208	3,558	3,558	3,558	0
Indonesia	859,378	1,046,198	1,163,173	1,272,387	1,309,911	2.9
Malaysia	358,795	442,740	563,666	552,456	620,894	12.4
Philippines	879,043	1,018,065	1,146,529	1,264,995	1,606,019	27.0
Singapore	3,891,902	5,481,720	6,791,398	7,489,205	7,869,152	5.1
Thailand	182,554	194,993	260,664	335,116	361,669	7.9
Vietnam‡	69,626	107,456	128,525	162,585	202,073	24.3
Subtotal	6,297,337	8,362,086	10,125,914	11,152,049	12,038,575	7.9
Australasia:						
Australia	1,205,248	1,247,172	1,374,197	1,531,739	1,651,747	7.8
Fiji	7,674	10,604	10,879	10,023	11,486	14.6
Kiribati	1,518	1,333	1,333	1,333	1,333	0
Nauru	48,271	48,353	48,353	54,004	54,004	0
New Hebrides	4,916	5,023	12,189	6,584	8,712	32.3
New Zealand	162,520	164,192	199,462	211,112	258,476	22.4
Papua New Guinea	14,550	15,329	16,217	16,718	21,022	25.7
Solomon Is.	629	1,008	1,746	2,018	1,822	-9.7
Tonga	9,644	13,722	14,180	20,663	23,549	14.0
Western Samoa	N.A.	N.A.	N.A.	714	927	29.8
Subtotal	1,454,970	1,506,734	1,678,556	1,854,908	2,033,078	9.6

SOURCE.--Lloyd's Register of Shipping Statistical Tables.

NOTE.--grt = gross registered tons; N.A. = not available.

*Excluding Panama, the 1978-79 change for Latin America is 13.3%.

†Excluding Liberia, the 1978-79 change for Sub-Saharan Africa is 7.1%.

‡Sum of D.P.R. Vietnam (12,011 grt) and Rep. Vietnam (57,615 grt).

TABLE 4F.--VESSELS LOST, BY COUNTRY (grt)

	1974	1975	1976	1977	1978
Australia	456 (3)	7,693 (4)	119 (1)	286 (1)	1,616 (1)
Argentina	617 (1)	6,646 (2)			2,527 (3)
Belgium	418 (1)	236 (2)			
Brazil	4,616 (1)		1,996 (1)		
Canada	22,898 (9)	1,146 (5)	3,758 (9)	9,524 (4)	1,167 (3)
Cyprus	52,069 (18)	29,873 (8)	78,217 (19)	42,736 (20)	83,473 (31)
Denmark	3,904 (6)	50,986 (8)	1,671 (6)	1,016 (3)	4,004 (12)
Finland	477 (1)			11,321 (1)	1,423 (2)
France	2,456 (4)	3,718 (2)	864 (4)	486 (2)	12,651 (3)
German D.R.			8,261 (2)	2,547 (1)	1,744 (1)
Germany, F.R.	24,228 (8)	7,199 (8)	4,303 (8)	2,516 (7)	44,131 (6)
Greece	119,282 (19)	146,152 (18)	127,103 (23)	144,569 (25)	782,291 (87)
Hong Kong	1,401 (1)				
India		1,348 (1)	10,979 (2)	10,620 (4)	38,308 (7)
Italy	74,062 (4)	6,278 (1)	5,456 (5)	1,388 (3)	17,211 (11)
Japan	77,623 (47)	66,641 (55)	80,617 (52)	63,862 (36)	39,011 (57)
Korea, Rep.	10,448 (13)	8,431 (15)	11,712 (10)	23,585 (17)	13,032 (17)
Liberia	105,357 (10)	249,125 (16)	352,771 (17)	291,626 (12)	205,550 (8)
Netherlands	171 (1)	1,803 (4)	2,730 (7)	1,271 (1)	4,145 (6)
Norway	52,288 (13)	6,219 (14)	10,662 (13)	88,185 (17)	16,763 (17)
Panama	107,885 (28)	154,353 (42)	217,157 (50)	198,910 (54)	223,867 (62)
Philippines	4,587 (4)	13,996 (6)	13,912 (3)	4,270 (3)	10,935 (11)
Poland	3,128 (4)	648 (1)			103 (1)
Portugal	7,405 (6)			6,275 (2)	
Singapore	40,802 (7)	34,348 (4)	19,447 (4)	40,560 (8)	21,754 (6)
Somalia	12,191 (2)	4,607 (1)			
Spain	3,998 (11)	9,377 (23)	71,422 (19)	7,312 (18)	8,824 (17)
Sweden	448 (2)	5,295 (4)	1,275 (1)	2,999 (1)	2,285 (1)
Turkey			11,547 (1)	149 (1)	1,875 (3)
USSR	18,566 (3)	11,688 (3)	9,500 (3)	5,626 (1)	
U.K.	25,672 (15)	62,739 (19)	4,138 (9)	40,321 (15)	11,447 (16)
U.S.A.	40,470 (16)	16,463 (11)	20,402 (11)	6,045 (20)	24,394 (21)
Yugoslavia	2,092 (1)			5,247 (2)	4,333 (2)
Others	49,643 (52)	88,253 (59)	86,090 (65)	59,875 (57)	131,949 (61)
Total	869,658 (311)	995,261 (336)	1,156,109 (345)	1,073,127 (336)	1,710,813 (473)

SOURCE.--Lloyd's Register of Shipping Statistical Tables, 1979.

NOTE.--grt = gross registered tons; no. of vessels shown in parentheses.

TABLE 5F.--SUBMARINE CABLE LINKS, BY SEA CROSSED

End Point		End Point		Length		Channels	
City*	Country	City*	Country	(km)	Use†	(No.)	Cable Name
		Atlantic, Northwest					
Bermuda	Bermuda	Mill Village	Canada	1,482	TZF	640	CANBER
Medway Harbour	Canada	Sue Wood Bay	Bermuda	1,463	TZFRD	640	CANBER
Clarenville	Canada	Terrenceville	Canada	102	TZFD	80	TAT-1
Terrenceville	Canada	Sydney Mines	Canada	502	TZFD	80	TAT-1
Clarenville	Canada	Sydney Mines	Canada	626	TZD	80	TAT-2
Hampden	Canada	Corner Brook	Canada	117	TZFD	24	ICECAN
Hampden	Canada	Corner Brook	Canada	117	TZFRD	80	CANTAT-1
Frederiksdal	Greenland	Hampden	Canada	1,576	TZFD	24	ICECAN
Manahawkin	U.S.A.	Bermuda	Bermuda	1,389	TZD	82	
Subtotal				7,374		1,730	
		Atlantic, Northeast					
Velbestad	Faeroe	Vestmannaeyjar	Iceland	746	TZF	30	SCOTICE
Hvitanes	Faeroe	Shetland Is.	U.K.	411	TZFD	480	
Penmarch	France	Casablanca	Morocco	1,926	TZFRD	640	
Vestmannaeyjar	Iceland	Frederiksdal	Greenland	1,552	TZFD	24	ICECAN
Lisbon	Portugal	Goonhilly	U.K.	1,761	TZD	640	
Bilbao	Spain	Goonhilly	U.K.	893	Z	480	
Bilbao	Spain	Goonhilly	U.K.	861	Z	1,380	
Gairloch	U.K.	Velbestad	Faeroe	530	TZF	26	SCOTICE
Subtotal				8,680		3,700	
		North Sea					
Fano	Denmark	Terschelling	Netherlands	339	TZFD	60	
Romo	Denmark	Oostmahorn	Netherlands	257	TZFRD	120	1
Romo	Denmark	Oostmahorn	Netherlands	259	TZFD	120	2
Hanstholm	Denmark	Kristiansand	Norway	122	TZFD	60	
Klitmoller	Denmark	Kristiansand	Norway	82	TZFRD	480	
Hirtshals	Denmark	Arendal	Norway	124	TZFD	60	
Uggerby	Denmark	Arendal	Norway	126	N.A.	2,700	
Fano	Denmark	Winterton	U.K.	541	TZFD	120	
Klitmoller	Denmark	Cayton Bay	U.K.	683	TZFD	1,260	
Fano	Denmark	Weybourne	U.K.	574	T	192	
Leer	Germany Fed.	Winterton	U.K.	465	TZR	120	1
Leer	Germany Fed.	Winterton	U.K.	461	TZFD	120	2

Wilhelmshaven	Germany Fed.	Winterton	U.K.	519	TZFD	1,260	3
Alkmaar	Netherlands	Lowestoft	U.K.	222	TZFD	N.A.	
Domburg	Netherlands	Aldeburgh	U.K.	152	Z	180	6
Domburg	Netherlands	Aldeburgh	U.K.	154	Z	1,260	7
Scheveningen	Netherlands	Lowestoft	U.K.	181	Z	60	1
Scheveningen	Netherlands	Lowestoft	U.K.	181	Z	60	2
Katwijk	Netherlands	Covehithe	U.K.	202	Z	120	1
Katwijk	Netherlands	Covehithe	U.K.	202	Z	480	2
Bergen	Norway	Aberdeen	U.K.	569	TZFD	36	
Kristiansand	Norway	Scarborough	U.K.	726	TZFRD	480	
Marske	U.K.	Goteborg	Sweden	945	TZFR	60	
Subtotal				8,086		9,408	
			Baltic Sea				
Molle Bugt	Denmark	Mielno	Poland	128	TZFD	60	
Helsingor	Denmark	Halsingborg	Sweden	6	ZF	42	
Saltholm	Denmark	Malmo	Sweden	9	TZFRD	900	
Mariehamn	Finland	Norrtalje	Sweden	63	Z	9	
Mariehamn	Finland	Norrtalje	Sweden	78	ZR	12	
Mariehamn	Finland	Norrtalje	Sweden	48	TZ	60	
Porkkala	Finland	Rohuneeme	USSR	56	TZR	11	
Stralsund	German Dem. Rep.	Malmo	Sweden	119	Z	12	
Stralsund	German Dem. Rep.	Malmo	Sweden	119	TZR	42	
Fehmarn	Germany Fed.	Malmo	Sweden	224	TZFD	480	
Fehmarn	Germany Fed.	Trelleborg	Sweden	200	TZFD	1,200	
Grossenbrode	Germany Fed.	Trelleborg	Sweden	202	TZFD	1,200	
Subtotal				1,252		4,028	
			Irish Sea				
Dublin	Ireland	Abergeirch	U.K.	117	Z	12	2
Dublin	Ireland	Abergeirch	U.K.	124	Z	12	3
Dublin	Ireland	Holyhead	U.K.	119	TZ	60	1
Dublin	Ireland	Holyhead	U.K.	117	TZ	60	2
Nefyn	U.K.	Dublin	Ireland	115	TZ	24	2
Nefyn	U.K.	Dublin	Ireland	117	TZ	24	3
Subtotal				709		192	
			English Channel				
Veurne	Belgium	Dover	U.K.	107	TZFD	3,900	
De Panne	Belgium	Dover	U.K.	89	TZFRD	420	6
Oostende	Belgium	Canterbury	U.K.	102	TZFRD	120	2

TABLE 5F. (continued)

End Point		End Point		Length		Channels	
City	Country	City	Country	(km)	Use†	(No.)	Cable Name
Oostende	Belgium	Broadstairs	U.K.	119	TZFRD	1,260	
Pirou	France	Jersey	U.K.	31	TZ	12	1
Pirou	France	Jersey	U.K.	33	TZ	12	
Calais	France	Dover	U.K.	37	TZFD	60	
Calais	France	Dover	U.K.	39	TZFRD	60	
Calais	France	Dover	U.K.	41	TZFD	60	
Boulogne	France	Dover	U.K.	46	TZFD	60	
Courseulles	France	Eastbourne	U.K.	191	TZFRD	2,875	
Domburg	Netherlands	Broadstairs	U.K.	354	TZFD	1,380	
Subtotal				1,189		10,219	
Atlantic, North							
Clarenville	Canada	Penmarch	France	4,063	TZD	48	TAT-2
Clarenville	Canada	Oban	U.K.	3,597	TZFD	48	TAT-1
Penmarch	France	Clarenville	Canada	4,091	TZD	48	TAT-2
Vendee	France	Tuckerton	U.S.A.	6,660	TZD	138	TAT-4
Vendee	France	Green Hill	U.S.A.	6,258	TZFRD	4,000	TAT-6
Cadiz	Spain	Green Hill	U.S.A.	6,410	TZD	845	TAT-5
Oban	U.K.	Clarenville	Canada	3,602	TZFRD	48	TAT-1
Oban	U.K.	Hampden	Canada	3,723	TZFRD	80	CANTAT-1
Bude	U.K.	Beaver Harbour	Canada	5,186	TZFRD	1,840	CANTAT-2
Widemouth	U.K.	Tuckerton	U.S.A.	6,517	TZD	138	TAT-3
Subtotal				50,107		7,233	
Atlantic, Western Central							
Bermuda	Bermuda	Tortola	Br. Virgin Is.	1,671	TZF	80	
Forteleza	Brazil	St. Thomas	U.S. Virgin Is.	4,071	TZFRD	640	
San Juan	Puerto Rico	W. Palm Beach	U.S.A.	2,069	TZD	50	
W. Palm Beach	U.S.A.	San Juan	Puerto Rico	2,104	TZD	50	
Vero Beach	U.S.A.	St. Thomas	U.S. Virgin Is.	2,184	TZD	142	
Jacksonville Beach	U.S.A.	St. Thomas	U.S. Virgin Is.	2,446	TZD	720	
Subtotal				14,545		1,682	
Caribbean Sea							
Grand Cayman	Cayman Is.	Kingston	Jamaica	693	ZF	120	
S. Domingo	Dominican Rep.	St. Thomas	U.S. Virgin Is.	715	TZD	128	

Kingston	Jamaica	Ft. Sherman	Panama Ca. Zn.	1,150	TZFRD	144	
Willemstad	Neth. Antilles	St. Maarten	Neth. Antilles	1,050	TZFRD	160	
St. Maarten	Neth. Antilles	St. Thomas	U.S. Virgin Is.	235	TZFRD	160	
Florida City	U.S.A.	Kingston	Jamaica	1,545	TZD	144	
Maiquetai	Venezuela	St. Thomas	U.S. Virgin Is.	1,009	TZFRD	83	
Subtotal				6,397		939	
		Straits of Florida					
Nassau	Bahamas	W. Palm Beach	U.S.A.	361	TZD	1,380	
Havana	Cuba	Key West	U.S.A.	220	TZD	24	
Havana	Cuba	Key West	U.S.A.	196	TCCT	8	2KZ-HVA
Havana	Cuba	Key West	U.S.A.	185	TCCT	8	3KZ-HVA
Havana	Cuba	Key West	U.S.A.	181	TCCT	8	4KZ-HVA
Key West	U.S.A.	Havana	Cuba	239	TZD	24	
Subtotal				1,382		1,452	
		Atlantic, Eastern Central					
Recife	Brazil	Las Palmas	Spain	5,050	TZFRD	160	
Sal	Cape Verde	Ascension	U.K.	3,145	TZFRD	360	SAT-1
Abidjan	Ivory Coast	Dakar	Senegal	2,550	TZFRD	480	
Casablanca	Morocco	Dakar	Senegal	2,711	X	640	
Lisbon	Portugal	Funchal	Portugal	1,148	TZ	120	CAM-1
Lisbon	Portugal	S. Cruz	Spain	1,363	TZFRD	360	SAT-1
Aguimes	Spain	Recife	Brazil	4,878	Z	160	BRACAN-1
S. Cruz	Spain	Sal	Cape Verde	1,585	TZFRD	360	SAT-1
Cadiz	Spain	Tenerife	Spain	1,396	Z	160	PENCAN-1
Las Palmas	Spain	Arrecife	Spain	346	Z	480	TRANSCAN
Cadiz	Spain	Las Palmas	Spain	1,365	Z	1,840	PENCAN-2
Las Palmas	Spain	Tenerife	Spain	111	Z	1,840	PENCAN-2
Cadiz	Spain	Las Palmas	Spain	1,369	Z	5,520	PENCAN-3
Aguimes	Spain	Camuri	Venezuela	5,999	Z	1,840	COLUMBUS
Subtotal				33,016		14,320	
		Red Sea					
Jiddah	Saudi Arabia	Port Sudan	Sudan	361	TCC	2	
Subtotal				361		2	
		Mediterranean Sea					
Algiers	Algeria	Pisa	Italy	1,107	TZ	480	
Larnaca	Cyprus	Beirut	Lebanon	217	Z	640	
Alexandria	Egypt	Catanzaro	Italy	1,682	TZ	480	

TABLE 5F. (continued)

End Point		End Point		Length		Channels	
City	Country	City	Country	(km)	Use†	(No.)	Cable Name
Alexandria	Egypt	Beirut	Lebanon	926	TZFRD	120	
Marseille	France	Algiers	Algeria	883	TZFRD	80	
Perpignan	France	Oran	Algeria	1,004	TZ	60	
Marseille	France	Algiers	Algeria	820	TZFRD	480	
Cannes	France	Ile Rousse	France	196	TZRD	96	
St. Raphael	France	La Foux	France	35	TZFRD	2,580	
Marseille	France	Tel Aviv	Israel	3,395	TZFRD	128	
Marseille	France	Beirut	Lebanon	3,400	TZFRD	120	
Perpignan	France	Tetouan	Morocco	1,404	TZFRD	96	
Perpignan	France	Bizerte	Tunisia	943	TZFRD	640	
Iraklion	Greece	Larnaca	Cyprus	963	Z	640	
Iraklion	Greece	Marseille	France	2,495	Z	640	
Iraklion	Greece	Athens	Greece	320	Z	1,840	
Khania	Greece	Siracusa	Italy	930	Z	60	
Lechaina	Greece	Calabria	Italy	528	Z	480	
Tel Aviv	Israel	Palo	Italy	2,713	TZFRD	1,380	TELPAL
Rome	Italy	Marseille	France	685	Z	3,440	MARPAL
Agrigenta	Italy	Tripoli	Libya	552	TZ	120	
Pozzallo	Italy	Malta	Malta	98	TZ	36	
Pantelleria	Italy	Kelibia	Tunisia	80	TZ	60	
Martil	Morocco	Marseille	France	1,472	X	2,580	
P.D. Mallorca	Spain	Algiers	Algeria	339	Z	480	
Barcelona	Spain	Pisa	Italy	796	Z	480	
Malaga	Spain	Rome	Italy	1,826	Z	640	MAT-1
Barcelona	Spain	Rome	Italy	950	Z	1,380	
Barcelona	Spain	P.D. Mallorca	Spain	339	Z	1,380	PENBAL-1
Valencia	Spain	P.D. Mallorca	Spain	302	Z	3,900	PENBAL-2
Bizerte	Tunisia	Marseille	France	865	TZ	128	
Bou-Ficha	Tunisia	Kelibia	Tunisia	109	Z	120	
Antalya	Turkey	Catania	Italy	2,006	TZFRD	480	
Subtotal				34,380		26,264	
		Adriatic Sea					
Durres	Albania	Brindisi	Italy	156	TZ	3	
Subtotal				156		3	

			Atlantic, Southeast				
Ascension	U.K.	Capetown	South Africa	4,804	TZFRD	360	SAT-1
Subtotal				4,804		360	
			Pacific, Northwest				
Agana	Guam	Ninomiya	Japan	2,656	TZFRD	138	TRANSPAC-1
Agana	Guam	Okinawa	Japan	2,528	TZFRD	845	TRANSPAC-2
Agana	Guam	Baler	Philippines	2,758	TZD	128	
Okinawa	Japan	Currimao	Philippines	1,341	TZFRD	1,600	OLUHO
Subtotal				9,283		2,711	
			Sea of Japan				
Nakhodka	USSR	Naoetsu	Japan	883	TZFRD	120	JASC
Subtotal				883		120	
			Yellow Sea				
Nanhui	China	Reihoku	Japan	872	TZFRD	480	
Subtotal				872		480	
			Pacific, Northeast				
Vancouver	Canada	Port Alberni	Canada	150	TZFRD	80	COMPAC
Port Angeles	U.S.A.	Ketchikan	U.S.A.	1,409	TZD	48	
Ketchikan	U.S.A.	Port Angeles	U.S.A.	1,370	TZD	48	
Subtotal				2,929		176	
			Pacific, North				
Port Alberni	Canada	Keawaula	U.S.A.	4,713	TZFRD	80	COMPAC
Subtotal				4,713		80	
			Pacific, Western Central				
Cairns	Australia	Madang	Papua N. Guin.	2,989	TZFRD	170	SEACOM
Tumon Bay	Guam	Hong Kong	Hong Kong	3,819	TZFRD	85	SEACOM
Madang	Papua N. Guin.	Tumon Bay	Guam	2,576	TZFRD	170	SEACOM
Wake	Wake Island	Agana	Guam	2,780	TZD	142	TRANSPAC-1
Subtotal				12,164		567	
			South China Sea				
Hong Kong	Hong Kong	K. Kinabalu	Malaysia	2,041	TZFRD	85	SEACOM
K. Kinabalu	Malaysia	Singapore	Singapore	1,606	TZFRD	85	SEACOM
Currimao	Philippines	Deep Water Bay	Hong Kong	880	TZFRD	1,840	OLUHO
Subtotal				4,527		2,010	

TABLE 5F. (continued)

End Point City	End Point Country	End Point City	End Point Country	Length (km)	Use†	Channels (No.)	Cable Name
			Coral Sea				
Cairns	Australia	Port Moresby	Papua N. Guin.	900	TZFRD	480	A-PNG
Subtotal				900		480	
			Pacific, Eastern Central				
Point Arena	U.S.A.	Hanauma Bay	U.S.A.	4,069	TZD	51	HAWAII-1
Hanauma Bay	U.S.A.	Point Arena	U.S.A.	4,093	TZD	51	HAWAII-1
S.L. Obispo	U.S.A.	Makaha	U.S.A.	4,413	TZD	143	HAWAII-2
Subtotal				12,575		245	
			Pacific, Central				
Midway	Midway Is.	Wake	Wake Island	2,052	TZD	142	TRANSPAC-1
Keawaula	U.S.A.	Suva	Fiji	5,691	TZFRD	80	COMPAC
Makaha	U.S.A.	Agana	Guam	6,760	TZFRD	845	TRANSPAC-2
Makaha	U.S.A.	Midway	Midway Is.	2,295	TZD	142	TRANSPAC-1
Subtotal				16,798		1,209	
			Tasman Sea				
Sydney	Australia	Auckland	New Zealand	2,204	TZFRD	480	TASMAN
Auckland	New Zealand	Sydney	Australia	2,358	TZFRD	85	COMPAC
Subtotal				4,562		565	
			Pacific, South				
Suva	Fiji	Auckland	New Zealand	2,334	TZFRD	85	COMPAC
Subtotal				2,334		85	
World total				244,978		90,260	

SOURCE.--International Telecommunication Union, List of Cables Forming the World Submarine Network, 19th ed. (Geneva: ITU, 1977).

*The cities given as end points are the largest settlements closest to landfall (where available).

†Following ITU classification: T = voice-frequency telegraphy; Z = telephony; F = facsimile transmission; V = television transmission; R = sound program transmission; D = data transmission; TCC = direct-current telegraphy; and X = T, Z, F, R, and D; N.A. = not available.

TABLE 6F.--SUBMARINE CABLE LINKS, BY COUNTRY

From	To		Year Laid	Length (km)	Channels (No.)	Use*
Albania:						
Durres	Brindisi	Italy	1940	156	3	TZ
Algeria:						
Algiers	Marseille	France	1957	883	80	TZFRD
Oran	Perpignan	France	1962	1,004	60	TZ
Algiers	Marseille	France	1972	820	480	TZFRD
Algiers	Pisa	Italy	1972	1,107	480	TZ
Algiers	P.D. Mallorca	Spain	1975	339	480	Z
Australia:						
Sydney	Auckland	New Zealand	1962	2,358	85	TZFRD
Sydney	Auckland	New Zealand	1975	2,204	480	TZFRD
Cairns	Port Moresby	Papua N. Guin.	1976	900	480	TZFRD
Cairns	Madang	Papua N. Guin.	1966	2,989	170	TZFRD
Bahamas:						
Nassau	W. Palm Beach	U.S.A.	1972	361	1,380	TZD
Belgium:						
Veurne	Dover	U.K.	1977	107	3,900	TZFD
De Panne	Dover	U.K.	1948	89	420	TZFRD
Oostende	Canterbury	U.K.	1958	102	120	TZFRD
Oostende	Broadstairs	U.K.	1971	119	1,260	TZFRD
Bermuda:						
Bermuda	Tortola	Br. Virgin Is.	1966	1,671	80	TZF
Sue Wood Bay	Medway Harbour	Canada	1971	1,463	640	TZFRD
Bermuda	Mill Village	Canada	1971	1,482	640	TZF
Bermuda	Manahawkin	U.S.A.	1961	1,389	82	TZD
Br. Virgin Is.:						
Tortola	Bermuda	Bermuda	1966	1,671	80	TZF
Brazil:						
Recife	Aguimes	Spain	1973	4,878	160	Z
Recife	Las Palmas	Spain	1972	5,050	160	TZFRD
Forteleza	St. Thomas	U.S. Virgin Is.	1977	4,071	640	TZFRD
Canada:						
Mill Village	Bermuda	Bermuda	1971	1,482	640	TZF
Medway Harbour	Sue Wood Bay	Bermuda	1971	1,463	640	TZFRD
Terrenceville	Clarenville	Canada	1956	102	80	TZFD
Sydney Mines	Terrenceville	Canada	1956	502	80	TZFD

TABLE 6F. (continued)

From	To		Year Laid	Length (km)	Channels (No.)	Use*
Sydney Mines	Clarenville	Canada	1959	626	80	TZD
Corner Brook	Hampden	Canada	1960	117	24	TZFD
Corner Brook	Hampden	Canada	1961	117	80	TZFRD
Port Alberni	Vancouver	Canada	1963	150	80	TZFRD
Clarenville	Penmarch	France	1959	4,091	48	TZD
Clarenville	Penmarch	France	1959	4,063	48	TZD
Hampden	Frederiksdal	Greenland	1962	1,576	24	TZFD
Clarenville	Oban	U.K.	1956	3,602	48	TZFRD
Hampden	Oban	U.K.	1961	3,723	80	TZFRD
Beaver Harbour	Bude	U.K.	1973	5,186	1,840	TZFRD
Clarenville	Oban	U.K.	1955	3,597	48	TZFD
Port Alberni	Keawaula	U.S.A.	1963	4,713	80	TZFRD
Cape Verde:						
Sal	S. Cruz	Spain	1969	1,585	360	TZFRD
Sal	Ascension	U.K.	1969	3,145	360	TZFRD
Cayman Is.:						
Grand Cayman	Kingston	Jamaica	1971	693	120	ZF
China:						
Nanhui	Reihoku	Japan	1976	872	480	TZFRD
Cuba:						
Havana	Key West	U.S.A.	1950	239	24	TZD
Havana	Key West	U.S.A.	1950	220	24	TZD
Havana	Key West	U.S.A.	1896	196	8	TCCT
Havana	Key West	U.S.A.	1899	185	8	TCCT
Havana	Key West	U.S.A.	1917	181	8	TCCT
Cyprus:						
Larnaca	Iraklion	Greece	1975	963	640	Z
Larnaca	Beirut	Lebanon	1975	217	640	Z
Denmark:						
Fano	Terschelling	Netherlands	1956	339	60	TZFD
Romo	Oostmahorn	Netherlands	1950	257	120	TZFRD
Romo	Oostmahorn	Netherlands	1950	259	120	TZFD
Hanstholm	Kristiansand	Norway	1955	122	60	TZFD
Klitmoller	Kristiansand	Norway	1967	82	480	TZFRD
Hirtshals	Arendal	Norway	1941	124	60	TZFD
Uggerby	Arendal	Norway	1977	126	2,700	N.A.

Molle Bugt	Mielno	Poland	1960	128	60	TZFD
Helsingor	Halsingborg	Sweden	1939	6	42	ZF
Saltholm	Malmo	Sweden	1951	9	900	TZFRD
Fano	Winterton	U.K.	1964	541	120	TZFD
Klitmoller	Cayton Bay	U.K.	1973	683	1,260	TZFD
Fano	Weybourne	U.K.	1950	574	192	T
Dominican Rep.:						
S. Domingo	St. Thomas	U.S. Virgin Is.	1968	715	128	TZD
Egypt:						
Alexandria	Catanzaro	Italy	1971	1,682	480	TZ
Alexandria	Beirut	Lebanon	1972	926	120	TZFRD
Faeroe:						
Velbestad	Vestmannaeyjar	Iceland	1961	746	30	TZF
Velbestad	Gairloch	U.K.	1961	530	26	TZF
Hvitanes	Shetland Is.	U.K.	1971	411	480	TZFD
Fiji:						
Suva	Auckland	New Zealand	1962	2,334	85	TZFRD
Suva	Keawaula	U.S.A.	1963	5,691	80	TZFRD
Finland:						
Mariehamn	Norrtalje	Sweden	1928	63	9	Z
Mariehamn	Norrtalje	Sweden	1938	78	12	ZR
Mariehamn	Norrtalje	Sweden	1951	48	60	TZ
Porkkala	Rohuneeme	USSR	1942	56	11	TZR
France:						
Marseille	Algiers	Algeria	1957	883	80	TZFRD
Perpignan	Oran	Algeria	1962	1,004	60	TZ
Marseille	Algiers	Algeria	1972	820	480	TZFRD
Penmarch	Clarenville	Canada	1959	4,063	48	TZD
Penmarch	Clarenville	Canada	1959	4,091	48	TZD
Ile Rousse	Cannes	France	1966	196	96	TZRD
La Foux	St. Raphael	France	1975	35	2,580	TZFRD
Marseille	Iraklion	Greece	1974	2,495	640	Z
Marseille	Tel Aviv	Israel	1968	3,395	128	TZFRD
Marseille	Rome	Italy	1976	685	3,440	Z
Marseille	Beirut	Lebanon	1970	3,400	120	TZFRD
Marseille	Martil	Morocco	1978	1,472	2,580	X
Perpignan	Tetouan	Morocco	1967	1,404	96	TZFRD
Penmarch	Casablanca	Morocco	1973	1,926	640	TZFRD
Marseille	Bizerte	Tunisia	1969	865	128	TZ
Perpignan	Bizerte	Tunisia	1975	943	640	TZFRD
Pirou	Jersey	U.K.	1938	31	12	TZ

TABLE 6F. (continued)

From	To		Year Laid	Length (km)	Channels (No.)	Use*
Pirou	Jersey	U.K.	1940	33	12	TZ
Calais	Dover	U.K.	1944	37	60	TZFD
Calais	Dover	U.K.	1945	39	60	TZFRD
Calais	Dover	U.K.	1968	41	60	TZFD
Boulogne	Dover	U.K.	1968	46	60	TZFD
Courseulles	Eastbourne	U.K.	1976	191	2,875	TZFRD
Vendee	Tuckerton	U.S.A.	1965	6,660	138	TZD
Vendee	Green Hill	U.S.A.	1976	6,258	4,000	TZFRD
German Dem. Rep.:						
Stralsund	Malmo	Sweden	1927	119	12	Z
Stralsund	Malmo	Sweden	1930	119	42	TZR
Germany Fed.:						
Fehmarn	Malmo	Sweden	1969	224	480	TZFD
Fehmarn	Trelleborg	Sweden	1975	200	1,200	TZFD
Grossenbrode	Trelleborg	Sweden	1978	202	1,200	TZFD
Leer	Winterton	U.K.	1963	465	120	TZR
Leer	Winterton	U.K.	1964	461	120	TZFD
Wilhelmshaven	Winterton	U.K.	1971	519	1,260	TZFD
Greece:						
Iraklion	Larnaca	Cyprus	1975	963	640	Z
Iraklion	Marseille	France	1974	2,495	640	Z
Athens	Iraklion	Greece	1975	320	1,840	Z
Khania	Siracusa	Italy	1962	930	60	Z
Lechaina	Calabria	Italy	1969	528	480	Z
Greenland:						
Frederiksdal	Hampden	Canada	1962	1,576	24	TZFD
Frederiksdal	Vestmannaeyjar	Iceland	1962	1,552	24	TZFD
Guam:						
Tumon Bay	Hong Kong	Hong Kong	1965	3,819	85	TZFRD
Agana	Ninomiya	Japan	1964	2,656	138	TZFRD
Agana	Okinawa	Japan	1976	2,528	845	TZFRD
Tumon Bay	Madang	Papua N. Guin.	1967	2,576	170	TZFRD
Agana	Baler	Philippines	1964	2,758	128	TZD
Agana	Makaha	U.S.A.	1975	6,760	845	TZFRD
Agana	Wake	Wake Island	1964	2,780	142	TZD

Hong Kong:						
Hong Kong	Tumon Bay	Guam	1965	3,819	85	TZFRD
Hong Kong	K. Kinabalu	Malaysia	1964	2,041	85	TZFRD
Deep Water Bay	Currimao	Philippines	1977	880	1,840	TZFRD
Iceland:						
Vestmannaeyjar	Velbestad	Faeroe	1961	746	30	TZF
Vestmannaeyjar	Frederiksdal	Greenland	1962	1,552	24	TZFD
Ireland:						
Dublin	Nefyn	U.K.	1937	115	24	TZ
Dublin	Nefyn	U.K.	1938	117	24	TZ
Dublin	Abergeirch	U.K.	1937	117	12	Z
Dublin	Abergeirch	U.K.	1938	124	12	Z
Dublin	Holyhead	U.K.	1947	119	60	TZ
Dublin	Holyhead	U.K.	1947	117	60	TZ
Israel:						
Tel Aviv	Marseille	France	1968	3,395	128	TZFRD
Tel Aviv	Palo	Italy	1975	2,713	1,380	TZFRD
Italy:						
Brindisi	Durres	Albania	1940	156	3	TZ
Pisa	Algiers	Algeria	1972	1,107	480	TZ
Catanzaro	Alexandria	Egypt	1971	1,682	480	TZ
Rome	Marseille	France	1976	685	3,440	Z
Siracusa	Khania	Greece	1962	930	60	Z
Calabria	Lechaina	Greece	1969	528	480	Z
Palo	Tel Aviv	Israel	1975	2,713	1,380	TZFRD
Agrigenta	Tripoli	Libya	1968	552	120	TZ
Pozzallo	Malta	Malta	1955	98	36	TZ
Pisa	Barcelona	Spain	1969	796	480	Z
Rome	Malaga	Spain	1970	1,826	640	Z
Rome	Barcelona	Spain	1974	950	1,380	Z
Pantelleria	Kelibia	Tunisia	1956	80	60	TZ
Catania	Antalya	Turkey	1976	2,006	480	TZFRD
Ivory Coast:						
Abidjan	Dakar	Senegal	1978	2,550	480	TZFRD
Jamaica:						
Kingston	Grand Cayman	Cayman Is.	1971	693	120	TZF
Kingston	Ft. Sherman	Panama Ca. Zn.	1963	1,150	144	TZFRD
Kingston	Florida City	U.S.A.	1963	1,545	144	TZD
Japan:						
Reihoku	Nanhui	China	1976	872	480	TZFRD
Ninomiya	Agana	Guam	1964	2,656	138	TZFRD

TABLE 6F. (continued)

From	To		Year Laid	Length (km)	Channels (No.)	Use*
Okinawa	Agana	Guam	1976	2,528	845	TZFRD
Okinawa	Currimao	Philippines	1977	1,341	1,600	TZFRD
Naoetsu	Nakhodka	USSR	1969	883	120	TZFRD
Lebanon:						
Beirut	Larnaca	Cyprus	1975	217	640	Z
Beirut	Alexandria	Egypt	1972	926	120	TZFRD
Beirut	Marseille	France	1970	3,400	120	TZFRD
Libya:						
Tripoli	Agrigenta	Italy	1968	552	120	TZ
Malaysia:						
K. Kinabalu	Hong Kong	Hong Kong	1964	2,041	85	TZFRD
K. Kinabalu	Singapore	Singapore	1964	1,606	85	TZFRD
Malta:						
Malta	Pozzallo	Italy	1955	98	36	TZ
Midway Is.:						
Midway	Makaha	U.S.A.	1964	2,295	142	TZD
Midway	Wake	Wake Island	1964	2,052	142	TZD
Morocco:						
Tetouan	Perpignan	France	1967	1,404	96	TZFRD
Casablanca	Penmarch	France	1973	1,926	640	TZFRD
Martil	Marseille	France	1978	1,472	2,580	X
Casablanca	Dakar	Senegal	1977	2,711	640	X
Neth. Antilles:						
St. Maarten	Willemstad	Neth. Antilles	1973	1,050	160	TZFRD
St. Maarten	St. Thomas	U.S. Virgin Is.	1973	235	160	TZFRD
Netherlands:						
Terschelling	Fano	Denmark	1956	339	60	TZFD
Oostmahorn	Romo	Denmark	1950	257	120	TZFRD
Oostmahorn	Romo	Denmark	1950	259	120	TZFD
Domburg	Broadstairs	U.K.	1974	354	1,380	TZFD
Alkmaar	Lowestoft	U.K.	1980	222	N.A.	TZFD
Domburg	Aldeburgh	U.K.	1947	152	180	Z
Domburg	Aldeburgh	U.K.	1972	154	1,260	Z
Scheveningen†	Lowestoft	U.K.	1954	181	60	Z
Scheveningen†	Lowestoft	U.K.	1954	181	60	Z

Katwijk	Covehithe	U.K.	1964	202	120	Z
Katwijk	Covehithe	U.K.	1968	202	480	Z
New Zealand:						
Auckland	Sydney	Australia	1975	2,204	480	TZFRD
Auckland	Sydney	Australia	1962	2,358	85	TZFRD
Auckland	Suva	Fiji	1962	2,334	85	TZFRD
Norway:						
Kristiansand	Hanstholm	Denmark	1955	122	60	TZFD
Kristiansand	Klitmoller	Denmark	1967	82	480	TZFRD
Arendal	Hirtshals	Denmark	1941	124	60	TZFD
Arendal	Uggerby	Denmark	1977	126	2,700	N.A.
Bergen	Aberdeen	U.K.	1954	569	36	TZFD
Kristiansand	Scarborough	U.K.	1968	726	480	TZFRD
Panama Ca. Zn.:						
Ft. Sherman	Kingston	Jamaica	1963	1,150	144	TZFRD
Papua N. Guin.:						
Port Moresby	Cairns	Australia	1976	900	480	TZFRD
Madang	Cairns	Australia	1966	2,989	170	TZFRD
Madang	Tumon Bay	Guam	1967	2,576	170	TZFRD
Philippines:						
Baler	Agana	Guam	1964	2,758	128	TZD
Currimao	Deep Water Bay	Hong Kong	1977	880	1,840	TZFRD
Currimao	Okinawa	Japan	1977	1,341	1,600	TZFRD
Poland:						
Mielno	Molle Bugt	Denmark	1960	128	60	TZFD
Portugal:						
Funchal	Lisbon	Portugal	1972	1,148	120	TZ
Lisbon	S. Cruz	Spain	1969	1,363	360	TZFRD
Lisbon	Goonhilly	U.K.	1969	1,761	640	TZD
Puerto Rico:						
San Juan	W. Palm Beach	U.S.A.	1959	2,104	50	TZD
San Juan	W. Palm Beach	U.S.A.	1960	2,069	50	TZD
Saudi Arabia:						
Jiddah	Port Sudan	Sudan	1923	361	2	TCC
Senegal:						
Dakar	Abidjan	Ivory Coast	1978	2,550	480	TZFRD
Dakar	Casablanca	Morocco	1977	2,711	640	X
Singapore:						
Singapore	K. Kinabalu	Malaysia	1964	1,606	85	TZFRD
South Africa:						
Capetown	Ascension	U.K.	1969	4,804	360	TZFRD

TABLE 6F. (continued)

From	To		Year Laid	Length (km)	Channels (No.)	Use*
Spain:						
P.D. Mallorca	Algiers	Algeria	1975	339	480	Z
Las Palmas	Recife	Brazil	1972	5,050	160	TZFRD
Aguimes	Recife	Brazil	1973	4,878	160	Z
S. Cruz	Sal	Cape Verde	1969	1,585	360	TZFRD
Barcelona	Pisa	Italy	1969	796	480	Z
Malaga	Rome	Italy	1970	1,826	640	Z
Barcelona	Rome	Italy	1974	950	1,380	Z
S. Cruz	Lisbon	Portugal	1969	1,363	360	TZFRD
Tenerife	Cadiz	Spain	1965	1,396	160	Z
Arrecife	Las Palmas	Spain	1971	346	480	Z
Las Palmas	Cadiz	Spain	1971	1,365	1,840	Z
P.D. Mallorca	Barcelona	Spain	1971	339	1,380	Z
Tenerife	Las Palmas	Spain	1972	111	1,840	Z
P.D. Mallorca	Valencia	Spain	1977	302	3,900	Z
Las Palmas	Cadiz	Spain	1978	1,369	5,520	Z
Bilbao	Goonhilly	U.K.	1970	893	480	Z
Bilbao	Goonhilly	U.K.	1975	861	1,380	Z
Cadiz	Green Hill	U.S.A.	1970	6,410	845	TZD
Aguimes	Camuri	Venezuela	1977	5,999	1,840	Z
Sudan:						
Port Sudan	Jiddah	Saudi Arabia	1923	361	2	TCC
Sweden:						
Halsingborg	Helsingor	Denmark	1939	6	42	ZF
Malmo	Saltholm	Denmark	1951	9	900	TZFRD
Norrtalje	Mariehamn	Finland	1928	63	9	Z
Norrtalje	Mariehamn	Finland	1938	78	12	ZR
Norrtalje	Mariehamn	Finland	1951	48	60	TZ
Malmo	Stralsund	German Dem. Rep.	1927	119	12	Z
Malmo	Stralsund	German Dem. Rep.	1930	119	42	TZR
Malmo	Fehmarn	Germany Fed.	1969	224	480	TZFD
Trelleborg	Fehmarn	Germany Fed.	1975	200	1,200	TZFD
Trelleborg	Grossenbrode	Germany Fed.	1978	202	1,200	TZFD
Goteborg	Marske	U.K.	1960	945	60	TZFR
Tunisia:						
Bizerte	Perpignan	France	1975	943	640	TZFRD

Bizerte	Marseille	France	1969	865	128	TZ
Kelibia	Pantelleria	Italy	1956	80	60	TZ
Kelibia	Bou-Ficha	Tunisia	1956	109	120	Z
Turkey:						
Antalya	Catania	Italy	1976	2,006	480	TZFRD
U.K.						
Dover	Veurne	Belgium	1977	107	3,900	TZFD
Dover	De Panne	Belgium	1948	89	420	TZFRD
Canterbury	Oostende	Belgium	1958	102	120	TZFRD
Broadstairs	Oostende	Belgium	1971	119	1,260	TZFRD
Oban	Clarenville	Canada	1955	3,597	48	TZFD
Oban	Clarenville	Canada	1956	3,602	48	TZFRD
Oban	Hampden	Canada	1961	3,723	80	TZFRD
Bude	Beaver Harbour	Canada	1973	5,186	1,840	TZFRD
Ascension	Sal	Cape Verde	1969	3,145	360	TZFRD
Winterton	Fano	Denmark	1964	541	120	TZFD
Cayton Bay	Klitmoller	Denmark	1973	683	1,260	TZFD
Weybourne	Fano	Denmark	1950	574	192	T
Shetland Is.	Hvitanes	Faeroe	1971	411	480	TZFD
Gairloch	Velbestad	Faeroe	1961	530	26	TZF
Jersey	Pirou	France	1938	31	12	TZ
Jersey	Pirou	France	1940	33	12	TZ
Dover	Calais	France	1944	37	60	TZFD
Dover	Calais	France	1945	39	60	TZFRD
Dover	Calais	France	1968	41	60	TZFD
Dover	Boulogne	France	1968	46	60	TZFD
Eastbourne	Courseulles	France	1976	191	2,875	TZFRD
Winterton	Leer	Germany Fed.	1963	465	120	TZR
Winterton	Leer	Germany Fed.	1964	461	120	TZFD
Winterton	Wilhelmshaven	Germany Fed.	1971	519	1,260	TZFD
Abergeirch	Dublin	Ireland	1937	117	12	Z
Abergeirch	Dublin	Ireland	1938	124	12	Z
Holyhead	Dublin	Ireland	1947	119	60	TZ
Holyhead	Dublin	Ireland	1947	117	60	TZ
Nefyn	Dublin	Ireland	1937	115	24	TZ
Nefyn	Dublin	Ireland	1938	117	24	TZ
Broadstairs	Domburg	Netherlands	1974	354	1,380	TZFD
Lowestoft	Alkmaar	Netherlands	1980	222	N.A.	TZFD
Aldeburgh	Domburg	Netherlands	1947	152	180	Z
Aldeburgh	Domburg	Netherlands	1972	154	1,260	Z
Lowestoft†	Scheveningen	Netherlands	1954	181	60	Z

TABLE 6F. (continued)

From	To		Year Laid	Length (km)	Channels (No.)	Use*
Lowestoft†	Scheveningen	Netherlands	1954	181	60	Z
Covehithe	Katwijk	Netherlands	1964	202	120	Z
Covehithe	Katwijk	Netherlands	1968	202	480	Z
Aberdeen	Bergen	Norway	1954	569	36	TZFD
Scarborough	Kristiansand	Norway	1968	726	480	TZFRD
Goonhilly	Lisbon	Portugal	1969	1,761	640	TZD
Ascension	Capetown	South Africa	1969	4,804	360	TZFRD
Goonhilly	Bilbao	Spain	1970	893	480	Z
Goonhilly	Bilbao	Spain	1975	861	1,380	Z
Marske	Goteborg	Sweden	1960	945	60	TZFR
Widemouth	Tuckerton	U.S.A.	1963	6,517	138	TZD
U.S. Virgin Is.:						
St. Thomas	Forteleza	Brazil	1977	4,071	640	TZFRD
St. Thomas	S. Domingo	Dominican Rep.	1968	715	128	TZD
St. Thomas	St. Maarten	Neth. Antilles	1973	235	160	TZFRD
St. Thomas	Vero Beach	U.S.A.	1964	2,184	142	TZD
St. Thomas	Jacksonville Beach	U.S.A.	1968	2,446	720	TZD
St. Thomas	Maiquetai	Venezuela	1966	1,009	83	TZFRD
U.S.A.:						
W. Palm Beach	Nassau	Bahamas	1972	361	1,380	TZD
Manahawkin	Bermuda	Bermuda	1961	1,389	82	TZD
Keawaula	Port Alberni	Canada	1963	4,713	80	TZFRD
Key West	Havana	Cuba	1950	220	24	TZD
Key West	Havana	Cuba	1896	196	8	TCCT
Key West	Havana	Cuba	1899	185	8	TCCT
Key West	Havana	Cuba	1917	181	8	TCCT
Key West	Havana	Cuba	1950	239	24	TZD

Keawaula	Suva	Fiji	1963	5,691	80	TZFRD
Tuckerton	Vendee	France	1965	6,660	138	TZD
Green Hill	Vendee	France	1976	6,258	4,000	TZFRD
Makaha	Agana	Guam	1975	6,760	845	TZFRD
Florida City	Kingston	Jamaica	1963	1,545	144	TZD
Makaha	Midway	Midway Is.	1964	2,295	142	TZD
W. Palm Beach	San Juan	Puerto Rico	1960	2,069	50	TZD
W. Palm Beach	San Juan	Puerto Rico	1959	2,104	50	TZD
Green Hill	Cadiz	Spain	1970	6,410	845	TZD
Tuckerton	Widemouth	U.K.	1963	6,517	138	TZD
Vero Beach	St. Thomas	U.S. Virgin Is.	1964	2,184	142	TZD
Jacksonville Beach	St. Thomas	U.S. Virgin Is.	1968	2,446	720	TZD
Ketchikan	Port Angeles	U.S.A.	1956	1,409	48	TZD
Port Angeles	Ketchikan	U.S.A.	1956	1,370	48	TZD
Hanauma Bay	Point Arena	U.S.A.	1957	4,069	51	TZD
Point Arena	Hanauma Bay	U.S.A.	1957	4,093	51	TZD
Makaha	S.L. Obispo	U.S.A.	1964	4,413	143	TZD
USSR:						
Rohuneeme	Porkkala	Finland	1942	56	11	TZR
Nakhodka	Naoetsu	Japan	1969	883	120	TZFRD
Venezuela:						
Camuri	Aguimes	Spain	1977	5,999	1,840	Z
Maiquetai	St. Thomas	U.S. Virgin Is.	1966	1,009	83	TZFRD
Wake Island:						
Wake	Agana	Guam	1964	2,780	142	TZD
Wake	Midway	Midway Is.	1964	2,052	142	TZD

SOURCE.--International Telecommunication Union, List of Cables Forming the World Submarine Network, 19th ed. (Geneva: ITU, 1977).

NOTE.--End points given are the largest settlements closest to the point of landfall. N.A. = not available.

*Use follows ITU classification (see table 5F).

†Two cables laid in the same year.

Appendix G

Tables, Marine Science and Technology

The three tables in this new Appendix to the *Ocean Yearbook* provide significant information on an important aspect of the marine-related, scientific capability of nearly 80 countries. The first table gives summary statistics of national surface fleets. The other tables provide details of 653 surface vessels and 81 submersibles. Because the distinction between research and survey vessels is not made by many, especially developing, countries, both types of vessels have been included for all countries. Due to variations among the sources consulted, particularly with regard to date and type of information gathered, the data are unavoidably incomplete. Future editions of the *Yearbook* will include refinements of these data, as well as complementary information on scientific personnel.

We wish to express our thanks to Desmond P. D. Scott and the staff of the Intergovernmental Oceanographic Commission of Unesco for supplying data on many vessels which were not included in the published sources.

THE EDITORS

TABLE 1G.--RESEARCH VESSEL FLEETS, BY COUNTRY

Country	Total				Average				
	Vessel	Gross Tonnage	Displacement (Tons)	Scientific Complement	Year	Length (M)	Horsepower	Crew	Berths
Algeria	1				1977.0 (1)	15 (1)	260 (1)	3 (1)	
Argentina	14				1979.0 (1)	38 (14)			
Australia	7	934 (2)	8,697 (6)	15 (3)	1968.0 (6)	61 (7)	2,883 (6)	9 (3)	
Bahamas	5	966 (1)				19 (5)			
Bahrain	1	320 (1)							21 (1)
Bangladesh*	3	705 (3)			1961.0 (1)	42 (1)			
Barbados	1					18 (1)	80 (1)		
Belgium*	2	404 (2)							
Belize	1					11 (1)	135 (1)		
Bermuda	5			6 (1)		12 (5)	300 (1)		
Brazil	28	98 (2)				26 (28)	63 (2)		
Canada	16	14,305 (16)	16,452 (10)	148 (15)	1962.3 (16)	45 (16)	1,993 (15)	26 (15)	
Cayman Is.*	4	3,531 (4)							
Chile	16					17 (15)			
China, P. R.*	5	6,855 (5)							
Colombia	7	1,324 (2)				25 (7)	359 (4)	35 (2)	49 (2)
Cuba	5					16 (5)	350 (1)		15 (1)
Cyprus	3	39 (1)				12 (3)	72 (1)	7 (1)	8 (1)
Denmark	4	1,083 (3)			1974.0 (2)	53 (4)	1,125 (4)		
Ecuador	16					23 (6)			
Egypt	4	692 (3)		18 (2)		21 (4)	456 (4)	13 (2)	12 (1)
Fiji*	1	105 (1)							
Finland*	8	3,156 (8)							
France	24	837 (5)	7,996 (10)	94 (8)	1968.4 (10)	29 (24)	792 (17)	13 (15)	12 (8)
German D. Rep.*	3	830 (3)							
Germany, Fed.	7	5,181 (6)	6,555 (6)	24 (1)	1959.0 (6)	50 (6)	1,330 (6)	49 (2)	
Greece	2	45 (1)				56 (1)			
Honduras	4					7 (4)	58 (4)		
Hong Kong	1	242 (1)	458 (1)	4 (1)	1951.0 (1)	36 (1)	385 (1)	17 (1)	
Iceland	1	449 (1)				41 (1)	1,000 (1)		
India	31	2,618 (18)		41 (3)		18 (26)	210 (21)	11 (21)	11 (12)
Indonesia	3	637 (3)		21 (1)	1963.0 (1)	46 (1)		47 (1)	
Iraq*	2	1,008 (2)							
Israel	2	139 (2)		9 (1)		25 (1)	250 (1)	8 (1)	6 (1)
Italy	18	184 (9)	2,287 (4)	67 (6)	1968.2 (6)	21 (17)	463 (16)	7 (16)	7 (6)
Jamaica	6	198 (2)			1970.7 (3)	15 (6)	258 (4)	5 (1)	9 (1)
Japan	10	6,016 (9)	4,062 (7)	36 (2)	1963.9 (9)	49 (10)	1,166 (8)	28 (8)	40 (6)

Kenya	4			4 (1)	1969.0 (1)	13 (4)		11 (1)	
Korea, Rep.	1		850 (1)		1971.0 (1)	44 (1)	1,500 (1)		
Kuwait*	2	297 (2)							
Lebanon	1					12 (1)	260 (1)		
Madagascar	1					12 (1)	5 (1)		6 (1)
Malaysia	2	269 (2)				26 (2)	513 (2)	12 (2)	17 (2)
Malta	1					7 (1)			
Mauritius	3	99 (1)				16 (3)	450 (1)		19 (1)
Mexico	1	283 (1)							
Monaco	2	69 (2)		5 (1)	1957.0 (1)	17 (2)	230 (2)	4 (2)	
Morocco	3					20 (3)	660 (2)		
Netherlands	3	2,035 (2)		32 (3)	1962.0 (3)	42 (3)	777 (3)	13 (3)	
Neth. Antilles	3					6 (3)	68 (1)	2 (1)	2 (1)
New Zealand	1	1,012 (1)		6 (1)	1973.0 (1)	75 (1)	1,600 (1)	21 (1)	
Norway	4	1,576 (2)	450 (1)	10 (1)	1971.0 (4)	60 (3)	1,300 (2)	9 (1)	
Panama	1	35 (1)		7 (1)		20 (1)	270 (1)	2 (1)	9 (1)
Philippines	12	2,960 (8)		35 (2)		34 (7)	308 (2)	19 (10)	
Poland*	3	587 (3)							
Puerto Rico	14	182 (3)		15 (1)		12 (14)	166 (11)	3 (4)	3 (2)
Romania*	1	659 (1)							
St. Kitts-Nevis*	1	256 (1)							
St. Vincent*	1	487 (1)							
Senegal*	1	178 (1)							
Seychelles	1					11 (1)	88 (1)		4 (1)
Singapore	3	196 (1)			1965.0 (1)	22 (2)	56 (1)		
South Africa	11	4,794 (10)	5,365 (7)	36 (9)	1962.1 (11)	35 (11)	479 (9)	13 (10)	
Spain	4			5 (1)		12 (3)	162 (2)	3 (1)	
Sweden*	2	1,185 (2)							
Syria	1					15 (1)			
Taiwan*	1	596 (1)							
Tanzania	1	498 (1)		8 (1)	1977.0 (1)	37 (1)	825 (1)	20 (1)	
Thailand	6	133 (3)	800 (1)	20 (3)	1960.0 (1)	27 (6)	380 (5)	14 (6)	4 (1)
Tunisia	2	90 (1)				17 (2)	285 (1)	11 (1)	
Turkey	7	239 (5)		20 (3)		15 (7)	176 (4)	4 (6)	
U.K.	48	36,076 (34)	30,403 (18)	153 (16)	1964.5 (43)	49 (46)	1,189 (19)	18 (12)	32 (11)
United Arab Emirates*	1	145 (1)							
U.S. Virgin Is.	3					6 (3)	63 (3)		
U.S.A.	140	13,348 (22)	82,689 (65)	1,133 (83)	1964.4 (83)	42 (140)	1,994 (63)	27 (73)	36 (37)
USSR	118	123,909 (44)	216,976 (78)	999 (18)	1963.7 (57)	73 (96)	2,336 (19)	54 (6)	88 (1)

TABLE 1G. (continued)

	Total				Average				
Country	Vessel	Gross Tonnage	Displacement (Tons)	Scientific Complement	Year	Length (M)	Horsepower	Crew	Berths
Venezuela	11	643 (4)			1963.3 (3)	17 (10)	153 (2)	6 (2)	14 (3)
Yugoslavia	7	416 (3)		23 (3)		16 (7)	140 (6)	6 (3)	3 (2)
World	694	246,153 (280)	384,040 (215)	3,234 (192)	1964.4 (275)	39 (606)	1,110 (287)	19 (237)	25 (104)

SOURCE.--Table 2G, unless otherwise indicated.

NOTE.--Surface research and survey vessels included; submersibles excluded. Numbers following in parentheses indicate the number of vessels for which values were reported.

*Aggregate figures from Lloyd's Register of Shipping Statistical Tables, 1979.

TABLE 2G.--SURFACE RESEARCH AND SURVEY VESSELS, BY COUNTRY

Institution and Vessel	Year*	Length (M)	Displacement (Tons)	Gross Tonnage	Hull† Type	Horsepower	Use‡	Scientists	Crew	Total Berths	Date of Information§
						Algeria					
Centre de Recherches Océanographiques et des Pêches (Algiers):											
"Georges Aime"	1977	15				260	O		3		05-77
						Argentina					
Centro de Investigación de Biología Marina [CIBIMA] (Buenos Aires):											
"Delfín"		9									
Dirección General de Pesca y Recursos Pesqueros (San Antonio Oeste):											
"El Austral"		65									
"IADO II"		5									
"IADO III"		17									
"Puerto Deseado"	1979	84									
Instituto Argentino de Oceanografía (Bahía Blanca):											
"La Tinta"		12									
Museo Provincial di Ciencias Naturales "Florentino Ameghino" (Santa Fé):											
"Cdro. Rivadavia"		52									
"Cormoran"		25									
"Goyena"		58									
"Islas Orcadas"		81									
"Petrel"		20									
"Thompson"		58									
Servicio de Hidrografía Naval (Buenos Aires):											
"Atlántico Sur"		36					O				
"Laurus"		16									
						Australia					
Commonwealth Scientific and Industrial Research Organization:											
"Courageous"	1975	28		287	1	548	F	5	8		00-77
"Kalinda"	1973	21	64		3		F	2	7		00-77
"Sprightly"	1973	44	870	647		1,900	OF	8	11		00-77
Royal Australian Navy Hydrographic Service:											
"Cook"		91	2,300			4,000					00-77
"Diamantina"	1969	92	2,127			5,500	O				00-77
"Kimbla"	1955	54	1,021			350					00-77
"Moresby"	1963	96	2,315			5,000					00-77
						Bahamas					
Forfar Field Station (Andros Island):											
N.A.		8					C				

TABLE 2G. (continued)

Institution and Vessel	Year*	Length (M)	Displacement (Tons)	Gross Tonnage	Hull† Type	Horsepower	Use‡	Scientists	Crew	Total Berths	Date of Information§
N.A.		8					C				
N.A.		9					C				
N.A.		9					C				
"San Andres"		63		966	1					21	
Bahrain											
Decca Survey Overseas Ltd.:											
"Decca Pilot"	1961	42		320							00-77
Barbados											
Bellairs Research Institute of McGill University (St. James):											
N.A.		18				80					
Belize											
Ministry of Trade and Industry (Belize City):											
"Panulirus Argus"		11				135					
Bermuda											
Bermuda Aquarium, Museum, and Zoo (Bermuda):											
N.A.		11									
N.A.		8									
Bermuda Biological Station for Research Inc. (St. George's):											
"Micmac"		12									
"Panuluris II"		20				300		6			
"Velella"		10									
Brazil											
Diretoria de Hidrografía e Navegação (Rio de Janeiro):											
"Orión"		24									
Instituto de Pesca (Sao Paulo):											
"Orión"		24									
Instituto de Pesca, Divisão de Pesca Maritima (Santos):											
"Cruz del Sur"		33					F				
"Diadorim"		24					F				
"Mestre Jeronimo"		36									
"Riobaldo"		24					F				
Programa de Pesquisa e Desenvolvimento Pesqueiro do Brasil (Brasilia):											
"Gilda Viana"		6		3		28					
"Serro Azul"		11		95		98					
Secretaria da Agricultura do Río Grande do Norte (Natal):											
"Ilha de Itamaracá I"		13									
"Investigador IV"		17									

Universidad Federal de Bahía (Salvador):										
"Agulmao"		10								
"Igarapesca"		15								
"Independencia"		25								
Universidad Federal de Río Grande do Sur, CECO (Porto Alegre):										
"Emilia"		15								
"Paiva Carvalho"		11								
"Prof. W. Besnard"		49					O			
"Veliger"		10								
Universidade de São Paulo, Instituto Oceanografico (São Paulo):										
"Stella Maris"		20								
Universidade do Río Grande, Base Oceanográfica Atlântica (Río Grande):										
"Alte Câmara"		63								
"Alte Saldanha"		93								
"Argus-Orion-Taurus"		45								
"Caravelas"		16								
"Itacurussa-Camocim"		16								
"Nogueira da Gama"		16								
"Paraibano-Río Branco"		16								
"Sirius-Canopus"		78								
Universidade Federal da Bahía, Instituto de Biologia (Salvador):										
N.A.		13								
Universidade Federal de Alagoas--CCBI (Maceió):										
"Investigador"		16								
Canada										
Bedford Institute of Oceanography, Department of Environment (Dartmouth):										
"Baffin"	1956	87	4,427	3,460		7,920	H	24	69	00-77
"Dawson"	1969	65	1,975	1,311		3,400	OH	13	30	00-77
"Hudson"	1963	90	4,793	3,721		7,500	OH	25	64	00-77
"Maxwell"	1962	35	276	262		700	H	7	14	00-77
Environment Canada:										
"A. P. Knight"	1956	22		140	3	304	F	3	6	00-77
"A. T. Cameron"	1958	54		753	1	1,000	F	10	25	00-77
"Caligus"	1967	17		41	5	240	F			00-77
"E. E. Prince"	1966	40	421	406	1	600	F	6	14	00-77
"G. B. Reed"	1962	54		753	1	1,000	F	10	25	00-77
"J. L. Hart"	1975	20		90	1	425	F	5	10	00-77
"Marinus"	1953	19	55	45	3	215	F	2	5	00-77
"Navicula"	1968	20	110	78	3	365	F	2	3	00-77
"Parizeau"	1967	65	1,993	1,314	1	3,400	O	13	32	00-77

TABLE 2G. (continued)

Institution and Vessel	Year*	Length (M)	Displacement (Tons)	Gross Tonnage	Hull† Type	Horsepower	Use‡	Scientists	Crew	Total Berths	Date of Information§
"Shamook"	1975	23		120	1	425	F	9	18		00-77
"Vector"	1967	40	557	516	1		X	9	15		00-77
"William J. Stewart"	1932	70	1,845	1,295	1	2,400	H	10	57		00-77
Chile											
Instituto de Fomento Pesquero [IFOP] (Santiago):											
"Lund"		15									
Instituto Hidrográfico de la Armada de Chile (Valparaiso):											
"Cynthia"		12									
Pontificia Universidad Católica de Chile (Santiago):											
"María Eugenia"		12									
Universidad Católica de Valparaiso (Valparaiso):											
"Yelcho"		65									
Universidad de Chile, Departamento de Biología (Santiago):											
N.A.		6									
Universidad de Chile, Sede Antofagasta (Antofagasta):											
"Sta. María Virginiall"											
"Stella Maris"		20									
Universidad de Chile, Sede Arica (Arica):											
"Guayita"		5									
"Pesquerita"		7									
Universidad de Chile, Sede Osorno (Osorno):											
N.A.		17									
"Tiberiades"		17									
Universidad de Concepción, Instituto de Biología (Concepción):											
"Lolito"		10									
Universidad del Norte, Centro de Investigaciones Marinas (Iquique):											
"Carlos Porter"		28									
"Fitz Roy"		16									
"Teararoa Rakei"		16									
Universidad del Norte, Departamento de Pesquerías (Antofagasta):											
"Caranco"		10									
Colombia											
Centro de Investigaciones Oceanográficas e Hidrográficas (Cartagena):											
"Arc. Quindio"		37		650	1	350			20	30	
"Arc. San Andres"		56		674	1	850			50	68	
"Batilancha"		10									
"CIOH-1"		6			1	150					

Inderena (Bogotá):										
"Inderenita"		54								
Universidad del Valle, Departamento de Biología (Cali):										
"Tigre"		6			85					
Universidad Tecnológica del Magdalena (Santa Marta):										
"Piscis II"		7								
					Cuba					
Centro de Investigaciones Marinas (Havana):										
(Due to receive 1980)		18			350				15	
Instituto de Oceanología (Havana):										
"Caribe"		15								
"Makaira"		19								
"Volga"		11								
"Xiphias"		18								
					Cyprus					
Department of Fisheries (Nicosia):										
N.A.		11		3		F				05-77
"F1"		10		3	72	F				05-77
"Triton"		16	39	1		F		7	8	05-77
					Denmark					
Wilh. Chr. Bech.:										
"Anne Bravo"		58	427		1,250					00-77
"Karen Bravo"		58	399		1,200					00-77
"Kirsten Bravo"	1974	60			1,500	S				00-77
"Martin Knudsen"	1974	37	257		550	O				00-77
					Ecuador					
Escuela Superior Politécnica del Litoral (Guayaquil):										
"Huayaipe"		26								
"Pinta"		8								
"Saint Jude"		10								
"Tohalli"		32								
Instituto Nacional de Pesca (Guayaquil):										
"Orión"		45								
"Rigel"		18								
					Egypt					
Institute of Oceanography and Fisheries (Alexandria):										
(Under construction, due 1979)		42	550	1	1,600				12	06-77
"Bahnasy"		13	22	3	50	C	12	18		06-77
"Faras-El-Bahr"		20	120	1	150	C	6	8		06-77

TABLE 2G. (continued)

Institution and Vessel	Year*	Length (M)	Displacement (Tons)	Gross Tonnage	Hull† Type	Horsepower	Use‡	Scientists	Crew	Total Berths	Date of Information§
University of Alexandria, Department of Oceanography (Alexandria):											
N.A.		8				25	C				06-77
France											
Centre National pour l'Exploitation des Océans:											
"Capricorne"	1969	47	710			1,200	OF	12	27		00-77
"Coriolis"	1963	38	450			700	OF	12	18		00-77
"Cryos"	1970	49	800				F	9	22		00-77
"Jean Charcot"	1965	74	2,100			3,360	O	22	48		00-77
"La Pelagia"	1965	32	242			800	OF	7	12		00-77
"Le Noroit"	1970	51	870			1,650	O	10	20		00-77
"Le Suroit"	1974	56	1,094	655		1,650	X	12	22		00-77
CERBOM (Nice):											
"St. Jacques"		10			3	24			1	14	04-77
"St. Maurice IV"		14			3	140			2	14	04-77
French Nuclear Research Centre, Biological Department:											
"Marara"	1973	43				1,200	B	10		28	00-77
Institut Scientifique et Technique des Pêches Maritimes (Nantes):											
"Ichthys"	1964	20		71	3	300	F		6	2	01-77
Laboratoire Arago (Banyuls-sur-Mer):											
"Néréis"		12		12		100			3	0	05-77
"Prof. Lacaze-Duthiers"		17		32		200			5	12	05-77
"Rufi"		6							2		05-77
Port of Bordeaux Authority:											
"Gardour"		33				1,500	H				00-77
STARESO de l'Université de Liège à Calvi (Calvi):											
"Recteur Dubuisson"		21		67			O				04-77
Station Marine d'Endoume et Centre d'Océanographie (Marseille):											
"Alciope"		19					O				09-77
"Antedon"		18					O				09-77
"Armanova"		10					O				09-77
Station Zoologique de Villefranche-sur-Mer (Villefranche):											
"Amphioxus"		7				18					10-75
"Catherine-Laurence"		21	130		3	240	O		4	14	10-75
"Korotneff"		20	110			360			6	10	10-75
"Sagitta"		7				15					10-75
Underwater Research and Study Group:											
"Triton"	1971	68	1,490								00-77

					Germany, Fed.						
Deutsches Hydrographisches Institut (Hamburg):											
"Atair"	1962	32	173	157		280					00-77
"Gauss"	1941	54	1,318	846		1,000					00-77
"Komet"	1969	68	1,595	1,252		3,800	H		42		00-77
"Meteor"	1964	82	3,054	2,615		2,000		24	55		00-77
"Suderoog"	1956	33	242	154		620					00-77
"Wega"	1962	32	173	157		280					00-77
Institut für Meereskunde:											
"Anton Dohrn"							H				00-77
					Greece						
Greek Atomic Energy Commission, Hydrobiology Laboratory (Athens):											
"Jenny"		56		45							05-77
University of Athens, Zoological Laboratory and Museum (Athens):											
"Mark IV"											05-77
					Honduras						
Laboratorio de Biología Marina (La Ceiba):											
N.A.		12			4	120					
N.A.		5			4	25					
N.A.		7			3	25					
N.A.		5			5	60					
					Hong Kong						
Agriculture and Fisheries Department:											
"Cape St. Mary"	1951	36	458	242		385	FH	4	17		00-77
					Iceland						
Fisheries Research:											
"Arni Fridriksson"		41		449	1	1,000					00-77
					India						
Central Institute of Fisheries Technology (Cochin):											
"Fishtech No. 1"		9		6		35	F		5		09-77
"Fishtech No. 2"		10		8		42	F		5		09-77
"Fishtech No. 4"		11		15		70	F		5		09-77
"Fishtech No. 5"		9		6		35	F		5		09-77
"Fishtech No. 6"		10		8		40	F		5		09-77
"Fishtech No. 7"		12		17		90	F		7	8	09-77
"Fishtech No. 8"		15		30		130	F		8	8	09-77
"Fishtech No. 9"		10		8		37	F		5		09-77
"Sindhukumari"		15		30		130	F		8	8	09-77
Central Marine Fisheries Research Institute (Cochin):											
(Under construction)		33					F				03-77

TABLE 2G. (continued)

Institution and Vessel	Year*	Length (M)	Displacement (Tons)	Gross Tonnage	Hull† Type	Horsepower	Use‡	Scientists	Crew	Total Berths	Date of Information§
"Cadalmin I"		13		186	3	93	C		7	2	03-77
"Cadalmin II"		13		186	3	93	C		7	2	03-77
"Chippi"		10				24	C				03-77
"Mantha"		10				24	C				03-77
Centre of Advanced Studies in Marine Biology (Parangipettai):											
N.A.		10					C				03-78
N.A.		10					C				03-78
Hydraulic Study Department (Calcutta):											
"Anusandhani"		52				1,160		6	44		05-77
Hydrobiological Research Station (Madras):											
N.A.		14									09-77
Marine Biological Research Station (Okha):											
"Gulf Shrimp"				30		87			6	6	04-77
"Indian Salmon"				30		87			6	6	04-77
"Ocean Perch"				58		200			8	8	04-77
"Rock Perch"				58		200			8	8	04-77
"Silver Pomfret"				30		87			6	6	04-77
"Varsha"		14				60			8	2	10-77
National Institute of Oceanography (Goa):											
"Gaveshani"		68		1,900	1	1,680		19	45	70	03-77
"Neendakara"		12			5						03-77
"Tarini"		15									03-77
Oil and Natural Gas Commission:											
N.A.		53						16	24		00-77
University of Cochin, Department of Marine Sciences (Cochin):											
N.A.		8					C				09-77
(Under construction)		13									09-77
Zoological Survey of India, Marine Biological Station (Madras):											
"Chotah Investigator"		10		12					6		10-77
Indonesia											
Janhidros:											
"Jalanidhi"	1963	46		432				21	47		
National Institute of Oceanology (Jakarta):											
"Mutiara"				15							09-77
"Samudra"				190							09-77

Israel											
Israel Oceanographic and Limnological Research Ltd. (Haifa):											
"Etziona"				19						6	05-77
"Shikmona"		25		120	1	250	X	9	8		05-77
Italy											
CNEN-EURATOM, Marine Contamination Laboratory (Fiascherino):											
"Odalisca"		18			3	185	C		4	4	03-77
Consiglio Nazionale delle Ricerche:											
"Bannock"	1963	63	1,258			3,003		26	28		00-77
"L. F. Marsili"	1969	55	817			1,000	O	22	26		00-77
"Luiciotta"	1973	31	128			660	F	6	8		00-77
ESPI--Research Centre for Fisheries and Marine Resources (Messina):											
"Centro Pesca"		14			3	115	FO	2	3		05-77
Institut de Recherche sur les Eaux (Roma):											
"Irsa-Mare"		13			3	320			2		
Institute of Marine Biology CNR (Venice):											
"Umberto d'Ancona"	1967	24	84	58	1	380	B	7	5	12	04-77
Institute of Zoology, Marine Biology Station (Messina):											
"Algesiro Matteo"	1970	29			3	470			6	16	05-77
"Colapesce"	1967	13			3	320	C		3	3	05-77
Laboratoire de Biologie Marine et de Peche (Fano):											
"Giannetto"		20		30	3	120	F		5		09-76
Laboratory for the Biological Improvement of Lagoons (Lesina):											
"Marina II"				9							06-77
Observatory for Experimental Geophysics (Trieste):											
"F. Vercelli"		20		48	3	150			6	4	07-75
Oceanographic Experimental Institute "Attilio Cerruti" (Taranto):											
"Attilio Cerruti"		14				260		4	1	5	05-77
University Center of Studies and Researches on Marine Biological Resources (Cesenatico):											
"Daphne"		13		15		180					05-77
Zoological Station and Aquarium (Napoli):											
"Federico Raffaele"		10		9	3	75			3		05-77
"Rinaldo Dohrn"		14		9	3	155			3		05-77
"Salvatore lo Bianco"		7		4	3	11			2		05-77
"San Gennaro"		5		2	3				2		
Jamaica											
Ministry of Agriculture, Fisheries Division (Kingston):											
"Albacore"	1964	11				100	X				
"Black Fin"	1976	22		100		365	X				
"Dolphin"	1972	22		98		365	X				

TABLE 2G. (continued)

Institution and Vessel	Year*	Length (M)	Displacement (Tons)	Gross Tonnage	Hull† Type	Horsepower	Use‡	Scientists	Crew	Total Berths	Date of Information§
University of the West Indies, Port Royal Marine Laboratory (Kingston):											
"Aurelia"		8									
"Caranx"		17			2	200	F		5	9	
"Pelagia"		8									
						Japan					
Fuyo Ocean Development and Engineering Co.:											
"Mawashio"	1971	31								20	00-77
Hakodate Marine Observatory:											
"Kofu Maru"	1962	47	319	346		650	O	16	25		00-77
Japanese Fisheries:											
"Soyo Maru"		51		494		800	F				00-77
Maritime Safety Agency:											
"Heiyo HM04"	1955	23	66	50		150	HO		11		00-77
"Kaiyo HM06"	1963	45	372	308		450	HO		23	31	00-77
"Meiyo HL03"	1962	45	478	260		700	HO		27	40	00-77
"Shoyo HL01"	1971	80	1,931	1,842		2,400	HO		49	73	00-77
"Takuyo HL02"	1956	62	728	773			HO		38	51	00-77
"Ten'Yo HM05"	1961	30	168	121		380	HO		16	25	00-77
Metal Mining Agency:											
"Hakurei Maru"	1974	77		1,822		3,800	G	20	35		00-77
						Kenya					
Fisheries Department (Mombasa):											
"Kenshore II"		10			2						
"Kisite"		9			3						
"Kongomile"		10			2						
"Shakwe"	1969	22						4	11		
						Korea, Rep.					
Ministry of Science and Technology:											
"Tam Yung"	1971	44	850			1,500	G				00-77
						Lebanon					
National Council for Scientific Research (Beirut):											
"Sita III"		12			1	260					12-76
						Madagascar					
Direction de l'Elevage et de la Pêche Maritime (Tananarive):											
"Makamby"		12			3	5				6	02-77

					Malaysia					
Fisheries Research Institute (Penang):										
"K. K. Aya"		29	184	3	700			11	20	03-77
"K. K. Jenahak"		23	85	3	325			12	13	03-77
					Malta					
University of Malta, Department of Biology (Msida):										
		7				C				03-75
					Mauritius					
Ministry of Fisheries (Port Louis):										
"Explorer"		11		3		C				09-77
"Investigator"		26	99	1	450	X			19	09-77
"Sphyrna"		10				C				09-77
					Mexico					
Instituto Nacional de Pesca (Campeche):										
"Onjuku"			283							
					Monaco					
Centre Scientifique de Monaco (Monaco):										
"Ramoge"		15	25	1	330			2		03-77
Musée Océanographique de Monaco (Monaco):										
"Winnaretta-Singer"	1957	18	44	3	130	FH	5	5		03-77
					Morocco					
Institut Scientifique des Pêches Maritimes (Casablanca):										
"El Idrissi"		35		3	1,000					06-77
"El Mounir"		18		3	320					06-77
Université Mohammed V, Département des Sciences de la Terre (Rabat):										
"Bn Majid"		7								05-77
					Netherlands					
Ministry of Transport and Public Works:										
"Cumulus"	1962	72	1,974		1,400	M	16	24		00-77
Netherlands Institute for Sea Research:										
"Aurelia TX-59"	1971	32			720		12	9		00-77
Tromso Marine Biological Station:										
"Asterias"	1953	21	61		210	H	4	5		00-77
					Neth. Antilles					
Caribbean Marine Biological Institute (Curacao):										
N.A.		5								
N.A.		6								
Department of Agriculture, Animal Husbandry, and Fisheries (Curacao):										
N.A.		8		4	68			2	2	

TABLE 2G. (continued)

Institution and Vessel	Year*	Length (M)	Displacement (Tons)	Gross Tonnage	Hull† Type	Horsepower	Use‡	Scientists	Crew	Total Berths	Date of Information§
						New Zealand					
Department of Scientific and Industrial Research (Wellington N.):											
"Tangaroa"	1973	75		1,012		1,600	O	6	21		00-77
						Norway					
Bergship A/S:											
"Arctic Surveyor"	1974	79		1,281		2,000					00-77
Geco A/S:											
"Longva 1"	1973	62					S				00-77
"Longva 2"	1975						S				00-77
Norwegian Defence Research Establishment:											
"Sverdrup"	1962	39	450	295		600	O	10	9		00-77
						Panama					
Smithsonian Tropical Research Institute (Balboa):											
"Benjamin"		20		35	1	270		7	2	9	
						Philippines					
Bureau of Coast and Geodetic Survey (Manila):											
"Arinya"		28		250			H		25		09-77
"Arlunya"		28		250			H		25		09-77
"Atyimba"		49		680			O		46		09-77
"Pathfinder"		50		1,057			O		37		09-77
Bureau of Fisheries and Aquatic Resources (Quezon City):											
"Researcher"		44		509			F	18	23		12-77
Institute of Fisheries Development and Research (Quezon City):											
"Albacore"		32		190		600	F	17	19		01-77
Silliman University (Dumaguete City):											
N.A.									2		05-77
N.A.									2		05-77
N.A.									4		05-77
Southeast Asian Fisheries Development Center [SEAFDEC] (Iloilo City):											
N.A.				4	4		B				04-77
N.A.				20	4		OF				04-77
Xavier University (Cagayan de Oro):											
"Dolphin"		10				15	C		2		05-77
						Puerto Rico					
Center for Energy and Environmental Research (Mayaguez):											
N.A.		6				40					
"Sultana"		13									

Commercial Fisheries Laboratory (Mayaguez):											
"Agustine Stahl"		11		12	3	130			2	2	
"Miguel A. Abreu"		13		10		184			2	4	
Departamento de Recursos Naturales (San Juan):											
"Bertram"		12				425					
"Julio G. Diaz" (on repair)		13				180					
"Mariner"		7				75					
N.A.		6									
N.A.		6				70					
N.A.		5				70					
"Proline"		5				70					
"Rachel Carson"		13				180					
Universidad de Puerto Rico, Departamento de Ciencias Marinas (Mayaguez):											
"Crawford"		38		160	1	400		15	6		
"Medusa"		18							2		
Seychelles											
Fisheries Division (Victoria):											
"Scyllarus"		11				88				4	08-77
Singapore											
Decca Survey S.E. Asia Pte. Ltd. (Singapore):											
"Decca Oceaneer"	1965	30		196			G				00-77
SEAFDEC, Marine Fisheries Research Department (Singapore):											
N.A.		13				56	C				11-77
University of Singapore, Zoology Department (Singapore):											
N.A.							C				03-76
South Africa											
Department of Industries:											
"Africana II"	1949	63	1,300	822		1,120	FO	4	31		00-77
"Benguela"	1968	44	546	486		1,190	FO	6	19		00-77
"Crustacea"	1959	23		99		165	CF	1	9		00-77
"Kuiseb"	1955	22		61		230	F	1	6		00-77
"Kunene"	1958	21	86	23		180	F	2	9		00-77
"Namib II"	1967	25		81			F	9	9		00-77
"Nautilus II"	1974	20		50		230	CF		9		00-77
"Sardinops"	1958	37	342	251	1	600	FO	4	16		00-77
"Trachurus"	1958	21	86	23		180	F	2	9		00-77
Navy:											
"Protea"	1972	80	2,800	2,898							00-77

TABLE 2G. (continued)

Institution and Vessel	Year*	Length (M)	Displacement (Tons)	Gross Tonnage	Hull† Type	Horsepower	Use‡	Scientists	Crew	Total Berths	Date of Information§
University of Cape Town:											
"Thomas B. Davie"	1965	29	205			420	O	7	13		00-77
						Spain					
Instituto Espanol de Oceanografia, Laboratorio del Mar Menor (Murcia):											
"La Torre"		9			3	49	C				01-77
Institute of Fishery Research (Barcelona):											
"Cornide de Saavedra"											01-77
Laboratorio Oceanográfico de Baleares (Palma de Mallorca):											
"Gavina II"		9					C				11-76
"Jafuda Cresques"		17			3	274		5	3		11-76
						Syria					
Centre for Marine Research, Latakia (Damascus):											
(Due to be available)		15									08-75
						Tanzania					
N.A.:											
"Kaskazi"	1977	37		498		825		8	20		00-79
						Thailand					
Phuket Marine Biological Center:											
"Pramong 8"		17		35					8	4	08-77
Phuket Marine Fisheries Station (Phuket):											
"Pramong 3"		21			3	250		4	10		03-77
"Pramong 10"		24			3	450		12	12		03-77
Royal Thai Navy:											
"Chandra"	1960	70	800			1,000	H	4	42		00-77
Songkhla Fisheries Station:											
"Songkhla I"		18		68		180			9		08-77
"Songkhla II"		9		30		22			4		08-77
						Tunisia					
INSTOP (Salammbo):											
"Hannoun"		21		90	1	285			11		12-76
"Mimoun"		12									12-76
						Turkey					
Ege University, Institute of Hydrobiology (Izmir):											
"Hippocampus"		16		18	3	80	B	8	4		03-77
"Nereis"		9			3	35	CB				03-77

Hydrobiological Research Institute (Istanbul):											
"Arar"		31		173	1	380		8	7		07-77
"Bulur"		12		16					3		07-77
"Gezer"		12		16					3		07-77
"Gorur"		12		16					3		07-77
Middle East Technical University, Marine Sciences Department (Ankara):											
"Kugu"		12				210	CO	4	2		05-77
U.K.											
British Arctic Survey:											
"Bransfield"	1970	99		4,816		6,400					00-77
"John Biscoe"	1956	67		1,584		1,350		34			00-77
Decca Survey Ltd."											
"Decca Engineer"	1966	48		620			G				00-77
"Decca Explorer"	1943	53		673			G				00-77
"Decca Mariner"	1965	46		449			G				00-77
"Decca Recorder"	1969	38		294			G				00-77
"Decca Scanner"	1968	52		1,019			G				00-77
"Decca Surveyor"	1963	48		568			G				00-77
"Decca Tracker"	1958	47		487			G				00-77
Department of Agriculture and Fisheries, Scotland (Aberdeen):											
"Clupea"	1968	32					F	5		16	00-77
"Explorer"	1956	62					F	5		38	00-77
"Goldseeker"	1967	15				110	F	2	4		00-77
"Mara"	1958	22				204		3	8		00-77
"Scotia"	1971	68	1,994				F	12		51	00-77
Gardline Shipping Ltd.:											
"Charterer"	1955	57		490		600				23	00-77
"Endurer"	1954	53		154		600			11	19	00-77
"Gardline Locator"	1975	52		698		960		12		26	00-77
"Isis Corer"	1945	20		69	3						00-77
"Researcher"		44		395		600					00-77
"Surveyor"	1946			491		700					00-77
"Tracker"	1953	53		431					11		00-77
Ministry of Agriculture, Fisheries and Food:											
"Arctic Privateer"	1968										00-77
"Cirolana"	1970	73	1,767	1,594		1,100	OM	11	35		00-77
"Corella"		42		460	1						00-77
Ministry of Defence:											
"Beagle"	1967	58	1,088				H			43	00-77
"Bulldog"	1967	58	1,088				H			43	00-77
"Endurance"	1968	93					H				00-77

TABLE 2G. (continued)

Institution and Vessel	Year*	Length (M)	Displacement (Tons)	Gross Tonnage	Hull† Type	Horsepower	Use‡	Scientists	Crew	Total Berths	Date of Information§
"Fawn"	1968	58	1,088				H			43	00-77
"Fox"	1967	58	1,088				H			43	00-77
"Hecate"	1965	80	2,800	2,898		2,000	O				00-77
"Hecla"	1965	80	2,800	2,898		2,000	O				00-77
"Herald"	1974	79		2,532							00-77
"Hydra"	1965	80	2,800	2,898		2,000	O				00-77
Natural Environment Research Council:											
"Challenger"	1971	55	1,440	998			B	9	25		00-77
"Discovery"	1962	80	2,800	2,665			O	21	45		00-77
"Edward Forbes"	1972	31	290	180			C	4	9		00-77
"Jane"	1970	16	60	41		220		4	4		00-77
"John Murry"	1967	41	595	441			CO	8	19		00-77
"Shackleton"	1971	61	1,658	1,103			O	13	32		00-77
Scottish Marine Biological Association (Oban):											
"Beaver Two"	1971	11									00-77
"Calanus"		23		72		235					00-77
University College of North Wales (Menai Bridge):											
"Prince Madog"		29		182	1	600		6	10		00-77
University College of Swansea (Swansea):											
"Ocean Crest"	1956	30	131		1						00-77
University College of Wales (Aberystwyth):											
"Kay BB"	1962	14		22	3	110				6	00-77
University of East Anglia (Norwich):											
"Envirocat"	1973	10			5		C				00-77
University of Liverpool (Liverpool):											
"Cuma"		20	71	53			B	4			00-77
University of Newcastle:											
"Bernica"	1973	15		47		110					00-77
Wimpey Laboratories Ltd.:											
"Wimpey Sealab"	1975	99	6,845	3,754		2,700	G				00-77
U.S. Virgin Is.											
Caribbean Research Institute, Ecological Research Center (St. Thomas):											
N.A.		7				155					
N.A.		5				10					
N.A.		5				25					

U.S.A.											
Alcoa Marine Corp.:											
"Alcoa Seaprobe"	1970	74	1,700		5			20	30		00-77
Cape Fear Technical Institute (Wilmington):											
"Advance II"		56					C				00-77
Columbia University, Lamont-Doherty Geological Observatory (Palisades):											
"Robert D. Conrad"	1962	64	1,370	1,425	1	1,000		20	7		00-78
"Vema"	1953	60	1,000			900		14	23		00-78
Decca Survey Systems, Inc. (Houston):											
"Decca Locator"	1975	32		192		1,340					00-77
"Decca Profiler"	1973	32		183		1,340					00-77
Duke University Marine Laboratory (Beaufort):											
"Eastward"	1964	36	610			640		15	15		00-78
"John de Wolf II"		19					C				00-77
Florida Department of Natural Resources (St. Petersburg):											
"Hernan Cortez"		22			3	500					
"Montezuma"		10			4	200					
Florida Institute for Oceanography (St. Petersburg):											
"Bellows"		20		80				10	3	13	
Florida Institute of Technology, Department of Oceanography and Ocean Engineering (Melbourne):											
"Jennie D"		13		19		640			1	6	
"Sea Hunter"		20				150			2	12	
Florida Institute of Technology, School of Applied Technology (Jensen Beach):											
"Aquarius"		20						12			
"Joie de Vivre"		14									
"LCM-6"		18									
"Waft"		17									
Florida State University, Department of Oceanography (Tallahassee):											
N.A.		5									
N.A.		6									
N.A.		7									
N.A.		7									
N.A.		9									
"Tursiops"	1953	20	95					5	3		00-77
General Oceanographics Inc.:											
"Seamark"		33		177						19	00-77
Gulf Coast Research Laboratory (Ocean Springs):											
"Gulf Researcher"		20									00-77
Harbor Branch Foundation (Fort Pierce):											
"Gosnold"	1962	30		166		250	0	6	8		00-77
"Johnson"		38					C				00-77

TABLE 2G. (continued)

Institution and Vessel	Year*	Length (M)	Displacement (Tons)	Gross Tonnage	Hull† Type	Horsepower	Use‡	Scientists	Crew	Total Berth	Date of Information§
Humboldt State University (Arcata):											
"Catalyst"		31					C				00-77
Johns Hopkins University Chesapeake Bay Institute (Baltimore):											
"Maury"	1950	20						4			00-78
"Ridgely Warfield"	1967	32	162					10	8		00-78
Marine Biological Laboratory (Woods Hole):											
"A E. Verrill"		20					C				00-77
Marine Environmental Sciences Consortium (Dauphin Island):											
"G. A. Rounsefell"		20									00-77
Marine Science Consortium (Wallops Island):											
"Annandale"		27					C				00-77
"Delaware Bay"		15					C				00-77
Massachusetts Institute of Technology, Sea Grant Program (Cambridge):											
"Edgerton"		20					C				00-77
Moss Landing Marine Laboratory (Moss Landing):											
"Oconostota"		30					C				00-77
National Oceanic and Atmospheric Administration [NOAA]:											
"Albatross IV"	1961	57	1,089				B	13	19	34	00-77
"Davidson"	1966	53	995				H	1	38	40	00-77
"Delaware II"	1967	47	680			1,000	F	8	15	24	00-77
"Discover"	1964	92	3,959			5,000	O	18	87		00-77
"David Starr Jordan"	1964	52	890			918	FO	13	16	35	00-77
"Fairweather"	1967	70	1,800			2,400	H	2	74	78	00-77
"Ferrel"	1968	41	360			750		2	16	18	00-77
"George B. Kelez"	1975	50	936				OB	8	24	32	00-77
"George M. Bowers"	1955	23	125			200	F	6	4	10	00-77
"Heck"	1966	27	190			800			10	10	00-77
"John N. Cobb"	1949	28	250			345	F	4	7	13	00-77
"Kingfish II"		13			4	440	C		1	4	02-79
"McArthur"	1965	53	995			800	H	1	38	40	00-77
"Miller Freman"	1967	65	1,782			2,150	F	9	23	35	00-77
"Mt. Mitchell"	1966	70	1,800			1,200	H	1	74	78	00-77
"Murre II"	1946	26	295		3	208	FB	6	3	9	00-77
"Oceanographer"	1964	92	3,959			5,000	O	18	87		00-77
"Onslow Bay"		15			3		C		1	4	
"Oregon"	1946	30	410			600	FB	4	7	15	00-77

"Oregon II"	1966	52	906	637	1	1,600	F	13	16	32	
"Peirce"	1962	50	760			800	H	2	36	40	00-77
"Rainier"	1967	70	1,800			1,200	H	2	74	78	00-77
"Researcher"	1968	85	2,772			3,200	O	13	67	84	00-77
"Rude"	1966	27	170			400			10	15	00-77
"Surveyor"	1960	89	3,150			3,200		9	90		00-77
"Townsend Cromwell"	1963	50	652			800	F	7	14		00-77
"Whiting"	1962	50	760			800	H	2	36	40	00-77
National Science Foundation, Office of Polar Programs:											
"Hero"	1968	38	640		3	760		8	12		00-77
Newfound Harbor Marine Institute at Seacamp (Big Pine Key):											
N.A.		8				110					
N.A.		8				110					
N.A.		8				110					
N.A.		8				110					
N.A.		8				110					
N.A.		8				110					
NOAA-AOML (Miami):											
"Virginia Key"		20		90	1		O		4	8	
Nova University Ocean Sciences Center (Dania):											
"Endless Seas"		22			5	740					
"Youngster III"		19					C				00-77
Occidental College, Department of Biology (Los Angeles):											
"Vantuna"		26					C				00-77
Old Dominion University, Institute of Oceanography (Norfolk):											
"Linwood Holton"		20					C				00-77
Oregon State University, School of Oceanography (Corvallis):											
"Cayuse"	1968	24	173	165			O	8			00-78
"Wecoma"	1975	54					O	14			00-78
Raytheon Company (Portsmouth):											
"Sub Sig II"	1975	36	200				X	15	6		00-77
Rideauship Delaware Corp.:											
"Yaquina"		55	865					17	19		00-77
Rutgers University (Piscataway):											
"Rutgers"		19					C				00-77
Sea Education Association (Woods Hole):											
"Westward"		30									00-77
Southeastern Massachusetts University (N. Dartmouth):											
"Corsair"	1943	20					C	2	2		00-77
Southern Maine Vocational Technical Institute (South Portland):											
"Aqualab III"		44					C				00-77

TABLE 2G. (continued)

Institution and Vessel	Year*	Length (M)	Displacement (Tons)	Gross Tonnage	Hull† Type	Horsepower	Use‡	Scientists	Crew	Total Berth	Date of Information§
State University of New York, Marine Sciences Research Center (Stony Brook):											
"Onrust"		17					C				00-77
Texas A&M University, Department of Oceanography (College Station):											
"Diaphus"		6						1	1		
"Gyre"	1973	53	950	297	1	850	0	18	10		
Texas Parks and Wildlife Department, Coastal Fisheries Branch (Austin):											
"Western Gulf"		22			1						
Tutor Hill Laboratory (New York):											
"Erline"		32									00-77
University of Alaska, Institute of Marine Science (Fairbanks):											
"Acona"	1962	61	154	197			0	9			00-78
University of California, Scripps Institution of Oceanography (La Jolla):											
"Alpha Helix"	1966	41	512			820	B	12	12		00-78
"Ellen B. Scripps"	1965	29	234			680	0	8	5		00-78
"Thomas Washington"	1965	64	1,362			1,000	0	25	19		00-78
University of Connecticut, Marine Science Institute (Groton):											
"T-441"		20					C				00-77
University of Delaware, College of Marine Studies (Lewes):											
"Cape Henlopen"	1975	37	165				0	12	7		00-78
University of Florida Marine Laboratory (Gainesville):											
"E. Lowe Pierce"		10			4	210			1		
University of Georgia, Skidaway Institute of Oceanography (Savannah):											
"Blue Fin"	1972	22						8	2		00-78
University of Hawaii, Hawaii Institute of Geophysics (Honolulu):											
"Kana Keoki"	1967	48					0	16			00-78
"Moana Wave"	1973	53	950	300			0	13			00-78
University of Maryland, Chesapeake Biological Laboratory (Solomons):											
"Aquarius"		20					C				00-77
"Orion"		17					C				00-77
University of Miami, Rosenstiel School of Marine and Atmospheric Science (Miami):											
"Calanus"	1970	20	97	83		300	0	6	2		00-78
"Columbus Iselin"	1972	58	794	260			0	13	6		00-78
"James M. Gilliss"	1962	64	1,422	965			0	17	22		00-78
"Orca"		14						12	1	0	
University of Michigan, Great Lakes and Marine Waters Center (Ann Arbor):											
"Laurentian"	1974	24					0	10			00-78
"Mysis"	1963	15					0	3			00-78

University of Rhode Island, Graduate School of Oceanography (Narragansett):										
"Endeavor"	1976	54					16			00-78
University of Southern California (Los Angeles):										
"Melville"	1979	75	2,075			O	31			00-78
"Sea Watch"	1964	20				O	20			00-78
"Velero IV"	1948	34	580	292		O	12			00-78
University of Texas (Austin):										
"Ida Green"	1965	40				S				00-78
"Longhorn"	1971	24				O	10			00-78
University of Washington, Department of Oceanography (Seattle):										
"Hoh"	1943	20	91			O	6			00-78
"Onar"	1954	20	95			O	6			00-78
"Thomas G. Thompson"	1965	68	1,362	1,151		O	19			00-78
U.S. Coast Guard:										
"Acushnet"	1968	65	1,745		3,000	O	9	71		00-77
"Evergreen"	1973	55	984	956		O	7	69		00-77
"Polar Star"	1976	120					10			00-77
U.S. Naval Oceanographic Office (Bay St. Louis):										
"Bartlett"	1966	64	1,300		2,400	O	15	26	41	
"Bowditch"	1958	139			8,500		28	60		00-77
"Chauvenet"	1968	120	4,000		3,600	H	57	69		
"De Steiguer"	1966	64	1,300		2,400	O	15	26	41	
"Dutton"	1958	139			8,500		28	60		00-77
"Harkness"	1968	120	4,000		3,600	H	57	69		
"Hayes"	1970	75	3,320		5,400	O	30	44	74	00-77
"Kane"	1965	87	2,600		7,200	O	26	48	74	
"Lynch"	1964	64	1,300		2,400	O	15	26	41	
"Michelson"	1958	139			8,500		28	60		00-77
"Silas Bent"	1964	87	2,600		7,200	O	26	48	74	
"Wilkes"	1969	87	2,600		7,200	O	27	48	74	
"Wyman"	1969	87	2,620		3,600		28	47	75	00-77
U.S. Naval Postgraduate School, Department of Oceanography (Monterey):										
"Acania"		38				C				00-77
U.S. Naval Research Laboratory (Washington):										
"Mizar"	1964	81		3,886			19	40		00-77
Virginia Institute of Marine Science (Gloucester Pt.):										
"Langley"		24				C				00-77
"Pathfinder"		17				C				00-77
"Retriever"		35				C				00-77
"Virginian Sea"		44				C				00-77

TABLE 2G. (continued)

Institution and Vessel	Year*	Length (M)	Displacement (Tons)	Gross Tonnage	Hull† Type	Horsepower	Use	Scientists	Crew	Total Berth	Date of Information§
Woods Hole Oceanographic Institution (Woods Hole):											
"Atlantis II"	1963	64	2,300	1,529				25	31		00-78
"Knorr"	1969	75	1,915			2,500		25	25		00-78
"Oceanus"	1975	54	962	298	1	2,800		13	12		00-78
USSR											
Academy of Sciences of the USSR:											
"A. Chirikov"	1964	90	2,674				O	70			00-77
"Academician Obruchev"	1954	18	79			150	O				00-77
"Akademik Korolev"											00-77
"Akademik Kurtchatov"		124		5,460		8,000	OM	81	85		00-77
"Akademil Sergei Korolev"	1971	180	21,250	17,114							00-77
"Altair"	1967	68	1,240								00-77
"Anadir"	1967	68	1,240								00-77
"Andrey Vilkitsky"	1964	90	2,674				O	70			00-77
"Andromeda"	1968	68	1,240								00-77
"Antares"	1968	68	1,240								00-77
"Anton Ktyda"	1969	68	1,240								00-77
"Argus"											00-77
"Artika"	1969	68	1,240								00-77
"Askold"	1970	68	1,240								00-77
"Berezan"	1970	68	1,240								00-77
"Boris Davidov"	1964	90	2,674				O	70			00-77
"Cheleken"		68	1,240								00-77
"Dolinsk"	1959	137		5,419							00-77
"Ekvator"		68	1,240				F				00-77
"Elton"		68	1,240								00-77
"F. Litke"	1964	90	2,674				O	70			00-77
"Izumrud"	1970			3,862							00-77
"Kildin"		68	1,240								00-77
"Kolguev"		68	1,240								00-77
"Krilon"		68	1,240								00-77
"Leonid Sobolev"							O				00-77
"Liman"		68	1,240								00-77
"Mars"		68	1,240								00-77

"Mikhail Lomonosov"	1957	102	5,960	3,897						00-77
"Morsoveic"		68	1,240							00-77
"Nerey"	1956	36		369			O			00-77
"Nikolai Zubov"	1964	90	2,674				O	70		00-77
"Novator"	1956	36		369						00-77
"Okean"		68	1,240							00-77
"Paleh"	1966	73		2,285						00-77
"Pelorus"		68	1,240							00-77
"Petrodvorets"	1938	78		1,965						00-77
"Professor Dobrynin"	1963	25	104			150	CO			00-77
"Rybachi"		68	1,240							00-77
"S. Chelyuskin"	1964	90	2,674				O	70		00-77
"Sejmen Dezhnev"	1964	90	2,674				O	70		00-77
"Sever"		68	1,240							00-77
"T. Bellinsgausen"	1964	90	2,674				O	70		00-77
"Taymyr"		68	1,240							00-77
"V. Golovnin"	1964	90	2,674				O	70		00-77
"Vega"		68	1,240							00-77
"Vityaz"	1948	109	5,710			3,000	O	73	65	00-77
"Vladimir Komarov"	1967	159	17,500				A			00-77
"Vladimir Obruchev"	1959	42		534						00-77
"Yug"	1957	39		348						00-77
"Yuri Gagarin"	1971	235	45,000	32,291			A			00-77
"Zapolara"		68	1,240							00-77
"Zarnitsa"	1957	39		348			H			00-77
"Zarya"	1975		580	333	3		O			00-77
"Zenith"							H			00-77
"Zvezda"	1957	39		348						00-77
Arctic and Antarctic Research Institute (Leningrad):										
"Baikal"	1964	111	6,900			3,400	O			00-77
"Balkhash"	1964	111	6,900			3,400	O			00-77
"Mikhail Somov"	1974		1,400							00-77
"Passat"										00-77
"Polyus"	1964	111	6,900			3,400	O			00-77
Arctic Marine Biology Institute:										
"Diana"										00-77
Far Eastern Scientific Center:										
"Kallisto"							G			00-77
"Mikhail Kalinin"										00-77
"Valerian Uryvayev"		55	1,130		1	875		15	40	00-77

TABLE 2G. (continued)

Institution and Vessel	Year*	Length (M)	Displacement (Tons)	Gross Tonnage	Hull† Type	Horsepower	Use‡	Scientists	Crew	Total Berth	Date of Information§
Maritime Hydrometeorological Service Administration:											
"Akademik Shirshov"	1968						M				00-77
Maritime Province Geological Administration:											
"Il'menit"	1974						G				00-77
Ministry of Fisheries:											
"A.I. Voyeyvkov"	1960	85	3,500	3,200			F	54	54		00-77
"Poisk"							F				00-77
"Yu.M. Shokalsky"	1959	85	3,500	3,200			F	54	54		00-77
"Zund"	1973	80		3,400			F	23		88	00-77
Moscow State University (Moscow):											
"Moskovsky Universitet"	1971							23			00-77
N.A.:											
"Nevelskoy"	1961	89	2,600								00-75
"Vladimir Vorobev"	1959	51	737						28		00-75
Navy:											
"A. Smirnov"		67	1,800			2,000					00-77
"Akademik Arkhangelsky"	1963	40		416			G				00-77
"Akademik Kovalevsky"	1949	38		284							00-77
"Akademik Vavilov"	1949	36		255							00-77
"Azimut"		55	800	1,276			H				00-77
"Deviator"		55	800	1,276			H				00-77
"Dmitri Laptev"		67	1,800			2,000					00-77
"Dmitri Ovstyn"		67	1,800			2,000					00-77
"Dmitri Sterlegov"		67	1,800			2,000					00-77
"E. Toll"		67	1,800			2,000					00-77
"Gidrolog"		55	800	1,276			H				00-77
"Gidrometr"		55	800	1,276			H				00-77
"Globus"		55	800	1,276			H				00-77
"Glubomer"		55	800	1,276			H				00-77
"Gorizont"		55	800	1,276			H				00-77
"Gradus"		55	800	1,276			H				00-77
"Kompas"		55	800	1,276			H				00-77
"MGLA"		39		299							00-77

"N. Kolomeytsev"		67	1,800			2,000					00-77
"N. Yevgenov"	1974	67	1,800			2,000					00-77
"Pamyat Merkuryia"		55	800	1,276			H				00-77
"Pielorus"	1962	55	800	1,276			H				00-77
"Rumb"		55	800	1,276			H				00-77
"S. Krakov"	1974	67	1,800			2,000					00-77
"Stefan Malygin"		67	1,800			2,000	H				00-77
"Tropik"		55	800	1,276			H				00-77
"V. Sukhotsky"	1974	67	1,800			2,000					00-77
"Vagach"		55	800	1,276			H				00-77
"Valerian Albanov"		67	1,800			2,000					00-77
"Vostok"		55	800	1,276			H				00-77
"Yug"		55	800	1,276			H				00-77
"Zenit"		55	800	1,276			H				00-77
P. N. Lebedev Institute:											
"Priboi"							M				00-77
"Professor Vize"							M				00-77
"Professor Zubov"							M				00-77
"Sergei Vavilov"	1954			3,561							00-77
Pacific Research Institute:											
"Akademik Berg"		85		3,165			F				00-77
"Akademik Knipovich"		85		3,165			F				00-77
"Poseidon"		85		3,165			F				00-77
"Professor Deryugin"	1968	85		3,165			F				00-77
Pacific Research Institute of Fisheries and Oceanography (Vladivostok):											
"Lakhtak"							F	23			00-77
Soviet Antarctic Expedition:											
"Ob"								23			00-77
Ukrainian Academy of Science (Sevastopol):											
"Academik Vernadsky"	1969	124									00-77
"Dmitry Mendeleyev"							O				00-77
Venezuela											
Dirección General de Desarrolla Pesquero (Caracas):											
"Golfo de Cariaco"	1952	20		74	1	250			8	13	
"Golfo de Paria"											
"Nemoto Maru"	1973	11		26	3	55			3	6	
Estación de Investigaciones Marinas de Margarita (Porlamar):											
"Dona Menca"		21									
"Dona Teresa"		14									
"La Salle"	1965	39		540						24	
"Mari Nieves"		14									

TABLE 2G. (continued)

Institution and Vessel	Year*	Length (M)	Displacement (Tons)	Gross Tonnage	Hull† Type	Horsepower	Use‡	Scientists	Crew	Total Berth	Date of Information§
Universidad del Zulia, Centro de Investigaciones Biológicas (Maracaibo):											
N.A.		7		3							
Universidad de Oriente, Instituto Oceanográfico (Cumano):											
"Dios Te Salve"		11									
"Guaiqueri II"		25									
"Yazmar"		12									
Yugoslavia											
Institute for Oceanography and Fisheries (Split):											
"Bios"		28		162	3	300		10	7		09-77
"Gira"		6									09-77
"Predvodnik"		19		54	3	220		5	5		09-77
Rudjer Boskovic Institute, Center for Marine Research (Zagreb):											
(Due to be available)		10				55	C			2	05-77
"Vila Velebita"		26		200	3	180		8	6		05-77
University of Ljubljana, Institute of Biology (Ljubljana):											
"PI-57"		10				32	C				06-77
Yugoslav Academy of Science and Arts, Biological Institute (Dubrovnik):											
"Baldo Kosic"		14				50				4	06-75

SOURCES.--Intergovernmental Oceanographic Commission, Unesco (unpublished marine research center reports); Jane's Ocean Technology, 1976-77; National Institute of Oceanography, India, and United Nations Environment Programme, Directory of Indian Ocean Marine Research Centres (1978); United Nations Environment Programme, Directory of Mediterranean Marine Research Centres (1977); The University-National Oceanographic Laboratory System (Woods Hole, Mass.), "List of Research Vessels Operated by UNOLS Member Institutions" (rev. 1/78) and "Other University Ships"; and USSR Academy of Sciences, Geological-Geophysical Atlas of the Indian Ocean (Moscow, 1975).

NOTE.--Vessels with a reported length of less than 5 meters have been omitted. N.A. = not available.

*Converted or built.

†Hull types: 1 = steel; 2 = ferro-cement; 3 = wood; 4 = fiberglass; 5 = aluminum.

‡Use: A = satellite/space program; B = biology; C = coastal work; F = fisheries research; G = geology; H = hydrology; L = limnology; M = meteorology; O = oceanography; S = seismology; and X = multipurpose.

§Due to the variation among sources, the date of publication of the printed source or the date the report was received by the IOC (if indicated) has been included for each vessel.

TABLE 3G.--SUBMERSIBLE RESEARCH AND SURVEY VESSELS, BY COUNTRY

Country and Operator	Vessel	Length (M)	Weight (Metric Tons)	Displacement (Long Tons)	Crew	Depth Capability (M)
Canada:						
Arctic Marine Ltd.	"Sea Otter"	4.1	2.8		3	460
Environment Canada	"Pisces IV"	6.0	10		3	2,000
Horton Maritime Explorations Ltd.	"Auguste Piccard"	28.5	167		6	700
	"Ben Franklin"	14.6	143		6	606
International Hydrodynamics Co.	"Pisces V"	5.8	10.7		3	2,000
	"Pisces IX"	5.8	10.7		3	2,000
Master Divers Ltd.	"Kittredge K-250"	3.2	1.0		1	76
Royal Navy	"SDL-1"	7.6	11.4		6	610
Colombia:						
Friendship S.A.	"DOWB"	5.5	9.1		2	
France:						
CNEXO	"Engin 3000 Cyana"	5.7	8.5		3	3,000
Compagnie Maritime d'Expertises	"Globule"				2	
	"Marco"					
	"Perry PC-12034"					
	"Deepstar 4000"	5.5	9		3	1,220
	"Argyronete"	27.8		255	10	600
	"Moana III"		9			450
	"Moana IV"					
	"Moana V"					
Groupe d'Intervention Sous la Mer (GISMER)	"Perry PC-4B Shelf-Diver"				4	
	"Archimède"	22.1	203.4		3	11,000
Intersub Ltd.	"Perry PC-8B"	5.6	5.5		2	244
	"Perry PC-1201"	6.1	7		2	305
	"Perry PC-1202"	9.6	14		5	305
Germany, Fed.:						
Bruker-Physik AG	"Mermaid I"					
	"Mermaid II"	5.2	6.8		2	300
	"Mermaid III"	7.2	11.5		4	250

TABLE 3G. (continued)

Country and Operator	Vessel	Length (M)	Weight (Metric Tons)	Displacement (Long Tons)	Crew	Depth Capability (M)
Italy:						
Sarda Estracione Lavovazione Prodotti Marini S.p.A.	"Antonio Magliúlo"	7.3	14.5			300
Sub Sea Oil Services	"Perry PC-5C"	5.6	5.5		2	365
	"Phoenix 66"					365
Japan:						
Maritime Safety Agency	"HU06 Shinkai"	16.5		91	4	600
Ocean Systems Japan Ltd.	"Hakuyo"	6.4		7	3	300
Norway:						
Fred Olsen Oceanics A/S	N.A.					500
	N.A.					1,000
Poland:						
Geological Inst., Warsaw	"Delphin II"		1.4		2	200
Taiwan:						
Kuo Feng Oceanic Development Co.	"Argus I Tours 64C"	6.9	14.0			300
U.K.:						
P & O Subsea Ltd.	"Perry PC-9"	7.9	10.2		2	412
	"Hyco Aquarius I"	4.2		5.0	2	335
Vickers Oceanics Ltd.	"Hyco Pisces I"	5.2	8		3	347
	"Hyco Pisces II"	6.2	11.9		3	732
	"Hyco Pisces III"	6.2	11.8		3	916
	"Hyco Pisces X"	6.1	10.6		3	750
	"Perry PC-15"	9.8	15		5	366
	"Perry PC-15D"	6.5	12.2		2	365
U.S.A.:						
Brown & Root Inc.	"Kittredge K-250"	3.2	1.0		1	76
	"Perry PC-9C"					
General Oceanographics	"Nekton Alpha"	4.6	2.1		2	300
	"Nekton Beta"	4.6	2.1		2	300
	"Nekton Gamma"	4.6	2.1		2	300
Harbor Branch Foundation Inc.	"Johnson Sea Link I"	7	9.6		4	330
	"Johnson Sea Link II"	7	9.6		4	660
International Underwater Contractors Inc.	"Perry PC-3B Techdiver"	6.7	3		2	183

	"Mermaid II"	5.2	6.3		2	305
	"Beaver Mk IV"	7.3	17		5	823
Lockheed Missiles & Space Co.	"Deep Quest"	12.2		50	5	2,520
Maui Divers of Hawaii	"Star II"	5.2	4.6		2	366
Naval Undersea Ctr.	"Makakai"	5.6	4.6	4.9	2	182
Ocean Systems Inc.	"Opsub"	5.4	4.7		2	610
Reynolds International Inc.	"Aluminaut"	15.5	81		3	4,550
Scripps Institution of Oceanography	"Star III"	5.2	4.5			600
Southwest Res. Inst.	"Nemo"	2.3	3.8		2	182
	"Deepview"	5.0	5.4		2	183
Texas A&M Univ.	"Perry PC-14C1 Diaphus"	6.0	4.6		2	365
U.S. Naval Weapons Ctr.	"Hikino"					10
U.S. Navy Submarine Development Group One	"NR-1"	41.6		400	7	
	"Trieste II"	23.9	84	303	3	6,060
	"Turtle"	7.9	26.5		3	1,980
	"Sea Cliff"	7.9	26.5		3	1,980
	"DSRV-1"	15		35	3	1,500
	"DSRV-2"	15		35	3	1,500
Vast Inc.	"Kittredge K-250"	3.2	1.0		1	76
Westinghouse Electric Corp.	"Deepstar 2000"	6.1	8.0		3	610
Woods Hole Oceanographic Inst.	"Alvin"	7.0		15	3	3,506
USSR:						
All-Union Scientific Res. Inst. of Sea Fishing & Oceanography	"Gvidon"			4.3	3	250
Academy of Sciences	"Hyco Pisces VI"	5.8	10.7		3	1,080
	"Hyco Pisces VII"	5.8	10.7		3	1,080
State Design Inst. of the Soviet Fishing Fleet	"Sever-2"	12		40	5	2,000
	"Thetis 2"					
Inst. for Res. for Marine Fishing and Oceanology	"OSA-3"			12	3	600
Pacific Fish Lab.	"Tinro 1"	11.0	40			300
	"Tinro 2"	7	10.5		2	405

SOURCE.--*Jane's Ocean Technology, 1976-77*.

NOTE.--CNEXO = Centre National pour l'Exploitation des Oceans; Ctr. = center; Inst. = institute; Res. = research; Univ. = university.

Appendix H

Military Activities

These seven tables detailing the configuration of the world's naval forces were prepared by Andrzej Karkoszka of SIPRI. Besides updating material found in *Ocean Yearbook 1* they complement Dr. Karkoszka's piece, "Naval Forces" (pp. 199–225), as well as the other essays on military activities.

THE EDITORS

TABLE 1H.--WORLD STOCK OF AIRCRAFT CARRIERS, BY COUNTRY

	1950	1955	1960	1965	1970	1976	1978
World total:							
Attack	44	43	30	29	25	21	21
Other	75	83	42	43	30	18	20
U.S.A.:							
Attack	27	19	14	16	15	15	15
Other	75	83	40	39	24	13	14
USSR							
Other						1	2
U.K.:							
Attack	12	16	8	5	4	1	1
Other			1	2	2	2	2
France:							
Attack	2	4	3	3	2	2	2
Other					1	1	
Australia:							
Attack	1	2	2	1	1	1	1
Other				1	1		
Canada:							
Attack	1	1	1	1			
Netherlands:							
Attack	1	1	1	1			
Spain:							
Other					1	1	1
Argentina:							
Attack			1	1	2	1	1
Brazil:							
Other			1	1	1	1	1
India:							
Attack				1	1	1	1

SOURCES.--J. L. Couhat, ed., Combat Fleets of the World 1976/77: Their Ships, Aircraft, and Armament (London: Arms & Armour, 1976); J. E. Moore, ed., Jane's Fighting Ships 1978-79 (London: Macdonald & Jane's, 1978); NATO's Fifteen Nations, Special Edition 1978, vol. 23; and G. Albrecht, comp., Weyers Flotten Taschenbuch 1977/78, Warships of the World (Munich: Bernard & Graefe Verlag, 1978).

NOTE.--Other = antisubmarine, amphibious helicopter assault, and utility.

TABLE 2H.--WORLD STOCK OF STRATEGIC SUBMARINES,* BY GROUPS OF COUNTRIES

	1955	1960	1965	1970	1976	1978
World total:						
Nucl.		7	42	70	104	121
Conv.		10	32	26	24	23
U.S.A.:						
Nucl.		3	33	41	41	41
Conv.						
Other NATO:						
Nucl.				5	8	8
Conv.				1	1	1
Total NATO:						
Nucl.		3	33	46	49	49
Conv.				1	1	1
USSR:						
Nucl.		4	9	24	55	72†
Conv.		10	32	25	23	22
Other WTO:						
Nucl.						
Conv.						
Total WTO:						
Nucl.		4	9	24	55	72†
Conv.		10	32	25	23	22

SOURCE.--See table 1H.

NOTE.--Nucl. = nuclear powered; Conv. = conventionally powered; WTO = Warsaw Treaty Organization.

*Equipped with medium- or long-range ballistic missiles.

†Two of these submarines are not operational, and 8 are of old design thus do not count under SALT I provisions.

TABLE 3H.--WORLD STOCK OF NUCLEAR-POWERED ATTACK SUBMARINES, BY COUNTRY

	1950	1955	1960	1965	1970	1976	1978
World total		1	15	48	108	157	167
U.S.A.		1	11	22	46	68	70
U.K.					4	9	10
USSR			4	26	58	80	87

SOURCE.--See table 1H.

TABLE 4H.--WORLD STOCK OF PATROL SUBMARINES,* BY GROUPS OF COUNTRIES

	1950	1955	1960	1965	1970	1976	1978
World total:							
Nucl.		1	15	48	108	157	167
Conv.	355	532	535	576	535	498	500
Developed:							
Nucl.		1	15	48	108	157	167
Conv.	351	527	502	516	459	365	346
U.S.A.:							
Nucl.		1	11	22	46	68	70
Conv.	194	190	158	139	52	11	9
Other NATO:							
Nucl.					4	9	10
Conv.	105	109	89	78	75	74	97
Total NATO:							
Nucl.		1	11	22	50	77	80
Conv.	299	299	247	217	127	85	106
USSR:							
Nucl.			4	26	58	80	87
Conv.	46	215	238	274	283	215	176
Other WTO:							
Nucl.							
Conv.					7	8	6
Total WTO:							
Nucl.			4	26	58	80	87
Conv.	46	215	238	274	290	223	182
Other Europe:							
Nucl.							
Conv.	3	9	12	18	27	34	32
Other developed:							
Nucl.							
Conv.	3	4	5	7	15	23	26
Total developing countries	4	5	33	60	76	133	154
Middle East			10	12	16	16	17
South Asia				1	8	11	12
East and Southeast Asia (excl. China)			2	12	14	18	18
Sub-Saharan Africa							
North Africa							2
Central America							
South America	4	5	9	13	11	28	29
China			12	22	27	60	76

SOURCE.--See table 1H.

NOTE.--Nucl. = nuclear powered; Conv. = conventionally powered; WTO = Warsaw Treaty Organization; Other Europe = Albania, Finland, Spain, Sweden, and Yugoslavia; Other developed = Australia, Japan, New Zealand, South Africa, and Taiwan; Middle East = Cyprus, Egypt, Iran, Iraq, Israel, Kuwait, Lebanon, Saudi Arabia, and Syria; South Asia = Bangladesh, India, Pakistan, and Sri Lanka; E. & S.E. Asia (excl. China) = Burma, Cambodia, Indonesia, Malaysia, Singapore, North Korea, South Korea, Thailand, the Philippines, and Vietnam; Sub-Saharan Africa = Cameroon, Ethiopia, Gabon, Ghana, Guinea, Ivory Coast, Kenya, Liberia, Madagascar, Mauritania, Nigeria, Senegal, Sudan, and Tanzania; North Africa = Algeria, Libya, Morocco, and Tunisia; Central America = Cuba, Dominican Republic, and Mexico; and South America = Argentina, Brazil, Chile, Colombia, Ecuador, Peru, Uruguay, and Venezuela.

*Post-World War II submarines displacing 700 tons or more.

†All conventionally powered.

TABLE 5H.--WORLD STOCK OF COASTAL SUBMARINES,* BY GROUPS OF COUNTRIES

	1950	1955	1960	1965	1970	1976	1978
World total	313	317	179	93	72	73	79
Developed	299	295	162	85	64	73	69
Developing	14	22	17	8	8		10
U.S.A.							
Other NATO		3	20	33	34	53	46
Total NATO		3	20	33	34	53	46
USSR	273	269	127	40	22	17	20
Other WTO							
Total WTO	273	269	127	40	22	17	20
Other Europe	26	23	15	12	8	3	
Other developed							3

SOURCE.--See table 1H.
NOTE.--WTO = Warsaw Treaty Organization.
*Submarines displacing less than 700 tons; all conventionally powered.

TABLE 6H.--WORLD STOCK OF MAJOR SURFACE WARSHIPS,* BY GROUPS OF COUNTRIES

	1950	1955	1960	1965	1970	1976	1978
World total:							
Miss.		2	18	111	191	383	512
Conv.	1,783	2,042	1,789	1,650	1,414	823	553
Developed:							
Miss.		2	18	111	189	347	445
Conv.	1,600	1,811	1,520	1,374	1,128	555	373
U.S.A.:							
Miss.		2	15	58	77	122	141
Conv.	817	835	704	654	478	84	81
Other NATO:							
Miss.				19	57	113	154
Conv.	520	582	402	330	294	185	109
Total NATO:							
Miss.		2	15	77	134	235	295
Conv.	1,337	1,417	1,106	984	772	269	190
USSR:							
Miss.			1	23	39	87	94
Conv.	150	256	260	228	206	187	97
Other WTO:							
Miss.					1	1	1
Conv.	5	7	14	14	9	4	3
Total WTO:							
Miss.			1	23	40	88	95
Conv.	155	263	274	242	215	191	100
Other Europe:							
Miss.			2	6	6	12	15
Conv.	67	70	75	69	71	34	27
Other developed:							
Miss.				5	9	12	40
Conv.	41	61	63	73	64	61	58

TABLE 6H. (continued)

	1950	1955	1960	1965	1970	1976	1978
Total developing countries:							
Miss.					2	36	67
Conv.	183	231	269	276	286	268	180
Middle East:							
Miss.					1	7	7
Conv.	18	20	15	18	21	14	10
South Asia:							
Miss.						4	7
Conv.	16	18	26	27	32	34	31
East and Southeast Asia (excl. China):							
Miss.					1	5	2†
Conv.	50	53	71	76	109	92	50‡
Sub-Saharan Africa:							
Miss.							
Conv.				1	1	2	2
North Africa:							
Miss.						1	1
Conv.				2	2	1	1
Central America:							
Miss.							
Conv.	30	30	29	29	22	34	29
South America:							
Miss.						9	32
Conv.	65	94	105	91	78	75	52
China:							
Miss.						13	18
Conv.	4	16	23	22	21	13	5

SOURCE.--See table 1H.

NOTE.--Miss. = missile armed (both SSMs and SAMs); Conv. = conventionally armed; WTO = Warsaw Treaty Organization. See table 4H for definitions of groups of countries.

*Cruisers, destroyers, frigates, and escorts (over 1,000 tons displacement).

†Taiwan's 22 missile-armed warships counted under "Other developed."

‡Taiwan's 11 conventionally armed warships counted under "Other developed."

TABLE 7H.--WORLD STOCK OF LIGHT NAVAL FORCES,* BY GROUPS OF COUNTRIES

	1950	1955	1960	1965	1970	1976	1978
World total:							
Miss.			5	141	281	529	691
Conv.	987	1,380	1,849	2,092	2,457	2,729	3,410
Developed:							
Miss.			5	112	188	244	299
Conv.	822	1,142	1,422	1,340	1,290	1,082	1,167
U.S.A.:							
Miss.					1	1	1
Conv.	147	120	35	18	35	24	26
Other NATO:							
Miss.					1	74	92
Conv.	190	267	230	233	241	198	231
Total NATO:							
Miss.					2	75	93
Conv.	337	387	265	251	276	222	257
USSR:							
Miss.			5	110	150	120	140
Conv.	395	516	769	653	600	424	512
Other WTO:							
Miss.				2	28	32	37
Conv.	16	54	141	191	164	188	145
Total WTO:							
Miss.			5	112	178	152	177
Conv.	411	570	910	844	764	612	657
Other Europe:							
Miss.					8	16	26
Conv.	60	117	180	194	196	198	184
Other developed:							
Miss.							3
Conv.	14	68	67	51	54	50	69
Total developing countries:							
Miss.				29	93	285	392
Conv.	156	238	427	752	1,141	1,647	2,243
Middle East:							
Miss.				3	32	58	62
Conv.	11	17	77	86	140	133	144
South Asia:							
Miss.						8	20
Conv.			1	9	7	34	60
E. & S.E. Asia (excl. China):							
Miss.				12	12	52	63
Conv.	55	120	149	260	403	423	726
Sub-Saharan Africa:							
Miss.						5	30
Conv.			5	20	54	106	179
North Africa:							
Miss.					16	13	25
Conv.			2	2	45	48	73
Central America:							
Miss.				12	18	23	26
Conv.	16	16	18	49	72	106	166
South America:							
Miss.						6	6
Conv.	42	32	26	47	38	87	125

TABLE 7H. (continued)

	1950	1955	1960	1965	1970	1976	1978
China:							
Miss.				2	15	120	160
Conv.			150	279	408	710	770

SOURCE.--See table 1H.

NOTE.--Miss. = missile armed; Conv. = conventionally armed; WTO = Warsaw Treaty Organization. See table 4H for definitions of groups of countries.

*This category includes corvettes, fast patrol boats, torpedo boats, and large and coastal patrol crafts (gun armed). Riverine craft are excluded. Corvettes are not included for the years 1950-76.

Appendix I

Tables, General Information

These tables have been added to the *Ocean Yearbook* in order to provide general cross-national data and ocean-related information which, while of general interest, does not fall within the domain of the other tabular appendices. In future editions, marine jurisdictional claims will be updated and the other tables will be expanded.

THE EDITORS

TABLE 11.--COUNTRY SUMMARIES

Country or Territory	Region*	Area (km^2)	Population: 1977 (Thousands)	Population: Annual Growth 1970-77 (%)	Gross National Product: 1977 (Millions $)	GNP Per Capita: 1977 ($)	GNP Per Capita: Annual Growth 1970-77 (%)
Afghanistan	7	647,500	14,304	2.2	3,150	220	2.7
Albania	4	28,749	2,545	2.5	1,670	660	4.1
Algeria	5	2,460,500	17,152	3.2	19,570	1,140	2.1
American Samoa	0	197	31	2.0	210	6,830	8.4
Angola	6	1,245,790	6,575	2.3	1,840	280	-3.4
Antigua	2	280	73	1.3	60	860	-3.7
Argentina	2	2,771,300	26,036	1.3	48,710	1,870	1.8
Australia	0	7,692,300	14,074	1.7	102,570	7,290	1.6
Austria	3	83,916	7,506	.2	48,390	6,450	3.8
Bahamas	2	11,396	213	2.7	520	2,450	-7.2
Bahrain	5	596	343	7.1	1,390	4,050	.2
Bangladesh	7	142,500	81,219	2.5	6,520	80	-.2
Barbados	2	430	248	.5	440	1,770	2.6
Belgium	3	30,562	9,845	.3	81,550	8,280	3.5
Belize	2	22,973	130	.9	100	750	4.7
Benin	6	115,773	3,229	2.9	670	210	.5
Bermuda	1	54	54	.5	460	8,520	2.4
Bhutan	7	46,600	1,231	2.3	110	90	-.3
Bolivia	2	1,098,160	5,154	2.7	2,460	480	2.9
Botswana	6	569,800	728	1.9	390	540	16.1
Brazil	2	8,521,100	116,100	2.9	163,880	1,410	6.7
British Virgin Is.	2	130	N.A.	N.A.	N.A.	N.A.	N.A.
Brunei	9	5,776	165	3.4	1,520	9,190	7.7
Bulgaria	4	111,852	8,835	.6	25,000	2,830	5.7
Burma	9	678,600	31,512	2.2	4,330	140	1.3
Burundi	6	28,490	4,156	1.9	550	130	.6
Cambodia	9	181,300	N.A.	N.A.	N.A.	N.A.	N.A.
Cameroon	6	47,400	7,882	2.2	3,280	420	1.0
Canada	1	9,971,500	23,320	1.2	194,660	8,350	3.4
Cape Verde	6	4,040	313	2.1	50	150	-2.1
Cayman Is.	2	260	N.A.	N.A.	N.A.	N.A.	N.A.
Central African Rep.	6	626,780	1,867	2.2	440	240	.9
Chad	6	1,284,640	4,221	2.2	560	130	-1.0
Chile	2	740,740	10,553	1.7	13,160	1,250	-1.8

China	8	9,600,000	902,337	1.6	372,800	410	4.5
Colombia	2	1,139,600	24,605	2.1	18,760	760	3.8
Comoros	6	2,170	370	3.8	70	180	-5.2
Congo	6	349,650	1,423	2.5	710	500	.8
Cook Is.	0	240	18	N.A.	N.A.	N.A.	N.A.
Costa Rica	2	51,000	2,061	2.5	2,870	1,390	3.2
Cuba	2	114,478	9,590	1.6	7,220	750	-1.2
Cyprus	5	9,251	644	.7	1,180	1,830	.3
Czechoslovakia	4	127,946	15,013	.7	63,640	4,240	4.3
Denmark	3	42,994	5,076	.4	46,520	9,160	2.3
Djibouti	6	23,310	300	3.5	130	450	-.3
Dominica	2	790	77	1.2	30	400	-4.1
Dominican Rep.	2	48,692	4,980	3.0	4,190	840	4.6
Ecuador	2	274,540	7,324	3.0	6,000	820	6.1
Egypt	5	1,000,258	37,796	2.1	12,950	340	5.2
El Salvador	2	21,400	4,256	3.1	2,510	590	2.1
Equatorial Guinea	6	27,972	338	2.2	N.A.	N.A.	N.A.
Ethiopia	6	1,178,450	30,245	2.5	3,280	110	.2
Faeroe Is.	3	1,340	43	N.A.	N.A.	N.A.	N.A.
Falkland Is.	2	12,168	2	N.A.	N.A.	N.A.	N.A.
Fiji	0	18,272	589	1.8	780	1,330	3.7
Finland	3	336,700	4,732	.4	29,300	6,190	2.8
France	3	551,670	53,051	.7	397,670	7,500	3.1
French Guiana	2	90,909	60	2.9	120	1,980	1.2
French Polynesia	0	4,000	145	3.3	690	4,770	1.8
Gabon	6	264,180	533	.9	1,700	3,190	6.5
Gambia	6	10,360	554	3.1	120	210	5.3
German D.R.	4	108,262	16,857	-.2	85,400	5,070	4.9
Germany, F.R.	3	248,640	61,418	.2	529,380	8,620	2.2
Ghana	6	238,280	19,634	3.0	3,940	370	-2.0
Gibraltar	3	6	29	N.A.	N.A.	N.A.	N.A.
Greece	3	132,608	9,231	.7	27,200	2,950	4.0
Greenland	1	2,175,600	50	1.0	310	6,280	6.3
Grenada	2	344	105	1.8	50	470	-3.2
Guadeloupe	2	1,779	319	.2	820	2,560	2.9
Guam	0	450	94	1.2	590	6,270	5.1
Guatemala	2	108,880	6,436	2.9	5,350	830	3.3
Guinea	6	246,050	4,989	3.0	1,000	200	2.5
Guinea-Bissau	6	36,260	747	2.0	130	180	-7.7
Guyana	2	214,970	817	2.0	430	520	.4
Haiti	2	27,713	4,749	1.7	1,090	230	2.1

TABLE 11. (continued)

Country or Territory	Region*	Area (km^2)	Population		Gross National Product		
						Per Capita	
			1977 (Thousands)	Annual Growth 1970-77 (%)	1977 (Millions $)	1977 ($)	Annual Growth 1970-77 (%)
Honduras	2	112,150	3,322	3.3	1,410	420	.0
Hong Kong	8	1,036	4,536	2.0	11,890	2,620	5.8
Hungary	4	92,981	10,628	.4	32,940	3,100	5.1
Iceland	3	102,952	223	1.3	1,710	7,690	3.0
India	7	3,136,500	631,726	2.1	100,180	160	1.1
Indonesia	9	1,906,240	133,505	1.8	42,680	320	5.7
Iran	5	1,647,240	34,782	3.0	N.A.	N.A.	N.A.
Iraq	5	445,480	11,803	3.4	18,490	1,570	7.1
Ireland	3	68,894	3,198	1.2	9,770	3,036	2.1
Israel	5	20,720	3,604	2.8	13,570	3,760	2.0
Italy	3	301,217	56,468	.7	199,270	3,530	2.0
Ivory Coast	6	323,750	7,463	6.0	5,710	770	1.1
Jamaica	2	11,422	2,101	1.7	2,230	1,060	-2.0
Japan	8	370,370	113,216	1.2	737,180	6,510	3.6
Jordan	5	96,089	2,888	3.3	1,960	2,270	6.5
Kenya	6	582,750	14,614	3.8	4,300	290	.9
Kiribati	0	684	55	1.6	40	700	6.0
Korea, D.P.R.	8	121,730	16,651	2.6	11,380	680	5.3
Korea, Rep.	8	98,400	35,953	2.0	35,150	980	7.6
Kuwait	5	16,058	1,137	6.2	14,420	12,690	-.9
Lao P.D.Rep.	9	236,804	3,200	N.A.	290	90	N.A.
Lebanon	5	10,360	2,939	2.5	N.A.	N.A.	N.A.
Lesotho	6	30,303	1,250	2.4	320	250	9.9
Liberia	6	111,370	1,684	3.4	740	410	1.1
Libya	5	1,758,610	2,636	4.1	17,189	6,520	-4.5
Liechtenstein	3	168	24	N.A.	N.A.	N.A.	N.A.
Luxembourg	3	2,590	357	.8	3,440	9,640	4.4
Macau	8	16	291	2.3	370	1,270	17.1
Madagascar	6	595,700	8,085	2.5	1,870	230	-2.7
Malawi	6	95,053	5,597	3.1	860	150	3.1
Malaysia	9	332,556	12,961	2.7	12,600	970	4.9
Maldives	7	298	140	4.0	20	140	N.A.
Mali	6	1,204,350	6,129	2.5	720	120	1.9
Malta	3	313	333	.3	620	1,870	11.9

Martinique	2	1,100	322	.1	1,120	3,470	5.7
Mauritania	6	1,085,210	1,503	2.7	410	270	-.1
Mauritius	6	1,856	906	1.3	670	740	6.9
Mexico	2	1,978,800	63,319	3.3	73,720	1,160	1.2
Monaco	3	2	25	N.A.	N.A.	N.A.	N.A.
Mongolia	8	1,564,619	1,530	3.0	1,330	870	1.6
Morocco	5	409,200	18,310	2.7	11,140	610	4.2
Mozambique	6	786,762	9,691	2.5	1,320	140	-4.3
Namibia	6	823,620	926	2.8	960	1,030	.8
Nauru	0	21	7	N.A.	N.A.	N.A.	N.A.
Nepal	7	141,400	13,322	2.2	1,450	110	2.4
Netherlands	3	33,929	13,864	.9	106,930	7,710	2.2
Netherlands Antilles	2	1,020	244	1.3	680	2,780	.5
New Caledonia	0	22,015	145	3.3	650	4,470	-5.9
New Hebrides	0	14,763	101	2.7	50	490	1.8
New Zealand	0	268,276	3,148	1.7	14,110	4,480	.9
Nicaragua	2	147,900	2,411	3.3	2,090	870	2.5
Niger	6	1,266,510	4,862	2.8	950	190	-1.8
Nigeria	6	924,630	78,982	2.6	40,540	510	4.4
Norway	3	323,750	4,034	.6	34,560	8,570	3.9
Oman	5	212,380	814	3.2	2,050	2,510	4.0
Pacific Is., Trust Territory	0	1,813	129	3.4	140	1,120	1.4
Pakistan	7	803,000	74,905	3.1	15,070	200	.8
Panama	2	75,650	1,771	3.1	2,120	1,200	-.1
Papua New Guinea	0	475,369	2,857	2.4	1,460	510	2.5
Paraguay	2	406,630	2,810	2.9	2,100	750	4.3
Peru	2	1,284,640	16,363	2.8	11,800	720	1.8
Philippines	9	300,440	44,473	2.7	20,410	460	3.7
Poland	4	312,354	34,724	1.0	114,280	3,290	6.3
Portugal	3	94,276	9,577	.8	17,580	1,840	3.1
Puerto Rico	2	8,891	3,303	2.8	8,090	2,450	.1
Qatar	5	10,360	215	10.3	2,440	11,370	-2.4
Reunion	6	2,512	499	1.8	1,450	2,900	-1.0
Rhodesia	6	391,090	6,683	3.3	3,070	460	-.1
Romania	4	237,503	21,648	.9	33,030	1,530	9.9
Rwanda	6	25,900	4,379	2.9	710	160	1.3
Samoa	0	2,849	154	1.2	N.A.	N.A.	N.A.
Sao Tome and Principe	6	964	82	1.3	40	460	8.0
Saudi Arabia	5	2,331,000	7,633	3.0	55,210	7,230	13.0
Senegal	6	196,840	5,240	2.6	1,980	380	.4

TABLE 11. (continued)

Country or Territory	Region*	Area (km^2)	Population: 1977 (Thousands)	Population: Annual Growth 1970-77 (%)	Gross National Product: 1977 (Millions $)	GNP Per Capita: 1977 ($)	GNP Per Capita: Annual Growth 1970-77 (%)
Seychelles	6	404	62	2.6	60	940	3.8
Sierra Leone	6	72,261	3,210	2.5	640	200	-1.3
Singapore	9	583	2,319	1.6	6,540	2,820	6.6
Solomon Is.	0	29,785	205	3.5	80	390	1.7
Somalia	6	637,140	3,660	2.3	430	120	-1.1
South Africa	6	1,222,480	26,952	2.7	37,640	1,400	1.1
Spain	3	505,050	36,298	1.0	118,170	3,260	3.6
Sri Lanka	7	65,500	14,097	1.7	2,290	160	1.3
St. Kitts-Nevis-Anguilla	2	389	50	.9	30	610	1.6
St. Lucia	2	616	118	2.3	70	560	.7
St. Vincent	2	389	103	2.3	30	320	-2.2
Sudan	6	2,504,530	16,919	2.6	5,650	330	2.5
Surinam	2	142,709	381	-1.1	710	1,870	6.3
Swaziland	6	17,364	511	2.5	270	530	5.6
Sweden	3	448,070	8,263	.4	77,200	9,340	1.2
Switzerland	3	41,440	6,327	.2	70,110	11,080	.1
Syria	5	186,480	7,835	3.3	6,700	860	6.1
Taiwan	8	32,260	16,793	2.0	19,800	1,180	5.5
Tanzania	6	939,652	16,363	3.0	3,440	210	2.1
Thailand	9	512,820	43,326	2.8	18,660	430	4.1
Togo	6	56,980	2,350	2.6	650	280	5.3
Tonga	0	997	92	1.9	40	400	.9
Trinidad and Tobago	2	5,128	1,118	1.2	2,930	2,620	1.5

Tunisia	5	164,206	5,866	2.0	4,940	840	6.5
Turkey	5	766,640	41,949	2.5	46,580	1,110	4.5
Tuvalu	0	26	6	N.A.	N.A.	N.A.	N.A.
Uganda	6	235,690	12,049	3.0	N.A.	N.A.	N.A.
U.K.	3	243,978	55,932	.1	254,100	4,540	1.6
United Arab Emirates	5	82,880	750	16.7	11,100	14,800	-3.6
Upper Volta	6	274,540	5,465	1.6	760	140	1.6
Uruguay	2	186,998	2,876	.2	4,170	1,450	1.3
U.S. Virgin Is.	2	342	98	3.6	510	5,190	.6
U.S.A.	1	9,363,396	216,729	.8	1,896,550	8,750	2.0
USSR	4	22,274,000	258,932	.9	861,210	3,330	4.4
Venezuela	2	911,680	13,513	3.4	35,480	2,630	3.2
Vietnam	9	329,707	50,647	3.1	N.A.	N.A.	N.A.
Wallis and Futuna	0	207	9	N.A.	N.A.	N.A.	N.A.
Western Sahara	5	266,770	N.A.	N.A.	N.A.	N.A.	N.A.
Yemen, Arab R.	5	194,250	4,982	1.9	2,540	510	N.A.
Yemen, Dem.	5	287,490	3,200	N.A.	290	90	N.A.
Yugoslavia	4	255,892	21,738	.9	45,600	2,100	5.1
Zaire	6	2,343,950	25,694	2.7	5,290	210	-1.4
Zambia	6	745,920	5,128	3.0	2,350	460	-.2
World total		135,830,000	4,147,013				
World mean†				1.8		1,919	3.2

SOURCES.--World Bank Atlas, 1979 (for population, GNP, GNP per capita, and related growth rates); National Basic Intelligence Factbook, July 1978 (Washington, D.C.: Government Printing Office, 1978) (for coastline and area); and Statesman's Year-Book, 1978-1979 (for population and area).

*Regions used in other tables of this volume: 0 = Australasia, 1 = Anglo-America, 2 = Latin America, 3 = Western Europe, 4 = Socialist Eastern Europe, 5 = Near East, 6 = Sub-Saharan Africa, 7 = South Asia, 8 = East Asia, 9 = Southeast Asia.

†The world mean is not an average of country values divided by the number of countries; it is a mean weighted by population and is independent of the number of countries.

TABLE 21.--MARINE JURISDICTIONAL CLAIMS, BY COUNTRY

Country or Territory	Coastline (km)	Coastline/ Area Ratio*	Ports (No.)		Jurisdictional Claims (nm)				Hypothetical† Area to 200 nm (Thousand km²)
			Major	Minor	Territorial Sea	Fishing	Economic	Other	
Albania	418	.0145	1	3	15				3.6
Algeria	1,183	.0005	9	8	12				40.0
Angola	1,600	.0013	3	15	20	200			147.0
Antigua	153	.5464	1	1	3				N.A.
Argentina‡	4,989	.0018	7	21	200				339.5
Australia§	25,760	.0033	12	N.A.	3	12			1,854.0
Bahamas	3,542	.3108	2	9	3	200			221.4
Bahrain	161	.2701	1	N.A.	3				1.5
Bangladesh	580	.0041	1	5	12	200	200		22.4
Barbados	97	.2256	1	2	12	200	200		48.8
Belgium	64	.0021	5	1	3	12			.8
Belize	386	.0168	4	4	3				9.0
Benin‖	121	.0010	1	1	200			100	7.9
Bermuda	103	1.9074	3	N.A.	3	200			N.A.
Brazil	7,491	.0009	8	23	200				924.0
Brunei	161	.0279	0	2	3				7.1
Bulgaria	354	.0032	3	3	12				9.6
Burma	3,060	.0045	4	6	200		200		148.6
Cambodia	443	.0024	2	5	12	200	200		16.2
Cameroon	402	.0085	1	3	50				4.5
Canada	90,908	.0091	19	300	12	200			1,370.0
Cape Verde	965	.2389	1	3	12	200	200		N.A.
Chile	6,425	.0087	10	20	3	200			667.3
China	14,500	.0015	10	180	12				281.0
Colombia	2,414	.0021	5	5	12	200	200		175.9
Comoros	340	.1567	0	1	12	200	200		N.A.
Congo	169	.0005	1	N.A.	30				7.2
Cook Is.	120	.5000	0	2	3				556.1
Costa Rica#	1,290	.0253	3	4	12	200		200	75.5
Cuba	3,735	.0326	8	44	12	200	200		105.8
Cyprus	537	.0580	3	6	12				29.0
Denmark	3,379	.0786	16	44	3	200			20.0
Djibouti	314	.0135	1	N.A.	12	200			N.A.
Dominica	148	.1873	0	2	3	12			N.A.
Dominican Rep.	1,288	.0265	5	17	6	200	200		78.4

Ecuador	2,237	.0081	3	11	200				338.0
Egypt**	2,450	.0024	3	8	12			18	50.6
El Salvador	307	.0143	2	1	200				26.8
Equatorial Guinea	296	.0106	2	3	12				82.6
Ethiopia††	1,094	.0009	2	N.A.	12				22.1
Faeroe Is.	764	.5701	0	1	3	200			N.A.
Falkland Is.	1,288	.1059	1	4	3				N.A.
Fiji	1,129	.0618	1	6	12				368.9
Finland	1,126	.0033	11	14	4	12			28.6
France	3,427	.0062	23	165	12	200			99.5
French Guiana	378	.0042	1	7	12	200			N.A.
French Polynesia	2,525	.6312	1	6	12	200			N.A.
Gabon	885	.0033	2	2	100	150			62.3
Gambia	80	.0077	1	N.A.	50				5.7
German D.R.	901	.0083	4	13	3				2.8
Germany, F.R.	1,488	.0060	10	11	3	200			11.9
Ghana	539	.0023	2	4	200				63.6
Gibraltar	12	2.0000	1	N.A.	3				N.A.
Greece	13,676	.1031	17	37	6				147.3
Greenland	44,087	.0203	7	16	3	200			N.A.
Grenada	121	.3517	1	1	12	200	200		N.A.
Guadeloupe	306	.1720	1	3	12	200			N.A.
Guatemala	400	.0037	2	3	12	200	200		28.9
Guinea	346	.0014	1	3	130				20.7
Guinea-Bissau	274	.0076	1	2	12	200	200		43.9
Guyana	459	.0021	1	3	3	200			38.0
Haiti	1,771	.0639	2	12	12	200	200		46.8
Honduras	820	.0073	3	9	12				58.6
Hong Kong	733	.7075	1	N.A.	3				N.A.
Iceland	4,988	.0484	4	50	4	200			252.8
India‡‡	7,003	.0022	9	79	12	200	200	300	587.6
Indonesia§§	54,716	.0287	10	70	12				1,577.3
Iran	3,180	.0019	7	6	12	50			45.4
Iraq	58	.0001	3	N.A.	12				.2
Ireland	1,448	.0210	6	38	3	200			110.9
Israel‖‖	273	.0132	3	5	6				6.8
Italy	4,996	.0166	16	22	12				161.0
Ivory Coast	515	.0016	2	3	12	200			30.5
Jamaica	1,022	.0895	3	10	12				86.8
Japan##	12,075	.0326	53	N.A.	12	200			1,126.0
Jordan	26	.0003	1	N.A.	3				N.A.

TABLE 2I. (continued)

Country or Territory	Coastline (km)	Coastline/ Area Ratio*	Ports (No.)		Jurisdictional Claims (nm)				Hypothetical† Area to 200 nm (Thousand km^2)
			Major	Minor	Territorial Sea	Fishing	Economic	Other	
Kenya	536	.0009	1	N.A.	12	200			34.4
Kiribati	1,143	1.6711	0	1	3	200			N.A.
Korea, D.P.R.***	2,495	.0205	6	26	12		200	50	37.8
Korea, Rep.	2,413	.0245	10	18	12				101.6
Kuwait	499	.0311	3	4	12				4.1
Lebanon	225	.0217	3	5	N.A.	6			6.6
Liberia	579	.0052	1	6	200				67.0
Libya†††	1,770	.0010	3	4	12				98.6
Macau	40	2.5000	1	N.A.	6	12			N.A.
Madagascar	4,828	.0081	4	N.A.	50	150			376.8
Malaysia	4,675	.0141	6	26	12				138.7
Maldives‡‡‡	644	2.1611	0	2		10	200		279.7
Malta	140	.4473	1	2	6	20			19.3
Martinique	290	.2636	1	5	12	200			N.A.
Mauritania	754	.0007	1	2	70	200			45.0
Mauritius	177	.0954	1	N.A.	12	200			345.0
Mexico	9,330	.0047	9	20	12	200	200		831.5
Monaco	4	2.0000	0	1	12				N.A.
Morocco	1,835	.0045	8	10	12	70			81.0
Mozambique	2,470	.0031	3	2	12		200		163.9
Namibia	1,489	.0018	2	N.A.	6	12			145.9
Nauru	24	1.1429	0	1	12				92.8
Netherlands	451	.0133	8	5	3	12			24.7
Netherlands Antilles	364	.3569	4	6	3	200			N.A.
New Caledonia	2,254	.1024	1	21	12	200			382.4
New Hebrides	2,528	.1712	0	2	3				179.9
New Zealand	15,134	.0564	3	N.A.	12	200	200		1,058.1
Nicaragua‡	910	.0062	3	7	3	200			46.6
Nigeria	853	.0009	5	10	30	200			61.5
Norway	5,832	.0180	9	69	4	200	200		590.5
Oman	2,092	.0099	1	3	12	200			163.8
Pakistan§§§	1,046	.0013	1	5	12	200	200	112	92.9
Panama‡	2,490	.0329	2	10	200				89.4
Papua New Guinea	5,152	.0108	5	8	12	200	200		684.2
Peru	2,414	.0019	7	20	200				229.4

Philippines‖‖‖	22,540	.0750	11	N.A.					551.4
Poland###	491	.0016	4	6	3	12		6	8.3
Portugal	1,793	.0190	6	34	12	200	200		517.4
Qatar	563	.0543	1	1	3				7.0
Reunion	201	.0800	1	N.A.	12				N.A.
Romania	225	.0009	5	2	12				9.3
Samoa	403	.1415	1	1	12				38.1
Sao Tome and Principe	209	.2168	1	N.A.	6	12			N.A.
Saudi Arabia**	2,510	.0011	3	6	12			18	54.9
Senegal	531	.0027	1	2	150	200	200		60.0
Seychelles	491	1.2153	0	1	3	12			N.A.
Sierra Leone	402	.0056	1	2	200				45.4
Singapore	193	.3310	3	2	3				.1
Solomon Is.	5,313	.1784	0	5	3	200			458.4
Somalia	3,025	.0047	3	N.A.	200				228.3
South Africa	2,881	.0024	8	N.A.	12	200			296.5
Spain	4,964	.0098	23	150	12	200	200		355.6
Sri Lanka****	1,340	.0205	3	9	12	200	200		150.9
St. Kitts-Nevis-Anguilla	193	.4961	0	3	3				N.A.
St. Lucia	158	.2565	1	1	3				N.A.
St. Vincent	84	.2159	1	1	3				N.A.
Sudan**	853	.0003	1	N.A.	12			18	26.7
Surinam	386	.0027	1	6	12	200	200		N.A.
Sweden	3,218	.0072	17	30	12	200		18	45.3
Syria**	193	.0010	3	2	12				N.A.
Taiwan	990	.0307	5	5	3	12			114.4
Tanzania	1,424	.0015	3	N.A.	50				65.1
Thailand	3,219	.0063	2	16	12				94.7
Togo	56	.0010	1	1	30	200			.3
Tonga††††	419	.4203	0	2					158.4
Trinidad and Tobago	362	.0706	3	6	12				22.4
Tunisia‡‡‡‡	1,408	.0086	4	8	12				25.0
Turkey§§§§	2,574	.0034	10	35	6	12			69.0
Tuvalu	24	.9231	0	1	3	200	200		211.5
U.K.	12,429	.0509	23	350	3	200			274.8
United Arab Emirates‖‖‖‖	1,448	.0175	3	1	3				17.3
Uruguay	660	.0035	4	6	200	200			34.8
U.S.A.	19,924	.0021	25	N.A.	3	200			2,222.0
USSR	46,670	.0021	52	116	12	200			1,309.5

TABLE 2I. (continued)

Country or Territory	Coastline (km)	Coastline/ Area Ratio*	Ports (No.)		Jurisdictional Claims (nm)				Hypothetical† Area to 200 nm (Thousand km²)
			Major	Minor	Territorial Sea	Fishing	Economic	Other	
Venezuela	2,800	.0031	6	17	12	200	200		106.6
Vietnam	3,444	.0104	9	23	12	200	200		210.6
Wallis and Futuna	129	.6232	0	2	12	200			71.9
Western Sahara	1,110	.0042	2	2	6	12			N.A.
Yemen Arab R.**	1,528	.0079	1	2	12			18	9.9
Yemen Dem.**	1,383	.0048	1	N.A.	12	200	200	18	160.5
Yugoslavia	3,935	.0154	9	24	10	12			15.3
Zaire	37	.0000	1	2	12				.3

SOURCES.--National Basic Intelligence Factbook, January 1980 (Washington, D.C.: Government Printing Office, 1978); Prescott's table 1 in this volume; and Sherwood E. Frezon, Summary of 1972 Oil and Gas Statistics for Onshore and Offshore Areas of 151 Countries, U.S. Geological Survey Professional Paper 885 (Washington, D.C.: Government Printing Office, 1974).

*Coastline divided by the land area (see table 1I).

†See Prescott and Frezon, sources above.

‡Claims continental shelf and superadjacent waters.

§Claims prawn and crayfish on continental shelf.

||100-nm mineral exploitation limit.

#Claims special competence over living resources to 200 nm.

**Claims an additional 6-nm necessary supervision zone.

††Claims sedentary fisheries to the limits of exploitation.

‡‡Claims an additional 100-nm fisheries conservation zone.

§§Claims archipelagic baselines and waters.

|| || Excludes areas occupied in 1967.

##3-nm territorial sea in international straits.

***50-nm military zone.

†††Gulf of Sidra sovereignty claim.

‡‡‡Territorial sea claimed within the area bounded by 7°9'N to 0°45'S and 72°30'E to 73°48'E.

§§§Claims the right to establish a 100-nm conservation zone.

|| || ||Claims archipelagic waters and territorial sea from archipelagic baselines to Treaty of Paris lines.

###Claims 3-nm contiguous zone beyond territorial sea.

****Claims pearling in the Gulf of Mannar.

††††Claims a polygonal territorial sea.

‡‡‡‡Claims a fishing zone to the 50-m isobath (12-65 nm in width).

§§§§Claims 12-nm territorial sea in the Black Sea.

|| || || || 12-nm territorial sea for Sharjah.

TABLE 31.--MARINE AREA SUMMARIES

Marine Area*	Area† (Thousand km²)	Percentage of Oceans	1977 Nominal Catch (Thousands of Metric Tons)
Atlantic and adjacent seas	109,708	34.4	N.A.
Arctic	7,336	2.0	N.A.
Hudson Bay	730	.2	N.A.
Atlantic, Northwest	5,207	1.4	3,018.6
Atlantic, Northeast	16,877	4.7	12,873.9
North Sea	427	.1	N.A.
Baltic Sea	382	.1	N.A.
Atlantic, Western Central	14,681	4.1	1,419.5
Caribbean Sea	2,517	.7	N.A.
Gulf of Mexico	1,508	.4	N.A.
Atlantic, Eastern Central	13,979	3.9	3,733.6
Mediterranean and Black Seas	2,980	.8	1,191.3
Mediterranean Sea	2,511	.7	N.A.
Black Sea	508	.1	N.A.
Atlantic, Southwest	17,756	4.9	1,125.3
Atlantic, Southeast	18,594	5.2	2,785.8
Atlantic, Antarctic	12,298	3.4	275.6
Indian and adjacent seas	72,307	20.0	N.A.
Indian, Western	30,198	8.4	2,319.3
Red Sea	453	.1	N.A.
Persian Gulf	230	.1	N.A.
Indian, Eastern	29,485	8.1	1,291.4
Andaman Sea	565	.2	N.A.
Indian, Antarctic	12,624	3.5	122.8
Pacific and adjacent seas	179,045	49.6	N.A.
Pacific, Northwest	20,476	5.6	18,607.7
Sea of Okhotsk	1,392	.4	N.A.
Sea of Japan	1,013	.3	N.A.
East China Sea	665	.2	N.A.
Yellow Sea	294	.1	N.A.
Pacific, Northeast	7,503	2.1	1,765.2
Bering Sea	2,262	.6	N.A.
Pacific, Western Central	33,530	9.3	5,761.4
South China Sea	2,975	.8	N.A.
Pacific, Eastern Central	57,467	15.9	1,992.9
Gulf of California	153	.0	N.A.
Pacific, Southwest	33,212	9.2	559.1
Pacific, Southeast	16,471	4.6	3,896.6
Pacific, Antarctic	10,386	2.9	3.4

SOURCES.--FAO, Yearbook of Fishery Statistics, 1977, vol. 44; and The World Almanac and Book of Facts, 1977.

*The designation of marine areas is adapted from FAO usage; see source cited.

†Because the areal extent of the various bodies are taken from different sources, the figures do not add to the subtotals given.

Contributors

Susan H. Anderson is coordinator for environmental projects for the law firm of Nossaman, Krueger, and Marsh and executive vice-president of the California Marine Parks and Harbors Association. Formerly she was marine recreation specialist for the Sea Grant Advisory Program at the University of Southern California. At the request of the Office of Marine Resources of the National Oceanic and Atmospheric Administration, she was coordinator for the First National Conference on Marine Recreation held in 1975 to develop recommendations for federal, state, and local actions to enhance marine recreational opportunities.

George F. Bass is professor of anthropology at Texas A&M University and serves as president of the Institute of Nautical Archaeology, which he founded in 1973. He received an M.A. in Near Eastern archaeology from the Johns Hopkins University and a Ph.D. in classical archaeology from the University of Pennsylvania, and has directed archaeological excavations both on land and under water in Turkey, Italy, and the United States. His fifth book on underwater archaeology is in press.

Elisabeth Mann Borgese is professor of political science at Dalhousie University; chairman of the Planning Council of the International Ocean Institute, Malta; and adviser to the Delegation of Austria at the Third United Nations Conference on the Law of the Sea. For some years she was a senior fellow at the Center for the Study of Democratic Institutions, Santa Barbara. She has written numerous books, monographs, and essays on international ocean affairs and marine resources management, including *The Ocean Regime* (1968), *The Drama of the Oceans* (1976), *The New International Economic Order and the Law of the Sea* (with Arvid Pardo, 1976), and *Seafarm* (1980).

Christopher Cragg is now based at the London School of Economics, completing a report for the British Social Science Research Council on the role of insurance in transporting oil and liquid natural gas to British ports. He has a philosophy, politics, and economics degree and a doctorate from Oxford. He has worked in the International Business Unit at the University of Manchester Institute of Science and Technology which provides consultancy and databank services to academics and business executives of all nationalities.

T. F. Gaskell is chief executive of Oil Industries' International Exploration and Production Forum, London. He has done research in geophysics at Cambridge University and holds a doctorate. He saw wartime service in underwater weapons and combined operations in London and Ceylon. He was chief scientist on HMS *Challenge* (1950–52) and chief petroleum physicist with the Anglo-Iranian Oil Co. in Iran and London (1946–73).

Norton Ginsburg is professor and chairman of the University of Chicago's Department of Geography. His interest in the oceans was stimulated during his tenure as academic dean at the Center for the Study of Democratic Institutions (1971–74). In 1972 he directed a Pacem in Maribus conference on the Mediterranean region. He is the author of several articles on ocean-related topics and has written and edited a number of papers and books on other subjects, for example, *An Atlas of Economic Development* (1961). He also has served as chairman of the Committee on Environment of the U.S. National Commission for Unesco and as consultant on environmental problems to SCOPE and Unesco.

Jozef Goldblat is senior member of the research staff at the Stockholm International Peace Research Institute (SIPRI). He has university degrees in international relations, law, and economics. Since the early sixties he has been involved, in different capacities including service for the United Nations, in disarmament negotiations in Geneva and New York. He also served on international control commissions in Korea and Vietnam. He has written numerous books, reports, articles, and brochures on truce supervision, the arms race, and disarmament problems. He has been published in Finland, France, India, Japan, Norway, Poland, Sweden, Switzerland, and the U.S.A.

S. J. Holt is professor of environmental studies at the University of California, Santa Cruz, and chairman of the Committee on Marine Mammals of IUCN. He has served as senior adviser on marine affairs in the Department of Fisheries, FAO (1979); conducted the FAO/UNEP Marine Mammal Project (Rome, 1974–76); served as Secretary of Unesco's Intergovernmental Oceanographic Commission (1971–72); and, as the UN senior regional adviser for ocean affairs, helped establish the International Ocean Institute, Malta, serving as its first director (1973–74). He has written extensively on fishery and resource management problems.

Bhupendra Jasani is a research fellow at the Stockholm International Peace Research Institute (SIPRI), which he joined in 1972. He received a Ph.D. and M.Sc. from the University of London and a B.Sc. from the University of Southampton. He was a member of the Medical Research Council (1958–72) and did research on the development of nuclear radiation detectors for medical and biological applications at the University College Hospital Medical School, London. He organized a SIPRI Symposium in 1973 on Nuclear Proliferation Problems. He has contributed to each *SIPRI Yearbook* since 1973 and has recently completed a book on the militarization of space, which will be published by SIPRI.

Andrzej Karkoszka is a research fellow at the Stockholm International Peace Research Institute (SIPRI) on leave of absence from the Polish Institute of International Affairs in Warsaw. He holds a Ph.D. in political science. He is the author of several articles and books on disarmament-related topics in Polish and English, including the SIPRI monograph, *Strategic Disarmament, Verification, and National Security,* and has contributed to the *SIPRI Yearbooks* 1978, 1979, and 1980.

George Kent is professor of political science and of urban and regional planning at the University of Hawaii. He received a doctorate from the University of Illinois. His primary concern is with the role of ocean and food politics in development. He is currently concentrating on the problem of the development of the Pacific islands, exploring ways in which planning can be deprofessionalized and localized.

Bruce D. Larkin is professor of politics and a participant in the Research Group in International Studies at the University of California, Santa Cruz. Following studies at the University of Chicago and Harvard University he joined the founding faculty of the University of California at Santa Cruz in 1965. In 1977–78 he was visiting research scholar at Tokyo University, and in 1980–81 he will return to Japan as a Fulbright Lecturer. He is a specialist on China and the international politics of the western Pacific. His major work is *China and Africa, 1949–1970* (1971).

George A. Levikov is currently with the Shipping Division of the UNCTAD Secretariat, Geneva. He graduated from Odessa Merchant Marine College and did graduate work at Moscow University, where he received a Candidate of Economics degree. He holds the rank of associate professor and has served as chief of a division of the State Planning and Scientific Research Institute of Maritime Transportation. As a member of Soviet delegations, he has participated in sessions of the UNCTAD Committee on Shipping, UNCTAD-V, and meetings of the Standing Commission on Transport of the Council of Mutual Economic Assistance. His monographs include *Sea Transport of Post-War Japan* (1969) and *International Merchant Marine Shipping* (1978).

Nikki Meith is a consultant for the United Nations Environment Programme on a variety of information projects, including the newsletter of the Regional Seas Programme, *The Siren.* She studied zoology and ecology at Duke University (B.A.), the University of North Carolina (M.A.), and most recently at the University of Ljubljana, Yugoslavia. She has published papers on organochlorine levels in deep sea fish, sublethal effects of DDT on barnacles, and the methodology of pollution studies.

Akio Miura is associate professor in algal cultivation and propagation at the Tokyo University of Fisheries. He graduated from the First Fisheries Institute (presently Tokyo University of Fisheries) in 1950 and received a doctorate from Tokyo Education University in 1977, presenting a dissertation on "Taxonomic Studies on *Porphyra* Cultivated in Japan."

Choon-ho Park is adjunct professor at the University of Hawaii Law School and research associate at the East-West Culture Learning Institute. He earned a doctorate at the Faculty of Law, University of Edinburgh (1971), following completion of a diploma in applied linguistics at the Faculty of Arts. He was a research associate in East Asian Legal Studies at Harvard Law School (1972–78), a fellow at the Woodrow Wilson International Center for Scholars (1973), and has taught at the Fletcher School of Law and Diplomacy. His publications include monographs and articles on the legal aspects of energy, fishing, and other uses of the sea in East Asia.

J. R. V. Prescott is reader in geography at the University of Melbourne and specializes in the study of international boundaries, both on land and sea. He received a Ph.D. from the University of London in 1961. His most recent books are *Boundaries and Frontiers* (1978), *Last of Lands—Antarctica* (with John Lovering, 1978), and *Australia's Continental Shelf* (1979).

S. J. R. Simpson is director of St. James's Research Group and Information Services, London. He was awarded a Ford Foundation Fellowship (1970–71) at the University of California, Berkeley, and subsequently served as adviser to several West African governments. He has been closely associated with North Sea oil development, having acted as consultant to oil service companies as well as undertaking research projects for the Economist Advisory Group. He is a member of the Panel of Advisors of the Intermediate Technology Development Group and has worked on projects for the Ministry of Overseas Development, the Department of Employment, and the London School of Hygiene and Tropical Medicine.

Susan Strange is Montague Burton Professor of International Relations at the London School of Economics. She has been a senior research fellow at the Royal Institute of International Affairs and has served as visiting professor at the University of Southern California (1978). She has written extensively on questions of international political economy and international monetary relations (*Sterling and British Policy*; *International Monetary Relations 1959–71*).

Peter S. Thacher is deputy executive director of the United Nations Environment Programme, Nairobi. He attended Dartmouth College and Yale University, from which he graduated in 1948. He has served as director of UNEP's Office in Geneva and director of Global Programmes as well as program director in the secretariat of the UN Conference on the Human Environment (Stockholm, 1972). Before joining the United Nations, he was a member of the U.S. Foreign Service and participated in conferences leading to the Treaty on Outer Space (1966), the Convention on Assistance and Return of Astronauts (1967), the Non-Proliferation Treaty (1968), and the Seabed Arms Control Treaty (1970). He is the author of numerous articles and reports in the field of environmental protection.

C. Vanderbilt is a marine scientist who has been responsible for day-to-day activities of the International Ocean Institute, Malta, since its founding in 1972.

Owen Wilkes is a research fellow at the Stockholm International Peace Research Institute (SIPRI) and was formerly on the staff of the International Peace Research Institute, Oslo (PRIO). He is engaged in a survey of foreign military presence and a study of the impact of developments in antisubmarine warfare technology on the strategic balance, with particular reference to the arctic.

Warren S. Wooster is professor of marine studies and fisheries at the University of Washington, Seattle. He has served as dean of the Rosenstiel School of Marine and Atmospheric Science of the University of Miami (1973–76), staff member of the Scripps Institution of Oceanography, director of investigations of the Peruvian Council of Hydrobiological Research (1957–58), director of Unesco's Office of Oceanography and secretary of the Intergovernmental Oceanographic Commission (1961–63), secretary and then president of the Scientific Committee on Oceanic Research (1964–72), and vice-president of the International Council for the Exploration of the Sea (1974–77). He is engaged in studies of the effects of changes in the physical environment on the abundance of marine fish; he also is concerned with the interactions of ocean science and ocean use.

Index